Heat and Mass Transfer

Heat and Mass Transfer

Kurt C. Rolle

University of Wisconsin–Platteville

Prentice Hall

Upper Saddle River, New Jersey   Columbus, Ohio

Library of Congress Cataloging-in-Publication Data

Rolle, Kurt C.
Heat and mass transfer/Kurt C. Rolle. — 1st ed.
p. cm.
Includes index.
ISBN 0-13-919309-X
1. Heat — Transmission. 2. Mass transfer. I. Title.
TJ260.R65 2000
621.402′2 — dc21 99-046326

Editor: Stephen Helba
Production Editor: Louise N. Sette
Production Supervision: York Production Services
Design Coordinator: Karrie Converse-Jones
Cover Designer: Jason Moore
Cover Art: © Jason Moore
Production Manager: Matthew Ottenweller
Marketing Manager: Chris Bracken

This book was set in Times Roman by York Graphic Services, Inc. and was printed and bound by Courier/Westford, Inc. The cover was printed by Phoenix Color Corp.

Printed in the United States of America

10 9 8 7 6 5 4 3 2 1

ISBN: 0-13-919309-X

Prentice-Hall International (UK) Limited, *London*
Prentice-Hall of Australia Pty. Limited, *Sydney*
Prentice-Hall of Canada, Inc., *Toronto*
Prentice-Hall Hispanoamericana, S. A., *Mexico*
Prentice-Hall of India Private Limited, *New Delhi*
Prentice-Hall of Japan, Inc., *Tokyo*
Prentice-Hall (Singapore) Asia Pte. Ltd., *Singapore*
Editora Prentice-Hall do Brasil, Ltda., *Rio de Janeiro*

To my parents, who gave me my start

To my wife
Joy

and to our children
Kurt, Loreli, Timothy, Heidi, Charity, and Sunshine

who have always been there for me

and to our grandchildren
who will inherit the earth and its environment

Preface

The purpose of this textbook is to present the fundamental concepts and ideas of heat transfer and mass transfer, and to demonstrate how they can be used in the design and analysis of a variety of engineering problems. The book has been written to be covered in a three-credit-hour junior or senior undergraduate engineering or engineering technology course. Numerous well-written and edifying books are addressed to the subject of heat transfer, so the reasons for writing yet another one need to be strong and apparent. These reasons can be listed and explained:

1. To present under one cover the subjects of heat and mass transfer and to show the complementary nature of these two subjects for many applications.
2. To present the subject matter so that the reader can grasp and be able to use the analytical and empirical methods that provide the engineer and technologist with necessary tools for assessing and solving many contemporary engineering problems.
3. To demonstrate the convenience of many numerical methods, made more accessible by the digital computer, in analyzing the more complicated situations involving heat and/or mass transfer.
4. To better show the relationships of heat and mass transfer to the disciplines of thermodynamics and fluid mechanics.
5. To provide a basic textbook that can be used in the new engineering and engineering technology courses in which design is to be integrated more intimately in the curricula.

Heat transfer is a field that has been extensively cultivated and studied. Each year it is flooded by books, articles, and papers that document the various problems and solutions that have been identified in heat transfer. As a consequence of this, a textbook writer seeking to harvest information proceeds with the problem of how to incorporate the latest findings into the content of the text. It often appears that the new information, to continue the metaphor, is a grasshopper plague rather than a nutrient for rich corn. The textbooks covering heat transfer have tended to become larger to accommodate the new information but, as a result, have become too large to be realistically covered in one three-credit-hour course, and many engineering schools can only offer one heat-transfer course at the undergraduate level. As a consequence, the engineering student studies from a book that contains much more material than can be discussed in the course.

This text has been composed so that, from the beginning, the student is introduced to the concepts through discussions and example problems that are "real world." This provides graphic, straightforward demonstrations that are not so abstract as to leave the reader questioning the relevancy of the subject matter. Nearly always these sorts of discussions provide the reader with experiences that will simultaneously address more than one of the five reasons listed for writing this book.

Previous authors have attempted to include diffusional and convective mass transfer in a heat-transfer text because these mechanisms are closely related to heat transfer in some very important applications. They both can be described mathematically in the same forms as conduction heat transfer and convective heat transfer, so it is comfortable and logical to include heat and mass transfer under one cover. The negligence in presenting mass transfer has been rationalized with the argument that it should be given a more serious and thorough study than it could receive as a part of heat transfer. In most engineering disciplines, with the exception of chemical engineering, mass transfer has not received such attention, so engineering students are not exposed to the subject of mass transfer. For instance, methods of predicting evaporation or drying of surfaces, moisture migration through solid materials such as building walls, and the wick action of alcohol in a heat pipe (all good mass-transfer design problems) are not seriously discussed, and concepts such as diffusivity, permeance, and Fick's Law are not explained. With an increasing emphasis on design content in undergraduate engineering programs, mass-transfer problems must be recognized in analyzing many of the complicated and constrained designs faced by engineers.

This book has been set down for engineering students so that they may design and analyze in a more professional manner the systems they may encounter in practice. It is not a design book, but it is intended to give a solid rational introduction to the scientific, mathematical, and empirical methods for treating heat- and mass-transfer phenomena in an applied context. The reader is expected to have a background that includes partial differential equations, thermodynamics, fluid mechanics, an introduction to electromagnetic radiation, and computer literary.

Many of the traditional closed-form problems and their solutions are presented, but the emphasis throughout the book is on applications of the concepts and methods to important engineering problems. As a result of this direction, the derivations of important equations or results are often outlined to give the reader a concise argument to show the rational basis for those results. The computer and its software are used only when more extended analysis warrants such use. Since the reader is presumed to be versed in the use of the computer, many of the problems can be easily handled with mathematical software packages. In addition, finite difference analysis, finite element analysis, and network analysis in radiation heat transfer can be more accessible with commercially available software or by individual algorithms. Readers are encouraged to make those advancements. Also, graphical techniques and approximation methods are introduced to demonstrate that there are still instances when reasonable estimates can be most expeditiously arrived at without computer elaborations. The emphasis is always on allowing students to understand the basic concepts of heat and mass transfer so that they can frame real-world problems in the appropriate setting. With the overwhelming amount of information that is now accessible and the variety of software packages available, engineers must understand how the information sources are relevant to their immediate problems or what the computing machines are doing to arrive at displayed results.

The method of presentation follows a standard chronological format: presenting and discussing a general concept or idea, defining the appropriate terminology, deriving or otherwise identifying governing equations or relationships that allow for analytical treatment of those concepts as related to actual physical situations, presenting some simple examples to reinforce the reader's understanding, detailing some long examples to convey the idea that more difficult problems can be treated, and finally developing one or two open-ended examples to demonstrate the difficulties and complications that are part of many actual engineering and technological problems. All of the settings are taken from the author's experiences in industry, education, consulting, and expert witnessing, and it is hoped that readers can see the inherent limiting features of the text and still be able to apply the concepts and methods to their specific designs and analyses.

Organization of the Text

The organization of the subject matter is traditional. Early on the reader is shown why heat and mass transfer are important in analyzing real engineering problems. In Chapter 1 an overview and review of thermodynamics are given, with the concepts of the boundary, system, surroundings, work, and heat sharpened for the student. The three modes of heat transfer are briefly introduced, and a short discussion of mass transfer is presented. Some mathematical preliminaries are reviewed as reminders to the reader of the impending textbook analyses, such as discussions of differential equation classifications and the boundary and initial conditions for evaluating them. A brief introduction to the unit systems and dimensions is given, including both English (foot/pound/degree Fahrenheit) and International System (SI) units. The engineer and technologists need to be able to converse in both of these systems even if there are social, economic, political, or personal preferences for one or the other. With heat transfer, where professional interactions with nonengineers can be a common occurrence, knowing the dual system of units is advantageous and desirable.

In Chapter 2 the various steady state conduction heat-transfer situations are presented: one-dimensional, multidimensional, and approximation methods. Early in the chapter Fourier's Law is introduced, and a comprehensive presentation is given for the thermal conductivity, including both empirical and analytic methods for determining this material property. The graphical solution to two-dimensional steady state conduction is developed in the section on shape factors. Some texts de-emphasize this method as being out of date, but the author has experienced an awakening of student understanding in the classical solutions to Laplace's equation for steady state conduction heat transfer when they are required to sketch, by hand, a temperature distribution and heat flow paths. The method also lends itself well to classroom examination questions.

In Chapter 3 transient conduction heat transfer is considered with convection or conduction boundary conditions. Lumped heat capacity systems, the semi-infinite solid, and the infinite plate are some of the types of problems considered in this chapter. A unique presentation is given for a straightforward method in deciding when a particular solid can be considered as "semi-infinite." The thermal probe method for determining the thermal conductivity of materials is presented because it is a good example of the utility of transient test methods. The finite difference techniques are used to demonstrate approaches to treating involved multidimensional heat-transfer problems. The finite element method is also briefly discussed.

Forced convection heat transfer is developed in Chapter 4 and free convection in Chapter 5. The derivation of the relationships used for predicting convection heat-transfer coefficients is outlined, beginning with a review of pertinent fluid dynamics concepts. Dimensional analysis is presented as a method for determining mathematical descriptions of complicated natural phenomena such as convection heat transfer. The subtleties of the development of the Reynolds–Coburn Analogy, which is the basis for empirical equations used in evaluating the Nusselt Number and convective heat-transfer coefficients, are presented without burying the reader in details. Experimental methods for determining the convective heat-transfer coefficients are presented to give readers an overview of the state of the art in convection heat-transfer analysis. Some novel problems that have direct engineering and technological applications are considered in both forced and free convection. Combined convection methods are then treated.

Radiation heat transfer is considered in two chapters. In Chapter 6 the concepts of electromagnetic radiation as a form of heat are presented. Black and gray bodies are introduced, and some discussions are given for the more involved but physically correct descriptive models of selective bodies, such as glass. The distinctions between opaque, transparent, and absorbing surfaces are sharpened. Shape or view factors are derived, and this leads to the applications and analysis of radiation heat-transfer systems. In Chapter 7 the concept of radiosity is introduced, and thermal circuits are presented and used to describe the interactions of radiation between two or more opaque surfaces. Also, the methods of treating gas radiation and radiation absorbing into a volume are presented. Applications to furnaces and incinerators are given as examples of radiation heat transfer applied to design problems.

In Chapter 8 the various mechanisms of mass transfer are discussed. Mixture analyses and mathematical descriptions of mass motions are reviewed, and then diffusional mass transfer is considered. Some methods of experimentally determining mass transfer are introduced, and Fick's Law is defined. Methods for measuring diffusivity and permeance, coefficients that provide quantitative methods for analyzing diffusion, are discussed. Then some of the common applications of mass transfer by diffusion are introduced and developed, including evaporation rates of water and other substances, water migration through solid materials, migration of chemicals into metals or other solids, and others.

Heat exchangers and heat pipes are important applications of heat transfer, and these are treated in Chapter 9. The concept of the *specific surface,* defined as the ratio of the heat exchanger contact surface to the heat exchanger volume, is presented as an important design and heat exchanger classification tool. The LMTD and effectiveness–NTU methods are developed for treating heat exchangers, and compact heat exchangers are considered. Heat pipes are utilized in many particular applications, from very large heat-transfer applications involving megawatts of heat to micro heat pipes involving milliwatts of power. The presentation here is one emphasizing their operation and limitations in the context of the developed heat-transfer topics.

Phase change heat transfer is another important subject that has been studied for a long time. In particular, boiling/condensing heat transfer has a large body of published information addressed to it. In Chapter 10 the accepted methods for treating boiling heat transfer are introduced along with the empirical methods. In addition, to continue the design flavor of this book, the phase changes of freezing and melting are involved in many important engineering problems. These phenomena are considered in the same chapter as the boiling heat-transfer material to give the reader an expanded experience of the subject of phase change heat transfer. The solid–liquid (freezing and melting) phase change has received considerable attention in the areas of, for instance, thermal storage systems, earth freezing–thawing actions on roads and foundations, and metal casting processes, Examples treating these and other applications are presented.

Contents

Appendices

Answers to Selected Problems

Index

1 INTRODUCTION

1–1 THE IMPORTANCE OF HEAT AND MASS TRANSFER

Heat transfer and mass transfer are terms that engineering students encounter in their studies and sometimes are required to understand in some detail. The terms *heat transfer* and *mass transfer* are self-explanatory to many students and vague to others. You expect to better understand what these terms mean after studying the subject matter of this text, and I hope that you will gain that benefit. Before we embark on the detailed study of the material, it is good and encouraging to reflect on why you should understand what heat and mass transfer are and what possible benefits can be derived from that knowledge. There are enough other subjects and concepts that you should understand to function as a competent scientist, engineer, technologist, or concerned citizen, and this introduction is intended to help you place the subject matter of this book in your personal priority list.

The modern technological civilization of which we are all a part relies to a profound extent on how it manages and uses its energy; energy that has been endowed and supplied through natural resources. The uses of energy can be identified as work, power, and heat, but in a real sense the work and power that are used finally degenerate into heat. Heat is the exchange of energy between "hot" and "cold" objects, and we understand the exchange to occur from the hot to the cold one. The science that explains and predicts how fast the energy exchange as heat occurs is called heat transfer. It is the science that integrates the various analytical and empirical tools to provide a forum, a body of knowledge, for designers, builders, operators, managers, and researchers to more accurately study heat as an energy exchange.

The recurring concern with energy and its conservation or saving by our societies requires to an important extent the understanding of the concepts of heat transfer and mass transfer. How much insulation to install in a building's ceiling or outside walls for given climatic conditions (shown in figure 1–1) or how much heat loss occurs through a modern insulated window (shown in figure 1–2) will determine how much heat (energy) is required to keep a house or building comfortable in winter or how much air conditioning is desired to keep that same structure comfortable in the summer. The design and operation of a steam generator, or boiler (shown in figure 1–3), requires an understanding of the heat transfer that occurs from the combustion of coal, gas, or oil to the water in the tubes on the sides of the steam generator. The design and construction of a radiator for an automobile engine to keep it cool when it is running (shown in figure 1–4) requires an understanding of the concepts of heat and mass transfer. Important other questions such as how to dissipate heat in electric power lines due to electrical resistance heating or to protect electrical cables from fires or high temperatures (figure 1–5), how to keep miniaturized computer circuits and systems cool as they function in their intended manners, how to properly specify the size of a heat pump or a heat pipe to supply heat to remote locations, or how to control or sense temperatures in an environment can be answered with heat-transfer analysis. Heat- and mass-transfer concepts are applicable and will need to be better understood with respect to sanitation and waste management, sterilization

Figure 1–1 Fiberglass is an effective and popular insulating material for reducing heat flow through building walls, floors, and roofs. It often is supplied in rolls, which can then be unrolled for installation in vertical walls, under floors, or above ceilings. Many times it is more convenient to install loose fiberglass by blowing it into areas as shown in this figure. (Reproduced with permission of Owens-Corning, Toledo, Ohio)

processes, and other hygienic processes. Also, in the various food handling and processing industries a grasp of heat and mass transfer will be mandatory for improving the design and construction of new equipment and processes as well as for more efficient operation of the existing ones.

In the field of metallurgy and material sciences many treatments and processes involve heat and/or mass transfer, and an understanding and appreciation of the concepts of heat and mass transfer are necessary if future advances in this field are to be accomplished. Appropriate methods of cooling aircraft engine bearings and power transmissions can best be designed and implemented only with a good understanding of the concepts of heat transfer and mass transfer.

Other direct applications of the ideas of heat and mass transfer are the mechanisms of frost formation in the ground and how deep to bury a water line so that the water will not freeze, how to construct a firm foundation or base to support oil pipelines in arctic climates without disturbing the environment, and how to prevent damage to highways or buildings due to frost formation. Storing nuclear wastes will finally require repositories that are selected after carefully considering heat and mass propagation through the medium surrounding such wastes (figure 1–6). Finally, if solar energy is to become the sustainable and economic energy supply it promises to be, the concepts of heat and mass transfer will need to be applied when designing and analyzing the required components and systems.

Mass transfer is the study of the movement of mass from one location to another. The reader may be able to think of many types of mass transfer: mechanical motion of mass from one place to another, bulk fluid motion due to a pressure difference in the fluid, and capillary action of fluid through restricted passageways. We will discuss the various modes of mass transfer in Chapter 8, but two particular types of mass transfer, *diffusional mass transfer* and *convective mass transfer* will be considered in some detail

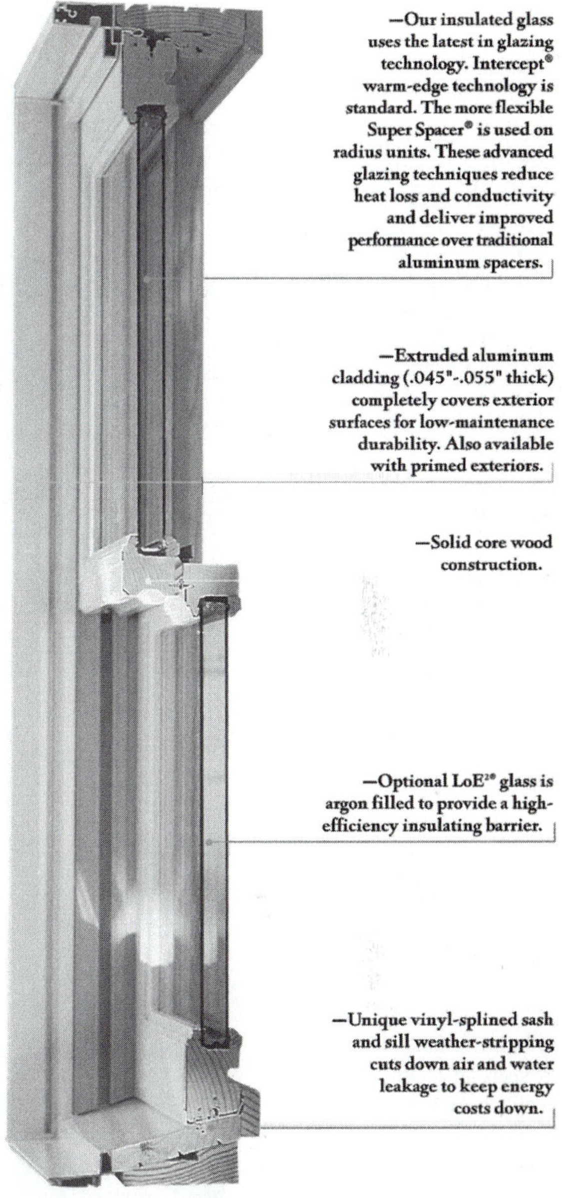

—Our insulated glass uses the latest in glazing technology. Intercept® warm-edge technology is standard. The more flexible Super Spacer® is used on radius units. These advanced glazing techniques reduce heat loss and conductivity and deliver improved performance over traditional aluminum spacers.
—Extruded aluminum cladding (.045"-.055" thick) completely covers exterior surfaces for low-maintenance durability. Also available with primed exteriors.
—Solid core wood construction.
—Optional LoE²® glass is argon filled to provide a high-efficiency insulating barrier.
—Unique vinyl-splined sash and sill weather-stripping cuts down air and water leakage to keep energy costs down.

Figure 1–2 Modern windows are designed to reduce heat flow without reducing the amount of sunlight that passes through. These high-tech windows are constructed of two or more parallel panes of glass with a sealed cavity separating the panes. The cavity may be a vacuum or filled with an inert gas such as argon. Often the inner surfaces of the glass panes are coated with a transparent material that increases the reflectivity of low-level radiation and thereby further reduces the total heat flow through the window. Notice the various characteristics cited on the right side, all of which reduce heat flow through the window/frame. (Reproduced with permission of Jeld-Wen, Inc., Klamath Falls, Oregon)

in this book. They are phenomena with similarities to heat transfer for analytic purposes, and they occur simultaneously with some heat-transfer processes. An understanding of mass transfer can help in analyzing many important processes such as drying, evaporation, humidification of air, or moisture movement through solid walls.

Mass transfer can also help in the analysis and design of disposal and burial tanks for toxic chemicals, radioactive wastes, and other substances. Even the degree of air pollution or contamination of fluorinated hydrocarbons, unburnt hydrocarbons in combustion, and acid rain can be better understood with the concepts of mass transfer. *Cryogenics,* which is the science of extremely low-temperature materials, has made possible the technologies for the production of extremely cold liquids, such as dry ice (liquid carbon dioxide), liquid oxygen (LOX), liquid nitrogen, and liquid hydrogen. Liquid

Figure 1–3 A modern steam generating unit for ultimately producing electrical power with steam turbines/electrical generating units is shown in cutaway in this figure. The walls are lined with tubes through which water passes and is boiled. The four nozzles on the left vertical side wall are used to spray a fuel, such as oil, gas, or pulverized coal, into the chamber where the fuel burns. The combustion of this fuel creates a hot fire ball in the chamber, and by radiation and convection heat transfer this heat is transferred to the water in the tubes. (Reproduced with permission of Combustion Engineering, Inc., Windsor, Conn.)

Figure 1–4 Many common devices can be classified as *heat exchangers* but are usually called by another name. Heat exchangers rely on heat to flow from one hot fluid stream to another, cooler fluid stream. In this figure is shown a radiator that is used to keep an internal combustion engine cool. It accomplishes this by cooling hot water that has passed through the engine and then passes through the radiator. Cool air passes around the radiator, thereby cooling the water and allowing this cooled water to recirculate through the engine again. Notice the scale to the right of the radiator to give some idea of the radiator's size. This radiator is used to cool tractor or combine engines but other radiators are used to cool automobile or truck engines. (Reproduced by permission of Young Radiator Company, Racine, Wisconsin)

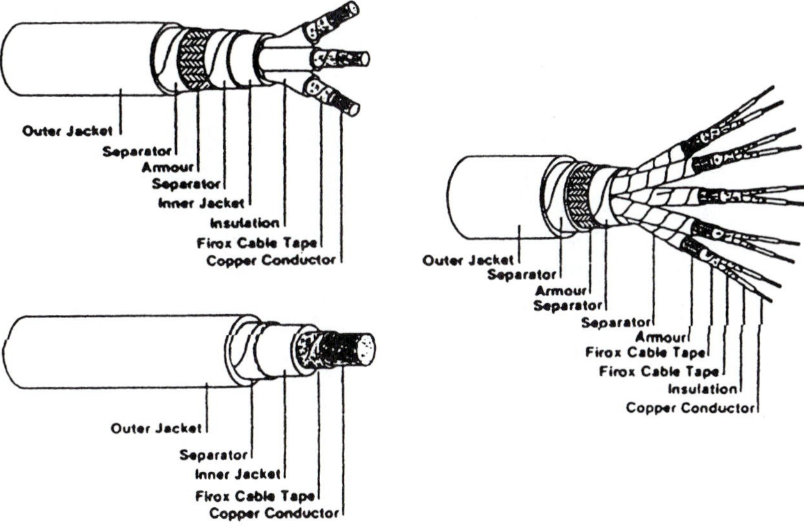

Figure 1–5 One purpose for insulating against heat flow is to prevent fire damage. In this figure is shown how electrical power cables can be wrapped in fire-retarding materials to prevent damage to the electrical cables. (Reproduced by permission of Cogebi, Inc., Dover, NH)

Figure 1–6 Technology produces much waste, which needs to be recycled, destroyed, or placed in regions where it will not cause damage to the environment. The disposal of nuclear waste products, which are classified as toxic chemicals because of their radioactivity, has been a major problem of the nuclear power industry and for government agencies. A proposed site for burial of much of the nuclear waste is Yucca Mountain, Nevada, shown in this figure. There is under construction a tunnel 800 feet below the mountain's surface approximately 25 or 30 feet wide and at least 5 miles long after entering the mountain base through a 2-mile-long entranceway. An understanding of heat and mass transfer is necessary to fully design and analyze this project and make it successful in containing the nuclear waste until it becomes safe. This may require a time span of 1000 to 1500 years.

oxygen and liquid hydrogen are important materials in the propulsion of rockets and space vehicles, but cryogenics relies at every turn on the concepts of heat and mass transfer.

Heat and mass transfer are mechanisms occurring throughout the natural environment and in the processes and devices of modern society. An understanding of those mechanisms and concepts of heat and mass transfer by scientists, engineers, technologists, and technicians is crucial if humankind is to come to realistic and long-term solutions to the problems arising from the demands by society and its perceived standards of living. It is equally important that the citizenry have an appreciation for the promises and limitations of technology due to the imposition of these heat- and mass-transfer mechanisms.

New Terms

A	area
a	acceleration
a,b	coefficients or constants
C	constant
c_p	specific heat at constant pressure
c_v	specific heat at constant volume
E	energy
$\dot{E}$	rate of energy change
e	energy per unit mass
$f(x)$	function of x
F	force
g	gravitational acceleration
g_c	32.17 ft · lbm/lbf · s^2
Hn	enthalpy
hn	enthalpy per unit mass
$\dot{\mathrm{Hn}}$	rate of change of enthalpy per unit time
h	convective heat transfer coefficient
KE	kinetic energy
ke	kinetic energy per unit mass
L	length
MW	molecular mass, molar mass, molecular weight
m	mass
$\dot{m}$	mass flow rate
N	number of moles
n	unit (=1) vector normal to a surface A
PE	potential energy
pe	potential energy per unit mass
p	pressure
Q	heat
$\dot{Q}$	heat transfer
$\dot{q}_A$	heat transfer per unit area
R	gas constant
R_u	universal gas constant
r	displacement, radius
T	temperature
t	time
U	internal or thermal energy
u	internal or thermal energy per unit mass
$\dot{U}$	rate of change of thermal energy per unit time
V	volume
$\dot{V}$	volume flow rate
V	velocity
v	specific volume
W	weight
Wk	work
$\dot{\mathrm{Wk}}$	power, rate of work per unit time
x	displacement, x coordinate
y	displacement, y coordinate
z	displacement, z coordinate
i	unit vector (=1) in the x direction
j	unit vector in the y direction
k	unit vector in the z direction
$\wp$	mass diffusivity
β	beta, relative humidity
κ	kappa, thermal conductivity
ρ	rho, density
σ	sigma, Stefan–Boltzmann constant
ψ	psi, mass fraction
Ω	omega, mole fraction
ω	omega, absolute humidity, humidity ratio
∇	Del operator, $= \mathbf{i}\dfrac{\partial}{\partial x} + \mathbf{j}\dfrac{\partial}{\partial y} + \mathbf{k}\dfrac{\partial}{\partial z}$ (in Cartesian coordinates)

1–2 DIMENSIONS AND UNITS IN HEAT AND MASS TRANSFER

The subjects of heat and mass transfer have developed through methods of empirical observations or experimentation. These methods involve watching physical events, recording and measuring such events, and extending the results with applied mathematics. The basic physical dimensions one can measure are length (L), mass (m), and time (t), and the *unit* is the name of the magnitude of those dimensions. Each of the basic dimensions has a different unit, and a unit system provides the framework for using accepted units. Two unit systems will be used in this text: the *International System* (SI) and the *English Engineering* (English) unit system. Since both of these systems of units are being used to a significant extent, the engineer or technologist may need to be conversant in both SI and English units. The units used for the basic dimensions in the two systems are given in table 1–1. Temperature, energy, and power are derived from the basic units, and these are also included in the table. Force is also included in the table because it is an important dimension. It can be derived from the three basic dimensions (L, m, and t) through Newton's Second Law of Motion,

$$F = Cma \tag{1–1}$$

where a is the acceleration of a mass, m, due to a force, F. The constant, C, provides the equality in Newton's law. In the SI system the Newton, the unit for force, is equal to a kilogram-meter per second squared (kg · m/s^2) and so C has a value of 1. It is therefore deleted from the equation so that equation (1–1) reads $F = ma$. In the English system the pound (lb) is used for both mass and force, so that the pound mass (lbm) is differentiated from the pound force, lbf. It is convenient to have 1 lbm equal to 1 lbf, but this can only be true when Ca in Equation (1–1) is equal to 1 lbf/lbm. It has been customary to use this criterion, where a = 32.17 ft/s^2, the nominal gravitational acceleration on the earth. From this, we get that C = 1 lbf · s^2/32.17 ft · lbm. It has been customary to use, instead of C, a different constant, g_c, which is the reciprocal of C, and then g_c = 32.17 ft · lbm/lbf · s^2. Equation (1–1) may then be written,

$$F = \frac{1}{g_c} ma \qquad \text{(English units only, } m \text{ in lbm)} \tag{1–2}$$

and

$$F = ma \qquad \text{(SI units, English units if } m \text{ in slugs)}$$

Sometimes in English units the **slug** is used for the mass unit. In this situation 1 slug = 1 lbf · s^2/ft, and then the constant C is identically 1 by definition and equation (1–1). The student should exercise care when using these units.

Table 1–1 Basic Dimensions and Units

Dimension	SI	English
Length	meter (m)	foot (ft)
Mass	kilogram (kg)	pound mass (lbm)
Force	Newton (N)	pound force (lbf)
Time	second (s)	second (s)
Temperature	Kelvin (K) or degree Celsius (°C)	Rankine (°R) or Fahrenheit (°F)
Energy	Joule (J)	foot-pound (ft · lbf) or British thermal unit (Btu)
Power	Watt (W)	Horsepower (hp) or Btu/s

The SI system utilizes prefixes to make the units more flexible over a wide range of values for dimensions and other quantities. As an example, the prefix *milli* represents 1/1000 or 10^{-3} (i.e., 1 mm = 1/1000 m). Similarly, 1 kg = 1000 g and 1 megawatt = 10^6 watts. The SI prefixes and their conversion factors are given in table 1–2.

Table 1–2 SI Prefixes

Prefix	Symbol	Amount	Multiple
giga	G	1 000 000 000	10^9
mega	M	1 000 000	10^6
kilo	k	1 000	10^3
hecto	h	100	10^2
deka	da	10	10
deci	d	0.1	10^{-1}
centi	c	0.01	10^{-2}
milli	m	0.001	10^{-3}
micro	μ	0.000 001	10^{-6}
nano	n	0.000 000 001	10^{-9}

There will be occasions when a conversion between the two unit systems will be necessary. The basic conversions between English and SI units are given in table 1–3. Other conversions may be found in the inside front cover.

Table 1–3 Conversion Factors Between SI and English Units

Unit	Multiply By	To Convert To
foot (ft)	0.3048	meter (m)
meter (m)	3.2808	foot (ft)
lbm	0.45359	kg
kg	2.2046	lbm
lbf	4.4484	Newton (N)
Newton (N)	0.2248	lbf
°Rankine	5/9	Kelvin
Kelvin	9/5	°Rankine
ft · lbf	1.356	Joule (J)
Joule (J)	0.737	ft · lbf
hp	745.7	W
W	0.001341	hp

1–3 THE CONCEPTS OF HEAT AND MASS TRANSFER

Heat transfer is the science that studies the rates at which heat occurs through temperature differences. The term *heat* arises in the study of thermodynamics, so it is useful to review the fundamental concepts of the science of thermodynamics. It is expected that you have already studied thermodynamics, so this will be a brief discussion of some of the important ideas and definitions that are pertinent to heat- and mass-transfer studies. If you feel that you know thermodynamics concepts, this section may be briefly reviewed and used as reference for the later discussions in this book.

The *thermodynamic system* is a volume in space separated from its *surroundings* by one or more bounding surfaces, called a *boundary*. In figure 1–7 is shown a simple thermodynamic system, not detailed, but presented to show that the boundary is really the

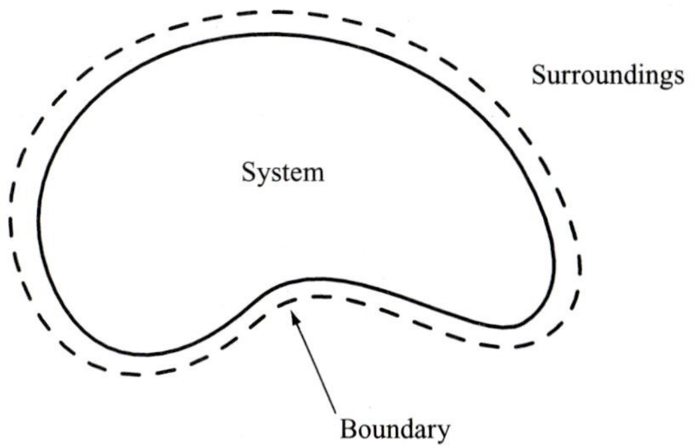

Figure 1–7 A diagram of a thermodynamic system.

surface needed to identify the system. Thermodynamics, and heat and mass transfer as well, begins when the boundary of the system is specified. As you may recall from thermodynamics, the boundary is not one unique, particular surface but is left to your choice. It is located wherever you, the analyzer, decide to place it. Often there is an optimum or most convenient location to define the boundary to most readily apply the principles and concepts of thermodynamics. Some of the guiding principles that are often used to select the boundary are

1. Select a boundary to encompass a system that is homogeneous, having uniform or constant conditions throughout, or whose properties can be determined.
2. Select a boundary that is physically well defined, such as a surface of a solid or liquid separating a gas.
3. Select a boundary where conditions such as the temperature, amount of heat, or amount of mass flow are known.

Once the boundary of a system has been decided upon, the system is described by its *properties*. For a thermodynamic system the properties would be the *mass, density, pressure, temperature, volume, weight, shape, energy, enthalpy, entropy,* and some others. Many of these properties are interrelated or dependent on each other. For instance, the mass (m), volume (V) and density (ρ) are related by a definition for density;

$$\rho = \lim_{V \to \delta V} \frac{m}{V} \tag{1–3}$$

where δV is the smallest volume of a system that can have an identified mass. Frequently we use the operational form,

$$m = \int \rho dV \tag{1–4}$$

and, for the average density in a volume V,

$$\rho = \frac{m}{V} \tag{1–5}$$

It is sometimes convenient to use the inverse of the density, the specific volume,

$$v = \frac{1}{\rho} = \lim_{m \to \delta m} \frac{V}{m} \tag{1–6}$$

where δm is the smallest identifiable mass of a system. The average specific volume would be

$$v = \frac{V}{m} \tag{1–7}$$

The weight of a system, W, is related to its mass by

$$W = mg \tag{1–8}$$

where g is the local gravitational acceleration. Comparing this result to Newton's Second Law of Motion, equation (1–1),

$$F = ma \qquad \textbf{(1–1, repeated)}$$

we can see that the weight corresponds to a force (the attraction between the system and the earth) and the local gravitational acceleration, g, corresponds to a potential acceleration, a.

Temperature is a system property that is the measure of the "hotness" of the system. The higher the temperature the hotter the system, and the lower the temperature, the less hot or cooler the system is. Temperature is an important property of systems in the analysis of heat and mass transfer. Thermal equilibrium occurs between two systems or bodies if their temperatures are the same. Thus, the concept of thermal equilibrium requires first understanding or at least recognizing temperature.

Energy is usually identified as *potential energy, kinetic energy,* or *internal energy.* Potential energy is associated with that energy due to gravitational attraction, or the system weight. For systems on or near the earth, or where g is nearly constant, we may write for the potential energy, *PE*,

$$\text{PE} = mgz = Wz \tag{1–9}$$

where z is a distance vector from some reference plane arbitrarily selected as having zero potential energy. As shown in figure 1–8, a system above the reference plane has positive potential energy and one below the plane has negative energy. One can visualize the

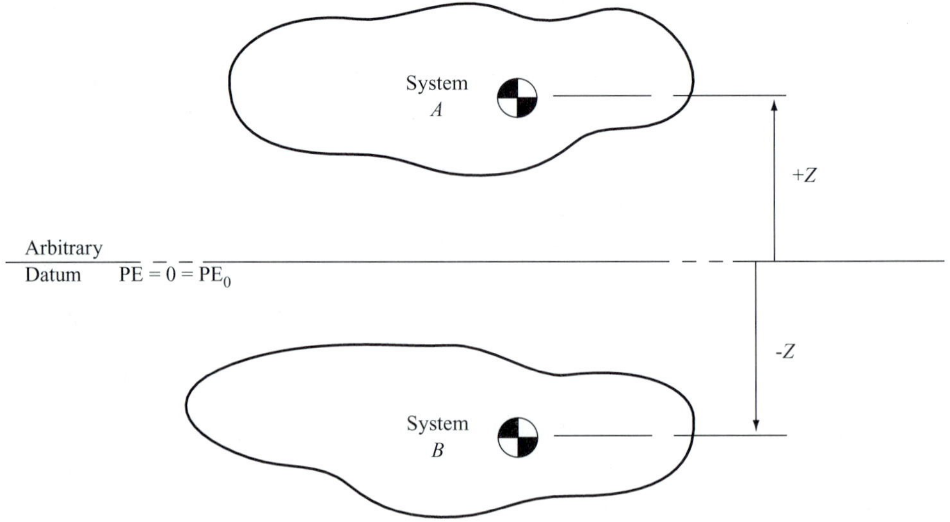

Figure 1–8 Diagram of method for determining gravitational potential energy.

kinetic energy, *KE*, to be associated with the motion or velocity of the system, and it is given by

$$\text{KE} = \frac{1}{2}m\mathbf{V}^2 \tag{1–10}$$

where **V** is the velocity vector of the system having kinetic energy in the amount KE.

The internal or thermal energy, U, is often associated with the temperature, T, of the system, although this can be erroneous thinking. It is true that systems having higher temperature or greater "hotness" have greater internal energy, but the internal energy is also a measure of the energy associated with the molecular, atomic, or nuclear forces or interactions of the submicroscopic particles making up any material. Internal energy can be viewed as an energy form composed of two separate parts: (1) energy due to hotness (or temperature) of the system, sometimes called the *sensible* energy, and (2) energy due to the system phase, whether solid, liquid, or vapor, sometimes called the *latent* energy. The sensible internal energy is related to the temperature by the definition for specific heat at constant volume, $c_v(T)$,

$$c_v = \frac{1}{m}\frac{\partial U}{\partial T} \tag{1–11}$$

For solids and liquids that are *incompressible* (meaning that their volume remains the same even if the shape changes or other conditions change), the partial derivative is an ordinary differential in equation (1–11), and then

$$dU = mc_v dT \tag{1–12}$$

Equation (1–11) is also descriptive of vapors provided the vapors behave according to the *Ideal Gas Law:*

$$pV = mRT = NR_uT$$

or

$$pv = RT \qquad \text{(per unit mass)} \tag{1–13}$$

where R is the gas constant, N is the number of moles of the vapor or substance, and R_u is the universal gas constant. The value for R_u may be taken as 8.31 kJ/kg · mol · K or 1544 ft · lbf/lbm · mol · °R. Thus, for a system that can be identified as either an incompressible substance or an ideal gas, the internal energy is obtained by integrating equation (1–12) over a temperature domain, T_o to T, so that

$$U(T) = U_0 + \int_{T_0}^{T} mc_v(T)dT \tag{1–14}$$

where U_0 is the internal energy at T_0. For convenience T_0 is often selected so that U_0 is zero, although such a condition is of itself arbitrary and subjective. Fortunately, the change in internal energy, ΔU, is more useful for most engineering analyses than is U and is

$$\Delta U = \int_{T_1}^{T_2} mc_v(T)dT \tag{1–15}$$

This result is independent of the arbitrary condition of U_0 at T_0. Often $c_v(T)$ is assumed constant so that $c_v(T) = c_v$, and the integration proceeds so that the internal energy or its change are linear functions of the temperature, T.

EXAMPLE 1–1 Determine the change in the internal energy of 3 kg of air when it is heated from 20°C to 120°C. Assume that $c_v(T)$ for air is given by the equation

$$c_v(T) = 0.683 + 0.0678 \times 10^{-3}T + 0.1658 \times 10^{-6}T^2 - 0.06789 \times 10^{-9}T^3 \frac{\text{kJ}}{\text{kg} \cdot \text{K}}$$

where temperature, T, is in Kelvin.

Solution Assuming that the air behaves as an ideal gas, we use equation (1–15) to evaluate the internal energy change,

$$\Delta U = \int_{T=293\text{K}}^{T=393\text{K}} mc_v(T)dT$$

After substituting for $c_v(T)$ and integrating, the result is

Answer $$\Delta U = 216.93 \text{ kJ}$$

Systems that change phase, from solid to liquid, liquid to vapor, or any of the other phase changes, require an accounting of the energy associated with such changes of phase. Since phase changes occur without temperature changes, the internal energy is affected by the latent part of the energy, or that energy associated with the phase. Thus, the change in internal energy associated with a system in proceeding from a temperature T_1 at phase i to a temperature T_2 at a phase j can be written

$$\Delta U = \int_{T_1}^{T_{ij}} mc_v(T)_i dT + \Delta U_{ij} + \int_{T_{ij}}^{T_2} mc_v(T)_j dT \qquad \textbf{(1–16)}$$

Here T_{ij} denotes the temperature at which the phase change from i to j occurs, ΔU_{ij} is the internal energy change of the material of the system as it proceeds from phase i to phase j (sometimes called the *latent internal energy*), and the specific heat terms are for the two phases respectively (i.e., $c_v(T)_i$ for phase i and $c_v(T)_j$ for phase j).

Another property of materials or systems that has many applications is the *enthalpy,* Hn, defined as

$$\text{Hn} = U + pV \qquad \textbf{(1–17)}$$

where p is the pressure. This pressure is equivalent in magnitude to the hydrostatic (normal) stress acting on a system at a discrete location. In the literature enthalpy is usually denoted as H (or h per unit of system mass). Here we denote it as Hn, or hn per unit of mass because we will use the letter h to denote the convective heat transfer coefficient, which is consistent with the heat transfer literature.

After a sufficient number of the properties of the system have been identified, the list of those properties is called the *state*, equivalent to the condition of the system. Thermodynamic analysis continues, after identifying the state of the system, by considering changes of the state, that is, what happens to the system as its energy changes from one state to another. The two methods by which the system's energy can change are through *work* and *heat*. *Work*, Wk, is the mechanical energy exchange defined as

$$\text{Wk} = \int d\text{Wk} = \int F dx \qquad \textbf{(1–18)}$$

where dWk is a differential amount of work. Work is also the energy exchange between a system and its surroundings associated with a force, F, acting through a displacement or distance, dx. A finite amount of work, as opposed to the differential amount, is the definite integral of equation (1–18) between x_1 and x_2. Thus, one must know how F varies with the vector displacement x, so the function $F = f(x)$ must be known. The case where the force F is constant gives a familiar result,

$$\text{Wk} = F\Delta x \qquad \text{(constant force)} \tag{1–19}$$

For our purposes we will often neglect any work done between the system and its surroundings, except the constant-pressure boundary work. This results from the following development: The boundary work is that work associated with a system's pressure causing its volume to change during a process, and this can be shown to be, from equation (1–18),

$$\text{Wk} = \int p dV \tag{1–20}$$

For constant-pressure processes, this result is

$$\text{Wk} = p\Delta V \tag{1–21}$$

The second method by which the energy of a system may change is *heat*. Heat is that energy exchange between a system and its surroundings which cannot be described as a force acting through a displacement. That is, heat and work are mutually exclusively defined energy exchanges. Heat, Q, can also be described as the energy exchange caused by a temperature difference; however, this then requires an acknowledgment of the *Second Law of Thermodynamics*. This law says, among other interpretations, that heat can only flow from a hot to a less-hot or cold system. It is an attempt to explain the known fact that heat will not flow from cold to hot, by itself.

Notice that heat does have a direction (from a hot system to a cold system), which is designated as positive (+) when *heating* a system and negative (−) when *cooling* a system. Likewise, work is associated with a direction and can be called positive when expanding the system boundary or flowing out of the system and negative when contracting the boundary or flowing into the system. The *First Law of Thermodynamics* is sometimes stated as the conservation of energy, written as

$$\boxed{\Delta E = Q - \text{Wk}} \tag{1–22}$$

and for systems that have energy changes associated only with internal energy (no kinetic or potential energy), this is

$$\Delta U = Q - \text{Wk} \tag{1–23}$$

For the system having boundary work and constant pressure processes only, the work is given by equation (1–21), and then equation (1–23) becomes

$$\Delta U + p\Delta V = Q \tag{1–24}$$

but the left side of this equation is the same as the change in enthalpy, ΔHn, for constant-pressure processes, so we can write this as

$$\Delta\text{Hn} = Q \tag{1–25}$$

This equation gives some clue as to why enthalpy is denoted as "Hn". It represents the "heat" for a constant-pressure process of a system having internal energy changes only. Also, it can be shown that for incompressible solids or liquids, or for ideal gases, the change in enthalpy is related to the temperature of the system through the definition of specific heat at constant pressure, $c_p(T)$,

$$c_p = \frac{1}{m}\frac{\partial \mathrm{Hn}}{\partial T} \tag{1–26}$$

For incompressible solids or liquids or ideal gases, the partial derivative is the same as an ordinary one, so we may write

$$d\mathrm{Hn} = mc_p dT \tag{1–27}$$

and a finite change in the enthalpy may be found by integrating equation (1–27) over the temperature domain T_1 to T_2. Referring to equation (1–16), which developed from the observation that materials can change phase, the enthalpy change of a material that proceeds through a phase change from phase i to phase j while experiencing a temperature change from T_1 to T_2 can be written

$$\Delta\mathrm{Hn} = m\int_{T_1}^{T_{ij}} c_p(T)_i dT + \Delta\mathrm{Hn}_{ij} + m\int_{T_{ij}}^{T_2} c_p(T)_j dT \tag{1–28}$$

where the notation is consistent with equation (1–16). The first and third terms on the right side are sometimes called the *sensible heat* of the material, and the middle term ($\Delta\mathrm{Hn}_{ij}$) is called the *latent heat* of the material. Notice that heat is here used to describe an enthalpy change, consistent with the restricted connection of equation (1–25).

Returning to equation (1–22), the conservation of energy statement, it can be seen that this equation could be written as a differential equation such as

$$dU = dQ - d\mathrm{Wk} \tag{1–29}$$

or,

$$\dot{U} = \dot{Q} - \dot{\mathrm{Wk}} \tag{1–30}$$

where the "dot" denotes a time derivative of the terms. Thus, $\dot{\mathrm{Wk}}$ is $d\mathrm{Wk}/dt$ or *power*, $\dot{U}$ is dU/dt, and $\dot{Q}$ is dQ/dt, or the *heat transfer*. For constant-pressure processes, equation (1–30) reduces to

$$\dot{\mathrm{Hn}} = \dot{Q} \tag{1–31}$$

and we will make considerable use of this result.

EXAMPLE 1–2

One hundred ft^3 of steam at atmospheric pressure of 14.7 $\mathrm{lbf/in}^2$ (psia) and 300°F is condensed and cooled to 80°F. Determine the amount of heat removed from the water during this cooling process. Assume that the specific heats at constant pressure are 1.00 Btu/lbm · °R for liquid water and 0.445 Btu/lbm · °R for steam as a vapor. Also assume that the latent heat of vaporization for water is 970 Btu/lbm and the density of steam at 14.7 psia, 300°F is 0.032 $\mathrm{lbm/ft}^3$.

Solution

We assume that the steam behaves as an ideal gas and the liquid water as an incompressible fluid. Then we can use equation (1–28) to evaluate the enthalpy change. Since the process is being carried out at constant pressure, we can then set the enthalpy change equal to the amount of heat by equation (1–25). The amount of water as steam is 100 ft^3, and the mass is found by using equation (1–5):

$$m = \rho V = \left(0.032 \frac{\text{lbm}}{\text{ft}^3}\right)(100 \text{ ft}^3) = 3.2 \text{ lbm}$$

The temperature domain is T_1 = 300°F + 460 = 760°R, T_{ij} = 212°F + 460 = 672°R, and T_2 = 80 + 460 = 540°R. Completing the evaluation from equation (1–28), we have

$$Q = (3.2 \text{ lbm})\left[\int_{760}^{672}\left(0.445 \frac{\text{Btu}}{\text{lbm}}\right)dT - \left(970 \frac{\text{Btu}}{\text{lbm}}\right) + \int_{672}^{540}\left(1 \frac{\text{Btu}}{\text{lbm}}\right)dT\right]$$

Answer $$= -3651.7 \text{ Btu}$$

Referring back to figure 1–7, note that the customary interpretation of the system includes the restriction that no mass be lost or gained (i.e., mass cannot cross over the boundary). Many times, however, the analyses must allow for mass movement. This is particularly true in mass-transfer studies, but also in many heat-transfer problems. The usual manner of considering mass motion is to define a *control volume*, or *open system,* which is a volume in space bounded by a boundary or *control surface* that allows mass to cross as well as heat and work. Control volumes are useful devices in many engineering problems, but they need to be treated somewhat differently than the system. The conservation of energy for a control volume as shown in figure 1–9 can be written as

$$\dot{Q} - \dot{\text{W}}\text{k} + \sum_{\text{in}} \dot{m}e - \sum_{\text{out}} \dot{m}e = \dot{E}_{\text{accumulated}} \tag{1–32}$$

The term $\sum_{\text{in}} \dot{m}e = \int_A \mathbf{V} \cdot \mathbf{n}e_{\text{in}}dA$ represents the rates of energy flows into the control volume per unit of time. The term $\mathbf{n}$ is the unit vector normal to the control surface at the inlet area A_{in}, the actual surface area of the entrance that allows for inflow.

Figure 1–9 Control volume diagram.

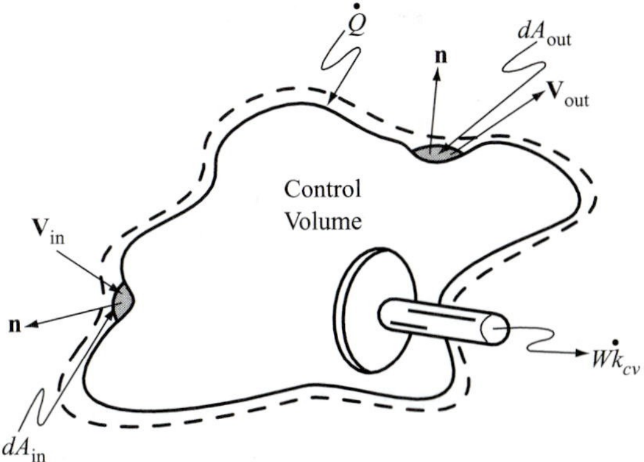

The same interpretation can be made for the term $\sum_{\text{out}} \dot{m}e$, which is the total energy of the material outflowing from the control volume. The term $\dot{E}_{\text{accumulated}}$ represents the energy accumulated within the control volume.

The conservation of mass may be written for the control volume as

$$\sum \dot{m}_{\text{in}} - \sum \dot{m}_{\text{out}} = \dot{m}_{\text{accumulated}} \tag{1–33}$$

where the mass flow rates are

$$\sum \dot{m}_{\text{in}} = \int_{A_{\text{in}}} V_{\text{in}} \cdot \mathbf{n} dA_{\text{in}} \quad \text{and} \quad \sum \dot{m}_{\text{out}} = \int_{A_{\text{out}}} V_{\text{out}} \cdot \mathbf{n} dA_{\text{out}}$$

If we recognize that there is work that needs to be accounted for in the inflows and the outflows of the control volume, and if only internal, kinetic, and potential energy forms are to be considered, then equation (1–32) can be revised to read

$$\dot{Q} - \dot{\text{W}}\text{k}_{cv} + \sum \dot{m}(\text{hn} + \text{ke} + \text{pe})_{\text{in}} - \sum \dot{m}(\text{hn} + \text{ke} + \text{pe})_{\text{out}} = \dot{E}_{\text{accumulated}} \tag{1–34}$$

In this equation the second term, $\dot{\text{W}}\text{k}_{cv}$ is sometimes referred to as the *shaft power.* This term is often used in engineering literature and refers to the fact that the majority of applications of control volumes to technological situations involve work that is obtained from or supplied to the control volume through a rotating shaft, thus the term *shaft power.* For instance, the pump or compressor requires work or power to function, so the shaft power would be supplied to the control volume. A steam turbine or internal combustion engine would operate as a heat engine and produce power or work that could only be extracted through the drive shaft or power shaft.

The *steady state, steady flow condition (SSSF)* is an important subset of control volume analyses. This condition occurs when the following conditions are met:

1. The state and rate of the mass inflowing and/or outflowing at any discrete location on the control surface do not vary with time. Thus the term *steady flow.*
2. The rates at which energy (as heat and work) crosses the control volume boundaries do not vary with time and the control volume does not move relative to the coordinate axis.
3. The state of the mass and energy within the control volume do not vary with time.

Items 2 and 3 constitute the steady state constraint. For the SSSF, equations (1–32) and (1–33) become

$$\dot{Q} - \dot{\text{W}}\text{k} = \sum \dot{m}e_{\text{out}} - \sum \dot{m}e_{\text{in}} \tag{1–35}$$

and

$$\sum \dot{m}_{\text{in}} = \sum \dot{m}_{\text{out}} \tag{1–36}$$

EXAMPLE 1–3 An automobile heater is required to provide 30 m^3/min of 30°C air to the passengers in the auto when the outside air temperature is −10°C. If a 50–50 coolant mixture of water and ethylene glycol is available at 85°C, determine the mass flow of coolant needed to heat the air if the coolant cannot leave the auto heater lower than 45°C.

Solution In figure 1–10 is shown a control volume for the heater, including some of the pertinent data that has been given for this problem. The mass flow rate of coolant will be determined from an application of the energy balance. We first assume that the heater is operating at SSSF conditions, that the coolant is an incompressible liquid, and that the air behaves as

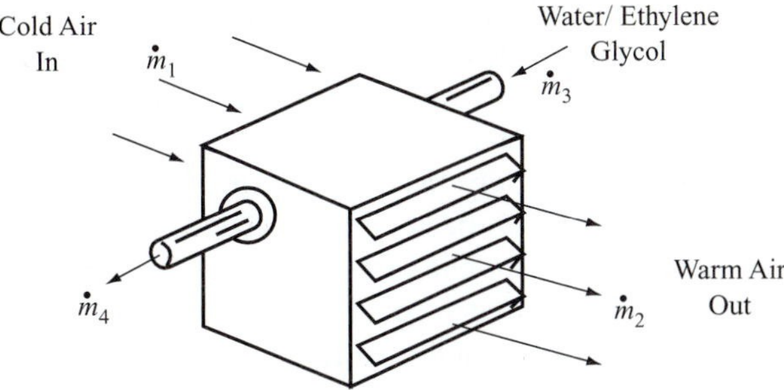

Figure 1–10 Schematic of an automobile heater.

an ideal gas. Also, there is no shaft power and it may be assumed that no heat transfer occurs external to the heater or its control volume. From equation (1–35) we then have, for the SSSF,

$$\sum \dot{m}_{in}(hn)_{in} = \sum \dot{m}_{out}(hn)_{out}$$

which becomes

$$\dot{m}_1(\text{hn})_1 + \dot{m}_3(\text{hn})_3 = \dot{m}_2(\text{hn})_2 + \dot{m}_4(\text{hn})_4$$

Assuming the air to be an ideal gas and the coolant an incompressible fluid, the enthalpies are $\text{hn}_1 = c_{p,\text{air}}T_1$, $\text{hn}_2 = c_{p,\text{air}}T_2$, $\text{hn}_3 = c_{p,\text{cool}}T_3$, and $\text{hn}_4 = c_{p,\text{cool}}T_4$. From appendix Tables B–3, B–4, and B–6 we find average specific heat values for ethylene glycol, air, and water, respectively. The specific heat of the mixture may be taken as the arithmetic average of the ethylene glycol and water. Then

$$c_{p,\text{cool}} = \frac{c_{p,\text{eg}}}{2} + \frac{c_{p,\text{water}}}{2} = \frac{2.385}{2} + \frac{4.18}{2} = 3.2825\frac{\text{kJ}}{\text{kg}\cdot\text{K}}$$

Since the air is assumed to be an ideal gas, its density is just p/RT,

$$\rho_{\text{air}} = \frac{p}{RT} = \frac{101\text{ kPa}}{\left(0.287\dfrac{\text{kN}\cdot\text{m}}{\text{kg}\cdot\text{K}}\right)(300\text{ K})} = 1.173\frac{\text{kg}}{\text{m}^3}$$

The mass flow of air is then

$$\dot{m}_{\text{air}} = \dot{m}_1 = \dot{m}_2 = \rho_{\text{air}}\dot{V}_{\text{air}} = \left(1.173\frac{\text{kg}}{\text{m}^3}\right)\left(30\frac{\text{m}^3}{\text{min}}\right) = 35.19\frac{\text{kg}}{\text{min}}$$

Then, using the energy balance, equation (1–35),

$$\dot{m}_{\text{air}}c_{p,\text{air}}(T_1 - T_2) = \dot{m}_{\text{cool}}c_{p,\text{cool}}(T_4 - T_3)$$

and

$$\dot{m}_{\text{cool}} = \dot{m}_{\text{air}}\left(\frac{c_{p,\text{air}}}{c_{p,\text{cool}}}\right)\left(\frac{T_1 - T_2}{T_4 - T_3}\right) = \left(35.19\frac{\text{kg}}{\text{min}}\right)\left(\frac{1.007}{3.2825}\right)\left(\frac{-10 - 30}{45 - 85}\right)$$

Answer

$$= 10.7955\frac{\text{kg}}{\text{min}}$$

In the study of heat and mass transfer, the work or power involved in control volume analyses is often negligible or zero. If there is no means of power extraction from rotating shafts, the work is clearly zero, but even the work that can be attributed to the volume change of the control volume is zero for the steady state situations.

Also, as this example shows, sometimes there are no heat losses or gains external to the system or control volume. The term *adiabatic* is used for those processes where no heat or heat transfer is involved. As it was analyzed, example 1–3 involves an adiabatic process. Heat was exchanged between the coolant and the air, so if the control volume had been defined as only the air-flow stream or only the coolant-flow stream, then the process would not have been adiabatic but rather one involving heat transfer. The automobile heater is an example of a class of devices called *heat exchangers*. Heat exchangers are important and very common devices that have a variety of applications, and in Chapter 9 they will be studied further.

Mixtures are composed of two or more distinct chemical species, an example being the water–ethylene glycol mixture of example 1–3. For engineering purposes it is convenient to categorize mixtures as one of two types: chemically reacting and chemically inert or nonreacting. For the purposes of this textbook, we will only consider the nonreacting mixtures. The analysis of mass transfers requires an understanding of mixtures and how to describe them.

The fundamental concept of mixtures is the qualitative and the quantitative description of the various components that make up the mixture. An identification of the types of chemical components will be presumed to be complete in this text so that the quantitative description of those components is of primary concern here. The mass analysis of a mixture is often the only information available to an engineer or technologist and it includes the mass fractions, ψ_i, of each component. The mass fraction ψ_i is defined as

$$\psi_i = \frac{m_i}{m_T} \tag{1–37}$$

where m_i is the mass of the ith component and m_T is the total mixture mass. It follows that

$$m_T = \sum m_i \tag{1–38}$$

and also

$$\sum \psi_i = 1.0 \tag{1–39}$$

In example 1–3 the mass fraction of water and the ethylene glycol in the coolant was 0.5.

It is often more convenient to have a molar analysis of a mixture. This includes the mole fraction, Ω_i, of each of the mixture components. The mole fraction is defined as

$$\Omega_i = \frac{N_i}{N_T} \tag{1–40}$$

where N_i is the number of moles of the ith component and N_T is the total number of moles in the mixture. It also follows that

$$N_T = \sum N_i \tag{1–41}$$

and

$$\sum \Omega_i = 1.0 \tag{1–42}$$

The relationship between the mass and mole analyses involves the molecular mass, MW_i, of the various components. The molecular mass represents the relative mass of a chemical species in one mole of that species. Thus,

$$m_i = MW_i N_i \tag{1–43}$$

and

$$m_T = \sum MW_i N_i$$

or

$$m_T = MW_m N_T \tag{1–44}$$

where MW_m is the mixture molecular mass, a convenient relative mass of a particular mixture. It is, from equations (1–44), that

$$MW_m = \frac{1}{N_T}\sum MW_i N_i \tag{1–45}$$

EXAMPLE 1–4 Determine the mole analysis of the coolant in the auto heater of example 1–3.

Solution The mass fractions for the coolant are given as

$\Psi_{water} = 0.5$, $m_{water} = 0.5$ kg in 1 kg of mixture

$\Psi_{eg} = 0.5$, $m_{eg} = 0.5$ kg in 1 kg of mixture

The molecular mass of water is nearly 18 kg/kg · mole or 18 lbm/lbm · mole and for ethylene glycol is 62.07. The number of moles of each component in 1 kg of mixture are, from Equation (1–43),

$$N_{water} = \frac{0.5\ \text{kg}}{18\ \dfrac{\text{kg}}{\text{kg}\cdot\text{mol}}} = 0.0278\ \text{kg}\cdot\text{mol}$$

$$N_{eg} = \frac{0.5\ \text{kg}}{62.07\ \dfrac{\text{kg}}{\text{kg}\cdot\text{mol}}} = 0.0081\ \text{kg}\cdot\text{mol}$$

The total moles of mixture, from equation 1–41, are 0.0278 + 0.0081 = 0.0359 kg · mol and the mole fractions are

Answer $\Omega_{water} = 0.774$ $\Omega_{eg} = 0.226$

There are occasions when knowing the mixture volume or the volume occupied by the various components in the mixture is necessary. A continuation of the mass and molar analysis of a mixture requires the model of how each component behaves, whether an ideal gas, a real gas, a liquid, or whatever. Essentially, the relationships between mass and volume are necessary, and the density gives such information. If a component is not incompressible, however, the density may vary considerably, depending on the temperature or the pressure to which that component is subjected. As a beginning toward treating this problem, if the mixture is assumed to be composed entirely of components that behave as ideal gases, some convenient results may be obtained. Using the ideal gas model (equation 1–13) and assuming that each component is free to occupy the total volume of the mixture, the pressure of each such component will vary, depending upon its mole fraction; that is,

$$p_i = \Omega_i p_T \tag{1–46}$$

where p_T is the mixture pressure, or total pressure acting on the mixture. The component pressure, p_i, is called the partial pressure of the ith component. A corollary to equation (1–46) is that the sum of the partial pressures is equal to the total mixture pressure,

$$\sum p_i = p_T \tag{1–47}$$

The partial pressure relationships of equations (1–46) and (1–47) were suggested by John Dalton in the early 1800s, and the result is often called Dalton's Law of Partial Pressures. Dalton suggested that it be used to analyze mixtures that do not behave as ideal gases; however, care must be exercised and sophisticated techniques are often required to use the partial pressure as an accurate mixture property. We will use the partial pressure in the analysis of diffusional mass transfer processes.

A particularly useful analysis of mixtures involves air and water vapor. The air and the water vapor (steam) behave, for technological purposes, as ideal gases. The water vapor acts to make the air more humid, and the term *relative humidity*, β, derives from the ratio of the partial pressure of the water vapor in the air to the saturation pressure of water at the air temperature. Thus,

$$\beta = \frac{p_{iw}}{p_g} \tag{1–48}$$

and for instance air at 80°F and 14.7 psia with a relative humidity of 50% would have a partial pressure of water vapor of 0.5 p_g, where p_g is the particular saturation pressure of water at, in this case, 80°F. From appendix table B–6, a steam table of thermodynamic properties, this pressure is approximately 0.506 psia, so the partial pressure of the water vapor is 0.253 psia. This information about the partial pressure then allows one to determine the mole fraction and the mass fraction of water vapor in a particular sample of air. It is the information that the engineer and technologist needs to determine moisture migration and, among other problems, the predicted occurrence of moisture condensation in air; dew, rain, fog, snow, and other phenomena.

The *dew point*, T_{dp}, represents the temperature to which a particular air–water vapor mixture must be cooled to in order that the relative humidity be 100%, or the point at which no more water can be present in the mixture as steam. For air at 50% relative humidity, 80°F, and 14.7 psia, the dew point is about 60°F, which is the saturation temperature of water at the pressure 0.253 psia as given in a steam table.

The traditional term used to describe the actual amount of water vapor present in air is the *absolute humidity*, ω, which is defined as

$$\omega = \frac{m_w}{m_{da}} \tag{1–49}$$

where m_w is the actual mass of water in a sample of air and m_{da} is the mass of the dry air, without the humidity, in the same sample. Notice that this term is not the mass fraction of water vapor in air but rather is a larger value because m_{da} is necessarily less that m_T, the total air mass of the sample of air. A good approximation for ω may be gotten from the equation

$$\omega = 0.622\frac{p_{iw}}{p_{da}} \tag{1–50}$$

which can be derived from equation (1–49) by assuming an ideal gas mixture. The dew point, the relative humidity, and the absolute humidity are conveniently related on appendix chart C–1, called a *psychrometric chart*. The dew point can be readily determined for a particular air–water vapor mixture by reading the *dry-bulb temperature* (the air temperature) at a point where the absolute humidity is the same as the prescribed air condition but the relative humidity is 100%. This requires moving to the left horizontally to the intersection point of the 100% relative humidity curve. Condensation of water (such as dew or rain) occurs if the particular air is cooled below the dew point. Then the final state will still be at 100% relative humidity, at a low (relatively speaking) temperature, and at a lower absolute humidity than the original air. The arithmetic difference between the two absolute humidity values (original minus final) will be the water condensing out of the air *per unit mass of dry air* if the air is cooled to that final lower temperature. We will refer back to these concepts in Chapter 8.

EXAMPLE 1–5 Determine the amount of water condensing out of air that is at 20°C, 80% relative humidity, and is then cooled to 5°C. Express the answer as a rate, kg/hr, and assume that 100 m^3/hr of the air is moving.

Solution The original condition of the air can be described by reading the values from the psychrometric chart, Appendix Chart C–1: $\beta = 80\%$, $T = 20°C$, $\omega_1 = 11.7$ g/kg_{da}, $T_{dp} = 16.5°C$.

The final state is

$$\beta = 100\%,\ T = (T_{dp}) = 5°C,\ \omega_2 = 5.4\ \text{g/kg}_{da}.$$

Assuming the air flow is measured at the original state, the air density (per unit mass dry air) is found, assuming 100 kPa of atmospheric pressure. The partial pressure of the water vapor in air is nearly the same as its saturation pressure at 20°C. From appendix table B–6 this is 2.339 kPa. Thus,

$$\rho = \frac{p}{RT} = \frac{\left(100 - 2.339\,\dfrac{\text{kN}}{\text{m}^2}\right)}{\left(0.287\,\dfrac{\text{kJ}}{\text{kg}\cdot\text{K}}\right)(293\ \text{K})} = 1.161\,\frac{\text{kg}}{\text{m}^3}$$

The mass flow rate of dry air is then determined from

$$\dot{m}_{da} = \rho\dot{V} = \left(1.161\,\frac{\text{kg}}{\text{m}^3}\right)\left(100\,\frac{\text{m}^3}{\text{hr}}\right) = 116.1\,\frac{\text{kg}}{\text{hr}}$$

and the amount of water condensing out per unit time is

$$\dot{m}_w = \dot{m}_{da}(\omega_1 - \omega_2) \tag{1–51}$$

The water condensing out is then

Answer

$$\dot{m}_w = \left(116.1\,\frac{\text{kg}}{\text{hr}}\right)\left(11.7\,\frac{\text{g}}{\text{kg}} - 5.4\,\frac{\text{g}}{\text{kg}}\right) = 731.43\,\frac{\text{g}}{\text{hr}}$$

1–4 THE MODES OF HEAT TRANSFER

Heat transfer occurs because of a temperature difference perceived at a system boundary, and this heat transfer always proceeds from the higher to the lower temperature. The three models that have been used to describe the various manners or modes of heat transfer are *conduction, convection,* and *radiation.*

Conduction heat transfer is the normal transfer of energy within solids. Conduction heat transfer also occurs through gases and liquids, but it then only predominates if the gases or liquids are stagnant or move slowly. The mathematical model for conduction heat transfer is Fourier's Law of Conduction,

$$\dot{Q}_x = -\kappa A \frac{\partial T}{\partial x} \tag{1–52}$$

where Q_x is the heat transfer in the x direction crossing the normal area A due to the temperature gradient $\partial T/\partial x$. This mechanism is illustrated in figure 1–11. The *thermal conductivity,* κ, is an index of the ability of a material to conduct heat due to a temperature gradient in that material, and equation (1–52) may be interpreted as the defining equation for thermal conductivity. The units for κ are energy–length per unit time–temperature–area. In SI units thermal conductivity is usually expressed in W/m · K (or W/m · °C) and in English units in btu/hr · ft · °R (or Btu/hr · ft · °F). Since equation (1–52) involves a temperature gradient instead of a temperature, the units of K and °C, or °R and °F, may be interchanged, and in the literature both forms of units have been used to report values for thermal conductivity. In appendix tables B–2, B–3, and B–4 list nominal values of thermal conductivity for some selected materials.

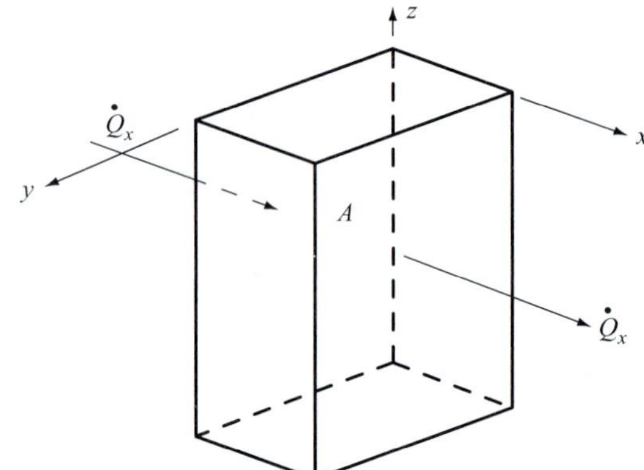

Figure 1–11 Conduction heat transfer in one direction.

Materials that conduct heat well, or rapidly, are called *conductors* and have high values of thermal conductivity; materials that do not conduct heat well are called *insulators* or *insulation* and have comparatively low values for thermal conductivity. For instance, Styrofoam is a good insulator or insulation material and it has a κ-value of approximately 0.029 W/m · K whereas copper, which is a good heat conductor, has a κ-value of approximately 400 W/m · K.

The range of κ-values extends over four orders of magnitude and even farther if more exotic materials are considered. It should be observed that these κ-values are only nominal values and could change significantly with temperature, pressure, or other conditions that may be out of the ordinary or other than listed in the table. In Chapter 2 we return to a further discussion of thermal conductivity and its potential variances.

Sometimes it is more convenient to consider the heat transfer per unit of area, $\dot{q}_{Ax}$, which is, from equation (1–52),

$$\dot{q}_{Ax} = \frac{\dot{Q}_x}{A} = -\kappa \frac{\partial T}{\partial x} \tag{1–53}$$

Notice that in equations (1–52) and (1–53) the heat transfer is defined in the x direction. For a regular three-dimensional coordinate system (such as Cartesian, cylindrical, or spherical coordinates) two additional equations will be necessary to define the heat transfer in the remaining two directions, y and z for instance, in the regular Cartesian system. The general form of Fourier's law is

$$\dot{q} = -\kappa \nabla T \tag{1–54}$$

where ∇ is the del or nabla operator and, in Cartesian coordinates,

$$\nabla T = \mathbf{i}\frac{\partial T}{\partial x} + \mathbf{j}\frac{\partial \mathbf{T}}{\partial \mathbf{y}} + \mathbf{k}\frac{\partial T}{\partial z} \tag{1–55}$$

where **i, j,** and **k** are called *unit vectors* in the x, y, and z directions, respectively, and signify that the del operation is a vector sum of three possible directions. Also, the thermal conductivity may vary with direction, and in Cartesian coordinates it will have three components: κ_x, κ_y, and κ_z. Thus, there are three components for conduction heat transfer in a material, and the total heat transfer can be treated as the vector sum of q_x, q_y, and q_z (in Cartesian coordinates), or

$$\dot{q} = \dot{q}_x + \dot{q}_y + \dot{q}_z \qquad \text{(vector sum)} \tag{1–56}$$

For one-directional heat transfer, or a material that has a temperature gradient in only one direction (say the x direction), at steady state equation (1–52) becomes

$$\dot{Q} = \dot{Q}_x = -\kappa A \frac{dT(x)}{dx} \tag{1–57}$$

If the area A and κ are constant and for a *steady state* (Q_x is constant) we then have

$$\dot{Q}_x = -\kappa A \frac{\Delta T}{\Delta x} \tag{1–58}$$

EXAMPLE 1–6 Determine the heat transfer per unit area through a sheet of 1-in.-thick plywood if one side of the plywood has a surface temperature of 100°F and the other side is 50°F.

Solution Assume that the plywood is in steady state and has constant thermal conductivity. Then we may use Equations (1–53) and (1–58). Referring to figure 1–12, we have $\Delta x = 1$ in. From

Figure 1–12 Heat flow by conduction through plywood. Example 1–6.

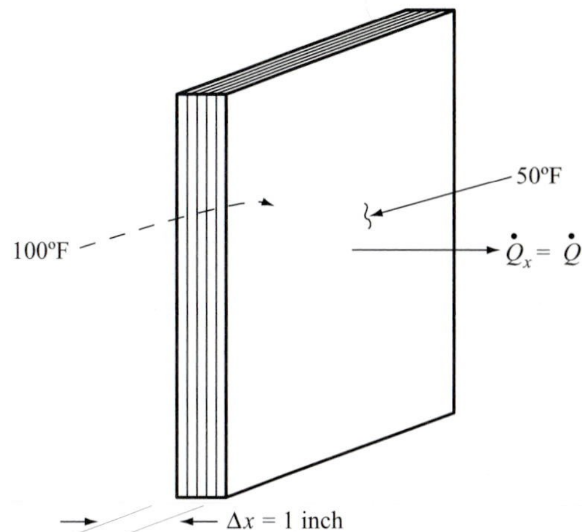

appendix table B–2 we find a thermal conductivity of 0.12 W/m · K for plywood. Using the conversion factor from appendix table B–1, $\kappa = 0.12 \times 6.9348 = 0.832$ Btu · in./hr · ft² · °F. Then

$$\dot{q}_A = \frac{\dot{Q}_x}{A} = -\kappa\frac{\Delta T}{\Delta x} = -\left(0.832\frac{\text{Btu}\cdot\text{in.}}{\text{hr}\cdot\text{ft}^2\cdot{}^\circ\text{F}}\right)\left(\frac{100^\circ\text{F} - 50^\circ\text{F}}{1\text{ in.}}\right)$$

Answer

$$\dot{q}_A = 41.6\frac{\text{Btu}}{\text{hr}\cdot\text{ft}^2}$$

The concept of conduction heat transfer will be extended and developed in the next two chapters using the basic ideas presented in this example.

Convection heat transfer is the mode of energy exchange that is associated with heat crossing a boundary or surface between a solid and a fluid. The fluid may be a gas or a liquid, and may be at low or high temperatures. Convection heat transfer is a macroscopic phenomenon and, when considered in detail, the model does not satisfy the thermodynamic definition of heat transfer. It is a model that is used with the idea of a boundary layer of a fluid at a solid surface so that there can be heat flow to (or from) the bulk fluid around the solid. Figure 1–13 shows a schematic of the phenomenon of convection heat transfer. The mathematical model for this mode of heat transfer is Newton's Law of Cooling,

$$\dot{Q} = hA(T_s - T_\infty) = hA\Delta T \qquad \textbf{(1–59)}$$

where T_s is the surface temperature, T_∞ is the ambient fluid temperature, and h is the convection heat transfer coefficient.

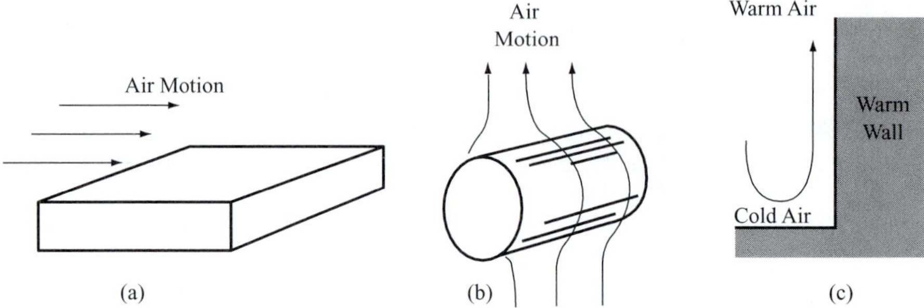

Figure 1–13 Examples of convection heat transfer: (a) between a flat plate and a fluid flowing past the plate, (b) between a cylinder or tube and a surrounding fluid, (c) between a vertical wall and natural convection of a fluid rising due to its density change.

The surface temperature is presumed to be higher than the ambient fluid temperature so that the surface is being "cooled"; however, the same equation may be used for "heating" of the surface when T_s is less than T_∞, and the ΔT term is then reversed, $\Delta T = (T_\infty - T_s)$. One should then call the equation Newton's Law of Heating. The simplicity of the equation masks the underlying complexity of convection as a model for heat transfer in the convection coefficient, h. If h is known or established to a reasonable degree, Newton's Law of Cooling/Heating is a simple and useful tool for engineering design. Unfortunately, the value for h under different and varying conditions is often hard to establish, and in Chapters 4 and 5 we discuss the various methods used for resolving this difficulty.

Convection heat transfer is usually classified as either *forced convection* or *free convection*. Free convection is also referred to as *natural convection* because it involves the convection of fluids that flow due solely to density changes induced by temperature changes. Without gravitational effects, or gravity, free convection is not possible. Forced convection, on the other hand, is present whenever a fluid is mechanically driven around a solid boundary (see figure 1–13a and b). Such effects as wind around building walls or around airplanes and autos are examples of forced convection conditions. The flow of a fluid through a ditch or channel or through a pipe or tube is also an example of forced convection. We will develop these concepts further in Chapter 4.

In table 1–4 are presented some approximate values for convection heat transfer coefficients for some common conditions. Again, recognize that the values in the table are only approximate values and should not be used as definitive results.

Table 1–4 Approximate Values for the Convection Heat Transfer Coefficient

Type of Convection	W/hr · m² · °C	Btu/hr · ft² · °F
Air at 6 ft/s (2 m/s) over a flat plate	12	2.1
Air at 100 ft/s (30 m/s) over a flat plate	75	13.2
Air flow past a 2-in. (5-cm)–diameter cylinder at 160 ft/s (50 m/s)	180	32
Water flow at 1 ft/s (0.3 m/s) in a 1-in. (2.5-cm)–diameter tube	3500	616
Natural Convection		
Vertical plate 1-ft (0.3-m) high in still air	4.5	0.79
Horizontal cylinder 2 in. (5 cm) in diameter in still air	6.5	1.14

EXAMPLE 1–7 A mercury-in-glass thermometer at 70°F is placed in water at 150°F. Estimate the rate of heat added to the thermometer per unit area if h is given as 160 Btu/hr · ft² · °F.

Solution Using Newton's Law of Cooling/Heating we can estimate the heat transfer per unit area as

$$\dot{q}_A = \frac{\dot{Q}}{A} = h(T_\infty - T_s) = \left(160 \frac{\text{Btu}}{\text{hr} \cdot \text{ft}^2 \cdot {}^\circ\text{F}}\right)(150 - 70^\circ\text{F})$$

Answer

$$= 12{,}800 \frac{\text{Btu}}{\text{hr} \cdot \text{ft}^2}$$

EXAMPLE 1–8 Estimate the heat transfer from a 4-m-high by 4-m-long vertical concrete retaining wall at 30°C into still air at 10°C.

Solution A rough approximate solution may be found by using table 1–4 to obtain a value for h of 4.5 W/m² · °C. Then, using Newton's Law of Cooling,

$$\dot{Q} = hA(T_w - T_\infty) = \left(4.5 \frac{\text{W}}{\text{m}^2 \cdot {}^\circ\text{C}}\right)(16\ \text{m}^2)(30 - 10^\circ\text{C})$$

Answer

$$= 1440\ \text{W}$$

The third model for describing heat transfer between systems is *radiation heat transfer*. This is the model that is used to explain observations of energy traveling through a distance without seeming to involve the intermediate volume. For instance, heat can be "felt" from an open fire even if one isn't directly at the fire. A fireplace transfers heat (energy) to a person or object some distance removed from the actual fire without heating (very well) the air in between the two. The solar energy coming from the sun, which we all feel and sense, has traveled some 32×10^6 km (93×10^6 mi) without depositing much energy in outer space between earth and the sun. Thus, heat can be transferred as an action over a distance, as opposed to intimate contact and transfer of energy such as we visualize for conduction and convection. Radiation heat transfer seems to operate best through a vacuum, and conduction cannot operate at all through that same vacuum. Radiation heat transfer is considered as a part of electromagnetic radiation that includes visible light, x-rays, γ-rays, ultraviolet, infrared, and even radio waves. It is customary to describe a *black body* as one that emits radiant heat described by the relationship

$$\dot{E}_b = A\sigma T^4 \qquad \text{(W or Btu/hr)} \tag{1–60}$$

where σ is the Stefan–Boltzmann Constant, 5.67×10^{-8} W/m$^2 \cdot$ K^4.

The temperature must be absolute, either Kelvin or degrees Rankine in equation (1–60) and in the subsequent developments and results from it. Then, if two *black bodies* radiate heat such that the smaller one can only "see" the other and not any other objects, such as a pan of water insulated from the ground and exposed to night sky, the net heat transfer between the two is given by

$$Q_{\text{net}} = A_1\sigma(T_1^4 - T_2^4) \tag{1–61}$$

Here the smaller body, 1, is assumed to have the higher temperature, but the same form of the equation can be used for radiation *to* the smaller body if the temperature terms are reversed. Thus, again we seek to describe heat as a positive quantity when flowing from hot to cold and not allow for negative heat transfers. Equations (1–60) and (1–61) are restrictive, and more general developments of the radiation concept are given in Chapters 6 and 7.

EXAMPLE 1–9 A metal casting at a temperature of 20°C is placed inside a heat-treat furnace. The furnace walls are at a temperature of 400°C and may be assumed to behave as a black body. If the casting is assumed to act as a black body and has an effective surface area of 0.4 m^2, determine the radiant and total heat transfer to the casting.

Solution The radiant heat transfer is the net amount to the casting, given by equation (1–61):

$$\dot{Q}_{\text{net}} = (0.4\ \text{m}^2)\left(5.67 \times 10^{-8}\frac{\text{W}}{\text{m}^2 \cdot \text{K}^4}\right)(673^4 - 293^4)$$

Answer $= 4.486$ kW

The total heat transfer is the amount emitted by the heat-treat furnace that reaches the casting. For the black body furnace walls this is given by equation (1–60):

$$\dot{Q} = E_b = A\sigma T^4 = (0.4\ \text{m}^2)\left(5.67 \times 10^{-8}\frac{\text{W}}{\text{m}^2 \cdot \text{K}^4}\right)(673^4)$$

Answer $= 4.653$ kW

Any real exchange of energy between two or more systems or surfaces involves more than just one of the three models of heat transfer which we have just considered, and the student should always be reminded of such complications. Yet, design and analysis of technological systems and an understanding and appreciation of the myriad of physical phenomena must begin with simple models. Heat transfer can be approached with the three models we have just considered, and in the following chapters they will all be developed further.

1–5 THE MODES OF MASS TRANSFER

Mass transfer occurs when a material crosses a system boundary. In general terms even the flow of a fluid through a pipe or tube could be considered to be mass transfer; however, the concept of mass transfer is usually considered in the context of mixtures of substances that change their mass or molar analysis (i.e., the fractions of the mixture components change). Also, for our purposes mass transfer will be limited to nonreacting mixtures and to the methods whereby a substance will migrate into or out of a mixture. There have been a number of suggested mechanisms to account for mass transfer, and the following have been compiled by Bird [2]:

1. Diffusion of substance caused by a concentration or density gradient of that substance at a boundary.
2. Diffusion due to a temperature gradient at a boundary.
3. Pressure diffusion due to a hydrostatic pressure gradient at a boundary.
4. Diffusion due to force fields acting on the mixture.
5. Mass transfer caused by forced convection (see section 1–4).
6. Mass transfer caused by free convection (see section 1–4).
7. Mass transfer caused by turbulence in a bulk fluid flow.
8. Interphase mass transfer or change of phase caused by a nonequilibrium condition at an interface of two or more phases of the material.

For the purposes of this text, we will consider only diffusion (item 1) and mass transfer caused by forced and free convection (items 5 and 6). Often we will refer to diffusion as mass transfer; however, you will need to recognize that diffusion may be mass transfer due to forced or free convection.

Examples of diffusion are to be found all around us. The evaporation of water into air, the migration of water or moisture through a concrete retaining wall, the vaporization of a flammable fuel in air such as in the combustion chamber of an auto engine or in the combustor of a gas turbine engine, the application of a preservative on a wood or metal surface, and the mixing of chlorine or other chemicals into water are instances where mass transfers are the crucial phenomena occurring. In all instances, the diffusion occurs because of a concentration gradient in the system and at some prescribed boundary.

A demonstration of the migration of a substance from a region of high concentration to one of low concentration can be carried out by placing a few crystals of potassium permanganate, $KMnO_4$, in a beaker of water. The $KMnO_4$ will begin to dissolve in the water and produce a dark purple color. This is the concentrated $KMnO_4$, and the color will fade out away from the crystals. The range of color then is from dark purple where the $KMnO_4$ is most concentrated to light and finally colorless where there is less and less concentrate of $KMnO_4$. The $KMnO_4$ color will slowly move away from the region where the crystals were placed, and this indicates that the concentration gradient drives the $KMnO_4$ to the region of low concentration. Fick suggested that diffusion of one substance A into another distinct substance B to produce a two-component mixture can be predicted by the equation

$$\dot{m}_{Ax} = -\rho \wp_{AB} \frac{\partial \psi_A}{\partial x} \tag{1–62}$$

where m_{Ax} is the mass migration of a substance A into a mixture of A and B per unit area, $kg/s \cdot m^2$ or $lbm/s \cdot ft^2$. The density, ρ, is defined as $\rho = \rho_A + \rho_B$ = total density of the mixture, and $\wp_{AB}$ is the diffusivity of a binary system or mixture. By definition, $\wp_{AB} = \wp_{BA}$. ψ_A is the mass fraction of substance A in the mixture, and the gradient $\partial\psi_A/\partial x$ is specified at a boundary where mass transfer is occurring. As we noted before, there will be other gradients or differences at the boundary, such as pressure and temperature, that may induce mass transfer, but consideration of those are beyond our purposes here.

The diffusivity, $\wp_{AB}$, is experimentally determined and also predicted, for certain substances, by kinetic theories of gases and liquids and other means [3]. In table 1–5 are given some values for the diffusivities of various types of substances diffusing into mixtures. The values for diffusivity extend over a wide range, depending on the phases of the migrating mass and the mixture. In Chapter 8 we will consider further the prediction of diffusivity and other parameters used to analyze mass transfer.

Table 1–5 Selected Experimental Values for Diffusivities

System	T (°C)	Diffusivity, $\wp_{AB}$ (cm^2/s)
Ammonia into air	20	0.236
Ar into O_2	20	0.20
CO_2 in N_2	25	0.165
Water (vapor) into air	20	0.256
Ethanol into water	25	0.90×10^{-5}*
He in SiO_2	20	2.4×10^{-10}
Hg in Pb	20	2.5×10^{-15}
Al in Cu	20	1.3×10^{-30}

Data abstracted from: Hirschfelder, J. O., C. F. Curtiss, and R. B. Bird, *Molecular Theory of Gases and Liquids,* Wiley, New York, 1954, p. 579; Johnson, P. A., and A. L. Babb, "Liquid Diffusion in Non-Electrolytes," *Chem. Rev.,* 56, 387–453, 1956; and Barrer, R. M., Diffusion in and Through Solids, Macmillan, New York, 1941.

*50% mole concentration of water–ethanol mixture.

EXAMPLE 1–10 Estimate the rate of evaporation of dry ice (solid CO_2) into air at 20°C if the mole fraction of CO_2 is 25% at 1 cm from the solid dry ice surface. Assume a diffusivity of 0.08 cm^2/s of CO_2 (vapor) in air at 20°C.

Solution The dry ice will sublimate, or have a solid to vapor phase change. Then the vapor CO_2 will migrate into the air and as a first approximation we can use Fick's law, equation 1–62, and assume that the CO_2 is an ideal gas when diffusing into air. The density for CO_2 then determined from the relationship

$$\rho_A = \frac{p_A}{RT} = \frac{\Omega_A p_T}{RT} = \frac{(0.25)\left(101\dfrac{\text{kN}}{\text{m}^2}\right)\left(44\dfrac{\text{kg}}{\text{kg}\cdot\text{mol}}\right)}{\left(8.315\dfrac{\text{kJ}}{\text{kg}\cdot\text{mol}}\right)(293\text{K})} = 0.456\frac{\text{kg}}{\text{m}^3}$$

and the diffusion is, using the mole fraction difference 0.25 − 1.00 over 1 cm as the concentration gradient,

Answer

$$\dot{m} = -\left(0.456\frac{\text{kg}}{\text{m}^3}\right)\left(0.08\frac{\text{cm}^2}{\text{s}}\right)\frac{(0.25 - 1.00)}{1\text{ cm}} = 0.0002736\frac{\text{kg}}{\text{m}^2 - \text{s}}$$

1–6 MATHEMATICAL PRELIMINARIES

The reader is presumed to have an understanding of the concepts of calculus through differential equations. Heat- and mass-transfer analyses often rely on differential equations to express general phenomena and to obtain solutions to particular problems.

As a short review let us consider a differential equation of independent variable x and dependent variable $y = y(x)$;

$$a_n\frac{d^ny}{dx^n} + a_{n-1}\frac{d^{n-1}y}{dy^{n-1}} + \cdots + a_1y(x) + a_0 = 0 \tag{1–61}$$

This is a linear, homogeneous ordinary differential equation of nth degree with constant coefficients. It is linear because the differentials of $y(x)$ are in elementary form; that is, there are no terms such as $(d^ny/dx^n)^m$ or x^nd^ny/dx^n. It is homogeneous because the right side of the equation $(=0)$ is not a function of x. It is ordinary because there are no partial derivatives and it is of nth degree because the highest order of the derivatives is n. In heat and mass transfer it turns out that a first- or second-order equation is most often used, and orders higher than two are rare. Also, the coefficients, $a_n, a_{n-1}, \ldots, a_1$, and a_0 are assumed to be constants.

The solution to such equations as (1–61) can be obtained using one of many different techniques: separation of variables, substitution, Fourier analysis, Laplace transforms, and others. For engineering analysis the crucial element in obtaining a useful and specific solution to a differential equation is the establishment of boundary values or boundary conditions (B.C.'s). A closed form solution to an nth-order ordinary differential equation can only be obtained if n boundary conditions are known. A boundary condition is the value of $y(x)$ at a prescribed boundary, or prescribed x. It could also be the value of one of the derivatives of $y(x)$ at a prescribed x. Finally, the boundary condition could be a separate external function that provides information of $y(x)$ or one or more of its derivatives at some boundary.

A common example of this, as we will see in later chapters, is the use of Newton's Law of Cooling/Heating at a solid surface or boundary of a heat-conducting material. Here the heat transfers at the boundary will be given by two separate modes or equations: convection and conduction. The governing equation for the conduction system is then said to be coupled to the surroundings. Coupling is used, in principle, to mathematically analyze two or more systems that physically interact through heat and mass transfer. It is used routinely in numerical and approximation methods, as we will see later.

Often in heat and mass transfer the properties, the heat transfer, or the mass transfer change with time. The problems are then called transient or non–steady state, as opposed to steady state problems, as indicated by equation (1–63). The properties are then functions of time, the general form of a differential equation of $y(t)$ is

$$b_n\frac{d^ny}{dt^n} + b_{n-1}\frac{d^{n-1}y}{dt^{n-1}} + \cdots + b_1\frac{dy}{dt} + b_0 = 0 \qquad \textbf{(1–64)}$$

and the solution to this equation is obtained after establishing specific values for $y(t)$ at known times, t. In heat and mass transfer only first-order differential equations in time are normally considered, so only one specified condition needs to be established with time. The initial condition (I.C.) is the most common that is prescribed, and it is the one that identifies $y(t)$ at $t = 0$, the initial condition. In principle, if the initial condition is not specified, another condition with respect to a positive nonzero time value may be established. The procedure is called explicit when the initial condition is specified, and implicit when another condition is specified. In numerical methods this distinction arises, and it can lead to different results, depending on the procedure.

The general heat- or mass-transfer problem involves three-dimensional space, so the differential equation (1–63) is a restricted, one-dimensional statement of the general case. Use of standard vector notation is a convenience in expressing the three-dimensional heat or mass transfer problems. Now, however, the differential equation is a partial differential equation and is written,

$$b_{ni}\nabla^n y_n + b_{(n-1)i}\nabla^{(n-1)}y_{(n-1)i} + \cdots + b_i\nabla y_i + b_{0i} = 0 \qquad \textbf{(1–65)}$$

where the operator, ∇, is the *del operator* introduced earlier, in equations (1–54) and (1–55):

$$\nabla = \mathbf{i}\frac{\partial}{\partial x} + \mathbf{j}\frac{\partial}{\partial y} + \mathbf{k}\frac{\partial}{\partial z} \qquad \textbf{(1–66)}$$

in Cartesian coordinates. The terms **i, j,** and **k** are unit vectors in the x, y, and z directions respectively. The special case of the del operator, ∇^2, is called the *Laplacian* and is

$$\nabla^2 = \frac{\partial^2}{\partial x^2} + \frac{\partial^2}{\partial y^2} + \frac{\partial^2}{\partial z^2} \tag{1–67}$$

Equation (1–65) is a more formidable equation than (1–63) and, recalling differential calculus, the solution to partial derivatives involves unknown functions of those variables held constant. A convenient method often used to avoid the problem of unknown functions is the *separation of variables,* which is based on the premise that a function of more than one variable, $F(x, y, z)$, can be written

$$F(x, y, z) = X(x) \cdot Y(y) \cdot Z(z) \tag{1–68}$$

It turns out that a partial differential equation such as that indicated in equation (1–65) for $F(x, y, z)$ can be reduced to a set of ordinary differential equations. For $F(x, y, z)$ of equation (1–68) in equation (1–65) the partial differential equation reduces to, at most, three ordinary differential equations: one with respect to x, one with respect to y, and one with respect to z. The crucial step in the method is the assumption of equation (1–68), and there should be no surprise when a final solution to $F(x, y, z)$ is in the form of equation (1–68).

For transient multidimensional heat- or mass-transfer problems, the fourth dimension, time, needs to be introduced, and then the separation of variables needs to be applied once more to give a functional form

$$F(x, y, z, t) = X(x) \cdot Y(y) \cdot Z(z) \cdot T(t) \tag{1–69}$$

The mathematical manipulations are escalated one step, without any additional mathematical profundity. The reader may recognize that a general mathematical statement, a *mathematical formulation*, in heat and mass transfer can usually begin with a second-order differential equation of the form

$$a\nabla^2 F(x, y, z, t) + b\nabla F(x, y, z, t) + c = 0 \tag{1–70}$$

with appropriate boundary conditions (B.C.'s) and initial conditions (I.C.'s). If all terms of equation (1–70) are included in an analysis and the indicated parameters are nonzero, then there will need to be six B.C.'s (two each of x, y, and z) and one initial or other time condition.

An attempt is often made to avoid such situations in normal engineering analysis by using orders of magnitude approximations. Thus, for instance, if moisture is migrating through a basement wall, the migration parallel to the wall might realistically be neglected and then the mass transfer problem becomes unidirectional and at most two-dimensional (in x and time). In this text the reader will see demonstrations of the use of mathematical formulations to address heat- and mass-transfer problems.

The student should recognize that the use of constant coefficients to describe the physical phenomena carried on in heat and mass transfer is probably a very crude approximation. In future years it may even be found that the use of constant coefficients (which means constant properties) is unacceptable because many of the system properties that impact on heat and mass transfer are strongly dependent on the variables of time, position, temperature, and others. Thus, in the future nonlinear and heterogenous (as opposed to homogeneous) differential equations will become more a part of engineering analyses.

Also, in future years more efforts will be made to better determine experimentally and analytically the diffusional and thermal properties of materials in order to address the concerns of accurate descriptions of heat and mass transfer. At various parts of this text the determination of thermal and diffusional properties are discussed.

1–7 ENGINEERING ANALYSIS OF HEAT AND MASS TRANSFER

Many technological and societal demands require that engineering analysis include heat and mass transfer. As we discussed in section 1–1, an understanding of heat- and mass-transfer phenomena will be needed to properly explain and to predict many physical processes. The subject can be rationally approached if we reflect on the types of problems and solutions to be encountered. In heat transfer there are essentially two classes of problems:

1. Given a temperature distribution, determine the heat transfer.
2. Given a rate of heating or of a heat transfer, determine the expected temperature distribution.

Steady state problems will be independent of time, and transient or non–steady state ones will be sensitive to time or a history of the situations and systems.

As we saw in section 1–6, to solve either a class 1 or a class 2 problem, boundary conditions and a time condition must be prescribed. Notice also that if heat transfer is by conduction, the temperature distributions will be for the conducting materials, if by convection the temperature distributions will be those of a flowing fluid or gas around some prescribed boundary, and if by radiation the temperature distributions will be surface temperatures of interacting bodies. As we will see later, radiation heat transfer can also account for differences in temperatures within particular media, such as gases and absorbing materials.

Diffusional mass transfer problems can also be classified into two classes:

1. Given concentration gradients, to determine the mass transfer.
2. Given the mass migration or mass transfer, to determine the various concentration gradients.

Here again, as with heat transfer concentration gradients or mass transfers do not change with time in the steady state problems and transient problems do involve time as an independent variable.

Mass transfers due to forced or free convection can also be analyzed by the classification method, but here the complications come thicker and faster, because fluid flow is involved and often turbulent, random conditions exist. Order-of-magnitude approximations are required to come to a reasonable formulation that can lead to tractable solutions.

1–8 SUMMARY

Heat transfer is a rate of energy transfer across a system boundary and is caused by a temperature difference that is received at the boundary. The three modes of heat transfer are conduction, convection, and radiation. Conduction heat transfer occurs in solids and wherever fluids are stagnant. It is the transfer of energy through molecular interactions at close range. The general equation used to describe conduction heat transfer is Fourier's Law of Conduction,

$$\dot{q}_A = -\mathbf{i}\kappa_x\frac{\partial T(x, y, z)}{\partial x} - \mathbf{j}\kappa_y\frac{\partial T(x, y, z)}{\partial y} - \mathbf{k}\kappa_z\frac{\partial T(x, y, z)}{\partial z} \tag{1–54}$$

and the one-dimensional, x-direction condition is

$$\dot{Q} = -\kappa_x A_x\frac{dT(x)}{dx} \tag{1–57}$$

For the situation when thermal conductivity, κ, the area, A_x, and the heat transfer are constant and one-dimensional,

$$\dot{Q} = -\kappa A\frac{\Delta T}{\Delta x} \tag{1–58}$$

Convection heat transfer is described as energy transfer between a solid surface and a fluid surrounding or flowing past that surface. This mode of heat transfer is described by Newton's Law of Cooling/Heating,

$$\dot{Q} = hA(\Delta T) \tag{1–59}$$

where ΔT is the temperature difference between the surface of the solid and the free stream or ambient, average surrounding temperature of the fluid. The area, A, is that of the surface and h is the convective heat transfer coefficient. Radiation heat transfer is the exchange of energy between two surfaces through thermo-electromagnetic radiation. A black body is a body or system having surfaces that emit thermal radiation in amounts predicted by the equation

$$\dot{E}_b = A\sigma T^4 \tag{1–60}$$

where T is the surface temperature in Kelvin (K) or degrees Rankine (°R) and σ is the Stefan–Boltzmann constant, $5.64 \times 10^{-8}\ \text{W/m}^2 \cdot \text{K}^4$. The net heat transfer between two black bodies, in which the smaller one can only "see" the larger one and no other surfaces, is given by

$$\dot{Q} = A_1\sigma(T_1^4 - T_2^4) \tag{1–61}$$

Mass transfer is the migration of mass from one location to another. The common interpretation of mass transfer is the condition in which one substance migrates into a volume occupied by different substance and thus creates a mixture. This mechanism is often called *diffusion,* and the mechanisms of mass transfer or diffusion can be listed as

1. Diffusion of a substance caused by a concentration or density gradient of that substance at a boundary.
2. Diffusion due to a temperature gradient at a boundary.
3. Pressure diffusion due to a hydrostatic pressure gradient at a boundary.
4. Diffusion due to force fields acting on the mixture.
5. Mass transfer caused by forced convection.
6. Mass transfer caused by free convection.
7. Mass transfer caused by turbulence in a bulk fluid flow.
8. Interphase mass transfer or change of phase caused by a nonequilibrium condition at an interface of two or more phases of the substance.

DISCUSSION QUESTIONS

Section 1–3

1–1 What is meant by a system? By a body?

1–2 What is the importance of a boundary in heat and mass transfer?

1–3 What is the difference between a boundary and a surface?

1–4 What is meant by system properties?

1–5 What is steady state?

1–6 What is a control volume?

1–7 What is heat transfer?

1–8 What is meant by reversible heat transfer?

Section 1–4

1–9 What are the three modes of heat transfer?

1–10 What is significant about the temperature difference when considering radiation heat transfer?

1–11 What mode of heat transfer does not satisfy the thermodynamic definition of heat transfer and why?

Section 1–5

1–12 What are the causes for diffusional mass transfer?

1–13 What is the difference between a concentration gradient and a density gradient?

PRACTICE PROBLEMS

Problems marked with an asterisk (*) may be more challenging than others.

Section 1–3

1–1 Ten lbm of CO_2 gas are cooled from 150°F to 80°F. If the specific heat, c_v, is given by

$$c_v = 0.323 - \frac{148}{T} + \frac{32{,}045}{T^2}$$

determine the internal energy change for the cooling process.

1–2 Two kg of water at 15°C are heated to 300°C at 101 kPa of pressure. Determine the amount of heat added during this process. Assume c_p is 4.18 kJ/kg · K for liquid water and 1.86 kJ/kg · K for steam. Also assume a value of 2258 kJ/kg for the latent heat of vaporization for water at 101 kPa.

1–3 One hundred lbm of air in a pressure tank are cooled from 180°F and 160 psia to 80°F when the surroundings are at 70°F. Determine the pressure at 80°F.

1–4 Steam is heated from 4 MPa, 480°C to 4 MPa, 640°C. Determine the heat added to the steam.

1–5 There are 300 kg of water heated from 20°C to 80°C. Determine the enthalpy change and the heat, Q.

1–6 Mercury is heated from 100°F to a vapor at 800°F. Assume hn_{fg} is 122 Btu/lbm at 676°F and use a c_p of 0.032 Btu/lbm · °F for liquid mercury and 0.015 for vapor. Determine the enthalpy change per unit mass and the heat per unit mass.

1–7 A refrigerator condenser is a heat exchanger that converts a refrigerant from a vapor to a liquid at constant pressure and approximately constant temperature. If ammonia is used as a refrigerant and 10 kg/s is condensed at 25°C when the surrounding temperature is 20°C, determine the heat transfer. (*Hint*: latent heat = hn_{fg} = heat of condensation.)

1–8 Ten m^3/s of air at 0°C, 80% relative humidity is heated to 25°C. Determine the amount of heat transfer required and the final relative humidity of the air. Determine the amount of water required to increase the final relative humidity to 50% at 25°C. How much additional heat is required to accomplish this humidification?

1–9 Air at 100°F, 70% relative humidity is to be conditioned to 75°F and 60% relative humidity. Determine the partial pressure of the water vapor in the air at both states, the amount of water removed per pound mass of dry air, and the lowest temperature to which the air must be cooled to accomplish this conditioning process.

1–10 One hundred kg of dry air is humidified by mixing with 1 kg of steam at 30°C. Determine the partial pressure of the dry air and the steam if the total pressure is 100 kPa.

***1–11** If for a real substance the enthalpy is a function of temperature and pressure, write an integral equation that expresses the enthalpy change per unit mass of the substance when the temperature and pressure change from T_1 to T_2 and p_1 to p_2 respectively. (*Hint:* Recall the total differential of a continuous function of two independent variables.)

Section 1–4

1–12 A styrofoam ice chest is 2 cm thick. If the inside of the chest is at 0°C and outside the chest it is 25°C, estimate the heat transfer by conduction through the styrofoam per unit area.

1–13 A concrete nuclear reactor containment wall has a temperature distribution, $T(x)$, given by the equation

$$T(x) = 800°\text{F} - 400\frac{°\text{F}}{\text{ft}^2}x^2 \qquad (x \text{ in feet})$$

Estimate the heat transfer through the wall per unit of area due to conduction at the center of the wall, where $x = 0.5$ ft.

1–14 A large 8-ft × 4-ft thermopane glass window loses 300 Btu/hr of heat when the inside temperature is 70°F. If the thermopane glass has an average thermal conductivity of 0.032 Btu/hr · ft · °F and is half an inch thick, estimate what the outside temperature of the window is.

1–15 A cast-iron frying pan 1/4 in. thick is used to prepare some food on a stove top. If the lower surface of the pan is at 600°F and the upper surface is 500°F, estimate the heat transfer through the pan per unit area.

1–16 Wind blows at 50 m/s around a 5-cm-diameter electric power line when the air temperature is −10°C. Estimate the heat loss of the power line if the surface temperature of the line is 5°C.

1–17 Water at 60°F flows through a copper tube of 1 in. inside diameter (ID) at approximately 1 ft/s. Determine the heat transfer to the water per foot of tube length if the inside surface temperature of the tube is 180°F.

1–18 An automobile is moving on a highway at 100 km/hr. Estimate the heat loss per unit area from the auto's roof, which is at 30°C, if the air temperature is 15°C.

1–19 Estimate the heat loss per unit area of a vertical south facing wall of a large office building when the air temperature is −10°F and the wall is at 5°F.

1–20 A well-clothed person walks into a large auditorium that is empty. If the auditorium walls are at an average temperature of 55°F and the average surface temperature of the person's clothes is 85°F, estimate the net radiation heat transfer between the person and the auditorium. An average person can be assumed to have a surface area of 19.4 ft^2 (1.8 m^2).

1–21 A mercury-in-glass thermometer reads an outdoor temperature of 20°C. If the sky and surroundings of the thermometer have an average surface temperature of 5°, estimate the net radiation from or to the thermometer.

1–22 The surface of the sun seems to be about 10,000°F. What would you guess the rate of heat emission from the sun to be for 1 sq. ft of area?

1–23 A radiation pyrometer is a device utilizing radiant heat to measure the temperature of a surface. Assume that a pyrometer has a surface area of 5 cm^2 and is at 20°C when directed toward a furnace opening having a temperature of 1100°C. Estimate the net rate of heat transfer toward the pyrometer if black body radiation is assumed.

Section 1–5

1–24 Predict the concentration of ammonia in air 5 mm from an interface of vapor ammonia and air if the interface area ia 1500 mm^2 and the evaporation rate of ammonia is found to be 0.002 g/min. Assume that the ammonia and air are at 20°C.

1–25 Liquid mercury is contained in a lead beaker as shown in figure 1–14. Estimate the amount of mercury that migrates by diffusion into the beaker after 48 hr if the concentration of mercury is 2% at a distance 0.01 cm into the beaker wall from the inside surface. Assume that the system is at 20°C and neglect evaporation to air.

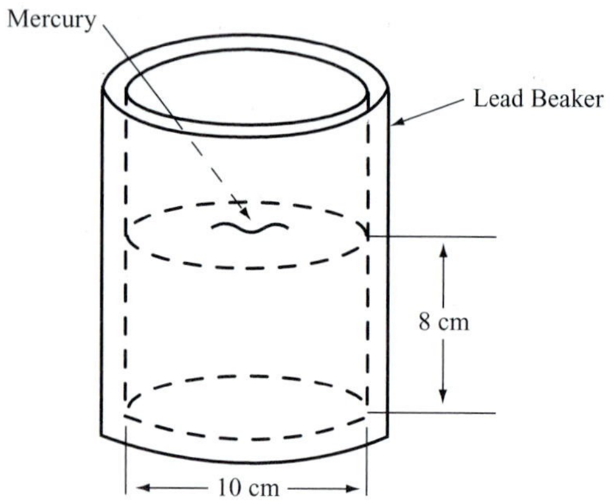

Figure 1–14 Beaker of problem 1–25.

1–26 A 4-ft-radius spherical container holds helium gas at 20 psia and 95°F. Estimate the amount of helium lost through diffusion in 24 hr if the sphere is made of SiO_2 (glass) and if it is assumed that at a point 1/8 in. into the container wall from the inside the concentration of helium is zero.

1–27 Water in a tightly closed flask evaporates at a rate of 0.00415 lbm/hr. Estimate the specific humidity of the air 1 in. above the liquid water–air surface if the dry-bulb temperature is 70°F. The water–air surface area is $3in^2$.

Section 1–6

Determine the number of boundary and/or initial or time conditions required to obtain a complete definite solution to the indicated differential equations.

1–28 The equation of two-dimensional conduction heat transfer with internal thermal energy generation,

$$\frac{\partial^2 T(x, y, t)}{\partial x^2} + \frac{\partial^2 T(x, y, t)}{\partial y^2} + \frac{g}{\kappa} = \left(\frac{1}{\alpha}\right)\frac{\partial T(x, y, t)}{\partial t}$$

1–29 An equation for steady state conduction heat transfer in one-dimensional cylindrical coordinates,

$$\frac{d^2 T(r)}{dr^2} + \left(\frac{1}{r}\right)\frac{dT(r)}{dr} = 0$$

1–30 A particular energy balance for two-dimensional convection heat transfer,

$$\rho c_p\left(u\frac{\partial T(x, y, t)}{\partial x} + v\frac{\partial T(x, y, t)}{\partial y}\right) = \kappa\frac{\partial^2 T(x, y, t)}{\partial y^2}$$

1–31 A particular momentum balance for fluid flow in the x-direction of Cartesian (x and y) coordinates with viscous and pressure forces,

$$\rho\left(u\frac{\partial u(x, y)}{\partial x} + v\frac{\partial v(x, y)}{\partial y}\right) = -\frac{\partial p}{\partial x} + \mu^2\frac{\partial^2 u(x, y)}{\partial y^2}$$

REFERENCES

The interested reader may want to investigate other publications for further discussions of those topics in this chapter. The following listing of references includes those publications that were used directly in writing this text and also some suggested advanced treatises. This is not a bibliography and is not intended to be a complete listing of applicable literature on the subjects.

Preface

1. Jakob, M., and G. A. Hawkins, *Elements of Heat Transfer*, 3rd edition, John Wiley & Sons, New York, 1957.
2. Incropera, F. P., and D. P. DeWitt, *Fundamentals of Heat and Mass Transfer,* John Wiley & Sons, New York, 1990.
3. Kreith, F., and M. S. Bohn, *Principles of Heat Transfer,* PWS, Boston, 1997.

Section 1–2

1. *Metric Practice Guide*, E380–74, American Society for Testing and Materials, Philadelphia, PA, 1974.

Section 1–3

1. Sonntag, R. E., and G. J. Van Wylen, *Introduction to Thermodynamics*, 3rd edition, John Wiley & Sons, New York, 1985.
2. Hatsopoulos, G. N., and J. H. Keenan, *Principles of General Thermodynamics*, John Wiley & Sons, New York, 1965.
3. Bejan, A., *Advanced Engineering Thermodynamics,* John Wiley & Sons, New York, 1988.
4. Moran, M. J., and H. N. Shapiro, *Fundamentals of Engineering Thermodynamics*, 3rd edition, John Wiley & Sons, New York, 1995.
5. Eringen, A. C., *Mechanics of Continua,* John Wiley & Sons, New York, 1967.
6. Truesdell, C., *Rational Thermodynamics*, McGraw-Hill, New York, 1969.

Section 1–4

1. Holman, J. P., *Heat Transfer*, 8th edition, McGraw-Hill, New York, 1995.
2. Brown, A. I., and S. M. Marco, *Introduction to Heat Transfer*, 3rd edition, McGraw-Hill, New York, 1958.
3. Jakob, M., and G. A. Hawkins, *Elements of Heat Transfer,* 3rd edition, John Wiley & Sons, New York, 1957.

Section 1–5

1. Jakob, M., and G. A. Hawkins, *Elements of Heat Transfer*, 3rd edition, John Wiley & Sons, New York, 1957.
2. Bird, R. B., W. E. Stewart, and E. N. Lightfoot, *Transport Phenomena*, John Wiley & Sons, New York, 1960.
3. Hirschfelder, J. O., C. F. Curtiss, and R. G. Bird, *Molecular Theory of Gases and Liquids*, John Wiley & Sons, New York, 1954.

Section 1–6

1. Wolf, H., *Heat Transfer*, Harper and Row, 1983.
2. Hildebrand, F. B., *Methods of Applied Mathematics*, 2nd edition, Prentice-Hall, 1965.
3. Jeffreys, H., *Cartesian Tensors*, Cambridge University Press, Cambridge, UK, 1965.

Section 1–7

1. Davies, G. A. O. (editor), *Mathematical Methods in Engineering*, Wiley Interscience, New York, 1984.
2. Kay, J. M., *An Introduction to Fluid Mechanics and Heat Transfer*, 2nd edition, Cambridge University Press, Cambridge, UK, 1968.

STEADY STATE CONDUCTION HEAT TRANSFER

2

In this chapter the reader is presented with the general problem of conduction heat transfer, beginning with a discussion of Fourier's Law of Conduction and thermal conductivity. Some of the methods of experimentally determining thermal conductivity are mentioned, and some analytical expressions for the dependency of thermal conductivity on temperature and other variables are presented. Then the general heat equation is derived from energy principles and the steady state condition is applied.

One-dimensional conduction heat transfer is considered at some length and the concepts of thermal resistance, electrical equivalances, and R-values are introduced. Steady state two-dimensional heat transfer is then developed, and some of the classical problems with their solutions are presented. The shape factor as an efficient tool for analysis is then developed and used in some applications. Graphical methods of analyzing two-dimensional problems are given and then numerical methods are presented. The use of finite differences is demonstrated and the finite element method is mentioned. Finally, some extended problems are presented to demonstrate the range of applications for conduction heat transfer, including some of the methods used to determine thermal properties.

New Terms

E_b bulk modulus of elasticity
$\dot{E}_{gen}$ energy generation
$\dot{e}_{gen}$ energy generation per unit volume
mm mass of one molecule
$\dot{Q}_r$ radial heat transfer
$\dot{q}_L$ heat transfer per unit length
R_T thermal resistance
R_{TL} thermal resistance per unit length
R_V thermal resistivity, R-value
R_{VL} radial thermal resistivity
r_{oc} critical radius of insulation
x_m mean distance between molecules
Γ molar diameter
$\mathcal{E}$ electrical potential
$\mathcal{I}$ electrical current
$\mathcal{R}_e$ electrical resistance per unit length
Lz Lorentz number
$\mathcal{C}$ electrical conductivity
ε_{fin} fin effectiveness

2–1 FOURIER'S LAW AND THERMAL CONDUCTIVITY

Heat transfer by conduction can be viewed as energy transfer due to molecular energy exchange. Conduction heat transfer requires a comparatively dense material to be effective, and solids are most common for this phenomenon. Even then, there are causes other than density that allow or retard conduction, such as the readiness for molecules to share energy with neighboring molecules. The engineer is primarily interested in conduction heat transfer at the macroscopic level, where temperature gradients at a boundary seem to be the parameter that most affects the heat transfer. We saw in Chapter 1 that Fourier's Law of Conduction is used to describe these actions through the equation

$$\dot{q} = -\kappa \nabla T \qquad \textbf{(1–54, repeated)}$$

where $\nabla T = \mathbf{i}\dfrac{\partial T}{\partial x} + \mathbf{j}\dfrac{\partial t}{\partial y} + \mathbf{k}\dfrac{\partial T}{\partial z}$ in Cartesian coordinates and **i, j,** and **k** are the unit vectors in the directions x, y, and z respectively. The conducted heat, $\dot{q}$, of equation (1–54) can be visualized as the resultant vector of three vector components, $\dot{q}_x$, $\dot{q}_y$, and $\dot{q}_z$, each which may be written

$$\dot{q}_x = -\kappa_x \frac{\partial T}{\partial x} \qquad \dot{q}_y = -\kappa_y \frac{\partial T}{\partial y} \quad \text{and} \quad \dot{q}_z = -\kappa_z \frac{\partial T}{\partial z}$$

In figure 2–1 these vector components are shown isometrically on regular rectangular and cylindrical coordinates. Notice that the three components, $\dot{q}_x$, $\dot{q}_y$, and $\dot{q}_z$ (or $\dot{q}_r$, $\dot{q}_\theta$, and $\dot{q}_z$) are each normal to the other two; they are orthogonal sets. We may also consider the thermal conductivity terms κ_x, κ_y, and κ_z to be thermal conductivities of the conducting medium in the three directions x, y, and z. An *isotropic* material, which is a material that has properties that do not vary with direction, has the thermal conductivities all equal, such that

$$\kappa_x = \kappa_y = \kappa_z = \kappa$$

For all of the applications of this text materials will be assumed to be isotropic unless otherwise stated. There are some important materials that are not isotropic (called anisotropic) which the reader should be aware of. These include most wood and wood products, composite materials, laminated plastics, and some synthetic materials. Thermal conductivity is a transport property that measures the ability of a medium to conduct heat, and it is defined by equation (1–54). It is often a function of the temperature, the pressure, or the material density, but to avoid extra complications, we will often assume thermal conductivity to be constant.

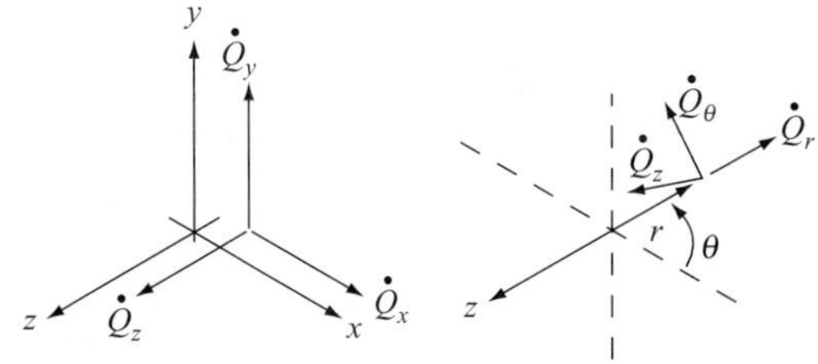

Figure 2–1 Conduction heat transfer as a vector quantity in rectangular and cylindrical coordinates.

It was mentioned in section 1–4 that materials conducting heat well or rapidly are called *conductors* and have a high value for thermal conductivity, whereas those materials that retard heat transfer are called *insulators* and have low values for thermal conductivity. In appendix table B–2 are listed thermal conductivity values for some solids. In appendix tables B–3 and B–4, the values for the thermal conductivities for some selected liquids and gases over ranges of temperature and pressure are presented. Notice that the thermal conductivities change with temperature, some dramatically and some not so much. Experimental data of thermal conductivity has been acquired over the years by using apparatus and equipment that measure the necessary properties (usually temperature) of conducting materials. The majority of the test equipment and experimentation relies on steady state one-dimensional heat transfer to accurately predict thermal conductivity.

The guarded hot plate, shown schematically in figure 2–2 from a diagram given in the American Society of Testing Materials (ASTM) Standards [1], has been a popular device to experimentally determine thermal conductivity of solids and some liquids. It is constructed so that a test material is surounded on all but two opposite faces with the same test material. Adequate insulation surrounds the device to assure one-dimensional heat transfer through the test material section. The guarded hot ring, the guarded hot box, the guarded ring, and the guarded tube are devices that use the same principles of design as those that are used in the guarded hot plate. These other devices are constructed for specific types of materials. The concentric cylinders shown in figure 2–3a and b are devices that were used

Figure 2–2 The guarded hot plate device. (From American Society for Testing and Materials (ASTM) Standard C 177-85 [1])

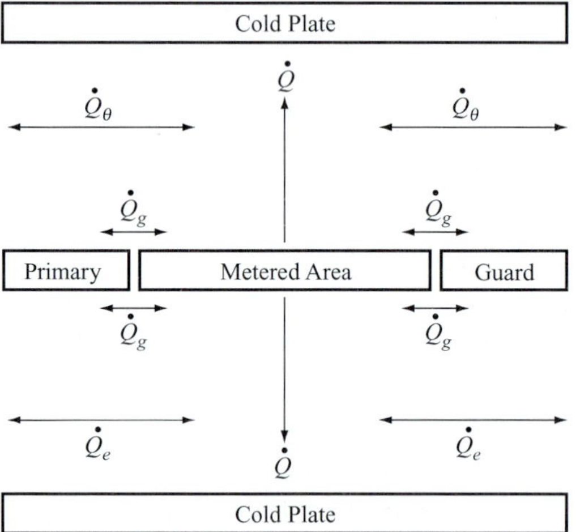

(a) Schematic of heat flow paths relative to the device

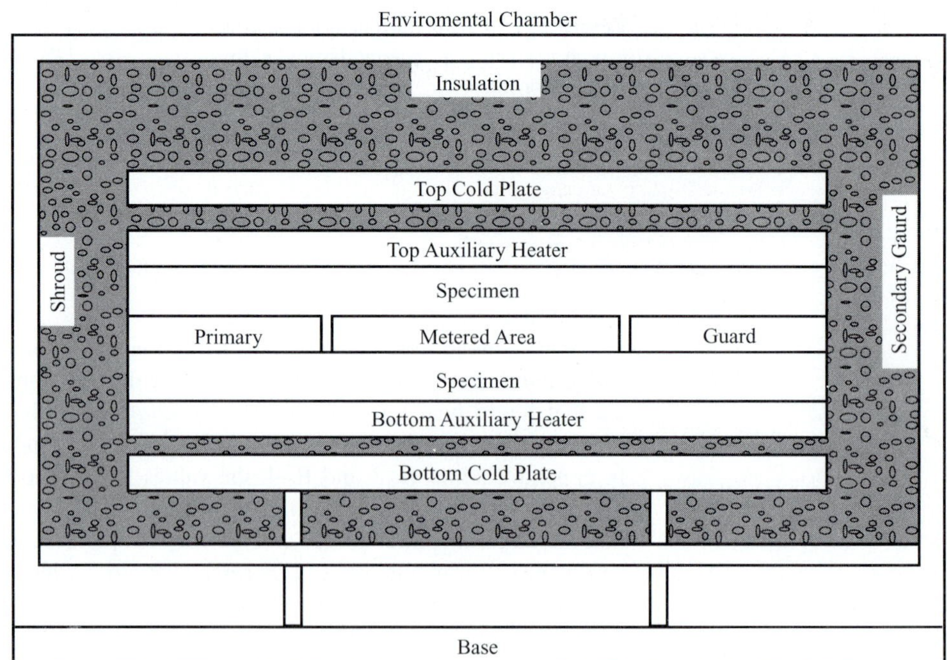

(b) Cross section of the device

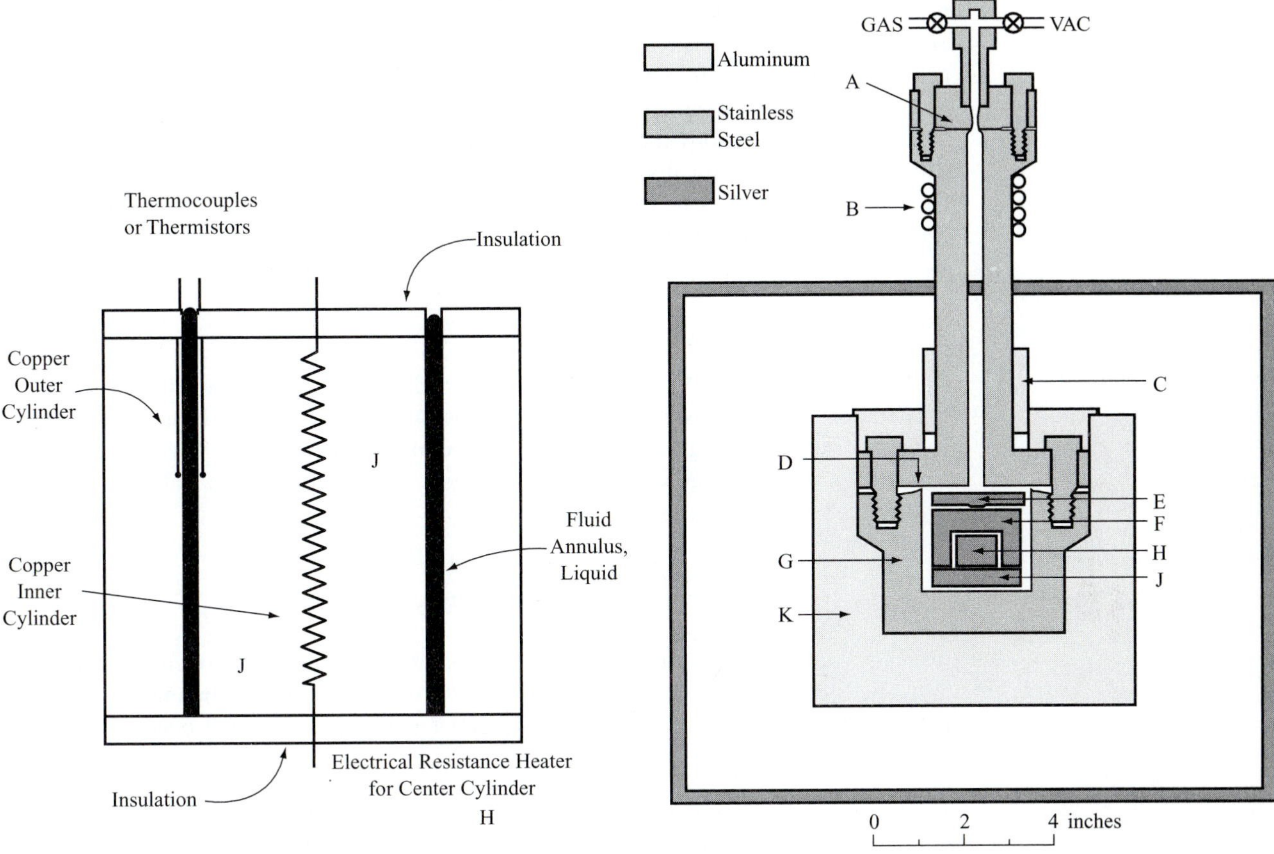

Thermal-conductivity apparatus.

A, Pressure seal for electric leads; B, cooling coil; C, furnace neck; D, Monel gasket; E, auxiliary guard; F, guard; G, pressure vessel; H, hotplate; J, coldplate; K, furnace.

(a) Arrangement for use with liquids.

The liquid is shown shaded. Heat flows radially from the heater H outward through a copper cylinder, J, through an annular liquid volume, through an outer copper cylinder. The thermal conductivity is so much greater for the copper than the liquid that thermocouples may be embedded in the copper near the liquid interface to monitor surface temperatures of the liquid.

(b) Arrangement for use with gases or vapors.

The gas or vapor is shown unshaded and its thermal conductivity is measured in the region between the hot plate (H) and the cold plate (J). The arrangement has a horizontal test section to minimize any convection effects. Vine [4] and others have used concentric arrangements similar to that used for liquids.

Figure 2–3 The concentric cylinder device.

by Bridgeman [2], Nuttall [3], and others for determining thermal conductivities of liquids or vapors. These devices also require insulation to prevent excessive heat transfer losses and are designed as to minimize convective effects in the test volume. We will discuss experimental determinations of thermal conductivity further in sections 2–7 and 3–5.

Analytic expressions for predicting the thermal conductivity of some materials have been developed. Specifically, from a consideration of gases that behave much like perfect gases, the *kinetic theory of gases* and a "rigid sphere molecule theory" the following equation has been suggested by Hirschfelder et al. [5],

$$\kappa = 8.328 \times 10^{-4}\sqrt{\frac{T}{\text{MW} \cdot \Gamma}} \quad (\text{W/cm} \cdot \text{K}) \tag{2–1}$$

where T is in kelvin, MW is the molar mass, and Γ is the molar diameter or "collision" diameter of a particular substance in angstroms, Å (1Å $= 10^{-10}$ m). Equation (2–1) is to be used for monatomic gases, such as helium, argon, and neon. For polyatomic gases such as nitrogen (N_2), oxygen (O_2), and hydrogen (H_2), a correction called the *Eucken Correction* needs to be applied, giving

$$\kappa_{\text{poly}} = \kappa\left(\frac{4c_v}{5R + 0.6}\right) \tag{2–2}$$

In table 2–1 are listed some parameters for helium, argon, and nitrogen that have been abstracted from Hirschfelder et al. [5]. Using the collision diameter in angstroms, c_v, and R from property tables, and recalling the molar masses, the thermal conductivity equations for these gases are

For helium $\quad \kappa = 0.8762 \times 10^{-4}\sqrt{T} \quad (\text{W/cm} \cdot {}^\circ\text{C})$ **(2–3)**
For argon $\quad \kappa = 0.09936 \times 10^{-4}\sqrt{T} \quad (\text{W/cm} \cdot {}^\circ\text{C})$ **(2–4)**
For nitrogen $\quad \kappa = 0.112 \times 10^{-4}\sqrt{T} \quad (\text{W/cm} \cdot {}^\circ\text{C})$ **(2–5)**

Also in table 2–1 are listed some values for thermal conductivity calculated from equations (2–3) through (2–5) and experimentally determined data. The thermal conductivity of mixtures of various gases can be expressed in analytic expressions such as for single substances, but additional complications arise.

There have been attempts made to write expressions for thermal conductivity of liquids. Hirschfelder et al. [5] is an excellent source for the interested reader to begin a study

Table 2–1 Kinetic Theory and Thermal Conductivity of some Gases

Gas	Molar Mass (MW)	Collision Diameter (Γ, Å)	Temperature (T, K)	Thermal Conductivity (κ, W/cm · °C) Calc.	Exper.*
He	4.00	2.18	100	0.000876	0.000730
			500	0.001959	0.002028
A	39.9	3.64	100	0.000099	0.000064
			500	0.000222	0.000264
N_2	28.02	3.75	100	0.000112	(0.000097)*
			500	0.000250	(0.000367)

Table condensed and revised from [5].

*Values in parentheses are approximate values.

of this particular aspect of heat transfer, and Bridgeman [2,6] used an heuristic argument to derive an equation for thermal conductivity of liquids, excluding liquid metals,

$$\kappa = 3.865 \times 10^{-23} \frac{\mathbf{v}_s}{x_m^2} \quad (\text{W/cm} \cdot {}^\circ\text{C, or W/cm} \cdot \text{K}) \tag{2–6}$$

where $\mathbf{v}_s$ is the sonic velocity (cm/s) in the medium and x_m is the mean distance of separation between molecules in cm. The mean distance of separation between molecules was computed by assuming a uniform cubical structure of the molecules which gives that

$$x_m = \sqrt[3]{\frac{\text{mm}}{\rho}} \tag{2–7}$$

where mm is the mass of one molecule in grams. Density has units of grams/cm^3 to give x_m centimeter units. The sonic velocity in a liquid or solid is usually written as

$$\mathbf{v}_s = \sqrt{\frac{E_b}{\rho}} \tag{2–8}$$

where E_b is the bulk modulus of elasticity and ρ is the density. The densities of liquids are normally inversely dependent on the temperature so that, using equations (2–6), (2–7), and (2–8) the thermal conductivity can be expected to be inversely proportional to the sixth power of temperature (see practice problem 2–3). In table 2–2 are listed some of the results given by Bridgeman for reference. These data may also be used with equation (2–6) to compare computed values of thermal conductivity to some experimentally measured values.

Table 2–2 Thermal Conductivity Parameters for Liquids

Liquids	**Temperature (T, K)**	**Sonic Velocity ($\mathbf{v}_s$, cm/s)**	**Mean Distance (x_m, cm $\times 10^7$)**	**Thermal Conductivity (κ, W/cm · °C)**	
				Calc.	**Exper.**
Methyl alcohol	303	1.13×10^5	0.408	0.0028	0.0021
Ethyl alcohol	303	1.14×10^5	0.459	0.0024	0.0018
Ether	303	0.92×10^5	0.560	0.0012	0.0014
Acetone	303	1.14×10^5	0.500	0.0019	0.0018
Carbon Bisulfide	303	1.18×10^5	0.466	0.0023	0.0016
Water	303	1.50×10^5	0.310	0.0065	0.0060

Table revised and converted by author from [6].

Solids have values for thermal conductivity that seem to be linearly proportional to temperature but, as with liquids, only if large temperature differences are involved. The Weidemann–Franz Law states that the ratio of the thermal conductivity to the electrical conductivity of a metal solid is proportional to the absolute temperature of the metal. Specifically,

$$\frac{\kappa}{C} = LzT \tag{2–9}$$

where C is the electrical conductivity (mohs/m or 1/ohm · m) Lz is called the *Lorentz number,* a constant of proportionality. A value of 2.43×10^{-8} volts^2/K^2 has been suggested for the Lorentz number; however, this value seems to be erroneous for very low temperatures or when the free electron theory of matter is not applicable.

An equation of the form

$$\kappa = \kappa_{T0} + a(T - T_0) \tag{2–10}$$

has been suggested for predicting thermal conductivities of solids and in appendix table B–8 are listed parameters κ_{T0} and a for various solids. It can be seen from the values of these parameters that thermal conductivity may increase or decrease with increasing temperature, depending on the particular material.

EXAMPLE 2–1 Estimate the thermal conductivity for solid sodium at its melting point of 97.9°C.

Solution

The thermal conductivity of the solid may be predicted by using equation 2–10 and data from appendix table B–8:

$$\kappa_{T0} = 142 \frac{\text{W}}{\text{m} \cdot \text{K}}$$

$$a = -0.01 \frac{\text{W}}{\text{m} \cdot \text{K}^2}$$

Then, for a temperature of 97.9°C,

Answer

$$\kappa = 142 \frac{\text{W}}{\text{m} \cdot \text{K}} - \left(0.01 \frac{\text{W}}{\text{m} \cdot \text{K}}\right)(370.9 - 200) = 140.29 \frac{\text{W}}{\text{m} \cdot \text{K}}$$

EXAMPLE 2–2 Estimate the thermal conductivity of glycerine at 300 K using equation (2–6) with equations (2–7) and (2–8). Then compare this result to the listed value given in appendix table B–3. The molecular weight (MW) of glycerine is 92 g/g-mol as given in any chemistry reference text. The mass of one molecule of glycerine may be approximated by the relationship

Solution

$$\text{mm} = \frac{\text{MW}}{\text{Avogadro's number}} = \frac{92 \dfrac{\text{g}}{\text{g} \cdot \text{mol}}}{6.022 \times 10^{23} \dfrac{\text{molecules}}{\text{g} \cdot \text{mol}}} = 15.277 \times 10^{-23} \frac{\text{g}}{\text{molecule}}$$

Using the density of glycerine from appendix table B–3,

$$x_m + \sqrt[3]{\frac{\text{mm}}{\rho}} = \sqrt{\frac{15.277 \times 10^{-23} \text{g/molecule}}{1260 \times 10^{-3} \text{ g/cm}^3}} = 0.495 \times 10^{-7} \text{ cm}$$

The bulk modulus of glycerine is 4.6×10^9 Pa, from appendix table B–3. Using equation (2–8), the sonic velocity is

$$\mathbf{v}_s = \sqrt{\frac{E_b}{\rho}} = \sqrt{\frac{4.6 \times 10^9 \text{ N/m}^2}{1260 \text{ kg/m}^3}} = 1.91 \times 10^3 \frac{\text{m}}{\text{s}} = 1.91 \times 10^5 \frac{\text{cm}}{\text{s}}$$

The thermal conductivity is then predicted from equation (2–6),

Answer

$$\kappa = (3.865 \times 10^{-23})\left(\frac{\mathbf{v}_s}{x_m^2}\right) = 3.014 \times 10^{-3} \frac{\text{W}}{\text{cm} \cdot \text{K}}$$

This compares favorably with the thermal conductivity given for liquid glycerine in appendix table B–3, of 2.87×10^{-3} W/cm · K.

2–2 THE GENERAL PROBLEM OF CONDUCTION HEAT TRANSFER

Let's now consider conduction heat transfer in a continuous medium such as a solid material or stagnant fluid having no flow. The conservation of energy of an element of such a medium can be expressed as

Rate of energy into the element = Rate of energy out of the element + Rate of accumulation of energy in the element **(2–11)**

For an element in a Cartesian coordinate space as shown in figure 2–4, the rates of energy into the element are the heat transfers into the element and the rates out of the element are those heat transfers out of the element (because heat transfer is energy that crosses a system boundary). The rate of accumulation of energy in the element may include an energy generation term. Such a term can be used as a mathematical device to account for such phenomena as radiation heat transfer absorbed in the element from afar; as a latent heat term to account for phase changes between a liquid and a solid, between a solid and a vapor, or between a vapor and a liquid; or as some viscous or internal frictional resistance such as electrical resistance in an electrical wire. We will denote the energy generation term as $\dot{E}_{\text{gen}}$.

Figure 2–4 Small element in a continuous medium subjected to conduction heat transfer.

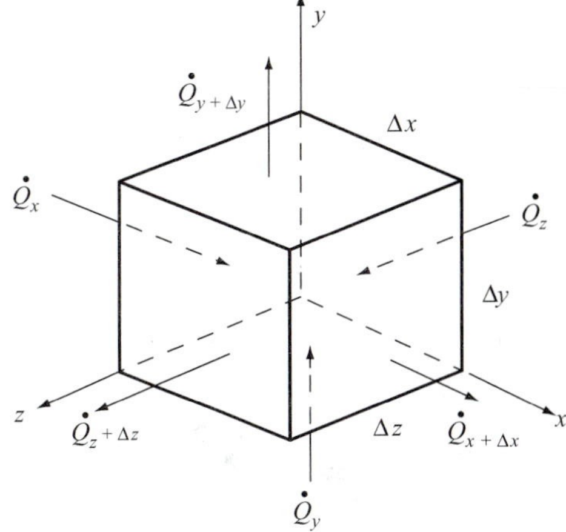

The accumulated energy term is accounted for as a rate of increase of sensible enthalpy in the element, $mc_p\frac{\partial T}{\partial t} = \rho V c_p \frac{\partial T}{\partial t}$. Entering these terms into the energy balance of equation (2–11) gives

$$\dot{Q}_x + \dot{Q}_y + \dot{Q}_x + \dot{E}_{\text{gen}} = \dot{Q}_{x+\Delta x} + \dot{Q}_{y+\Delta y} + \dot{Q}_{z+\Delta z} + \rho V c_p \frac{\partial T}{\partial t} \tag{2–12}$$

Using Fourier's Law of Conduction to describe the heat transfer terms, this equation becomes, for isotropic materials so that thermal conductivity is independent of direction,

$$-\kappa A_x \frac{\partial T}{\partial x} - \kappa A_y \frac{\partial T}{\partial t} - \kappa A_z \frac{\partial T}{\partial t} + \dot{E}_{\text{gen}} = -\kappa A_{x+\Delta x}\left[\frac{\partial T}{\partial t}\right]_{x+\Delta x} - \kappa A_{y+\Delta y}\left[\frac{\partial T}{\partial t}\right]_{y+\Delta y}$$
$$- \kappa A_{z+\Delta z}\left[\frac{\partial T}{\partial t}\right]_{z+\Delta z} + \rho V c_p \frac{\partial T}{\partial t} \tag{2–13}$$

For the element in cartesian coordinates, we have $V = \Delta x \cdot \Delta y \cdot \Delta z$, $A_x = A_{x+\Delta x} = \Delta y \cdot \Delta z$, $A_y = A_{y+\Delta y} = \Delta x \cdot \Delta z$, and $A_z = A_{z+\Delta z} = \Delta x \cdot \Delta y$. Substituting these relationships into equation (2–13) gives us

$$-\kappa\Delta y\cdot\Delta z\left[\frac{\partial T}{\partial x}\right]_x - \kappa\Delta x\cdot\Delta z\left[\frac{\partial T}{\partial y}\right]_y - \kappa\Delta x\cdot\Delta y\left[\frac{\partial T}{\partial z}\right]_z + \dot{E}_{gen} = -\kappa\Delta y\cdot\Delta z\left[\frac{\partial T}{\partial x}\right]_{x+\Delta x} - \kappa\Delta x\cdot\Delta z\left[\frac{\partial T}{\partial y}\right]_{y+\Delta y}$$
$$- \kappa\Delta x\cdot\Delta y\left[\frac{\partial T}{\partial z}\right]_{z+\Delta z} + \rho V c_p\frac{\partial T}{\partial t} \qquad \textbf{(2–14)}$$

If the conduction heat transfer terms on the right side of the equation are combined on the left side and each term is divided by the volume element, $\Delta x\cdot\Delta y\cdot\Delta z$, we obtain

$$\frac{\left\{\kappa\left[\frac{\partial T}{\partial x}\right]_{x+\Delta x} - \kappa\left[\frac{\partial T}{\partial x}\right]_x\right\}}{\Delta x} + \frac{\left\{\kappa\left[\frac{\partial T}{\partial y}\right]_{y+\Delta y} - \kappa\left[\frac{\partial T}{\partial y}\right]_y\right\}}{\Delta y} + \frac{\left\{\kappa\left[\frac{\partial T}{\partial z}\right]_{z+\Delta z} - \kappa\left[\frac{\partial T}{\partial z}\right]_z\right\}}{\Delta z} + \dot{e}_{gen} = \rho c_p\frac{\partial T}{\partial t} \qquad \textbf{(2–15)}$$

If the volume is now reduced, in the limit, to zero (i.e., taking Δx, Δy, and Δz to zero), we get from calculus

$$\frac{\partial}{\partial x}\left[\kappa\frac{\partial T}{\partial x}\right] + \frac{\partial}{\partial y}\left[\kappa\frac{\partial T}{\partial y}\right] + \frac{\partial}{\partial z}\left[\kappa\frac{\partial T}{\partial z}\right] + \dot{e}_{gen} = \rho c_p\frac{\partial T}{\partial t} \qquad \textbf{(2–16)}$$

This equation represents the starting point for analysis of conduction heat transfer within the regular Cartesian coordinate system. The energy generation term, $\dot{e}_{gen}$, is the energy generation per unit of volume or generation density. It is correctly defineable only to a smallest volume, δV, of a smallest mass, δm. *For many conduction heat-transfer problems the energy generation is zero, and then the governing equation is*

$$\frac{\partial}{\partial x}\left[\kappa\frac{\partial T}{\partial x}\right] + \frac{\partial}{\partial y}\left[\kappa\frac{\partial T}{\partial y}\right] + \frac{\partial}{\partial z}\left[\kappa\frac{\partial T}{\partial z}\right] = \rho c_p\frac{\partial T}{\partial t} \qquad \textbf{(2–17)}$$

and this equation is sometimes referred to as the *Heat Equation.* It is applicable for those situations when non–steady state or transient conditions occur, with no energy generations.

For *steady state conditions* the time rate of change of temperature is zero, and then the heat equation becomes

$$\frac{\partial}{\partial x}\left[\kappa\frac{\partial T}{\partial x}\right] + \frac{\partial}{\partial y}\left[\kappa\frac{\partial T}{\partial y}\right] + \frac{\partial}{\partial z}\left[\kappa\frac{\partial T}{\partial z}\right] = 0 \qquad \textbf{(2–18)}$$

and this equation is sometimes referred to as *Laplace's Equation.* It has been extensively studied and applied to a variety of physical situations, and we use it here for the general case of one-, two-, or three-dimensional steady state conduction heat transfer without energy generation.

The energy balance equations that are equivalent to the heat equation in cylindrical and spherical coordinates are also useful for many applications. They are, in cylindrical coordinates,

$$\frac{1}{r}\frac{\partial}{\partial r}\left[\kappa r\frac{\partial T}{\partial r}\right] + \frac{1}{r^2}\frac{\partial}{\partial \theta}\left[\kappa\frac{\partial T}{\partial \theta}\right] + \frac{\partial}{\partial z}\left[\kappa\frac{\partial T}{\partial z}\right] = \rho c_p\frac{\partial T}{\partial t} \qquad \textbf{(2–19)}$$

and, in spherical coordinates,

$$\frac{1}{r^2}\frac{\partial}{\partial r}\left[\kappa r^2\frac{\partial T}{\partial r}\right] + \frac{1}{r^2\sin\theta}\frac{\partial}{\partial \theta}\left[\kappa\sin\theta\frac{\partial T}{\partial \theta}\right] + \frac{1}{r^2\sin^2\theta}\frac{\partial}{\partial \phi}\left[\kappa\frac{\partial T}{\partial \phi}\right] = \rho c_p\frac{\partial T}{\partial t} \qquad \textbf{(2–20)}$$

The student may want to derive these two general equations, using methods similar to the earlier derivation of equation (2–17).

EXAMPLE 2–3 Define the volume element for conduction heat transfer with energy generation in cylindrical coordinates.

Solution: Cylindrical coordinates are usually denoted as r (radial), θ (circumferential), and z (axial). In figure 2–5 is shown the physical situation for a regular cylindrical coordinate system. The volume element is then included on the space and it has sides of Δr, $r\Delta\theta$, and Δz. The volume element is, as a first approximation,

Answer

$$\Delta V = (\Delta r)(r\Delta\theta)(\Delta z)$$

Figure 2–5 Cylindrical coordinate system with a unit element.

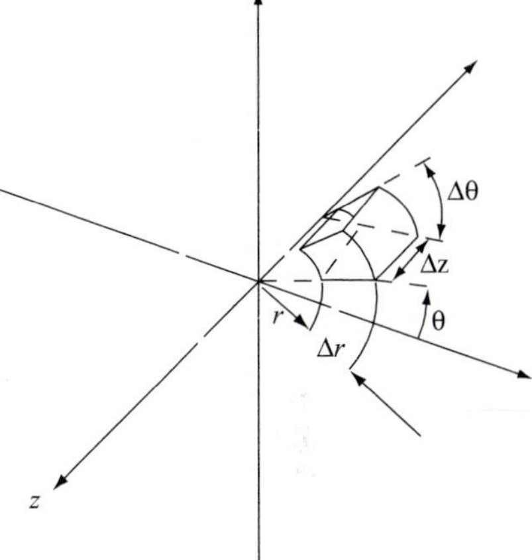

Notice in this example that heat transfer can occur in three directions that are each normal or perpendicular to each other. That is, the heat transfers may occur across the face areas of the element in each of the three coordinate directions, and these areas are normal or orthogonal to each other. Also, the cylindrical coordinate system is an orthogonal one and the heat transfers in the three directions r, θ, and z are always orthogonal to each other at any point in the space.

EXAMPLE 2–4 Determine the temperature distribution in an insulated 10-cm-diameter by 3-m-long copper rod heated at one end to 600°C and cooled to 30°C at the other end. Assume the heat transfer is only along the axial direction and the thermal conductivity of copper is a constant, as listed in appendix table B–2.

Solution: We assume one-dimensional heat transfer in the z direction of a cylindrical coordinate system. In figure 2–6 is shown the schematic of the physical situation of the rod conducting

Figure 2–6 Rod conducting heat axially due to a temperature distribution.

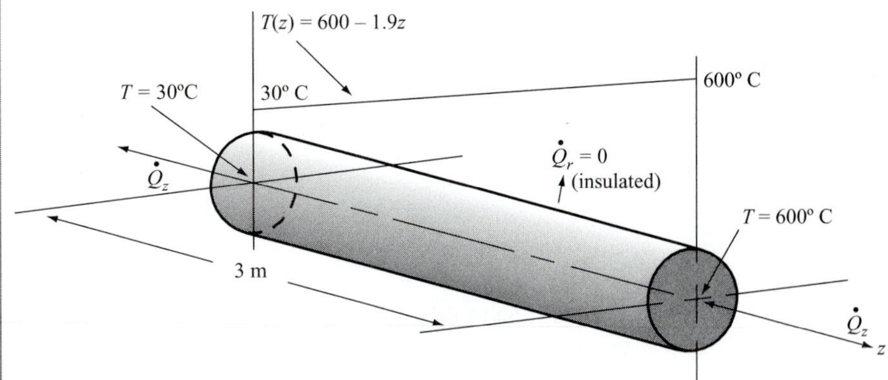

heat axially. From equation (2–19) we eliminate all terms that contain derivatives other than those with respect to z. The result is

$$\frac{\partial}{\partial z}\left[\kappa\frac{\partial T}{\partial z}\right] = 0$$

Since the thermal conductivity is constant, this equation becomes

$$\kappa\frac{\partial^2 T}{\partial z^2} = 0$$

or

$$\frac{\partial^2 T}{\partial z^2} = 0$$

Also, the temperature is only changing in the z direction and is therefore only a function of z. We write that $T = T(z)$ and the partial derivatives are the same as ordinary derivatives. The governing equation is then

$$\frac{d^2 T}{dz^2} = 0 \qquad 0 \leq z \leq 3\text{ m}$$

which is a second-order ordinary differential equation.

To obtain a solution to this equation requires that two boundary conditions be identified:

B.C. 1	$T(z) = 600°\text{C}$	at $z = 0$
B.C. 2	$T(z) = 30°\text{C}$	at $z = 3\text{ m} = 300\text{ cm}$

The governing equation can be integrated twice after separating variables, to give

$$T(z) = C_1 z + C_2$$

Using B.C. 1 gives $T(0) = 600°\text{C} = C_2$, and from B.C. 2, $T(300\text{ cm}) = 30°\text{C} = C_1(300\text{ cm}) + 600°\text{C}$, so that

$$C_1 = (30°\text{C} - 600°\text{C})/300\text{ cm} = -1.9°\text{C/cm}$$

The temperature distribution is

Answer

$$T(z) = -1.9z + 600°\text{C}$$

This result is shown graphically in the schematic of figure 2–6 as a superimposed temperature distribution.

2–3 STEADY STATE ONE-DIMENSIONAL HEAT TRANSFER

Steady state one-dimensional heat transfer is an idealization that is used to describe many important actual heat-transfer processes. It is probably the single most used concept in all of heat transfer for engineering analysis and the design of technological devices and processes. The governing equation for one-dimensional steady state conduction heat transfer without energy generation is, in Cartesian coordinates,

$$\frac{d}{dx}\kappa\frac{dT}{dx} = 0 \tag{2–21}$$

and if thermal conductivity is taken as a constant, this becomes

$$\kappa\frac{d^2T}{dx^2} = 0 \tag{2–22}$$

The solution to this equation requires two specified boundary conditions. Typically the temperature values at two positions (x-coordinate points) and/or the heat transfer, $\dot{Q}_x$, at one of the boundaries are the boundary conditions. Consider, for instance, a slab or

Figure 2–7 Steady state one-dimensional heat transfer through a slab or wall.

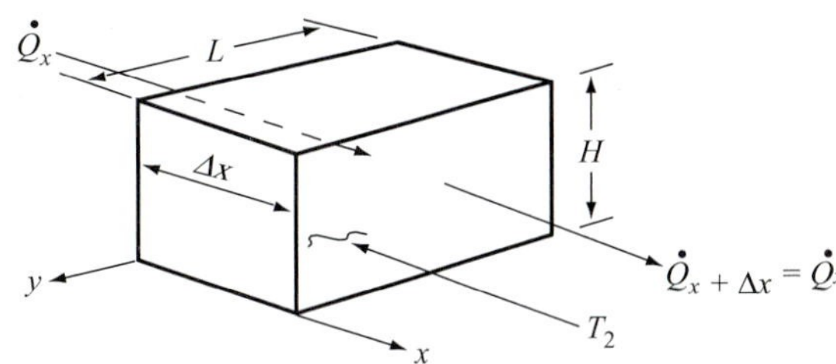

wall whose length (L) and height (H) are much greater than the thickness, Δx, as shown in figure 2–7. For constant thermal conductivity we use equation (2–22) and the mathematical formulation,

$$\kappa \frac{d^2T}{dx^2} = 0 \qquad 0 \leq x \leq \Delta x$$

$$\begin{array}{lll} \text{B.C. 1} & T(x) = T_1 & \text{at } x = 0 \\ \text{B.C. 2} & T(x) = T_2 & \text{at } x = \Delta x \end{array} \tag{2–23}$$

Then, integrating the governing equation twice gives

$$T(x) = C_1 x + C_2$$

which is a description of a linear temperature distribution in x. Using the first boundary condition gives

$$T(x) = T_1 = C_1(0) + C_2 = C_2$$

and the second boundary condition gives

$$T(x) = T_2 = C_1(\Delta x) + C_2 = C_1(\Delta x) + T_1$$

and

$$C_1 = \frac{(T_2 - T_1)}{\Delta x}$$

so that

$$T(x) = \left(\frac{T_2 - T_1}{\Delta x}\right)x + T_1 \tag{2–24}$$

If the heat transfer is directed as shown in figure 2–7 we may write, from Fourier's Law,

$$\dot{Q} = -\kappa \frac{dT}{dx} \tag{2–25}$$

From equation (2–24),

$$\frac{dT}{dx} = \frac{T_2 - T_1}{\Delta x}$$

so

$$\dot{Q} = -\kappa A\left(\frac{T_2 - T_1}{\Delta x}\right) = -\kappa HL\left(\frac{T_2 - T_1}{\Delta x}\right) \tag{2–26}$$

The temperature T_2 must be less than T_1 if heat transfer is to occur in the direction shown. Equation (2–26) is the starting point for many further studies involving one dimensional heat transfer by conduction. It can be revised by defining *thermal resistance*, R_T,

$$R_T = \frac{\Delta x}{\kappa A} \tag{2–27}$$

We may then write equation (2–26) as

$$\dot{Q} = \frac{\Delta T}{R_T} \tag{2–28}$$

where ΔT is the absolute value of the temperature difference across the slab or wall. In figure 2–7 ΔT would be $T_1 - T_2$. Often it is more convenient to consider heat transfer per unit of area normal to the heat transfer itself, q_A

$$\dot{q}_A = \frac{\dot{Q}}{A} = -\kappa\frac{T_2 - T_1}{\Delta x} = \frac{\Delta T}{R_T A} \tag{2–29}$$

The product in the denominator of this equation, $R_T A = \Delta x/\kappa$, is the *thermal resistivity* of the material and it will be denoted by R_V. It can be seen that the thermal resistivity R_V is

$$R_V = \frac{\Delta x}{\kappa} \tag{2–30}$$

in x-direction heat transfer, and

$$\dot{q}_A = \frac{\Delta T}{R_V} \tag{2–31}$$

A term often used in building construction and in heating and air conditioning applications is the *R-value,* defined as

$$1 \text{ R-value} = 1 \text{ hr} \cdot \text{ft}^2 \cdot {}^\circ\text{R/Btu} = 1 \text{ hr} \cdot \text{ft}^2 \cdot {}^\circ\text{F/Btu}$$

Using conversion factors from appendix table B–1 we have

$$1 \text{ R-value} = 0.176 \text{ m}^2 \cdot {}^\circ\text{C/W}$$

EXAMPLE 2–5 A concrete retaining wall is exposed on one side to 100°F air and on the other side to 55°F earth. If the retaining wall is 16 in. thick, 12 ft high, and 40 ft long, determine the heat transfer through the wall, the heat transfer per square foot of surface area, the temperature distribution through the wall, and the wall R-value. Assume a value of 9.7 Btu · in./hr · ft² · °F for the thermal conductivity of concrete.

Solution We assume steady state conditions, that the earth is uniformly at 55°F on one side of the wall, and that the air is uniformly at 100°F on the exposed side of the wall. The heat transfer may be predicted by equation (2–26)

$$\dot{Q} = -\kappa\frac{T_2 - T_1}{\Delta x} = -\left(9.7\frac{\text{Btu} \cdot \text{in.}}{\text{hr} \cdot \text{ft}^2 \cdot {}^\circ\text{F}}\right)(12 \times 40 \text{ ft}^2)\left(\frac{55^\circ\text{F} - 100^\circ\text{F}}{16 \text{ in.}}\right)$$

Answer $$\dot{Q} = 13{,}095 \frac{\text{Btu}}{\text{hr}}$$

The heat transfer per unit area is

Answer $$\dot{q}_A = \frac{\dot{Q}}{A} = \frac{13.095 \dfrac{\text{Btu}}{\text{hr}}}{12 \times 14 \text{ ft}^2} = 27.28 \frac{\text{Btu}}{\text{ft}^2 \cdot \text{hr}}$$

The temperature distribution is linear because there is no energy generation and the distribution is given by the equation and shown in figure 2–8.

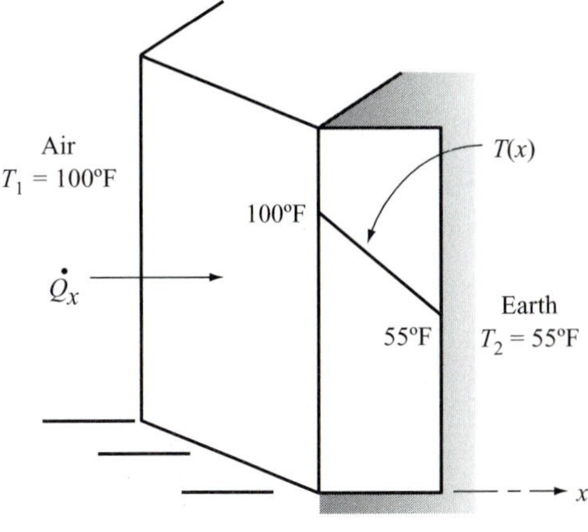

Figure 2–8 Cross section of retaining wall of example 2–5.

Answer

$$T(x) = T_1 - \left(\frac{T_1 - T_2}{\Delta x}\right)x = 100°\text{F} - \frac{45°\text{F}}{16\text{ in.}}x = 100°\text{F} - 2.815x$$

The R-value is

$$R_V = \frac{\Delta x}{\kappa} = \frac{16\text{ in.}}{9.7\dfrac{\text{Btu}\cdot\text{in.}}{\text{hr}\cdot\text{ft}^2\cdot°\text{F}}} = 1.649\,\frac{\text{hr}\cdot\text{ft}^2\cdot°\text{F}}{\text{Btu}} = 1.649$$

The convenience in defining thermal resistance and thermal resistivity (or R-value) becomes apparent when heat transfer occurs through more than one homogenous material. Thermal resistance, heat transfer, and temperature differences can be compared to the similar phenomena of electrical current through a conducting material. The concept of Ohm's Law, $\mathcal{E} = \mathcal{I}\mathcal{R}$ describes the relationship between current $\mathcal{I}$ (in amperes), electric potential $\mathcal{E}$ (in voltage), and electrical resistance $\mathcal{R}$ (in ohms). In figure 2–9a is shown a resistor having resistance of $\mathcal{R}$ ohms across which an electric potential is impressed. The result is a current $\mathcal{I}$. Analogous to this process is heat transfer Q_x due to a temperature difference ΔT, with a thermal resistance R_T. This process is illustrated in figure 2–9b, and from a comparison of the two diagrams in the figure the heat transfer is analogous to electrical current, the temperature difference is analogous to electrical potential, and the two resistances are physically comparable. Electrical circuits may be used to simulate heat-transfer processes through more than one material and in somewhat complex paths. Such modeling has been the essence of the analog computers that have been used to predict heat transfers or temperature distributions in complicated processes. We use it for helping to understand the heat transfer processes involving two or more homogeneous materials and use examples to demonstrate the electrical analogy.

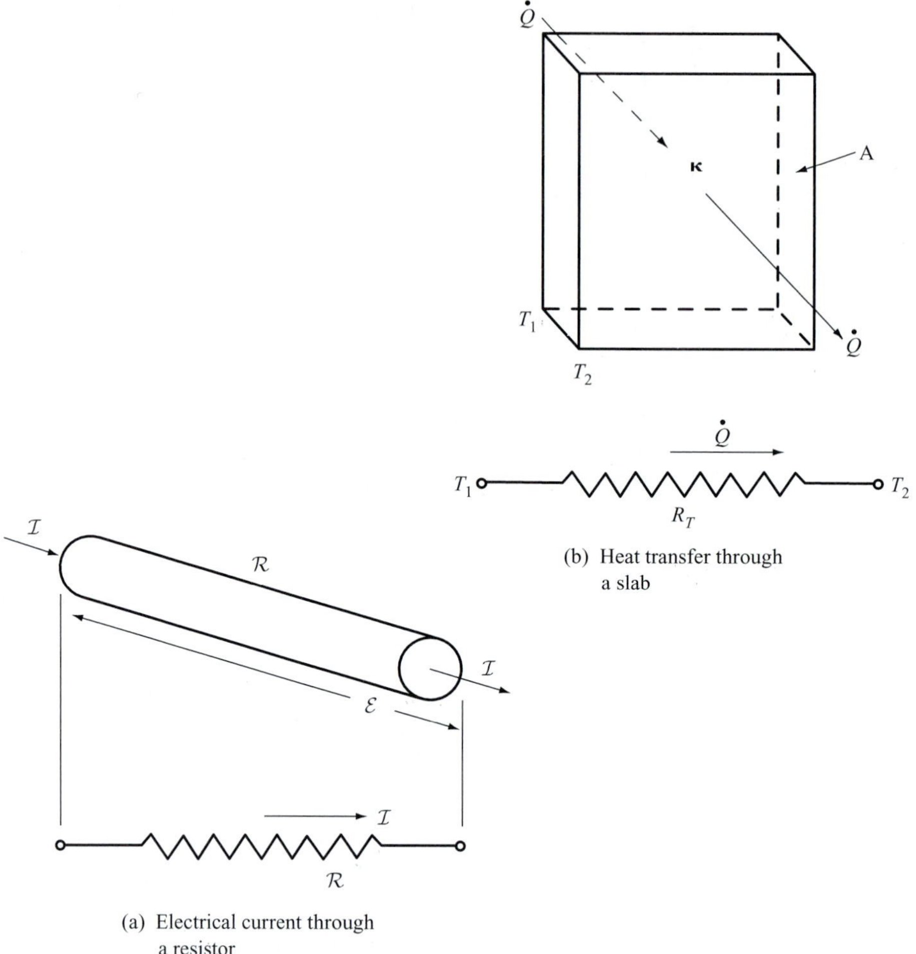

Figure 2–9 Heat transfer/electrical analogy.

EXAMPLE 2–6 The cross section of a refrigerator wall, shown in figure 2–10, exposes the various materials making up that wall. The inside surface temperature of the wall is 0°C and the outside is 28°C. Determine the temperature distribution through the refrigerator wall and the expected heat transfer per unit of area, which is a heat gain to the refrigerator.

Solution The refrigerator cabinet is a three-dimensional structure (figure 2–10a). However, we visualize the typical wall cross-section (figure 2–10b) as a one-dimensional system. The heat transfer problem can then be approximated as one-dimensional, steady state with no energy generation. For the wall section shown in figure 2–10 we write

$$\dot{q}_A = \frac{\Delta T_{\text{overall}}}{\Sigma R_V} = \frac{\Delta T_{\text{overall}}}{R_{V1} + R_{V2} + R_{V3}}$$

Note that, because the heat transfer occurs in series through three different materials, the thermal resistivity summation is

$$\sum R_V = R_{V1} + R_{V2} + R_{V3}$$

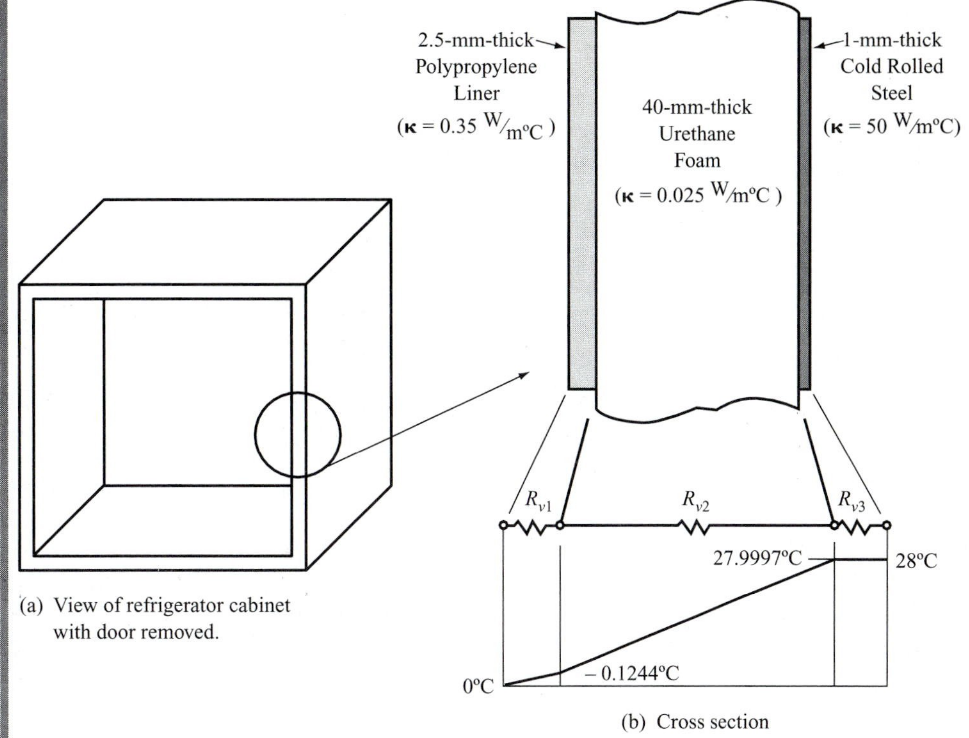

(a) View of refrigerator cabinet with door removed.

(b) Cross section

Figure 2–10 Cross section of a typical refrigerator wall, example 2–6.

For the polyethylene liner,

$$R_{V1} = \frac{\Delta x_1}{\kappa_1} = \frac{2.5 \text{ mm}}{0.35 \dfrac{\text{W}}{\text{m} \cdot {}^\circ\text{C}}} = 0.00714 \frac{\text{m}^2 \cdot {}^\circ\text{C}}{\text{W}}$$

for the urethane foam,

$$R_{V2} = \frac{\Delta x_2}{\kappa_2} = \frac{40}{0.025} = 1.6 \frac{\text{m}^2 \cdot {}^\circ\text{C}}{\text{W}}$$

and for the steel cabinet,

$$R_{V3} = \frac{1 \text{ mm}}{50 \dfrac{\text{W}}{\text{m}^2 \cdot {}^\circ\text{C}}} = 0.00002 \frac{\text{m}^2 \cdot {}^\circ\text{C}}{\text{W}}$$

The thermal resistivity summation is then

$$R_{V1} + R_{V2} + R_{V3} = 1.60716 \text{ m}^2 \cdot {}^\circ\text{C/W}$$

The overall temperature difference, $\Delta T_{\text{overall}} = T_o - T_1$, is 28°C and the expected heat loss per square meter of refrigeration surface area is

Answer

$$\dot{q}_A = \frac{28^\circ\text{C}}{1.60716 \dfrac{\text{m}^2 \cdot {}^\circ\text{C}}{\text{W}}} = 17.422 \frac{\text{W}}{\text{m}^2}$$

Since the heat transfer progresses through the successive layers of the three wall materials we can also write,

$$\dot{q}_A = \frac{\Delta T_1}{R_{V1}} = \frac{\Delta T_2}{R_{V2}} = \frac{\Delta T_3}{R_{V3}} = 17.422\,\frac{\text{W}}{\text{m}^2}$$

and

$$\Delta T_1 = 17.422\,\frac{\text{W}}{\text{m}^2} \times R_{V1} = 0.1244°\text{C}$$

$$\Delta T_2 = 17.422 \times R_{V2} = 27.8753°\text{C}$$
$$\Delta T_3 = 17.422 \times R_{V3} = 0.0003°\text{C}$$

Thus, the temperature of the outside surface of the polypropylene liner corresponds to the inside surface of the urethane foam,

$$T_1 = T_i + \Delta T_1 = 0.1244°\text{C}$$

The temperature of the outside of the urethane foam is

$$T_2 = T_1 + \Delta T_2 = 27.9997°\text{C}$$

and the outside of the steel, as a check, is

$$T_o = T_2 + \Delta T_3 = 28.0000°\text{C}$$

which agrees with the known value of the outside temperature. In figure 2–10b this temperature distribution is overlayed on the sketch for the wall section.

After reading through this example you can see that thermal resistivity (or thermal resistance!) increases when heat transfer is forced to occur through multiple layers of materials in series or in succession. The thermal resistances, or thermal resistivities, are added arithmetically. Also, as we saw in Chapter 1, convective heat transfer occurs between solid boundary surfaces and a surrounding fluid or vapor. This mechanism has been described by the convection equation (1–59)

$$\dot{Q} = hA\Delta T \qquad \textbf{(1–59, repeated)}$$

where ΔT is the temperature difference of the surface, T_s, and the ambient or bulk fluid temperature, T_o. This equation can be revised to conform to the thermal resistance concept by writing it as

$$\dot{Q} = \frac{\Delta T}{R_T}$$

where

$$R_T = \frac{1}{hA} \qquad \textbf{(2–32)}$$

Notice that the thermal resistivity for convection heat transfer can also be written

$$R_V = \frac{1}{h} \qquad \textbf{(2–33)}$$

Thus, in example 2–6 we could include thermal resistivity at the inside wall surface and at the outside surface if the wall surface temperatures had not been specified. In situations where the surface temperature is not known, the surrounding fluid temperature and the convective heat transfer coefficient must be given.

Sometimes the engineer or technologist encounters a situation where heat transfer occurs simultaneously across two or more dissimilar materials. In figure 2–11 is sketched

Figure 2–11 The concept of parallel and series heat transfer.

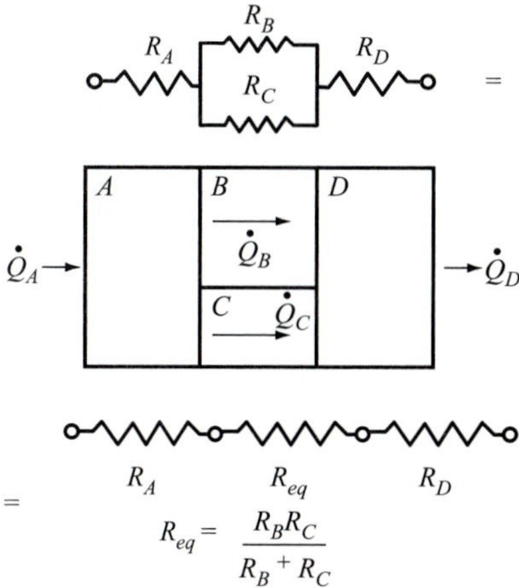

a simple version of this situation along with an analogous electrical circuit. Notice that elements B and C are in a parallel arrangement regarding heat transfer and for the electrical analogy. When this sort of condition arises, it is best to avoid the use of thermal resistivity and work only with the thermal resistances, as we shall now see. For the system of figure 2–11 the heat transfers can be written

$$\dot{Q}_A = \dot{Q}_B + \dot{Q}_C = \dot{Q}_D$$

or, using thermal resistances again,

$$\frac{\Delta T_A}{R_{TA}} = \frac{\Delta T_B}{R_{TB}} + \frac{\Delta T_C}{R_{TC}} = \frac{\Delta T_D}{R_{TD}}$$

But, since B and C are parallel, the temperature differences ΔT_B and ΔT_C are equal to each other, denoted ΔT_{eq} and then

$$\frac{\Delta T_A}{R_{TA}} = \frac{\Delta T_{\text{eq}}}{R_{TB}} + \frac{\Delta T_{\text{eq}}}{R_{TC}} = \frac{\Delta T_D}{R_{TD}}$$

Adding the two middle terms together gives

$$\frac{\Delta T_A}{R_{TA}} = \frac{\Delta T_{\text{eq}}}{R_{T\text{eq}}} = \frac{\Delta T_D}{R_{TD}} \tag{2–34}$$

where

$$R_{T\text{eq}} = \frac{R_{TB}R_{TC}}{R_{TB} + R_{TC}}$$

This result is indicated in figure 2–11 as an electrical analogy.

EXAMPLE 2–7 Consider a thermopane window constructed of two parallel panes of regular window glass but separated by a sealed air gap. The window frame is an aluminum casing as shown in figure 2–12. Determine the thermal resistance of the window itself (not including the convective resistance), the heat loss through the window, and the temperature distribution across the window.

Solution We can see from figure 2–12 that the heat transfer occurs through the window panes in series and in parallel with the aluminum frame. Then, for the process of heat loss, we can

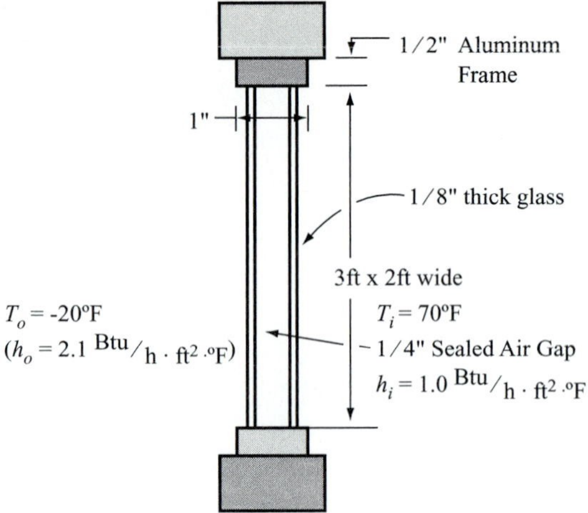

Figure 2–12 Aluminum frame window, example 2–7.

visualize an electrical analogy as shown in figure 2–13. Using a result similar to that of equation (2–34), we have for the window resistance, R_{Teq},

$$R_{Teq} = \frac{(2R_{Tg} + R_{Tair})R_{Taluminum}}{2R_{Tg} + R_{Tair} + R_{Taluminum}}$$

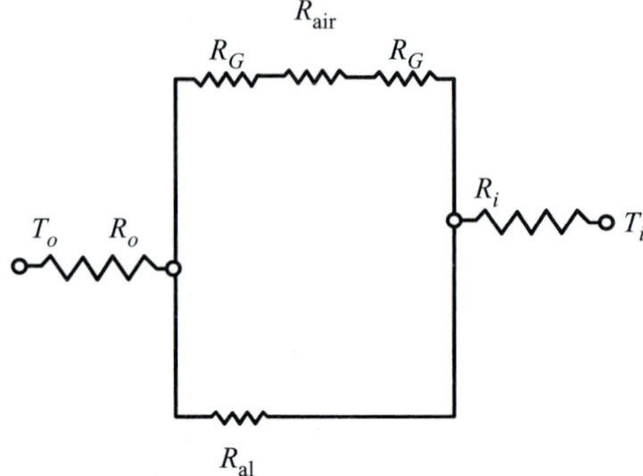

Figure 2–13 Electrical analogy for heat loss in window of figure 2–12.

If we use thermal conductivity values listed in appendix table B–2 we get

$$R_{Tg} = \frac{\Delta x_g}{A_g \kappa_g} = \frac{0.125 \text{ in.}}{6 \text{ ft}^2 \times 4.7087 \dfrac{\text{Btu} \cdot \text{in.}}{\text{hr} \cdot \text{ft}^2 \cdot {}^\circ\text{F}}} = 0.0044 \frac{\text{hr} \cdot {}^\circ\text{F}}{\text{Btu}}$$

$$R_{Tair} = \frac{0.25 \text{ in.}}{6 \text{ ft}^2 \times 0.1803 \dfrac{\text{Btu} \cdot \text{in.}}{\text{hr} \cdot \text{ft}^2 \cdot {}^\circ\text{F}}} = 0.231 \frac{\text{hr} \cdot {}^\circ\text{F}}{\text{Btu}}$$

and $$R_{Taluminum} = \frac{1 \text{ in.}}{(6.4236 \text{ in}^2 - 6 \text{ in}^2)\left(1657.4 \dfrac{\text{Btu} \cdot \text{in.}}{\text{hr} \cdot \text{ft}^2 \cdot {}^\circ\text{F}}\right)} = 0.00142 \frac{\text{hr} \cdot {}^\circ\text{F}}{\text{Btu}}$$

The equivalent resistance is then

Answer $$R_{Teq} = 0.0041 \frac{\text{hr} \cdot {}^\circ\text{F}}{\text{Btu}} = \text{Thermal resistance of the window}$$

and the overall thermal resistance, including the convective effects, is

$$\sum R_T = R_{To} + R_{Teq} + R_{Ti} = \frac{1}{h_o A_o} + R_{Teq} + \frac{1}{h_i A_i}$$

$$= \frac{1}{6.4236\ \text{ft}^2 \times 2.1\ \dfrac{\text{Btu}}{\text{hr}\cdot\text{ft}^2\cdot{}^\circ\text{F}}} + 0.0041\ \frac{\text{hr}\cdot{}^\circ\text{F}}{\text{Btu}} + \frac{1}{6.4236\ \text{ft}^2 \times 1.0\ \dfrac{\text{Btu}}{\text{hr}\cdot\text{ft}^2\cdot{}^\circ\text{F}}}$$

$$= 0.0741 + 0.0041 + 0.1557 = 0.231\ \frac{\text{hr}\cdot{}^\circ\text{F}}{\text{Btu}}$$

The heat loss through the window and its frame is then

Answer

$$\dot{Q} = \frac{\Delta T_{\text{overall}}}{\sum R_T} = \frac{70^\circ\text{F} - (-20^\circ\text{F})}{0.231\dfrac{\text{hr}\cdot{}^\circ\text{F}}{\text{Btu}}} = 389.4\ \frac{\text{Btu}}{\text{hr}}$$

The temperature distributions are linear through each of the window components and are nonlinear through the boundary layers at either side of the window. We will discuss the methods used to predict the temperature distribution through the boundary layer in Chapters 4 and 5, but for now we sketch in a nonlinear temperature distribution in figure 2–14. The outer and inner surface temperatures of the glass and window frame can be predicted from equation (2–30);

$$\dot{Q} = \frac{\Delta T}{\sum R_t} = 389.4\ \frac{\text{Btu}}{\text{hr}} = \frac{\Delta T_o}{R_{To}} = \frac{\Delta T_i}{R_{Ti}}$$

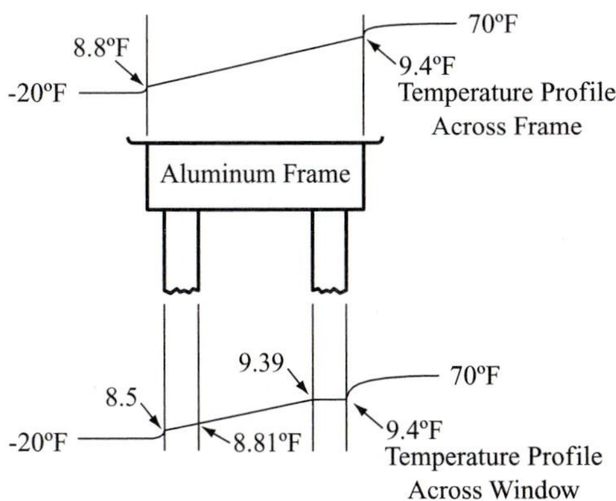

Figure 2–14 Temperature distribution through window of example 2–7.

Since $R_{To} = 0.0741$ hr · °F/Btu and $R_{Ti} = 0.1557$ hr · °F/Btu, the convective temperature differences ΔT_o and ΔT_i can be determined. They are 28.8°F for the outer convection and 60.6°F for the inner. Then

$$T_{so} = T_o + \Delta T_o = 8.8^\circ\text{F}$$

Similarly, the inside surface temperature is found:

$$T_{si} = T_i + \Delta T_i = 9.4^\circ\text{F}$$

These temperatures are indicated in figure 2–14. The glass–air–glass circuit has a heat transfer given by

$$\dot{Q}_g = \frac{\Delta T_{\text{overall}}}{2R_{Tg} + R_{T\text{air}}} = \frac{9.4^\circ\text{F} - 8.8^\circ\text{F}}{\left(2 \times 0.0044 + 0.231\ \dfrac{\text{hr}\cdot{}^\circ\text{F}}{\text{Btu}}\right)} = 2.5\ \frac{\text{Btu}}{\text{hr}}$$

Then, for the temperature distribution for the glass we note that

$$\dot{Q}_g = \frac{\Delta T_g}{R_{Tg}} = \frac{\Delta T_{\text{air}}}{R_{T\text{air}}} = 2.5\,\frac{\text{Btu}}{\text{hr}}$$

The temperature differences are

$$\Delta T_g = 2.5 \times 0.0044 = 0.01°\text{F}$$
$$\Delta T_{\text{air}} = 0.58°\text{F}$$

and the temperature of the inside of the outer pane of glass is 8.8°F + 0.01°F = 8.81°F. The outside temperature of the inside pane of glass is 9.4°F − 0.01°F = 9.39°F. These results are indicated in figure 2–14.

Notice that the aluminum frame is an excellent conductor of heat and has a heat transfer given by

$$\dot{Q}_{\text{aluminum}} = 389.3 - 2.5 = 386.8\frac{\text{Btu}}{\text{hr}}$$

Thus, in an actual window assembly as shown in figure 2–12 a large portion of the window glass will have a greater temperature drop than predicted by the preceding analysis. As a result there will be heat transfers in the two dimensions of the window plane and the inside temperature of the window glass will be higher than the inside temperature of the aluminum frame. The frame, being such a good heat conductor, acts as a short in an electrical circuit with the effect that a disproportionate amount of heat transfer occurs. As a consequence of this effect, many window frames are constructed of a material, such as wood, with better insulating qualities.

There are many important applications where heat transfer is best analyzed in cylindrical or in spherical coordinates. Water or steam flowing through a pipe, refrigerants flowing through condenser or evaporator tubes, and high-temperature gases flowing through nozzles or diffusers are only three of many applications where cylindrical coordinates are convenient in analyzing the heat-transfer processes. In section 2–2 the heat conduction equation (2–19) was given in cylindrical coordinates, and in example 2–4 the steady state conduction in the axial or z direction was considered. Here we consider steady state conduction heat transfer in the radial or r direction. Equation 2–19 becomes, for the one-dimensional condition, including energy generation,

$$\frac{1}{r}\frac{d}{dr}\kappa r\frac{dT}{dr} + \dot{e}_{\text{gen}} = 0 \tag{2–35}$$

and for no energy generation,

$$\frac{1}{r}\frac{d}{dr}\kappa r\frac{dT}{dr} = 0 \tag{2–36}$$

For constant thermal conductivity conditions, equation (2–36) is

$$\frac{d}{dr}r\frac{dT}{dr} = 0 \tag{2–37}$$

Typically, the radial conduction heat transfer problems involve at least two radii, an inner and an outer radius, as indicated in figure 2–15. The two radii, r_i and r_o are most often the actual physical surfaces and the location of prescribed boundary conditions. The

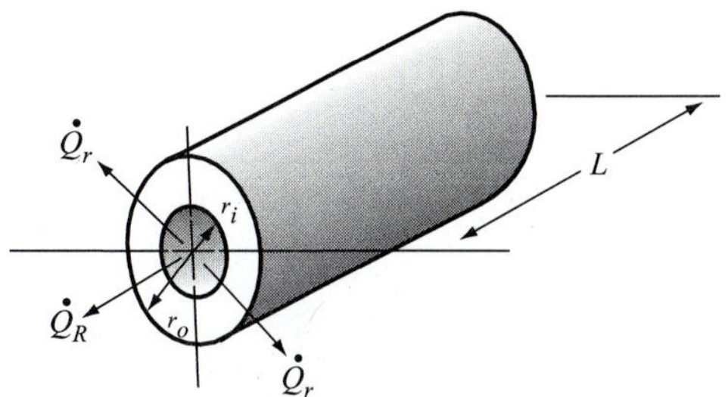

Figure 2–15 One-dimensional radial conduction heat transfer.

solution to equations (2–35), (2–36), or (2–37) can only be completed with two boundary conditions, and these are customarily the temperatures or the heat transfers at the two radii. There are exceptions, such as boundary conditions at the centerline axis for solid cylinders with energy generation or some prescribed intermediate condition between the inner and outer boundaries. Typically the heat transfers at either r_i or r_o would be convective heat transfer, but they could be conduction to another surrounding material or radiation heat exchange with some other body. Consider the boundary conditions for equation (2–37) and refer to Figure 2–15:

$$\begin{aligned} &\text{B.C. 1} \quad T(r) = T_i \quad \text{at } r = r_i \\ &\text{B.C. 2} \quad T(r) = T_o \quad \text{at } r = r_o \end{aligned} \qquad \textbf{(2–38)}$$

Then, solving equation (2–37) by integrating once,

$$r\frac{dt}{dr} = C_1 \qquad \textbf{(2–39)}$$

and separating variables to obtain

$$\int dT = C_1 \int \frac{dr}{r} \qquad \textbf{(2–40)}$$

we may integrate the indefinite integral, obtain a second constant of integration, and apply the boundary conditions. A direct method is to take the definite integral of both sides at the two boundary conditions:

$$\int_{T_i}^{T_o} dT - C_1 \int_{r_i}^{r_o} \frac{dr}{r} \quad \text{and} \quad T_o - T_i = C_1 ln\frac{r_o}{r_i}$$

The constant C_1 may be explicitly determined by recalling the definition of radial conduction heat transfer

$$\dot{Q}_r = -\kappa A_r \frac{\partial T}{\partial r} = -\kappa A_r \frac{dT}{dr}$$

For $T(r)$, and $A_r = 2\pi rL$ (referring to figure 2–15). Then,

$$\dot{Q}_r = -2\pi\kappa rL\frac{dT}{dr}$$

but we have in equation (2–39) that

$$C_1 = r\frac{dT}{dr}$$

so, comparing these results,

$$C_1 = -\frac{\dot{Q}_r}{2\pi\kappa L}$$

The general equation for one-dimensional steady state heat transfer as specified by equation (2–37) with the prescribed boundary conditions of equation (2–38) is

$$T_o - T_i = -\frac{\dot{Q}_r}{2\pi\kappa L}\ln\frac{r_o}{r_i} \quad \text{or} \quad \dot{Q}_r = \frac{2\pi\kappa L}{ln\,(r_o/r_i)}(T_i - T_o) \tag{2–41}$$

This important result gives the total conduction heat transfer, in kW, Btu/hr, or associated units. It is often convenient to use the thermal resistance concept and write equation (2–41) as

$$\dot{Q}_r = \frac{\Delta T}{R_T} \tag{2–42}$$

where

$$R_T = \frac{\ln(r_o/r_i)}{2\pi\kappa L} \tag{2–43}$$

The thermal resistance, as we saw in rectangular coordinate systems, allows for straightforward analysis of those systems involving more than one homogenous material. Those same methods are applicable in radial heat transfer situations as indicated by equations (2–42) and (2–43). Also, it is frequently convenient to consider heat transfer per unit of length of a cylindrical object. From equation (2–41) we may write

$$\dot{q}_L = \frac{\dot{Q}_r}{L} = -\frac{2\pi\kappa}{\ln(r_o/r_i)}(T_o - T_i) \tag{2–44}$$

where q_L is the radial heat transfer per unit of length, in W/m, Btu/hr · ft, or similar units. The radial thermal resistivity may be defined as

$$R_{VL} = \frac{ln\ r_o/r_i}{2\pi\kappa} \tag{2–45}$$

and then

$$\dot{q}_L = \frac{\Delta T}{R_{VL}} \tag{2–46}$$

EXAMPLE 2–8 A 2-in.-outside-diameter (OD) by 1.5-in.-inside-diameter (ID) wrought iron water line is buried in the ground. If water flow through the pipe is such that the inside surface of the pipe is 60°F and the outer surface is 50°F where it is surrounded by earth, determine the heat transfer or heat loss per unit length of the pipe and the temperature distribution through the pipe.

Solution The heat loss per unit length of the pipe is determined from equation (2–44). From appendix table B–2 the thermal conductivity for wrought iron is 29.5 Btu/hr · ft · °F so that

Answer

$$\dot{q}_L = \frac{2\pi\left(29.5\ \frac{\text{Btu}}{\text{hr}\cdot\text{ft}\cdot{}^\circ\text{F}}\right)}{\ln\frac{1}{0.75}}(60^\circ\text{F} - 50^\circ\text{F}) = 6443\ \frac{\text{Btu}}{\text{hr}\cdot\text{ft}}$$

The temperature distribution through the pipe is found from the equation

$$dT = -\frac{\dot{q}_L}{2\pi\kappa}\frac{dr}{r} \quad \text{or} \quad T(r) = -\frac{\dot{q}_L}{2\pi\kappa}\ln r + C_1$$

Using the boundary condition that $T = 60°\text{F}$ at $r = 0.75$ in.,

$$C_1 = 60°\text{F} + \frac{\dot{q}_L}{2\pi\kappa}\ln(0.75)$$

Then

$$T(r) = 60°\text{F} - \frac{\dot{q}_L}{2\pi\kappa}\ln\left(\frac{r}{0.75}\right)$$

This equation demonstrates the nonlinear temperature distribution through the pipe, a result that is due to the radial heat transfer of the pipe.

For those conditions where convection at the boundary of the conducting system is included in the analysis, the thermal resistance for such convection as

$$R_{TL} = \frac{1}{2\pi rhL} \tag{2–47}$$

and the radial thermal resistivity (radial R-value) is

$$R_{VL} = \frac{1}{2\pi rh} \tag{2–48}$$

EXAMPLE 2–9 A 10-cm-OD by 6-cm-ID steam line delivers superheated steam at 1000°C. The line is steel wrapped with 10 cm of asbestos and 1 cm of plaster over the asbestos. The cross section of the line is shown in figure 2–16. If the convective heat transfer coefficients are 2500 W/m^2 · °C for the steam and 7.0 W/m^2 · °C for air surrounding the steam line, determine the outside surface temperature and the temperature distribution of the line when heat losses are 710 W/m.

Solution The heat loss may be described by a variation of equation (2–46),

$$\dot{q}_L = \frac{\Delta T_{\text{overall}}}{\Sigma R_{VL}}$$

We use the thermal resistivity concept and refer to the electrical analogy in figure 2–16 to obtain

$$\sum R_{VL} = R_{VL,\text{steam}} + R_{VL,\text{steel}} + R_{VL,\text{asbestos}} + R_{VL,\text{plaster}} + R_{VL,\text{air}}$$

where we have

$$\sum R_{VL} = \frac{1}{2\pi r_1 h_{\text{steam}}} + \frac{\ln(r_2/r_1)}{2\pi\kappa_{\text{steel}}} + \frac{\ln(r_3/r_2)}{2\pi\kappa_{\text{asbestos}}} + \frac{\ln(r_4/r_3)}{2\pi\kappa_{\text{plaster}}} + \frac{1}{2\pi r_4 h_{\text{air}}}$$

The thermal conductivity values for the steel, the asbestos, and the plaster are found from appendix table B–2. Substituting these values into the preceding equation gives

$$\sum R_{VL} = 0.002122 + 0.001891 + 1.12083 + 0.0954 + 0.142103 = 1.362346\ \frac{\text{m} \cdot °\text{C}}{\text{W}}$$

Figure 2–16 Steam line and temperature distribution for example 2–9.

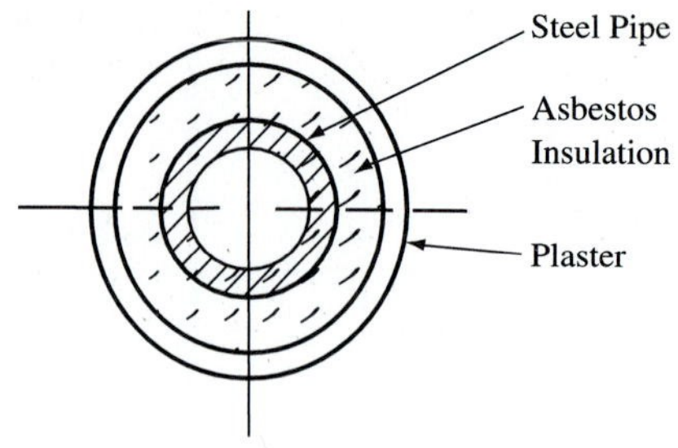

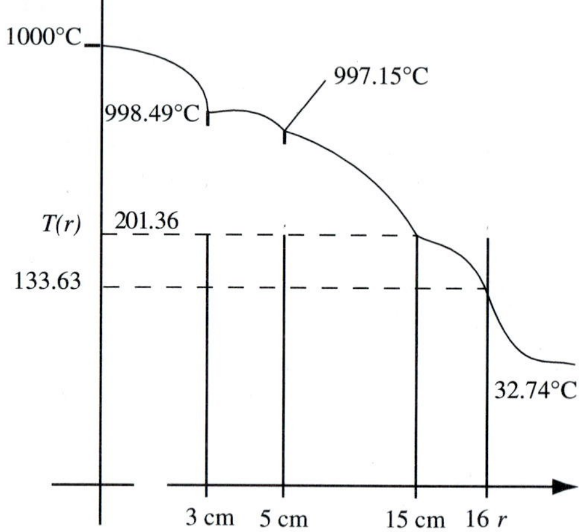

Then, because the heat loss is expected to be 710 W/m, we obtain the overall temperature difference $\Delta T_{\text{overall}}$:

$$\Delta T_{\text{overall}} = \dot{q}_L \sum R_{VL} = \left(710\,\frac{\text{W}}{\text{m}}\right)\left(1.362346\,\frac{\text{m}\cdot{}^\circ\text{C}}{\text{W}}\right) = 967.26^\circ\text{C}$$

Similarly, because the heat transfer progresses in series through the various components,

$$\begin{aligned}
\Delta T_{\text{steam}} &= (710\text{ W/m})\ (0.002122\text{ m}\cdot{}^\circ\text{C/W}) = 1.51^\circ\text{C} \\
\Delta T_{\text{steel}} &= (710)\ (0.001891) = 1.34^\circ\text{C} \\
\Delta T_{\text{asbestos}} &= (710)\ (1.12083) = 795.79^\circ\text{C} \\
\Delta T_{\text{plaster}} &= (710)\ (0.09540) = 67.73^\circ\text{C} \\
\Delta T_{\text{air}} &= (710)\ (0.142103) = 100.89^\circ\text{C}
\end{aligned}$$

and the interface temperatures are directly determined:

$$\begin{aligned}
T_1 &= T_{\text{steam}} - \Delta T_{\text{steam}} = 1000^\circ\text{C} - 1.51^\circ\text{C} = 998.49^\circ\text{C} \\
T_2 &= T_1 - \Delta T_{\text{steel}} = 998.49^\circ\text{C} - 1.34^\circ\text{C} = 997.15^\circ\text{C} \\
T_3 &= T_2 - \Delta T_{\text{asbestos}} = 997.15 - 795.79 = 201.36^\circ\text{C} \\
T_4 &= T_3 - \Delta T_{\text{plaster}} = 201.36 - 67.73 = 133.63^\circ\text{C}
\end{aligned}$$

and

$$T_5 = T_{\text{air}} = T_4 - \Delta T_{\text{air}} = 133.63 - 100.89 = 32.74^\circ\text{C}$$

Answer These results are summarized on figure 2–16. Notice that the outer surface temperature is $T_4 = 133.63^\circ\text{C}$.

There are some important applications of radial one-dimensional heat transfer with energy generation, described by equation (2–35), which we will consider in section 2–7.

Some heat-transfer processes are best analyzed in the spherical coordinate system. For instance, many storage containers and tanks are approximately spherical to take advantage of structural strength and can lose or gain energy through heat transfers. Also, many sources of energy, such as the sun and fireballs inside a furnace, are sometimes approximated as spherical steady state heat conductors. Even some rectangular or irregular-shaped objects, when buried or surrounded by a poor heat conductor, can behave approximately as spherical energy sources. We consider here the steady state condition with no energy generation. In equation (2–20) heat conduction in spherical coordinates is set down. For steady state radial heat conduction with no energy generation, this equation becomes

$$\frac{1}{r^2}\frac{d}{dr}\kappa r^2\frac{dT}{dr} = 0 \tag{2–49}$$

and for constant thermal conductivity it is

$$\frac{1}{r^2}\frac{d}{dr}r^2\frac{dt}{dr} = 0 \tag{2–50}$$

The solution of either of these two equations requires two specified boundary conditions, and these are usually temperature values or heat transfers at radii. These radii normally coincide with physical surfaces of some container or other heat conducting material. Fourier's Law for one-dimensional radial heat conduction in spherical coordinates is written

$$\dot{Q}_r = -\kappa A_r\frac{\partial T}{\partial r} = -\kappa A_r\frac{dt}{dr}$$

where $A_r = 4\pi r^2$, the surface area of a sphere. Using this equation with the following boundary conditions:

B.C. 1	$T(r) = T_i$	at $r = r_i$
B.C. 2	$T(r) = T_o$	at $r = r_o$

and after separating variables, we obtain

$$\int_{T_i}^{T_o} dT = -\int_{r_i}^{r_o}\frac{\dot{Q}_r}{4\pi\kappa}\frac{dr}{r^2} \tag{2–51}$$

Performing the integration gives the temperature distribution through the material,

$$T(r) = T_i + \frac{\dot{Q}_r}{4\pi\kappa}\left(\frac{1}{r} - \frac{1}{r_i}\right) \tag{2–52}$$

which shows that the temperature is related in a hyperbolic form, $f(r^{-1})$, to the radius. A characteristic temperature distribution through a material for heat transfer outward in a spherical system is shown in figure 2–17. The heat transfer density, or heat transfer per unit of spherical area, $\dot{q}_A$, is given by

$$\dot{q}_A = \frac{\dot{Q}_r}{4\pi\kappa r^2} \tag{2–53}$$

which shows that the intensity of heat transfer (heat transfer per unit area) is inversely related to the square of distance from the origin (the Inverse-Square Law). This sort of relationship occurs in the theory of gravity, radiation heat transfer (as we shall see later in this text), and electrostatic field theory among others.

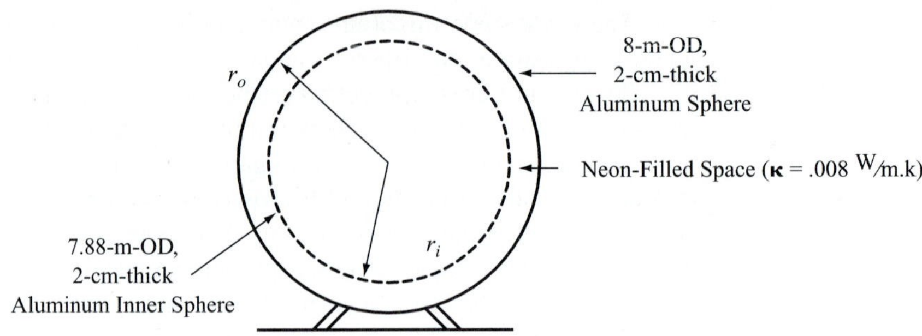

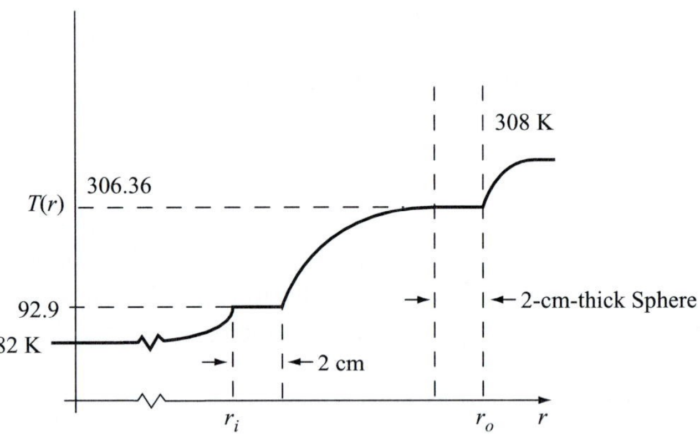

Figure 2–17 Spherical tank with radial heat transfer.

EXAMPLE 2–10 A spherical steel tank is used to store liquid oxygen at 82 K (oxygen boils at 90.188 K, 1 atm). The tank is shown in cross section in figure 2–17 and is double walled with neon gas in the space between the two walls. Assume the tank is designed to function properly when surrounding temperatures reach as high as 35°C and with convective heat transfer coefficient of 25 W/m^2 · °C for air. Then determine the temperature distribution through the tank and the maximum heat gain of the oxygen if the convective heat transfer coefficient is estimated as 4 W/m^2 · °C for the oxygen.

Solution From equation (2–52) we may write an expression for the radial heat transfer, $\dot{Q}_r$,

$$\dot{Q}_r = \frac{4\pi\kappa}{\left(\frac{1}{r_o} - \frac{1}{r_i}\right)}(T_o - T_i) \tag{2–54}$$

or

$$\dot{Q}_r = \frac{\Delta T}{R_{Tr}}$$

where R_{Tr} is the spherical thermal resistance given by

$$R_{Tr} = \frac{\left(\frac{1}{r_i} - \frac{1}{r_o}\right)}{4\pi\kappa} \quad \text{(for conducting media)} \tag{2–55}$$

$$R_{Tr} = \frac{1}{4\pi r^2 h} \quad \text{(for convecting media)}$$

Then the various thermal resistances may be computed. For the oxygen,

$$R_{T,\text{oxygen}} = \frac{1}{4\pi(3.92\ \text{m}^2)\left(4\dfrac{\text{W}}{\text{m}^2\cdot{}^\circ\text{C}}\right)} = 0.001295\ \frac{{}^\circ\text{C}}{\text{W}}$$

for the inner shell, with conductivity value for aluminum listed in appendix table B–2,

$$R_{T,\text{aluminum}} = \frac{\left(\dfrac{1}{3.92\ \text{m}} - \dfrac{1}{3.94\ \text{m}}\right)}{4\pi\left(236\ \dfrac{\text{W}}{\text{m}\cdot{}^\circ\text{C}}\right)} = 0.0000004\ \frac{{}^\circ\text{C}}{\text{W}}$$

for the neon, assuming it is stagnant,

$$R_{T,\text{neon}} = \frac{\left(\dfrac{1}{3.94\ \text{m}} - \dfrac{1}{3.98\ \text{m}}\right)}{4\pi\left(0.008\ \dfrac{\text{W}}{\text{m}\cdot{}^\circ\text{C}}\right)} = 0.02537\ \frac{{}^\circ\text{C}}{\text{W}}$$

for the outer shell,

$$R_{T,\text{aluminum}} = \frac{\left(\dfrac{1}{3.98\ \text{m}} - \dfrac{1}{4.00\ \text{m}}\right)}{4\pi\left(236\ \dfrac{\text{W}}{\text{m}\cdot{}^\circ\text{C}}\right)} = 0.0000004\ \frac{{}^\circ\text{C}}{\text{W}}$$

and for the air

$$R_{R,\text{air}} = \frac{1}{4\pi(16\ \text{m}^2)\left(25\ \dfrac{\text{W}}{\text{m}^2\cdot{}^\circ\text{C}}\right)} = 0.000199\ \frac{{}^\circ\text{C}}{\text{W}}$$

The total thermal resistance is

$$\sum R_T = 0.02686\ \frac{{}^\circ\text{C}}{\text{W}}$$

Then, with the overall temperature difference of 308 − 82 = 226 K we obtain for the heat gain to the oxygen,

Answer

$$\dot{Q}_r = \frac{\Delta T_{\text{overall}}}{\Sigma R_T} = \frac{226\ \text{K}}{0.02686\ \dfrac{\text{K}}{\text{W}}} = 8414\ \text{W} = 8.414\ \text{kW}$$

The temperatures at the various interfaces can be computed using the same sort of method as we demonstrated for rectangular and cylindrical coordinates. We compute for the inner surface, and referring to figure 2–17,

$$T_1 = T_{\text{oxygen}} + \Delta T_{\text{oxygen}} = T_{\text{oxygen}} + \dot{Q}_r R_{T,\text{oxygen}} = 82\ \text{K} + (8414\ \text{W})(0.001295)$$
$$= 92.9\ \text{K} = -172.212^\circ\text{C}$$

Similarly,

$$T_2 = T_1 + \Delta T_{\text{al},i} = 92.9 + 0.0033 = 92.9\ \text{K}$$
$$T_3 = T_2 + \Delta T_{\text{ne}} = 92.9 + 213.46 = 306.36\ \text{K}$$
$$T_4 = T_3 + \Delta T_{\text{al},o} = 306.36 + 1.64 = 308\ \text{K}$$

The temperature distributions through each of the elements can be described by the hyperbolic relationship of equation (2–52). These characteristics and the interface temperatures are given in figure 2–17.

2–4 STEADY STATE TWO-DIMENSIONAL HEAT TRANSFER

Conduction heat transfer was considered in section 2–3 as a one-dimensional phenomenon, and although this is a convenient and clear approach to heat transfer, the actual conduction through materials is more accurately described by two- or three-dimensional heat transfer. In many engineering or technological problems the one-dimensional heat transfer approximation model gives results and interpretations that are in substantial error. Thus, engineers and technologists must understand the significant divergences between one- and two-dimensional heat transfer.

Conduction heat transfer in two dimensions through a solid material with the restrictions of steady state and no energy generation can be described by Laplace's Equation (2–18)

$$\frac{\partial}{\partial x}\kappa\frac{\partial T}{\partial x} + \frac{\partial}{\partial y}\kappa\frac{\partial T}{\partial y} = 0 \qquad \textbf{(2–18, repeated)}$$

For a constant thermal conductivity this becomes

$$\frac{\partial^2 T}{\partial x^2} + \frac{\partial^2 T}{\partial y^2} = 0 \qquad \textbf{(2–56)}$$

and the solution for the temperature, $T(x, y)$, requires four boundary conditions. Also, there is a fundamental problem of the nature of partial derivatives that makes any analysis using equation (2–56) complicated. One manner of overcoming the difficulty is through the *method of separation of variables,* which begins with the premise that, for the function $T(x, y)$ in equations (2–16) or (2–56), a function of two or more variables can be separated into distinct functions. Thus for the temperature as a function of x and y we write

$$T(x, y) = X(x) \cdot Y(y) \qquad \textbf{(2–57)}$$

Notice in this separation that the total function is specified as a product of (in this instance) two functions, $X(x)$ and $Y(y)$, which are individually functions of only one variable. Then, by using the chain rule of calculus we find

$$\frac{\partial T(x, y)}{\partial x} = Y(y) \cdot \frac{dX(x)}{dx} \quad \text{and} \quad \frac{\partial T(x, y)}{\partial y} = X(x) \cdot \frac{dY(y)}{dy} \qquad \textbf{(2–58)}$$

Also,

$$\frac{\partial^2 T(x, y)}{\partial x^2} = Y(y) \cdot \frac{d^2X(x)}{dx^2} \quad \text{and} \quad \frac{\partial^2 T(x, y)}{\partial y^2} = X(x) \cdot \frac{d^2Y(y)}{dy^2}$$

Substituting results (2–58) into equation (2–56) gives

$$Y(y)\frac{d^2X(x)}{dx^2} + X(x)\frac{d^2Y(y)}{dy^2} = 0$$

Now, separating all of the X's from the Y's by separating variables, gives

$$-\frac{1}{X(x)} \cdot \frac{d^2X(x)}{dx^2} = \frac{1}{Y(y)} \cdot \frac{d^2Y(y)}{dy^2} = p^2 \qquad \textbf{(2–59)}$$

Equation (2–59) is now two ordinary differential equations of second order and where p^2 is a constant. It is convenient to write the constant as p^2 instead of p because of the form of the solution to the equations. Notice also that the negative sign attached to the expression containing the $X(x)$ function could just as well be attached to the expression for the $Y(y)$ function. Thus the negative sign is arbitrary but must be assigned to one or the other function. The application of equation (2–59) to a two-dimensional heat transfer situation requires that the four boundary conditions be such that two of them specify conditions of $X(x)$ and the other two specify conditions of $Y(y)$.

From the analysis of differential equations, it is found that one set of solutions to equation (2–59) can take the form

$$T(x,y) = (a \sin px \pm b \sin px)(ce^{-py} \pm de^{py}) \tag{2–60}$$

where a, b, c, and d are constants. Substituting this general equation, after taking partial derivatives with respect to x and then y, back into equations (2–16) or (2–59) proves this contention. The reader may want to verify that equation (2–60) represents a solution to equation (2–56) (see problem 2–26). Now that we have a general solution for two-dimensional heat transfer as described by equation (2–56), we should ask if there are other solutions to the equation, and therefore to the description of the temperature distribution. It is implied that the constants a, b, c, and d can be determined explicitly through the boundary conditions. Because of the harmonic nature with which the sine and cosine functions can be treated (see appendix A–7), we can write the solution such that $p = \pi/L$ where L is the interval or range of x. Also, such terms as $\sin 2\pi x/L$, $\sin 3\pi x/L, \ldots$, $\cos 2\pi x/L$, $\cos 3\pi x/L, \ldots$, $e^{-2\pi y/L}$, $e^{2\pi y/L}$, $e^{-3\pi y/L}$, $e^{3\pi y/L}$, and others having arguments of $n\pi x/L$ or $n\pi y/L$, for sine, cosine, and the exponential functions may be included in a more comprehensive general solution. Thus, a general form of the solutions to equation (2–16) could be written, noting that $p_n = n\pi/L$,

$$T(x,y) = \sum_{n=0}^{\infty} (a_n \sin p_n x + b_n \cos p_n x)(c_n e^{-p_n y} \pm d_n e^{p_n y}) \tag{2–61}$$

where the constants a_n, b_n, c_n, and d_n are still to be evaluated. The final form will depend on the various specific boundary conditions and any qualitative expectations of temperature of heat transfers for a particular situation. We need to consider examples to demonstrate the analytic methods further.

The first problem on conduction heat transfer described in detail by J. B. J. Fourier (1768–1830), after whom *Fourier's Law of Conduction* was named, was the semi-infinite solid wall. We consider an application of that problem.

EXAMPLE 2–11 A high concrete wall is exposed to cold air such that its outer vertical surfaces are at 0°F. The wall is 3 ft thick and is supported by a gravel bed. The temperature distribution of the horizontal surface between the wall and the gravel bed is described by the equation $T(x, 0) = 50°\text{F} \sin \pi x/L$. Determine the temperature distribution through the wall and the heat transfer from the wall to the air.

Solution The wall cross section is shown in figure 2–18a with the base temperature distribution superimposed on the section. The heat transfer through the wall is two-dimensional, and if we assume steady state conditions and constant thermal conductivity we may then use equation (2–56)

$$\frac{\partial^2 T}{\partial x^2} + \frac{\partial^2 T}{\partial y^2} = 0 \qquad 0 < x < L, 0 < y$$

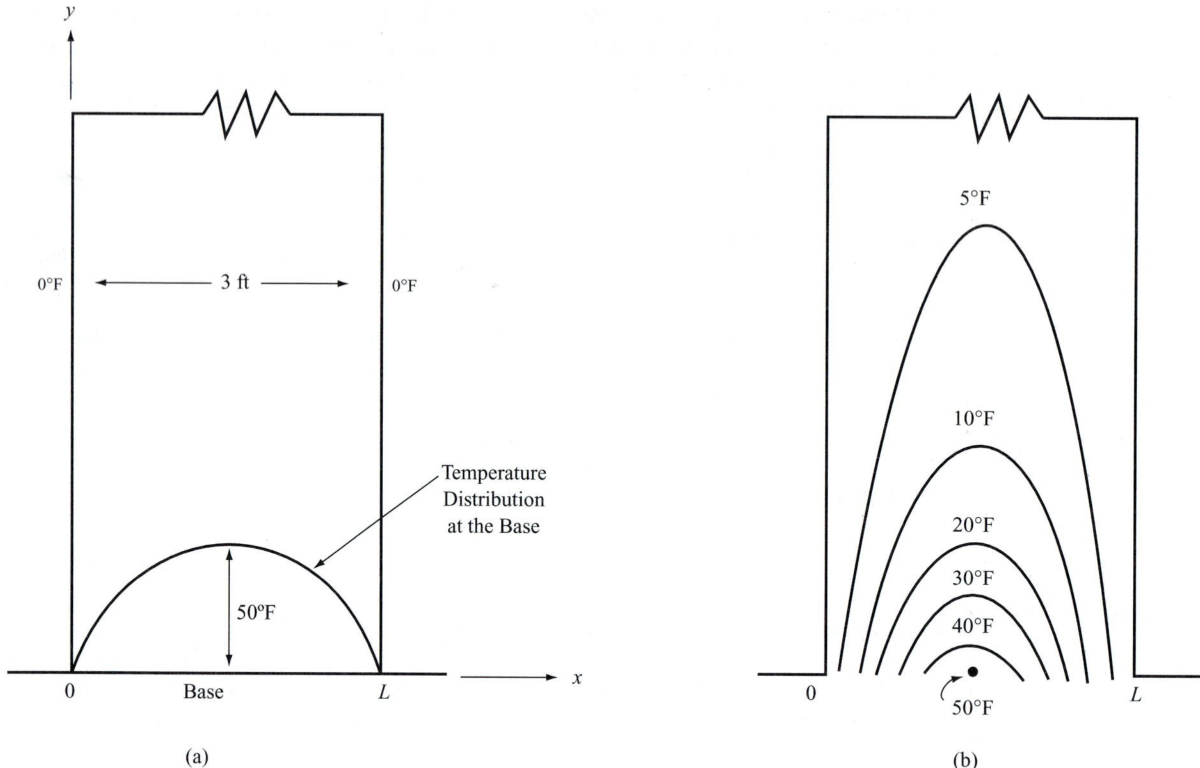

Figure 2–18 Concrete wall with steady state temperature distribution as predicted in example 2–11.

with the following boundary conditions:

B.C. 1	$T(x, y) = 0°F$	at $y > 0, x = 0$
B.C. 2	$T(x, y) = 0°F$	at $y > 0, x = L$
B.C. 3	$T(x, y) = 0°F$	as $y \to \infty, 0 < x < L$
B.C. 4	$T(x, y) = 50°F \sin \pi x/L$	at $y = 0, 0 < x < L$

The general solution to equation (2–56) is given by equation (2–61)

$$T(x, y) = \sum_{n=0}^{\infty} (a_n \sin p_n x + b_n \cos p_n x)(c_n e^{-p_n y} \pm d_n e^{p_n y})$$

An application of B.C. 1 requires that $b_n = 0$; B.C. 3 indicates that d_n must be zero because the exponential term e^{py} increases without bound as y increases toward an infinite value. Thus, the general form of the solution is

$$T(x, y) = \sum_{n=0}^{\infty} A_n e^{-n\pi y/L} \sin \frac{n\pi x}{L}$$

Since $\sin (n\pi x/L) = 0$ for $n = 0$, we may write the summation from $n = 1$ to ∞,

$$T(x, y) = \sum_{n=1}^{\infty} A_n e^{-n\pi y/L} \sin \frac{n\pi x}{L} \tag{2–62}$$

Notice that $A_n = a_n c_n$ and is still a constant to be determined. The form of the solution to T(*x, y*) given in equation (2–62) suggests a sum of sine functions and is a special case of *Fourier Series.* (see [7]–[10] and appendix A–8). The constants, or *Fourier Coefficients*, A_n, are determined from the equation

$$A_n + \frac{2}{L}\int_0^L T(x, 0)\sin\frac{n\pi x}{L}dx \tag{2–63}$$

where *T(x, 0)* is the temperature distribution evaluated at $y = 0$, B.C. 4. Thus, $T(x, 0) = 50°\text{F}\sin \pi x/L$, and with this substitution,

$$A_n = \frac{100°\text{F}}{L}\int_0^L\left(\sin\frac{\pi x}{L}\right)\left(\sin\frac{n\pi x}{L}\right)dx$$

Using some calculus of integration, we get

$$A_1 = 50°\text{F} \quad \text{and} \quad A_n = 0 \qquad \text{for } n > 1$$

Therefore, the solution can be written as

$$T(x, y) = 50°\text{F}(e^{-\pi y/L})\sin\frac{\pi x}{L}$$

This result is graphically shown in figure 2–18b, where the 50°F, 40°F, 30°F, 20°F, 10°F, and 5°F isotherms are plotted.

Notice that the 50°F is only at the bottom center, where $x = L/2$ and $y = 0$. The heat transfer through the wall can be determined by first obtaining the temperature gradient in the *y* direction at $y = 0$, the base, and then applying Fourier's law across the base:

$$\dot{Q}_y = -\kappa A_y\frac{\partial T}{\partial y} = -\kappa W\int_0^L\frac{\partial T}{\partial y}dx$$

where *W* is the length of the wall. We can set this to 1 ft. The gradient $\partial T/\partial y$ at $y = 0$ is

$$\frac{\partial T}{\partial y} = -50°\text{F}\left(\frac{\pi}{L}\right)e^{-\pi(0)/L}\sin\frac{\pi x}{L} = -50°\text{F}\left(\frac{\pi}{L}\right)\sin\frac{\pi x}{L}$$

Then, where $\kappa = 0.809$ Btu/hr · ft · °F from appendix table B–2,

Answer $$\dot{Q}_y = -50°\text{F}\left(\kappa\frac{\pi}{L}\right)\int_0^L\sin\frac{\pi x}{L}dx = 80.9\frac{\text{Btu}}{\text{hr}}$$

This example indicates solutions to more general heat-transfer processes. For instance, the situation of a high wall, or very tall or "long" extension that is surrounded on both sides by a particular constant temperature, call it T_0, and where the base temperature distribution is sinusoidal with an amplitude of $T_A - T_0$ can be written

$$T(x,y) = (T_A - T_0)e^{-\pi y/L}\sin\frac{\pi x}{L} + T_0 \tag{2–64}$$

where *L, y*, and *x* are parameters as in example 2–11. The solution can also be used for the reverse situation, where heat-transfer progresses from the surrounding to the wall and down through the base. A further extension can be made by noticing that the temperature distribution of the base can be written as any particular function of *x, T*(*x*, 0), at $y = 0$, and then for the problem framed by the equation and associated boundary conditions,

$$\frac{\partial^2 t}{\partial x^2} + \frac{\partial^2 T}{\partial y^2} = 0 \qquad 0 < y, 0 < x < L$$

$$\begin{aligned} T(x,y) &= T_0 && \text{at } y > 0, x = 0 \\ T(x,y) &= T_0 && \text{at } y > 0, x = L \\ T(x,y) &= T_0 && \text{as } y \to \infty, 0 < x < L \\ T(x,y) &= T(x,0) && \text{at } y = 0, 0 < x < L \end{aligned}$$

the temperature distribution is

$$T(x,y) = \sum_{n=1}^{\infty} A_n e^{-n\pi y/L} \sin\frac{n\pi x}{L} + \mathrm{T}_0 \qquad \textbf{(2–65)}$$

This result indicates that the temperature distribution is determined by an infinite series. When the summation is carried out, many infinite series converge or approach a finite value; other series do not approach a finite value but approach an infinite value. It is important that the solution converge to a finite value for physical and for mathematical reasons. Also, there are situations when a temperature distribution can be described and the heat transfer is undefined, or does not converge to a finite value. We see this in the next example.

EXAMPLE 2–12 For the wall of example 2–11, determine the temperature distribution in the wall if the base temperature is 50°F over the full area. Then discuss the heat flow through the wall.

Solution The temperature, $T(x, y)$, is given by equation (2–62) and the coefficient A_n is determined from equation (2–63) with $T(x, 0) = 50°\text{F}$. The result is

$$T(x, y) = \frac{100°\text{F}}{\pi} \sum_{n=1}^{\infty} (1 - \cos n\pi)\left(\frac{1}{n}\right) e^{-n\pi y/L} \sin\frac{n\pi x}{L} \qquad \textbf{(2–66)}$$

The temperature distribution described by this equation can be approximated by a summation over a finite number of integers and the temperature profiles are indicated in figure 2–19b. It can be shown (see Carslaw and Jaeger [7], pp. 164, 431–433) that equation (2–66) reduces to

$$T(x, y) = \frac{100°\text{F}}{\pi} \tan^{-1}\left[\left(\frac{\sin(\pi x/l)}{\sinh(\pi y/L)}\right)\right]$$

where $\sinh \pi y/L$ is the hyperbolic sine function defined in appendix A–5. Isotherms are defined as lines of constant temperature, so, if $T(x, y)$ a constant, we get

$$\sin\frac{\pi x}{L} = C \sinh\frac{\pi y}{L} \qquad \textbf{(2–67)}$$

where

$$C = \tan\frac{T\pi}{100°\text{F}}$$

It can also be demonstrated that the directional lines of heat transfer may be described by

$$\cosh\frac{\pi y}{L} + \cos\frac{\pi x}{L} = K\left(\cosh\frac{\pi y}{L} - \cos\frac{\pi x}{L}\right) \qquad \textbf{(2–68)}$$

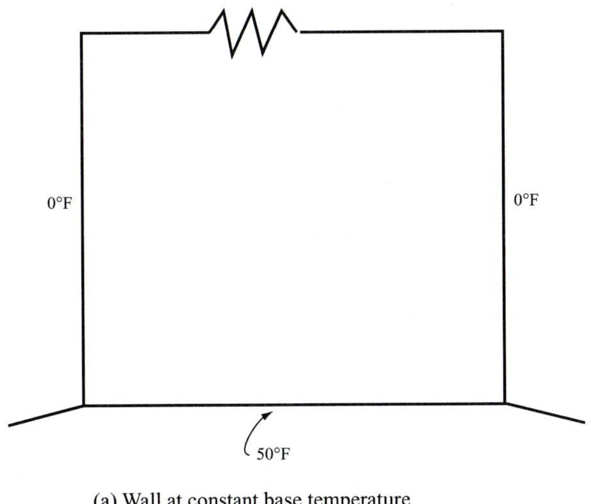

(a) Wall at constant base temperature

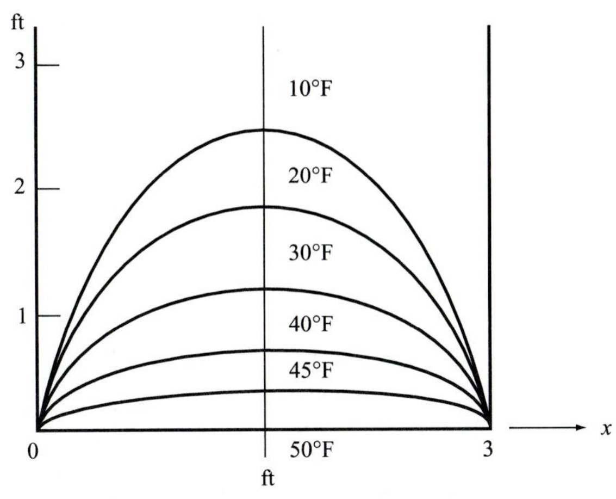

(b) Temperature distribution

Figure 2–19 Example 2–12.

where K is a constant and cosh $(\pi y/L)$ is the hyperbolic cosine function defined in appendix A–5. The isotherms and heat flow lines, given by equations (2–67) and (2–68), are shown in figure 2–19 for selected values. For instance, at a temperature of 40°F the constant in equation (2–67),

$$C = \tan(40\pi/100) = \tan 72° = 3.0777$$
$$\text{at } T = 30°\text{F}, C = \tan(30\pi/100) = \tan 54° = 1.3764$$
$$\text{at } T = 20°\text{F}, C = \tan(20\pi/100) = \tan 36° = 0.7265$$
$$\text{at } T = 10°\text{F}, C = \tan(10\pi/100) = \tan 18° = 0.3249$$
$$\text{at } T = 5°\text{F}, C = \tan(5\pi/100) = \tan 9° = 0.1584$$

These isotherms are plotted in figure 2–19b along with the heat-transfer directional lines, computed from equation (2–68). The heat-flow lines delineate paths of amounts of heat transfer, and it can be seen that these lines diverge or spread apart from the base. Therefore, the heat transfer per unit of area decreases away from the base and is most intense at the base. The heat transfer can be considered by using the method of example 2–11; however, the temperature gradient is undefined at the corners A and B in figure 2–19a. That is, the temperature has a discontinuity at $x = 0$, $y = 0$ and at $x = L$, $y = 0$ where the temperature is 0°F on one surface up to the corner and is 50°F along the base. This is an example of a large potential heat transfer created by a large temperature gradient. It is therefore an unrealistic assumption to insist on a constant temperature across a base as shown in figure 2–19.

In the next example the effect of a shorter wall or extension is considered. The temperature distribution is similar to that for t he infinite wall, but with some subtle differences in the mathematical analysis.

EXAMPLE 2–13 Consider a fin extending out from an internal combustion engine block, as shown in figure 2–20. Determine the temperature distribution through the fin if the surfaces of the fin are at 100°C and the base, where the fin is attached to the engine block surface, has a temperature distribution given by $T(x, y) = 100°\text{C} + 100°\text{C} \sin \pi y/L$. Determine the heat transfer through the fin.

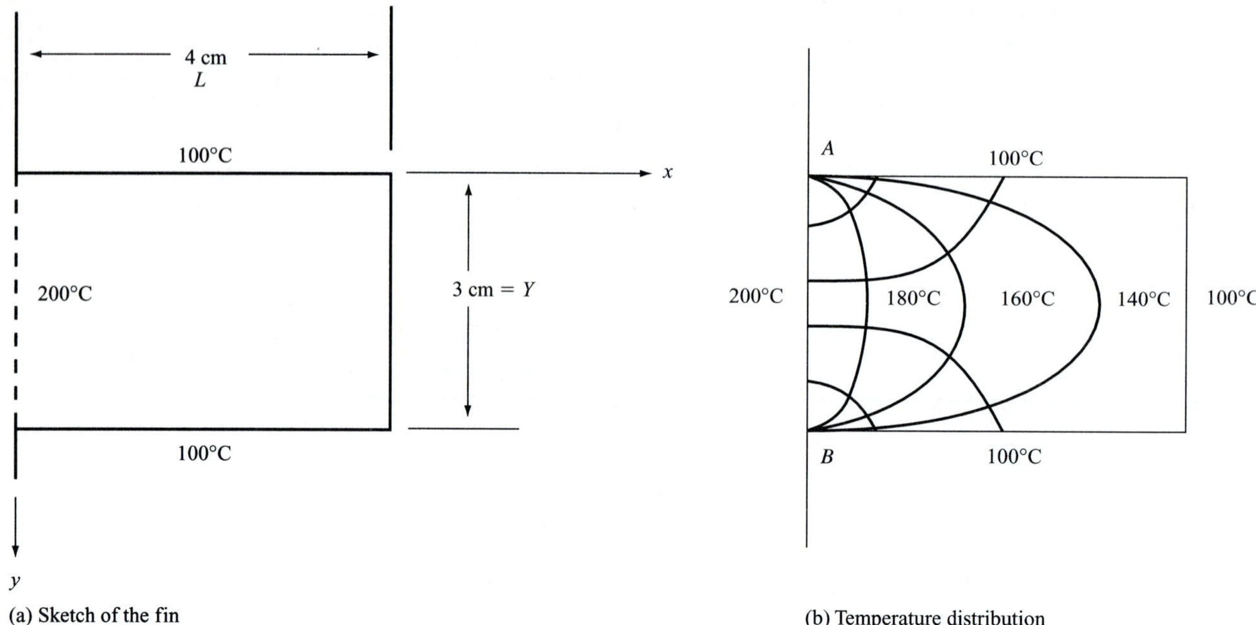

Figure 2–20 Heat transfer through a short fin.

Solution The fin is not infinitely long, so the previous analysis needs to be revised to consider the situation given in figure 2–20a. The algebraic manipulations are reduced if we define a temperature $T'(x, y) = T(x, y) - 100°\text{C}$. The governing equation is,

$$\frac{\partial^2 T}{\partial x^2} + \frac{\partial^2 T}{\partial y^2} = 0 \qquad 0 < x < L = 4\text{ cm}, 0 < y < Y = 3\text{ cm}$$

B.C. 1	$T'(x, y) = 0\ (T = 100°\text{C})$	at $y = 0, 0 < x < L$
B.C. 2	$T'(x, y) = 0°\text{C}$	at $y = Y, 0 < x < L$
B.C. 3	$T'(x, y) = 0°\text{C}$	at $0 < y < Y, x = L$
B.C. 4	$T'(x, y) = 100°\text{C} \sin \pi y/Y$	at $0 < y < Y, x = 0$

By using the methods just discussed it can be expected that a general form of a solution can be written as

$$T'(x, y) = \sum_{n=0}^{\infty} (a_n \sin p_n y + b_n \cos p_n y)(c_n e^{-p_n x} \pm d_n e^{p_n x})$$

Since the fin is not infinitely long, it is more convenient to write for the solution

$$T'(x, y) = \sum_{n=0}^{\infty} (a_n \sin p_n y + b_n \cos p_n y)(g_n \sinh p_n x + f_n \cosh p_n x)$$

where the hyperbolic functions, $\sinh p_n x$ and $\cosh p_n x$, replace the exponential functions and are defined as

$$\sinh p_n y = \frac{1}{2}(e^{p_n y} + e^{-p_n y})$$

$$\cosh p_n y = \frac{1}{2}(e^{p_n y} + e^{-p_n y})$$

and g_n and f_n are constants. An application of B.C. 1 requires that $b_n = 0$ and, from B.C. 3, $f_n = 0$ because $\cosh p_n x$ is not zero for any n. Further, because the temperature profile in the x direction at $y = Y/2$ (centerline) drops from 200°C at $x = 0$ to 100°C at $x = L$, the argument for the sinh term needs to be $n\pi(L - x)/Y$ instead of $n\pi x/Y$. The solution is then written, where $p_n = n\pi/Y$ and $A_n = g_n a_n$,

$$T'(x, y) = \sum_{n=1}^{\infty} A_n[\sinh p_n(L - x)]\sin p_n y$$

From B.C. 4, we have

$$100°\text{C} \sin\frac{n\pi}{Y} = \sum_{n=1}^{\infty} A_n \sinh p_n L \sin p_n y$$

Applying the definition of the Fourier Coefficient of equation (2–63) and recognizing that $A_n \sinh p_n L$ is a constant gives

$$A_n \sinh p_n L = \frac{1}{2}\int_0^Y 100°\text{C} \sin\frac{\pi y}{Y} \sin\frac{n\pi y}{Y} dy$$

$$= \frac{200°\text{C}}{Y}\int_0^Y \sin\frac{\pi y}{Y} \sin\frac{n\pi y}{Y} dy$$

similar to the results from example 2–11

$$A_n \sinh p_n L = 100°\text{C} \quad \text{or} \quad A_n = \frac{100°\text{C}}{\sinh p_n L}$$

and $A_n = 0$ for $n > 0$.

Therefore the solution can be written as

$$T(x, y) = 100°\text{C} + \frac{100°\text{C}}{\sinh p_n L} \sinh p_n(L - x) \sin p_n y$$

or

$$T(x, y) = 100°\text{C} + \frac{100°\text{C}}{\sinh(\pi L/Y)} \sinh\frac{\pi}{L}(L - x)\sin\frac{\pi y}{Y} \tag{2–69}$$

Isotherms of 120°C, 140°C, 160°C, and 180°C, which were predicted by using equation (2–69), are shown in figure 2–20b. Notice how similar is the temperature distribution of the infinitely long wall and the short wall. The heat transfer through the extension may be determined by the same method used in example 2–11. First the temperature gradient in the y direction is found at the base, where x is zero:

$$\frac{\partial T(0, y)}{\partial x} = -\frac{100°\text{C}\pi}{Y \sinh\frac{\pi L}{Y}} \cosh\frac{\pi}{Y}(L - 0) \sin\frac{\pi y}{Y} = -\frac{100°\text{C}\pi}{Y \sinh(\pi L/Y)} \cosh\frac{\pi L}{Y} \sin\frac{\pi y}{Y}$$

The heat transfer per unit of area is

$$\dot{q}_A = -\kappa\frac{\partial T(0, y)}{\partial x}$$

and over the full base

$$\dot{Q}_y = \int_0^Y \dot{q}_A \, dy \qquad \text{(assuming unit depth)}$$

We carry through with the calculation

$$\dot{Q}_y = \frac{100°\text{C}\kappa\pi}{Y\tanh\frac{\pi L}{Y}}\int_0^Y \sin\frac{\pi y}{Y}\,dy$$

In this example $L = 4$ cm, $Y = 3$ cm, and $\kappa = 40$ W/m · °C,

Answer

$$\dot{Q}_y = 10{,}894.9\,\frac{\text{W}}{\text{m}} = 108.94\,\frac{\text{W}}{\text{cm}}$$

The results from this example can be generalized to problems that are similar but with different temperature distributions at the base, $x = 0$. For the general problem of a finite length (L) wall, fin, or extension of constant width (Y) having a temperature T_0 at its three exposed surfaces and having temperature distribution $T(x, y)$ at the base where $x = 0$, the temperature distribution can be written as

$$T(x, y) = T_0 + \sum_{n=0}^{\infty} \frac{A_n}{\sinh\,(n\pi L/Y)} \sin\frac{n\pi y}{Y} \sinh\frac{n\pi}{Y}(L - x) \tag{2–70}$$

where A_n is determined from the relationship

$$A_n = \frac{2}{Y}\int_0^y T(x, 0)\sin\frac{n\pi y}{Y}dy \tag{2–63, repeated}$$

Other configurations involving two- or three-dimensional conduction heat transfer, which are treated in rectilinear, Cartesian coordinates, have been studied and some of the results are compiled by Carslaw and Jaeger [7] and Arpaci [11], among others. Many engineering problems, however, involve objects and heat transfers that are best described in cylindrical or spherical coordinate systems. We consider one of those types of problems in the following example.

EXAMPLE 2–14 A ceramic ($\kappa = 1.25$ Btu/hr · ft · °F) plug of uniform diameter 1 in. and length of $^5\!/_8$ in. (0.625 in.) is to be fired in a furnace. The plug is placed on end as shown in figure 2–21, is at 400°F on that end, and is at 500°F on the other surfaces. The plug is at steady state, and the temperature distribution through the plug is to be determined.

Solution

The plug, resting on one end, is experiencing axisymmetric conduction heat transfer. That is, heat transfer is occurring radially and longitudinally but not angularly around the plug. The governing equation for conduction heat transfer with constant thermal conductivity is then

$$\frac{\partial^2 T(r, z)}{\partial r^2} + \frac{1}{r}\frac{\partial T(r, z)}{\partial r} + \frac{\partial^2 T(r, z)}{\partial z^2} = 0$$

It is convenient to assume that the temperature function can be separated such that

$$T(r, z) = R(r) \cdot Z(z) \tag{2–71}$$

Then, as we saw with rectangular coordinates, the separation of variables technique provides two ordinary differential equations,

$$\frac{d^2 Z(z)}{dz^2} = p^2 Z(z) \tag{2–72}$$

$$\frac{d^2 R(r)}{dr^2} + \frac{1}{r}\frac{dR(r)}{dr} = -p^2 R(r) \tag{2–73}$$

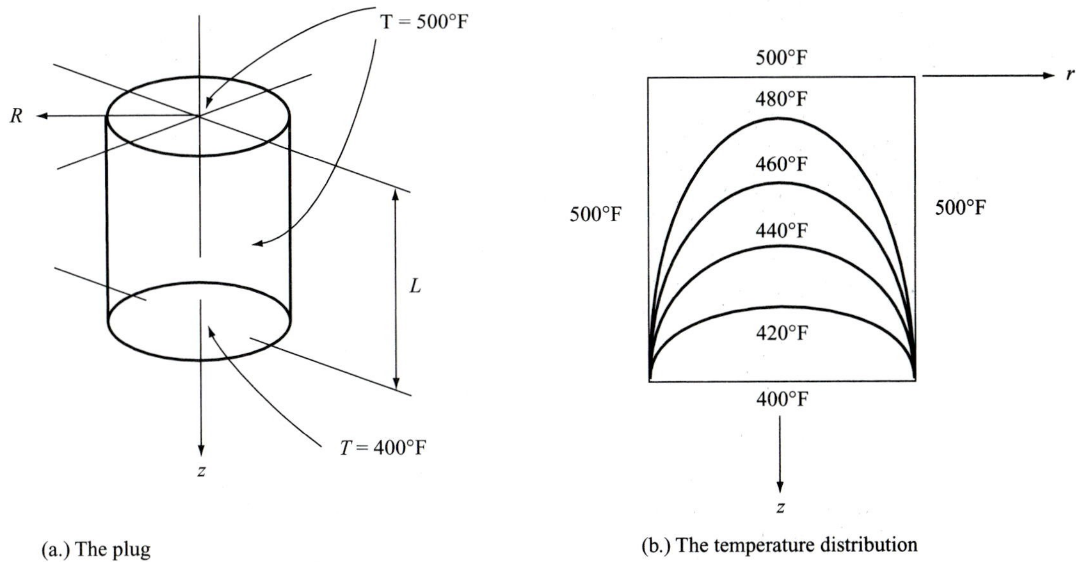

Figure 2–21 Ceramic plug fired in a kiln, example 2–14.

The solution to equation (2–72) may be written as

$$Z(z) = a \sinh pz + b \cosh pz$$

Equation (2–73) is a particular second order differential equation called *Bessel's Equation* of zero order (see appendix A–10). The solution to Bessel's Equation of zero order is, for *R(r)*,

$$R(r) = cJ_0(pr) + dY_0(pr)$$

where $J_0(pr)$ = Bessel Function of the first kind and zero order having argument *pr*
$Y_0(pr)$ = Bessel Function of the second kind and zero order having argument *pr*

In appendix table A10–1 are given some selected values for these two special functions for an argument $x(=pr)$ and in figure A10–2 these functions are graphically displayed. The reader may want to refer to that information to get a better understanding of the character of the Bessel Functions. If we define a temperature function

$$T'(r, z) = T(r, z) - 500°F \tag{2–74}$$

then we may write boundary conditions as

B.C. 1	$T'(r, z) = 0$	at $0 < r < R, z = 0$
B.C. 2	$T'(r, z) = 0$	at $r = R, 0 < z < L$
B.C. 3	$T'(r, z) =$ finite	at $r = 0, 0 < z < L$
B.C. 4	$T'(r, z) = -100°F$	at $0 < r < R, z = L$

The general solution is

$$T'(r, z) = (a \sinh pz + b \cosh pz)(cJ_0pr + dY_0pr) \tag{2–75}$$

and from B.C. 1 the constant $b = 0$. Also, from B.C. 3 the constant $d = 0$ because $Y_0(0) \to \infty$. Then the solution is

$$T'(r, z) = AJ_o(pr) \sinh pz$$

From B.C. 2 we have

$$0 = AJ_0(pR) \sinh pz$$

which can only be satisfied, for all z, if

$$J_0(pR) = 0$$

From a study of the character of the Bessel Function J_0 it is found from appendix table A10–1 that zero values occur at the following arguments:

$$\begin{aligned} pR &= 2.4048 = p_1R \\ pR &= 5.5201 = p_2R \\ pR &= 8.6537 = p_3R \\ pR &= 11.7915 = p_4R \\ pR &= 14.9309 = p_5R \end{aligned}$$

and so on. Further arguments for zero values of the Bessel Function can be numerically determined. It can be determined that the difference between succeeding arguments approaches π (pi) and even at the fourth increment above, $p_5R - p_4R = 3.1394$, the difference is nearly pi. Thus, $p_6R \approx 14.9309 + 3.14159 = 18.07249$.

The complete solution to the temperature distribution through the plug may then be written

$$T'(r, z) = \sum_{n=1}^{\infty} A_n J_0(pr) \sinh p_n z \tag{2–76}$$

where $p_n = x_n/R$ and x_n is the argument of $J_0(x_n) = 0$. From B.C. 4 we also must have that

$$-100°\text{F} = \sum_{n=1}^{\infty} A_n J_0 (pr) \sinh p_n L \tag{2–77}$$

It can be shown that, based on an analysis similar to the Fourier Coefficients, that

$$A_n \sinh p_n L = \frac{2(-100°\text{F})}{R^2 J_1^2(p_n R)} \int_0^R r J_0(p_n r)\, dr \tag{2–78}$$

where $J_1(p_n R)$ is the Bessel Function of the first kind and first order. Further (see appendix A–10),

$$\int_0^R r J_0(p_n r)\, dr = \frac{R}{p_n} J_1(p_n r)$$

so that the solution is

$$T'(r, z) = \frac{-200°\text{F}}{R} \sum_{n=1}^{\infty} \frac{J_0(p_n r) \sinh p_n z}{p_n[\sinh(p_n L)][J_1(p_n R)]} \tag{2–79}$$

The isotherms of 420°F, 440°F, 460°F, and 480°F are shown in figure 2–21b. The heat transfer cannot be explicitly determined because of the discontinuity in the temperature at the outer rim of the base, where $r = R$ and $z = 0$.

The general solution for the temperature distribution in a cylinder as shown in figure 2–21, where one end and the cylindrical surface are at a constant temperature, T_0, while the remaining end surface is at T_f can be written, from equation (2–79),

$$T(r, z) = T_0 + (T_f - T_0)\left(\frac{2}{R}\right) \sum_{n=1}^{\infty} \frac{[J_0(p_n r)][\sinh(p_n z)]}{(p_n)[\sinh(p_n L)][J_1(p_n R)]} \tag{2–80}$$

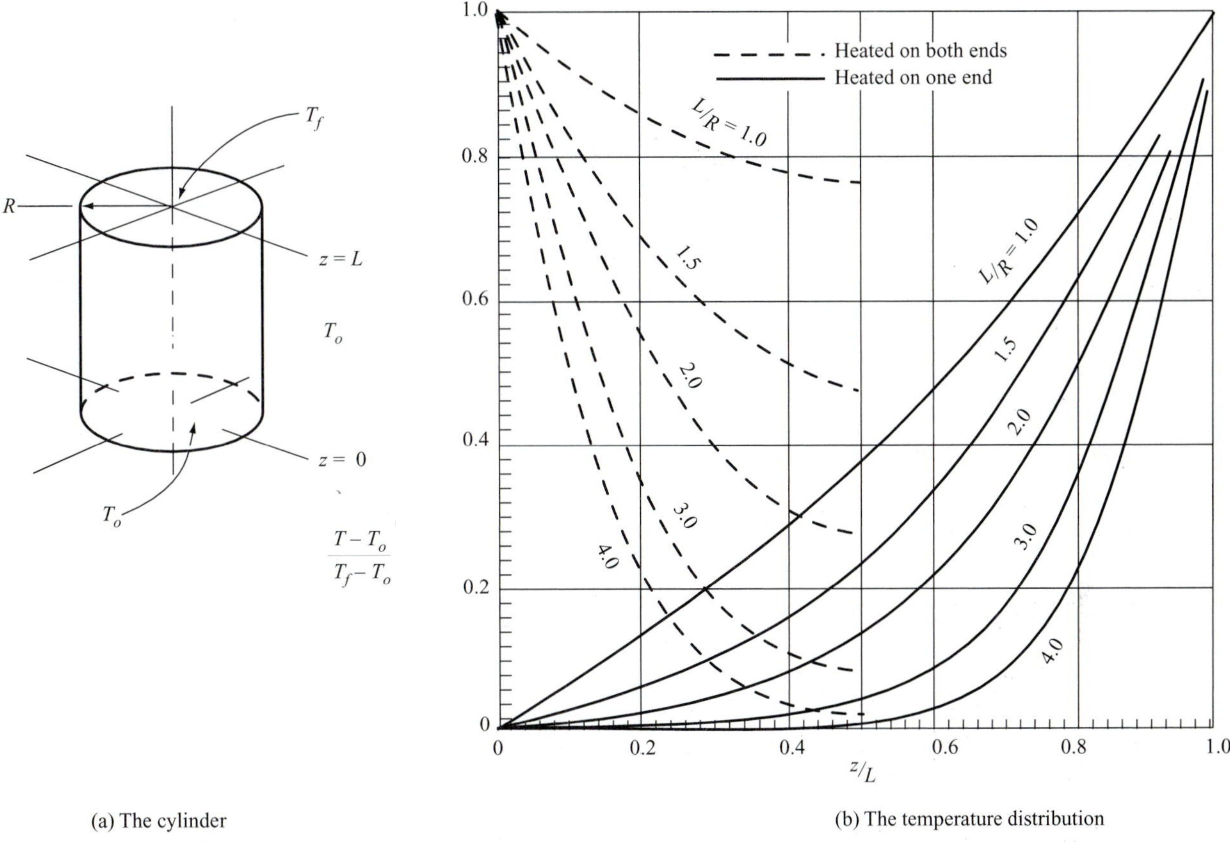

(a) The cylinder

(b) The temperature distribution

Figure 2–22 Centerline temperature in a solid cylinder.

This result can be used to aid in determining numerical solutions to the problem of a cylindrical plug with constant temperature, T_0, on all surfaces except one end, which is at T_f as shown in figure 2–22a. The temperature along the axis, at $r = 0$, is determined from equation (2–80) but with the simplification that $J_0(p_n 0) = 1$ so

$$T(0, z) = T_0 + (T_f - T_0)\left(\frac{2}{R}\right)\sum_{n=1}^{\infty} \frac{\sinh(p_n z)}{(p_n)[\sinh(p_n L)][J_1(p_n R)]}$$

This result is graphically presented in figure 2–22b, and the result can be extended, by using a principle of superposition, to the problem of a cylinder having both ends at different constant temperatures than the cylindrical surface. The following examples demonstrate the use of the data from figure 2–22.

EXAMPLE 2.15 A wood dowel 4 cm in diameter and 4 cm long is exposed to 200°C at one end while its other surfaces are at 30°C. What is its temperature at the center of the axis.

Solution The length to radius ratio, L/R, is 2, and at the center of the dowel $z/L = 0.5$. From figure 2–22b we have

$$\frac{T - T_0}{T_f - T_0} \approx 0.14$$

Since $T_0 = 30°C$ and $T_f = 200°C$, then

Answer
$$T = (0.14)\,(T_f - T_0) + T_0 = 53.8°C$$

EXAMPLE 2–16 For the wood dowel of example 2–15, what is the center temperature if the cylindrical surface is warmed to 50°C while one end is still at 30°C and the other end is at 200°C.

Solution We may use a principle of superposition to solve for the centerline temperature. First we recognize that the dowel temperature can be modeled as two separate dowel temperature distributions: one dowel having surface temperatures of 50°C, except one end at 30°C, and another dowel at 50°C with an end (opposite end) at 200°C. For the 30°C-end dowel

$$T_{30} - T_0 = (0.14)\,(30°\text{C} - 50°\text{C}) = -2.8°\text{C}$$

and for the 200°C-end dowel,

$$T_{200} - T_0 = (0.14)\,(200 - 50) = 21°\text{C}$$

The superposition principle can be applied by adding the two solutions; that is

$$(T_c - T_0) = (T_{200} - T_0) + (T_{30} - T_0)$$

Then

Answer

$$T = 21°\text{C} - 2.8°\text{C} + 50°\text{C} = 68.2°\text{C}$$

Notice that the graph in figure 2–22 can also be used to determine the temperature distribution along the centerline axis for a given L/R ratio where z is the independent variable. Further, the same graph can be used for temperature distribution along the centerline axis where the ends are heated, or cooled, to the same end temperature by using the dashed line. This would make the superposition problem demonstrated in example 2–16 somewhat shorter.

Other conduction heat transfer problems involving cylindrical and spherical coordinate systems have been considered, and some of the results have been compiled by Carslaw and Jaeger [7], Arpaci [11], and others. In most of the problems where analytic solutions have been obtained, the heat transfer occurs with simple geometric shapes and isothermal boundary conditions. Complicated geometric objects and nonisothermal boundary conditions can create mathematical complexities that prevent setting down complete solutions. The topics in the next section address some of the more straightforward results of those problems that have been solved, and then methods or techniques for obtaining approximate solutions to more complicated (but more realistic) problems are shown.

2–5 SHAPE FACTOR METHODS

In the preceding discussions of conduction heat transfer it was shown that some two-dimensional heat-transfer problems can be analyzed by mathematical methods. Many of the complicated derivations for problems more involved than those considered are beyond the purposes of this text; however, the engineer or technologist should be able to use the results from those studies. In particular, the general problem of conduction heat transfer between two isothermal surfaces, whether one-, two-, or three-dimensional, sketched in figure 2–23 can be treated by means of the *shape factor, S,* defined by,

$$\dot{Q} = S\kappa\Delta T \tag{2–81}$$

where κ is the thermal conductivity (presumed to be a constant) and ΔT is the temperature difference between the two surfaces. In table 2–3 are listed some of the most common configurations where shape factors have been determined and can assist in analyzing heat transfer processes. Notice that some cases are one-dimensional, some are two-dimensional, and others are three-dimensional, and in all cases the use of equation (2–81) is implied. A limitation of this method is that temperature distributions cannot be determined.

Figure 2–23 General configuration of conduction heat transfer between two isothermal surfaces.

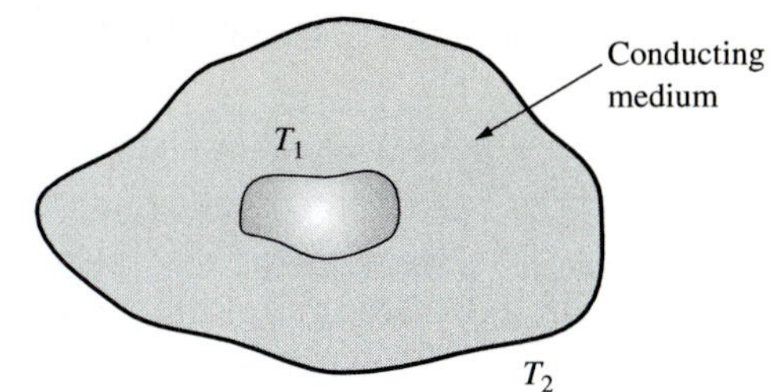

Table 2–3 Conduction Shape Factors

System	Sketch	Shape Factor	Assumption
1. Plane wall, one-dimensional heat transfer		$\frac{A}{Y} = \frac{WL}{Y}$	$W \gg Y, L \gg Y$
2. Annular cylinder		$\frac{2\pi L}{\ln r_o/r_i}$	$L \gg r_o$
3. Spherical shell	Cross Section	$\frac{4\pi r_o r_i}{r_o - r_i}$	
4. Square bar with center hole		$\frac{2\pi L}{\ln(0.54W/r)}$	$L \gg W$
5. Cylindrical bar with eccentric hole		$\frac{2\pi L}{\cosh^{-1}\left[\frac{r_1^2 + r_2^2 - Y^2}{2r_1r_2}\right]}$	$L \gg r_2$
6. Cylinder in a semi-infinite medium, placed vertically and one end flush with surface		$\frac{2\pi L}{\ln(2L/r)}$	$L \gg 2r$

Table 2–3 (continued) Conduction Shape Factors

System	Sketch	Shape Factor	Assumption
7. Rectangular parallelopiped surrounded by semi-infinite medium	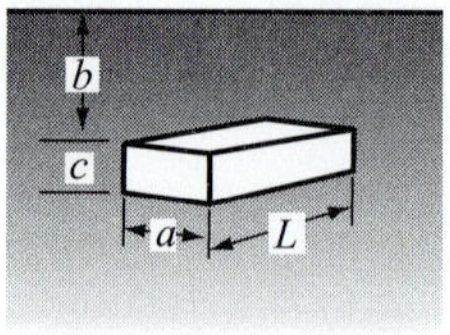	$\dfrac{1.685L}{\left[\log\left(1+\dfrac{b}{a}\right)\right]^{0.59}}\left(\dfrac{c}{b}\right)^{0.078}$	
8. Cylinder surrounded by semi-infinite medium	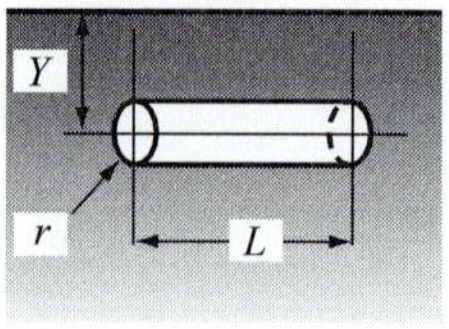	$\dfrac{2\pi L}{\cosh^{-1}\dfrac{Y}{r}}$	$L \gg r$
		$\dfrac{2\pi L}{\ln(2Y/r)}$	$L \gg r, Y > 3r$
		$\dfrac{2\pi L}{\ln\dfrac{L}{r}\left(1-\dfrac{\ln(L/2Y)}{\ln(L/r)}\right)}$	$Y \gg r, L \gg Y$
9. Sphere surrounded by infinite medium		$4\pi r$	
10. Sphere surrounded by semi-infinite medium	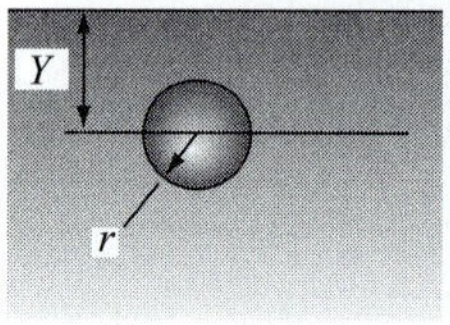	$\dfrac{4\pi r}{\left(1-\dfrac{r}{2Y}\right)}$	
11. Conduction between two isothermal parallel cylinders surrounded by an infinite medium	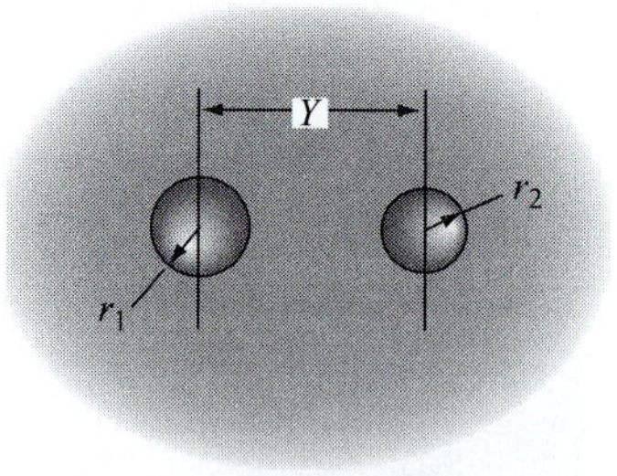	$\dfrac{2\pi}{\cosh^{-1}\left(\dfrac{L^2-1-r^2}{2r}\right)}$	$r = r_1/r_2, L = Y/r_2$
12. Conduction between two isothermal spheres in an infinite medium	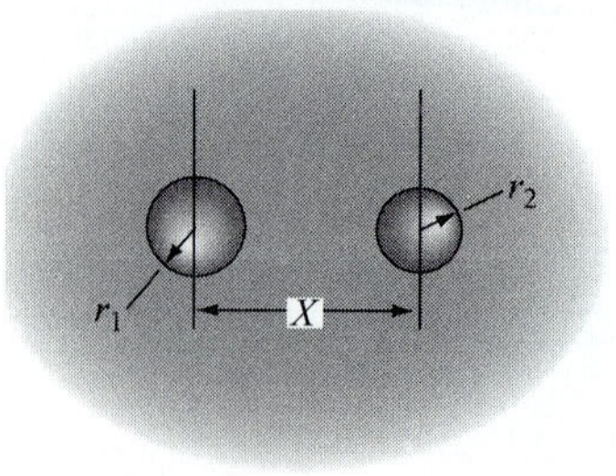	$\dfrac{4\pi}{\dfrac{r_2}{r_1}\left(1-\dfrac{(r_1/X)^4}{1-(r_2/X)^2}\right)-\dfrac{2r_2}{X}}$	$x > 5r_{max}$

Table 2–3 (continued) Conduction Shape Factors

System	Sketch	Shape Factor	Assumption
13. Thin rectangular plate surrounded by a semi-infinite medium, parallel to surface		$\dfrac{\pi W}{\ln(4W/L)}$	$Y = 0, W > L$
		$\dfrac{2\pi W}{\ln(4W/L)}$	$Y \gg W, W > L$
		$\dfrac{2\pi W}{\ln(2\pi Y/L)}$	$Y > 2W, W \gg L$
14. Conduction between two parallel disks surrounded by an infinite medium	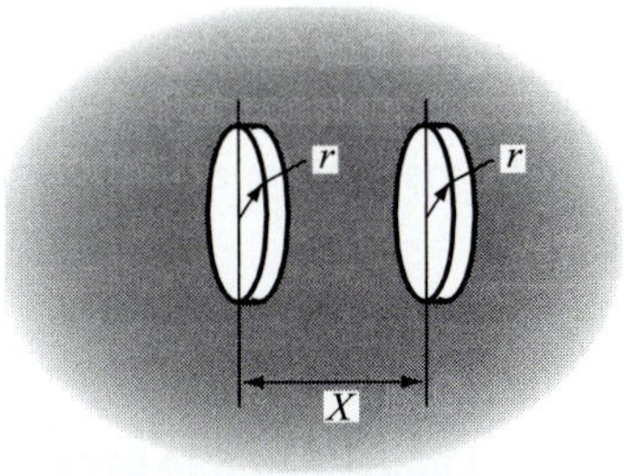	$\dfrac{4\pi}{2\left[\dfrac{\pi}{2} - \tan^{-1}\left(\dfrac{r}{X}\right)\right]}$	$x > 5r$
15. Thin disk surrounded by semi-infinite medium and parallel to surface		$4r$	$Y = 0$
		$8r$	$Y \gg 2r$
16. Sphere surrounded by semi-infinite medium with adiabatic surface.	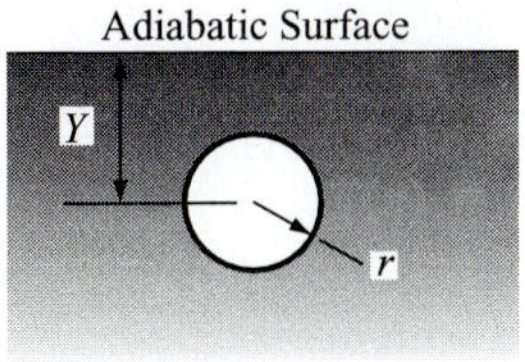 Adiabatic Surface	$\dfrac{4\pi r}{1 + (r/2Y)}$	
17. Edge with conduction between one line and opposite two surfaces	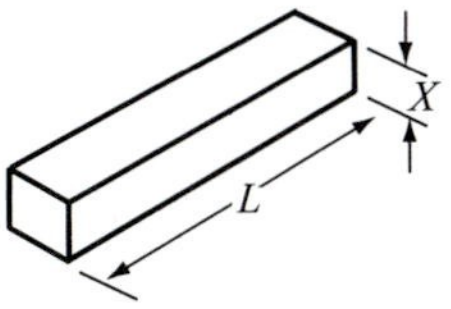	$0.559\ L$	$L \gg X$
18. Square cube with conduction between one corner and opposite three surfaces	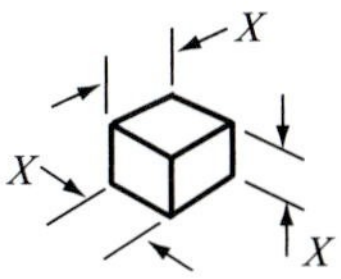	$0.15\ X$	

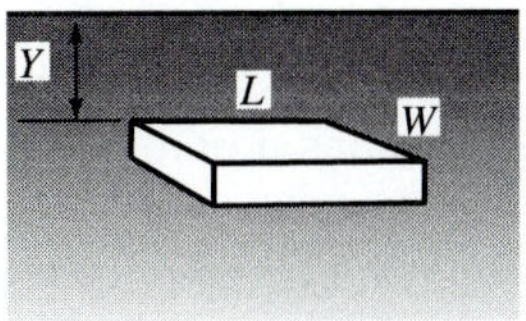

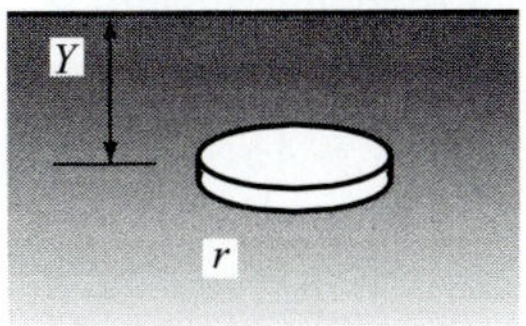

EXAMPLE 2–17 A steel spherical container 3 ft in diameter is buried in a landfill. The container is filled with a chemical that keeps the outer surface of the barrel at 100°F while the earth's surface is at 50°F. Determine the heat transfer from the container if it is buried under 3 ft of earth. Then compare this heat loss to the loss that would occur if the container were buried under 200 ft of earth and the earth has a nominal temperature of 55°F.

Solution The heat transfer may be predicted by using equation (2–81) with a shape factor of system 10 from table 2–3

$$S = \frac{4\pi r}{1 - \dfrac{r}{2Y}} = \frac{4\pi(1.5\text{ ft})}{\left(1 - \dfrac{1.5\text{ ft}}{2(3\text{ ft})}\right)} - 25.13\text{ ft}$$

From appendix table B–2 a thermal conductivity value for soil may be used, which is 0.301 Btu/hr · ft · °R. Then

Answer
$$\dot{Q} = S\kappa\Delta T = (25.13\text{ ft})\left(0.301\frac{\text{Btu}}{\text{hr}\cdot\text{ft}\cdot{}^\circ\text{F}}\right)(100^\circ\text{F} - 50^\circ\text{F}) = 378.3\frac{\text{Btu}}{\text{hr}}$$

If the container were buried under 200 ft of earth, the heat loss would be similar to that of a sphere buried in an infinite medium, system 9 of table 2–3. Comparing the shape factors of systems 10 and 9,

$$S_9 = 4\pi r = 4\pi(1.5\text{ ft}) = 18.85\text{ ft}$$

$$S_{10} = \frac{4\pi r}{1 - \dfrac{r}{2Y}} = 18.92\text{ ft} \qquad \text{for } Y = 200\text{ ft}$$

We assume that the earth above the container, which is between the container and ground level, is at 55°F so that the heat transfer acts as if it were from a sphere in an infinite medium. The heat loss is

Answer
$$\dot{Q} = S\kappa\Delta T = (18.85\text{ ft})\left(0.301\frac{\text{Btu}}{\text{hr}\cdot\text{ft}\cdot{}^\circ\text{F}}\right)(100^\circ\text{F} - 50^\circ\text{F}) = 255.3\frac{\text{Btu}}{\text{hr}}$$

Thus, burying the container in earth at greater depth tends to reduce the heat loss.

EXAMPLE 2–18 Estimate the heat gain of a thick-walled ice chest shown in figure 2–24. Assume the outer surface is 35°C and the inner surface is −5°C. The ice chest is made of styrofoam with a thermal conductivity value of 0.029 W/m · °C.

Solution The ice chest may be visualized as 6 plane walls of the system 1 type (table 2–3), 12 edges of the system 17 type, and 8 corners of the system 18 type.

$$S_{\text{corner}} = 8 \times 0.15L = 8 \times 0.15 \times 0.05\text{ m} = 0.06\text{ m}$$

$$S_{\text{edge}} = 4 \times 0.559 \times (04\text{ m} + 0.4\text{ m} + 0.5\text{ m}) = 2.9068\text{ m}$$

$$S_{\text{walls}} = \frac{1}{0.05\text{ m}}[(4 \times 0.5 \times 0.4\text{ m}^2) + (2 \times 0.4 \times 0.4\text{ m}^2)] = 22.4\text{ m}$$

Since these components all have the same temperature difference, we may add them all up and then use equation (2–81). The result is

$$S_{\text{total}} = 25.3068\text{ m}$$

and

Answer
$$\dot{Q} = S_{\text{total}}\,\kappa\Delta T = (25.3068\text{ m})\left(0.029\frac{\text{W}}{\text{m}\cdot{}^\circ\text{C}}\right)(35^\circ\text{C} - 15^\circ\text{C}) = 29.36\text{ W}$$

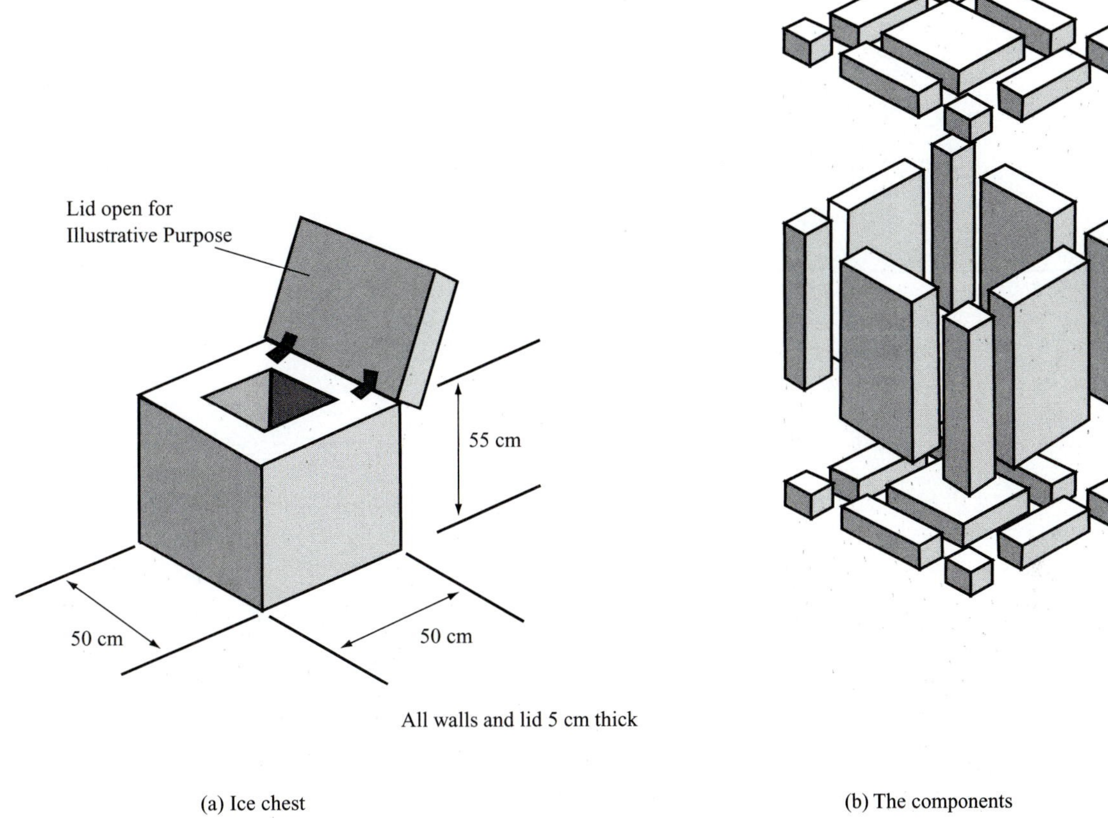

Figure 2–24 Shape factors to determine heat gain in an ice chest.

There are other shape factors than those given in table 2–3 that can be found in the literature (see, for example [12]); however, many systems will not have identified shape factors.

For many systems and problems a good approximation for heat transfer may be obtained by using a graphical method that uses for its authority the separation of variables with the Laplace Equation and the orthogonal characteristic of the isotherms and heat-flow lines. The method, shown in figure 2–25, is restricted to two-dimensional conduction heat transfer in rectangular coordinates with isothermal boundary conditions. It requires that isotherms be sketched approximately parallel to the inner and outer surfaces, which are at temperatures T_1 and T_2 respectively. The number of isotherms that are constructed is arbitrary, and the student will find that a correct judgment of this number will be developed after attempting to solve some problems. In figure 2–25, 4 isotherms between the inner and the outer surfaces are identified, and they are spaced unequally. This unequal spacing was made because freehand heat-flow lines need to be constructed perpendicular to the isotherms and spaced such that a grid of approximate squares results, as indicated in figure 2–25b. That is, $\Delta x \approx \Delta y$, where Δx is the spacing of the isotherms and Δy is the spacing of the heat-flow paths.

It can be seen that the geometry requires that the heat-flow paths diverge outward and so Δy increases. As a consequence of this divergence the corresponding Δx must also increase, and this is reflected in the unequal spacing of the isotherms. It requires some

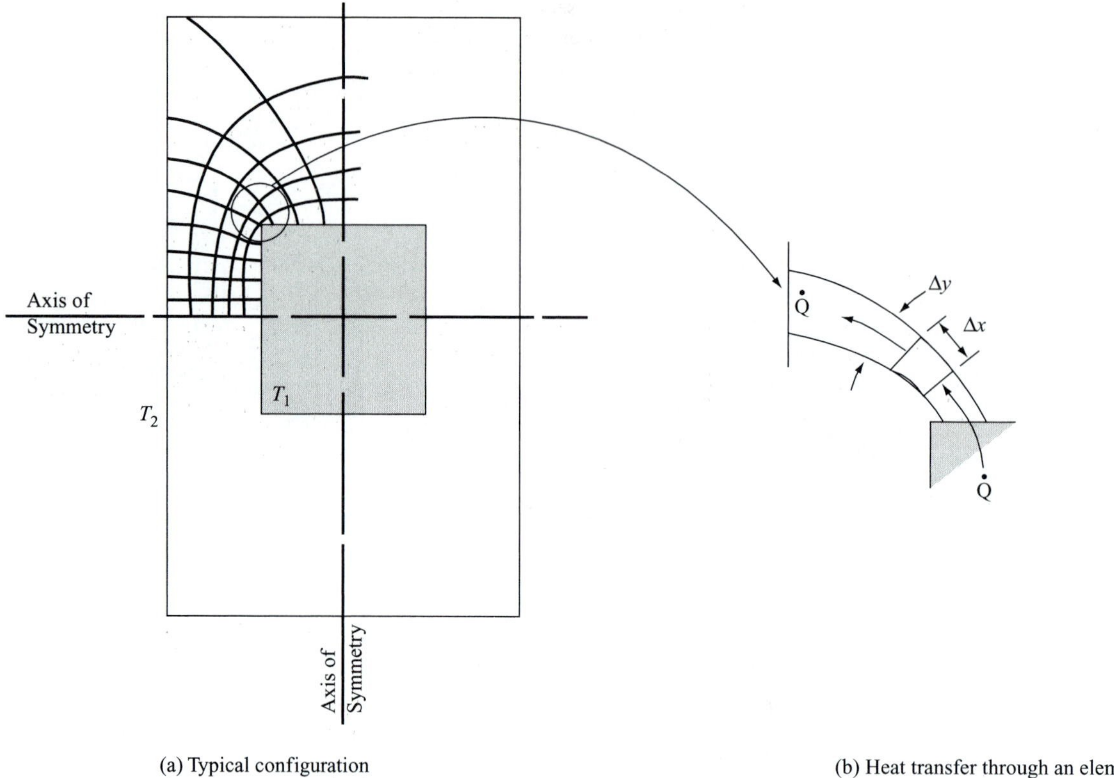

Figure 2–25 Graphical method of analyzing two-dimensional conduction heat transfer.

practice to complete a reasonably accurate sketch such as shown in figure 2–25, but with a soft pencil, a good eraser, and some paper, the student can gain a competency in solving nearly all problems in two-dimensional, Cartesian coordinates, conduction heat transfer with constant thermal conductivity. After constructing the isotherm/heat flow grid the heat transfer through a particular path such as that shown in figure 2–25b is, where L is the depth,

$$\dot{Q}_{\text{path}} = \kappa L \Delta y \frac{\Delta t}{\Delta x} \approx \kappa L \Delta T$$

since $\Delta x \approx \Delta y$. Also,

$$\Delta T = \frac{\Delta T_{\text{overall}}}{N} = \frac{T_1 - T_2}{N}$$

where N is the number of elements through which the heat transfer must pass between T_1 and T_2. Notice that the heat transfer is two-dimensional but the heat transfer Q_{path} is constrained to the identified path and heat transfer cannot occur between adjacent heat-flow paths. Therefore, the lines of geometric symmetry that are sketched in figure 2–25 are heat-flow lines because heat cannot cross lines of symmetry. From Chapter 1, recall that lines or surfaces that do not allow heat transfer are called adiabatic surfaces, or just *adiabats*. Lines of symmetry and heat flow lines are adiabats. Then the total heat transfer between the surface at T_1 and the surface at T_2 is the sum of the various Q_{path} terms, or

$$\dot{Q} = \kappa L M \Delta T = \kappa L \frac{M}{N} \Delta T_{overall} \tag{2–82}$$

where M is the number of heat flow paths. Notice that LM/N is the same as the shape factor in equation (2–81), for this special case. For the example in figure 2–25 the number of elements, N, is 5, and the number of paths, M, is 36 (= 4 × 9), so $M/N = 7.2$ (unitless). We saw in examples 2–11, 2–12, and 2–13 that mathematical analysis can be used to obtain the same sorts of isotherm/ heat-flow grids in Cartesian coordinates as those shown here. It is suggested that for many problems an adequate representation of a solution can be obtained by using freehand graphical techniques. The graphical method provides a quick and approximate solution that includes, along with the predicted heat transfer, the temperature distribution in the conducting medium. Techniques can also be applied to obtain an approximate solution to the heat transfer when mathematical difficulties arise, such as in example 2–12, where a temperature gradient is undefined at a corner.

EXAMPLE 2–19 Estimate the heat transfer through the fin of example 2–12.

Solution By referring to example 2–12, Figure 2–19, the temperature distribution can be visualized. Without this information a freehand sketch can be approximated by noting that the temperature is constant at the base and must gradually reduce to the boundary temperature of 0°F as y → ∞. The identical temperature distribution as in figure 2–19 should result. The heat-flow lines are then sketched perpendicular to the isotherms and spaced such that square grids are obtained. The solution to this problem is shown in figure 2–26b. The number of heat flow paths is 14 (= 7 × 2) and the number of elements is 6. Then, for unit depth (or L = 1),

Answer

$$\dot{Q} = \kappa \frac{M}{N} \Delta T_{overall} = \left(0.809 \frac{\text{Btu}}{\text{hr} \cdot \text{ft} \cdot {}^\circ\text{F}}\right)\left(\frac{14}{6}\right)[50^\circ\text{F} - (-5^\circ\text{F})] = 94.64 \frac{\text{Btu}}{\text{hr}}$$

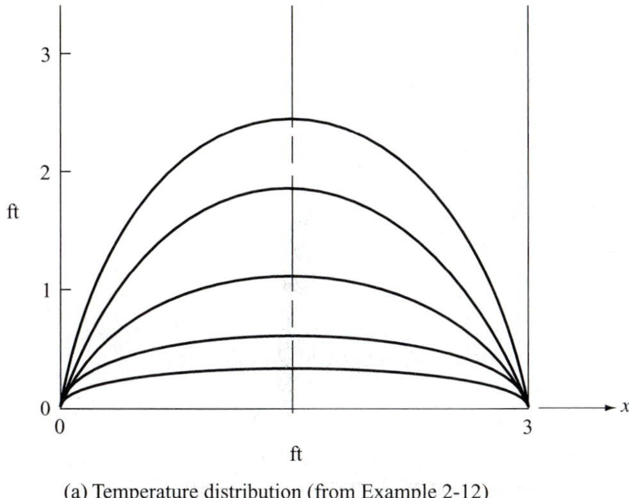

(a) Temperature distribution (from Example 2-12)

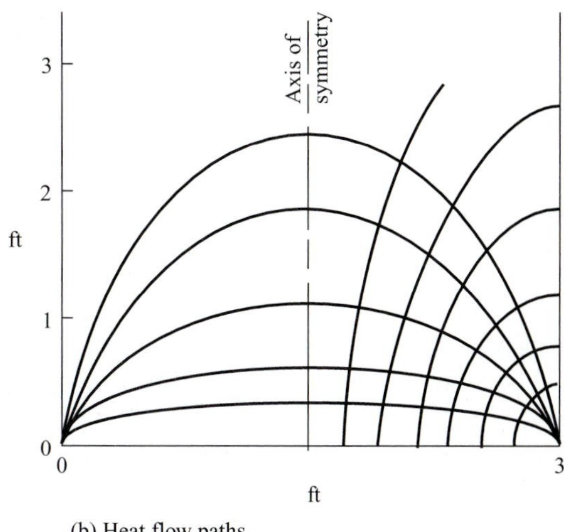

(b) Heat flow paths

Figure 2–26 Graphical solution to heat transfer through a concrete wall with uniform temperature at its base.

Notice that the temperature discontinuity at the bottom corners is troublesome, even with the sketch. There is a valid question whether one or two more heat paths should be included at each corner. Surely the answer obtained here is correct to within no more than $\pm 10\%$ of the true amount.

2–6 NUMERICAL METHODS OF ANALYSIS

A numerical method for analyzing conduction heat transfer that yields approximate solutions for temperatures and/or heat transfers in one-, two-, or three-dimensional regions is the *Finite Difference Technique.* The method, which is the basis of nearly all of the commercially available computer software programs involving conduction heat transfer, can be used for complicated systems and for situations when thermal conductivity varies. It can also be a convenient method for non–steady state heat transfer problems, as we will see in the next chapter. Finite differences uses the approximations for temperature gradients,

$$\frac{\partial T}{\partial x} \approx \frac{\Delta T}{\Delta x} \quad \text{and} \quad \frac{\partial T}{\partial y} \approx \frac{\Delta T}{\Delta y} \tag{2–83}$$

which are finite temperature gradients to study heat transfer occurring in materials. Consider, for instance, the situation of a temperature distribution $T(x, y)$ and the resulting temperature gradient in the plane of constant y as shown in figure 2–27. The temperature gradient, $\partial T/\partial x$, is approximated by equation (2–83a),

$$\frac{\partial T(x, y)}{\partial x} \approx \frac{[T(n + 1, y) - T(n - 1, y)]}{2\Delta x} \tag{2–84}$$

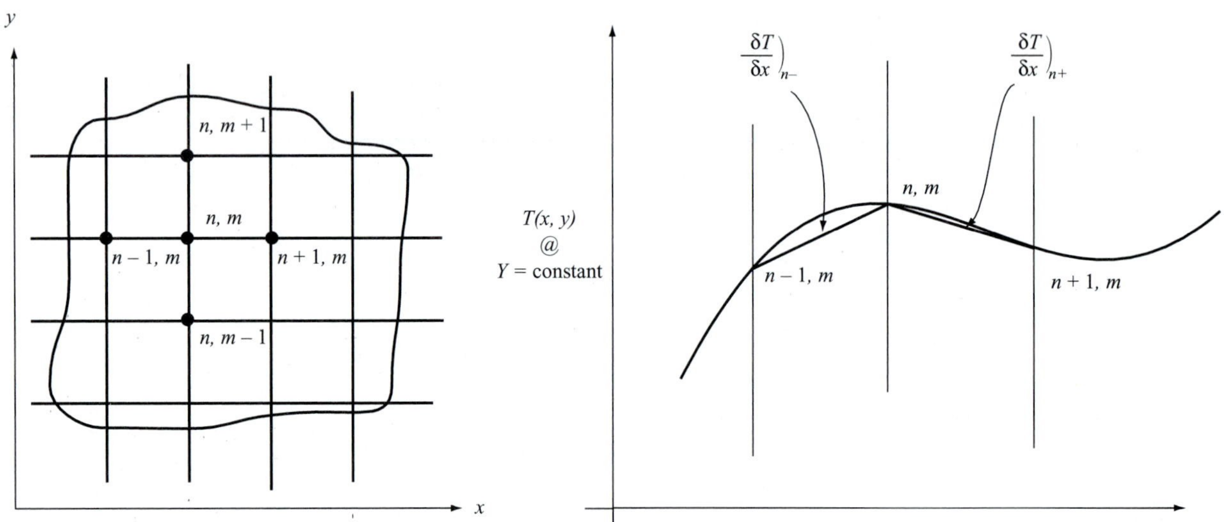

Figure 2–27 Finite difference approximations of temperature gradients.

Two other identifiable approximate temperature gradients can be associated with the exact temperature distribution at point n. These are the approximations to the left partial derivative,

$$\frac{\partial T(n + 0, y)}{\partial x} \approx \frac{[T(n, y) - T(n - 1, y)]}{\Delta x} \tag{2–85}$$

and the right partial derivative,

$$\frac{\partial T(n-0,y)}{\partial x} \approx \frac{[T(n+1,y) - T(n,y)]}{\Delta x} \tag{2–86}$$

The combination of equations (2–85) and (2–86) yields equation 2–84. The second partial derivatives are best approximated by referring to figure 2–27 and using equations (2–85) and (2–86) to obtain

$$\begin{aligned}\frac{\partial^2 T(n,y)}{\partial x^2} &\approx \frac{[\partial T(n-0,y)/\partial x] - [\partial T(n+0,y)/\partial x]}{\Delta x}\\ &\approx \frac{[T(n+1,y) - 2T(n,y) + T(n-1,y)]}{\Delta x^2}\end{aligned} \tag{2–87}$$

Similarly, the second partial derivative of $T(x, y)$ with respect to y becomes

$$\frac{\partial^2 T(x,m)}{\partial y^2} \approx \frac{[T(x,m+1) - 2T(x,m) + T(x,m-1)]}{\Delta y^2} \tag{2–88}$$

From these results the steady state two-dimensional conduction heat-transfer equation (2–56) for constant thermal conductivity is

$$\begin{aligned}\frac{\partial^2 T(x,y)}{\partial x^2} &\approx \frac{[T(n+1),y) - 2T(n,y) + T(n-1,y)]}{\Delta x^2}\\ &+ \frac{[T(x,m+1) - 2T(x,m) + T(x,m-1)]}{\Delta y^2}\end{aligned}$$

The power of the finite difference techniques is best utilized in the two-dimensional situations, and in figure 2–28 such a configuration is shown. Notice in the figure that the conducting medium has been divided into a grid. The intersection of the grid lines represents a node point, or node, that has a neighborhood surrounding it with properties of the node. In figure 2–28 it is shown that these neighborhoods are defined as squares, where

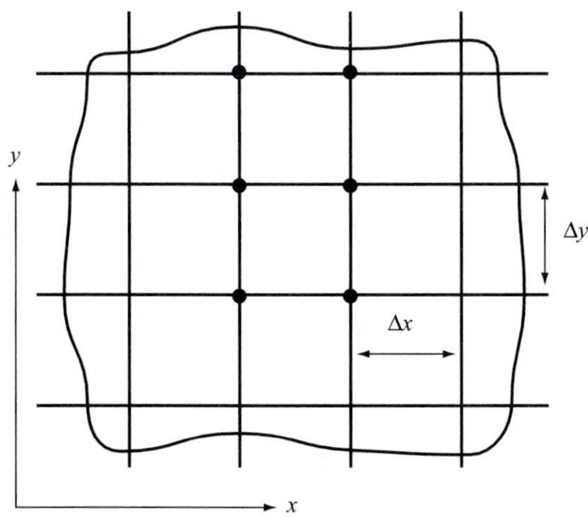

(a) A conducting system

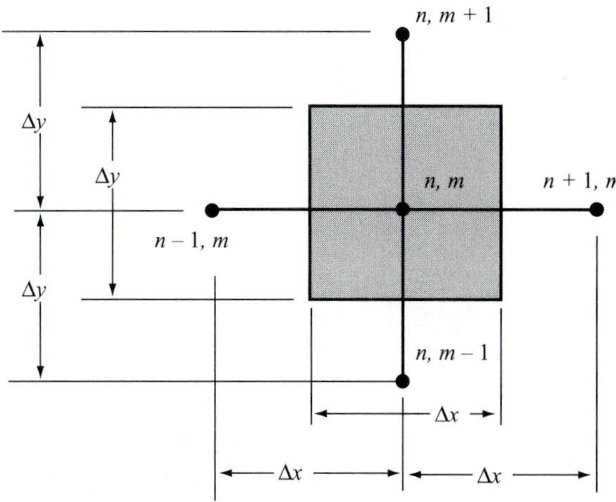

(b) Typical node and its neighborhood

Figure 2–28 Finite difference technique with node points defined on a conducting medium.

$\Delta y = \Delta x$. This restriction on the geometry is applied to decrease the algebraic complexities, but the general situation where $\Delta y \neq \Delta x$ could also be considered. The student needs to recognize that more care is needed in the algebraic manipulations when the nodes are not squares.

The finite difference technique proceeds with an energy balance for each node and its neighborhood. The boundary of each node, midway to the adjacent nodes, represents a true thermodynamic system boundary, and that is where the heat transfer can be identified. Thus for a typical node the energy balance, assuming steady state conditions, is

$$\dot{Q}_{\text{other}} - \kappa \sum \frac{\Delta T}{\Delta x} - \kappa \sum \frac{\Delta T}{\Delta y} = 0 \tag{2–89}$$

where $\dot{Q}_{\text{other}}$ is the nonconduction heat transfer across the node boundary (positive is heat transfer to the node, negative is out of the node) or any energy generation phenomenon that may be occurring in the node. For an inner node surrounded by other similar nodes and with no energy generation, the $\dot{Q}_{\text{other}}$ term is zero. The energy balance of equation (2–89) needs to be applied to all of the nodes of a conducting medium, and these result in a set of linear equations. This set of equations can be displayed as a matrix

$$[M]\{T\} = \{\dot{Q}\} \tag{2–90}$$

Where $[M]$ is the matrix formed by the coefficients a_{ij} of the terms containing temperatures T_i in the node energy balances. The notation $\{T\}$ is the matrix notation for the temperatures of all of the nodes, called a column vector in matrix analysis, and $\{\dot{Q}\}$ is the boundary nonconduction heat-transfers energy generation quantities. For our purposes in heat transfer the matrix $[M]$ will be a square nonsingular matrix having the same number of rows (m) and columns (n), and for such matrices it can be shown that

$$\{T\} = [M]^{-1}\{\dot{Q}\} \tag{2–91}$$

where $[M]^{-1}$ is the inverse of the matrix $[M]$. Thus, the solution to the various node temperatures, $\{T\}$, can be obtained by determining the inverse of the matrix $[M]$.

There are many "matrix inverse" packages, some with spreadsheet software, for a computer to facilitate these operations. It is not the purpose of this text to consider all of them but the reader who is interested in more details and discussions of advanced treatments of numerical methods, or computational heat transfer methods, should consult, for instance, Gerald and Wheatley [13], Shih [14], or Jaluria and Torrance [15]. The purpose here is to give the student an appreciation of the actual processes that occur in machine (computer) calculations to obtain answers. Two general methods are used to solve for a set of linear equations: iterative and elimination. With the next two examples we will demonstrate two iterative methods: Relaxation and Gauss–Siedel.

EXAMPLE 2–20 Estimate the heat transfer through the square fin shown in figure 2–29a. Neglect heat transfer or temperature differences in the z direction.

Solution In figure 2–29b is shown a simple 4-node grid for using finite differences to determine the fin temperature and from,those temperatures the heat transfer may be approximated. Each node is numbered and the energy balance is applied as shown in figure 2–30. The thermal conductivity is taken as 39 W/m · °C for matrix form as cast iron from appendix table B–2. The node equations can be written in

$$[M]\{T\} = \{\dot{Q}\}$$

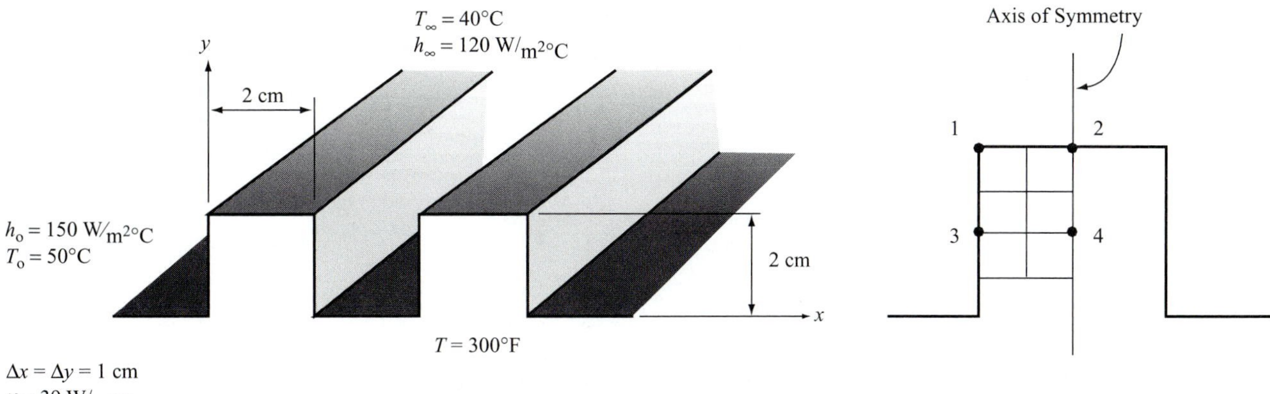

Figure 2–29 Square cast-iron fin with two-dimensional heat-transfer analysis (example 2–20).

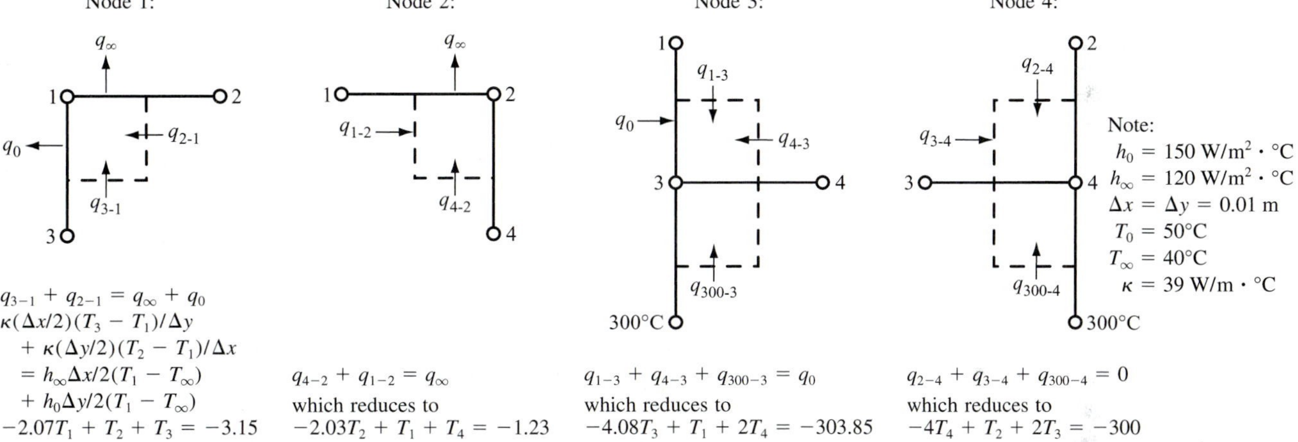

Figure 2–30 Node equations for square fin of example 2–20.

where

$$[M] = \begin{bmatrix} -2.069 & 1.0 & 1.0 & 0.0 \\ 1.0 & -2.038 & 0.0 & 1.0 \\ 1.0 & 0.0 & -4.063 & 2.0 \\ 0.0 & 1.0 & 2.0 & -4.0 \end{bmatrix} = \begin{bmatrix} a_{11}\ a_{12}\ a_{13}\ a_{14} \\ a_{21}\ a_{22}\ a_{23}\ a_{24} \\ a_{31}\ a_{32}\ a_{33}\ a_{34} \\ a_{41}\ a_{42}\ a_{43}\ a_{44} \end{bmatrix}$$

$$\{T\} = \begin{bmatrix} T_1 \\ T_2 \\ T_3 \\ T_4 \end{bmatrix} \quad \text{and} \quad \{Q\} = \begin{bmatrix} -3.077 \\ -1.538 \\ -303.077 \\ -300.0 \end{bmatrix}$$

A computational method called the *Relaxation Method* can be used to solve for the four node temperatures. The relaxation method involves the following step-by-step procedure:

1. Write each node equation so that one side is zero. For the set of equations in figure 2 – 30, this means that the terms on the right side are all transposed to the left. The resulting equations for this example are given in table 2–4.

Table 2–4 Set of Node Equations of Square-Fin Problem of Example 2–20

1: $-2.07T_1 + T_2 + T_3 + 3.15 = r_1$
2: $-2.03T_2 + T_1 + T_4 + 1.23 = r_2$
3: $-4.08T_3 + T_1 + 2T_4 + 303.85 = r_3$
4: $-4T_4 + T_2 + 2T_3 + 300 = r_4$

Note: When the correct node temperatures are used in these equations, the residuals r_1, r_2, r_3, and r_4 are all zero.

2. Assume values for each of the node temperatures. Some judgment needs to be made in the initial assumptions of these temperatures to avoid large numbers of computations.
3. Compute the value of each node equation, using the assumed node temperature values. This result is called a *residual* (r), and when all the node temperatures are "correct," the residuals are all zeros.
4. If the residual is not zero for any node equation, then the nonzero residual having the largest absolute value is "relaxed" to zero by adjusting the node temperature. The node temperature is determined from the equation

$$T_i^k = T_i^{k-1} - r_i/a_{ii} \tag{2–92}$$

where k denotes the kth relaxation step and $k + 1$ is the next step. The residuals of those nodes affected by the adjusted node temperature are then computed.
5. Steps 3 and 4 are continued until all of the residuals are zero or nearly zero. Some judgment is required on deciding how close to zero the residuals must be; however, nonzero residuals less than 2% of the node temperature is usually sufficient for engineering purposes.

In table 2– 5 is shown the hand computations for the node temperatures using initial values for the node temperatures of $T_1 = 220°\text{C}$, $T_2 = 240°\text{C}$, $T_3 = 250°\text{C}$, and $T_4 = 270°\text{C}$. The residuals at the initial conditions are $r_1 = 37.8°\text{C}$, $r_2 = 4.0°\text{C}$, $r_3 = 43.9°\text{C}$, and $r_4 = -40°\text{C}$. The temperature of node 3 is relaxed for the first step since the residual is greatest for this node. The new node temperature, T_3, after the first step (k = 1) is

$$\begin{aligned} T_3^{(1)} &= T_3^{(0)} - r_3/a_{33} = 250°\text{C} - (43.9°\text{C})/(-4.08) \\ &= 260.8°\text{C} \end{aligned}$$

Node temperatures 1 and 4 are affected by the change in temperature of node 3, and the new residuals of those nodes are $r_1 = 48.6°\text{C}$ and $r_4 = -18.4°\text{C}$. The second relaxation step proceeds with node 1, and this continues through the 17th relaxation step, as indicated in table 2–5. After the 17th step, the residuals are all less than 2% of any of the node temperatures; however, further relaxation steps could be performed if desired.

The heat transfer through the fin can be approximated by the heat transfer across the lower boundaries of the nodes 3 and 4. This can be set down as

$$\dot{Q}_{\text{fin}} = 2\kappa\left[\left(\frac{\Delta x}{2\Delta y}\right)(300°\text{C} - T_3) + \left(\frac{\Delta x}{2\Delta y}\right)(300°\text{C} - T_4)\right]$$

Answer

$$= \left(39\,\frac{\text{W}}{\text{m}\cdot°\text{C}}\right)[(300°\text{C} - 274.9°\text{C})] + (300°\text{C} - 279.1°\text{C})] = 1794.0\,\frac{\text{W}}{\text{m}}$$

Table 2–5 Relaxation Method for Solution to Example Problem 2–18 (T in °C, r the residual in °C)

Relax Step	Node Relaxed	T_1	r_1	T_2	r_2	T_3	r_3	T_4	r_4
0	(initial)	220	37.8	240	4.0	250	43.9	270	−40.0
1	3		48.6			260.8	0.0		−18.4
2	1	243.5	0.0		27.5		23.5		
3	2		13.5	253.5	0.0				−4.9
4	3		19.3			266.6	0.0		6.6
5	1	252.8	0.0		9.3		9.3		
6	2		4.6	258.1	0.0				15.8
7	4				4.0		17.3	274.0	0.0
8	3		8.8			270.8	0.0		8.5
9	1	257.1	0.0		8.3		4.3		
10	4				10.4		8.5	276.1	0.0
11	2		5.1	263.2	0.0				5.1
12	3		7.2			272.9	0.0		9.3
13	4				2.3		4.7	278.4	0.0
14	1	260.6	0.0		5.8		8.2		
15	3		2.0			274.9	0.0		4.0
16	2		4.9	266.1	0.0				6.9
17	4				1.3		2.6	279.1	0.0
Solution		260.6°C		266.1°C		274.9°C		279.1°C	

An alternate solution can be obtained be computing the convective heat transfer around the outer surface of the fin. This can be written as

Alternate Answer

$$\dot{Q}_{\text{fin}} = 2\left[h_0 \Delta y (T_3 - 50°\text{C}) + h_0\left(\frac{\Delta y}{2}\right)(T_1 - 50°\text{C}) + h_0\left(\frac{\Delta y}{2}\right)(300°\text{C} - 50°\text{C})\right]$$

$$+ 2\left[h_\infty\left(\frac{\Delta x}{2}\right)(T_1 - 40°\text{C}) + h_\infty\left(\frac{\Delta x}{2}\right)(T_2 - 40°\text{C})\right] = 1901.6\,\frac{\text{W}}{\text{m}}$$

The agreement between these two results indicates that the solution to the node temperatures is reasonable but further relaxing the set of node equations could provide closer agreement.

The relaxation method for determining the node temperatures in a finite difference model for steady state conduction heat transfer is tractable for hand calculations when the number of nodes is not large. In example 2–20 it was demonstrated how a system of four nodes can be analyzed with relaxations; however, if there had been many more nodes, the procedures would have become tedious and impractical. The relaxation method has been demonstrated because it is, in principle, the same procedure that is used for the *Gauss–Siedel Iteration Method,* which is a powerful and useful method for solving sets of linear equations. Also, the Gauss–Siedel Iteration Method can be adapted for use with a computer whereas the relaxation method does not lend itself to the computer.

The Gauss–Siedel Iteration Method is particularly useful for systems where the matrix of the coefficients, [M], is nonsingular and diagonally dominant. By diagonally dominant is meant that the coefficients a_{ii} have absolute values greater than other a_{ij} coefficients; that is, $|a_{11}| \geq |a_{12}|$, $|a_{11}| \geq |a_{13}|$, and so on, and in general $|a_{ii}| \geq |a_{ij}|$ for $j \neq i$.

In finite difference modeling of conduction heat transfer, the sets of mode temperature equations naturally tend to provide diagonally dominant coefficient matrices. The Gauss–Siedel Iteration Method can be outlined by the following steps:

1. Arrange the node equations so that the matrix of the coefficients is diagonally dominant.
2. Write the node equations in an explicit form to solve for the node temperature. As an example, in example problem 2–20 the equation for node 2 should be written

$$T_2 = (-1.538 - T_1 - T_4)/(-2.038)$$

The general form for the ith node temperature is

$$T_i^{k+1} = \frac{1}{a_{ii}}\dot{Q}_i - \frac{1}{a_{ii}}\sum_{j=1}^{i-1} a_{ij}T_i^{k+1} - \frac{1}{a_{ii}}\sum_{j=i}^{N} a_{ij}T_i^k \tag{2–93}$$

and this can be rearranged to read, for N nodes

$$T_i^{k+1} = T_i^k + \frac{w}{a_{ii}}\left[\dot{Q}_i - \sum_{j=1}^{i-1} a_{ij}T_i^{k+1} - \sum_{j=i}^{N} a_{ij}T_i^k\right] \tag{2–94}$$

where the superscript k denotes the temperature at the kth iteration, $k + 1$ at the $k + 1$–th iteration and so on. The term w is the overrelaxation factor. If the relaxation method is employed, $w = 1$, and for the Gauss–Siedel Iteration Method it can be shown that w must be less than 2.0. The least number of required iterations to obtain an acceptable solution by the Gauss–Siedel method occurs with w approximately equal to 1.25.
3. Assume initial values for the node temperatures T_i at $k = 0$.
4. Calculate new values for the node temperatures, $T_i^{(k+1)}$, using equation (2–94). Notice that the newest node temperature values are used for all calculations. Since the nodes before the ith node are iterated first, these new temperature values are used in the calculation, whereas the nodes after the ith node have the older temperature values.
5. The iteration proceeds until every node temperature remains essentially constant or converges to a unique value. A check such as

$$\left[T_i^{(k+1)} - T_i^{(k)}\right] \le \varepsilon \tag{2–95}$$

where ε is some small value for a convergence criterion, is a method to halt the iteration procedure.

EXAMPLE 2–21 For the fin of example 2–20, refine the prediction for the temperature distribution by using a greater number of nodes.

Solution The temperature distribution through the fin can be predicted to a greater precision if the node grid is made finer. The node grid shown in figure 2–31 results by reducing the node spacing to one half, or $\Delta x = \Delta y = 0.5$ cm. There are 12 nodes, the node equations from an energy balance of the nodes are given in table 2–6a, and this is an excessive number for relaxation methods. The equivalent set of equations formed for Gauss–Siedel iteration and using an overrelaxation factor $w = 1.2$ is given in table 2–6b. Using an $\varepsilon = 1.0$ kelvin for convergence criteria, and beginning the process with assumed initial temperature values given in table 2–7 for $k = 0$, the iteration proceeds through 7 iterations to arrive at the solution displayed in the $k = 7$ row of the table. The results given in table 2–7 may be generated by hand calculations or with a software program and a computer. Comparison of these results to the same fin with a 4-node relaxation model as given in table 2–5 shows good agreement.

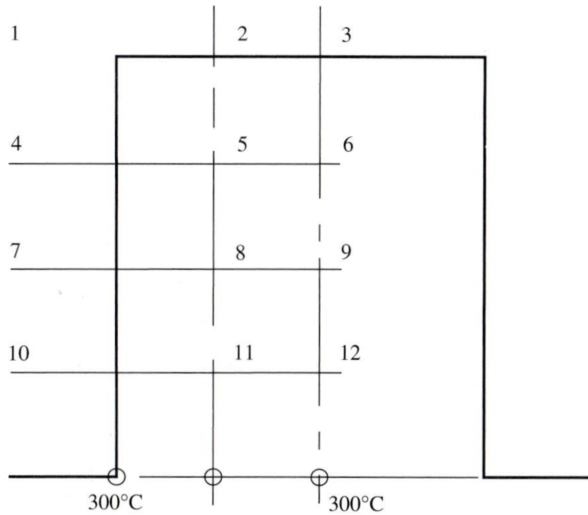

Figure 2–31 Square fin with twelve nodes, revised from example 2–20.

Table 2–6a

Node	Node Equation
1	$-2.069T_1 + T_2 + T_4 = -1.577$
2	$0.5T_1 - 2.015T_2 + 0.5T_3 + T_5 = -0.615$
3	$T_2 - 2.015T_3 + T_6 = -0.615$
4	$0.5T_1 - 2.019T_4 + T_5 + 0.5T_7 = -0.962$
5	$T_2 + T_4 - 4T_5 + T_6 + T_8 = 0$
6	$T_3 + 2T_5 - 4T_6 + T_9 = 0$
7	$0.5T_4 - 2.019T_7 + T_8 + 0.5T_{10} = -0.962$
8	$T_5 + T_7 - 4T_8 + T_9 + T_{11} = 0$
9	$0.5T_6 + T_8 - 2T_9 + 0.5T_{12} = 0$
10	$0.5T_7 - 2.019T_{10} + T_{11} = -150.962$
11	$T_8 + T_{10} - 4T_{11} + T_{12} = -300.0$
12	$0.5T_9 + T_{11} - 2T_{12} = -150.0$

Table 2–6b Implicit Form of Node Equations

$$T_1 = T_1 + 1.2(1.577 + T_2 + T_4 - 2.069T_1)/2.069$$
$$T_2 = T_2 + 1.2(0.615 + 0.5T_1 + 0.5T_3 + T_5 - 2.015T_2)/2.105$$
$$T_3 = T_3 + 1.2(0.615 + T_2 + T_6 - 2.015T_3)/2.015$$
$$T_4 = T_4 + 1.2(0.962 + 0.5T_1 + T_5 + 0.5T_7 - 2.019T_4)/2.019$$
$$T_5 = T_5 + 1.2(T_2 + T_4 + T_6 + T_8 - 4T_5)/4$$
$$T_6 = T_6 + 1.2(T_3 + 2T_5 + T_9 - 4T_6)/4$$
$$T_7 = T_7 + 1.2(0.962 + 0.5T_4 + T_8 + 0.5T_{10} - 2.019T_7)/2.019$$
$$T_8 = T_8 + 1.2(T_5 + T_7 + T_9 + T_{11} - 4T_8)/4$$
$$T_9 = T_9 + 1.2(T_6 + 2T_8 + T_{12} - 4T_9)/4$$
$$T_{10} = T_{10} + 1.2(150.962 + 0.5T_7 + T_{11} - 2.019T_{10})/2.019$$
$$T_{11} = T_{11} + 1.2(300 + T_8 + T_{10} + T_{12} - 4T_{11})/4$$
$$T_{12} = T_{12} + 1.2(300 + T_9 + 2T_{11} - 4T_{12})/4$$

Table 2–7 Gauss–Siedel Iteration of Example 2.21

	Node Temperatures (°C)											
k	T_1	T_2	T_3	T_4	T_5	T_6	T_7	T_8	T_9	T_{10}	T_{11}	T_{12}
0	220	230	240	240	250	260	250	260	270	260	270	280
1	229.5	243.1	252.0	243.7	252.0	264.2	264.8	265.0	268.3	276.9	282.6	284.1
2	237.4	247.5	254.7	250.9	260.9	259.9	262.0	269.1	271.0	280.2	283.5	284.6
3	242.5	254.3	255.6	255.4	259.4	261.6	267.3	270.5	272.0	281.6	284.3	285.3
4	248.0	253.9	256.2	256.8	261.0	262.7	267.9	271.5	272.9	282.0	284.8	285.7
5	247.5	255.0	257.4	257.5	261.8	263.6	268.7	272.2	273.5	282.4	285.1	286.0
6	248.6	255.9	258.3	258.4	262.7	264.4	269.3	272.7	274.0	282.7	285.4	286.2
7	249.5	256.8	259.1	259.2	263.4	265.1	269.8	273.2	274.5	283.0	285.6	286.5

Iteration stopped. The solution to the temperatures is given by the results of iteration 7.

The elimination methods do not use iterations to obtain solutions but rather substitutions and reduction of equations. The *Gauss Elimination Method* is a popular and useful method of solving for sets of linear equations. A difficulty associated with this method occurs when certain operations may involve division by zero, and these can be avoided with special techniques. Many of the computer software packages that include provisions for conduction heat-transfer analyses are structured with Gauss elimination and partial pivoting. It may be used to solve for the finite difference analyses of this section with the computer. When the computer is called upon, your most important task is writing down the node equations in a matrix form of equation (2–90). In examples 2–20 and 2–21 we saw how node equations can be obtained by applying an energy balance to the node. In table 2–8 are compiled some of the nodes often encountered in rectangular Cartesian coordinates.

The finite difference method can also be used for cylindrical or spherical coordinate systems and for three-dimensional coordinates. The three-dimensional node neighborhood elements in these systems are defined in table 2–9.

It should be recognized that the three-dimensional problems require more node equations, and these equations will often have more terms and be more challenging to write than the two-dimensional node equations. The elements shown in table 2–9 may be an aid in visualizing the three-dimensional problems and in writing the node equations.

Sometimes it may be more convenient to use the thermal resistance concept when considering the three-dimensional problems. Recall from section 2–3 that the thermal resistance for conduction heat transfer in the x direction is

$$R_{Tx} = \Delta x / kA \qquad \textbf{(2–27, repeated)}$$

so that

$$\dot{Q}_x = \frac{\Delta T}{R_{Tx}} \qquad \textbf{(2–28, repeated)}$$

Also, for convection heat transfer the thermal resistance is

$$R_{T0} = 1/h_0 A \qquad \textbf{(2–32, repeated)}$$

so that

$$\dot{Q}_{\text{convection}} = \frac{\Delta T}{R_{T0}}$$

Table 2–8 Some Typical Nodes, Node Neighborhoods, and Resulting Node Equations (steady state conduction with constant thermal conductivity)

I. Internal Node

$$-4\,T_{m,n} + T_{m,n+1} + T_{m,n-1} + T_{m+1,n} + T_{m-1,n} = 0 \qquad (\text{for } \Delta x = \Delta y)$$

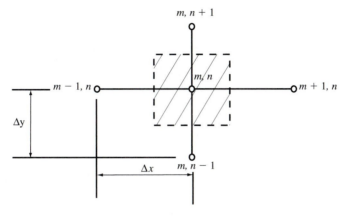

II. Boundary Node with Convection

$$-(2h\Delta y/\kappa + 4)T_{m,n} + 2T_{m+1,n} + T_{m,n+1} + T_{m,n-1} = -[2h\Delta y/\kappa]T_{\infty} \qquad (\text{for } \Delta x = \Delta y)$$

III. Boundary Node with Specified Heat Transfer per Unit Area, $\dot{q}$ ($\dot{q} = 0$ for insulated boundary or axis of symmetry)

$$-4T_{m,n} + 2T_{m+1,n} + T_{m,n-1} + T_{m,n+1} = 2\dot{q}\Delta y/\kappa \qquad (\text{for } \Delta x = \Delta y)$$

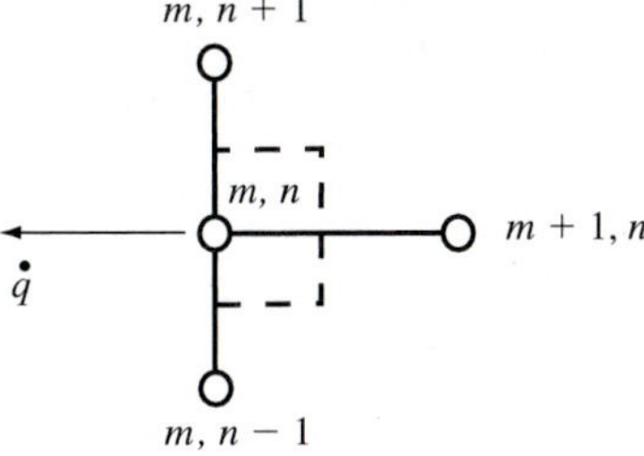

IV. Outer Corner Node with Convection

$$-[(2h\Delta y/\kappa) + 2]T_{m,n} + T_{m,n-1} + T_{m-1,n} = -2h\Delta y/\kappa T_0 \qquad (\text{for } \Delta x = \Delta y)$$

Table 2–8 (continued) Some Typical Nodes, Node Neighborhoods, and Resulting Node Equations (steady state conduction with constant thermal conductivity)

V. Inner Corner Node with Convection

$$-(6 + 2h\Delta x/\kappa)T_{m,n} + T_{m,n-1} + T_{m+1,n} + 2T_{m,n+1} + 2T_{m-1,n} = -[2h\Delta x/\kappa]T_0 \qquad (\text{for } \Delta x = \Delta y)$$

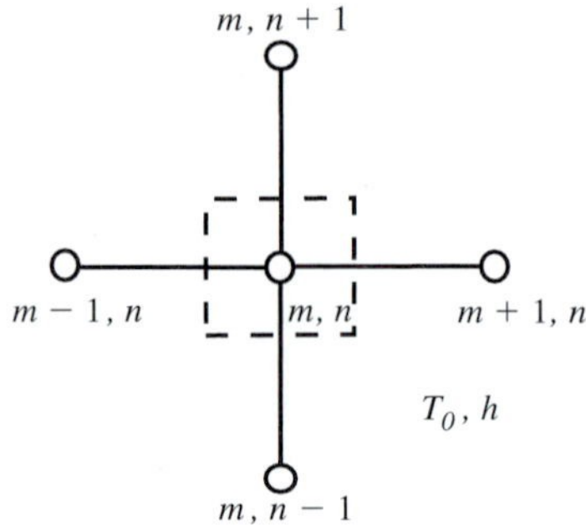

VI. Internal Node Near Irregular Boundary

$$-2\left(\frac{\Delta x}{\Delta a} + \frac{\Delta y}{\Delta b}\right)T_{m,n} + \frac{2\Delta x}{\Delta a\left(\dfrac{\Delta a}{\Delta x + 1}\right)}T_a + \left[\frac{2}{\left(\dfrac{\Delta b}{\Delta y} + 1\right)}\right]T_{m,n-1} + \frac{2}{\left(\dfrac{\Delta a}{\Delta x} + 1\right)}T_{m+1,n} + \frac{2\Delta y}{\Delta b\left(\dfrac{\Delta b}{\Delta y} + 1\right)}T_b = 0$$

(for $\Delta x = \Delta y$)

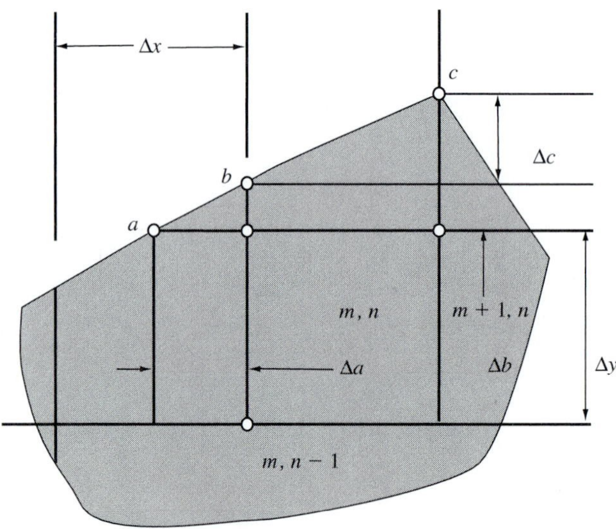

VII. Surface Node *b*, on Irregular Boundary

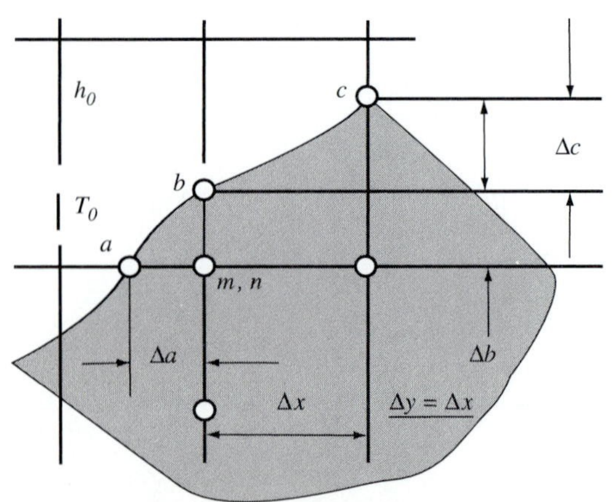

$$-\left[\frac{\Delta b}{\sqrt{(\Delta a)^2 + (\Delta b)^2}} + \frac{\Delta b}{\sqrt{(\Delta c)^2 + (\Delta y)^2}} + \frac{\Delta a + \Delta x}{\Delta b} + \left(\sqrt{\Delta c^2 + \Delta y^2} + \sqrt{(\Delta a)^2 + (\Delta b)^2}\right)\frac{h}{\kappa}\right]T_b + \frac{\Delta b}{\sqrt{(\Delta a)^2 + (\Delta b)^2}}T_a$$

$$+ \frac{\Delta b}{\sqrt{(\Delta c)^2 + (\Delta y)^2}}T_c + \frac{\Delta a + \Delta x}{\Delta b}T_{m,n} =$$

$$-\frac{h}{\kappa}\left(\sqrt{\Delta c^2 + \Delta y^2} + \sqrt{\Delta a^2 + \Delta b^2}\right)T_0$$

Table 2–9 Node Neighborhoods in Rectangular, Cylindrical, and Spherical Coordinates

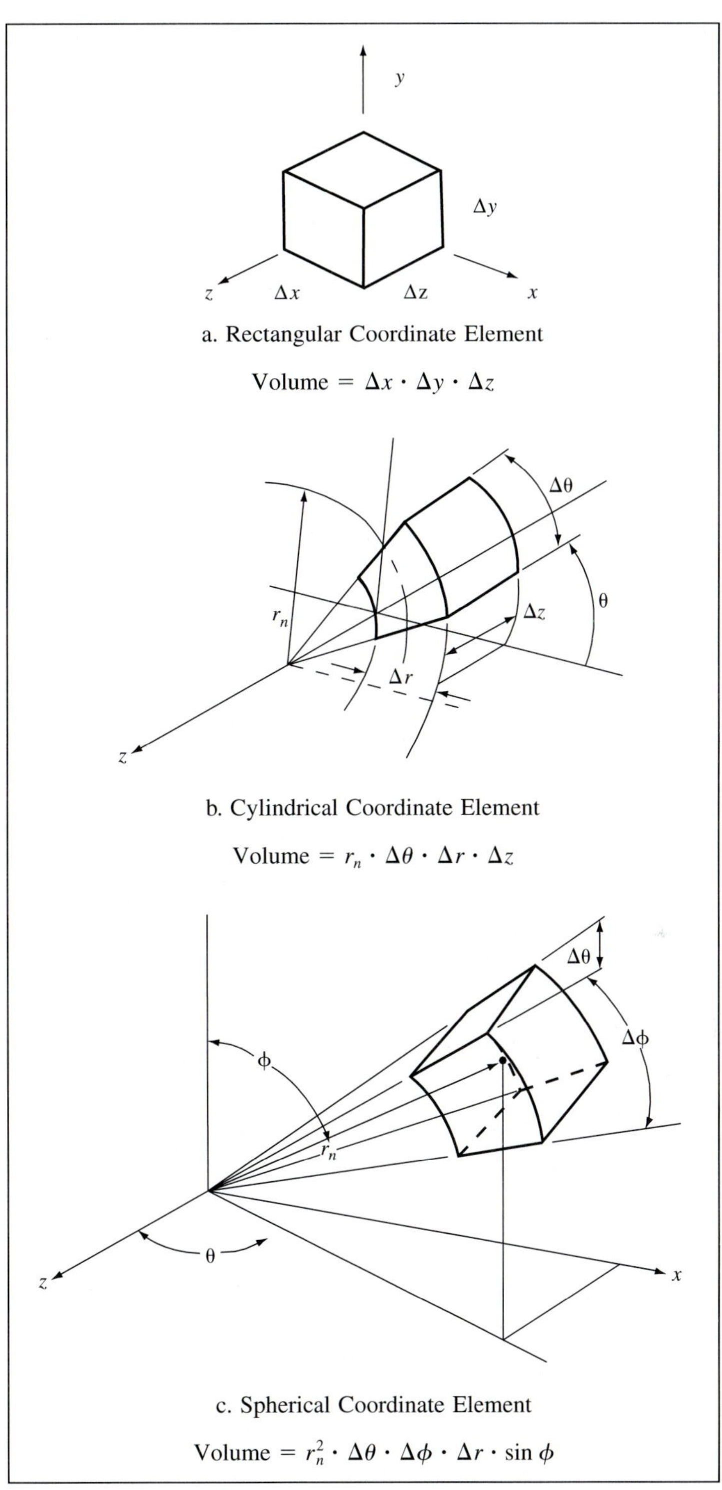

Table 2–10 Node Element Models with Heat Transfer in Three Dimensions in the Three Common Coordinate Systems

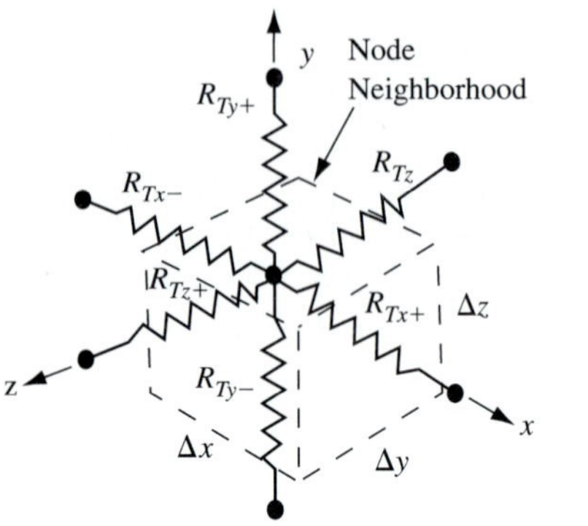

	Thermal Resistance	**Conduction**	**Convection**
a. Node in Rectangular Cartesian Coordinates	R_{Tx}	$\Delta x/(\kappa\Delta y \cdot \Delta z)$	$1/(h\Delta y \cdot \Delta z)$
	R_{Ty}	$\Delta y/(\kappa\Delta x \cdot \Delta z)$	$1/(h\Delta x \cdot \Delta z)$
	R_{Tz}	$\Delta z/(\kappa\Delta x \cdot \Delta y)$	$1/(h\Delta x \cdot \Delta y)$
b. Node in Cylindrical Coordinates	R_{Tr}	$\Delta r/[\kappa(r_n \pm \Delta r/2)\,(\Delta\theta)\,(\Delta z)]$	$1/[h(r_n \pm \Delta r/2)(\Delta\theta)\,(\Delta z)]$
	$R_{T\theta}$	$r_n(\Delta\theta)/(\kappa\Delta r \cdot \Delta z)$	$1/(h\Delta r \cdot \Delta z)$
	R_{Tz}	$(\Delta z)/(\kappa r_n \cdot \Delta\theta \cdot \Delta r)$	$1/hr_n\Delta\theta \cdot \Delta r$
c. Node in Spherical Coordinates	R_{Tr}	$\dfrac{(\Delta r)}{\left(r_n \pm \Delta\dfrac{r}{2}\right)^2(\Delta\phi)(\Delta\theta)\kappa\sin\theta}$	$\dfrac{1}{h\left(r_n \pm \dfrac{\Delta r}{2}\right)^2(\Delta\theta)\,(\Delta\phi)}$
	$R_{T\theta}$	$\dfrac{(\Delta\phi)\sin\theta}{\kappa(\Delta\theta)\,(\Delta r)}$	$\dfrac{1}{hr_n(\Delta\phi)\,(\Delta r)}$
	$R_{T\phi}$	$\dfrac{(\Delta\theta)}{\kappa(\Delta r)(\Delta\theta)\sin(\phi \pm \Delta\phi)}$	$\dfrac{1}{hr_n(\Delta\theta)\,(\Delta r)}$

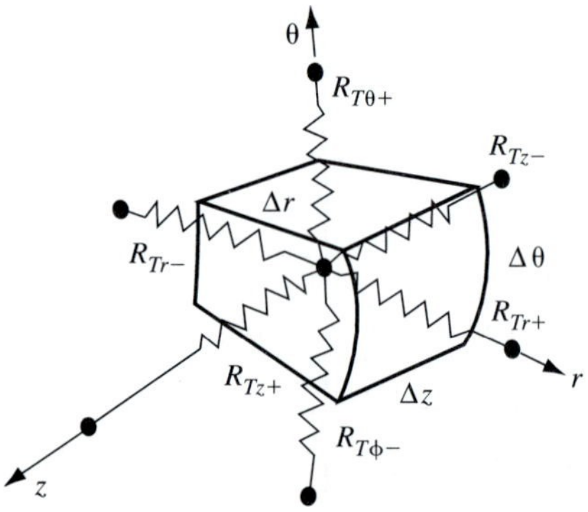

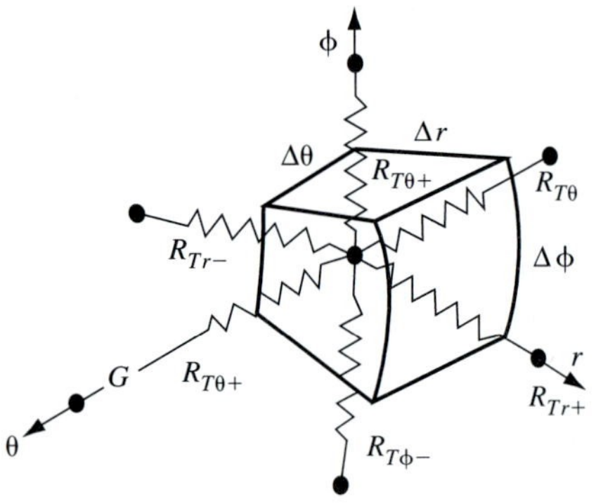

The energy balance equation (2–89) can be written with thermal resistances instead of the conduction terms as

$$\dot{Q}_{\text{other}} = \sum \frac{\Delta T_i}{R_{Ti}} = 0 \tag{2–96}$$

where ΔT_i and R_{Ti} can be in all three coordinate directions.

In table 2–10 are presented models of the nodes where thermal resistances are used to visualize the conduction or convection heat transfers. In these models it should be remembered that the thermal resistances could be replaced by the usual conduction or convection heat transfer terms. Using the resistance elements indicated in table 2–10 for the various terms in the summation of equation (2–96) gives the node equation for each of the identified nodes of a particular system. The following example demonstrates the methods used to analyze a typical three-dimensional problem. It is framed in cylindrical coordinates to further demonstrate the treatment required for analysis in those coordinates.

EXAMPLE 2–22 A large oak timber, 30 cm in diameter, is a horizontal structural member in a building. As a study of potential fire damage to the timber, consider a high-temperature gas rising up past the timber over a 40-cm-wide path as shown in figure 2–32. It is expected that the boundary conditions for the timber outer surface are symmetrical in the axial direction, and the temperature of the gases passing around the timber is written as $T(\theta, z)$. The convective heat transfer coefficient is $h(\theta, z)$ to indicate the variation of this property angularly and axially over the timber surface. Write the node equations for nodes 13, 40, and 41, shown shaded in figure 2–32b, for a finite difference analysis. Also write the equations that can be used to describe the conduction heat transfer in the axial, radial, and angular directions through the timber at nodes 13 and 40.

Solution In figure 2–32b is shown a proposed finite difference node grid with the neighborhoods also indicated. Since the boundary conditions are expected to be symmetrical in the axial direction, there will be a line of symmetry midway across the 40-cm-wide gas path. In figure 2–32b it is indicated that the gas path is 20 cm wide with an adiabatic surface at one end. Considering node 13, there will be thermal resistances axially, radially, and angularly as shown in figure 2–33b, which are

$$R_{Tr+} = \frac{1}{h(\theta_{13}, z_{13}) \cdot r_{13} \cdot \Delta\theta \cdot \dfrac{\Delta z}{2}} = \frac{1}{h(-22.5°, 0)\ (0.15\text{ m})\ (\pi/8)\ (0.05\text{ m})} = \frac{339.5}{h(-22.5°, 0)}$$

$$R_{Tr-} = \frac{r}{\kappa\left(r_{13} - \dfrac{\Delta r}{2}\right)(\Delta\theta)\left(\dfrac{\Delta z}{2}\right)} = \frac{1}{k(0.125\text{ m})\ (\pi/8)\ (0.05\text{ m})} = \frac{407.4}{k}$$

$$R_{T\theta+} = R_{T\theta-} = \frac{r_{13}(\Delta\theta)}{\kappa(\Delta r/2)\ (\Delta z/2)} = \frac{(0.15)\ (\pi/8)}{\kappa\left(\dfrac{0.05}{2}\right)\left(\dfrac{0.1}{2}\right)} = \frac{47.1}{k}$$

$$R_{Tz+} = \frac{\Delta z}{\kappa r_{13}(\Delta\theta)\ (\Delta r/2)} = \frac{0.1}{\kappa(0.15)\ (\pi/8)\ (0.025)} = \frac{67.9}{k}$$

where κ is the thermal conductivity of the oak. The energy balance for node 13 can be written as

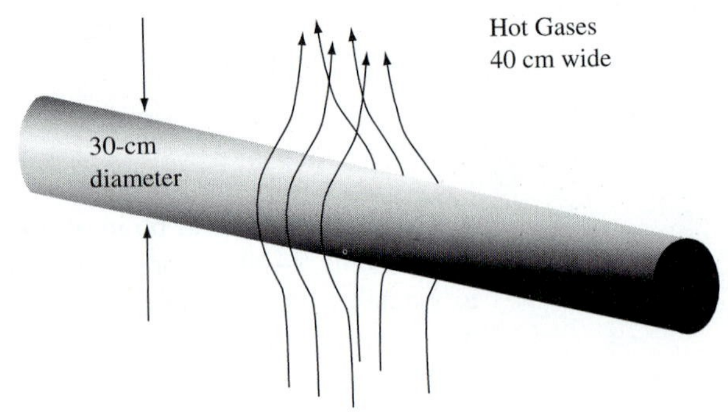

(a) Sketch of the system.

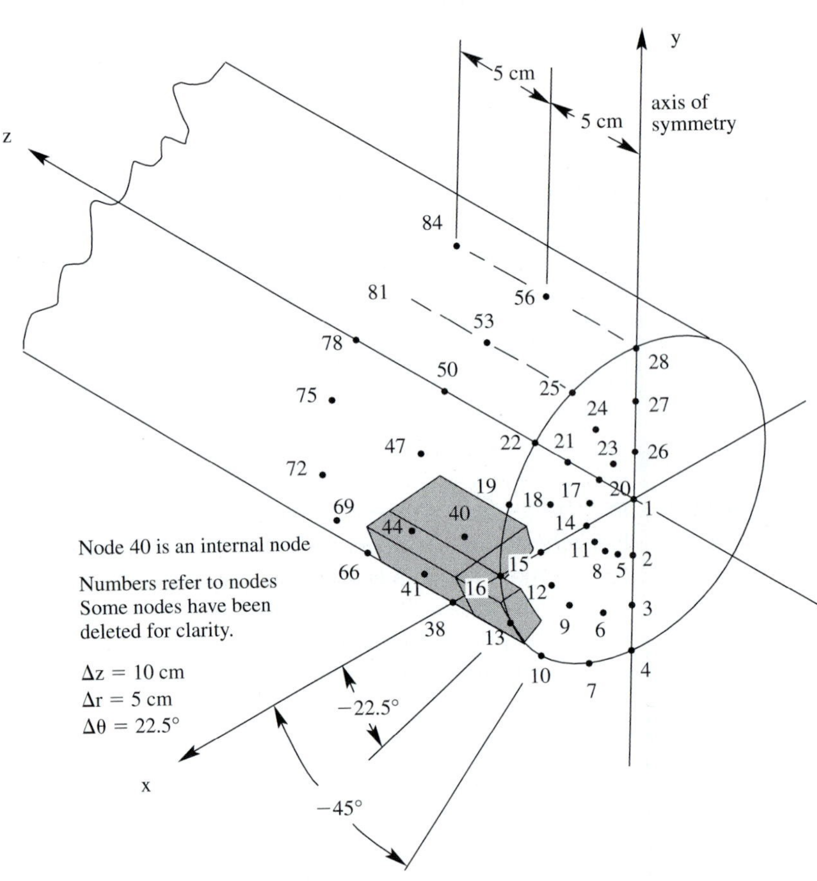

(b) Finite difference model. Numbers refer to nodes. Some nodes are deleted for clarity.

Figure 2–32 Oak beam subjected to hot gases.

$$\frac{T(-22.5^\circ) - T_{13}}{R_{Tr+}} + \frac{T_{41} - T_{13}}{R_{Tz+}} + \frac{T_{16} - T_{13}}{R_{T\theta+}} + \frac{T_{10} - T_{13}}{R_{T\theta-}} + \frac{T_{12} - T_{13}}{R_{Tr-}} = 0 \quad \textbf{(2–97)}$$

and the reader can see that this equation reduces to a linear one with node temperatures as the independent variables. This node equation, combined with the complete set of node equations for the model of the oak beam, can then be used to solve for the individual node temperatures. The heat transfers at node 13 are the terms of equation (2–97); that is,

Figure 2–33 Individual nodes for a finite difference analysis of an oak beam subjected to hot gases.

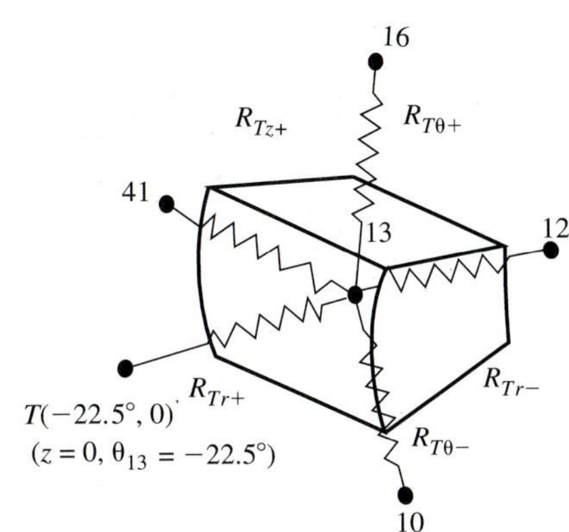

$\Delta\theta = 22.5° = \frac{\pi}{8}$ rad

$r_{13} = 15$ cm $= 0.15$ m

$\Delta z = 10$ cm $= 0.1$ m

$\Delta r = 5$ cm $= 0.05$ m

(a) Node 13 and its neighborhood

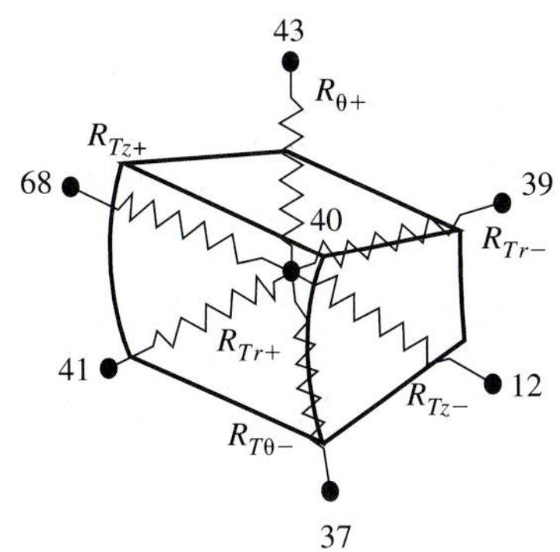

$r_{40} = 0.1$ m

$\Delta\theta = \pi/8$

$\Delta z = 0.1$ m

$\Delta r = 0.05$ m

(b) Node 40 and its neighborhood

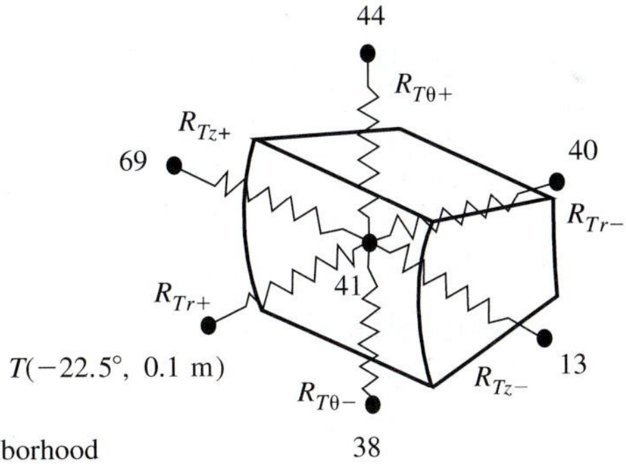

$\Delta\theta = \pi/8$

$r_{41} = 0.15$ m

$\Delta z = 0.1$ m

$\Delta r = 0.05$ m

(c) Node 41 and its neighborhood

$\dot{Q}_z = (T_{41} - T_{13})/R_{Tz+}$	is the heat transfer from node 41 to node 13, due to conduction
$\dot{Q}_r = (T_{12} - T_{13})/R_{Tr-}$	is that heat transfer from 12 to 13
$\dot{Q}_r = T(-22.5°, 0) - T_{13}$	is convection heat transfer from the hot gases, having temperature $T(-22.5°, 0)$, to node 13
$\dot{Q}_\theta = (T_{10} - T_{13})/R_{T\theta-}$	is conduction from 10 to 13
$\dot{Q}_\theta = (T_{16} - T_{13})/R_{T\theta+}$	is conduction from 16 to 13

These heat transfers can be determined after the individual node temperatures are computed.

With an analysis similar to that used for node 13 and referring to figure 2–33b, the node equation for 40 may be written as

$$\frac{T_{39}-T_{40}}{R_{Tr-}}+\frac{T_{41}-T_{40}}{R_{Tr+}}+\frac{T_{12}-T_{40}}{R_{Tz-}}+\frac{T_{68}-T_{40}}{R_{Tz+}}+\frac{T_{43}-T_{40}}{R_{T\theta+}}+\frac{T_{37}-T_{40}}{R_{T\theta-}}=0 \quad \textbf{(2–98)}$$

where the thermal resistances are for node 40. They are

$$R_{Tr-}=\frac{(\Delta r)}{\kappa\left(r_{40}-\dfrac{\Delta r}{2}\right)(\Delta\theta)\,(\Delta z)}=\frac{(0.05)}{\kappa(0.075)\,(\pi/8)\,(0.1)}=\frac{16.98}{\kappa}$$

$$R_{Tr+}=\frac{(0.05)}{\kappa(0.125)\,(\pi/8)\,(0.1)}=\frac{10.2}{\kappa}$$

$$R_{T\theta+}=R_{T\theta-}=\frac{r_{40}(\Delta\theta)}{\kappa(\Delta r)\,(\Delta z)}=\frac{7.85}{\kappa}$$

$$R_{Tz+}=R_{Tz-}=\frac{(\Delta z)}{\kappa r_{40}(\Delta\theta)\,(\Delta r)}=\frac{50.9}{\kappa}$$

Equation (2–98) also reduces to a linear one with node temperatures as the independent variables. The heat transfers associated with node 40 are the terms

$$\begin{aligned}
Q_r &= (T_{39}-T_{40})/R_{Tr-} && \text{from node 39 to node 40}\\
Q_r &= (T_{41}-T_{40})/R_{Tr+} && \text{from node 41 to 40}\\
Q_z &= (T_{12}-T_{40})/R_{Tz-} && \text{from 12 to 40}\\
Q_z &= (T_{68}-T_{40})/R_{Tz+} && \text{from 60 to 40}\\
Q_\theta &= (T_{37}-T_{40})/R_{T\theta-} && \text{from 37 to 40}\\
Q_\theta &= (T_{43}-T_{40})/R_{T\theta+} && \text{from 43 to 40}
\end{aligned}$$

Finally, referring to figure 2–33c for node 41 and its neighborhood, the node equation is

$$\frac{T_{40}-T_{41}}{R_{Tr-}}+\frac{T(-22.5^\circ,0.1\text{ m})-T_{41}}{R_{Tr+}}+\frac{T_{38}-T_{40}}{R_{T\theta-}}+\frac{T_{44}-T_{40}}{R_{T\theta+}}+\frac{T_{13}-T_{40}}{R_{Tz+}}+\frac{T_{69}-T_{40}}{R_{Tz-}}=0 \quad \textbf{(2–99)}$$

The thermal resistances for node 41 are

$$R_{Tr+}=\frac{1}{h(\theta_{41},z_{41})\,(r_{41}(\Delta\theta)(\Delta z)}=\frac{1}{h(-22.5^\circ,0.1\text{ m})\,(0.15\text{ m})\,(\pi/8)\,(0.1\text{ m})}=\frac{169.8}{h(-22.5^\circ,0.1\text{ m})}$$

$$R_{Tr-}=\frac{(\Delta r)}{\kappa\left(r_{41}-\dfrac{\Delta r}{2}\right)(\Delta\theta)\,(\Delta z)}=\frac{(0.05)}{\kappa(0.125)(\pi/8)\,(0.1)}=\frac{50.9}{\kappa}$$

$$R_{T\theta+}=R_{T\theta-}=\frac{r_{41}(\Delta\theta)}{\kappa(\Delta r/2)\,(\Delta z)}=\frac{23.56}{\kappa}$$

$$R_{Tz-}=R_{Tz+}=\frac{(\Delta z)}{\kappa r_{41}(\Delta\theta)\,(\Delta r/2)}=\frac{67.9}{\kappa}$$

Similar node equations can be written for the remaining nodes for a complete analysis of the heat transfers occurring in and around the oak beam. At node 38, for instance, the heat transfer terms would be the same as for node 41 but with different node temperatures. At node 38 the adjacent nodes would be 44, 37, 35, 10, 66, and $T(-45°, 0.1\text{ m})$ for the hot gases. Also, at node 38, the convective heat transfer coefficient should be identified as $h(-45°, 0.1\text{ m})$.

The four examples shown to demonstrate the methods of analysis using finite differences to write node equations and thereby provide a means of determining temperature distributions did not include direct discussions of the accuracy of the solutions. The finite difference method is an approximation method, and by using smaller differences (achieved by decreasing the size of the node neighborhoods) the exact solution should be obtained or at least approached. Decreasing the node sizes also increases the number of nodes and node equations required to analyze the same region. As a consequence, the computational efforts will increase with decreasing node size. Some techniques are available that provide some assurances that a reasonable solution has been obtained for the temperature distribution. Three of these are

1. Compare results of the finite difference method to analytic results for the same or nearly the same problem.
2. Repeat the numerical solution using smaller nodes until the solution set of the latest attempt agrees closely with the immediately preceding attempt. This method implies that the solution will converge to the correct one.
3. Use heat transfer calculations to check the temperature distribution of the region. An application of this method was shown in example 2–20, where the conduction heat transfer at the base of the fin was compared to convection heat transfer at the outer surfaces of the same fin.

A decision regarding whether two or more solutions are in reasonable agreement always requires judgment and reflection on other undefined aspects that can impact on the problem. For instance, in example problem 2–20 the solution to the heat transfer through the fin was determined by two separate means: determining the convection heat transfer at the outer surface (1901.6 W/m) and determining the conduction/convection heat transfer at the base (1794.0 W/m). The difference between these two solutions is roughly 107.6 W/m or 6%. As we will see later, the convective heat transfer coefficients can have differences of at least 10% and even thermal conductivity of the materials can have disparities of 12% or more. These, among other complications, can render uncertainties of at least 10%, so the analysis of problem 2–18 had results that are in good agreement.

The finite element method has been used to solve many conduction heat-transfer processes, and for extended systems it is usually superior to the finite difference method. The finite element method adapts well to one-, two-, or three-dimensional problems with complicated geometric shapes or boundary conditions. Many computer software packages can perform the necessary mathematical and numerical operations and can also generate the grid nodes, or finite elements, automatically. With the power of the finite element method in conjunction with the digital computer many of the difficult conduction heat-transfer problems are tractable. For a further discussion of the finite element method the reader is referred to, for instance, Kikuchi [16].

2-7 APPLICATIONS OF STEADY STATE HEAT TRANSFER

In this section we will consider five applications of steady state heat transfer: fins and extended surfaces, guarded-hot-plate devices for experimentally determining thermal conductivity, thermal contact resistance phenomena, critical thickness of insulation, and heat generation in an electrical wire. These discussions and analyses are intended to demonstrate the usefulness of the concepts of heat transfer and are not intended to represent a complete list of those applications. The reader should also see that these applications can be developed to a greater extent and sophistication than the discussions given here would indicate; that is, all of these applications can be interpreted as open ended problems with complications beyond those included here.

Fins Engineers and technologists often encounter heat transfer problems that require, as a solution, high rates of heat transfer. Cooling, heating, and drying are three phenomena that the reader can reflect on and relate to heat transfers. Convection heat transfer is often a limiting factor in these phenomena, and as a consequence, the designer must increase heat transfer by one of three methods: increase the temperature difference in the boundary layer of convection, increase the convective heat-transfer coefficent h, or increase the surface area for convection.

A successful method of increasing the surface area for convection is through the use of *fins.* A fin is a section of material extending out from a base surface as shown in figure 2–34. If the fin has a cross section that is uniform and rectangular, as indicated in figure 2–34a, it is called a *square fin* and is probably the simplest for mathematical analysis. You should recognize that there are all sorts of different fins, some of which are not even called fins, but affect heat transfers in much the same way—increasing convection by increasing the surface area. Figure 2–35 shows three other types of fins—circumferential, rod, and tapered—and figure 2–36 shows some configurations of devices incorporating these fins, including the frying pan with its fin (the handle). You can probably think of many more fins.

Figure 2–34 The square fin.

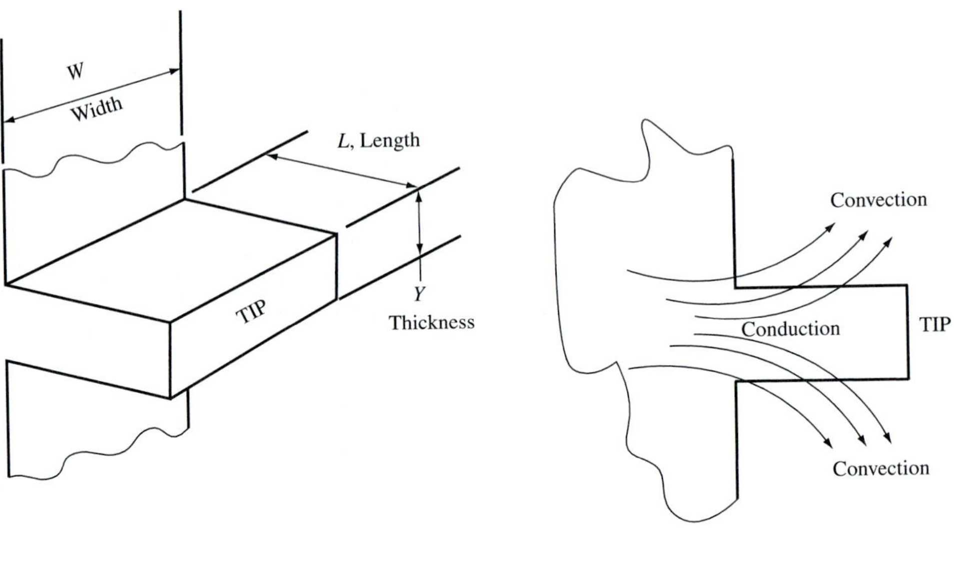

(a) Sketch of the fin

(b) Heat transfer through a fin

It is instructive to consider the square fin and the heat transfer mechanisms occurring with it. Notice, in figure 2–34b, that the heat transfer is a combination of conduction and convection along the fin and by analyzing a differential element of the square fin, shown in figure 2–37, one can predict the fin temperature and the heat transfers. For steady state conditions on the element ΔV, the conservation of energy is

$$\dot{Q}_x = \dot{Q}_{x+\Delta x} + \dot{Q}_{\text{convection}} \tag{2–100}$$

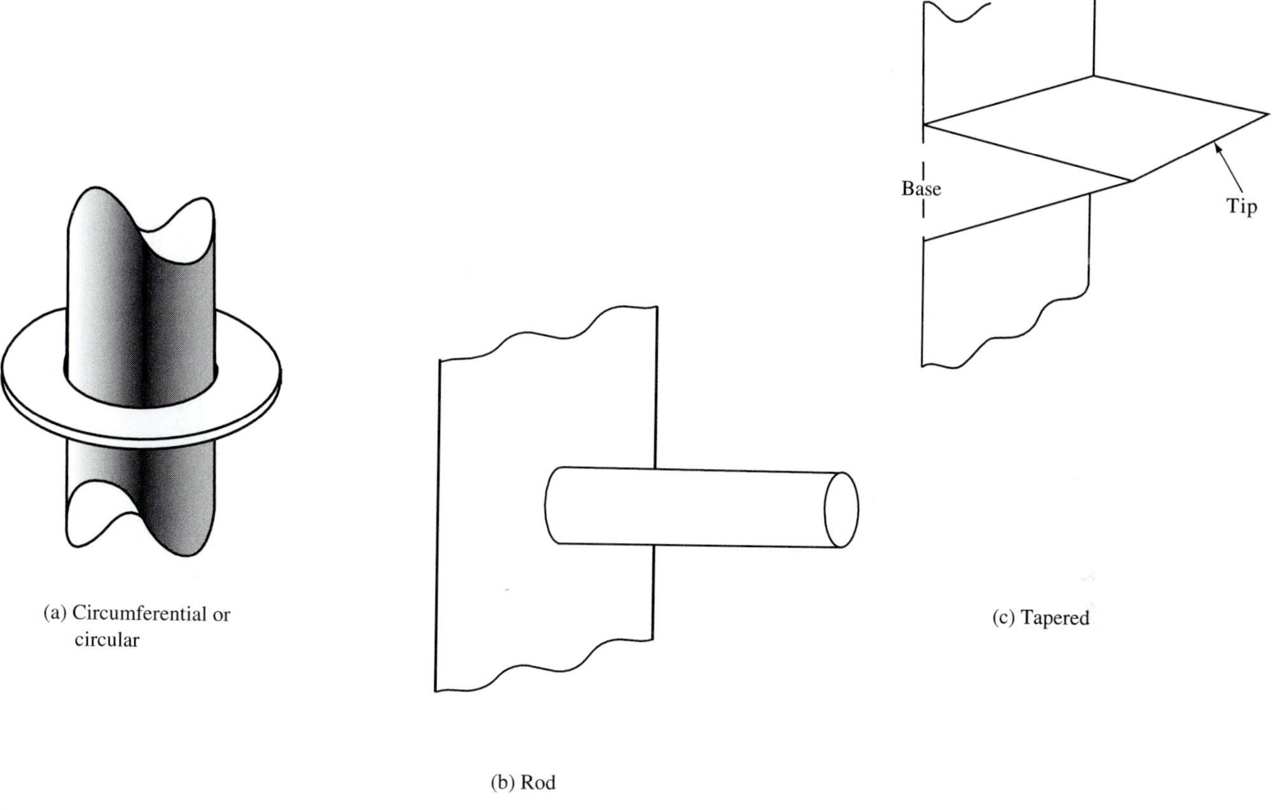

(a) Circumferential or circular

(b) Rod

(c) Tapered

Figure 2–35 Some types of fins.

where $\dot{Q}_x$ is the conduction heat transfer into ΔV at x, $\dot{Q}_{x+\Delta x}$ is the conduction heat transfer out of ΔV at $x + \Delta x$, and $\dot{Q}_{\text{convection}}$ is the convection heat transfer out of ΔV across the surface A_s. Recalling equation (1–52) for conduction, we have

$$\dot{Q}_x = -\kappa A_x \left[\frac{\partial T(x, y)}{\partial x}\right]_x \quad \text{and} \quad \dot{Q}_{x+\Delta x} = -\kappa A_{x+\Delta x} \left[\frac{\partial T(x, y)}{\partial x}\right]_{x+\Delta x}$$

Recalling equation (1–57) for convection, the convection becomes

$$\dot{Q}_{\text{convection}} = hA_s[T(x, y) - T_\infty]$$

For the square fin $A_x = A_{x+\Delta x}$. Then, assuming thermal conductivity is constant, noting that $A_s = P\Delta x$ where P is the perimeter $(2W + 2Y)$, and performing the necessary algebraic manipulations to equation (2–100) gives

$$\frac{\kappa A}{Ph\Delta x}\left\{\left[\frac{\partial T(x, y)}{\partial x}\right]_{x+\Delta x} - \left[\frac{\partial T(x, y)}{\partial x}\right]_x\right\} = T(x, y) - T_\infty$$

If Δx is now taken to zero, this equation becomes

$$\frac{\kappa A}{Ph}\left[\frac{\partial^2 T(x, y)}{\partial x^2}\right] = T(x, y) - T_\infty$$

If the temperature distribution $T(x, y)$ is assumed to be uniform at any particular point x, then $T(x, y) = T(x)$, and if the cross-sectional area is assumed to be constant, as for a square fin, then this equation becomes

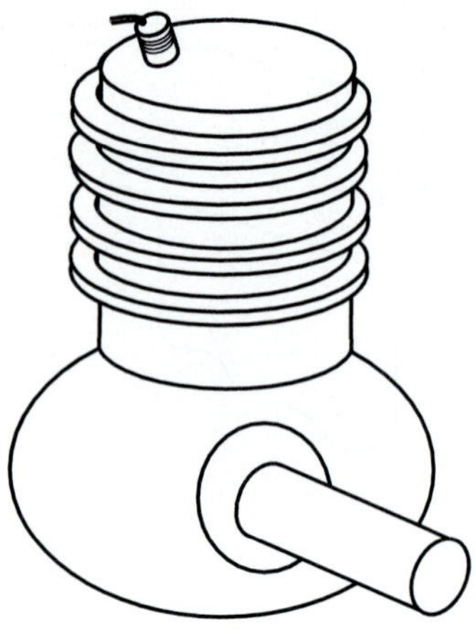

Engine/Compressor

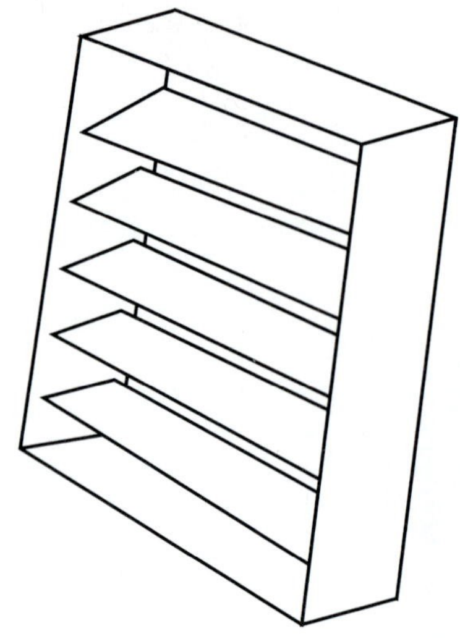

Radiator/Heat Exchanger

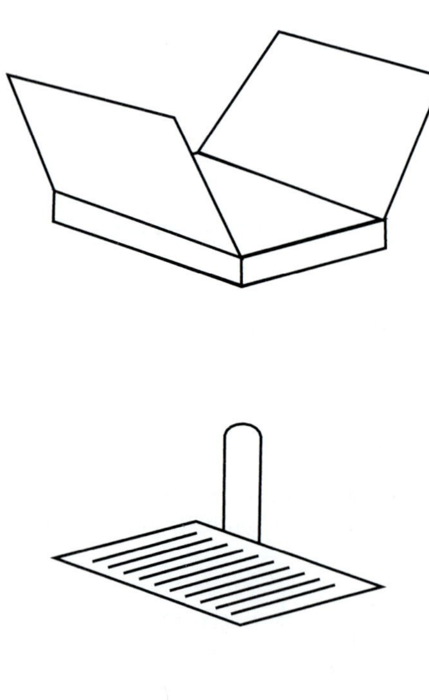

Electrical Components
Fins for Cooling

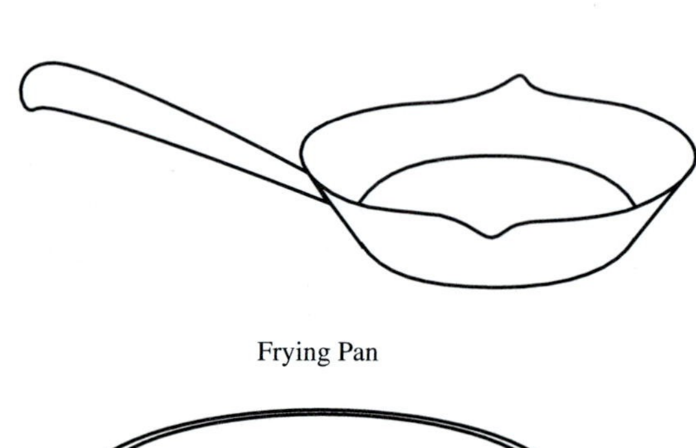

Frying Pan

Handles, Kettle

Figure 2–36 Devices with their fins.

Figure 2–37 Differential element of the square fin.

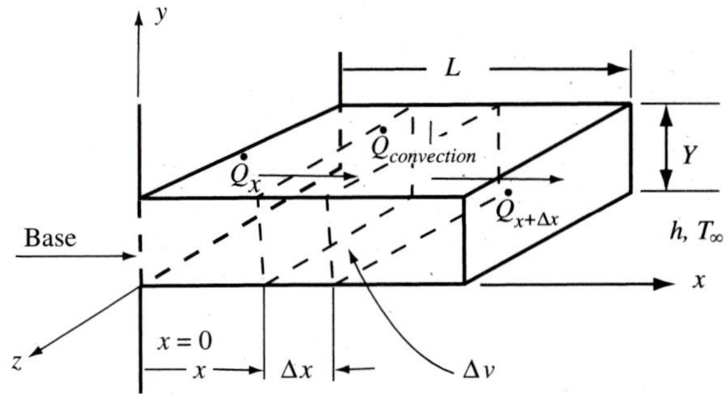

$$\frac{\kappa A}{Ph}\left[\frac{d^2T(x)}{dx^2}\right] = T(x) - T_\infty \quad \textbf{(2–101)}$$

It is convenient, for mathematical manipulations, to define

$$m^2 = \frac{Ph}{\kappa A} \quad \text{and} \quad \theta(x) = T(x) - T_\infty$$

which then allows equation (2–101) to be written as

$$\frac{d^2\theta(x)}{dx^2} = m^2\theta(x) \quad \textbf{(2–102)}$$

The solution to equation (2–102) can be written as

$$\theta(x) = C_1e^{-mx} + C_2e^{mx} \quad \textbf{(2–103)}$$

A boundary condition may be written as

B.C. 1 $\quad \theta(x) = \theta_0 = T(0) - T_\infty \quad$ at $x = 0$, the base

Three square-fin cases are customarily considered to obtain a second boundary condition and thereby evaluate the constants, C_1 and C_2:

Case I. Very long fin, or a fin that is very thin; that is, $Y \ll L$ so that the additional boundary condition is

B.C. 2(I) $\quad T(x) = T_\infty$ or $\theta(x) = 0 \quad$ as $x \to \infty$

Case II. Fin with adiabatic tip. The boundary condition is then

B.C. 2(II) $\quad \dfrac{dT(x)}{dx} = 0$ or $\dfrac{d\theta(x)}{dx} = 0 \quad$ at $x = L$

Case III. Fin with convection heat transfer at the tip so that a convective coefficient h_L at the tip can be identified. The second boundary condition is then

B.C. 2(III) $\quad \dot{Q}_x = -\kappa\dfrac{dT(x)}{dx} = h_L[T(x) - T_\infty] \quad$ at $x = L$

or $\quad \dot{Q}_x = -\kappa\dfrac{d\theta(x)}{dx} = h_L\theta_L \quad$ at $x = L$

Applying B.C. 1 and B.C. 2(I) to equation (2–103) gives the temperature distribution for a case I fin,

$$\theta(x) = \theta_0e^{-mx} \quad \text{(case I)} \quad \textbf{(2–104)}$$

This relationship results because B.C. 2(I) requires that C_2 be zero. The temperature distribution is plotted in the graph of figure 2–38a.

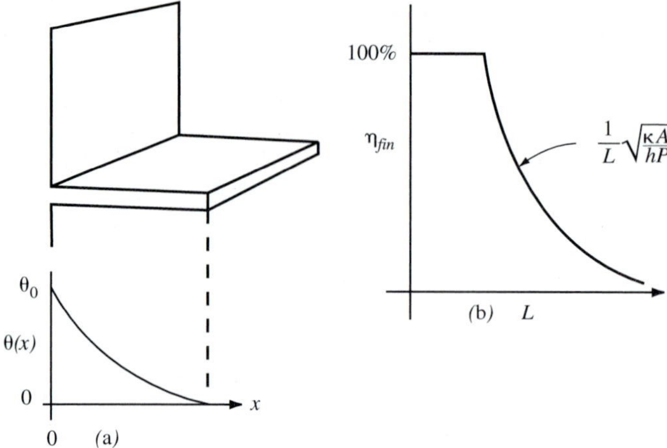

Figure 2–38 Long square fin, its temperature, and its fin efficiency as length varies.

For fins described by case II conditions, the temperature distribution is, after applying B.C. 1 and B.C. 2(II) to equation (2–103),

$$\theta(x) = \theta_0 \frac{\cosh[m(L - x)]}{\cosh mL}$$

where the hyperbolic function, introduced previously in section 2–4, example problem 2–12, is defined as

$$\cosh mx = \frac{1}{2}(e^{-mx} + e^{mx})$$

For case III fins, the temperature distribution can then be written

$$\theta(x) = \theta_0 \left(\frac{\cosh[m(L - x)] + \frac{h_L}{m\kappa}\sinh[m(L - x)]}{\cosh mL + \frac{h_L}{m\kappa}\sinh mL} \right)$$

The hyperbolic sine function is defined as

$$\sinh mx = \frac{1}{2}(e^{mx} - e^{-mx})$$

The heat transfer through the square fins, $\dot{Q}_{\text{fin}}$, can be determined conveniently by one of two methods. In one method the amount of heat flowing through a fin is determined by the equivalent amount of heat that passes through the fin base, as indicated in figure 2–39a. This heat flow is evaluated from the temperature gradient at the base, where $x = 0$, and Fourier's Law:

$$\dot{Q}_{\text{fin}} = \dot{Q}_{\text{conduction/base}} - \kappa A_0 \left[\frac{dT(x)}{dx} \right]_{x=0} = -\kappa A_0 \left[\frac{d\theta(x)}{dx} \right]_{x=0} \qquad \textbf{(2–105)}$$

Equation (2–105) is applicable for any fin.

Another method that can be used to evaluate the fin heat transfer is to determine the overall convection heat transfer at the fin surfaces as indicated in figure 2–39b. This is done by the integration

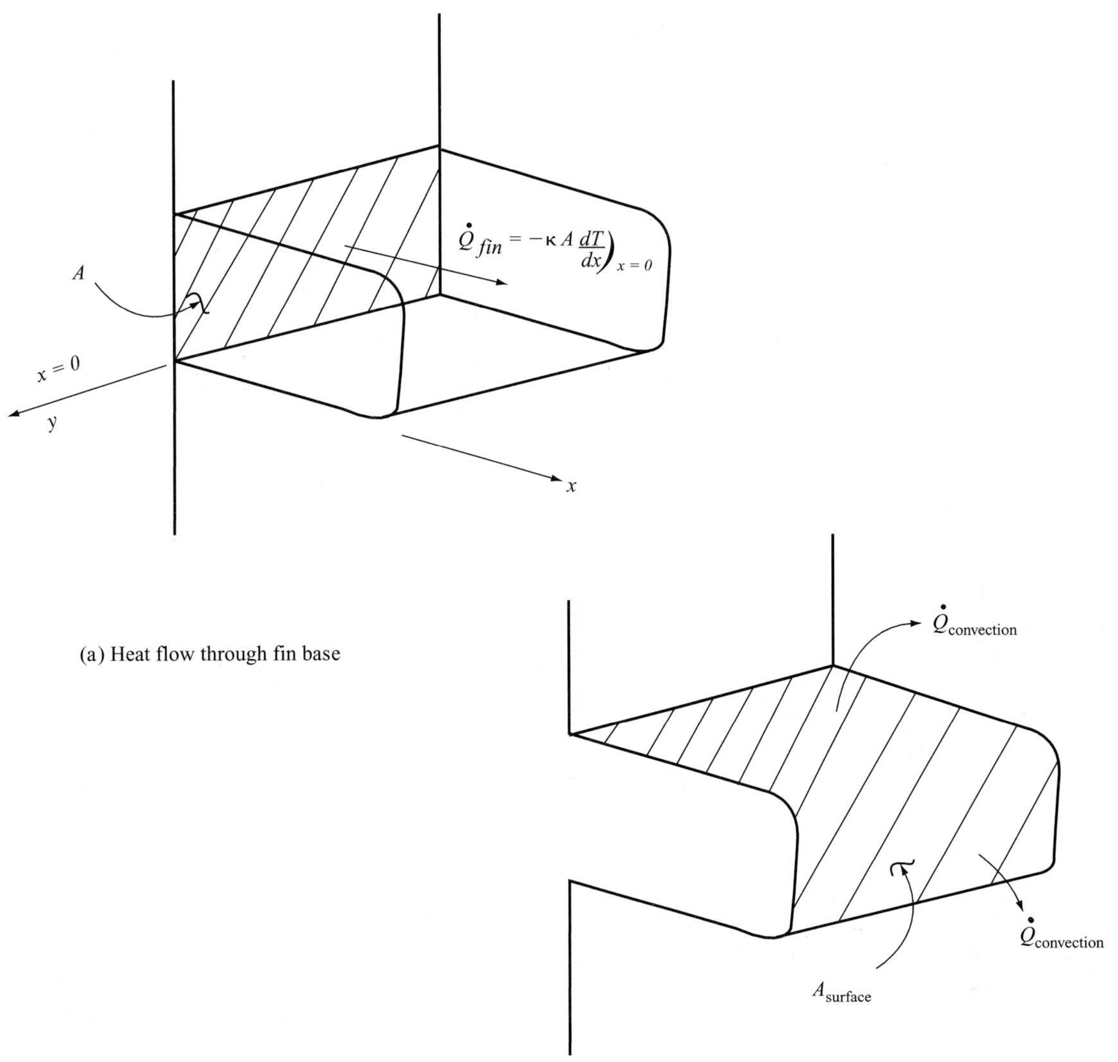

Figure 2–39 Methods for determining total heat flow through fins.

$$\dot{Q}_{\text{fin}} = \dot{Q}_{\text{convection}} = \int_0^L hP[T(x) - T_\infty]dx = \int_0^L hP\theta(x)dx \tag{2–106}$$

Equation (2–106) is applicable for any square fin having the same cross-sectional area over its entire length L.

For case I fins, the heat transfer can be found by using equations (2–104) and (2–105), noting that the fin cross-sectional area at the base, A_0, is the same throughout the fin, A.

$$\dot{Q}_{\text{fin}} = -\kappa A_0 \frac{d}{dx}[\theta_0 e^{-mx}]_{x=0} = \kappa A \theta_0 m e^{-m(0)} = \kappa A m \theta_0 = \theta_0 \sqrt{hP\kappa A}$$

The same result could be gotten by using equations (2–104) and (2–106),

$$\dot{Q}_{\text{fin}} = \int_0^{L=\infty} hP\theta_0 e^{-mx}dx = hP\theta_0\left(\frac{-1}{m}\right)[e^{-mx}]_0^\infty = \frac{hP}{m}\theta_0 = \theta_0\sqrt{hP\kappa A}$$

The derivation of the expressions for the fin heat transfers of cases II and III is left as exercises for the student (see practice problems 2–61 and 2–62). The fin efficiency is customarily defined as

$$\eta_{\text{fin}} = \frac{\dot{Q}_{\text{fin}}}{\dot{Q}_0} \tag{2–107}$$

where $\dot{Q}_0$ is the heat transfer from a fin that would be at the base temperature T_0 throughout its length. This is an idealization for any fin but is useful for comparing various size and types of fins. We have, discounting any convection at the tip,

$$\dot{Q}_0 = \theta_0 \int_0^L h(x)P(x)dx \tag{2–108}$$

For square fins with constant convective coefficient and neglecting the fin tip area, this becomes

$$\dot{Q}_0 = hPL\theta_0 \qquad \text{(square fins)}$$

For case I fins, the fin efficiency reduces to

$$\eta_{\text{fin}} = \frac{\theta_0\sqrt{hP\kappa a}}{hPL\theta_0} = \frac{1}{L}\sqrt{\frac{\kappa A}{hP}} \tag{2–109}$$

In Figure 2–38b the fin efficiency curve for the very long fin as given by equation (2–109) is sketched. Notice that the efficiency increases with shorter length L and that the value of Q_{fin} becomes greater than 100% for a short enough length. The fin efficiency cannot be greater than 100%, so equation (2–109) is not applicable for short fins, as we already knew. The sketch of the fin efficiency in figure 2–38 illustrates the characteristic behavior of all fins; that is, the fin efficiency is 100% when the fin length is zero, will remain high for short fins, and will decrease in a hyperbolic manner to lower values as the fin length increases. Thus, a dichotomy between efficiency and fin heat transfer occurs: Short fins have high fin efficiencies, but long fins allow greater amounts of heat transfer. The engineer and the technologist will continually face similar trade-offs in design and analysis.

The efficiency of a square fin with uniform convective heat transfer coefficient, $h_L = h$, can be shown to be

$$\eta_{\text{fin}} = \frac{1}{L}\sqrt{\frac{\kappa A}{hP}}\left(\frac{\sinh mL + \dfrac{h}{m\kappa}\cosh mL}{\cosh mL + \dfrac{h}{m\kappa}\cosh mL}\right) \tag{2–110}$$

and in figure 2–40 this relationship is shown graphically. The tapered fin efficiency relationship is also included on this graph. Also, in figure 2–41 are shown the fin efficiency relationships for circumferential fins of various sizes. Thus, by knowing some of the fin parameters, one can determine the fin efficiency and compute the fin heat transfer $\dot{Q}_{\text{fin}}$ by using equation (2–107), in the form

$$\dot{Q}_{\text{fin}} = \eta_{\text{fin}}\dot{Q}_0 \tag{2–107, revised}$$

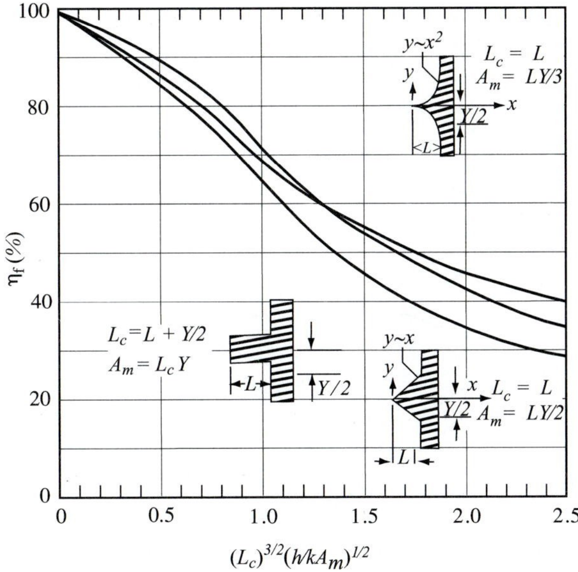

Figure 2–40 Fin efficiency for square and tapered fins [17].

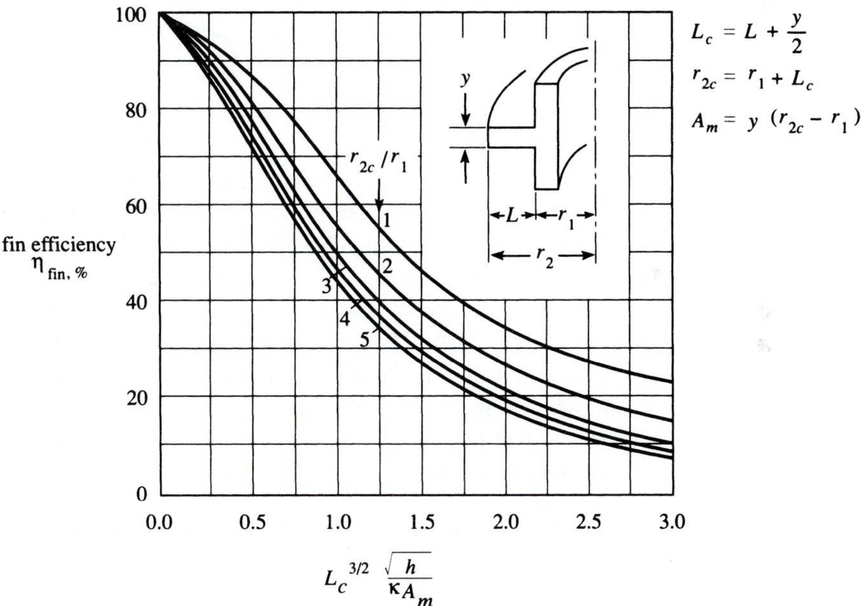

Figure 2–41 Fin efficiencies for circumferential fins [17].

Another important concept that needs to be considered when selecting fins is that of the fin spacing. Figure 2–42 shows multiple fins attached to a plane wall or base. The spacing between the fins $(S - Y)$ must be sufficient to allow each fin to perform as a single fin. If fins are not spaced far enough apart, the fluid (gas or liquid) surrounding the fin may not be able to adequately convect heat, radiation effects may be significant, or solids or foreign matter may become embedded between the fins. Any of these situations could decrease the effectiveness of the individual fins, and as a general rule, fins should be used only if the parameter $P\kappa/hA$ is of the order of magnitude of 10 or more. Then, if multiple fins are needed, the spacing $(S - Y)$ should be at least 3 times the fin thickness Y in order that adequate convection or radiation heat transfer occur. A parameter that

Figure 2–42 Multiple-fin situation.

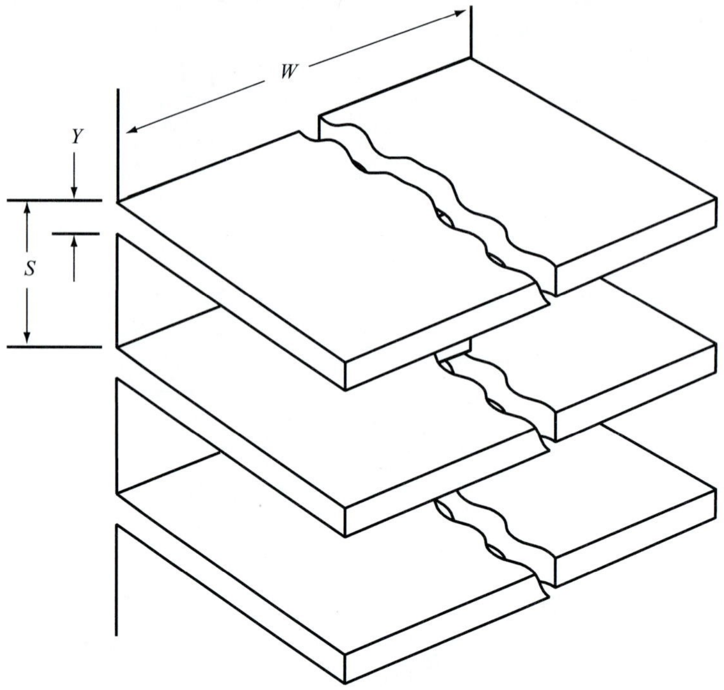

compares the effectiveness of multiple fins is the *fin effectiveness,* ε_{fin}, defined as the ratio of the heat transfer from the finned wall to the heat transfer if the wall and fin were at the base temperature. The fin effectiveness may be expressed as

$$\varepsilon_{\text{fin}} = \frac{\dot{Q}_{\text{fin}} + \dot{Q}_{\text{base}}}{\dot{Q}_0 + \dot{Q}_{\text{base}}} = \frac{\dot{Q}_T}{\dot{Q}_0 + \dot{Q}_{\text{base}}} = \frac{\dot{Q}_T}{hA_T\theta_0} \tag{2–111}$$

where $\dot{Q}_T$ is the total heat transfer of one fin and the base between one pair of fins. The heat transfer $\dot{Q}_0$ is the ideal heat transfer from a fin, plus from the area between two adjacent fins. The surface area of one fin plus the area of the base between one pair of fins is designated as A_T. It can then be shown that the fin effectiveness is related to the fin efficiency by the equation

$$\varepsilon_{\text{fin}} = 1 - \left[\frac{A_{\text{fin}}}{A_T} - \eta_{\text{fin}}\left(\frac{A_{\text{fin}}}{A_T}\right)\right] \tag{2–112}$$

where A_{fin} is the surface area of one fin. The fin effectiveness is sometimes called the area-weighted fin efficiency.

EXAMPLE 2–23 A series of five square aluminum fins are used to cool a microelectronic circuit board. The micro–circuit board is attached to one side of an aluminum plate and the fins are attached to the other side, as shown in figure 2–43. Determine the expected amount of heat transfer that the fins can dissipate if the circuit board has a temperature of 80°C, the air surrounding the fins is 20°C, and the convective heat transfer coefficient is 200 W/m^2 · °C.

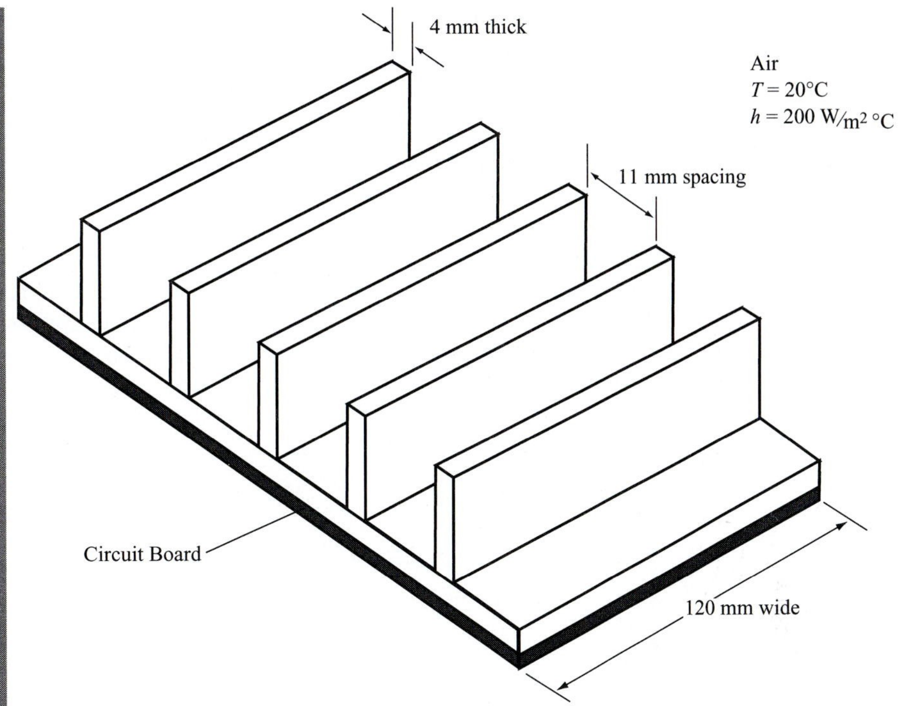

Figure 2–43 Electronic circuit cooling fins, example 2–23.

Solution

We assume that the fin base temperature is the same as the circuit board, 80°C. The heat transfer dissipated by each fin can then be determined from equation (2–111). The total heat transfer will be five times this amount. Referring to figure 2–43, we have

$$L_c = L + (Y/2) = 20 \text{ mm} + (4 \text{ mm}/2) = 22 \text{ mm} = 0.022 \text{ m}$$
$$A_m = YL_c = 4 \times 22 \text{ mm} = 88 \text{ mm}^2 = 88 \times 10^{-6} \text{ m}^2$$

and using $\kappa = 236$ W/m · K for aluminum,

$$L_c^{3/2}\left(\frac{h}{\kappa A_m}\right)^{1/2} = (0.022)^{3/2}\left(\frac{200}{(236)(88 \times 10^{-6})}\right)^{1/2} = 0.32$$

The fin efficiency is approximately 0.90 (90%) from figure 2–40. The surface areas are

$$A_f = 2 \times 120 \text{ mm} \times 20 \text{ mm} + 160 \text{ mm} \times 4 \text{ mm} = 5440 \text{ mm}^2$$
$$= 0.00544 \text{ m}^2$$
$$A_b = 11 \text{ mm} \times 120 \text{ mm} = 1320 \text{ mm}^2 = 0.00132 \text{ m}^2$$
$$A_T = a_f + A_b = 0.00676 \text{ m}^2$$

From equation (2–112), the fin effectiveness is

$$\varepsilon_{\text{fin}} = 1 - [0.00544/0.00676 - (0.9)\ (0.0544/0.00676)] = .9195$$

and the heat transfer per fin plus its surrounding base, using equation (2–111), is

$$\dot{Q}_T = \varepsilon_{\text{fin}} h A_T \theta_0 = (0.9195)\left(200 \frac{\text{W}}{\text{m}^2 \cdot °\text{C}}\right)(0.00676 \text{ m}^2)(80°\text{C} - 20°\text{C}) = 74.6 \text{ W}$$

The total heat transfer dissipated by the five fins is

Answer

$$\dot{Q}_{T,\text{fins}} = 373 \text{ W}$$

Occasionally fins will be attached at both ends. This requires some care in the analysis and often means returning to the fundamental concepts we have been discussing. The following example illustrates a method of considering these special fins.

EXAMPLE 2–24 A circular rod is attached at both ends to isothermal surfaces, at temperatures T_1 and T_2 as shown in figure 2–44a. Air, having temperature T_∞, surrounds the rod, and the convective heat transfer coefficient h describes the conditions around the rod. Determine the steady state temperature distribution through the rod, assuming the temperature is only a function of the axial position, x, and plot the two following conditions: (1) where $T_\infty > T_1 \geq T_2$ to describe heating of the rod and surfaces, and (2) where $T_\infty < T_1 \leq T_2$ describing the cooling of the rod and the surfaces.

Figure 2–44

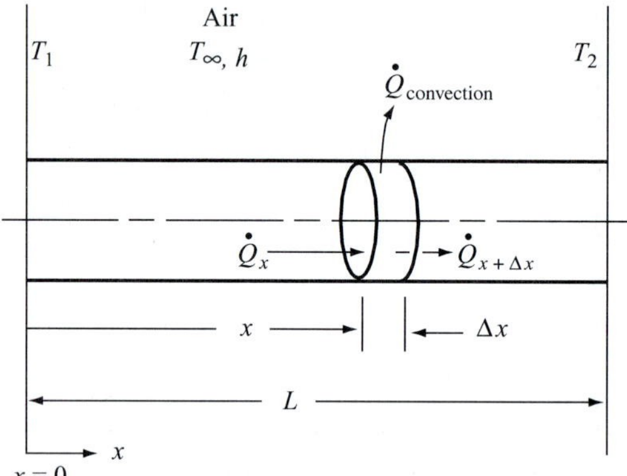

(a) Rod attached at both ends to isothermal surfaces

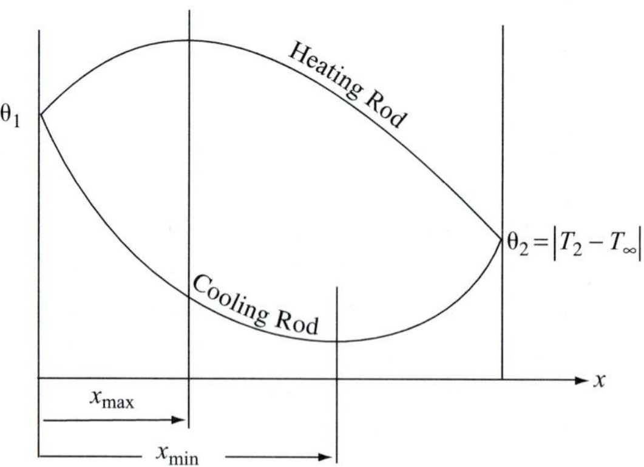

(b) Temperature distributions through the rod

Solution In figure 2–44a is shown an element of the rod, and its energy balance is

$$\dot{Q}_x = \dot{Q}_{x+\Delta x} + \dot{Q}_{convection}$$

Recalling Fourier's Law for Conduction and the convection heat-transfer relationship, this balance becomes

$$-\kappa A\left[\frac{dT(x)}{dx}\right]_x = -\kappa A\left[\frac{dT(x)}{dx}\right]_{x+\Delta x} + hP\Delta x[T(x) - T_\infty]$$

rearranging the parameters and terms and taking $\Delta x \to 0$ gives

$$\frac{d^2T(x)}{dx^2} = \frac{hP}{\kappa A}[T(x) - T_\infty] = m^2\theta(x) = \frac{d^2\theta(x)}{dx^2}$$

where $m^2 = hP/\kappa A$ and $\theta(x) = T(x) - T_\infty$. We know the solution can be written

$$\theta(x) = C_1e^{-mx} + C_2e^{mx} \tag{2–113}$$

and two boundary conditions can be identified:

B.C. 1 $\quad \theta(x) = T_1 - T_\infty = \theta_1 \quad$ at $x = 0$
B.C. 2 $\quad \theta(x) = T_2 - T_\infty = \theta_2 \quad$ at $x = L$

Applying these two boundary conditions to the solution, equation (2–113), and with some algebraic manipulations, and the solution can be written

$$\theta(x) = \left(\frac{1}{e^{2mL} - 1}\right)[(\theta_1e^{2mL} - 2\theta_1 - \theta_2e^{mL})e^{-mx} + (\theta_2e^{mL} - \theta_1)e^{mx}] \tag{2–114}$$

We may arbitrarily specify either of the surfaces to have the higher temperature. If T_1 is greater than T_2, then the temperature distribution takes the general form shown in figure 2–44b. The upper curve describes the expected temperature distribution if the surrounding air temperature is greater than that of either of the surfaces and the rod is heated by the air. The maximum $\theta(x)$, θ_{max}, is located at a coordinate point, x_{max}, where the temperature gradient is zero. This situation coincides with the solution to the position x when the first derivative of the function $\theta(x)$ of equation (2–114) is set to zero; that is,

$$\frac{d\theta(x)}{dx} = 0 = \left(\frac{1}{e^{2mL} - 1}\right)[-m(\theta_1e^{2mL} - 2\theta_1 - \theta_2e^{mL})e^{-mx} + m(\theta_2e^{mL} - \theta_1)e^{mx}]$$

Solving for x,

$$x_{max} = \frac{1}{2m}\ln\left(\frac{\theta_1e^{2mL} - 2\theta_1 - \theta_2e^{mL}}{\theta_2e^{mL} - \theta_1}\right) \tag{2–115}$$

The situation where the surfaces are both at higher temperatures than the air and the rod is being cooled by the air gives a rod temperature distribution indicated by the lower curve of figure 2–48b. Equation (2–114) applies to either condition (for rod heating $\theta(x)$ is negative and for rod cooling $\theta(x)$ is positive).

Thus, equation (2–115) can also be used to determine the location of the minimum rod temperature.

The performance of irregular fins, ones that do not lend themselves well to mathematical analysis, is often determined by experiment. With the advent of the computer there is also a trend to apply more numerical methods to fins, which we saw in section 2–6. Further discussion of fins and their performances is given by Kern and Kraus [18].

Guarded-Hot-Plate Devices

Determination of the thermal conductivity of conducting materials is necessary for any complete analysis of conduction heat-transfer processes. Although it would be convenient to analytically predict thermal conductivity of materials, the experimental values of thermal conductivity are the basis upon which all information about this property has been accumulated, including the analytical theories. Thus, the experimental determination of thermal conductivity is a fundamental charge of engineers, scientists, and technologists for the advancement of heat transfer. An important method that has been used in the past and continues to be used to determine thermal conductivity of materials is the *guarded-hot-plate method.* This method is a direct application of one-dimensional steady state heat transfer and requires a device called a guarded hot plate as shown schematically in figure 2–2.

The fundamental operation of the device includes applying heat to one side of a flat plate made of a material having the unknown thermal conductivity. The temperature is simultaneously measured on both sides of the plate after a steady state test condition has been achieved, and the thermal conductivity is then computed through the use of equation (2–26). It can be seen in figure 2–2 that the guarded-hot-plate device includes insulation to prevent unnecessary heat losses, symmetrical application of heat transfers in the horizontal and vertical planes, and "guard plates" (identified as *primary guards* and *secondary guards*). Also, during the course of experiments the auxiliary heaters can provide thermal conditions such that the gap and edge heat fluxes are minimized. These items assure that steady state one-dimensional conduction heat transfer can be adequately approximated in an actual experiment.

EXAMPLE 2–25

A test material is subjected to a guarded hot plate experiment and the data shown in table 2–11 are obtained. Determine the thermal conductivity of the material.

Solution

The thermal conductivity for the specimen material can be computed from equation (2–26). Referring to table 2–11 and the diagram of the testing device, we see that the specimens have a thickness of 1 cm and an area 110 cm² (11 cm × 10 cm) for one-dimensional conduction. The temperature difference between the faces of the specimen may be considered as the average of the thermocouple readings of number 2 minus 1 and number 3 minus 4. Then

$$\Delta T = \frac{1}{2}(2.885 - 2.618 + 2.926 - 2.663\ \text{mV})\left(\frac{20°\text{C}}{\text{mV}}\right) = 5.3°\text{C}$$

The energy supplied to the electric heater is the electrical power, 0.31 amps × 62.44 volts = 19.3564 W. This power may be expected to represent the heat transfer through both specimens because steady state is assumed and also it is expected that the experiment was conducted in a fashion such that one-dimensional heat transfer occurred through each specimen. From equation (2–26) we find

Answer

$$\kappa = \frac{\dot{Q}\Delta x}{A\Delta T} = \frac{(19.3564\ \text{W})(0.02\ \text{m})}{(0.011\ \text{m}^2)(5.3°\text{C})} = 6.640\frac{\text{W}}{\text{m}\cdot°\text{C}} = 6.640\frac{\text{W}}{\text{m}\cdot\text{K}}$$

Thermal Contact Resistance

Until now the conduction has been assumed to be occurring in solids or continuous media. The materials have been assumed to be homogenous and isotropic and if there where more than one distinct material involved in heat transfer, as in section 2–3 where composite parallel and series assemblies were considered, the heat transfer was assumed to

Table 2–11 (example 2–25) Thermal Data for Experimental Determination of Thermal Conductivity

Run	Heater Data		Thermocouple Data (millivolts, mV)					
	A, amps	*V*, volts	1	2	3	4	5	6
1	0.312	62.40	2.618	2.885	2.928	2.664	2.632	2.896
2	0.310	62.47	2.619	2.884	2.925	2.666	2.633	2.895
3	0.308	62.45	2.617	2.886	2.925	2.659	2.628	2.891
Ave.	0.310	62.44	2.618	2.885	2.926	2.663	2.631	2.894

Thermocouple conversion factor 20°/mV

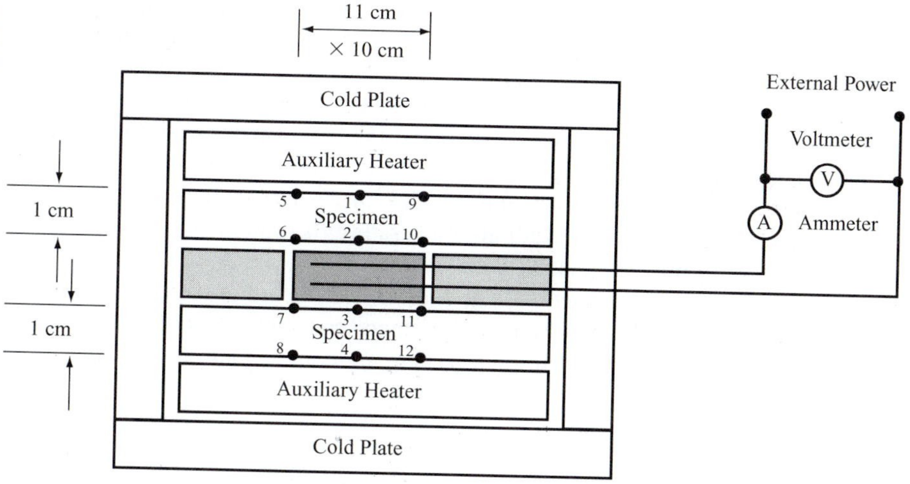

Diagram of testing device

occur between adjacent, different materials in an unhindered manner. A more accurate model of heat transfer between two systems includes the idea of *thermal contact resistance.* In figure 2–45 is shown the premise of such a phenomenon, where two systems, seeming to be in physical contact, are not contacting as intimately as one would expect. The physical gaps and cavities between the two surfaces will be filled with a fluid; most likely atmospheric air. Sometimes the fluid will become trapped in the gaps and reach a thermal equilibrium with the surrounding surfaces.

At a macroscopic scale, which we are using in heat transfer, it is useful to model these imperfections in the contacting surfaces as a convective medium having a perceptible and measurable temperature difference across it. Thus, as indicated in figure 2–45, the temperature profile across the contacting surfaces of systems *A* and *B* includes a temperature drop, ΔT_{TC}, and the heat transfer across these two adjoining surfaces is

$$\dot{Q}_{TC} = h_{TC} A \Delta T_{TC} \tag{2-116}$$

where h_{TC} is the thermal contact coefficient. It is effectively a convection heat transfer coefficient, and the term $1/h_{TC}A$ is called the thermal contact resistance, R_{TC}.

It is reasonable to expect that the thermal contact resistance will be less if the contacting surfaces are smooth or if there is external pressure applied to both surfaces so that

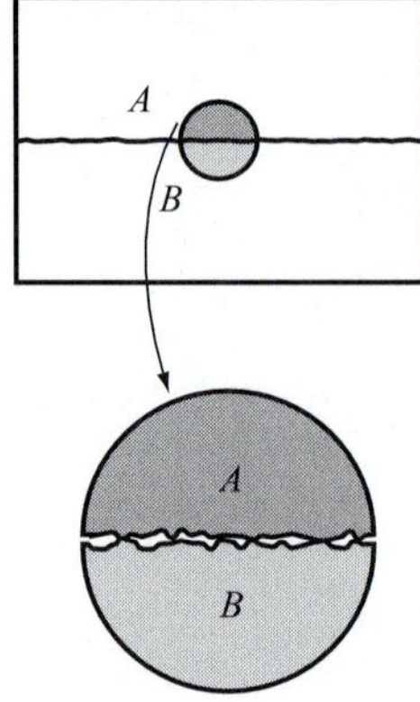

Figure 2–45 Thermal contact resistance.

they are more closely in contact. These sorts of results have been deduced from past experimental data. In table 2–12 are listed some representative thermal contact per unit of area ($R_{TC} \cdot A$) resistances associated with characteristic systems. At this time the information on thermal contact resistance is sparse and the mechanisms that impact on predicting thermal contact resistance are not well understood. Requirements for precise temperature measurements, however, will indicate that thermal contact resistance must be better understood because it occurs with all temperature measurements. In addition, any time two surfaces are joined or abutted with heat transfer expected to occur across that boundary, thermal contact resistance will be present and the engineer or designer must recognize the phenomenon. Thermal contact resistance can sometimes be reduced by applying a grease or oil to the abutted surfaces. There are commercial products that are intended for such an application.

Table 2–12 Thermal Contact Resistance for Some Surfaces

	Roughness (rms)				$R_{TC} \cdot$ A	
Surface	**(in. × 10^6)**	**(μm)**	**Temperature (°C)**	**Pressure (Atm)**	**hr · ft^2 · °F/Btu**	**m^2 · °C/W**
416 Stainless	100	2.54	90–200	2–30	0.0015	0.000264
304 Stainless	45	1.14	20	40–70	0.003	0.000528
Aluminum	100	2.54	150	10–25	0.0005	0.000088
	10	0.25	150	10–25	0.0001	0.000018
Copper	50	1.27	20	12–200	0.00004	0.000007
	150	3.81	20	10–50	0.0001	0.000018
Mild steel	120	3.0	20	20	0.0022	0.000394
with oxide	120	3.0	20	40	0.001	0.000204
coating, 10^{-10} m [21]	120	3.0	20	60	0.0008	0.000139

EXAMPLE 2–26 A test was conducted to determine the approximate thermal contact resistance that occurs between concrete blocks and styrofoam. The test used a concrete block wall with a 4-ft × 8-ft sheet of 1-in. styrofoam as the specimen and a test cell called a guarded hot box (based on the same principle as the guarded hot plate) was then used in conjunction with the test specimen. A set of thermocouples were placed at critical locations on the surfaces and the following data was collected:

Concrete Blocks: 8-in. thick, $\kappa = 9.71$ Btu · in./hr · ft^2 · °F
Exposed surface temperature = 100°F

Styrofoam: 1-in. thick, $\kappa = 0.201$ Btu · in./hr · ft^2 · °F
Exposed surface temperature = 50°F

Rate of heat transfer across appropriate unit area of specimen = 6.289 Btu/hr · ft^2

Solution

In figure 2–46 is shown the typical cross section of the specimen. Using equation (2–26) for steady state one-dimensional conduction, we determine the contact surface temperatures for the concrete and the styrofoam:

$$T_{s2} = T_3 + \frac{\dot{Q}(\Delta x)_s}{\kappa} = 50°\text{F} + \frac{(6.289 \text{ Btu/hr} \cdot \text{ft}^2)\ (1 \text{ in.})}{(0.201 \text{ Btu} \cdot \text{in./hr} \cdot \text{ft}^2 \cdot °\text{F})} = 81.289°\text{F}$$

$$T_{c2} = T_1 - \frac{\dot{Q}(\Delta x)_c}{\kappa} = 100°\text{F} - \frac{(6.289 \text{ Btu/hr} \cdot \text{ft}^2)\ (8 \text{ in.})}{(9.71 \text{ Btu} \cdot \text{in./hr} \cdot \text{ft}^2 \cdot °\text{F})} = 94.82°\text{F}.$$

The temperature drop across the contact boundary is then $T_{c2} - T_{s2} = 13.531°\text{F} = \Delta T_{TC}$ and the contact resistance per unit of area, $R_{TC}A$, can be determined from the relationship

Answer

$$R_{TC}A = \frac{\Delta T_{TC}}{\dot{q}_A} = \frac{13.531°\text{F}}{6.289 \dfrac{\text{Btu}}{\text{hr} \cdot \text{ft}^2}} = 2.152 \frac{\text{hr} \cdot \text{ft}^2 \cdot °\text{F}}{\text{Btu}}$$

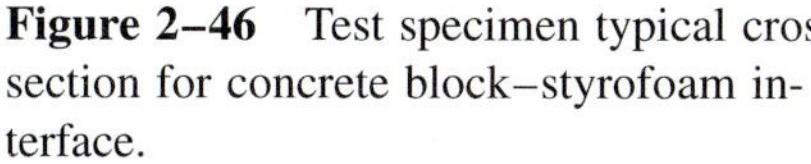

Figure 2–46 Test specimen typical cross section for concrete block–styrofoam interface.

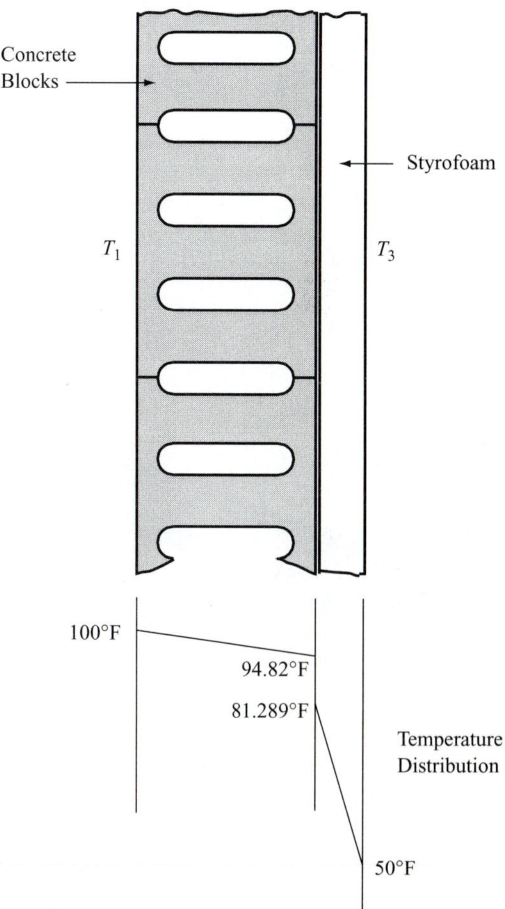

Further information and discussions of thermal contact resistances can be found in [19], [20], and [21].

Critical Thickness of Insulation

Insulation is used to increase thermal resistance or decrease thermal conductance. Thus, the engineer expects to use more insulation to provide greater thermal resistance and a lesser or thinner layer to provide a reduced thermal resistance. In general this situation can be expected, but, on some occasions the reverse phenomena occurs; that is, the use of more insulation provides less thermal resistance and greater heat transfers. This can happen when the surface area of convection is simultaneously increased with increased insulation material, such as when insulation is wrapped around a cylindrical surface. The situation is indicated in Figure 2–47a. Notice that the condition is one of conduction–convection heat transfer through a radial element, given by

$$\dot{q}_L = \frac{T_i - T_o}{\Sigma R_{\text{TL}}} \tag{2–117}$$

where

$$\sum R_{\text{TL}} = \frac{\ln(r_o/r_i)}{2\pi\kappa} + \frac{1}{2\pi r_o h}$$

If r_o is allowed to vary, as would happen if insulation were wrapped around a cylindrical surface, then equation (2–117) would predict the heat transfer per unit length for any outer radius, r_o, and $\dot{q}_L$ is then a function of r_o. Taking the first derivative of $\dot{q}_L$ with respect to r_o, from equation (2–117), and setting that result to zero gives the condition of maximum heat transfer. That is,

$$\frac{d\dot{q}_L}{dr} = 0 = \frac{2\pi(T_i - T_o)\left(\dfrac{1}{\kappa r_o} - \dfrac{1}{h r_o^2}\right)}{\left[\dfrac{\ln(r_o/r_i)}{\kappa} + \dfrac{1}{h_o r_o}\right]^2}$$

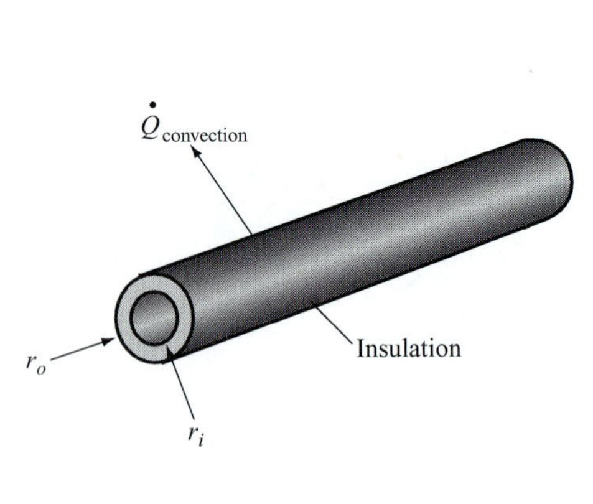

(a)

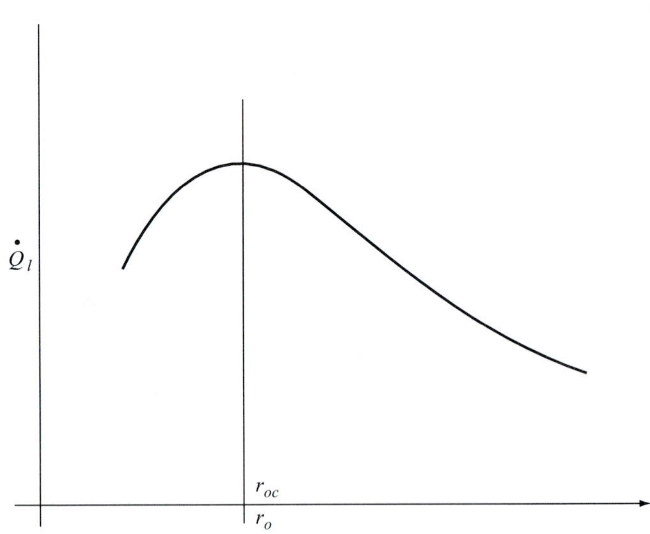

(b) Typical relationship of heat transfer to outer radius

Figure 2–47 Critical thickness of insulation.

This result implies that the term

$$\frac{1}{\kappa r_o} - \frac{1}{h_o r_o} = 0 \quad \text{or} \quad r_o = \frac{\kappa}{h_o} = r_{oc} \tag{2–118}$$

Thus, if κ/h_o (the ratio of the conductivity to the convection heat transfer coefficient at the outer radius r_o, is greater than r_o, one should expect an increase in heat transfer when more insulation is wrapped around a cylindrical surface. If, by applying this added layer of insulation, the radius is increased to an amount such that r_o is now greater than κ/h_o, then any further insulation will decrease heat transfer. Typical results of $\dot{q}_L$ vs r_o are shown in figure 2–47b. The radius r_{oc} is called the *critical radius.*

EXAMPLE 2–27 A 2-cm-diameter water line has an outer surface temperature of 100°C. If the line is surrounded by 10°C air having $h_o = 2.1\ \text{W/m}^2 \cdot °\text{C}$, determine the heat transfer per meter of line. Then, determine the thickness of asbestos insulation ($\kappa = 0.17\ \text{W/m} \cdot °\text{C}$) needed to provide insulating qualities to the water line. Determine the heat transfer per meter of line for this condition.

Solution The water line without insulation is

Answer

$$\dot{q}_L = 2\pi r_o h_o (T_s - T_o) = 2\pi(0.01\ \text{m})\left(2.1\frac{\text{W}}{\text{m}^2 \cdot °\text{C}}\right)(100 - 10°\text{C}) = 11.9\ \frac{\text{W}}{\text{m}}$$

The critical radius is, from equation (2–118)

$$r_{oc} = \frac{\kappa}{h_o} = \frac{0.17\dfrac{\text{W}}{\text{m} \cdot °\text{C}}}{2.1\dfrac{\text{W}}{\text{m}^2 \cdot °\text{C}}} = 0.081\ \text{m} = 8.1\ \text{cm}$$

The asbestos must be at least 7.1 cm (=8.1 cm − 1 cm) thick before it provides insulating qualities. At the critical radius the heat transfer is determined from the equation

$$\dot{q}_L - \frac{2\pi(T_s - T_o)}{\dfrac{\ln(r_{oc}/r_i)}{\kappa} + \dfrac{1}{h_o}}$$

where r_i = radius of the line, 1 cm. Then

Answer

$$\dot{q}_L = 31.1\frac{\text{W}}{\text{m}}$$

Notice that the 8 cm of asbestos insulation increases the heat transfer by 19.2 W/m of water line. Adding further insulation would reduce the heat transfer.

Heat Generation in a Conducting Material

Materials that conduct heat can also generate energy. At least three cases can be cited where this phenomena occurs: nuclear reactor rods, electrical conduction through wires and other components, and materials passing through a phase change. Here we will consider the problem of conduction of electricity through a cylindrical rod or wire. Other problems of internal energy generation can be handled with a similar mathematical analysis. Consider a cylindrical wire conducting electric energy as indicated in Figure 2–48.

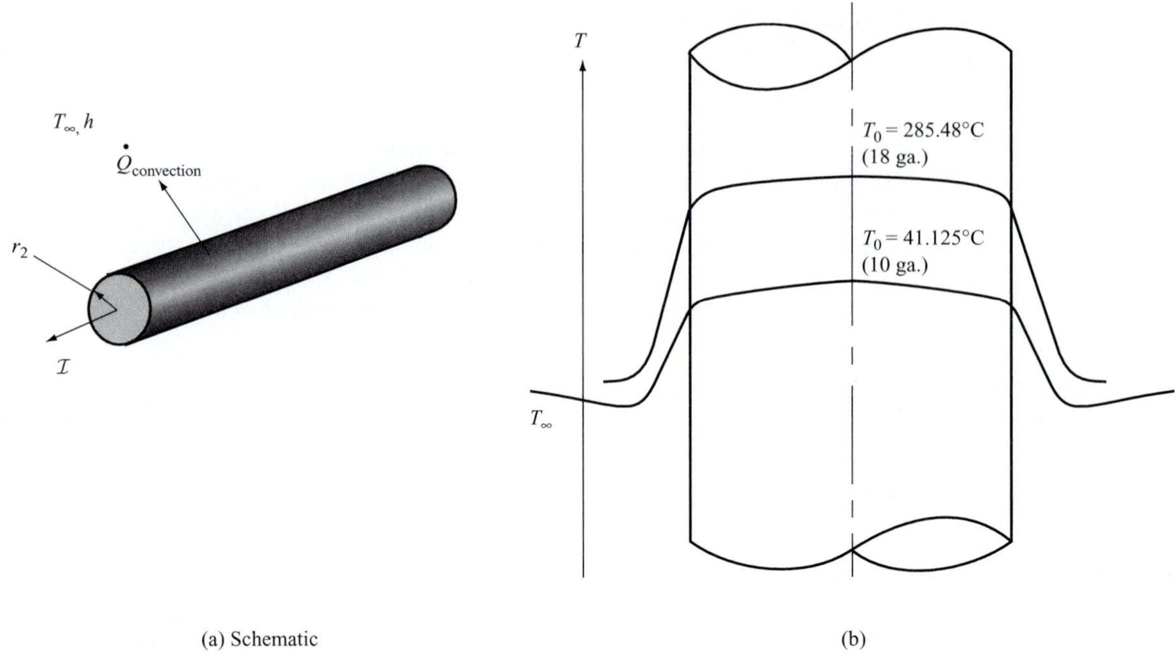

Figure 2–48 Copper electrical wire temperature distribution.

The wire has internal electrical resistance $\mathcal{R}_e$ per unit of length of wire. The internal energy generation due to the electrical current $\mathcal{I}$ passing through the wire is

$$\dot{e}_L = \mathcal{I}^2\mathcal{R}_e\left(\frac{\mathrm{W}}{\mathrm{m}}\right) \tag{2–119}$$

The volumetric energy generation would be

$$\dot{e}_{\text{gen}} = \frac{1}{\pi r_o^2}\mathcal{I}^2\mathcal{R}_e \tag{2–120}$$

where r_0 is the outside radius of the wire. The governing equation for this problem can be written

$$\frac{d}{dr}\left[r\frac{dT(r)}{dr}\right] + \frac{r}{\kappa}\dot{e}_{\text{gen}} = 0 \qquad r_o \geq r \geq 0 \tag{2–121}$$

with various possible boundary conditions. We list some of these:

B.C. 1a.	$\dot{e}_L = \dot{e}_{\text{gen}}(\pi r_o^2) = -2\pi\kappa r\dfrac{dT(r)}{dr}$	at $r = r_o$
B.C. 1b.	$-\kappa\dfrac{dT(r)}{dr} = h_o(T_o - T_\infty)$	at $r = r_o$
B.C. 1c.	$\dot{e}_L = 2\pi r_o h_o[T_o - t_\infty]$	at $r = r_o$
B.C. 2a.	$T(r) = T_o$	at $r = r_o$
B.C. 2b.	$T(r) = T_c$ (center temperature)	at $r = 0$

Equation 2–121 can be solved directly by separating the variables,

$$r\frac{dT(r)}{dr} = -\left(\frac{\dot{e}_{\text{gen}}}{2\kappa}\right)r^2 + C_1$$

Using B.C. 1a gives $C_1 = 0$. Separating variables once more and integrating gives

$$T(r) = -\left(\frac{\dot{e}_{\text{gen}}}{4\kappa}\right)r^2 + C_2$$

Using B.C. 2b allows for C_2 to be evaluated. Then

$$T(r) = -\left(\frac{\dot{e}_{\text{gen}}}{4\kappa}\right)r^2 + T_C \qquad \textbf{(2–122)}$$

If B.C. 1c and B.C. 2a are used to evaluate the constants C_1 and C_2, then the temperature distribution through the wire can be expressed as

$$T(r) = T_\infty + \dot{e}_{\text{gen}}\left(\frac{r_o}{2h_o} + \frac{1}{4\kappa}(r_o^2 - r^2)\right) \qquad \textbf{(2–123)}$$

It is instructive to consider how the wire temperature depends on the various parameters included in equations (2–122) and (2–123). In the next example we see how the temperature of an electrical wire can be affected by the wire size. In appendix table B–7 are listed the resistance per unit length, $\mathcal{R}_e$, of common copper that is often used for electrical wiring for use in the analysis.

EXAMPLE 2–28 Determine the temperature distribution through a 10-gauge copper wire conducting 2 amperes of electric current. Assume that the convective coefficient, *h,* is 100 W/m² · °C. Then compare these temperatures to another 18-gauge copper wire conducting 2 amperes with the same convective coefficient. Assume the surrounding temperature is 25°C.

Solution The temperature distribution can be obtained by using either equation (2–122) or (2–123). The thermal conductivity of copper may be taken as 400 W/m · °C, from appendix table B–1, and the electrical resistance, $\mathcal{R}_e$, is 0.9989 Ω/ft (3.277 m) for 10-gauge wire and 6.385 Ω/ft (20.948 Ω/m) for 18-gauge wire from appendix table B–7. The wire diameters are 101.9 mils (=0.1019 in.) for 10-gauge wire and 0.0403 in for 18-gauge wire. The radii are then 0.001294 m for 10-gauge wire and 0.000512 m for 18-gauge wire. Substituting into equation (2–123), for 10-gauge wire we have

$$T(r) = 25°\text{C} + \frac{(2\text{ amps})^2(3.277\ \Omega/\text{m})}{(\pi)\,(0.001294^2)}\left(\frac{0.001294\text{ m}}{2\times 100\,\dfrac{\text{W}}{\text{m}^2\cdot°\text{C}}} + \frac{1}{\dfrac{1}{4}\times 400\,\dfrac{\text{W}}{\text{m}\cdot°\text{C}}}(0.001294^2\text{ m}^2 - r^2)\right)$$

$$= 25°\text{C} + 16.122°\text{C} + 1557.4\,(0.001294^2\text{ m}^2 - r^2)$$

$$= 41.122°\text{C} + 1557.4(0.001294^2\text{ m}^2 - r^2) \qquad \textbf{(2–124)}$$

The difference between the center temperature, T_c, of 41.122°C where $r = 0$, and the outside surface, where $r = 0.001294$ m, is only 0.003°C. Within the precision limits of standard temperature measuring devices, the wire temperature can be assumed to be uniform throughout. The temperature distribution, however, is shown somewhat exaggerated in figure 2–48. Also, for 18-gauge-copper wire the temperature distribution is

$$T(r) = 285.467°\text{C} + 0.0167°\text{C} - 63590.6\,r^2 \qquad \textbf{(2–125)}$$

with a centerline temperature of 285.4837°C and an outer surface temperature of 285.467°C. Again, this indicates that the wire temperature is effectively constant, but an exaggerated graph of this temperature distribution is shown in figure 2 – 48. Notice how much greater than the temperature of the 10-gauge wire is the temperature of the 18-gauge wire. This can be explained in many ways; for instance, the energy generation per unit volume of wire is greater for the 18-gauge wire than for the 10-gauge wire. Also, the convective surface area, the outer wire surface, is less for the 18-gauge wire than for the 10-gauge wire.

2–8 SUMMARY

The general equation for Fourier's Law of Heat Conduction is

$$\dot{q} = -\kappa \Delta T \tag{1–54}$$

where κ is the thermal conductivity. Thermal conductivity may be predicted for monatomic ideal gases from the equation

$$\kappa = 8.328 \times 10^{-4}\sqrt{\frac{T}{\text{MW} \cdot \Gamma}} \tag{2–1}$$

and for diatomic ideal gases,

$$\kappa_{\text{poly}} = \kappa\left(\frac{4c_v}{5R + 0.6}\right) \tag{2–2}$$

For liquids, thermal conductivity may be predicted from

$$\kappa = 3.865 \times 10^{-23}\frac{\mathbf{V}}{x_m^2} \tag{2–6}$$

and for solids,

$$\kappa = \kappa_{T0} + a(T - T_0) \tag{2–10}$$

Thermal conductivity may be taken as a constant if the temperature does not vary or change too much.

The general form of the energy balance of a material conducting heat and with an energy generation, in Cartesian coordinates, is

$$\frac{\partial}{\partial x}\left[\kappa\frac{\partial T}{\partial x}\right] + \frac{\partial}{\partial y}\left[\kappa\frac{\partial T}{\partial y}\right] + \frac{\partial}{\partial z}\left[\kappa\frac{\partial T}{\partial z}\right] + \dot{e}_{\text{gen}} = \rho c_p\frac{\partial T}{\partial t} \tag{2–16}$$

For no energy generation the equation is

$$\frac{\partial}{\partial x}\left[\kappa\frac{\partial T}{\partial x}\right] + \frac{\partial}{\partial y}\left[\kappa\frac{\partial T}{\partial y}\right] + \frac{\partial}{\partial z}\left[\kappa\frac{\partial T}{\partial z}\right] = \rho c_p\frac{\partial T}{\partial t} \tag{2–17}$$

And for steady state conditions,

$$\frac{\partial}{\partial x}\left[\kappa\frac{\partial T}{\partial x}\right] + \frac{\partial}{\partial y}\left[\kappa\frac{\partial T}{\partial y}\right] + \frac{\partial}{\partial z}\left[\kappa\frac{\partial T}{\partial z}\right] = 0 \tag{2–18}$$

For one-dimensional steady state conduction heat transfer and no energy generation, in rectangular coordinates

$$\dot{Q} = -\kappa A\frac{\Delta T}{\Delta x} \tag{2–26}$$

in cylindrical coordinates

$$\dot{Q} = \frac{2\pi\kappa L\Delta T}{\ln(r_o/r_i)} = \frac{2\pi\kappa L\Delta T}{\ln(D_o/D_i)} \tag{2–41}$$

and in spherical coordinates

$$\dot{Q} = -\frac{4\pi\kappa\Delta T}{\left(\dfrac{1}{r_o} - \dfrac{1}{r_i}\right)} \tag{2–54}$$

Thermal resistance, R_T, is

$R_T = \dfrac{\Delta x}{\kappa A}$ for conduction in the x direction

$R_T = \dfrac{1}{hA}$ for convection in the x direction

$R_T = \dfrac{\ln(r_o/r_i)}{2\pi\kappa L}$ for conduction in the r direction, cylindrical coordinates

$R_T = \dfrac{1}{2\pi rhL}$ for convection in the r direction, cylindrical coordinates

$R_T = \dfrac{\dfrac{1}{r_i} - \dfrac{1}{r_o}}{4\pi\kappa}$ for conduction in the r direction, spherical coordinates

$R_T = \dfrac{1}{4\pi r^2 h}$ for convection in the r direction, spherical coordinates

For two-dimensional steady state conduction heat transfer, no energy generation and constant thermal conductivity,

$$\nabla^2 T(x, y) = 0 \tag{2–56}$$

and the general solution may be written

$$T(x, y) = \sum_{n=0}^{\infty} (a_n \sin p_n x + b_n \cos p_n x)(c_n e^{-p_n y} \pm d_n e^{p_n y}) \tag{2–61}$$

For two-dimensional conduction heat transfer in cylindrical coordinates,

$$\frac{\partial^2 T(r, z)}{\partial r^2} + \left(\frac{1}{r}\right)\frac{\partial T(r, z)}{\partial r} + \frac{\partial^2 T(r, z)}{\partial z^2} = 0$$

with the general solution

$$T'(r, z) = (a \sinh pz + b \cosh pz)(cJ_0 pr + dY_0 pr) \tag{2–75}$$

For steady state conduction heat transfer occurring between two isothermal surfaces, the shape factor, S, is defined as

$$\dot{Q} = S\kappa\Delta T$$

Graphical methods may be used to obtain approximate temperature distributions and heat transfers through two-dimensional systems having two isothermal boundaries. Isotherms and heat flow paths are constructed perpendicularly and so that the grid is composed of approximate squares. Then

$$\dot{Q} = L\frac{M}{N}\kappa\Delta T_{\text{overall}} \tag{2–82}$$

where M = number of heat flow lines and N = number of temperature steps between the overall temperature difference, $\Delta T_{\text{overall}}$.

Finite difference temperature gradients are approximated by the equations

$$\partial T(x, y)/\partial x \approx [T(n + 1, y) - T(n - 1, y)]/2\Delta x \tag{2–84}$$

$$\partial T(x, y)/\partial x \approx [T(n, y) - T(n - 1, y)]/\Delta x \tag{2–85}$$

$$\partial T(x, y)/\partial x \approx [T(n + 1, y) - T(n, y)]/\Delta x \tag{2–86}$$

Also,

$$\partial^2 T(x, y)/\partial x^2 \approx [T(n + 1, y) - 2T(n, y) + T(n - 1, y)]/\Delta x^2 \tag{2–87}$$

For a typical finite difference node the energy balance for two-dimensional problems can be written as

$$\dot{Q}_{\text{other}} - \kappa\sum\frac{\Delta T}{\Delta x} - \kappa\sum\frac{\Delta T}{\Delta y} = 0 \tag{2–89}$$

which gives n equations for n nodes and an $n \times n$ matrix such that

$$\{\dot{Q}\} = [M]\{T\} \tag{2–90}$$

For two- or three-dimensional steady state conduction, one can also use the resistance concepts to write

$$\dot{Q}_{\text{other}} = \sum\frac{\Delta T_i}{R_{Ti}} \tag{2–96}$$

Fins

Fins and extended surfaces increase the area for convection heat transfer with conducting media. For square fins that are very long in proportion to the thickness, the temperature distribution may be approximated by the equation

$$T(x) = T_\infty + (T_0 - T_\infty)e^{-mx} \tag{2–104}$$

and the fin heat transfer by the equation

$$\dot{Q}_{\text{fin}} = -\sqrt{hP\kappa A}(T_0 - T_\infty)$$

where $m^2 = Ph/\kappa A$ and P is the perimeter of the fin.

Fin efficiency, η_{fin}, is defined as

$$\eta_{\text{fin}} = \frac{\dot{Q}_{\text{fin}}}{\dot{Q}_0} \tag{2–107}$$

where

$$\dot{Q}_0 = (T_0 - T_\infty)\int_0^L h(x)P(x)dx \tag{2–108}$$

For multiple fins, having uniform spacing between each fin, the fin effectiveness is defined as

$$\varepsilon_{\text{fin}} = \frac{\dot{Q}_T}{\dot{Q}_{\text{base}} + \dot{Q}_0} \tag{2–111}$$

where $\dot{Q}_T = \dot{Q}_{\text{base}} + \dot{Q}_{\text{fin}}$. Also,

$$\varepsilon_{\text{fin}} = 1 - [A_{\text{fin}}/A_T - \eta_{\text{fin}}(A_{\text{fin}}/A_T)] \tag{2–112}$$

Thermal Contact Resistance

Thermal contact resistance, R_{TC}, is defined as

$$\dot{Q}_{\text{TC}} = \frac{\Delta T_{\text{TC}}}{R_{\text{TC}}} \tag{2–116}$$

where ΔT_{TC} is the temperature difference at a boundary between two surfaces that are in physical contact.

Critical Thickness of Insulation

For a cylinder, such as a tube, pipe of wire, that is insulated with an annular layer of insulation, the heat loss from the cylinder can be increased rather than decreased if the condition

$$\frac{\kappa}{h_0} > r_{\text{oc}} \tag{2–118}$$

where r_{oc} is the critical radius or outside radius of the insulation.

Heat Generation in a Conducting Material

For a cylindrical wire conducting electric current the governing equation is

$$\frac{d}{dr} r \frac{dT(r)}{dr} + \frac{r}{\kappa}\dot{e}_{\text{gen}} = 0 \tag{2–121}$$

where

$$\dot{e}_{\text{gen}} = \frac{1}{2\pi r_o^2}\mathcal{I}^2\mathcal{R}_e \qquad (\text{W/m}^3\text{), or Btu/hr} \cdot \text{ft}^2) \tag{2–120}$$

The solution for the boundary conditions $T = T_o$ at $r = r_0$ and $\dot{e}_{\text{gen}} = 2\pi h_o r_o(T_o - T_\infty)$ at $r = r_o$ is

$$T(r) = T_\infty + \dot{e}_{\text{gen}}\left[\frac{r_o}{2h_o} + \frac{1}{4\kappa}\right][r_o^2 - r^2] \tag{2–123}$$

DISCUSSION QUESTIONS

Section 2–1

2–1 Describe the physical significance of thermal conductivity.

2–2 Why is thermal conductivity affected by temperature?

2–3 Why is thermal conductivity not affected to a significant extent by material density?

Section 2–2

2–4 Why are heat of vaporization, heat of fusion, and heat of sublimation accounted as energy generation in the usual derivation of energy balance equations?

Section 2–3

2–5 Why are heat transfers and electrical conduction similar?

2–6 Describe the difference between thermal resistance, thermal conductivity, thermal resistivity, and R-value.

Section 2–4

2–7 Why do solutions for temperature distributions in heat conduction problems need to converge?

2–8 Why is the heat conduction in a fin not able to be determined for the case where the base temperature is constant, as in figure 2–9.

2–9 What is meant by an isotherm?

2–10 What is meant by a heat-flow line?

Section 2–5

2–11 What is a shape factor?

2–12 Why should isotherms and heat-flow lines be orthogonal or perpendicular to each other?

Section 2–6

2–13 Can you identify a physical situation when the partial derivatives from the left and right (see equations 2–85 and 2–86) are not the same?

Section 2–7

2–14 Can you explain when fins may not be advantageous in increasing the heat transfer at a surface?

2–15 Why should thermal contact resistance be of concern to an engineer?

PRACTICE PROBLEMS

Problems marked with an asterisk (*) may be more difficult for some students.

Section 2–1

2–1 Compare the value for thermal conductivity of helium at 20°C using equation (2–3) and the value from appendix table B–4.

2–2 Predict the thermal conductivity for neon gas at 200°F. Use a value of 3.9 Å for the collision diameter for neon.

2–3 Show that thermal conductivity is proportional to temperature to the 1/6 power for a liquid according to Bridgeman's Equation (2–6).

2–4 Predict a value for thermal conductivity of liquid ammonia at 300 K. Use the equation (2–6) suggested by Bridgeman.

2–5 Plot the value for thermal conductivity of copper as a function of temperature as given by equation (2–10). Plot the values over a range of temperatures from −40°F to 160°F.

2–6 Estimate the thermal conductivity of platinum at −100°C if its electrical conductivity is 6×10^7 mhos/m, based on the Wiedemann–Franz Law. *Note:* 1 mho = 1 amp/volt, 1 amp = 1 coulomb/s, 1 W = 1 J/s, and 1 J = 1 volt · coulomb.

2–7 Calculate the thermal conductivity of carbon bisulfide using equation (2–6) and compare this result to the listed value given in table 2–2.

Section 2–2

2–8 Estimate the temperature distribution in a stainless steel rod, 1 in. in diameter, that is 1 yard long with 3 in. of one end submerged in water at 40°F and the other end held by a person. Assume the person's skin temperature is 82°F, the temperature in the rod is uniform at any point in the rod, and steady state conditions are present.

***2–9** Derive the general energy equation for conduction heat transfer through a homogenous, isotropic medium in cylindrical coordinates, equation (2–19).

***2–10** Derive the general energy equation for conduction heat transfer through a homogenous, isotropic medium in spherical coordinates, equation (2–20).

2–11 Determine a relationship for the volume element in spherical coordinates.

Section 2–3

2–12 An ice-storage facility uses sawdust as an insulator. If the outside walls are 2-ft-thick sawdust and the sideboard thermal conductivity is neglected, determine the R-value of the walls. If the inside temperature is 25°F and the outside temperature is 85°F, estimate the heat gain of the storage facility per square foot of outside wall.

2–13 The combustion chamber of an internal combustion engine is at 800°C when fuel is burned in the chamber. If the engine is made of cast iron with an average thickness of 6.5 cm between the combustion chamber and the outside surface, estimate the heat transfer per unit of area if the outside surface temperature is 50°C and the outside air temperature is 30°C.

2–14 Triple-pane window glass has been used in some building construction. Triple-pane glass is a set of three glass panels, each separated by a sealed air gap as shown in figure 2–49. Estimate the R-value for triple-pane windows and compare this to the R-value for single-pane glass. Note that the air within the gap is sealed and cannot move so that it acts as a conducting media only.

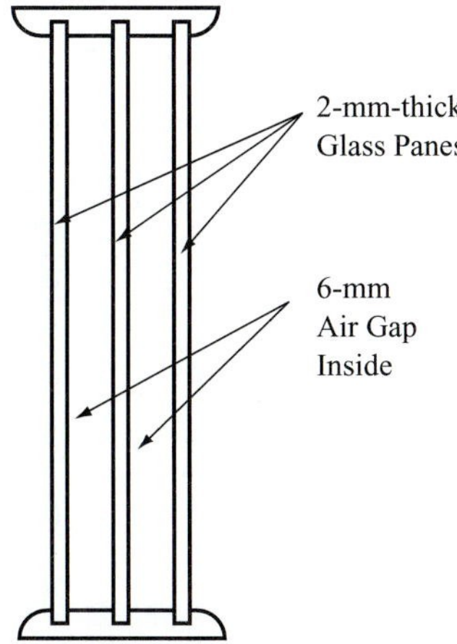

Figure 2–49 Triple-pane window.

2–15 For the outside wall shown in figure 2–50, determine the R-value, the heat transfer through the wall per unit area, and the temperature distribution through the wall if the outside surface temperature is 36°C and the inside surface temperature is 15°C.

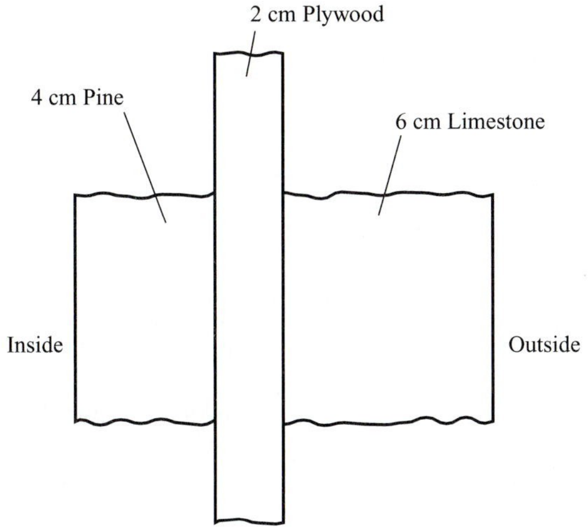

Figure 2–50 Outside wall.

2–16 Determine the heat transfer per foot of length through a copper tube having an outside diameter of 2 in. and an inside diameter of 1.5 in. The pipe contains 180°F ammonia and is surrounded by 80°F air.

2–17 A steam line is insulated with 15 cm of rock wool. The steam line is a 5-cm-OD iron pipe with a 5-mm-thick wall. Estimate the heat loss through the pipe per meter length if steam at 120°C is in the line and the surrounding temperature is 20°C. Also determine the temperature distribution through the pipe and insulation.

2–18 Evaporator tubes in a refrigerator are constructed of 1-in.-OD aluminum tubing with $^1/_8$-in.-thick walls. The air surrounding the tubing is at 25°F, and the refrigerant in the evaporator is at 15°F. Estimate the heat transfer to the refrigerant over 1 ft of length.

2–19 Teflon tubing of 4 cm OD and 2.7 cm ID conducts 1.9 W/m outward when the outside temperature is 80°C. Estimate the inside temperature of the tubing. Also predict the thermal resistance per unit of length.

2–20 A spherical glass flask, 4 m in diameter with a 15-mm-thick wall, is used to heat grape juice. During the heating process the outside surface temperature of the flask is 82°C and the inside surface is 80°C. Estimate the thermal resistance of the flask, the heat transfer through the flask if it is assumed that only the bottom half is heated, and the temperature distribution through the flask wall.

2–21 A Styrofoam spherical container having a 1-in.-thick wall and 2-ft diameter holds dry ice (solid carbon dioxide) at −85°F. If the outside temperature is 60°F, estimate the heat gain in the container and establish the temperature distribution through the 1-in. wall.

2–22 Determine the overall thermal resistance per unit area for the wall shown in figure 2–51. Exclude the thermal resistance due to convection heat transfer in the analysis. Then, if the heat transfer is expected to be 190 W/m^2 and the exposed brick surface is 10°C, estimate the temperature distribution through the wall.

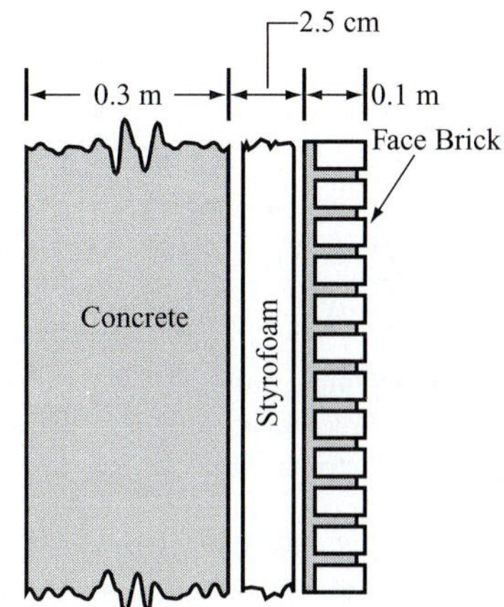

Figure 2–51 Structural wall.

2–23 Determine the thermal resistance per unit of length of the tubing shown in figure 2–52. Then predict the heat transfer through the tubing if the inside ambient temperature is $-10°C$ and the outside is 20°C.

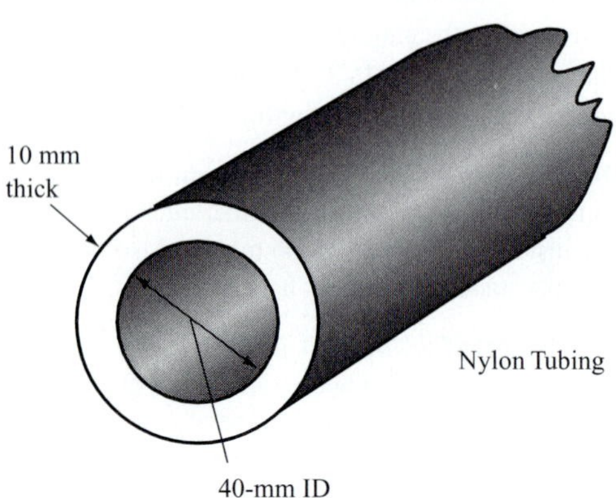

Figure 2–52 Tubing.

***2–24** Determine the heat transfer through the wall of example problem 2–5 if the thermal conductivity is affected by temperature through the relationship

$$\kappa = 9.2 + 0.007T \frac{\text{Btu} \cdot \text{in.}}{\text{hr} \cdot \text{ft}^2 \cdot °\text{F}} \quad (T \text{ in } °\text{F})$$

***2–25** Determine the temperature distribution through a slab if $\kappa = aT^{0.001}$, T is in K, and a is a constant. Then compare this to the case where $\kappa = a$.

Section 2–4

2–26 Show that $T(x, y) = (a \sin px + b \cos px)(ce^{-py} + de^{py})$ satisfies the equation

$$\frac{\partial^2 T(x, y)}{\partial x^2} + \frac{\partial^2 T(x, y)}{\partial y^2} = 0$$

2–27 For the wall of example problem 2–11, determine the heat transfer in the y direction at 3 ft above the base. Then plot the temperature distribution at this level.

2–28 Write the governing equation and the necessary boundary conditions for the problem of a tapered wall as shown in figure 2–53.

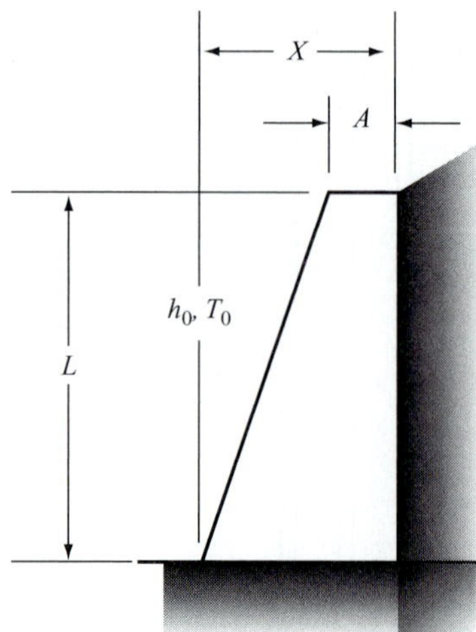

Figure 2–53 Tapered wall with heat transfer.

2–29 Write the governing equation and the necessary boundary conditions for the problem of a heat exchanger tube as shown in figure 2–54.

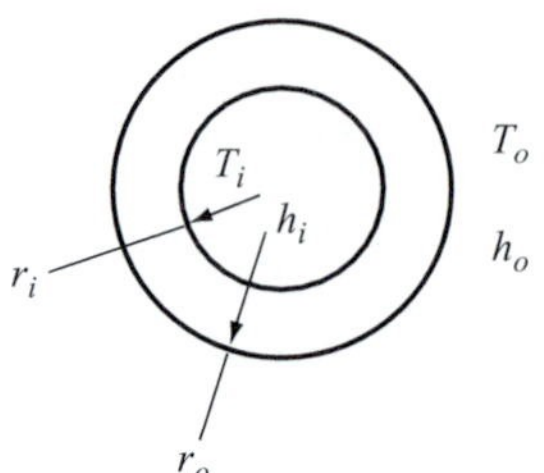

Figure 2–54 Heat exchanger tube.

2–30 Write the governing equation and the necessary boundary conditions for the problem of a spherical concrete shell as sketched in figure 2–55.

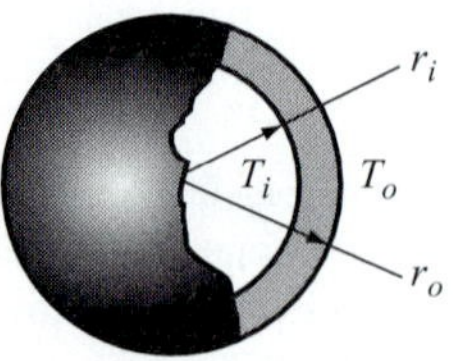

Figure 2–55 Spherical thick-walled shell.

2–31 Determine the Fourier Coefficient A_n for the problem resulting in a temperature distribution of

$$T(x, y) = \sum_{n=0}^{\infty} A_n e^{-n\pi y/L} \sin n\pi x/L$$

involving a boundary temperature distribution given by

$$T(x, 0) = \cos \pi x/L$$

for $0 < x < L$.

2–32 Determine the Fourier Coefficient A_n for the problem involving a boundary temperature distribution given by

$$T(x, 0) = T_0\left(1 - \frac{x}{L}\right)$$

and where the solution to the temperature distribution is

$$T(x, y) = \sum_{n=0}^{\infty} A_n e^{-n\pi y/L} \sin n\pi x/L$$

2–33 Plot the Bessel Function of the first kind of zero and first order, J_0 and J_1, for arguments from 0 to 10.

2–34 Plot the Bessel Function of the second kind of zero and first order, Y_0 and Y_1, for arguments from 0 to 10.

2–35 A silicon rod 20 cm in diameter and 30 cm long is exposed to a high temperature at one end so that the end is at 400°C while the remaining surfaces are at 60°C. Estimate the centerline temperature distribution through the rod.

2–36 A Teflon rod 6 in. in diameter and 2 ft long is at 230°F. It is then exposed at one end to cool air so that that end reaches 80°F while the cylindrical surface cools to 150°F. The other end remains at 230°F at steady state. Determine the expected temperature distribution.

Section 2–5

2–37 A water line of 2-in. diameter is buried horizontally 4 ft deep in earth. Estimate the heat loss per foot from the water line if water at 50°F flows through the line and the outside temperature of the line is assumed to be 50°F. The surface temperature of the earth is −20°F.

2–38 A chimney is constructed of square concrete blocks with a round flue as shown in figure 2–56. Estimate the heat loss through the cement blocks per meter of chimney if the outer surface temperature is −10°C and the inner surface temperature is 150°C.

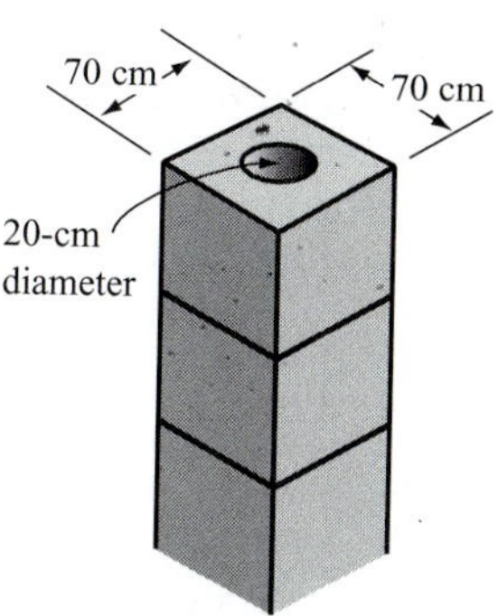

Figure 2–56 Chimney and flue.

2–39 Nuclear waste is placed in drums 50 cm in diameter by 100 cm long and buried in sand. Water lines are buried adjacent to the drums to keep them cool. The suggested typical arrangement is shown in figure 2–57. Estimate the heat transfer between a drum and the water line.

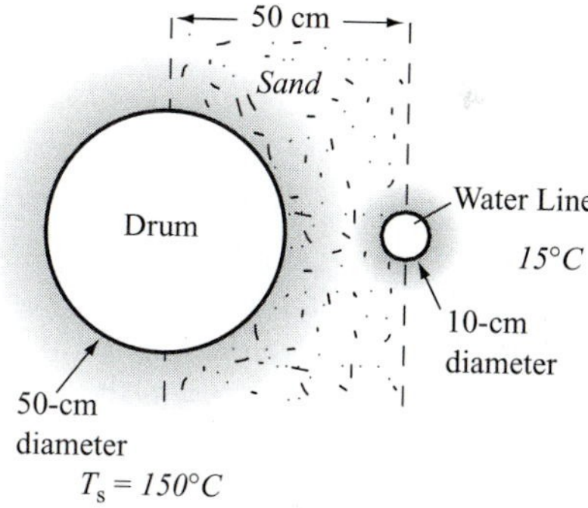

Figure 2–57 Nuclear waste drums.

2–40 Steel pins are driven into asphalt pavement as shown in figure 2–58. Estimate the heat transfer between a pin when it is at 60°F and the surface when it is at 110°F.

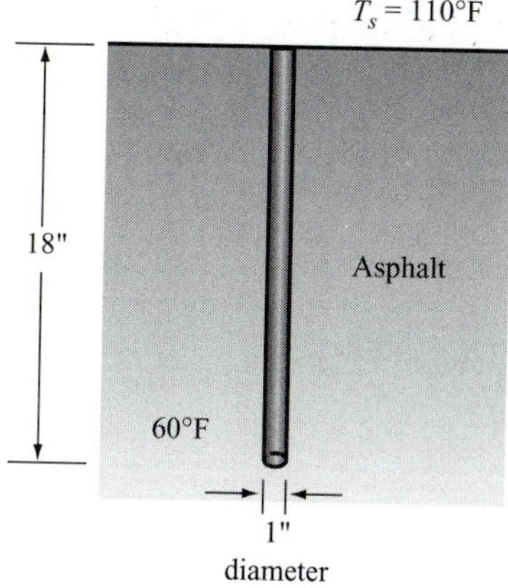

Figure 2–58 Steel pins in asphalt.

2–41 A heat treat furnace sketched in figure 2–59 has an inside surface temperature of 1200°C and an outside surface temperature of 60°C. If the walls are assumed to be homogenous with thermal properties the same as asbestos, estimate the heat transfer from the walls, excluding the door.

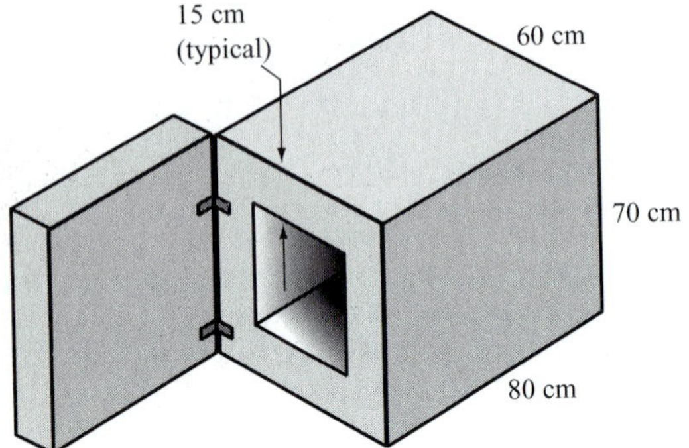

Figure 2–59 Heat-treat furnace.

2–42 A small refrigerator freezer, 16 in. × 16 in. × 18 in. outer dimensions, has an inside surface temperature of 10°F and an outside surface temperature of 80°F. If the walls are uniformly 3 in. thick, homogeneous, and with thermal properties the same as styrofoam, estimate the heat transfer through the walls and door of the refrigerator.

2–43 Using graphical methods, estimate the temperature distribution and the heat transfer per meter depth between the two surfaces at the corner shown in figure 2–60. Notice that the scale is 1 to 8.

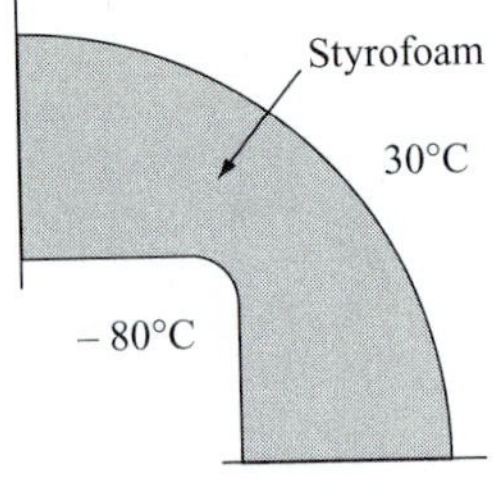

Scale: 1 to 8

Figure 2–60 Heat transfer at a corner.

2–44 Using graphical methods, estimate the temperature distribution through the phenolic disk surrounding the silicon chip sketched in figure 2–61. Then estimate the heat transfer per millimeter of depth.

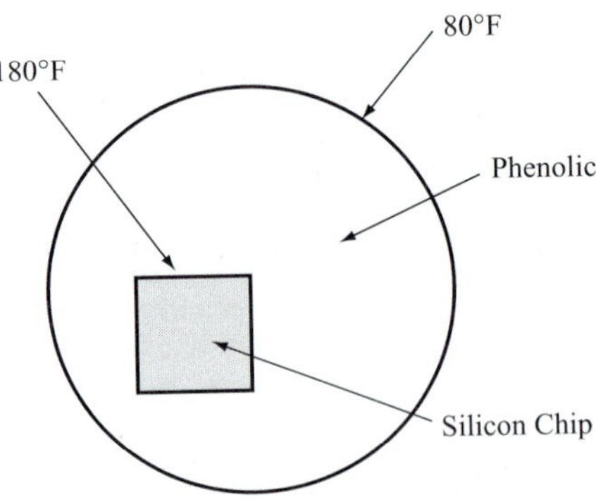

Scale: 100 to 1

Figure 2–61 Silicon chip embedded in phenolic.

2–45 Using graphical techniques, estimate the temperature distribution through the earth around the electrical power line shown in figure 2–62. Then estimate the heat transfer necessary between the line and the ground surface for steady state conditions. Express your answer in W/m.

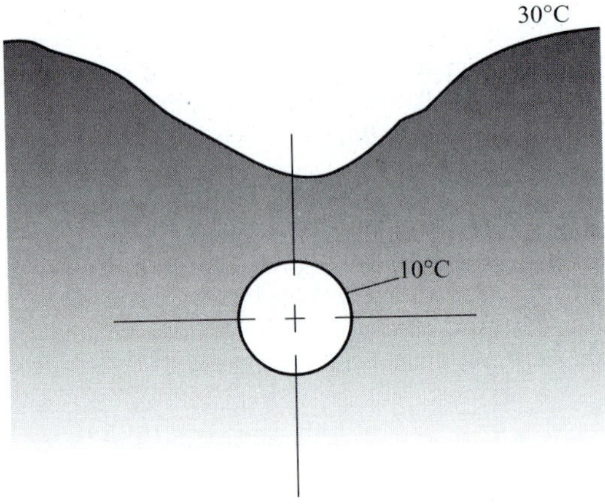

Figure 2–62 Buried high-power line.

***2–46** Using graphical techniques, estimate the temperature distribution through the cast iron engine block and head shown in figure 2–63.

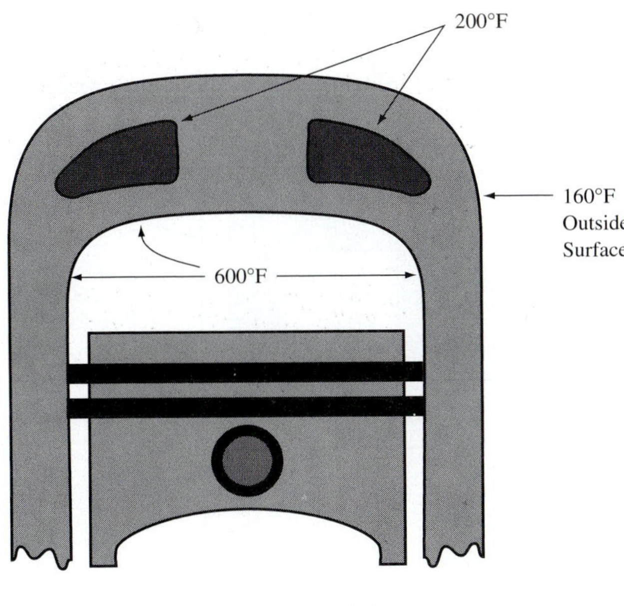

Figure 2–63 Sketch of a gasoline engine.

***2–47** A Bunsen burner is used to heat a block of steel. The surfaces of the steel may be taken as 50°C except on the bottom where the burner is heating the block. In figure 2–64 is shown the temperature profile at the bottom surface and the overall configuration of the heating process. Using graphical techniques, estimate the temperature profile through the block and the heat transfer through the block.

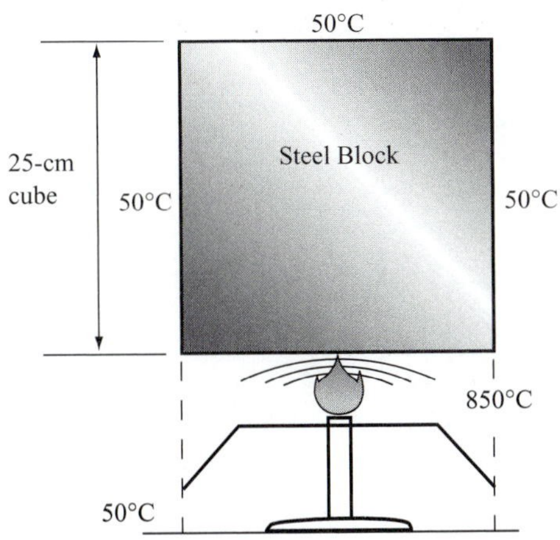

Figure 2–64 Bunsen burner and steel block.

Section 2–6

2–48 Estimate the heat transfer from the fin shown in figure 2–65. Write the necessary node equations and then solve for the temperatures by the relaxation method or matrix inversion. Assume the fin is aluminum.

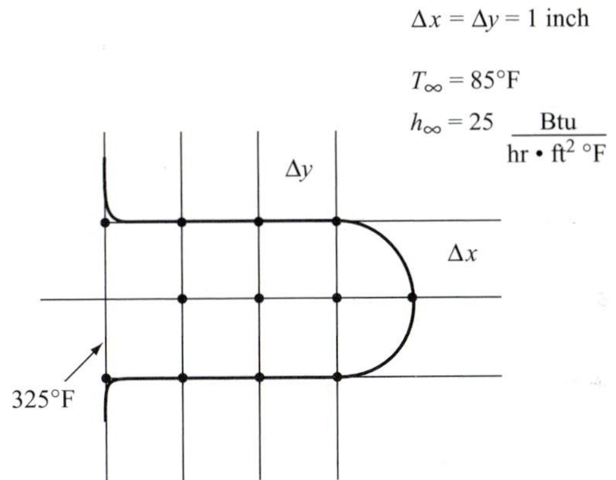

Figure 2–65 Fin heat transfer.

2–49 Write the node equations for the model of heat transfer through the compressor housing section shown in Figure 2–66. Then solve for the node temperatures by the relaxation method or by matrix inversion methods. Assume still air is in the four circumferential slots.

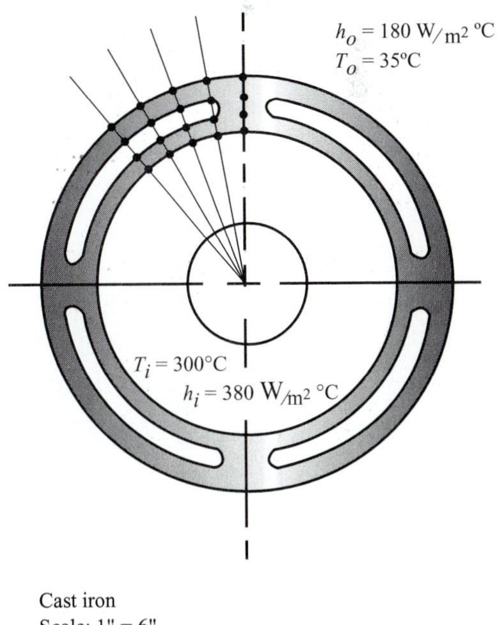

Figure 2–66 Compressor housing section

***2–50** Write the node equations for describing heat transfer through the buried waste shown schematically in figure 2–67. Notice that energy generation occurs due to a pyrolytic reaction of the waste (slow chemical reaction) and that there are boundaries that require reference to table 2–6.

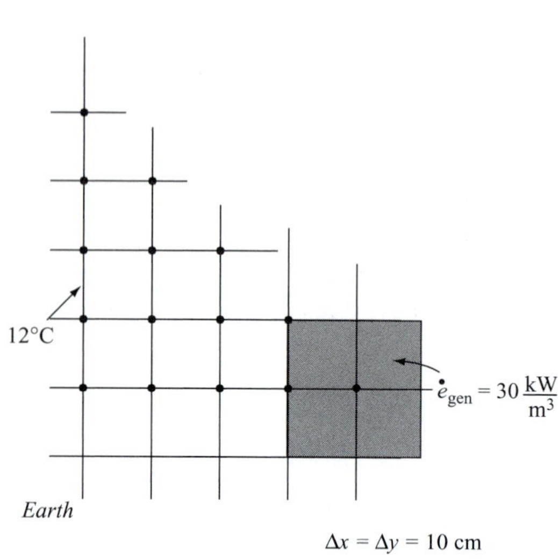

Figure 2–67 Buried waste mass.

***2–51** Write node equations for determining the temperature distribution through the cast iron lathe slide shown in figure 2–68. Notice that the sliding surface is assumed to be adiabatic and that there are irregular boundary profiles.

***2–52** A concrete chimney flue is surrounded by a Styrofoam insulator as shown in figure 2–69. Construct an appropriate grid model and then write the node equations needed to determine the temperature distribution.

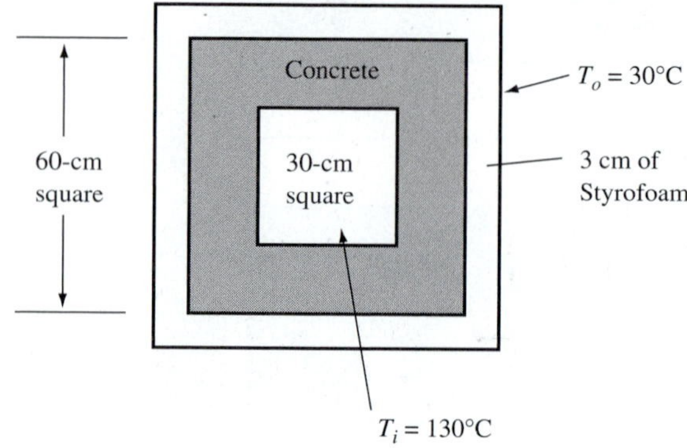

Figure 2–69 Chimney flue.

***2–53** Consider the chimney flue of figure 2–69. If the Styrofoam is removed and the outer boundary condition is the same, write the necessary node equations and solve for the node temperatures. What is the heat transfer through the chimney flue?

***2–54** Write the node equations for nodes 1 and 2 of the model of the oak beam sketched in figure 2–32.

2–55 In figure 2–70 is shown a section of a large granite surface plate used for precision measurements. A person touches the surface and thereby induces heat transfer through the plate. Neglecting any radiation involved, write the node equations for nodes 1, 5, and 12 of the node model of the plate shown in the figure. Assume steady state conditions and that the plate is 65°F beyond the nodes indicated in the figure.

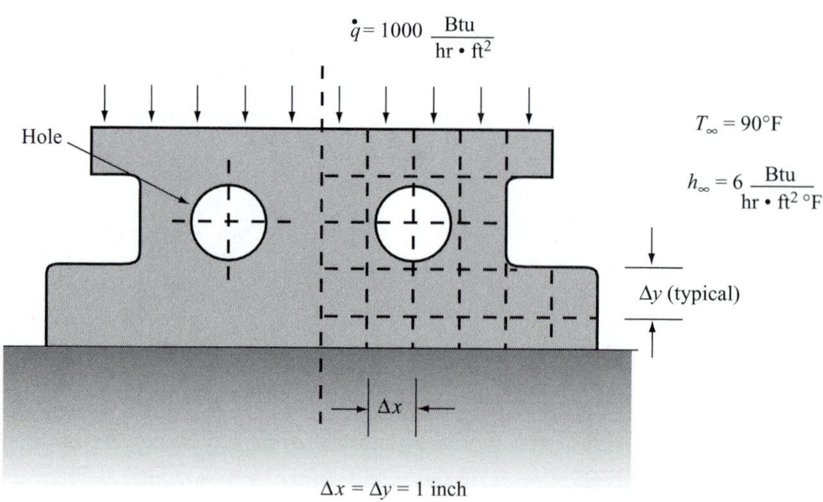

Figure 2–68 Lathe slide.

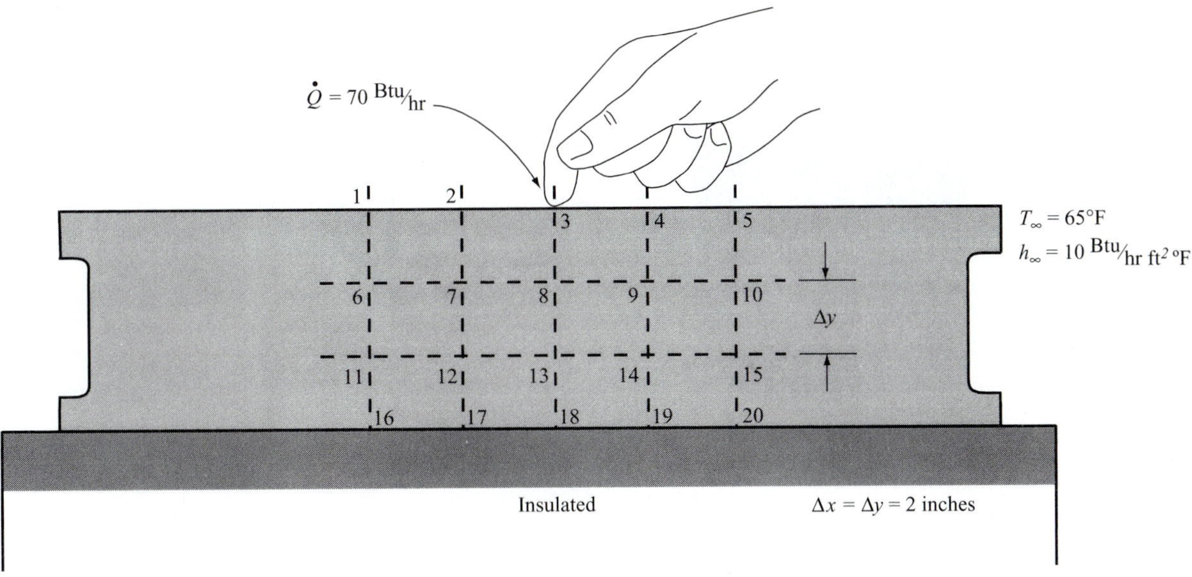

Figure 2–70 Surface plate.

***2–56** Write the complete set of node equations for the surface plate shown in figure 2–70 and estimate the temperatures and the heat transfer through the plate (see problem 2–55).

***2–57** A plutonium nuclear fuel rod, shown in figure 2–71, has energy generation in the amount of 3000 Btu/s · ft³. For the grid node model shown, write the node equations and solve for the temperatures. Assume $\kappa = 10 \frac{\text{W}}{\text{m} \cdot \text{K}}$.

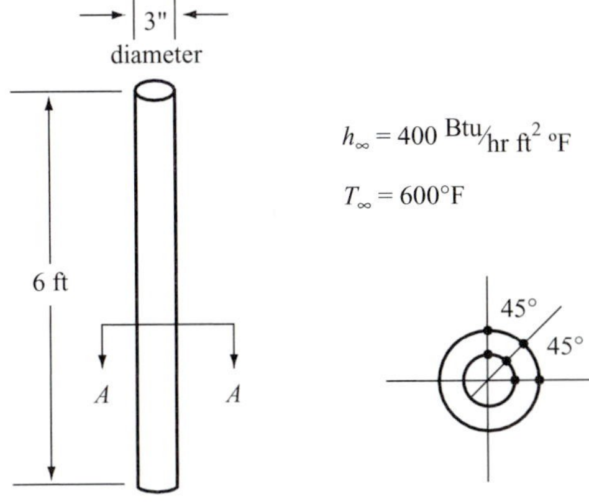

Figure 2–71 Plutonium fuel rod.

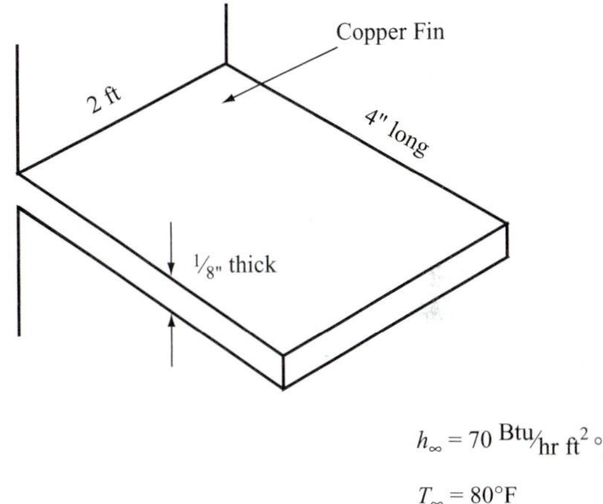

Figure 2–72 Problem 2–58.

Section 2–7

2–58 Determine the heat transfer and fin efficiency for a copper fin shown in figure 2–72. The fin can be assumed to be very long and its base temperature taken as 200°F.

2–59 A square brass fin, 30 cm wide, 1 cm thick, and 5 cm long is surrounded by air at 27°C and $h = 300$ W/m² · °C. The base temperature of the fin is 170°C. Determine the fin tip temperature, the fin heat transfer, and the fin efficiency.

2–60 A square aluminum fin having base temperature of 100°C 5-mm thick, and 5-cm length is surrounded by water at 40°C. Using h of 400 W/m² · °C compare the heat transfer of the fin predicted by the three conditions:

(a) Very long fin.

(b) Adiabatic tip.

(c) Uniform convection heat transfer over fin, including tip.

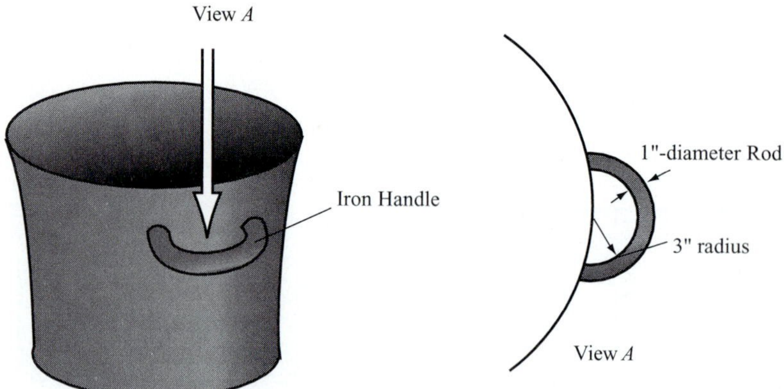

Figure 2–73 Cooking pot handle.

***2–61** Show that the fin heat transfer for a fin that is square and has an adiabatic tip is

$$\dot{Q}_{\text{fin}} = \theta_0 \sqrt{hP\kappa A} \tanh mL$$

***2–62** Show that the fin heat transfer for a fin that is square and has fin-tip convective coefficient h_L can be written

$$\dot{Q}_{\text{fin}} = \theta_0 \sqrt{hP\kappa A} \left(\frac{\sinh mL + \dfrac{h_L}{\kappa m} \cosh mL}{\cosh mL + \dfrac{h_L}{\kappa m} \sinh mL} \right)$$

***2–63** Derive an expression for the heat transfer from a tapered fin having base of Y thickness, L length, κ thermal conductivity, h_0 convective coefficient, and T_0 base temperature. The surrounding fluid temperature is T_∞. See figure 2–75.

***2–64** Show that the fin effectiveness, ε_{fin}, is related to the fin efficiency, η_{fin}, by equation (2–112)

$$\varepsilon_{\text{fin}} = 1 - \left(\frac{A_{\text{fin}}}{A_T} - \eta_{\text{fin}} \frac{A_{\text{fin}}}{A_T} \right)$$

2–65 A circumferential steel fin is 8 cm long, is 3 mm thick, and is on a 20-cm-diameter rod. The surrounding air temperature is 20°C and $h = 35$ W/m^2 · °C. The surface temperature of the rod is 300°C. Determine
(a) Fin efficiency.
(b) Heat transfer from the fin.

2–66 A brass rod 1 cm in diameter and 30 cm long protrudes from a brass surface at 150°C. The rod is surrounded by air at 10°C with $h = 10$ W/m^2 · °C. Determine the heat transfer through the rod.

2–67 A circumferential cast-iron fin attached to a compressor housing is 1 in. thick, 3 in. long, and 3 in. in diameter, and the convection heat transfer coefficient is 16 Btu/hr · ft^2 · °F. If the base temperature is 160°F and the surrounding air is 80°F, determine the fin efficiency and the heat transfer through the fin.

2–68 A handle on a cooking pot, as shown in figure 2–73, can be modeled as a rod fin with an adiabatic tip at the furthest section from the attachment points. For the handle shown in the figure, determine the temperature distribution and the heat transfer through the handle if the pot surface temperature is 190°F, the surrounding air temperature is 90°F and $h = 160$ Btu/hr · ft^2 · °F.

2–69 An aluminum fin is attached at both ends in a compact heat exchanger as shown in figure 2–74. For the situation shown in the figure, determine the temperature distribution and the heat transfer through the fin. Notice that the analysis requires using the governing equation $d^2\theta(x)/dx^2 = m^2\theta(x)$ with appropriate boundary conditions to determine the temperature distribution.

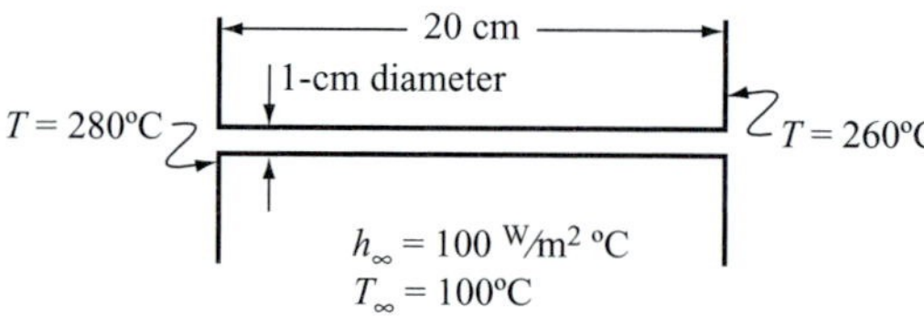

Figure 2–74 Compact heat exchanger fin.

***2–70** For the tapered aluminum fin shown in figure 2–75, determine the fin efficiency and the heat transfer through the fin.

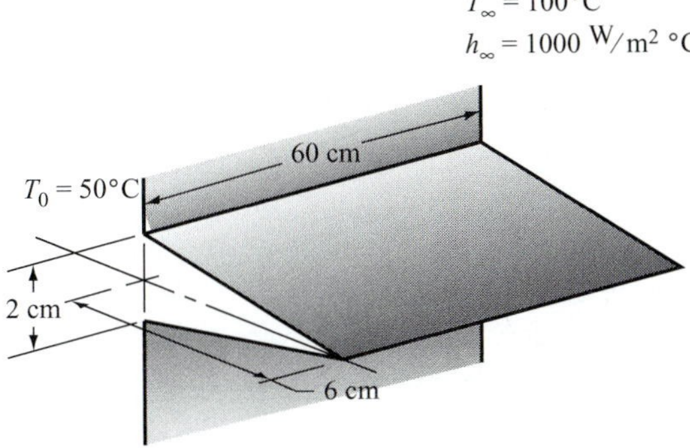

Figure 2–75 Tapered fin.

2–71 Determine the expected temperature drop at the contact between two 304 stainless steel parts if the overall temperature drop across the two parts is 100°C, as indicated in figure 2–76.

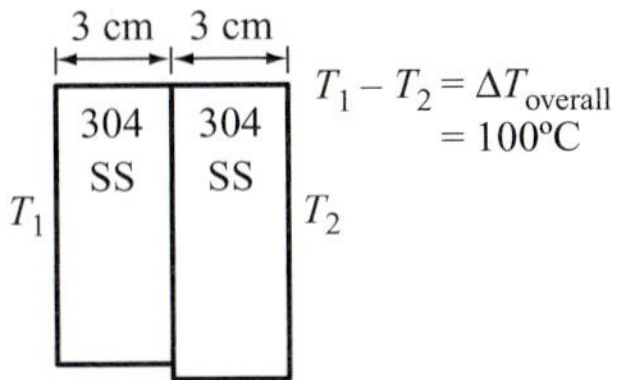

Figure 2–76 Heat transfer at contact surface.

2–72 A weldment made of mild steel is bolted to another mild steel surface. The contact pressure is estimated as 20 atm. and the expected heat transfer between the two parts is 300 Btu/hr · in^2. Estimate the temperature drop at the contact due to thermal contact resistance.

2–73 For example 2–26, estimate the temperature drop at the contact surface if the heat transfer is reduced to 3 Btu/hr · ft^2.

2–74 A guarded-hot-plate test results in the following data. Estimate the thermal conductivity of the test material. The heater surface is 10 cm × 10 cm.

Thermal Conductivity Data

Test No.	Heater Data: *A*, amps	Heater Data: *V*, volts	Thermocouple Data (millivolts, mV): 1	Thermocouple Data (millivolts, mV): 2
1	0.05	8.6	2.669	2.775
2	0.055	8.4	2.672	2.780
3	0.049	8.8	2.662	2.771

Thermocouple conversion: 22°/mV

2 cm
Auxiliary Heater
Heater
Auxiliary Heater
1
2

Diagram of testing device

2–75 A steam line has an outer surface diameter of 3 cm and temperature of 160°C. If the line is surrounded by air at 25°C and $h = 3.0$ W/m^2 · °C, determine the heat transfer per meter of line. Then determine the thickness of asbestos insulation needed to provide insulating qualities to the steam line.

2–76 Electric power lines require convective cooling from the surrounding air to prevent excessive temperatures in the wire. If a 1-in.-diameter line is wrapped with nylon to *increase* heat transfer with the surroundings, how much nylon can be wrapped around the wire before it begins to act as an insulation. Assume $h = 5$ Btu/hr · ft^2 · °F.

2–77 Estimate the temperature distribution through a bare 16-gauge copper wire conducting 1.5 amps of electric current if the surrounding air is at 10°C and $h = 65$ W/m^2 · °C.

2–78 Aluminum wire has resistivity of $0.286 \times 10^{-7}\ \Omega \cdot$ m where resistivity is defined as resistance(ohms) × area/length. Determine the temperature distribution through an aluminum wire of $\frac{1}{4}$-in. diameter carrying 200 A of current if it is surrounded by air at 80°F and with $h = 200$ Btu/hr · ft^2 · °F.

***2–79** Determine the temperature distribution through a uranium slab, shown in figure 2–77. Assume 4500 Btu/min · ft^3 of energy is generated internal to the slab and the slab is surrounded by water at 190°F with $h = 450$ Btu/hr · ft^2 · °F.

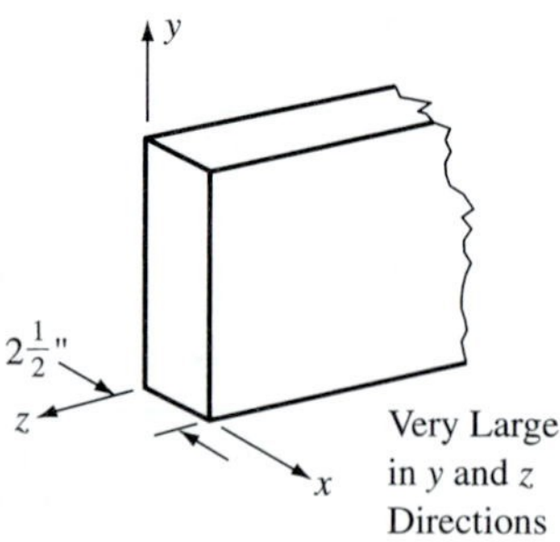

Figure 2–77 Uranium slab.

Hint: Use of the equation $d^2T/dx^2 + \dot{e}_{gen}\kappa = 0$ with appropriate boundary condition must be made to obtain the temperature distribution function. Use $\kappa = 38$ W/m · K for uranium.

2–80 Plutonium plates of 6-cm thickness generate 60 kW/m^3 of energy. It is exposed on one surface to pressurized water which cannot be more than 280°C. The other surface is well insulated. What must the convective heat transfer coefficient be at the exposed surface. *Hint:* Use of the governing equation $d^2T/dx^2 + e_{gen}/\kappa = 0$ plus appropriate boundary conditions must be made to obtain the temperature function. The maximum temperature is 315°C and use $\kappa = 10$ W/m · K for plutonium.

REFERENCES

1. American Society of Testing Materials (ASTM), Test Method for Steady State Thermal Transmission Properties by Means of the Guarded Hot Plate, *ASTM Standards,* Vol. 04.06, C177–85, 1989.
2. Bridgeman, P. W., The Thermal Conductivity of Liquids Under Pressure, *Proc. Am. Acad. Arts Sci.* 59, 141–169, 1923.
3. Nuttall, R. L., and D. C. Ginnings, Thermal Conductivity of Nitrogen from 50°C to 500°C and 1 to 100 Atmospheres, *J. Res Natl Bureau of Standards,* 58, 5, May 1957.
4. Vine, R. G., Measurements of Thermal Conductivity of Gases at High Temperatures, *Trans. ASME,* 82C, 48, 1960.
5. Hirschfelder, J. O., C. F. Curtiss, and R. B. Bird, *Molecular Theory of Gases and Liquids,* John Wiley & Sons, 1966.
6. Bridgeman, P. W., The Thermal Conductivity of Liquids, *Proc. Natl. Acad. Sci. U.S.A.* 9, 341–345, 1923.
7. Carslaw, H. S., and J. C. Jaeger, *Conduction of Heat in Solids,* 2nd edition, Oxford, 1965.
8. Stroud, K. A., *Fourier Series and Harmonic Analysis,* Stanley Thornes Publishers, Ltd., 1984.
9. Churchill, R. V., *Fourier Series and Boundary Value Problems,* McGraw-Hill, New York, 1941.
10. Tolstov, G. P., *Fourier Series* (Trans. R. A. Silverman), Dover Publications, New York, 1962.
11. Arpaci, V. S., *Conduction Heat Transfer,* Addison-Wesley, Reading, MA, 1966.
12. Hahne, E., and U. Grigull, Formfaktor und Formweiderstand der Stationaren Mehrdimensionalen Warmleitung, *Int. J. Heat Mass Transfer,* 18, 751, 1975.
13. Gerald, C. F., and P. O. Wheatley, *Applied Numerical Analysis,* 3rd edition, Addison-Wesley, Reading, MA, 1984.
14. Shih, T. M., *Numerical Heat Transfer,* Hemisphere Publishing Co., New York, 1984.
15. Jaluria, Y., and K. E. Torrance, *Computational Heat Transfer* Hemisphere Publishing Co., New York, 1986.
16. Kikuchi, N., *Finite Element Methods in Mechanics,* Cambridge Univ. Press, UK, 1986.
17. Gardner, K. A., Efficiency of Extended Surfaces, *Trans. ASME,* 67, 621–631, 1945.
18. Kern, D. Q., and A. D. Kraus, *Extended Surface Heat Transfer,* McGraw-Hill, New York, 1980.
19. Fletcher, L. S., Recent Developments in Contact Conduction Heat Transfer, *J. Heat Transfer,* 110, 4(b), 1059, November 1988.
20. Salerno, L. J., A. L. Spivak, and P. Kittel, Thermal Conductance of Pressed Copper Contacts at Liquid Helium Temperatures, *NASA Tech Brief ARC-11572,* 10, 5, National Aeronautics and Space Administration, U.S. Government Printing Office, Washington, DC, 1984.
21. Mian, M. N., F. R. Al-Astrabadi, P. W. O'Callaghan, and S. D. Probert, Thermal Resistance of Pressed Contacts Between Steel Surfaces: Influence of Oxide Films, *J. Mech. Eng. Sci.,* IMechE, 1979.

TRANSIENT CONDUCTION HEAT TRANSFER

In this chapter the general problems of transient or unsteady state heat transfer are considered. Attention is directed to those problems involving conduction heat transfer, where the conducting medium's temperature is changing in time. Boundary conditions are often considered as constant or uniform after some initial state where a jump discontinuity occurs in the boundary condition. The development of the material in this chapter begins with the concept of general energy balances, governing equations, and typical boundary conditions. The property of thermal diffusivity is introduced as a direct outcome of the manipulation of the governing differential equations. The development continues with lumped heat capacity systems and the use of the time constant as a parameter to compare simple transient heat-transfer systems. Some well-known one-dimensional transient heat-transfer problems are considered, such as the semi-infinite solid subjected to a abrupt change in a uniform boundary condition at an instant and the problem of a line source of thermal energy in an infinite medium. The line source of thermal energy represents a description of an important test method for determining thermal conductivity of materials.

The conduction of heat in slabs, cylinders, and spheres that are subjected to sudden changes in the boundary condition is presented and extended to two- and three-dimensional cases. Some energy methods are used to provide the reader with greater versatility. In particular, some systems with internal energy generation are considered. A few examples of transient heat transfer applications are presented, such as heat-treatment processes in metals and earth or ground temperature seasonal fluctuations due to climatic changes. Finite difference techniques are developed to show how the power of digital computers may be used to obtain approximate solutions to a wide variety of transient heat-transfer problems. Finally, graphical methods such as the Schmidt Plot are demonstrated to give the reader an appreciation for some of the numerical approximations carried out in finite difference techniques and also to show the student that some problems can be analyzed with very elementary tools and engineering intuition.

New Terms

t_c time constant

α thermal diffusivity

3–1 GENERAL PROBLEMS IN TRANSIENT CONDUCTION

Heat-transfer processes tend to change the energy of the system or systems involved. For instance, heat transfers out of a system will tend to decrease the system's energy and heat transfers into the system will increase the energy. If the system is a solid or fluid and does not experience a phase change during heat transfer, the energy changes can often be associated with temperature changes. In systems where the material undergoes phase changes, the energy change can often be carried out without a temperature change, as we saw in

Chapter 1. Here we consider systems that do not experience phase changes, unless otherwise noted. Any heat transfer therefore tends to change the system's temperature over some time period. For the steady state situations, considered in Chapter 2, the system temperatures were unchanged in time, and here we consider the conditions where the system temperatures do change in time. We call this the *transient* or *unsteady state* condition and there are two reasons for a system to have an unsteady state condition: (1) the heat transfers into the system are not balanced by the heat transfers out of the system, and (2) the heat transfers at the boundary or boundaries suddenly change. The fundamental concept of conservation of energy demonstrates these possibilities and if we write this for the system shown in figure 3–1, we have

$$\dot{Q}_{\text{in}} + \dot{E}_{\text{gen}} = \dot{Q}_{\text{out}} + \frac{d\text{Hn}}{dt} \tag{3–1}$$

Figure 3–1 Transient heat-transfer system.

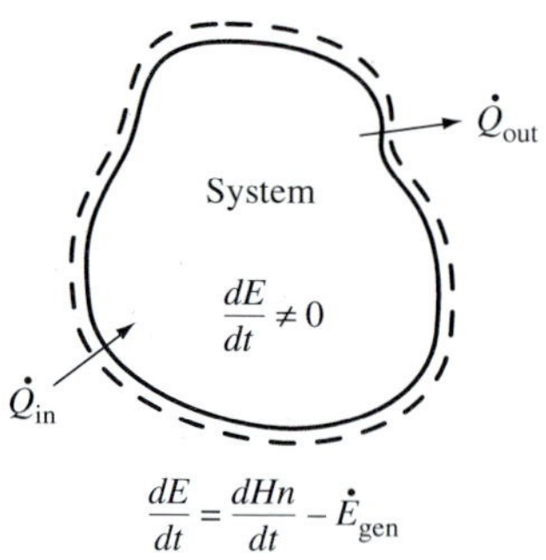

Transient conditions are a result of an imbalance or nonequilibrium of a system with its surroundings, and the actions involved in these transient processes tend to bring the system to a steady state or a condition where it is in equilibrium with its surroundings. This means that the temperature or temperatures of a system will tend to become equal to the temperatures of the boundaries or, in the case of convective boundary conditions, the temperature of the surrounding fluid. In the case of radiation heat transfer impacting at the boundary, the system temperature will tend toward the temperature of those surrounding, radiating surfaces.

EXAMPLE 3–1 The heat transfer into a 10-kg block of ice is 400 W. Determine the rate of change of energy of the ice and the rate of change of the ice temperature, assuming that the heat transfer by conduction through the ice is so rapid that its temperature may be considered to be uniform at any instant.

Solution From equation (3–1) we have $\dot{Q}_{\text{in}} = 400$ W, $\dot{Q}_{\text{out}} = 0$, and

Answer

$$\frac{dE}{dt} = \dot{Q}_{\text{in}} = 400 \text{ W}$$

Also, we assume that the energy change may be written,

$$\frac{dE}{dt} = \rho V c_p \frac{dT}{dt} = m c_p \frac{dT}{dt} = 400 \text{ W}$$

Answer

The rate of change of temperature is then

$$\frac{dT}{dt} = \frac{400\text{ W}}{mc_p} = \frac{400\text{ W}}{(10\text{ kg})\left(2.04\frac{\text{kJ}}{\text{kg}\cdot\text{K}}\right)} = 0.0196\frac{\text{K}}{\text{s}} = 0.0196\frac{°\text{C}}{\text{s}}$$

In Chapter 2 we developed the equations for conduction in a homogeneous isotropic material. For the transient or unsteady conditions this equation is written, in Cartesian coordinates,

$$\frac{\partial^2 T(x, y, z, t)}{\partial x^2} + \frac{\partial^2 T(x, y, z, t)}{\partial y^2} + \frac{\partial^2 T(x, y, z, t)}{\partial z^2} + \frac{\dot{e}_{\text{gen}}}{\kappa} = \frac{\rho c_p}{\kappa}\frac{\partial T(x, y, z, t)}{\partial t} \tag{3–2}$$

and the term $\kappa/\rho c_p$ is called the *thermal diffusivity,* α,

$$\alpha = \frac{\kappa}{\rho c_p} \tag{3–3}$$

Equation (3–2) can then be written

$$\frac{\partial^2 T}{\partial x^2} + \frac{\partial^2 T}{\partial y^2} + \frac{\partial^2 T}{\partial z^2} + \frac{\dot{e}_{\text{gen}}}{\kappa} = \frac{1}{\alpha}\frac{\partial T}{\partial t} \tag{3–4}$$

Thermal diffusivity is often compared to a "thermal inertia" of a material. It gives a measure of the ratio of conduction heat transfer through a material to its capability to store thermal energy. The larger the value of thermal diffusivity, the greater is the rate of change of temperature likely to be for a given amount of conduction heat transfer. Materials having low thermal diffusivity values are good materials for thermal storage applications. In table 3–1 are listed some typical values for thermal diffusivity for some common materials. Also, in appendix tables B–2, B–3, and B–4 are more extensive listings of thermal properties needed to determine the thermal diffusivity. We will use thermal diffusivity in the following sections.

Table 3–1 Typical Values for Thermal Diffusivity

Material	Temperature °F	Temperature °C	Thermal Diffusivity × 10^6 ft²/s	Thermal Diffusivity × 10^6 m²/s
Water (liquid)	70	20	1.5	0.10
Ice	32	0	12.8	1.10
Granite (stone)	70	20	14.7	1.37
Steel (1.0% carbon)	70	20	126.17	11.72
Copper	70	20	1209.22	112.34
Aluminum	70	20	906.1	84.18
Wood (oak)	70	20	1.38	0.13
Asbestos (loose)	32	0	3.77	0.35
Air	70	20	238.5	22.16
Carbon dioxide	70	20	114.1	10.6

The general differential equation (3–4) that describes the situations of transient conduction heat transfer requires six separate boundary conditions plus one initial condition or one condition describing the state of the system and its boundary at some instant in time. For one-dimensional transient problems only two boundary conditions and the one initial condition are required. In the next section we consider the elementary problem of systems whose temperatures are only dependent on time.

3–2 LUMPED HEAT CAPACITY SYSTEMS

Consider a system that is a good conductor of heat, so good that heat conduction can occur over an infinitesimal temperature gradient. In this situation we can say that the system temperature is uniform throughout and is independent of the position in the system. For the transient condition, we may write that the system temperature is only a function of time; that is,

$$T(x, y, z, t) = T(t) \qquad (\kappa \to \infty, \text{ or very small heat conducting system}) \tag{3–5}$$

Also, for a very small system whose volume does not allow for any temperature gradients we may again say that the system temperature is a function of time alone, as given by equation (3–5).

A method of deciding when a system can be expected to satisfy one of these requirements is to compare the heat transfer at the boundary of the system to the conduction heat transfer through the system. The boundary heat transfer is often convection, so we may write that

$$\dot{Q}_{\text{convection}} \propto hA\Delta T$$

and the conduction heat transfer as

$$\dot{Q}_{\text{conduction}} \propto \kappa A \frac{\Delta T}{\Delta x}$$

The ratio of the boundary heat transfer to the conduction heat transfer, called the *Biot Number,* Bi, is then

$$\text{Bi} = \frac{\dot{Q}_{\text{boundary}}}{\dot{Q}_{\text{conduction}}} = \frac{hA\Delta T\Delta x}{\kappa A\Delta T} = \frac{h\Delta x}{\kappa} \tag{3–6}$$

where Δx is some "characteristic" length, often defined as

$$\Delta x = \frac{\text{System volume}}{\text{Boundary surface area}} = \frac{V_{\text{system}}}{A_{\text{surface}}} = \frac{V_s}{A_s} \tag{3–7}$$

If the system is a regular shape, such as a sphere, a cylinder, a cube, a parallelopiped, a wall, or so on, the definition of (3–7) is adequate. If the system is extremely irregular, then subjective judgments must be made, which are beyond the scope of this book.

As we will see in succeeding sections, if the Biot Number is less than about 0.1 (unitless), we may assume a uniform temperature as given in equation (3–5). We call such systems *lumped heat capacity* and they are the single most important class of transient heat-transfer systems. The assumptions and approximations required for this system are often not accurate, but the lumped heat capacity analysis often provides a good first approximation of the actual behavior of a transient heat conducting system. In figure 3–2 is shown a typical lumped heat capacity system. Writing the energy balance for the system gives

$$\dot{Q}_{\text{boundary}} = -\frac{dE}{dt} \tag{3–8}$$

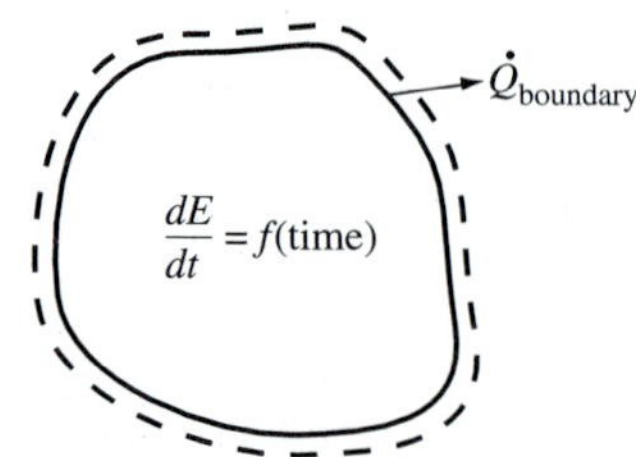

Figure 3–2 Lumped heat capacity system.

which describes the situation of a system that has a boundary heat loss causing a decrease in the system energy. The negative sign in the rate of energy change term, dE/dt, reflects this condition.

Notice that we could also consider the situation where boundary heat transfer is into the system and the rate of energy change of the system is then positive. As we can see, equation (3–8) describes the heating or cooling processes of lumped heat capacity systems. Often the boundary heat transfer is convection, so we may write that

$$\dot{Q}_{\text{boundary}} = hA_s[T(t) - T_\infty]$$

where T_∞ is the surrounding temperature. The energy rate change is

$$\frac{dE}{dt} = \rho V_s c_p \frac{dT(t)}{dt}$$

Equation (3–8) then becomes

$$hA_s[T(t) - T_\infty] = -\rho V_s c_p \frac{dT(t)}{dt} \tag{3–8a}$$

This equation may be rewritten to simplify the mathematical manipulations by defining $\theta(t) = T(t) - T_\infty$. From this definition we also have that $\dfrac{d\theta(t)}{dt} = \dfrac{dT(t)}{dt}$ and then equation (3–8) is

$$hA_s\theta(t) = -\rho V_s c_p \frac{d\theta(t)}{dt} \tag{3–9}$$

for $t \geq 0$. We may identify an initial condition such that

$$T(t) = T_i \quad \text{or} \quad \theta(t) = \theta_i = T_i - T_\infty \qquad \text{at } t = 0$$

where T_i is an initial temperature at time zero. The system at time zero is suddenly subjected to the convective boundary condition for all t greater than zero, described by equation (3–9). This equation can be solved directly upon separating the variables to obtain

$$\int_{\theta_i}^{\theta} \frac{d\theta(t)}{\theta(t)} = -\frac{hA_s}{\rho V_s c_p}\int_0^t dt = -\frac{hA_s}{m_s c_p}\int_0^t dt$$

Here m_s is the mass of the system, its density times volume. Integrating this equation gives

$$\ln \theta(t) - \ln \theta_i = -\frac{hA_s}{\rho V_s c_p} t = -\frac{hA_s}{m_s c_p} t$$

or, written in exponential form,

$$\frac{\theta(t)}{\theta_i} = e^{-\frac{hA_s}{m_s c_p} t} \tag{3–10}$$

The function $\theta(t)/\theta_i$ is plotted in the graph of figure 3–3, and it shows that θ eventually reaches zero or that the temperature $T(t)$ eventually equals that of the surroundings, T_∞, but requires an infinite time period to reach that equilibrium. Realistically, a lumped heat capacity system will reach a temperature equilibrium with its surroundings in a finite time

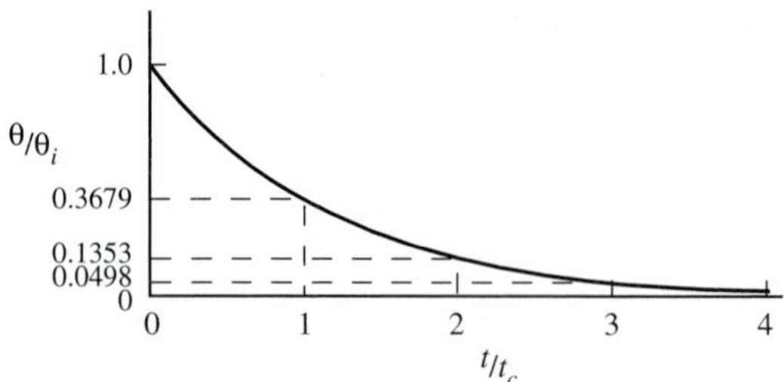

Figure 3–3 Temperature of a lumped heat capacity system subject to a convective heat transfer at its boundary at $t > 0$.

period; however, that period may be quite extended as indicated by equation (3–10) and its result. It is sometimes convenient to define the quantity $\rho V c_p / h A_s$ as the time constant, t_c,

$$\text{Time constant, } t_c = \frac{\rho V_s c_p}{h A_s} = \frac{m_s c_p}{h A_s} \tag{3–11}$$

so that equation (3–10) is then

$$\frac{\theta(t)}{\theta_i} = e^{-t/t_c}$$

For the instant when the time t expired is one time constant, the temperature function

$$\frac{\theta(t)}{\theta_i} = e^{-1} = 0.3679...$$

This result means that the temperature difference between the system and its surroundings has been reduced by 63.21% or that the difference is 36.79% of the initial condition. In the graph of figure 3–3 the temperature function at one, two, three, and four time constants is indicated. After a time period of four time constants, the temperature difference between the system and its surroundings is 1.83% of the initial difference.

The time constant is often used to describe the response of systems to transient heat transfer. Thermometers, thermocouples, and thermistors are often described partly by their time constant because these devices are expected to sense and measure their surroundings' temperature.

EXAMPLE 3–2 A mercury-in-glass thermometer 3 mm in diameter by 20 cm long at 15°C is placed in water at 75°C, as shown in figure 3–4. Determine the time constant of the thermometer and the time when the thermometer indicates within 5% of the actual water temperature. Assume that $h = 180\ \text{W/m}^2 \cdot °\text{C}$ for the water/thermometer and that the thermometer has the same properties as glass.

Solution The Biot Number is first determined. Using a value of $1.4\ \text{W/m}^2 \cdot °\text{C}$ for the thermal conductivity of glass, from appendix table B–2, gives

$$\text{Biot Number} = \text{Bi} = \frac{hV_s}{\kappa A_s} = \frac{\left(180 \dfrac{\text{W}}{\text{m}^2 \cdot \text{K}}\right)[\pi \times (0.0015\ \text{m})^2 \times (0.2\ \text{m})]}{\left(1.4 \dfrac{\text{W}}{\text{m} \cdot \text{K}}\right)[\pi \times (0.003\ \text{m}) \times (0.2\ \text{m})]} = 0.096$$

Thus, since Bi is less than 0.1, we may use the lumped heat capacity approximation and the time constant is given by equation (3–11),

Figure 3–4 Mercury-in-glass thermometer.

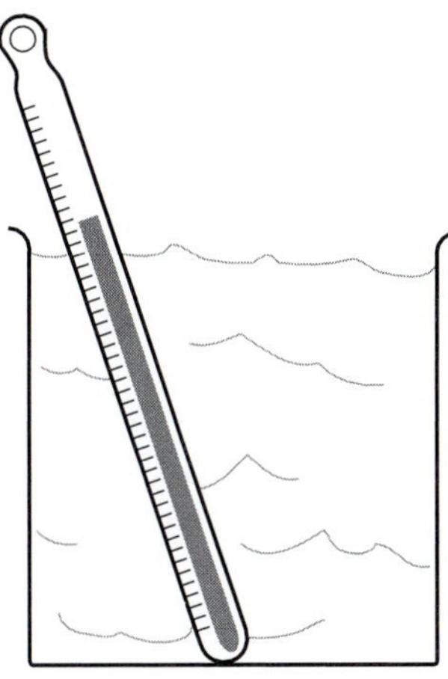

$$t_c = \frac{\rho V_s c_p}{hA_s} = \frac{\left(2500\,\frac{\text{kg}}{\text{m}^3}\right)[\pi \times (0.0015\text{ m})^2 \times (0.2\text{ m})]\left(750\,\frac{\text{kJ}}{\text{kg}\cdot\text{K}}\right)}{\left(180\,\frac{\text{W}}{\text{m}^2\cdot\text{K}}\right)[\pi \times (0.003\text{ m}) \times (0.2\text{ m})]} = 7.8125\text{ s}$$

If we interpret that the thermometer must indicate within 5% of the water temperature to mean that 5% of 75°C is 3.75°C so that the thermometer indicates 71.25°C (75°C − 3.75°C), then the time for this to happen is predicted by equation (3–10),

$$\frac{71.25°\text{C} - 75°\text{C}}{15°\text{C} - 75°\text{C}} = e^{-t_c/7.8125\text{ s}} = 0.0625$$

which gives that

Answer

$$t = 21.7\text{ s}$$

Therefore, the thermometer requires nearly 22 seconds to have an indicated temperature within 5% of the actual value.

EXAMPLE 3–3 Common bricks are heated in an oven, called a kiln, to dry and to harden them. A brick is 4 in. by 8 in. by 2 in. with a convective heat-transfer coefficient of 0.8 Btu/hr · ft² · °F in the kiln. The brick is placed in the kiln at 70°F and the kiln temperature is 640°F. Estimate the time constant and the time needed for the brick to reach 500°F.

Solution We first check the Biot Number to see if it is less than 0.1. If so then we can use the lumped heat capacity method for the brick.

$$\text{Bi} = \frac{hV_s}{\kappa A_s} = \frac{\left(0.8\,\frac{\text{Btu}}{\text{hr}\cdot\text{ft}^2\cdot°\text{F}}\right)(4 \times 8 \times 2\text{ in}^3)}{(2 \times 4 \times 2\text{ in}^2 + 8 \times 12\text{ in}^2)\left(12\,\frac{\text{in.}}{\text{ft}}\right)} = 0.094$$

Therefore, since Bi is less than 0.1, we are justified in using lumped heat capacity. The time constant is

Answer

$$t_c = \frac{\rho V_s c_p}{hA_s} = \frac{\left(99.75\,\frac{\text{lbm}}{\text{ft}^3}\right)(64\text{ in}^3)\left(0.20\,\frac{\text{Btu}}{\text{lbm}\cdot°\text{F}}\right)}{\left(0.8\,\frac{\text{Btu}}{\text{hr}\cdot\text{ft}^2\cdot°\text{F}}\right)(112\text{ in}^2)\left(12\,\frac{\text{in.}}{\text{ft}}\right)} = 1.1875\text{ hr} = 4275\text{ s}$$

From equation (3–10) the time for the brick to reach 500°F is determined.

$$\frac{500°\text{F} - 640°\text{F}}{70°\text{F} - 640°\text{F}} = e^{-t/4275\text{ s}}$$

so that

Answer $$t = 6002 \text{ s} = 1.67 \text{ hr}$$

Sometimes it is important to know how much heat can be transferred between a system and its surroundings; that is, we may want to determine the heat, Q, transferred over some time period, t, where

$$Q = \int_0^t \dot{Q}\, dt \tag{3–12}$$

This quantity is also equal to the change of the system energy, given by $mc_p\,[T(t) - T_i]$. For the configuration shown in figure 3–2, we write

$$Q = -mc_p[T(t) - T_i] \tag{3–13}$$

and, since $T(t)$ can be predicted from equation (3–10), equation (3–13) becomes

$$Q = mc_p(T_i - T_\infty)(1 - e^{-t/t_c}) \tag{3–14}$$

If the term $mc_p(T_i - T_\infty)$ is defined as Q_o and interpreted as the maximum energy that can be transferred between the system and its surroundings, then equation (3–14) may be rewritten as

$$\frac{Q}{Q_o} = 1 - e^{-t/t_c} \tag{3–15}$$

and this equation demonstrates that Q approaches Q_o exponentially but requires an infinite time to accommodate the heat exchange. In figure 3–5 is shown the function Q/Q_o predicted by equation (3–15), which is the compliment of the system temperature function given by equation (3–10).

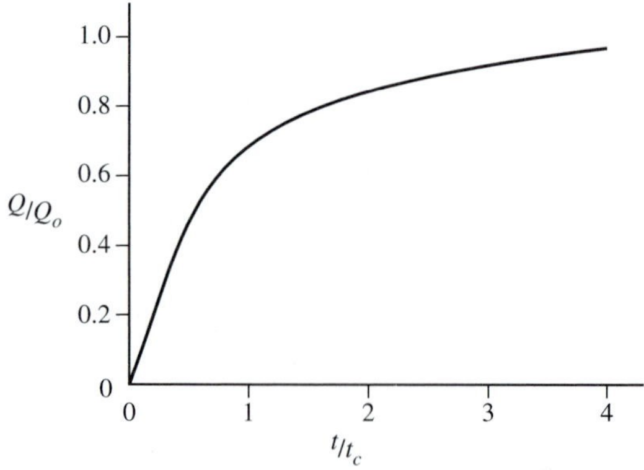

Figure 3–5 Thermal energy exchange of a lumped heat capacity system.

The reader should observe that many systems can be accurately modeled as lumped heat capacity ones, if the system Biot Number is less than about 0.1. As we shall see in the following sections if the Biot Number is greater than 0.1, other methods should be used to analyze the heat transfers and predict the system temperatures. As stated at the beginning of this section, lumped heat capacity analysis is useful for providing a beginning for more precise analyses of those heat-conducting systems where there is a transient or unsteady state condition.

3–3 ONE-DIMENSIONAL TRANSIENT HEAT TRANSFER

There are many situations where transient heat transfer effectively occurs in only one dimension. As an example, the sketch shown in figure 3–6 indicates a solid that has an extremely large boundary. The boundary is so large that it can be considered as extending infinitely in its plane. Heat transfer at the boundary is one-dimensional, uniformly distributed along the boundary surface, and the heat transfer through the solid will also be one-dimensional. The configuration shown in figure 3–6 is identified as a semi-infinite solid because it extends to infinity along the boundary and also because it extends to infinity along the x direction or away from its boundary surface. We will now consider some specific situations of one-dimensional transient heat transfer.

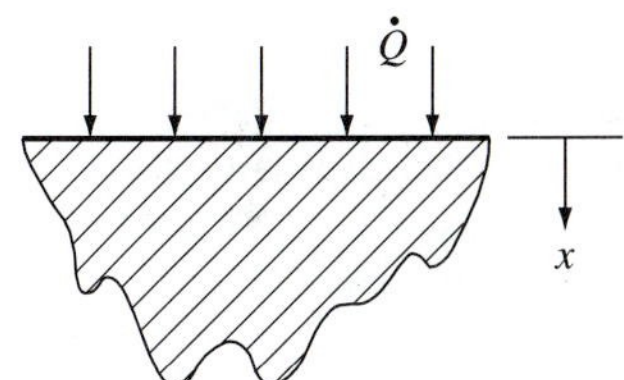

Figure 3–6 Example of transient heat transfer in one dimension.

Semi-Infinite Solid at T_i Subject to a Boundary Temperature T_∞

A semi-infinite solid, as shown in figure 3–6, subject to a heat transfer at the boundary, has a temperature distribution given as $T(x, t)$. For the specific condition where its initial temperature is uniformly T_i throughout and suddenly (beginning at some time and continuing) the solid has a temperature T_∞ at its boundary, the mathematical problem can be given as

$$\frac{\partial^2 T(x, t)}{\partial x^2} = \frac{1}{\alpha}\frac{\partial T(x, t)}{\partial t} \qquad x \geq 0, t > 0 \tag{3–16}$$

with the initial condition at $t = 0$,

$$\text{I.C.} \qquad T(x, t) = T_i \qquad x \geq 0, t = 0$$

and boundary conditions

$$\begin{aligned} &\text{B.C. 1} \qquad T(x, t) = T_B \qquad x = 0, t > 0 \\ &\text{B.C. 2} \qquad T(x, t) = T_i \qquad x \to \infty, t > 0 \end{aligned}$$

The physical situation of this transient problem is indicated in figure 3–7, and there are actual conditions that are approximated by this model. The situation where a large solid, such as a steel plate, is suddenly subjected to a stream of cold water or a hot gas

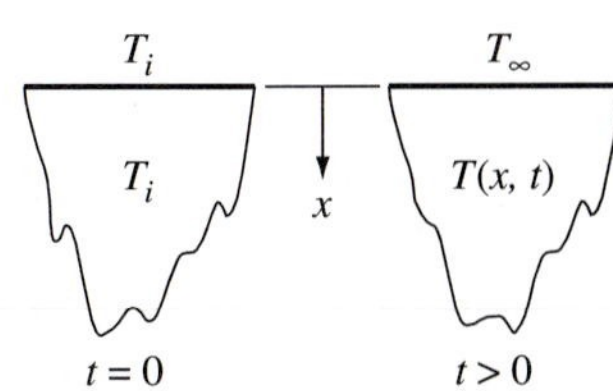

Figure 3–7 Semi-infinite solid subject to instantaneous boundary temperature change from T_i to T_B.

can often be treated with this model. As a first approximation, one may use this model to describe the heat transfer between the earth and the sun at sunrise and at sunset. The reader can identify other equally applicable situations. It can be shown (see Carslaw and Jaeger [1] and Hill and Dewynne [2]) that a solution to equation (3–16) with its associated initial and boundary conditions, by using Laplace transform techniques, is

$$T(x, t) = T_B + (T_i - T_B)\operatorname{erf}(x/2\sqrt{\alpha t}) \tag{3–17}$$

where the *error function, erf,* is defined as

$$\operatorname{erf}(z) = \frac{2}{\sqrt{\pi}} \int_0^z e^{-\eta^2} d\eta \tag{3–18}$$

The variable η is called a dummy variable to allow for the integration. In figure 3–8 is shown the temperature distribution of equation (3–17), presented in a dimensionless form that is equal to the error function of the argument $x/2\sqrt{\alpha t}$. In appendix table A–9 are listed the values for erf(z) for its argument z. Also, in appendix table A–9 is given an approximation to facilitate the use of the computer to predict the error function.

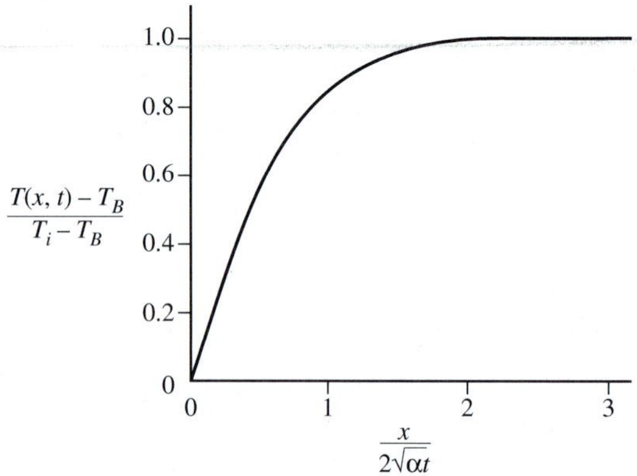

Figure 3–8 Dimensionless temperature distribution in a semi-infinite solid subject to boundary temperature, T_∞.

Notice that erf(3.6) = 1.000000, which means that the temperature of the infinite solid is unchanged at a depth x_{un} such that

$$x_{\text{un}} \geq 3.6 \times 2\sqrt{\alpha t} = 7.2\sqrt{\alpha t} \tag{3–19}$$

Thus, the region of changing temperature increases with time but is finite and definite, and any solid may be treated as a semi-infinite solid, as long as the solid's temperature is not affected by other boundary conditions at the depth, x_{un}.

The conduction heat transfer can be predicted at any location in the infinite solid and at any time by using Fourier's Law of Heat Transfer,

$$\frac{\dot{Q}}{A} = \dot{q}_A = -\kappa \frac{\partial T(x, t)}{\partial x}$$

In this specific problem, referring to equation (3–17),

$$\frac{\partial T(x, t)}{\partial x} = (T_i - T_B) \frac{2}{\sqrt{\pi}} e^{-x^2/4\alpha t} \frac{\partial}{\partial x}\left(\frac{x}{2\sqrt{\alpha t}}\right)$$

or

$$\frac{\partial T(x, t)}{\partial x} = \frac{(T_i - T_B)}{\sqrt{\pi \alpha t}} e^{-x^2/4\alpha t}$$

Substituting this into Fourier's Law of Conduction gives

$$\dot{q}_A = -\kappa(T_i - T_B)\frac{1}{\sqrt{\pi\alpha t}}e^{-x^2/4\alpha t} \tag{3–20}$$

This equation describes the conduction heat transfer in the solid at any location x and at any time t.

EXAMPLE 3–4 A large marble surface plate 3 ft by 3 ft by 1 ft thick is at 70°F when its top surface is suddenly sprayed with 50°F water. Determine the time before the temperature is 65°F $\frac{1}{2}$ in. below the surface. Also determine the surface heat transfer per unit area at this time.

Solution The marble plate is not a semi-infinite solid; however, we may use that model and then check to see if the plate's thickness, 1 ft, is sufficient to accommodate the depth x_{un}. The dimensionless temperature is

$$\frac{T(x, t) - T_B}{T_i - T_B} = \frac{65°F - 50°F}{70°F - 50°F} = 0.75$$

From equation (3–17)

$$0.75 = \text{erf}\,(x/2\sqrt{\alpha t})$$

The argument of an error function of 0.75 is approximately 0.82 from appendix table A–9, or by using the equation in appendix A–9. Then

$$0.82 = \frac{x}{2\sqrt{\alpha t}} \quad \text{or} \quad t = \frac{x^2}{4\alpha(0.82)^2}$$

Using a value of 12×10^{-7} m²/s = 1.29×10^{-5} ft²/s for thermal diffusivity of marble gives

Answer

$$t = \frac{\left(\frac{0.5}{12}\text{ft}^2\right)}{4\left(1.29 \times 10^{-5}\frac{\text{ft}^2}{\text{s}}\right)(0.82)^2} = 50.0 \text{ s}$$

Checking for the depth where the marble temperature will be unchanged, from equation (3–19)

$$x_{un} = 7.2\sqrt{\alpha t} = 7.2\sqrt{(1.29 \times 10^{-5})(50.0)} = 0.18 \text{ ft}$$

Since the marble plate is 1 ft thick, we are justified in approximating a solution based on a semi-infinite model. The heat transfer at the boundary is predicted by equation (3–20),

$$\dot{q}_A = -\kappa(T_i - T_B)\frac{1}{\sqrt{\pi\alpha t}}e^{-x^2/4\alpha t}$$

At the boundary $x = 0$ so that the heat transfer is

$$\dot{q}_A = -\kappa(T_i - T_B)\frac{1}{\sqrt{\pi\alpha t}}$$

Using $\kappa = 2.5$ W/m · °C = 1.445 Btu/hr · ft · °F for marble,

$$\dot{q}_A = \left(1.445\frac{\text{Btu}}{\text{hr}\cdot\text{ft}\cdot°\text{F}}\right)(70°\text{F} - 50°\text{F})\frac{1}{\sqrt{\pi(1.29 \times 10^{-5}\text{ ft}^2/\text{s})(50.0\text{ s})}}$$

Answer

$$= -642.0\frac{\text{Btu}}{\text{hr}\cdot\text{ft}^2}$$

The negative sign indicates that the heat transfer is up from the marble plate (negative direction in figure 3–7), instead of down into the plate.

Semi-Infinite Solid at T_i with Boundary Heat Transfer $\dot{q}_B$

The semi-infinite solid at a uniform temperature T_i that is suddenly subjected to a boundary heat transfer of $\dot{q}_B$ continuously is similar to the immediately preceding problem. In figure 3–9 is shown the physical situation, and the mathematical problem is analyzed with the governing equation

$$\frac{\partial^2 T(x, t)}{\partial x^2} = \frac{1}{\alpha}\frac{\partial T(x, t)}{\partial t} \qquad x \geq 0, t > 0 \tag{3–21}$$

with the initial condition

$$\text{I.C.} \qquad T(x, t) = T_i \qquad x \geq 0, t = 0$$

and the boundary conditions

$$\text{B.C. 1} \qquad -\kappa\frac{\partial T(x, t)}{\partial x} = \dot{q}_B \qquad x = 0, t > 0$$

$$\text{B.C. 2} \qquad T(x, t) = T_i \qquad x \to \infty, t > 0$$

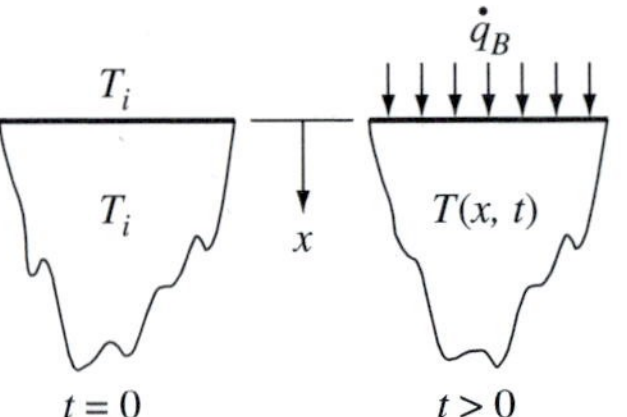

Figure 3–9 Semi-infinite solid at T_i subjected to boundary heat transfer $\dot{q}_B$.

The solution to this problem may be obtained using techniques similar to the previous problem where the boundary temperature was a constant. The result for this case can be written

$$T(x, t) = T_i + \frac{1}{\kappa}\left\{2\dot{q}_B\sqrt{\frac{\alpha t}{\pi}}e^{-x^2/4\alpha t} - \dot{q}_B x\left[1 - \operatorname{erf}(x/2\sqrt{\alpha t})\right]\right\} \tag{3–22}$$

Notice that the heat transfer at the boundary, at $x = 0$, is constant and the temperature at the boundary will change with time. For the previous problem, the boundary temperature was constant and the heat transfer at the boundary changed. The conduction heat transfer in the solid at any location and time may be obtained from Fourier's Law of Conduction. This requires that the temperature gradient, $\partial T(x, t)/\partial t$, be determined, and this is left as an exercise for the student to prove (see practice problem 3–19).

EXAMPLE 3–5 At 10 o'clock A.M. the sun appears from behind a cloud bank and 600 W/m^2 of direct solar radiation strikes the earth perpendicularly. Assuming that the earth's temperature is 10°C, that the radiation is constant for 1 hr, and that the earth's thermal conductivity is 0.52 W/m · °C, determine the surface temperature of the earth and the earth's temperature at a depth of 2 cm at 10:40 A.M. Assume the earth's density is 2000 kg/m^3 and its specific heat is 1700 J/kg · °C.

Solution The earth's temperatures can be predicted from equation (3–22) because this problem satisfies all of the prescribed boundary and initial conditions of equation (3–21). At the earth's surface we have $x = 0$, $T_i = 10$°C, and $\dot{q}_B = 600$ W/m^2, and at 10:40 A.M. the time $t = 40$ min or 2400 s. The thermal diffusivity can be found from equation (3–3),

$$\alpha = \frac{\kappa}{\rho c_p} = \frac{\left(0.52\ \frac{\text{W}}{\text{m}\cdot\text{K}}\right)}{\left(2000\ \frac{\text{kg}}{\text{m}^3}\right)\left(1700\ \frac{\text{J}}{\text{kg}\cdot\text{K}}\right)} = 1.53 \times 10^{-7}\ \frac{\text{m}^2}{\text{s}}$$

The surface temperature or boundary temperature at 10:40 A.M., using equation (3–22) with $t = 40$ min $= 2400$ s, is

Answer $$T_s = T_B = T(0, 2400\ \text{s}) = 34.9°\text{C}$$

The temperature of the earth at a depth of 2 cm is found from the same equation, except now $x = 2$ cm $= 0.02$ m. The result is

Answer $$T(0.02\ \text{m}, 2400\ \text{s}) = 18.36°\text{C}$$

Semi-Infinite Solid Subjected to Convection Heat Transfer

The semi-infinite solid with a convective heat transfer at its boundary is often a more accurate model of heat transfer between a large object and its surroundings than the previous two models. While this discussion considers the solid, a stagnant fluid, having no convective currents, could also be considered. The physical situation described by this model is sketched in figure 3–10 and the governing differential equation for the solid is the same as for the preceding two cases. Only the boundary condition at $x = 0$ changes. We may write for this model,

$$\frac{\partial^2 T(x,t)}{\partial x^2} = \frac{1}{\alpha}\frac{\partial T(x,t)}{\partial t} \qquad x \geq 0, t > 0 \tag{3–23}$$

with the initial condition $T(x, t) = T_i$, $x \geq 0$, $t = 0$, and the two boundary conditions

$$\text{B.C. 1} \qquad -\kappa\frac{\partial T(x,t)}{\partial x} = h_\infty[T(x,t) - T_\infty] \qquad x = 0, t > 0$$

$$\text{B.C. 2} \qquad \frac{\partial T(x,t)}{\partial x} = 0 \qquad x \rightarrow \infty, t > 0$$

Figure 3–10 Semi-infinite solid at T_i subjected to convective heat transfer $\dot{q}_B$.

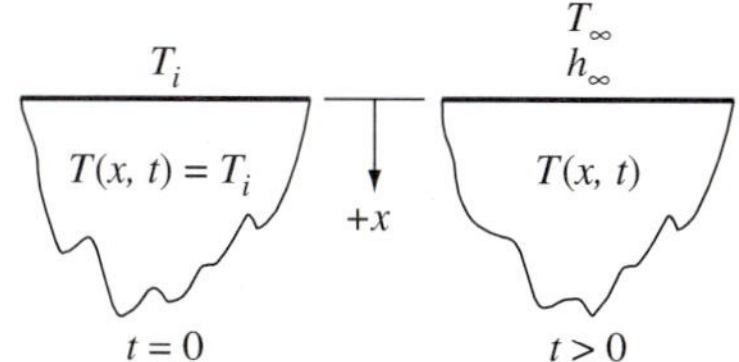

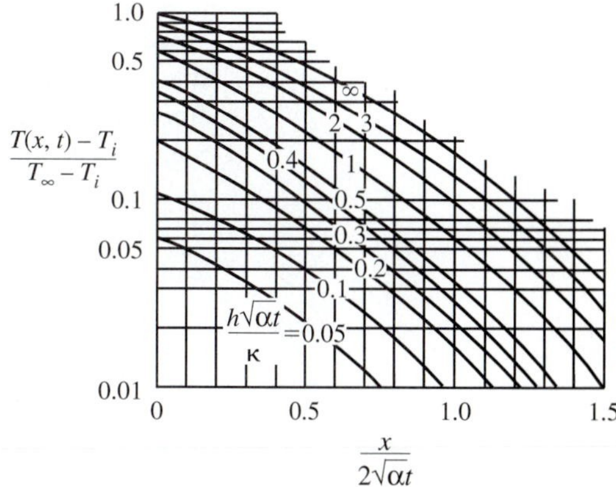

A solution to this mathematical problem can be obtained using an approach similar to the previous two cases for the semi-infinite solid. A solution has been shown by Schneider [3] to be,

$$T(x, t) = T_i + (T_\infty - T_i)\left\{1 - \operatorname{erf}\left(\frac{x}{2\sqrt{\alpha t}}\right) - e^{\left(\frac{hx}{\kappa} + \frac{h^2\alpha t}{\kappa^2}\right)}\left[1 - \operatorname{erf}\left(\frac{x}{2\sqrt{\alpha t}} + \frac{h\sqrt{\alpha t}}{\kappa}\right)\right]\right\} \quad \textbf{(3–24)}$$

and in figure 3–10 is shown a graphical solution of this equation. It can be seen from figure 3–10 or from equation (3–24) that the temperature distribution through the semi-infinite solid has the same general character for the convective boundary heat transfer as for the constant boundary temperature or the constant boundary heat transfer. The surface temperature will change with time and the depth of penetration of the heat transfer increases with time, but in all three of these cases of semi-infinite solids the temperature remains unchanged at a sufficient distance from the boundary. The conduction heat transfer through the solid can be predicted from Fourier's Law of Conduction, upon determining the first partial derivative of equation (3–24) with respect to x. This development is left as an exercise for the student (see problem 3–24).

Infinite Flat Plate Subjected to Convection Heat Transfer

An infinite plate is one having two identifiable infinite boundaries that are parallel to each other, as shown in figure 3–11. For the transient heat transfer problem of an infinite plate at an initial temperature T_i throughout suddenly subjected to a convective heat transfer on both surfaces, a mathematical statement is, referring to figure 3–11,

$$\frac{\partial^2 T(x, t)}{\partial x^2} = \frac{1}{\alpha}\frac{\partial T(x, t)}{\partial t} \qquad L \geq |x| \geq 0, t > 0 \quad \textbf{(3–25)}$$

with initial condition $T(x, t) = T_i$, $L \geq x \geq -L$, $t = 0$, and

B.C. 1 $$-\kappa\frac{\partial T(x, t)}{\partial x} = h_\infty[T(x, t) - T_\infty] \qquad x = \pm L, t > 0$$

B.C. 2 $$\frac{\partial T(x, t)}{\partial x} = 0 \quad \text{(a symmetry condition)} \qquad x = 0, t > 0$$

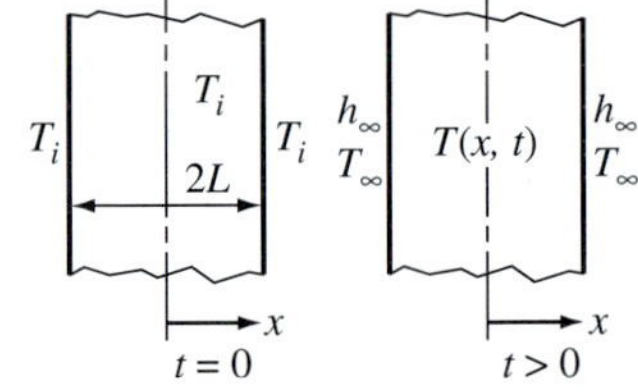

Figure 3–11 Infinite flat plate at T_i subjected to convective boundary heat transfer.

Since the temperature distribution is a function of x and t, from the separation of variables technique the solution is given as a temperature function dependent on time and another distinct temperature function of position. For the parameters, $\theta_0 = T(0, t) - T_\infty$, the centerline temperature, and $\theta_i = T_i - T_\infty$, the initial temperature, $\theta(x, t) = T(x, t) - T_\infty$, and the *Fourier Number,* $\text{Fo} = \alpha t/L^2$.

A solution to this problem has been obtained and is discussed by Schneider [3]. The result is

$$\frac{\theta(x, t)}{\theta_i} = \sum_{n=1}^{\infty} C_n e^{-\xi_n^2 \text{Fo}} \cos\frac{\xi_n x}{L} \quad \textbf{(3–26)}$$

where $$C_n = \frac{4\sin\xi_n}{2\xi_n + \sin 2\xi_n} \quad \textbf{(3–27)}$$

The terms ξ_n are called *eigenvalues* and are positive roots of the transcendental equation

$$\xi_n \tan \xi_n = \text{Bi} = \frac{h_\infty L}{\kappa} \tag{3–28}$$

In appendix table A–11–1 the first five roots of equation 3–28 and the coefficient, C_n, of equation (3–27) are listed for various Biot Numbers. Heisler [4] has shown that a sufficiently accurate (but approximate) solution can be obtained for Fo ≥ 0.2 (representing an elapsed time period well beyond the initial condition at $t = 0$) if only the first term of the infinite series of equation (3–26) is used. Then,

$$\frac{\theta(x, t)}{\theta_i} = C_1 e^{-\xi_1^2 \text{Fo}} \cos \frac{\xi_1 x}{L} \qquad \text{for Fo} > 0.2 \tag{3–29}$$

Heisler [4] and Grober et al. [5] have presented graphical charts or solutions to this problem, and these are given in figure 3–12a and b. These graphical solutions are sometimes referred to as *Heisler Charts.* The ratio θ_0/θ_i is the time-dependent part of equation (3–29), namely

$$\frac{\theta_0(t)}{\theta_i} = C_1 e^{-\xi_1^2 \text{Fo}} \tag{3–30}$$

and this is presented in figure 3–12a. The spatial or position-dependent part of equation (3–29),

$$\frac{\theta(x)}{\theta_0} = \cos \frac{\xi_1 x}{L} \tag{3–31}$$

is presented in figure 3–12b. Equation (3–29) could be written

$$\frac{\theta(x, t)}{\theta_i} = \left(\frac{\theta(x)}{\theta_0}\right)\left(\frac{\theta_0(t)}{\theta_i}\right) \tag{3–29, revised}$$

The graphical solution given in figure 3–12a, applicable only if the Fourier Number is greater than 0.2, implies that the temperature history is not determined for the period immediately following the sudden boundary condition change. Heisler [4] has presented charts that can be used to analyze the transient heat transfers when $0 \leq \text{Fo} \leq 0.2$, or equation (3–26) can be used for such an analysis. If the Fourier Number is greater than about 0.002, then the first five terms, determined from the information given in appendix table A–11–1, of the infinite series of equation (3–26) should give sufficiently accurate results for the temperature distribution.

Sometimes it is important to know how much thermal energy has been exchanged between the plate and its surroundings. This energy can be described as the enthalpy of the plate and applying the conservation of energy, equation (1–27), we get

$$\Delta H = Q$$

For the plate we may also write for the enthalpy change

$$\Delta H = \int_{\text{volume}} \rho c_p [T(x, t) - T_\infty] dV \tag{3–32}$$

Since the plate is visualized as extending to infinity along its faces we write for the differential volume, $dV = Adx$. The enthalpy change per unit area is then, from equation (3–32),

$$\Delta H = 2\,A \int_0^L \rho c_p [T(x, t) - T_\infty] dx = Q \tag{3–33}$$

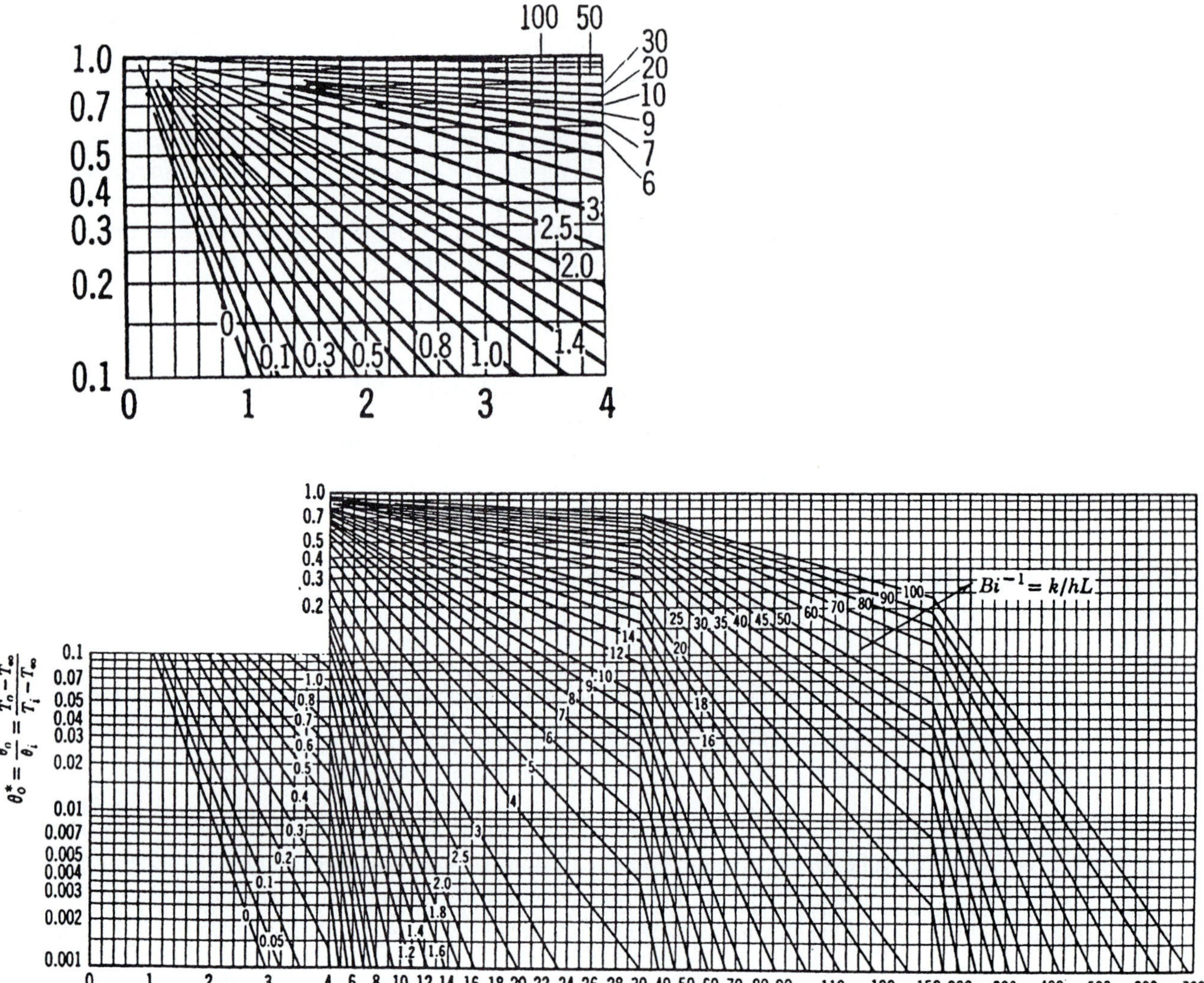

(a) Center line time history

Figure 3–12 Graphical solution to the infinite plate subjected to convective heating. (By permission of American Society of Mechanical Engineers, New York, NY)

The maximum amount of thermal energy exchange is

$$\Delta H_0 = Q_0 = 2\rho c_p(T_i - T_\infty)A\int_0^L dx$$
$$= \rho c_p(T_i - T_\infty)2\,AL \tag{3–34}$$

If equation (3–33) is divided by (3–34), we find that

$$\frac{Q}{Q_0} = \frac{1}{L}\int_0^L \frac{\theta(x,t)}{\theta_i}dx \tag{3–35}$$

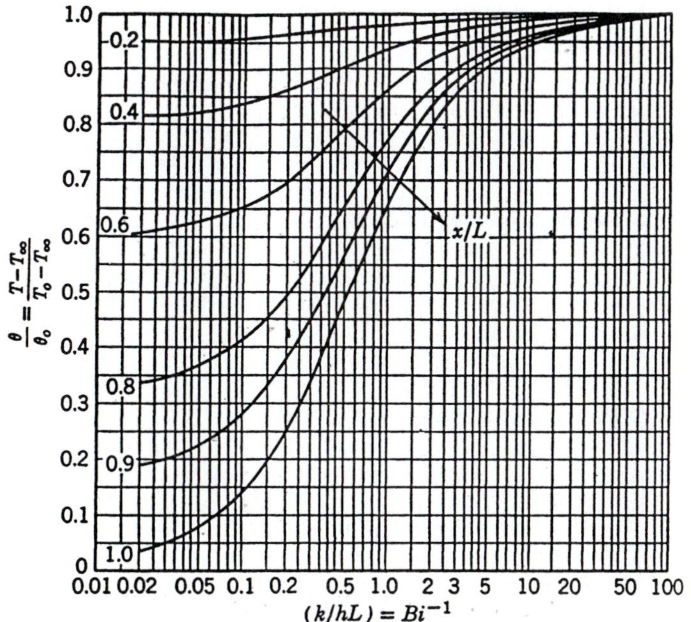

(b) Temperature distribution

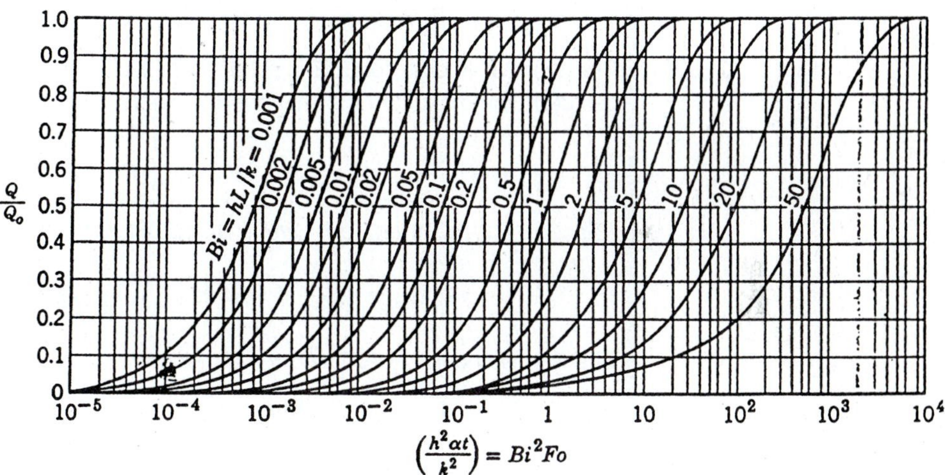

(c) Thermal energy transfer function

Using the result from equation (3–29) to complete the integration of equation (3–35), we obtain

$$\frac{Q}{Q_0} = 1 - C_1 e^{-\xi_1^2 \, \mathrm{Fo}} \frac{\sin \xi_1}{\xi_1} \qquad (\mathrm{Fo} > 0.2) \tag{3–36}$$

The graphical form of this result has been given by Grober [5] and is presented in figure 3–12c.

EXAMPLE 3–6 Large sheets of 1-in.-thick glass are taken out of an oven at 400°F and placed vertically to allow for air cooling. If the air is at 80°F and the convective heat transfer coefficient is 3 Btu/hr · ft² · °F, determine the center temperature after 20 min of cooling and the amount of thermal energy loss per unit area of the glass. Assume the following properties for glass,

$$\kappa = 0.45 \text{ Btu/hr} \cdot \text{ft} \cdot {}^\circ\text{F}$$
$$c_p = 0.2 \text{ Btu/lbm} \cdot {}^\circ\text{F}$$
$$\alpha = 3.66 \times 10^{-6} \text{ ft}^2\text{/s}$$

The Biot Number, hL/κ, is

$$\text{Bi} = \frac{h_\infty L}{\kappa} = \frac{\left(3\dfrac{\text{Btu}}{\text{hr} \cdot \text{ft}^2 \cdot {}^\circ\text{F}}\right)\left(\dfrac{1}{24}\text{ ft}\right)}{\left(0.45\dfrac{\text{Btu}}{\text{hr} \cdot \text{ft} \cdot {}^\circ\text{F}}\right)} = 0.277$$

The Fourier Number, $\alpha t/L^2$, is,

$$\text{Fo} = \frac{\alpha t}{L^2} = \frac{\left(3.66 \times 10^{-6}\dfrac{\text{ft}^2}{\text{s}}\right)(1200 \text{ s})}{\left(\dfrac{1}{24}\text{ ft}\right)^2} = 2.53$$

The quantity θ_0/θ_i is read from figure 3–12(a) to be approximately 0.56. Using the definitions for θ_o and θ_i, the centerline temperature is determined,

Answer
$$T(0, 20 \text{ min}) = T_\infty + (400^\circ\text{F} - 80^\circ\text{F})(0.56) = 260^\circ\text{F}$$

The thermal energy lost during the 20 min can be predicted from the results of figure 3–12(c). The maximum thermal energy loss is

$$Q_{A0} = \frac{Q_0}{A} = 2\, L\rho c_p \theta_i$$

Since $\alpha = \kappa/\rho c_p$, then $\rho c_p = \kappa/\alpha = 34.153 \text{ Btu/ft}^3 \cdot {}^\circ\text{F}$, and then

$$Q_{A0} = 2\left(\frac{1}{24}\text{ ft}\right)\left(34.153\frac{\text{Btu}}{\text{ft}^3 \cdot {}^\circ\text{F}}\right)(320^\circ\text{F}) = 910.75\frac{\text{Btu}}{\text{ft}^2}$$

The Biot Number is 0.277 and the parameter

$$\frac{h^2 \alpha t}{\kappa^2} = \frac{\left(3\dfrac{\text{Btu}}{\text{hr} \cdot \text{ft}^2 \cdot {}^\circ\text{F}}\right)^2\left(3.66 \times 10^{-6}\dfrac{\text{ft}^2}{\text{s}}\right)(1200 \text{ s})}{\left(0.45\dfrac{\text{Btu}}{\text{hr} \cdot \text{ft} \cdot {}^\circ\text{F}}\right)^2} = 0.1952$$

Then, using figure 3–12c, we read that

$$\frac{Q_A}{Q_{A0}} = \frac{Q}{Q_0} \approx 0.5$$

Answer
$$Q_A = (0.5)Q_{A0} = 455.4 \text{ Btu/ft}^2$$

EXAMPLE 3–7 A large plate surrounds a cryogenic tank with a vacuum between the two, as shown in figure 3–13. The plate is 2 cm thick, is constructed of 40% nickel steel and is at a temperature of 10°C when it is suddenly subjected to air at 30°C with convective coefficient of 200 W/m^2 · °C against its outer surface. Estimate the time before the inner surface, facing the vacuum, reaches 20°C and find the temperature at the outer surface at this time.

Figure 3–13 Section of cryogenic tank.

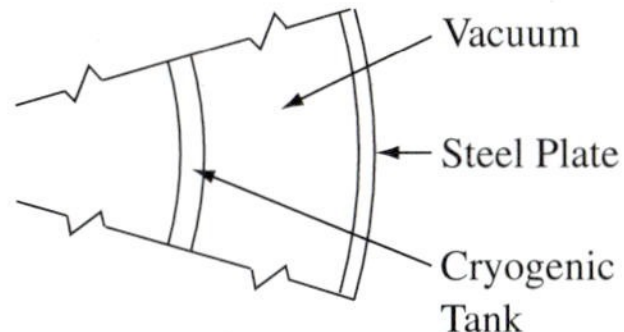

Assume that no radiation heat transfer occurs across the vacuum so that the inner surface of the plate is an adiabatic surface.

Solution

The plate may be considered as an infinite plate having thickness of 4 cm because there is no heat transfer across its inner surface. From B.C. 2, the centerline of an infinite plate is an adiabatic surface, due to symmetry of the temperature distribution. Thus, although the plate is actually only 2 cm thick, it has a temperature distribution predicted by a plate 4 cm thick having symmetrical boundary conditions on both sides. Using our prevision notation, the temperature, $T_{inner} = T_0$. The properties of 40% nickel steel are found in appendix table B–2, and they are $\kappa = 10\ \text{W/m} \cdot {}^\circ\text{C}$, $\alpha = 0.279 \times 10^{-5}\ \text{m}^2/\text{s}$, $c_p = 460\ \text{J/kg} \cdot \text{K}$, and $\rho = 8169\ \text{kg/m}^3$. The temperature function is

$$\frac{\theta_0}{\theta_i} = \frac{20^\circ\text{C} - 30^\circ\text{C}}{10^\circ\text{C} - 30^\circ\text{C}} = 0.5$$

and the inverse Biot Number is

$$\kappa/hL = (10\ \text{W/m} \cdot {}^\circ\text{C})/(200\ \text{W/m}^2 \cdot {}^\circ\text{C})(0.02\ \text{m}) = 2.5$$

The Fourier Number is read from figure 3–12a as, approximately, 2.2. The time is then computed from

Answer

$$t = (2.2)(L)^2/\alpha = (2.2)(0.02\ \text{m})^2/(0.279 \times 10^{-5}\ \text{m}^2/\text{s}) = 315.4\ \text{s} = 5.257\ \text{min}$$

The outside surface temperature is found from the result of figure 3–12b, $\theta/\theta_0 \approx 0.83$ at $x/L = 1.0$. Then

Answer

$$T_{outside} = T_\infty + (0.83)(T_0 - T_\infty) = 21.7^\circ\text{C}$$

Notice that the temperature at any other point in the plate may be predicted from the result of figure 3–12b using appropriate x/L values or by using equation (3–31), a cosine function. Further, the analyses using the Heisler Charts can be extended to include the boundary condition, or physical condition, where the surfaces of the plate immediately, at $t > 0$, change to the surrounding temperature, T_∞. This can be done by assuming an infinite Biot Number or infinite convective heat transfer coefficient. Thus, the inverse Biot Number, Bi^{-1}, would be zero for an infinite Biot Number, and in figure 3–12a this can be read. The temperature distribution through the plate for an infinite Biot Number can be approximated in figure 3–12b by using an extrapolation of the x/L curves to $\text{Bi}^{-1} \approx 0.01$. The analytic expression for the temperature distribution through the infinite plate can also be used for this special situation where the plate surfaces are immediately the same as the surrounding temperature. Here, from equation (3–28), $\xi_1 = \pi/2$ for a Biot Number of infinity. Then, using equation (3–27), $C_1 = 4/\pi$ and equation (3–29) becomes

$$\frac{\theta(x, t)}{\theta_i} = \frac{4}{\pi} e^{-\frac{\pi^2}{4}\text{Fo}} \cos\frac{\pi x}{2L} \qquad (\text{for Fo} \geq 0.2,\ -L \leq x \leq L)$$

Infinite Cylinder Subjected to Convection Heat Transfer

Long cylinders or rods can often be approximated as infinite cylinders, and heat transfer can then be considered as one-dimensional if the heat transfer at the ends are neglected. For the infinite cylinder subjected to a convection heat transfer suddenly imposed at the outer surface, shown in figure 3–14, the mathematical formulation may be written

$$\frac{\partial^2 T(r,t)}{\partial r^2} + \frac{1}{r}\frac{\partial T(r,t)}{\partial r} = \frac{1}{\alpha}\frac{\partial T(r,t)}{\partial t} \qquad r_0 \geq r \geq 0,\ t > 0 \tag{3–37}$$

with initial condition $T(r, t) = T_i$, $r_0 \geq r \geq 0$, $t = 0$, and boundary conditions

$$\text{B.C. 1} \qquad -\kappa\frac{\partial T(r,t)}{\partial r} = h_\infty[T(r,t) - T_\infty] \qquad r = r_0,\ t > 0$$

$$\text{B.C. 2} \qquad \frac{\partial T(r,t)}{\partial r} = 0 \qquad r = 0,\ t > 0$$

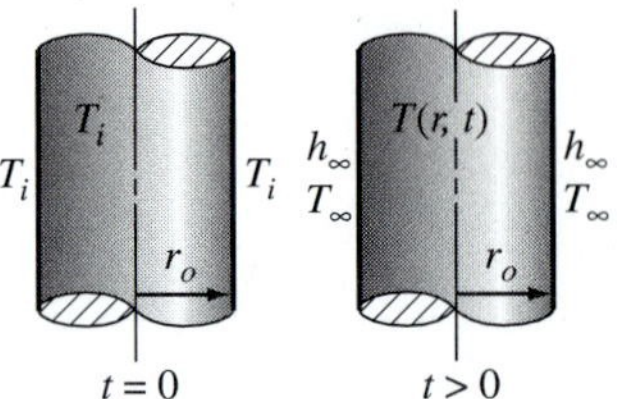

Figure 3–14 Infinite cylinder subjected to convection heat transfer.

The solution to this problem has been obtained and is discussed by Schneider [3]. The result is

$$\frac{\theta(r,t)}{\theta_i} = \sum_{n=1}^{\infty} C_n e^{-\beta_n^2 \mathrm{Fo}_c} J_0\left(\frac{\beta_n r}{r_0}\right) \tag{3–38}$$

where $\theta(r, t) = T(r, t) - T_\infty$, $\theta_i = T_i - T_\infty$, $\mathrm{Fo}_c = \alpha t/r_0^2$, and

$$C_n = \frac{2J_0(\beta_n)}{\beta_n[J_0^2(\beta_n) + J_1^2(\beta_n)]} \tag{3–39}$$

J_1 is the Bessel Function of the first kind, first order, and J_0 is the Bessel Function of the first kind, zero order. The Bessel Functions are tabulated in appendix table A–10. The argument β_n is the eigenvalue or positive roots of the transcendental equation

$$\beta_n \frac{J_1(\beta_n)}{J_0(\beta_n)} = \frac{h_\infty r_0}{\kappa} = \mathrm{Bi}_c \tag{3–40}$$

and the first five roots, β_n, of equation (3–40) are tabulated in appendix table A–11–2. Heisler [4] has shown that for $\mathrm{Fo}_c = \alpha t/r_0^2 \geq 0.2$ the solution of equation (3–34) can be approximated by the first term of the infinite series; that is,

$$\frac{\theta(r,t)}{\theta_i} = C_1 e^{-\beta_1^2 \mathrm{Fo}_c} J_0\left(\beta_1 \frac{r}{r_0}\right) \tag{3–41}$$

for $\mathrm{Fo}_c \geq 0.2$. For the solution to the short-term conditions where $\mathrm{Fo}_c < 0.2$, additional terms need to be included in the series solution of equation (3–38). If the Fourier Number is, say, greater than 0.002, then the first four terms of the infinite series should give an adequate answer to the temperature distribution of equation (3–38). The centerline temperature history, θ_0/θ_i, is the time dependent part of equation (3–41), namely

$$\frac{\theta_0(t)}{\theta_i} = C_1 e^{-\beta_1^2 \mathrm{Fo}_c} \tag{3–42}$$

and this is presented in graphical form in figure 3–15a. The part of the temperature function that is dependent on position is

$$\frac{\theta(r)}{\theta_0} = J_0\left(\beta_1 \frac{r}{r_0}\right) \tag{3–43}$$

and this function is presented in figure 3–15b. An energy balance can be applied to the infinite cylinder, as was done for the infinite plate, and the result is (for $\mathrm{Fo}_c \geq 0.2$)

$$\frac{Q}{Q_0} = 1 - 2C_1 e^{-\beta_1^2 \mathrm{Fo}_c} \frac{J_0(\beta_1)}{\beta_1} \tag{3–44}$$

This result is given in graphical form in figure 3–15c. The charts presented in figure 3–15 are sometimes referred to as Heisler Charts in the same way that the charts in figure 3–12 are called Heisler Charts. The student needs to understand which physical condition is being studied, whether an infinite plate or an infinite cylinder. Also, the special boundary condition where the surface of the cylinder suddenly changes to the surrounding temperature, represented by an infinite Biot Number and infinite convective heat-transfer coefficient can be treated with the charts, using an inverse Biot Number of zero.

EXAMPLE 3–8 Steel rods 2 cm in diameter at 300°C are quenched in oil at 100°C. Assuming that the surface temperatures of the rods reach 140°C within 75 s after being immersed in the oil, estimate the convective heat transfer coefficient between the oil and the rods.

Solution The properties of steel are gotten from appendix table B–2. We will use

$$\kappa = 19 \text{ W/m} \cdot {}^\circ\text{C}$$
$$\rho = 7963.3 \text{ kg/m}^3$$
$$c_p = 460 \text{ J/kg} \cdot \text{K}$$
$$\alpha = 0.526 \times 10^{-5} \text{ m}^2\text{/s}$$

The temperature functions are

$$\theta_i = T_i - T_\infty = 300^\circ\text{C} - 100^\circ\text{C} = 200^\circ\text{C}$$
$$\theta_s = T(r_0, t) - T_\infty = 140^\circ\text{C} - 100^\circ\text{C} = 40^\circ\text{C}$$

and

$$\frac{\theta_B}{\theta_i} = \frac{40^\circ\text{C}}{200^\circ\text{C}} = 0.2$$

The Fourier Number is

$$\frac{\alpha t}{r_0^2} = \frac{\left(0.526 \times 10^{-5} \frac{\text{m}^2}{\text{s}}\right)(75 \text{ s})}{(0.01 \text{ m})^2} = 3.945$$

and by graphical trial and error using the two graphs, figure 3–15a and b, the value for h may be determined. Assuming a value of 0.9 for θ_B/θ_0 gives κ/hr_0 of approximately 4.5 from figure 3–15b at $r/r_0 = 1.0$. Using this value for κ/hr_0, the value for $\theta_0\theta_i$ is found to be

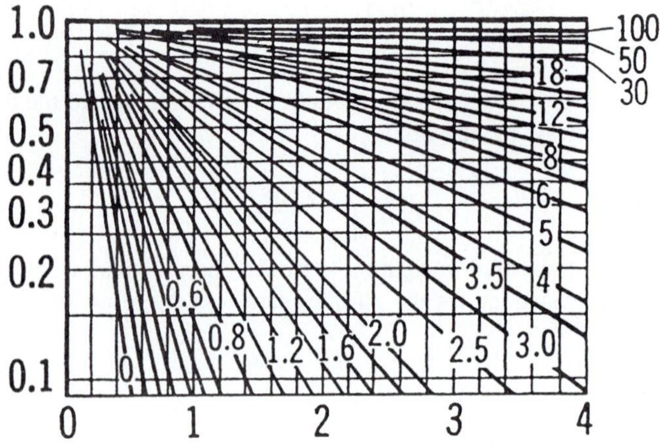

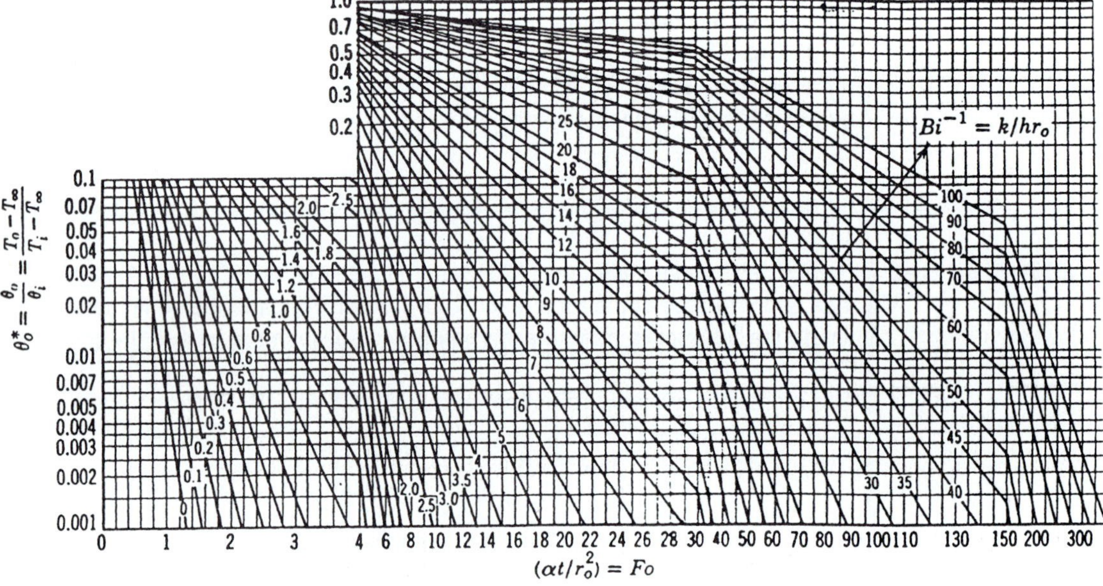

(a) Center temperature history

Figure 3–15 Graphical solutions to the temperature and thermal energy exchange for an infinite cylinder with a surrounding fluid. (By permission of American Society of Mechanical Engineers, New York, NY)

approximately 0.22 from figure 3–15a. Since this result was arrived at by iteration, we check to see if the value of θ_s/θ_i is 0.2:

$$\theta_s/\theta_i = [\theta_0/\theta_i][\theta_s/\theta_0] = [0.22][0.9] = 0.198 \approx 0.2$$

so that the assumed value $\theta_s/\theta_0 = 0.9$ was reasonable. Therefore, using the inverse Biot Number

$$\kappa/hr_0 = 4.5$$

gives

Answer

$$h = \kappa/4.5r_0 = (19\ \text{W/m} \cdot {}^\circ\text{C})/(4.5)(0.01\ \text{m}) = 422.2\ \text{W/m}^2 \cdot {}^\circ\text{C}$$

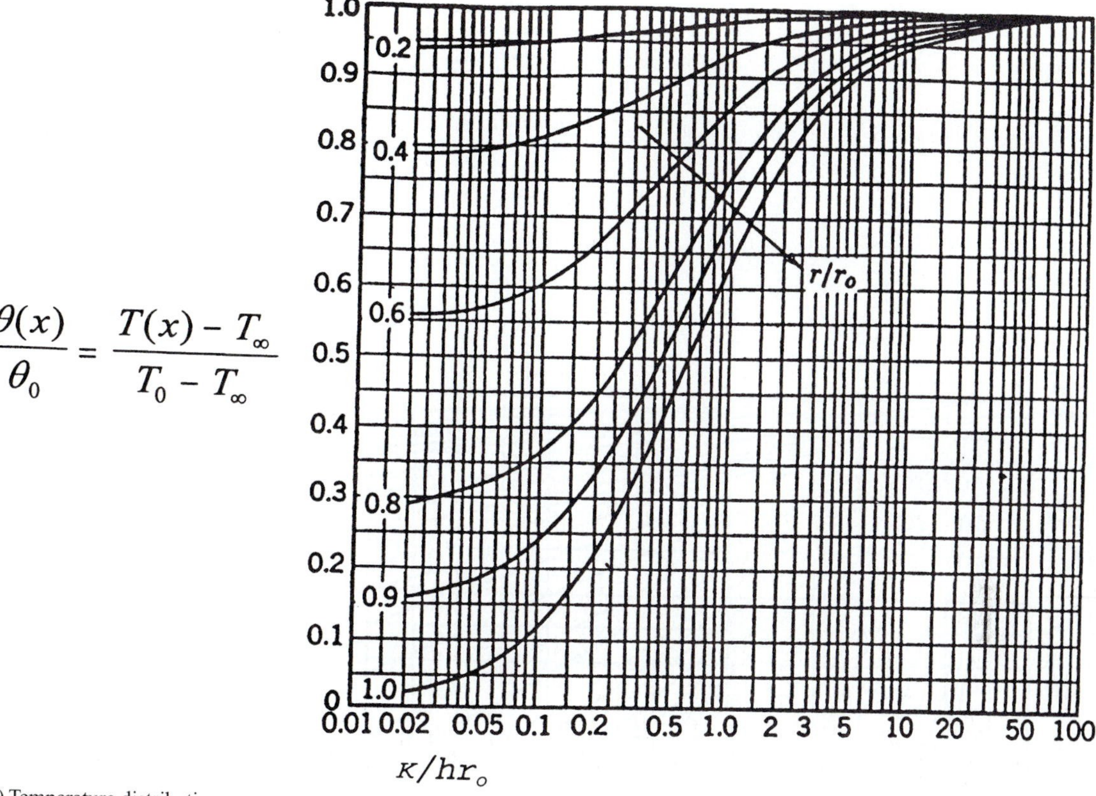

(b) Temperature distribution

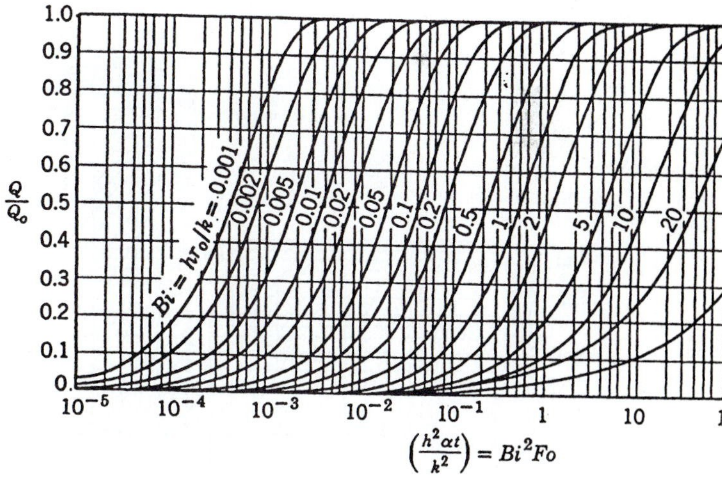

(c) Thermal energy transfer function

Sphere Subjected in Convection Heat Transfer at Its Surface

The problem of solid spheres at a temperature T_i, suddenly subjected to convection heat transfer at its surface and shown in figure 3–16 can be mathematically formulated as

$$\frac{\partial^2 T(r,t)}{\partial r^2} + \frac{2}{r}\frac{\partial T(r,t)}{\partial r} = \frac{1}{\alpha}\frac{\partial T(r,t)}{\partial t} \qquad \text{for } r_0 \geq r \geq 0, t > 0 \tag{3–45}$$

with initial condition

$$T(r,t) = T_i \qquad r_0 \geq r \geq 0, t = 0$$

Figure 3–16 Solid sphere subjected to convection heat transfer.

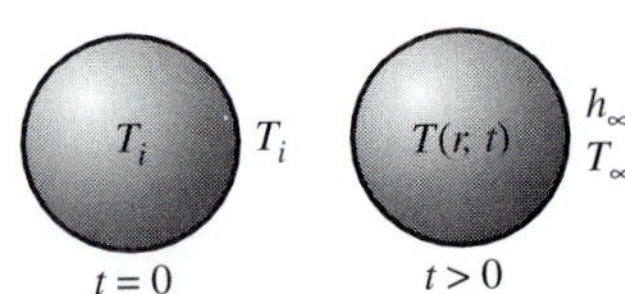

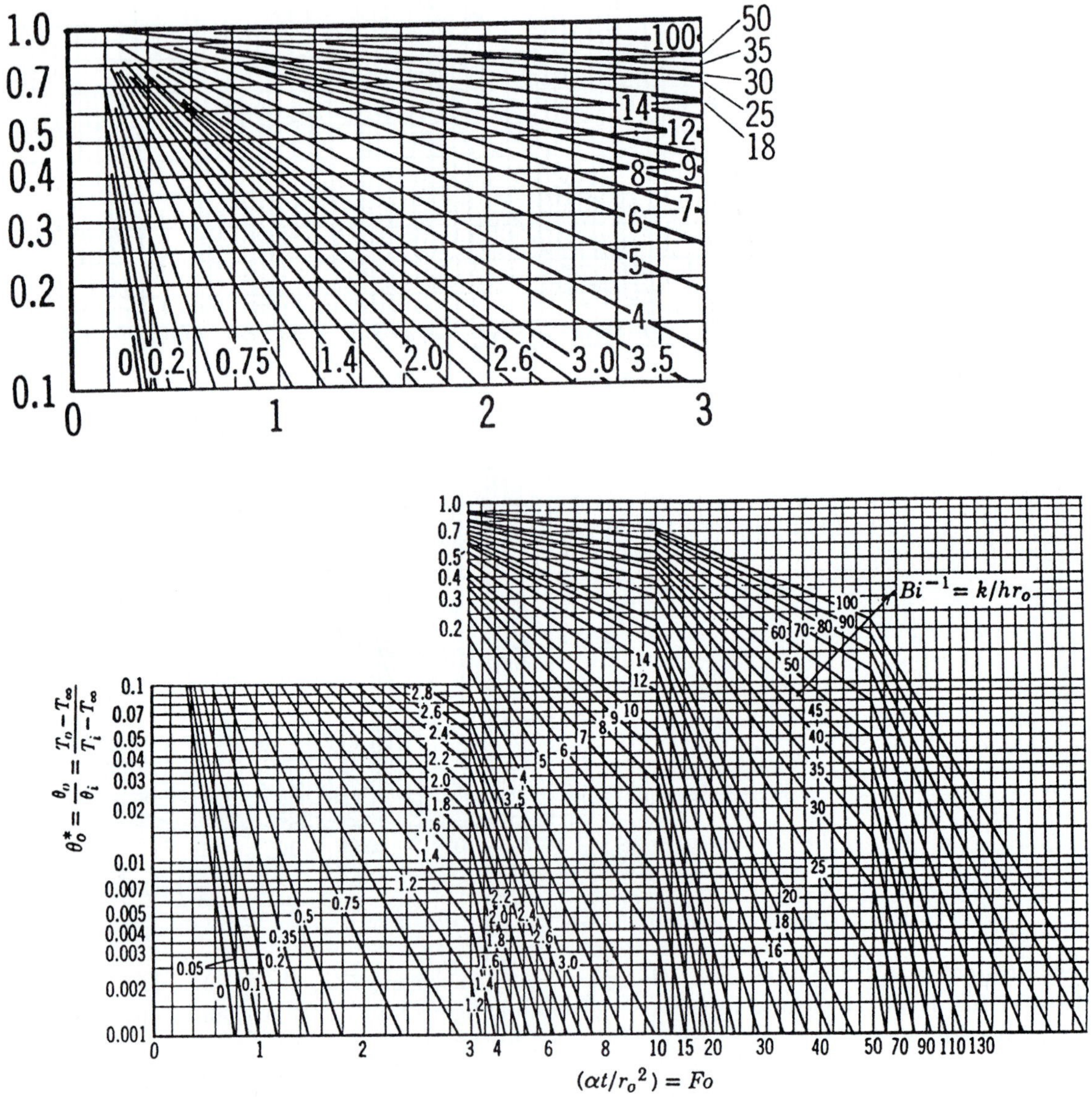

(a) Center temperature history

Figure 3–17 Solid sphere subjected to convection heat transfer. (By permission of American Society of Mechanical Engineers, New York, NY) and boundary conditions

$$\text{B.C. 1} \qquad -\kappa \frac{\partial T(r,t)}{\partial r} = h_\infty[T(r,t) - T_\infty] \qquad r = r_0, t > 0$$

$$\text{B.C. 2} \qquad \frac{\partial T(r,t)}{\partial r} = 0 \qquad r = 0, t > 0$$

The solution to equation (3–45) can be written as

$$\frac{\theta(r,t)}{\theta_i} = \sum_{n=1}^{\infty} C_n e^{-\zeta_n^2 \text{Fo}_r} \frac{r_0}{\zeta_n r} \sin\left(\zeta_n \frac{r}{r_0}\right) \tag{3–46}$$

where

$$\text{Fo}_r = \frac{\alpha t}{r_0^2} \tag{3–47}$$

$$C_n = \frac{4(\sin \zeta_n - \zeta_n \cos \zeta_n)}{2\zeta_n - \sin 2\zeta_n} \tag{3–48}$$

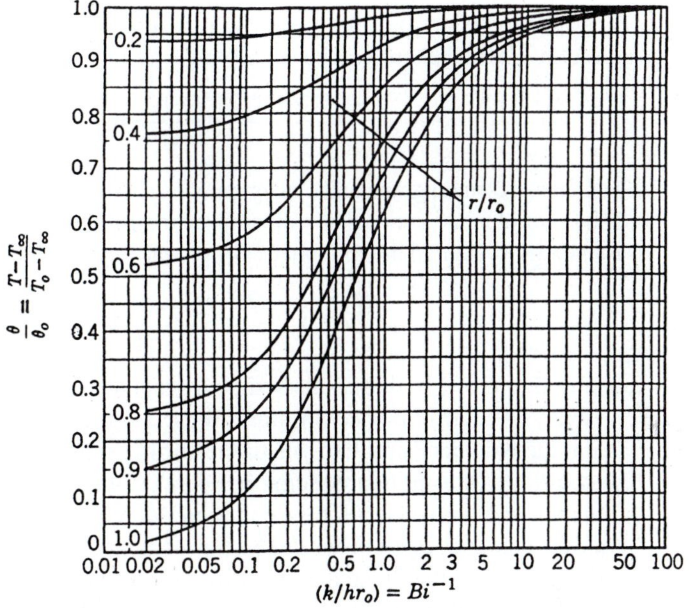

(b) Temperature distribution

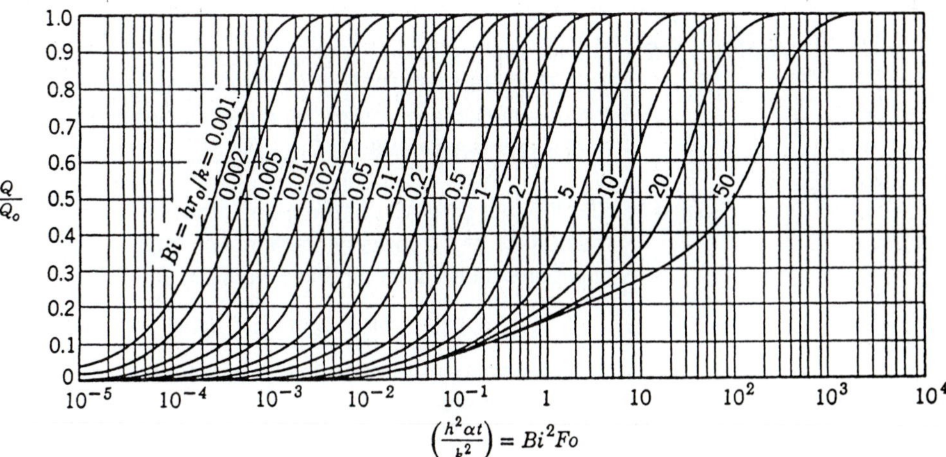

(c) Thermal energy transfer function

and ζ_n are the positive roots of the transcendental equation

$$1 - \zeta_n \cot \zeta_n = \text{Bi} = \frac{h_\infty r_0}{\kappa} \tag{3–49}$$

The first four roots of equation (3–49) and C_n are tabulated in appendix table A–11.3. Heisler [4] has shown that for $\text{Fo}_r \geq 0.2$ the solution of equation (3–46) can be reasonably approximated by using the first term of the infinite series so that

$$\frac{\theta(r, t)}{\theta_i} = C_1 e^{-\zeta_1^2 \text{Fo}_r} \frac{r_0}{\zeta_1 r} \sin\left(\zeta_1 \frac{r}{r_0}\right) \tag{3–50}$$

and the center temperature history can be identified as the time dependent portion of equation (3–50),

$$\frac{\theta_0(t)}{\theta_i} = C_1 e^{-\zeta_1^2 \text{Fo}_r} \tag{3–51}$$

The function is graphically displayed in figure 3–17a. The radial temperature distribution through a solid sphere can be represented by the remaining part of equation (3–50)

$$\frac{\theta(r)}{\theta_0} = \frac{r_0}{\zeta_1 r} \sin\left(\zeta_1 \frac{r}{r_0}\right) \tag{3–52}$$

which is presented graphically in figure 3–17b. An energy balance can be applied to the solid sphere to obtain a ratio of the amount of thermal energy exchange between it and its surroundings to the maximum amount of thermal energy which can be exchanged. This result is

$$\frac{Q}{Q_0} = 1 - \frac{3C_1}{\zeta_1^3} e^{-\zeta_1^2 \text{Fo}_r} (\sin \zeta_1 - \zeta_1 \cos \zeta_1) \tag{3–53}$$

and in figure 3–17c this result is shown graphically. The method of using these results in an analysis of a solid sphere follows those for the infinite plate and the infinite cylinder. Also, the boundary problem where the surface temperature of the sphere changes immediately to that of the surrounding temperature can be included by using an infinite Biot Number and zero inverse Biot Number. More complete solutions of the various heat-transfer configurations discussed in this section as well as other regular geometrical objects with prescribed boundary conditions have been presented by Heisler [4], Grober, et al. [5], and Schneider [6].

EXAMPLE 3–9 Granite rocks of 3-cm diameter are used as thermal storage material in a rock bed storage. The rocks are 10°C when they are suddenly subjected to hot air at 50°C. Assuming that the surfaces of the rocks immediately become 50°C, determine the time before the center of the rocks are 30°C.

Solution The thermal diffusivity of the granite rock is, from table 3–1, $\alpha = 1.37 \times 10^{-6}$ m²/s. Since the surface of the rocks immediately changes to the surrounding air temperature, the Biot Number will be infinite and its inverse is zero. From equation (3–49) we have

$$1 - \zeta_n \cot \zeta_n = \text{Bi} = \infty \quad \text{and} \quad \zeta_n \cot \zeta_n = -\infty$$

For the cotangent function this requires that $\zeta_1 = \pi$. Using equation (3–48), the constant C_1 can be evaluated:

$$C_1 = \frac{4(\sin \zeta_1 - \zeta_1 \cos \zeta_1)}{2\zeta_1 - \sin \zeta_1} = \frac{4(\sin \pi - \pi \cos \pi)}{2\pi - \sin \pi} = 2$$

The values for C_1 and ζ_1 can be verified from appendix table A11–3, which lists the values for a Biot Number of 100. Using equation (3–51)

$$\frac{\theta_0(t)}{\theta_i} = C_1 e^{-\zeta_1^2 \,\mathrm{Fo}_r} = \frac{T_0 - T_\infty}{T_i - T_\infty} = \frac{30°\mathrm{C} - 50°\mathrm{C}}{10°\mathrm{C} - 50°\mathrm{C}} = 0.5 = 2e^{-\pi^2 \,\mathrm{Fo}_r}$$

and

$$\mathrm{Fo}_r = \alpha t / r_0^2 = 0.21069$$

The time is then

Answer

$$t = 0.21069 \frac{r_0^2}{\alpha} = 138.4 \text{ s}$$

A similar result can be obtained by using the chart in figure 3–17a, where, for an inverse Biot Number of zero and θ_0/θ_i of 0.5, the Fourier Number, Fo_r, is approximately 0.2. This is in substantial agreement with our result using the analytic expressions.

3–4 TWO-DIMENSIONAL TRANSIENT HEAT TRANSFER

Many engineering problems involve heat transfer in two or three dimensions. For the transient heat-transfer problems, the analysis of these multidimensional cases can proceed with the separation of variables technique applied to the conduction heat-transfer equation. In Cartesian coordinates this is equation (3–4), and in many cases where regular geometric configurations are involved and the boundary conditions are identified, solutions have been obtained. Carslaw and Jaeger [1], Hill and Dewynne [2], and Schneider [3] have considered many of the situations where rigorous mathematical analysis can be applied to obtain solutions. In section 3–3 we considered various configurations and boundary condition problems, and here we will extend the use of the solutions to those problems. Consider, for instance, the long cylindrical rod shown in figure (3–18), which has heat transfer along its surfaces. At the end of the rod the heat transfer will be radial (on the cylindrical surface) and axial. The heat transfer is therefore two-dimensional in cylindrical coordinates and requires the application of the conduction equation in the form

$$\frac{\partial T(r,z,t)}{\partial r^2} + \frac{1}{r}\frac{\partial T(r,z,t)}{\partial r} + \frac{\partial T(r,z,t)}{\partial z^2} = \frac{1}{\alpha}\frac{\partial T(r,z,t)}{\partial t}$$

Figure 3–18 Heat transfer at the end of a long cylindrical rod.

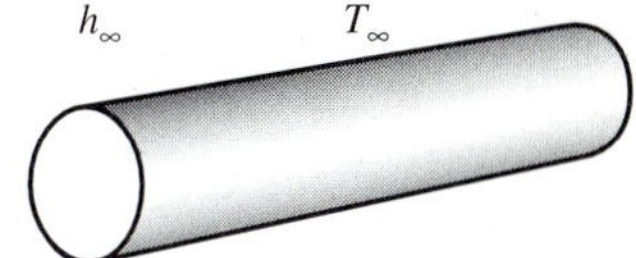

From the separation of variables techniques, the product solution can be written

$$\theta(r,z,t) = \theta_{\text{semi-infinite solid}}(z,t) \cdot \theta_{\text{infinite cylinder}}(r,t) \qquad \textbf{(3–54)}$$

If the boundary and initial conditions are identical to those for the semi-infinite solids and infinite cylinders from section 3–3, each of these two components of the solution can be determined. Here we have defined the function

$$\theta(r,z,t) = T(r,z,t) - T_\infty \qquad \textbf{(3–55)}$$

Many other regular geometric configurations having regular boundary and initial conditions suggest the application of other one-dimensional solutions as components of a product solution for multidimensional problems.

The thermal energy exchange between a system and its surrounding can also be extended from the one-dimensional case to those involving two or three dimensions. For instance, it can be shown (see Langston [7]) that in two-dimensional cases the thermal energy exchange function, Q/Q_0, defined in section 3–3 can be written

$$\frac{Q}{Q_0} = \left[\frac{Q}{Q_0}\right]_i + \left[\frac{Q}{Q_0}\right]_j \left(1 - \left[\frac{Q}{Q_0}\right]_i\right) \tag{3–56}$$

and for three dimensional cases,

$$\frac{Q}{Q_0} = \left[\frac{Q}{Q_0}\right]_i + \left[\frac{Q}{Q_0}\right]_j \left(1 - \left[\frac{Q}{Q_0}\right]_i\right) + \left[\frac{Q}{Q_0}\right]_k \left(1 - \left[\frac{Q}{Q_0}\right]_i\right)\left(1 - \left[\frac{Q}{Q_0}\right]_j\right) \tag{3–57}$$

where $[Q/Q_0]_i$, $[Q/Q_0]_j$, and $[Q/Q_0]_k$ denote the thermal energy exchange functions in the three coordinate directions of the coordinate system discussed in section 3–3. These terms can be interchanged in the two defining equations (3–56) and (3–57) if there would be a convenience in such a manipulation.

EXAMPLE 3–10 Sausages are taken out of a freezer and left to thaw until they are at 70°F. They are 1.5 in. in diameter and 7 in. long, and are boiled in water at 200°F where the convection heat transfer coefficient is 100 Btu/hr · ft² · °F. Estimate the center temperature of the sausages after 18 min of boiling, the temperature distribution through each sausage, and estimate their thermal energy increase. Assume that the sausages have the same properties as water and are cylindrical (see figure 3–19).

Figure 3–19 Boiling sausage.

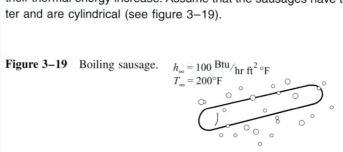

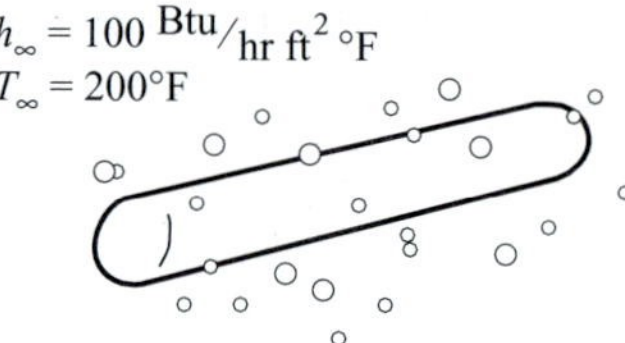

Solution The center temperature of each sausage may be determined from the equation

$$\frac{\theta_0}{\theta_i} = \left[\frac{\theta_0}{\theta_i}\right]_{\text{plate}} \cdot \left[\frac{\theta_0}{\theta_i}\right]_{\text{cylinder}} \tag{3–58}$$

We will use the following properties for the sausages (water at 70°F = 21.4°C):

$$\kappa = 0.64 \text{ W/m} \cdot \text{°C} = 0.37 \text{ Btu/hr} \cdot \text{ft} \cdot \text{°F}$$
$$c_p = 4180 \text{ J/kg} \cdot \text{°C} = 1.0 \text{ Btu/lbm} \cdot \text{°F}$$
$$\rho = 1000 \text{ kg/m}^3 = 62.4 \text{ lbm/ft}^3$$
$$\alpha = \kappa/\rho c_p$$

$$= \frac{\left(0.37 \dfrac{\text{Btu}}{\text{hr} \cdot \text{ft} \cdot \text{°F}}\right)}{\left(62.4 \dfrac{\text{lbm}}{\text{ft}^3}\right)\left(1.0 \dfrac{\text{Btu}}{\text{lbm} \cdot \text{°F}}\right)\left(3600 \dfrac{\text{s}}{\text{hr}}\right)} = 16 \times 10^{-7} \frac{\text{ft}^2}{\text{s}}$$

The dimensionless parameters for an infinite cylinder and an infinite plate can be determined:

$$\text{Bi}_{\text{cylinder}} = \frac{h_\infty r_0}{\kappa} = \frac{\left(100\,\dfrac{\text{Btu}}{\text{hr}\cdot\text{ft}^2\cdot{}^\circ\text{F}}\right)\left(\dfrac{0.75}{12}\,\text{ft}\right)}{\left(0.37\,\dfrac{\text{Btu}}{\text{hr}\cdot\text{ft}\cdot{}^\circ\text{F}}\right)} = 16.9$$

The inverse Bi is then 0.059. The Fourier Number is

$$\text{Fo}_{\text{cylinder}} = \frac{\alpha t}{r_0^2} = \frac{(16\times10^{-7}\,\text{ft}^2/\text{s})(1080\,\text{s})}{(0.0625\,\text{ft})^2} = 0.442$$

From figure 3–15a we read that $[\theta_0/\theta_i]_{\text{cylinder}} \approx 0.12$. For an infinite plate

$$\text{Bi}_{\text{plate}} = \frac{h_\infty L}{\kappa} = \frac{(100)(3.5/12)}{(0.37)} = 78.7$$

The inverse Bi is then 0.0126, and the plate Fourier Number is

$$\text{Fo} = \frac{\alpha t}{L^2} = \frac{(16\times10^{-7}\,\text{ft}^2/\text{s})(1080\,\text{s})}{(0.29167\,\text{ft})^2} = 0.0203$$

Since the Heisler Charts do not apply for values less than about 0.2, we must either use the analytic expression of equation (3–26) with the first five terms of the infinite series included in the summation or estimate a value (recognizing the rough extrapolation in figure 3–12a that must be made). From the Heisler Chart (figure 3–12a) we estimate

$$\left[\frac{\theta_0}{\theta_i}\right]_{\text{plate}} \approx 1.0$$

This result implies that the sausages are long enough that the center (axially) temperature is unchanged due to heat transfers at the ends. As a check we may use equation (3–26) and the parameters from appendix table A11–1. Interpolating between Bi = 50 and 100:

$$\begin{aligned}
\xi_1 &= 1.5487 & C_1 &= 1.2729\\
\xi_2 &= 4.6464 & C_2 &= -0.4234\\
\xi_3 &= 7.7445 & C_3 &= 0.2529\\
\xi_4 &= 10.8428 & C_4 &= -0.1794\\
\xi_5 &= 13.9421 & C_5 &= 0.1383
\end{aligned}$$

Then, using equation (3–26) with the first five terms:

$$\frac{\theta(x,t)}{\theta_i} = C_1 e^{-\xi_1^2\,\text{Fo}}\cos\xi_1\frac{x}{L} + C_2 e^{-\xi_2^2\,\text{Fo}}\cos\xi^2\frac{x}{L} + C_3 e^{-\xi_3^2\,\text{Fo}}\cos\xi_3\frac{x}{L} + C_4 e^{-\xi_4^2\,\text{Fo}}\cos\xi_4\frac{x}{L} + C_5 e^{-\xi_5^2\,\text{Fo}}\cos\xi_5\frac{x}{L}$$

Substituting values into this equation gives

$$\begin{aligned}
\frac{\theta(x,t)}{\theta_i} &= 1.2729e^{-1.5487^2(0.0203)}\cos 1.5487\frac{x}{L} - 0.4234e^{-4.6464^2(0.0203)}\cos(4.6464)\frac{x}{L}\\
&\quad + 0.2529e^{-7.7445^2(0.0203)}\cos 7.7445\frac{x}{L} - 0.1794e^{-10.8428^2(0.0203)}\cos 10.8428\frac{x}{L}\\
&\quad + 0.1383e^{-13.9421^2(0.0203)}\cos 13.9421\frac{x}{L}
\end{aligned}$$

At the axial center $x = 0$, so this equation reduces to

$$\frac{\theta_0}{\theta_i} = 1.2729e^{-0.049} - 0.4234e^{-0.438} + 0.2529e^{-1.217} - 0.1794e^{-2.387} + 0.1383e^{-3.946}$$
$$= 0.9999$$

This result is nearly identical to that obtained from the Heisler Chart (1.0), so the extra effort in using the analytic method is probably not necessary in this example. The center temperature is determined from

$$\frac{\theta_0}{\theta_i} = \left[\frac{\theta_0}{\theta_i}\right]_{\text{plate}} \cdot \left[\frac{\theta_0}{\theta_i}\right]_{\text{cylinder}} \approx (1.0)(0.12) = 0.12 = \frac{T_0 - T_\infty}{T_i - T_\infty}$$

Answer and

$$T_0 = 200°\text{F} + (0.12)(70°\text{F} - 200°\text{F}) = 184.4°\text{F}$$

Using figure 3–12b for an inverse Biot Number of 0.0126, the axial temperature distribution through each sausage at the centerline can be determined. These results are presented in table 3–2.

Table 3–2

x/L	x, in.	θ/θ_i	$T/(x, 0, 18 \text{ min})$, °F
0.0	0.0	1.00	184.4
0.2	0.7	0.95	185.18
0.4	1.4	0.82	187.2
0.6	2.1	0.59	190.7
0.8	2.8	0.33	194.85
0.9	3.15	0.17	197.35
1.0	3.5	0.02	199.69

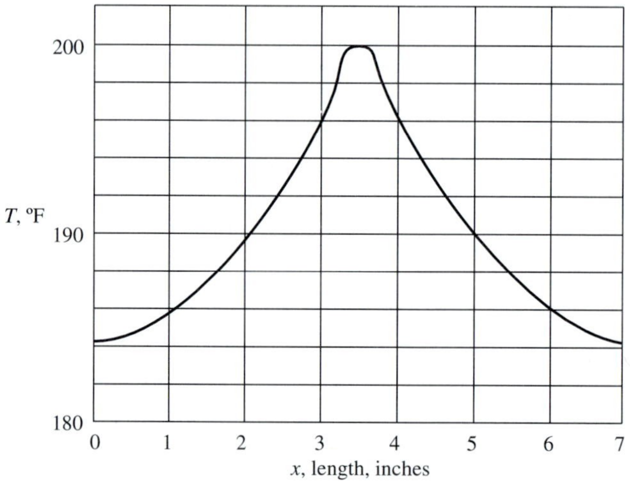

Similar distributions can be obtained in the axial and radial directions at any other location.

The thermal energy increase of each sausage can be found by using equation (3–35). We first compute the parameters $\text{Bi}^2_{\text{plate}}\,(\alpha t/L^2)$ and $\text{Bi}^2_{\text{cyl}}\,(\alpha t/r_0^2)$ and use the corresponding charts to determine the dimensionless temperature ratios. For the plate,

$$\text{Bi}^2_{\text{plate}}\left(\frac{\alpha t}{L^2}\right) = 126.05$$

and, from figure 3–12c,

$$Q/Q_0 \approx 0.20$$

For the cylinder,

$$\mathrm{Bi}^2_{\text{cylinder}}\left(\frac{\alpha t}{r_0^2}\right) = 129.95$$

and, from figure 3–15c,

$$Q/Q_0 \approx 0.38$$

From equation (3–56) we find

$$Q/Q_0 = (0.38) + (0.20)(1 - 0.38) = 0.504$$

so that

$$Q = \frac{Q}{Q_0} \cdot Q_0 = \frac{Q}{Q_0} \cdot \rho c_p V \theta_i$$

Answer

$$= (0.504)\left(62.4\,\frac{\text{lbm}}{\text{ft}^3}\right)\left(1\,\frac{\text{Btu}}{\text{lbm}\cdot{}^\circ\text{F}}\right)\left(\pi \cdot 0.0625^2\,\text{ft}^2 \cdot \frac{7}{12}\,\text{ft}\right)(130^\circ\text{F}) = 29.3\ \text{Btu}$$

In the following example we consider a three-dimensional transient heat-transfer problem that uses the analytic expressions for the analysis.

EXAMPLE 3–11 Bricks at 25°C are fired in a kiln that is at 400°C (see figure 3–20). The bricks are rectangular parallelepipeds of size 10 cm by 18 cm by 36 cm, and the convection heat transfer coefficient is 380 W/m^2 · °C. Determine the time required for the center of a brick to reach 350°C, and determine the temperature of a corner at this time. Use the following properties for brick:

$$\kappa = 18.5\ \text{W/m}\cdot{}^\circ\text{C}$$
$$\rho = 3000\ \text{kg/m}^3$$
$$c_p = 800\ \text{J/kg}\cdot{}^\circ\text{C}$$
$$\alpha = \kappa/\rho c_p = 7.7 \times 10^{-6}\ \text{m}^2/\text{s}$$

Figure 3–20 Brick fired in a kiln.

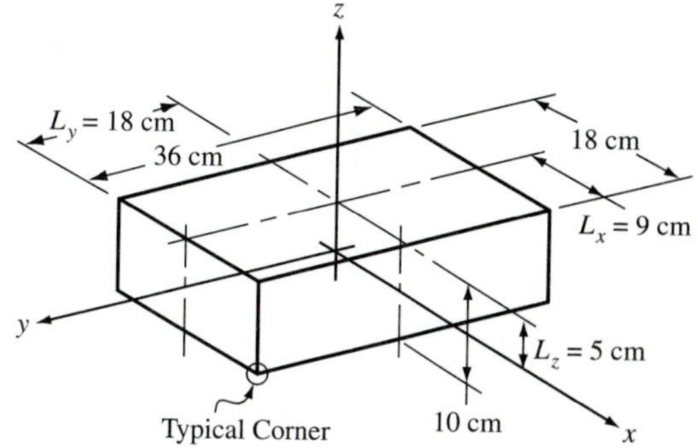

Solution In this example the temperature is known and the time is to be determined. We may use an equation of the form

$$\frac{\theta(x, y, z, t)}{\theta_i} = \left[\frac{\theta(x, t)}{\theta_i}\right]_{x,\text{plate}} \cdot \left[\frac{\theta(y, t)}{\theta_i}\right]_{y,\text{plate}} \cdot \left[\frac{\theta(z, t)}{\theta_i}\right]_{z,\text{plate}} \qquad \textbf{(3–59)}$$

where, for this problem, at the center of the brick

$$\frac{\theta(x, y, z, t)}{\theta_i} = \frac{\theta_0(t)}{\theta_i} = \left[\frac{\theta_0(t)}{\theta_i}\right]_{x,\text{plate}} \cdot \left[\frac{\theta_0(t)}{\theta_i}\right]_{y,\text{plate}} \cdot \left[\frac{\theta_0(t)}{\theta_i}\right]_{z,\text{plate}} = \frac{350°\text{C} - 400°\text{C}}{25°\text{C} - 400°\text{C}}$$
$$= 0.133$$

By a trial and error method using the graphical solutions for infinite plates as given in figure 3–12a, the time may be determined. Instead of this approach, we will use the analytic expressions for the infinite plate given by equation (3–26) and approximated by the first-term solution, given by equation (3–30). Since the solution of equation (3–29) applies to $\alpha t/L^2$ of greater than 0.2, we will need a final check to see if this criterion is satisfied. Using equation (3–30) we have, for the x direction,

$$\text{Bi}_x = hL_x/\kappa = (380\ \text{W/m}^2 \cdot °\text{C})(0.09\ \text{m})/(18.5\ \text{W/m} \cdot °\text{C}) = 1.8486$$

where $L_x = 9$ cm $= 0.09$ m. Then

$$\left[\frac{\theta_0(t)}{\theta_i}\right]_{x,\text{plate}} = C_1 e^{-\xi_1^2\,\text{Fo}} = C_1 e^{-\xi_1^2 \frac{\alpha t}{L^2}} = 1.170 e^{-10.37\times 10^{-4}\,t}$$

where C_1 and ξ_1 are interpolated from appendix table A–11–1, at $Bi_x = 1.8486$. Similarly, for the y direction we get

$$\text{Bi}_y = hL_y/\kappa = (380)(0.18)/(18.5) = 3.697$$

and, again interpolating from appendix table A–11–1 to determine C_1 and ξ_1 at $\text{Bi}_y = 3.697$, with $L_y = 18$ cm $= 0.18$ m,

$$\left[\frac{\theta_0(t)}{\theta_i}\right] = C_1 e^{-\xi_1^2\,\text{Fo}} = 1.223 e^{-3.67\times 10^{-4}\,t}$$

In the z direction we get, for $L_z = 5$ cm $= 0.05$ m

$$\text{Bi}_z = hL_z/\kappa = (380)(0.05)/(18.5) = 1.027$$

and

$$\left[\frac{\theta_0(t)}{\theta}\right]_{z,\text{plate}} = 1.121 e^{-23.1\times 10^{-4} t}$$

Combining these into equation (3–59) at $x = y = z = 0$ yields

$$\frac{\theta_0}{\theta_i} = 1.604 e^{-37.14\times 10^{-4}\,t} = 0.133$$

The time can then be directly calculated:

Answer

$$t = \frac{1\ \text{s}}{37.14 \times 10^{-4}} \ln\left(\frac{0.133}{1.604}\right) = 670.4\ \text{s} = 11.17\ \text{min}$$

The minimum Fourier Number is that for the infinite plate in the y direction:

$$\text{Fo} = \frac{\alpha t}{L_y^2} = \frac{(7.7 \times 10^{-6}\ \text{m}^2/\text{s})(670.4\ \text{s})}{(0.18\ \text{m})^2} = 0.16\ (\text{minimum})$$

The use of the approximation, equation (3–30), for $\text{Fo} \geq 0.2$ is not quite satisfied, so we should include the first five terms of the series solution for the general relationship of equation (3–26). For purposes of illustration, however, we will continue with this approximation. The corner temperature may be found from the equation

$$\frac{\theta_{\text{corner}}}{\theta_i} = \left(\frac{\theta_{x=L_x}}{\theta_i}\right) \cdot \left(\frac{\theta_{y=L_y}}{\theta_i}\right) \cdot \left(\frac{\theta_{z=L_z}}{\theta_i}\right) \tag{3–60}$$

since, at the corner $x = L_x$, $y = L_y$, and $z = L_z$. The individual terms can be determined using a graphical technique from figure 3–12b or from the analytic expressions, equation (3–31). We will use the analytic approach. In the x direction

$$\frac{\theta_{x=L_x}}{\theta_i} = \cos \xi_{x1} \frac{L_x}{L_x} = \cos 1.044 = \cos 59.84° = 0.5024$$

where ξ_{x1} is interpolated from appendix table A–11–1 at $\text{Bi}_x = 1.8486$. Similarly, in the y and the z directions we find that

$$\frac{\theta_{y=L_y}}{\theta_i} = \cos 1.24297 = \cos 71.2° = 0.322$$

and

$$\frac{\theta_{z=L_z}}{\theta_i} = \cos 0.866 = 0.648$$

Then

$$\frac{\theta_{\text{corner}}}{\theta_i} = (0.5024)(0.322)(0.648) = 0.105$$

and

Answer

$$T_{\text{corner}} = T_\infty + \frac{\theta_{\text{corner}}}{\theta_i}(T_i - T_\infty) = 400°\text{C} + (0.105)(350°\text{C} - 400°\text{C}) = 395.8°\text{C}$$

In the final example of this section we compare solutions using the infinite plate/infinite cylinder combination to solutions using the semi-infinite solid/infinite cylinder combination.

EXAMPLE 3–12 Long carbon steel bars 4-in. in diameter by 8 ft long are placed in a uniform stack to cool, as shown in figure 3–21. The bars are 400°F when stacked, and the ends are exposed to air at 100°F with a convection coefficient of 40 Btu/hr · ft^2 · °F. The cylindrical surfaces may be assumed to be exposed to an average air temperature of 100°F with convection coefficient of 35 Btu/hr · ft^2 · °F. Estimate the temperature 6 in. in from either end of the bars after $\frac{1}{2}$ hr of cooling.

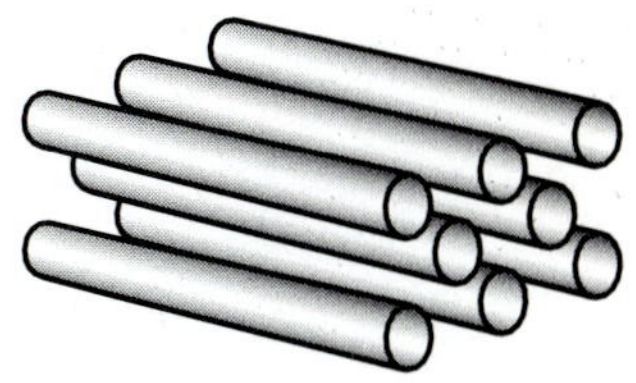

Figure 3–21 Example 3–12, carbon steel bars stacked for cooling.

Solution

The properties of the carbon steel are taken to be

$$\begin{aligned} \kappa &= 60.5\ \text{W/m} \cdot °\text{C} = 34.96\ \text{Btu/hr} \cdot \text{ft} \cdot °\text{F} \\ c_p &= 434\ \text{J/kg} \cdot °\text{C} = 0.1037\ \text{Btu/lbm} \cdot °\text{F} \\ \rho &= 7854\ \text{kg/m}^3 = 490\ \text{lbm/ft}^3 \\ \alpha &= 17.7 \times 10^{-6}\ \text{m}^2/\text{s} = 1.9 \times 10^{-4}\ \text{ft}^2/\text{s} \end{aligned}$$

The temperature 6 in. in from the end may be estimated by using the product solution of the infinite plate and infinite cylinder combination:

$$\frac{\theta(r, z, t)}{\theta_i} = \left[\frac{\theta_0(t)}{\theta_i}\right]_{\text{cylinder}} \cdot \left[\frac{\theta_0(t)}{\theta_i}\right]_{\text{plate}} \cdot \left[\frac{\theta(r)}{\theta_0}\right]_{\text{cylinder}} \cdot \left[\frac{\theta(z)}{\theta_i}\right]_{\text{plate}} \tag{3–61}$$

Since the bars are long, they may be approximated as semi-infinite solids and the product solution

$$\frac{\theta(r, z, t)}{\theta_i} = \left[\frac{\theta_0(t)}{\theta_i}\right]_{\text{cylinder}} \cdot \left[\frac{\theta(t)}{\theta_i}\right]_{\text{semi-inf., } z=6 \text{ in.}} \cdot \left[\frac{\theta(r)}{\theta_0}\right]_{\text{cylinder}} \cdot \left[\frac{\theta(z)}{\theta}\right]_{\text{semi-inf., } z=6 \text{ in.}} \tag{3–62}$$

can also be used. The Biot Number for the infinite cylinder is

$$\text{Bi}_{\text{cyl}} = hr_0/\kappa = (35\ \text{Btu/hr} \cdot \text{ft}^2 \cdot {}^\circ\text{F})(2/12\ \text{ft})/(34.96\ \text{Btu/hr} \cdot \text{ft} \cdot {}^\circ\text{F}) = 0.167$$

and its inverse is 5.99. The Fourier Number,

$$\text{Fo}_c = \frac{\alpha t}{r_0^2} = \frac{(1.9 \times 10^{-4}\ \text{ft}^2/\text{s})}{\left(\dfrac{2}{12}\ \text{ft}\right)^2} = 12.3$$

and from figure 3–15a,

$$[\theta_0/\theta_i]_{\text{inf. cyl}} \approx 0.02$$

Since we are determining the centerline temperature, we have $[\theta(z)/\theta_0]_{\text{inf. plt}} = 1.0$. For the method using the infinite plate in combination with the infinite cylinder, we find for the Biot Number,

$$\text{Bi} = \frac{h_\infty L}{\kappa} = \frac{\left(40\ \dfrac{\text{Btu}}{\text{hr} \cdot \text{ft}^2 \cdot {}^\circ\text{F}}\right)(4\ \text{ft})}{\left(34.96\ \dfrac{\text{Btu}}{\text{hr} \cdot \text{ft} \cdot {}^\circ\text{F}}\right)} = 4.58$$

and its inverse is 0.22. The Fourier Number, Fo, is

$$\text{Fo} = \frac{\alpha t}{L^2} = \frac{(1.9 \times 10^{-4})(1800)}{(4)^2} = 0.0214$$

Since the Fourier Number is less than about 0.2, we should use the analytic series solution with at least the first five terms of the infinite series. For purposes of illustration, however, we will use the approximation using the Heisler Charts. From figure 3–12a,

$$[\theta_0/\theta_i]_{\text{inf. plt}} \approx 0.95$$

Also, from figure 3–12b, where $x/L = 3.5\ \text{ft}/4\ \text{ft} = 0.875$

$$[\theta(x)/\theta_0]_{\text{inf. plt}} \approx 0.44$$

Using equation (3–61),

$$\theta/\theta_i = (0.02)(0.95)(1.0)(0.44) = 0.00836$$

so that the centerline temperature of each bar 6 in. in from either end after one-half hour of cooling is

Solution

$$T = 100^\circ\text{F} + (0.00836)(400^\circ\text{F} - 100^\circ\text{F}) = 102.5^\circ\text{F}$$

We may also obtain a similar result if we use the semi-infinite solid with convective heat transfer at the boundary in combination with the infinite cylinder. For the semi-infinite solid, we use the results given in figure 3–10 with the parameters

$$h_\infty \frac{\sqrt{\alpha t}}{\kappa} = \left(40 \frac{\text{Btu}}{\text{hr} \cdot \text{ft}^2 \cdot {}^\circ\text{F}}\right) \frac{\sqrt{\left(1.9 \times 10^{-4} \frac{\text{ft}^2}{\text{s}}\right)(1800 \text{ s})}}{34.96 \frac{\text{Btu}}{\text{hr} \cdot \text{ft} \cdot {}^\circ\text{F}}} = 0.669$$

and

$$\frac{x}{2\sqrt{\alpha t}} = \frac{(0.5 \text{ ft})}{2\sqrt{(1.9 \times 10^{-4})(1800)}} = 0.427$$

Then, from figure 3–10,

$$\frac{T - T_\infty}{T_i - T_\infty} \approx 0.19$$

Using equation (3–62), noting that the fourth term is included in the second term,

$$\theta/\theta_i = (0.02)(0.19)(1.0) = 0.0038$$

Answer and

$$T = 100^\circ\text{F} + (0.0038)(400^\circ\text{F} - 100^\circ\text{F}) = 101.14^\circ\text{F}$$

which is in agreement with the previous answer using the infinite plate and cylinder.

The analysis of many of the two- and three-dimensional transient heat-transfer problems can proceed in an orderly fashion by writing equations for the combinations of temperature functions as demonstrated by these last three examples and by equations such as (3–54), (3–58), (3–59), (3–60), (3–61), and (3–62).

3–5 APPLICATIONS TO SOLIDS

Transient conduction heat transfer occurs often in solids and should be an important part of any complete analysis of engineering designs. Applications to specific situations are many and varied, and the reader can think about some of them. Here we consider three specific problems involving transient conduction heat transfer: (1) cooling rates in metals to control metallurgical properties; (2) earth temperatures affected by daily or seasonal temperature changes to study, for instance, thermal blankets to provide for more efficient and acceptable earth shelter homes; and (3) line sources of thermal energy to predict thermal conductivity and to predict behavior of buried electrical power cables.

Cooling Rates in Metals

Metals, iron and steel in particular, can exhibit large differences in material properties if they are heated and subsequently cooled. The hardness, tensile strength, and ductility of metals seem to be affected by the cooling rate of a metal. That is, if the metal is heated to a high temperature and then cooled at a prescribed rate, these mechanical properties will be different. In figure 3–22 is shown hardenability information of five specific steels. The purpose of presenting the figure is to demonstrate that the cooling rate affects the hardness of steel in a profound way. The cooling rate has usually been determined experimentally by measuring temperatures with thermocouples or thermistors embedded in material samples. A commonly used test for the purposes of finding cooling rate versus hardness, ductility, or tensile strength of steels is the end-quench (or Jominy) test.

In figure 3–23 is shown the mechanical arrangement for the Jominy test. A Jominy bar of 1-in. (25-mm) diameter and 4-in. (100-mm) length with a flange is heated to 1300°F (700°C) and placed in a vertical position as shown in figure 3–23, and a stream of water at a temperature of 75°F (24°C) is directed to the bottom of the Jominy bar. From our previous discussions of heat transfer you can see that the cooling rate at the quenched end

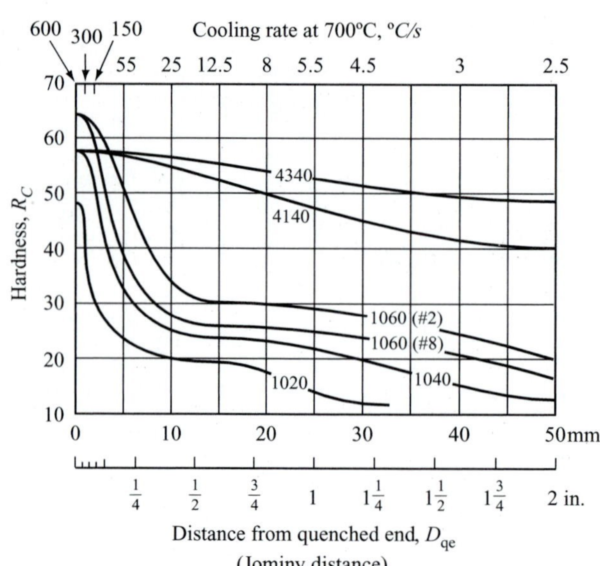

Figure 3–22 Hardenability of steels as affected by the cooling rate and the type of steel. The distance from the quenched end is indicated in figure 3–23.

Steel Number	Components in Steel Other Than Iron (Fe)					Grain Size
	Carbon (C)	Manganese (Mn)	Nickel (Ni)	Chromium (Cr)	Molybdenum (Mo)	
1020	0.20	0.90	0.01	—	—	8
1040	0.40	0.89	0.01	0.01	—	8
1060	0.60	0.81	0.02	—	—	2 and 8
4140	0.40	0.79	0.01	1.01	0.22	8
4340	0.40	0.75	1.71	0.77	0.30	8

will be greatest and will diminish away from the quenched end. The distance from the quenched end is called the Jominy distance, D_{qe}, and is used as a parameter in figures 3–22 and 3–23. Notice that the information in these two figures suggests that the cooling rate is a function of the Jominy distance. We know that the cooling rate must also be a function of temperature and time from beginning of quench. We will use the results from the previous discussions to predict a cooling rate history at various locations on a standard Jominy bar.

EXAMPLE 3–13 Predict the cooling rate at the surface of a Jominy bar at six locations, $D_{qe} = 1$ mm, 5 mm, 10 mm, 20 mm, 30 mm, and 50 mm for a Jominy test. Assume $h = 10\ \text{W/m}^2 \cdot {}^\circ\text{C}$ along the cylindrical surface of the bar exposed to air at 24°C and that the water keeps the quenched end of the bar at 24°C for the complete test.

Solution We will predict the temperatures of the surface of the Jominy bar by using the semi-infinite solid with constant surface temperature in combination with the solution for an infinite cylinder. The solution for the semi-infinite solid with constant boundary temperature is given by equation (3–17),

$$\frac{\theta_{\text{semi-inf. solid}}}{\theta_i} = \text{erf}\left(\frac{x}{2\sqrt{\alpha t}}\right) = \frac{T(x, t) - T_B}{T_i - T_B} \qquad \textbf{(3–17, repeated)}$$

Figure 3–23 Jominy end-quench test.

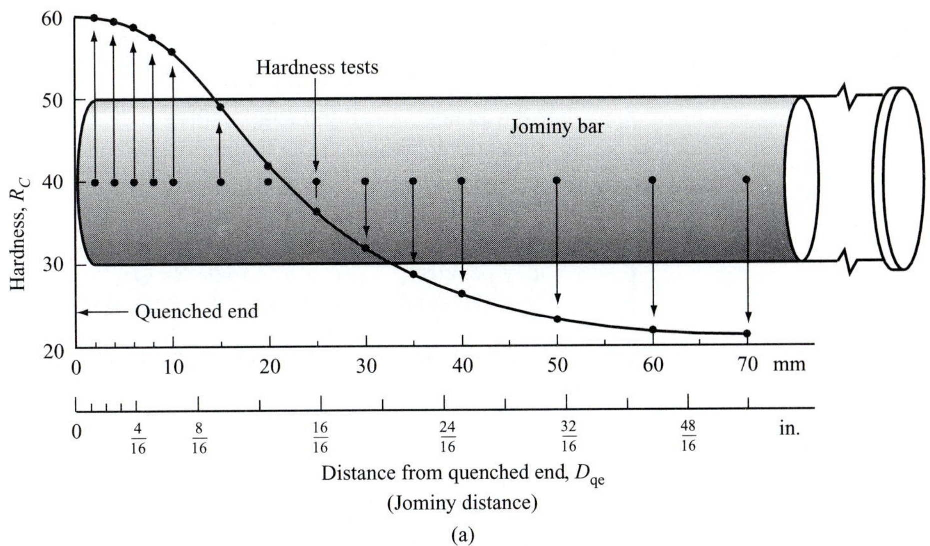

(a)

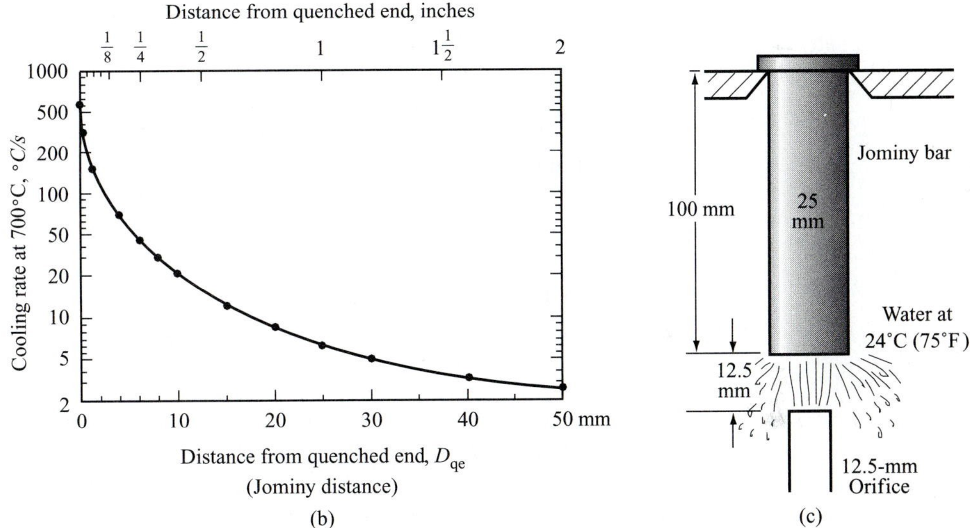

(b) (c)

For the infinite cylinder we may use equation (3–41),

$$\frac{\theta(r, t)}{\theta_i} = C_1 e^{-\beta_1^2 \, \mathrm{Fo}_c} J_0\left(\beta_1 \frac{r}{r_0}\right) \qquad \textbf{(3–41, repeated)}$$

where $Fo_c = \alpha t / r_0^2$ and is required to be greater than 0.2. The surface cooling rates require that $r = r_0$. For this problem we assume the properties of steel are $\rho = 7854\ \mathrm{kg/m^3}$, $c_p = 434\ \mathrm{J/kg \cdot {}^\circ C}$, $\kappa = 60.5\ \mathrm{W/m \cdot {}^\circ C}$, and $\alpha = 17.7 \times 10^{-6}\ \mathrm{m^2/s}$. The Fourier number, Fo_c, is then

$$\mathrm{Fo}_c = (17.7 \times 10^{-6})t/(0.0125\ \mathrm{m})^2 = 1.416t > 0.2$$

Thus, time is restricted to values greater than $0.2/1.416 = 0.141$ s.

The Biot Number for the infinite cylinder portion of this problem,

$$\text{Bi} = hr_0/\kappa = (10\ \text{W/m}^2 \cdot {}^\circ\text{C})(0.0125\ \text{m})/(60.5\ \text{W/m} \cdot {}^\circ\text{C}) = 0.002$$

so that the lumped heat capacity method may be used, rather than the infinite cylinder. Thus, we can use equation (3–10) and write

$$\frac{\theta(t)}{\theta_i} = e^{-t/t_c}$$

where $t_c = \rho c_p V/hA_s = \rho(c_p \cdot \pi r_0^2 L)/(h \cdot 2\pi r_0 L) = \rho c_p r_0/2h$
$= (7854\ \text{kg/m}^3)(434\ \text{J/kg} \cdot {}^\circ\text{C})(0.0125\ \text{m})/(2)(10\ \text{J/m}^2 \cdot {}^\circ\text{C s}) = 2130\ \text{s}$

Combining the semi-infinite solid with the lumped heat capacity for an infinite cylinder gives

$$\frac{T(x,t) - T_B}{T_i - T_B} = e^{-t/t_c}\text{erf}\left(\frac{x}{2\sqrt{\alpha t}}\right) \tag{3–63}$$

Notice that the temperature is a function of x and time, t, only (not r) because of the lumped heat capacity assumption. Thus, there is no significant difference in the bar temperature in a radial direction. The cooling rate is the first derivative with respect to time of equation (3–63),

$$\frac{\partial T(x,t)}{\partial t} = (T_i - T_B)\left[-e^{-t/t_c}\text{erf}\left(\frac{x}{2\sqrt{\alpha t}}\right) + e^{-t/t_c}\frac{\partial}{\partial t}\text{erf}\left(\frac{x}{2\sqrt{\alpha t}}\right)\right] \tag{3–64}$$

From appendix table A–9, $\partial/\partial z\ \text{erf}\ z = 2\ e^{-z^2}/\sqrt{\pi}$, or

$$\frac{\partial}{\partial t}\text{erf}(z) = \frac{\partial}{\partial z}\frac{\partial}{\partial t}\text{erf}(z)$$

so that the cooling rate is

$$\frac{\partial T(x,t)}{\partial t} = (T_i - T_\infty)e^{-t/t_c}\left[-\frac{1}{t_c}\text{erf}\left(\frac{x}{2\sqrt{\alpha t}}\right) - \frac{x}{2t^{3/2}\sqrt{\alpha\pi}}e^{-x^2/4\alpha t}\right] \tag{3–65}$$

Using $T_i = 700^\circ\text{C}$, $T_\infty = 24^\circ\text{C}$, and $\alpha = 17.7 \times 10^{-6}\ \text{m}^2/\text{s}$ with the variables x and t gives results listed in table 3–3. Also, the maximum cooling rates are included in this table, and

Table 3–3 Cooling Rate for Jominy Bar (°C/s)

Time, t (s)	Jominy Distance, D_{qe} (m)						
	0.001	**0.005**	**0.010**	**0.020**	**0.030**	**0.050**	**Maximum**
0.25	342.8	442.4	13.03	0.317	0.317	0.317	655.9 (at 2.975 mm)
0.50	124.6	316.8	76.3	0.349	0.317	0.317	327.4 (at 4.208 mm)
1.00	44.71	159.3	110.6	3.494	0.317	0.317	163.7 (at 5.95 mm)
2.00	15.93	67.3	79.3	19.24	1.128	0.317	81.93 (at 8.422 mm)
4.00	5.657	25.99	39.9	27.84	7.387	0.317	41.03 (at 11.92 mm)
8.00	2.0087	9.61	16.87	19.91	12.46	1.512	20.58 (at 16.89 mm)
15.00	0.7814	3.84	7.13	10.82	10.21	3.983	11.042 (at 23.21 mm)
30.00	0.5098	2.52	4.869	8.459	10.06	8.008	10.142 (at 33.0 mm)
60.00	0.0993	0.498	0.9795	1.8268	2.45	2.850	2.850 (at 47.4 mm)

this data as obtained by taking the first derivative of the cooling rate with respect to the Jominy distance and setting this result to zero:

$$\frac{\partial}{\partial x}\frac{\partial T}{\partial t} = (T_i - T_B)e^{-t/t_c}e^{-x^2/4\alpha t}\left(\frac{-1}{t_c\sqrt{\alpha\pi t}} - \frac{1}{2t^{3/2}\sqrt{\alpha\pi}} + \frac{x}{4t^{5/2}\alpha^{3/2}\sqrt{\pi}}\right) = 0 \qquad \textbf{(3–66)}$$

Since the term $(T_i - T_\infty)e^{-t/t_c}$ is not likely to be zero, the bracket term must be zero,

$$-\frac{1}{t_c\sqrt{\alpha\pi t}} - \frac{1}{2t^{3/2}\sqrt{\alpha\pi}} + \frac{x}{4t^{5/2}\alpha^{3/2}\sqrt{\pi}} = 0 \qquad \textbf{(3–67)}$$

with the resulting criterion, where $\alpha = 17.7 \times 10^{-6}$ and the time constant is 2130 s, that

$$\frac{1}{2130} + \frac{1}{2t} = \frac{x^2}{4} \times 17.7 \times 10^{-6} \qquad \textbf{(3–68)}$$

The Jominy distance at maximum cooling rate for any time can be determined from equation (3–68) and is indicated in the last column of table 3–3. The cooling rates were then obtained at these Jominy distances.

The cooling rates presented in table 3–3 are also given in graphical form in figure 3–24, and a comparison of these results to the data presented in figures 3–22 and 3–23 shows reasonable close agreement, particularly if the maximum cooling rates are used as the comparison. Unfortunately, the information about the cooling rate curve shown in figure 3–23 does not indicate that the curve describes cooling rates at different times; rather, it seems to imply that the cooling rate decreases as the Jominy distance increases. In the actual solution given in table 3–3 and in figure 3–24, the wave nature of the heat transfer is reflected in the behavior of the cooling rate. Notice that at any one particular location or Jominy distance the cooling rate increases, reaches a maximum, and then decreases with time. Also, the cooling rate is zero at the quenched end because of the boundary condition that the temperature be equal to the water temperature for all time after $t = 0$. The cooling rate at an instant in time in the Jominy bar increases from the quenched end, reaches a maximum, and then decreases again. It is not clear how the Jominy test can be extended in a quantitative manner beyond its intended function as a comparative test for iron and steel hardenability.

Cooling rates are also of concern in other ways in metallurgical processes. For instance, the casting of metals in sand molds can result in varied cooling rates in the casting. In centrifugal casting of cast iron or steel, where molten material is poured into a cylindrical tube, capped at both ends, and spun at a high rate of speed, the inside and outside diameters of the final casting, a tubing, are cooled at significantly different rates, and the metallurgical properties are noticeably different as well. Forging of metals in dies and welding of two or more separate parts are also processes that can result in varied metallurgical properties of the final assembly and attributable to varied cooling rates. In many of these and other applications the boundary and/or initial conditions are such that analysis must involve many approximations. In addition, cooling rates for the manufacture of glass and certain plastics can profoundly affect the final products.

Seasonal Earth Temperature Changes

The earth is subjected to changing seasons, changing weather patterns, and daily changes due to the sun's rise and fall each day. All of these variations tend to change the temperature of the earth itself. The earth's temperature is most affected near the surface and remains less affected at greater depths, which is expected and has been experimentally verified in a large number of situations. As a first approximation to provide some insight into this behavior and to analyze some simple engineering design situations, let us consider

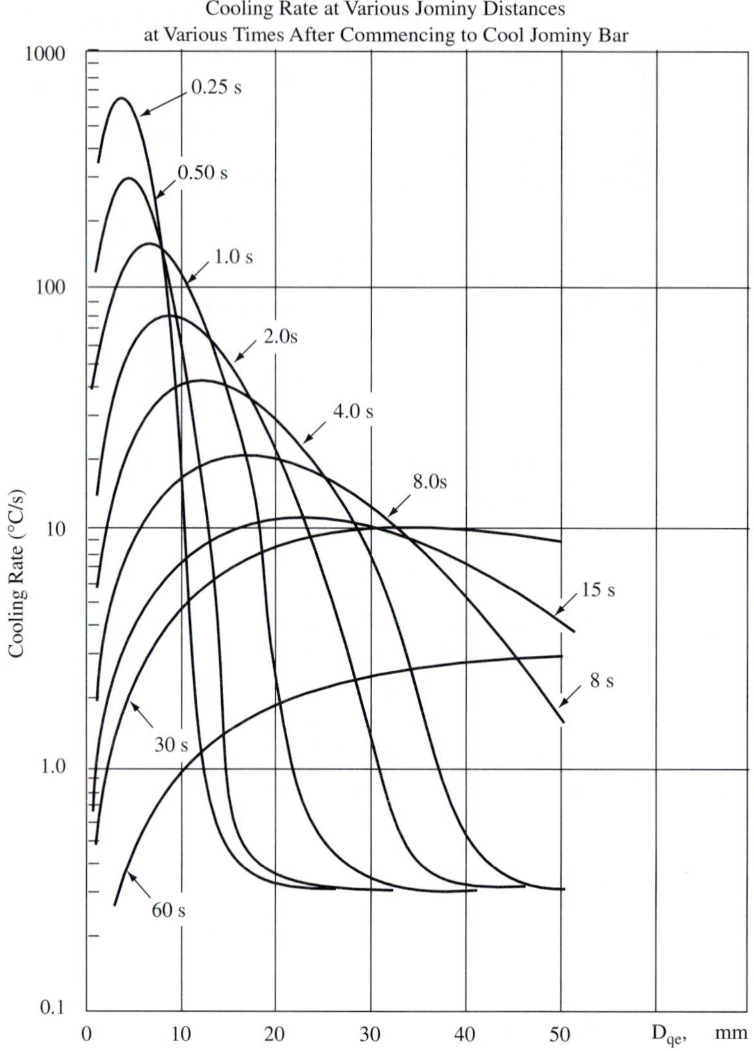

Figure 3–24 Cooling rates in a Jominy bar.

the earth to be a homogeneous, semi-infinite solid having constant thermal conductivity, density, and specific heat. Let us also assume that the surface temperature varies as a periodic function. We frame the mathematical problem for this situation as

$$\frac{\partial^2 T(x, t)}{\partial x^2} = \frac{1}{\alpha}\frac{\partial T(x, t)}{\partial t} \qquad 0 < x < \infty \tag{3–69}$$

B.C. 1 $\quad T(0, t) = A\cos(\omega t) \quad$ at $x = 0$

B.C. 2 $\quad \dfrac{\partial T}{\partial t} \to 0 \quad$ as $x \to \infty$

I.C. $\quad T(x, 0) = 0 \quad$ at $t = 0$

Carslaw and Jaeger [1] presented a derivation and discussion of this problem and showed that a solution may be written as

$$T(x, t) = Ae^{-x\sqrt{\frac{\omega}{2\alpha}}}\cos\left(\omega t - x\sqrt{\frac{\omega}{2\alpha}}\right) \tag{3–70}$$

for the temperature at any location in the earth. In equation (3–69) and the solution (3–70), the time t is to be interpreted as measured from noon if the variation is a daily variation due to the rise and fall of the sun ($t = 0$ at noon). That is, the surface temperature of the earth should be greatest at solar noon, when the sun is directly overhead. For the case where seasonal variations are being studied, the time t should be measured from some midsummer day, such as July 24 or so. The time would most conveniently be expressed in the units of seconds or minutes if daily variations are considered, but units of days would be appropriate for seasonal changes. Climatic changes have been much studied, but these phenomena do not lend themselves to straightforward scientific and mathematical analysis and are beyond the purposes of this textbook.

From the results given by equation (3–70) it can be seen that ω represents a frequency and for one cycle we may write $\omega t = 2\pi$. Then the period, or time for one complete cycle is

$$t = 2\pi/\omega \tag{3–71}$$

For daily changes the frequency, ω, would be $2\pi/24$ hr or 1 cycle per day. For seasonal changes ω would be expressed as $2\pi/365$ days or 1 cycle per year. In general, the seasonal changes and daily changes in temperature will be of a similar magnitude, although the seasonal temperature fluctuations are the greater. In B.C. 1 of equation (3–69), the magnitude or amplitude of the temperature changes is denoted as A. From the solution of equation (3–70) it can be seen that the amplitude of the temperature changes in the earth decreases with depth, since the term $e^{-x\sqrt{\frac{\omega}{2\alpha}}}$ becomes progressively smaller with increasing values of x, the depth. Also, the amplitude diminishes as the frequency is greater; that is, the daily fluctuations in temperatures occur approximately 365 times more often than do the yearly fluctuations. As a consequence, the seasonal temperature changes cause earth temperatures to change to a greater depth than do the daily changes by an amount of about 19, since

$$\frac{\sqrt{\omega_{\text{daily}}}}{\sqrt{\omega_{\text{seasonal}}}} = \sqrt{365} \approx 19$$

Notice that the temperature "waves" predicted by equation (3–70) at greater depths are further out of phase with the surface conditions. This can be seen by the magnitude of the phase-shift term $x\sqrt{\dfrac{\omega}{2\alpha}}$ in the argument for the cosine function of equation (3–70). Again, the daily oscillations of temperature will be further out of phase than the yearly oscillations.

EXAMPLE 3–14

Estimate the depth below the earth's surface at which the temperature will vary $\pm 2°C$ if the seasonal changes are given by the equation

$$T(0, t) = 20°\text{C} \cos \omega t$$

Assume the earth's average surface temperature is zero and that $t = 0$ on July 24. Also predict when the temperature is maximized and when it is minimized at this depth.

Solution

From equations (3–69) and (3–70), we can identify the terms $A = 20°C$ and $\omega = 2\pi/365$ rad/day, and we assume for the earth $\alpha = 1.01 \times 10^{-6}$ m^2/s. Using equation (3–70) we may directly solve for the depth, x, where

$$T(x, t) = 2°\text{C} = 20°\text{C}\, e^{-x\sqrt{\frac{\omega}{2\alpha}}}$$

Solving for x,

Answer

$$x = \ln\left[\frac{2°\text{C}}{20°\text{C}}\right]\sqrt{\frac{2\alpha}{\omega}} = \ln(0.1)\sqrt{\frac{2 \times 1.01 \times 10^{-6}\,\dfrac{\text{m}^2}{\text{s}}}{2\pi\,\dfrac{1\text{ cycle}}{365\text{ days} \times 24\text{ hr/day} \times 3600\text{ s/hr}}}} = 7.33\text{ m}$$

so that, at a depth of 7.33 m the yearly or seasonal temperature change will be ±2°C when the surface temperature change is ±20°C. The day at which the maximum temperature can be expected to occur at the 7.33-m depth is predicted from equation (3–70). Using, for the maximum temperature,

$$\cos\left(\omega t_{max} - x\sqrt{\frac{\omega}{2\alpha}}\right) = 1$$

Then

$$\omega t_{max} - x\sqrt{\frac{\omega}{2\alpha}} = 0$$

and

$$t_{max} = \frac{x}{\sqrt{2\alpha\omega}} = 133.8\text{ days}$$

Using a calendar, we see that the earth temperature (for this particular example) maximizes 133.8 days after July 24, or roughly in the evening of December 5. The minimum temperature occurs when

$$\cos(\omega t_{min} - x\sqrt{\omega/2\alpha}) = -1$$

so that

$$(\omega t_{min} - x\sqrt{\omega/2\alpha}) = \pi$$

or

$$t_{min} = 316.6\text{ days}$$

This will roughly occur in the early afternoon of June 6.

This example should serve to show the qualitative behavior of the earth's cooling and heating. From equations (3–69) and (3–70) we may generalize by normalizing the temperature response. From equation (3–70) we write $T(x, t)A = \xi = x(\sqrt{\omega/2\alpha}$ and then

$$f(\xi, t) = e^{-\xi}\cos(t - \xi) \tag{3–72}$$

to demonstrate the time lag and amplitude decrease of the temperature waves. The function $f(\xi t)$ is plotted against t for various values of ξ in figure 3–25, and this data demonstrates how the relative magnitude of the amplitude declines from a value of 1.0 at the earth's surface, $\xi = 0$, to approximately 0.1 when $\xi = 3\pi/4$. The phase lag can also be visualized from the data of figure 3–24.

The actual heat transfers that occur in the earth are much more complex than the simple approximations we have been discussing here. The earth's thermal conductivity, density, and specific heat vary with location, time, and temperature. There are also energy sources within the earth, so there is a prevailing energy flow as heat outward from the earth's core to the surface. Climatic conditions are so irregular that any simple harmonic function as suggested by B.C. 1 of equation (3–69) is far from describing these phenomena, and even the daily and seasonal variations can only be roughly approximated by mathematical functions. Earth that becomes saturated with moisture or other liquids and areas where fissures penetrate deeply into the earth can create conditions where convection heat transfer may be as great as conduction. Earthquakes, floods, and hurricanes are only some of the happenings that can complicate any analysis of heat transfer near or on the earth's

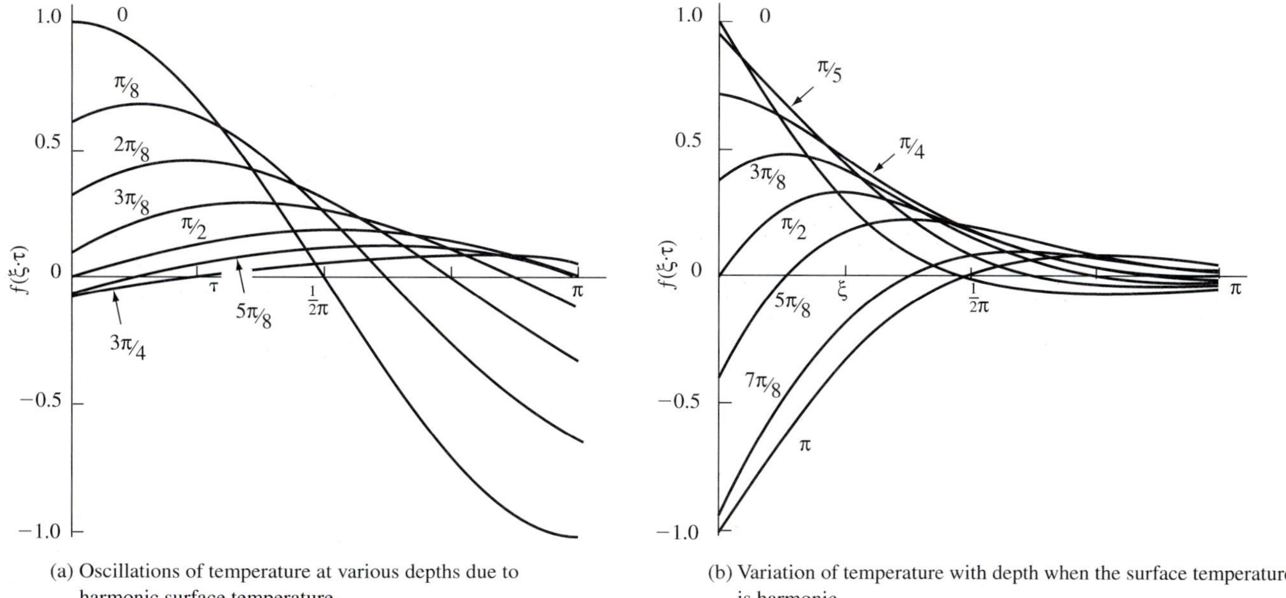

(a) Oscillations of temperature at various depths due to harmonic surface temperature

(b) Variation of temperature with depth when the surface temperature is harmonic

Figure 3–25 Temperature distribution in a semi-infinite medium subjected to an oscillating surface temperature and described by equation 3–72.

surface. The engineer and technologist should recognize, however, that the qualitative behavior of the materials is often the best available information in a design or application.

A particular example of this situation is the design of earth shelter buildings. These structures are intended to utilize the moderating effects of the earth to reduce the amount of cooling required in the warm seasons and to reduce the amount of heating necessary in the cold seasons. The earth, as we have already discussed, remains at a constant 45°F (7°C) at a depth of about 20 ft (6 m) below the surface. By proper design, analysis, and care in construction of such buildings much of the energy demands for heating and cooling are provided by the earth and substantial reductions in operating costs can be realized. One particular adaptation of the concepts we have considered is shown in figure 3–26. The structure is surrounded by ground except, in this particular example, one open side. For the normal earth shelter structure the earth moderates the inside temperature of the structure so that the ground temperature is roughly 2 to 3 months out of phase with the seasonal temperatures.

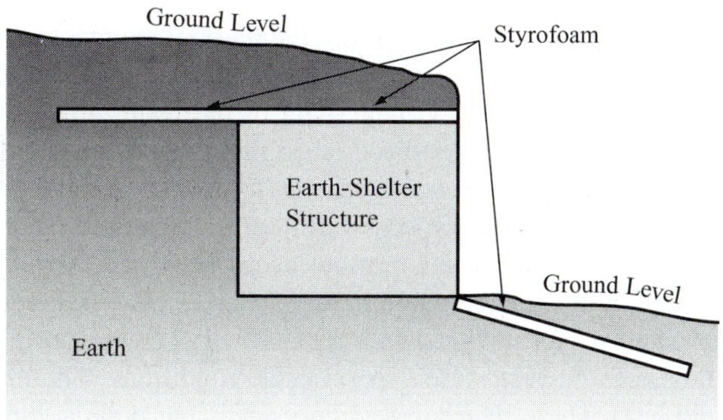

Figure 3–26 Schematic of thermal blanket for earth-shelter structure.

The configuration suggested in figure 3–26 includes a thermal blanket or layer of insulation buried 0.6 m (2 ft) deep, and this layer reduces the heat transfer more than does ground without insulation. As a result, the earth's temperature variation lags the seasonal changes by even more than is expected. With proper design and insulation, the earth temperature at 2- or 3-m depths can lag or be out of phase with surface variations by as much as 6 months. This would represent the ideal situation; coolest ground temperatures with warmest surface temperatures and warmest ground with coolest surface.

Line Source of Thermal Energy in an Infinite Solid

We consider another example of transient heat transfer involving an infinite solid at a uniform temperature, T_i, throughout. Suddenly a line source of energy at temperature T_0 is inserted into the solid as shown in figure 3–27. The energy is provided at a constant rate $\dot{e}_{\text{gen}}$ (power/unit length), and the temperature of the solid begins to increase nearest the line source. Since the line source of thermal energy is a line, there is no mass associated with the source. Also, the problem of a line sink of thermal energy, where the line is at a lower temperature than the solid, is mathematically the same problem.

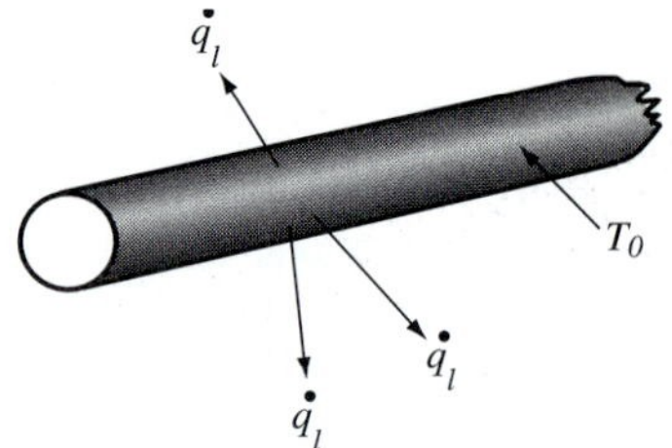

Figure 3–27 Line source of thermal energy in an infinite solid.

Two important applications that we will consider, however, involve sources of thermal energy and so that is how we will continue the development. These two applications are the buried electrical power cable and the thermal probe device. Buried electrical power cables conduct electricity and thereby represent a line source of power from the dissipated power, $\mathcal{I}^2\mathcal{R}$. The ground surrounding these cables must conduct heat away from the cable or the temperature of the cable will become excessively high. As we have discussed earlier, increasing the temperature of a conducting metal often increases the electrical resistance of that metal and so, if a power cable becomes hot from insufficient ground cooling, its resistance rises, increases the electrical resistance, and then increases the dissipated power. The result is an even greater rise in the cable temperature until finally the cable may melt. It is therefore important that a steady state condition be achieved in buried electrical power cables.

A thermal probe device, shown in figure 3–28, is a tube encasing one or more thermocouples, thermistors, or other temperature sensors, and a source of thermal energy. The source is most often a resistance electric heater. The probe is inserted into a solid or other material, the power is supplied to the thermal source, and simultaneously the temperature sensor data is recorded. If the temperatures are measured continuously during some heating period, the thermal conductivity of the material surrounding the probe can be determined. We saw in earlier discussions that the thermal conductivity may be experimentally determined by steady state methods. Using the thermal probe, data can be collected in a transient heat-transfer process and in much less time. Shorter test times are not only more convenient but can result in more accurate data. Often materials contain moisture or air pockets, or are amalgamated materials, and convection heat transfer and/or mass transfers can occur if thermal gradients are applied over extended time periods, such as would be required for steady state experimental conditions. Thermal probes can often function in rapid tests where thermal gradients occur but data is gotten quickly before other effects can begin to complicate the process.

Figure 3–28 Thermal probe device.

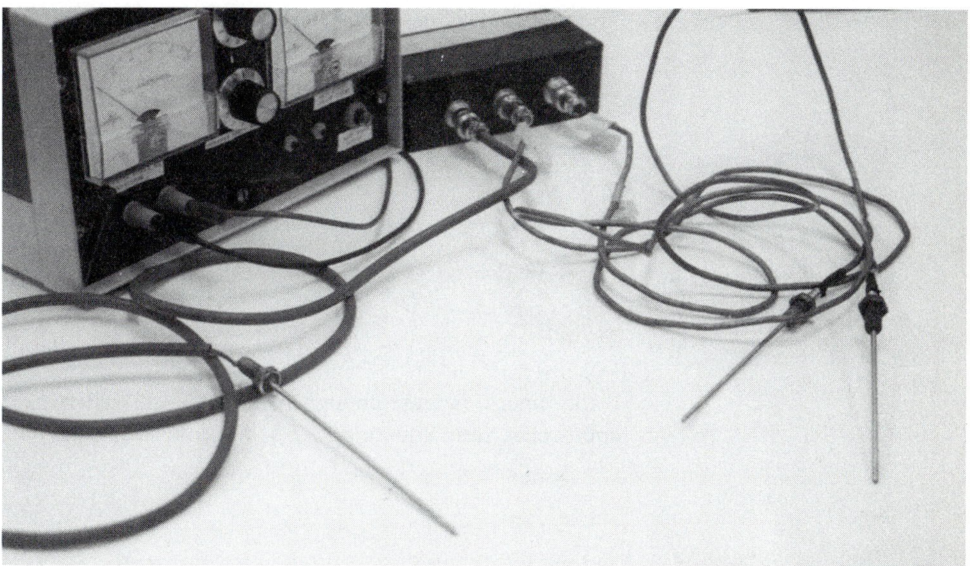

(a) Photograph of a setup with three thermal probes. Each probe is 3 mm in diameter and 130 mm long, but the size of these devices can be as small as 2 mm × 10 mm (or less) and as large as 50 mm × 600 mm (or larger)

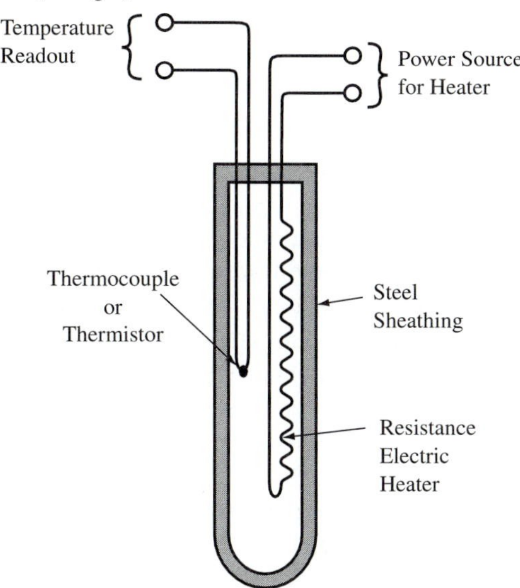

(b) Schematic of a typical thermal probe

The governing equation for the line source of thermal energy in an infinite medium may be written, for the infinite medium,

$$\frac{1}{r}\frac{\partial T(r,t)}{\partial r} + \alpha\frac{\partial^2 T(r,t)}{\partial r^2} = \frac{\partial T(r,t)}{\partial t} \qquad 0 < r < \infty, t > 0$$

with the initial condition

$$T(r,0) = T_i \qquad t = 0, r > 0 \tag{3–73}$$

and boundary conditions, assuming a finite diameter source but with no mass or specific heat associated with the source,

$$\text{B.C. 1} \qquad -2\pi kr\frac{\partial T(r,t)}{\partial r} = \dot{e}_{\text{gen}} \qquad r = r_o \text{ (radius of source)}, t > 0$$

$$\text{B.C. 2} \qquad \partial T(r,t)/\partial r \to 0 \qquad \text{as } r \to \infty, t > 0$$

The solution to this problem has been given by Carslaw and Jaeger [1] as

$$T(r, t) = T_0 + \frac{\dot{e}_{gen}}{4\pi\kappa}\int_{r^2/4\alpha t}^{\infty} \frac{1}{z}e^{-z}\,dz \tag{3–74}$$

where the integral can be written

$$\int_{r^2/4\alpha t}^{\infty} \frac{1}{z}e^{-z}\,dz = -\ln\left(\frac{r^2}{4\alpha t}\right) - \gamma + \Delta z \tag{3–75}$$

The term $\gamma = (0.5772157\ldots)$ is called the Euler–Mascheroni Constant and

$$\Delta z = -\sum_{n=1}^{\infty}\frac{(-1)^n}{n \cdot n!}\left(\frac{r^2}{4\alpha t}\right)^n = \frac{r^2}{4\alpha t} - \frac{1}{4}\left(\frac{r^2}{4\alpha t}\right)^2 + \frac{1}{18}\left(\frac{r^2}{4\alpha t}\right)^3 - \frac{1}{96}\left(\frac{r^2}{4\alpha t}\right)^4 + \cdots \tag{3–76}$$

If the time, t, is large enough then, since the value for Δz predicted by equation (3–76) approaches zero, equation (3–75) becomes

$$\int_{r^2/4\alpha t}^{\infty} \frac{1}{z}e^{-z}\,dz = -\ln\left(\frac{r^2}{4\alpha t}\right) - \gamma \tag{3–77}$$

and the temperature is then predicted from

$$T(r, t) = T_0 - \dot{e}_{gen}\frac{1}{4\pi\kappa}\left(\ln\frac{r^2}{4\alpha t} + \gamma\right) \tag{3–78}$$

Also, the temperature at any one particular location in the infinite medium changes from an initial value, T_1, to another value, T_2, as time elapses according to

$$T_2 - T_1 = \frac{\dot{e}_{gen}}{4\pi\kappa}\ln\left(\frac{t_2}{t_1}\right) \tag{3–79}$$

which comes from equation (3–78). Equation (3–79) is used to calculate thermal conductivity of materials from time and temperature data taken with a thermal probe. It is applicable only for a select time frame. The time t_1 must be greater than some minimum elapsed time (considering $t = 0$ as the initial condition stated in equation 3–73) and t_2 must be less than some maximum amount. The conditions are mentioned by McGaw [8] and Rolle and Wetzel [9], and it can be shown that

$$t_1 > 10r_p^2/\alpha \tag{3–80}$$

and

$$t_2 < 40r_{max}^2/\alpha \tag{3–81}$$

where r_{max} represents the radial distance from the line source of thermal energy at which the solid or surrounding material may have a boundary or discontinuity and r_p is the radius of the line source of thermal energy. Geometrically a line has zero radius; however, there are physical and arithmetic problems in allowing t_2 to approach zero below the criterion of equation (3–81). Also, infinite solids are mathematically convenient but there are physical constraints that prohibit use of data collected beyond a time suggested by equation (3–81).

Notice that equation (3–79) predicts that T_2 can increase without bound as time elapses. If some limit on time is not imposed, a buried power cable, for instance, will eventually reach a high enough temperature (according to equation 3–79) to melt a cable. Power cables, however, are buried at some finite depth, so the earth's temperature may rise around the cable, and eventually these temperature rises will occur at the earth's surface. At such a time the boundary conditions at the surface (convection heat transfer) will be such that steady state conditions will begin to be established. We have discussed steady state heat transfer in Chapter 2. Other applications, such as buried nuclear or hazardous wastes at great depths, encourage extreme temperatures in the earth surrounding those materials, as indicated by equation (3–79).

EXAMPLE 3–15 A buried electric power cable of 0.460-in. (1.2-cm) diameter carries 800 A of current at 440 V. Estimate the maximum temperature expected in the wire if the ground temperature is normally 45°F, the cable is buried 3 ft (0.9 m) deep, and the electrical insulation surrounding the cable is neglected.

Solution The power cable, shown in cross section in figure 3–29, is buried 3 ft below the surface, so we may estimate that the transient conditions of the line source of thermal energy in an infinite solid and steady state conditions begin when the ground temperature begins to change at the surface. Thus, we use equation (3–81) with $r_{max} = 3$ ft to predict t_2. The temperature predicted from equation (3–74) can be visualized as a wave moving from the source outward in a radial direction. When the wave first reaches a location where the material can no longer be considered as a homogeneous and continuous medium the transient conditions begin to be replaced by one of steady state. We use, for the properties of the earth,

$$\kappa = 0.52 \text{ Btu/h} \cdot \text{ft} \cdot {}^\circ\text{F}$$
$$\rho = 94 \text{ lbm/ft}^3$$
$$c_p = 0.19 \text{ Btu/lbm} \cdot {}^\circ\text{F}$$

so that

$$\alpha = \kappa/\rho c_p = 0.029 \text{ ft}^2/\text{h}$$

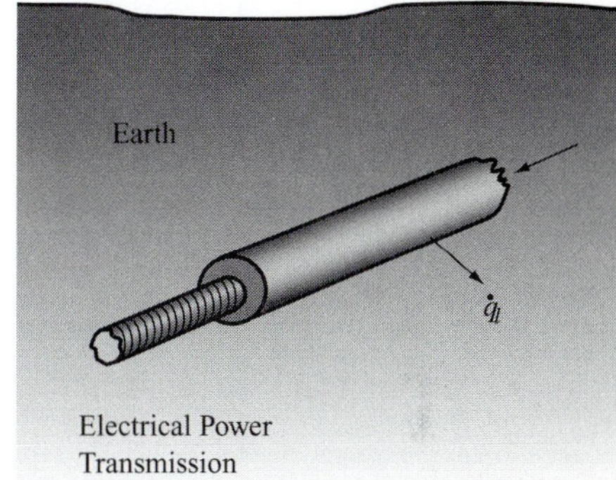

Figure 3–29 Buried electrical power cable.

From equation (3–81) we predict the maximum t_2 at $r_{max} = 3$ ft,

$$t_2 = 40(r_{max})^2/\alpha = (40)(3\text{ ft})^2/(0.029 \text{ ft}^2/\text{hr}) = 12{,}414 \text{ hr}$$

The equilibrium temperature, or maximum cable temperature when the cable reaches steady conditions, can be assumed to occur at the time of 12,414 hr. Using equation (3–78) with $T_1 = T_0 = 45$°F and

$$\dot{e}_{gen} = \mathcal{I}^2\mathcal{R}_L = (800 \text{ amps})^2(0.00004901 \text{ ohms/ft}) = 31.4\,\frac{\text{W}}{\text{ft}} = 107\,\frac{\text{Btu}}{\text{hr} \cdot \text{ft}}$$

where $\mathcal{R}_L$ is read from appendix table B–7. We have

$$T = 45^\circ\text{F} - \frac{107\,\dfrac{\text{Btu}}{\text{hr}\cdot\text{ft}}}{4\pi\left(0.52\,\dfrac{\text{Btu}}{\text{hr}\cdot\text{ft}\cdot{}^\circ\text{F}}\right)}\left[\ln\frac{\left(\dfrac{0.23}{12}\text{ft}\right)^2}{4 \times 0.029\,\dfrac{\text{ft}^2}{\text{hr}} \times 12{,}414 \text{ hr}} + 0.5772\ldots\right]$$

Answer

$$= 284^\circ\text{F}$$

In this example it is shown how a transient condition may become steady state. Notice the long time period before the temperature wave strikes the earth's surface, roughly 1.4 years. If the cable were buried deeper, the time would be even longer and the cable temperature higher.

EXAMPLE 3–16 A thermal probe is used to measure thermal conductivity of a granite rock. A 5.01-mm-diameter hole, 25 cm deep, has been drilled in the rock, and a 5.00-mm-diameter by 25-cm-long probe inserted into the hole. The data shown in table 3–4 has been obtained from a test where the probe was supplied electrical power and the time, temperature, and voltage across the probe heater were simultaneously recorded. The heater resistance was 100.0 ohms, and it was assumed to extend the full length of the probe. Determine the thermal conductivity of the granite rock.

Table 3–4 Thermal Probe Test Data

Time, t, s	Temperature °C	Voltage across Heater
0	2.14	103.4
30	3.68	103.4
60	7.43	103.4
90	12.13	103.4
120	15.47	103.4
150	18.05	103.4
180	20.17	103.3
210	21.96	103.3
240	23.51	103.3
270	24.87	103.3
300	26.09	103.2
330	27.20	103.2
360	28.21	103.1
390	29.14	103.0
420	29.99	102.9
450	30.79	102.9
480	31.54	102.8
510	32.25	102.7
540	32.91	102.6
570	33.52	102.5
600	34.11	102.5

Solution The thermal conductivity is predicted from equation (3–79) in the form

$$\kappa = \frac{\dot{e}_{gen}}{4\pi(T_2 - T_1)} \ln\left(\frac{t_2}{t_1}\right)$$

From the data of table 3–4 we assume the average voltage is 103 volts and the thermal energy per unit length is

$$\dot{e}_{gen} = \frac{\mathcal{E}^2}{\mathcal{R}L} = \frac{103^2}{(100 \text{ ohms})(0.25 \text{ m})} = 424.36 \frac{\text{W}}{\text{m}}$$

Since there is an abundance of data in table 3–4 we may approximate the solution by using $t_1 = 60$ s and $t_2 = 540$ s. Then $T_1 = 7.43$°C and $T_2 = 32.91$°C, and the thermal conductivity is

Answer

$$\kappa = \frac{424.36 \text{ W/m}}{4\pi(T_2 - T_1)} \ln\left(\frac{t_2}{t_1}\right) = 2.912 \frac{\text{W}}{\text{m} \cdot \text{°C}}$$

It is prudent to check that the minimum test time is less than the 60 s used in the previous calculation. From equation (3–80)

$$t_1 > 10\, r_p^2/\alpha = (10)(0.0025\text{ m})^2/(\alpha)$$

The thermal diffusivity is computed from the property values given in appendix table B–2,

$$\alpha = \kappa/\rho c_p = 2.79\text{ W/m}\cdot{}^\circ\text{C}/(2630\text{ kg/m}^3)(775\text{ J/kg}\cdot{}^\circ\text{C})$$
$$= 1.369 \times 10^{-6}\text{ m}^2/\text{s}$$

so that

$$t_1 > 44\text{ s}$$

Therefore we were justified in using the data recorded at 60 s, and the thermal conductivity can be assumed to be 2.912 W/m · °C for the granite rock.

3–6 NUMERICAL METHODS OF ANALYSIS

Many times transient heat transfer occurs with complications that make a straightforward mathematical analysis difficult. Even if boundary and initial conditions can be identified, there are often not convenient analytic solutions to a particular situation. In these instances a numerical method can often provide adequate tools for analyzing a problem or a design. An extension of the finite difference techniques developed in section 2–6 can be made to include transient problems. The finite difference for temperature changes with time is defined as

$$\frac{\partial T}{\partial t} \approx \frac{\Delta T}{\Delta t} = \frac{T_n^{\Delta t} - T_n}{\Delta t} \tag{3–82}$$

where T_n is the temperature of some node, n, at an initial time or at the beginning of a time interval, Δt, and $T_n^{\Delta t}$ is the temperature of that same node after an elapsed time Δt. The finite difference or change of enthalpy of a heat conducting material that is not experiencing a phase change is

$$\frac{\partial \text{Hn}}{\partial t} = \dot{\text{H}}\text{n} = \rho V c_p \frac{\partial T}{\partial t} \approx \rho V c_p \frac{\Delta T}{\Delta t} \tag{3–83}$$

The finite difference technique applied to a transient heat-transfer process proceeds by the following pattern:

1. Identify the nodes and their neighborhoods, and determine how long a time period t_p is needed to complete the analysis.
2. Apply an energy balance to each node and its neighborhood. At this point the energy balance should include a term for the enthalpy change of the node neighborhood. In our discussions of steady state conduction, the heat transfers into the neighborhood were equal in magnitude to the heat transfers leaving that neighborhood. Here the heat transfers are not equal, but their difference is the change in enthalpy; that is,

$$\sum_{\text{in}} \dot{Q} - \sum_{\text{out}} \dot{Q} = \dot{\text{H}}\text{n} \tag{3–84}$$

3. Arrange each node/neighborhood energy balance equation (3–84) explicitly for $T_n^{\Delta t}$; that is,

$T_n^{\Delta t}$ = some function of heat transfer terms and the properties ρ, c_p, and V

$$T_n^{\Delta t} = T_n + \frac{\Delta t}{\rho V c_p}\left(\sum_{\text{in}} \dot{Q} - \sum_{\text{out}} \dot{Q}\right) \tag{3–85}$$

4. After identifying each node temperature equation, the calculations may proceed by hand or by machine. There are two common methods of proceeding with the calculations, and they determine the precise form of equation (3–85): the *explicit method* and the *implicit method.* In the explicit method, the heat transfer terms are determined by using the initial temperatures or the temperatures at the beginning of the time interval Δt for all nodes, including the nth node. Thus, the right side of equation (3–85) will be a function of the initial temperatures and not the final node temperature, $T_n^{\Delta t}$.

In the implicit method the heat transfer terms are determined by using the final temperatures for all nodes, including the nth node. In this way equation (3–85) is more conveniently written as

$$T_n = T_n^{\Delta t} + \frac{\Delta t}{\rho V c_p}\left(\sum_{\text{out}} \dot{Q} - \sum_{\text{in}} \dot{Q}\right) \tag{3–86}$$

and, if the initial or beginning temperature is known, the set of equations (3–86) are more conveniently treated as a matrix and the temperatures solved through the matrix methods. The form of equation (3–86) suggests that "earlier" temperatures, T_n's, can be solved directly if "later" temperatures, $T_n^{\Delta t}$'s, are known. For this reason the implicit method is sometimes referred to as the backward-difference method and the explicit method as the forward-difference method.

The calculational procedure for the explicit method requires that the time interval, Δt, be defined and the node temperatures all be known. Also, the node neighborhood volume, V, must be defined. As a general procedure, the computations begin at the initial condition, a prescribed time interval is used, and the node temperature $T_n^{\Delta t}$ is determined for every node of the system. These new temperatures are then used as initial conditions for a second set of computations using the same time interval Δt, and the next set of node temperatures is obtained. This sort of routine proceeds until the time period has expired; that is, until t_p (the time period of the full analysis) $\leq N\Delta t$, where N is the number of time periods.

The selection of a time interval, Δt, will determine the amount of computational time is required to reach a solution. If the time interval is large, the number of computations is reduced; however, if a detailed history of the transient conditions is desired, then smaller intervals should be selected. There is an upper limit to the time interval, a *critical time interval,* Δt_c, which cannot be exceeded. This critical time interval can be determined by noting that the general form of the energy balance equation (3–85) can be written

$$T_n^{\Delta t} = T_n + \frac{\Delta t}{\rho V c_p} \sum \frac{1}{R_{mn}}(T_m - T_n) \tag{3–87}$$

if the heat transfer is assumed to all be positive, or into the element, so that $\sum Q_{\text{out}} = 0$.

Here we denote T_m as the temperatures of the nodes adjacent to node n, and R_{mn} is the thermal resistance between node n and its adjacent nodes. The thermal resistances are the same as those listed in table 2–8 for the various coordinate systems. Rearranging equation (3–87) to read

$$T_n^{\Delta t} = T_n\left(1 - \sum \frac{\Delta t}{\rho V c_p R_{mn}}\right) + T_m \sum \frac{\Delta t}{\rho V c_p R_{mn}} \tag{3–88}$$

it can be seen that the coefficient for the node temperature T_n, namely the coefficient $1 - \sum (\Delta t/\rho V c_p R_{mn})$, must be greater than zero. If it is less than zero, then a situation can occur where all of the adjacent node temperatures, T_m, are the same or larger than T_n and yet $T_n^{\Delta t}$ is less than T_n. This would allow a node to cool down without any external work and violates the Second Law of Thermodynamics. It also violates your intuition!

EXAMPLE 3–17 Determine the maximum time interval allowed for transient heat-transfer analysis of the node shown in figure 3–30. Assume the thermal diffusivity is 10×10^{-6} m²/s.

Figure 3–30 Two-dimensional node for example 3–17.

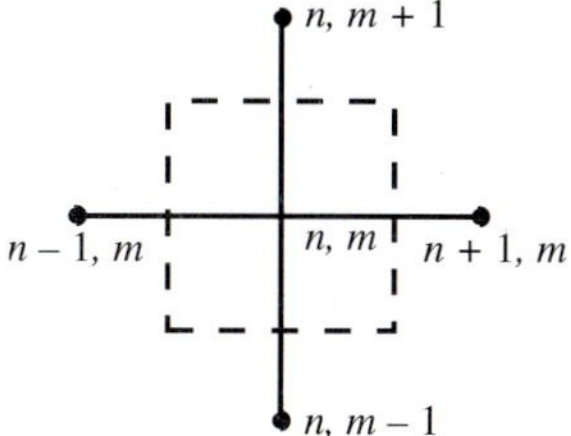

Solution The energy balance for the node is written, assuming all conduction heat-transfer terms are positive, as

$$\kappa\Delta y\frac{T_{n-1,m} - T_{n,m}}{\Delta x} + \kappa\Delta y\frac{T_{n+1,m} - T_{n,m}}{\Delta x} + \kappa\Delta x\frac{T_{n,m-1} - T_{n,m}}{\Delta y} + \kappa\Delta x\frac{T_{n,m+1} - T_{n,m}}{\Delta y} = \rho c_p(\Delta x \cdot \Delta y)\frac{T_{n,m}^{\Delta t} - T_{n,m}}{\Delta t}$$

Using the information that $\Delta x = \Delta y$ and that $\kappa/\rho c_p = \alpha$, this equation can be written

$$T_{n,m}^{\Delta t} = T_{n,m}\left(1 - 4\alpha\frac{\Delta t}{\Delta x^2}\right) + \alpha\frac{\Delta t}{\Delta x^2}(T_{n-1,m} + T_{n+1,m} + T_{n,m-1} + T_{n,m+1})$$

and the term

$$1 - 4\alpha\frac{\Delta t}{\Delta x^2} \geq 0 \quad \text{or} \quad 4\alpha\frac{\Delta t}{\Delta x^2} \leq 1$$

where Δt can be interpreted as the critical time interval, or the maximum time interval allowed for numerical analysis. Then

Answer

$$\Delta t_c = \frac{\Delta x^2}{4\alpha} = \frac{(0.01\ \text{m})^2}{4 \times 10^{-5}\ \text{m}^2/\text{s}} = 2.5\ \text{s}$$

Notice in this last example that the node was defined in two dimensions. If the node is defined in one dimension, so that the node temperatures $T_{n,m+1}$ and $T_{n,m-1}$ are the same as $T_{n,m}$ then the energy balance equation reduces to

$$T_{n,m}^{\Delta t} = T_{n,m}\left(1 - 2\alpha\frac{\Delta t}{\Delta x^2}\right) + \alpha\frac{\Delta t}{\Delta x^2}(T_{n-1,m} - T_{n+1,m})$$

and the critical time interval is then

$$\Delta t_c = \frac{\Delta x^2}{2\alpha} \tag{3–89}$$

Also, if the node would have been defined in three-dimensional space, then the critical time interval is

$$\Delta t_c = \frac{\Delta x^3}{6\alpha} \tag{3–90}$$

These critical time intervals apply only for internal nodes where there is only conduction heat transfer. At boundary nodes or where there are convection or radiation heat transfers or where there are energy generation terms, the critical time intervals are determined from energy balances, which may be more involved than the ones given in example 3–17. In all of these analyses, the crucial step involves making sure that the coefficient for the node temperature is greater than zero, as we discussed. In a system composed of many nodes the time interval must be the same for all of the nodes, and it must be less than or equal to the *least* critical time interval.

EXAMPLE 3–18 For the finite difference nodes used to analyze transient heat transfer through a semi-infinite solid shown in figure 3–31, determine the maximum allowable time step, Δt.

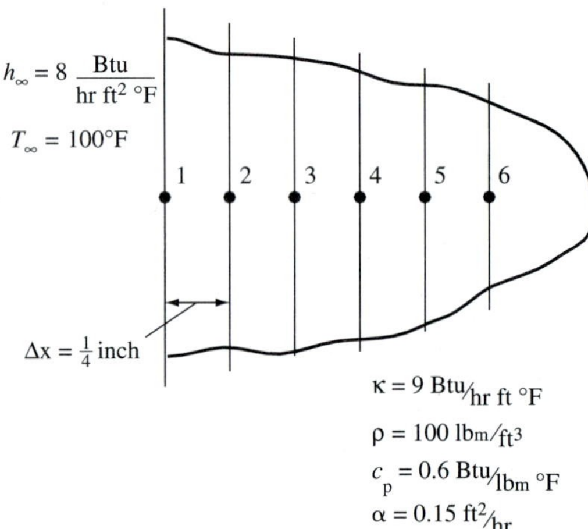

Figure 3–31 Semi-infinite solid for example 3–18.

Solution Referring to figure 3–30, the first five nodes are identified, with node 1 being the boundary node and nodes 2, 3, 4, 5, ... being internal nodes. The energy balance for any of the internal nodes, writing the energy balance so that all heat transfers are positive, is

$$\kappa \frac{T_{n+1} - T_n}{\Delta x} + \kappa \frac{T_{n-1} - T_n}{\Delta x} = \rho c_p \Delta x \frac{T_n^{\Delta t} - T_n}{\Delta t}$$

or

$$T_n^{\Delta t} = T_n\left(1 - 2\alpha \frac{\Delta t_c}{\Delta x^2}\right) + \alpha \frac{\Delta t_c}{\Delta x^2}(T_{n+1} + T_{n-1})$$

which gives the restriction that

$$\Delta t_c \leq \frac{\Delta x^2}{2\alpha} = \frac{(^1\!/_{48}\ \text{ft})^2}{2 \times 0.15\ \text{ft}^2/\text{hr}} = 1.45 \times 10^{-3}\ \text{hr} = 5.2\ \text{s}$$

For boundary node 1 the energy balance is

$$\kappa\frac{T_2 - T_1}{\Delta x} + h(T_\infty - T_1) = \rho c_p \frac{\Delta x}{2\Delta t}(T_1^{\Delta t} - T_1)$$

so that

$$T_1^{\Delta t} = T_1\left\{1 - \left[\frac{\alpha}{(\Delta x)^2} + \left(\frac{h}{\rho c_p \Delta x}\right)\right]2\Delta t\right\} + \frac{2\alpha\Delta t T_2}{(\Delta x)^2} + \frac{2h\Delta t T_\infty}{\rho c_p \Delta x}$$

with the maximum time interval limited by the criteria to

$$1 - [\alpha/(\Delta x)^2 + h/\rho c_p \Delta x]2\Delta t \geq 0$$

Therefore, the critical time interval for boundary node 1 is

$$\Delta t_c = 1/\left[\frac{2\alpha}{(\Delta x)^2} + \frac{2h}{\rho c_p \Delta x}\right]$$

$$= 1/[(2)(0.15\ \text{ft}^2/\text{hr})/(^1\!/_{48}\ \text{ft})^2$$

$$+ (2)(8\ \text{Btu/hr}\cdot\text{ft}^2\cdot{}^\circ\text{F})/(100\ \text{lbm/ft}^3 \times 0.6\ \text{Btu/lbm}\cdot{}^\circ\text{F} \times (^1\!/_{48}\ \text{ft})]$$

$$= 0.00142\ \text{hr} = 5.1\ \text{s}$$

When the two critical time interval values, that for the internal nodes and that for the boundary node are compared, the boundary node has a shorter time, so that will be the maximum allowable time interval.

In the *implicit* method of solving for the node temperatures, the energy balance of each node, from equation (3–86), is written so that the heat transfers are identified with the new node temperatures, $T_n^{\Delta t}$ $T_m^{\Delta t}$, and so on. The node temperature equation, using thermal resistances, for the *implicit* method would be

$$T_n^{\Delta t} = T_n + \frac{\Delta t}{\rho V c_p}\sum\frac{1}{R_{mn}}(T_m^{\Delta t} - T_n^{\Delta t}) \qquad \textbf{(3-91)}$$

The implicit method provides a set of node temperature equations that must be solved simultaneously as was discussed in section 2–6. The implicit method requires techniques for solving matrices or some other numerical methods. As a consequence, the explicit method appears to be more straightforward; however, the node temperature equation (3–88) can be written

$$T_n = T_n^{\Delta t}\left(1 + \frac{\Delta t}{\rho V c_p}\sum\frac{1}{R_{mn}}\right) - \frac{\Delta t}{\rho V c_p}\sum\frac{1}{R_{mn}}T_m^{\Delta t} \qquad \textbf{(3–92)}$$

and this equation shows that the time interval has no maximum restriction placed upon it. The coefficient of the node's latest temperature, $T_n^{\Delta t}$, will always be positive, and therefore there will not be situations where the Second Law of Thermodynamics is violated, so unless a detailed history of the transient conditions are desired larger time intervals may be used in the computations.

Because of the manner of writing equation (3–92), which solves for an earlier node temperature, the implicit method is really a backward-difference method, as we noted earlier. The explicit method is then a forward-difference method, to distinguish its manner of directly predicting the future temperatures. Thus, the implicit method is convenient if some specified final temperature distribution is known and an initial condition is to be determined.

The numerical methods discussed here can be used to solve for any transient conduction heat-transfer problem, and the only limitations are in the degree of precision and accuracy required in the analysis. Often the computations produce results that are far removed from the correct one, so convergence to more accurate results is a major concern. A good background in the analytic solutions discussed earlier can often be the only way to judge whether a numerical solution for some complicated system is approximately correct. With care and experience, all of the problems discussed in this chapter can be handled with these numerical methods. Further, problems involving variable thermal conductivities, density, or specific heats may often only be analyzed to any degree of satisfaction with these numerical methods.

There are many software packages commercially available for using machine computations to analyze transient heat transfer situations. The student should be aware of the capabilities and the limitations of these devices; however, a discussion of these are beyond the purposes of this text.

3–7 GRAPHICAL METHODS OF ANALYSIS

Before the computer became a convenient mechanism for using numerical methods, graphical methods were often used to obtain solutions to transient heat-transfer problems. The methods are particularly suited to one-dimensional problems with convection or conduction boundary conditions and can be interpreted as the graphical equivalent to the finite difference techniques discussed in section 3–6. In figure 3–32 is shown the graphical technique for determining the temperature distribution in a semi-infinite solid that is initially at a temperature T_i except at its boundary, which is taken as T_1. The temperature at node 2, T_2, changes to $T_2^{\Delta t_1}$ after one time interval Δt. The value for $T_2^{\Delta t_1}$ is determined by constructing line A' from T_1 to T_3. This construction corresponds to the mathematical analogy of linearly averaging between two temperatures, T_1 and T_3, to determine T_2. Such a technique is the essential part of the numerical and the graphical techniques. The second time interval for the solid in figure 3–32 results in a change in the temperature of node 3, T_3, to $T_3^{\Delta t2}$. This is determined by construction of the line B′ between nodes 2 and 4.

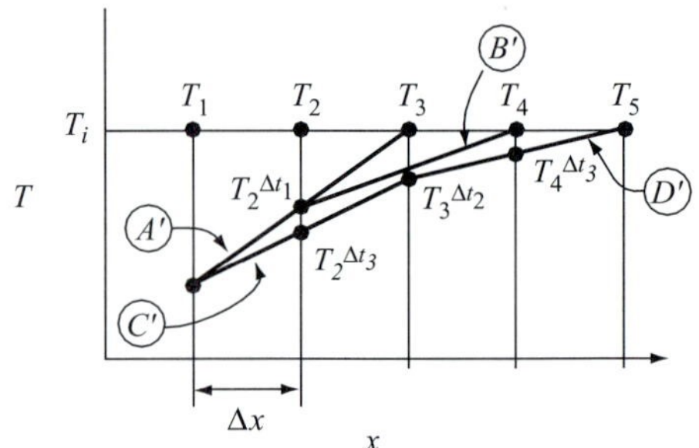

Figure 3–32 Graphical method of solving for temperature distribution in a semi-infinite solid.

Notice that a node temperature change is obtained by constructing a straight line between two nodes adjacent to the node under analysis. Thus, node 2 remained unchanged during the second time interval because the straight line between nodes 1 and 3 passes through the temperature of node 2, $T_2^{\Delta t1} = T_2^{\Delta t2}$. The third time interval provides changes in the temperatures of nodes 2 and 4, temperatures that are determined by construction of lines C′ and D′. The fourth, fifth, and succeeding intervals can be considered and the node temperatures determined further and further into the solid. In figure 3–32 the analysis has stopped after four intervals, and the temperature distribution is approximated by the node temperatures T_1, T_2, T_2, T_4, and T_5.

The graphical analysis is based on the same sort of energy balance as the numerical methods of section 3–6. For one-dimensional conduction heat transfer of an interval using the explicit method of predicting node temperatures, the time interval applying to *internal nodes* was restricted by the criterion, from equation (3–89),

$$\frac{\Delta x^2}{2\alpha\Delta t} \leq 1.0$$

and the least amount of time steps, or maximum time step, for a given Δx and α we use

$$\Delta t_{\text{max}} = \frac{\Delta x^2}{2\alpha} \tag{3–93}$$

for *all internal node temperatures* in one-dimensional analysis.

The graphical method can be used for any one-dimensional solid with a known initial temperature distribution and with convection or conduction boundary conditions. For a convection boundary condition where

$$\kappa\frac{\partial T}{\partial x} = h(T_w - T_\infty) \quad \text{or} \quad \kappa\frac{\partial T}{\partial x} = h(T_\infty - T_w) \tag{3–94}$$

where T_w is the temperature of the solid at the boundary and T_∞ is the bulk fluid temperature, the slope of the temperature distribution at the boundary is

$$\left[\frac{\partial T}{\partial x}\right]_{\text{wall}} = \frac{h}{\kappa}(T_w - T_\infty) \quad \text{or} \quad \left[\frac{\partial T}{\partial x}\right]_{\text{wall}} = \frac{h}{\kappa}(T_\infty - T_w) \tag{3–95}$$

In the graphical method this information is used to handle the convection boundary condition in the manner demonstrated in figure 3–33. The node separation, Δx, is defined and the term κ/h is used to define the node point ∞ in the fluid where the temperature is T_∞. Line a is then drawn between ∞ and node m. The temperature T_n is then directly determined on this line. Succeeding time intervals are handled like the above example of the solid with a constant boundary temperature, except here the fluid temperature T_∞ is constant at the node ∞.

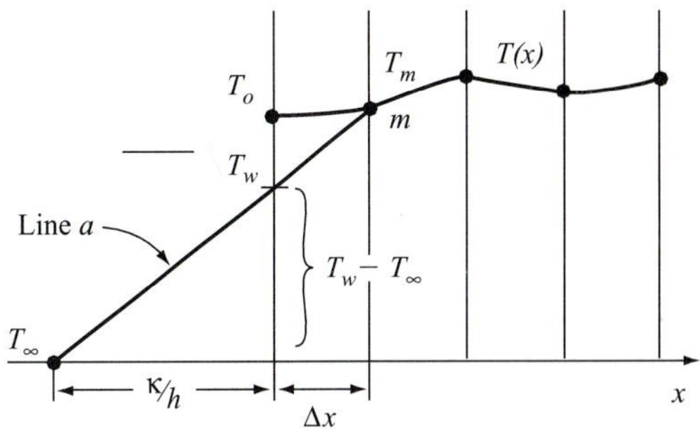

Figure 3–33 Typical handling of a transient heat transfer with convection boundary condition using graphical methods.

EXAMPLE 3–19 A concrete containment wall 2 ft thick has an outer surface temperature of 70°F and an inner temperature of 130°F with the temperature linearly distributed through the wall between these two values. It is then subjected to air on both inner and outer surfaces that is at 50°F with a convective heat transfer coefficient of 10 Btu/hr · ft² · °F. Determine the temperature distribution through the wall after 8 hr, using 3-in. increments between the nodes and assuming the following properties for concrete:

$$\kappa = 0.82 \text{ Btu/hr} \cdot \text{ft} \cdot {}^\circ\text{F}$$
$$c_p = 0.21 \text{ Btu/lbm} \cdot {}^\circ\text{F}$$
$$\rho = 143 \text{ lbm/ft}^3$$
$$\alpha = 0.027 \text{ ft}^2\text{/hr}$$

Solution

We have $\Delta x = 3$ in. $= 0.25$ ft and then, from equation (3–93)

$$\Delta t = \frac{\Delta x^2}{2\alpha} = \frac{(0.25 \text{ ft})^2}{2 \times 0.0027 \text{ ft}^2\text{/hr}} = 1.157 \text{ hr} = 4167 \text{ s}$$

Also the term $\kappa/h = 0.082$ ft and the graphical analysis can proceed as shown in figure 3–34. The initial temperature distribution is indicated on the figure. Seven time intervals were used to reach the 8-hr time span. The actual time for seven intervals is 8.099 hr, which is slightly in excess of the 8 hr. The final temperature distribution is indicated in the figure.

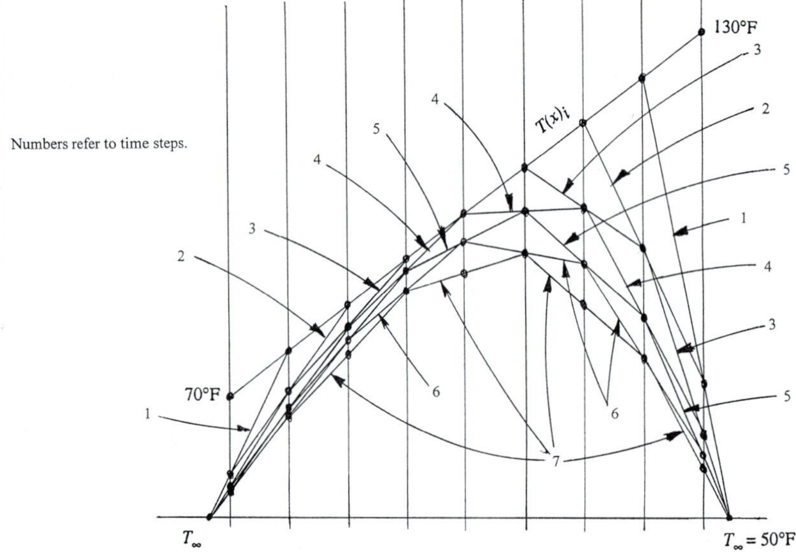

Figure 3–34 Graphical analysis of a concrete containment wall.

The problem of two solid heat conducting materials placed together can be analyzed conveniently by the graphical method, provided that one-dimensional heat transfer can be assumed. In this situation the boundary condition is

$$\kappa_A \left[\frac{\partial T}{\partial x} \right]_A = \kappa_B \left[\frac{\partial T}{\partial x} \right]_B \tag{3–96}$$

and this can be approximated as

$$\kappa_A \left[\frac{\Delta T}{\Delta x} \right]_A = \kappa_B \left[\frac{\Delta T}{\Delta x} \right]_B \tag{3–97}$$

or, for $\Delta x_A = \Delta x_B$,

$$\frac{\Delta T_A}{\Delta T_B} = \frac{\kappa_B}{\kappa_A} \tag{3–98}$$

For the initial condition where two infinite solids at different uniform temperatures are placed together, the boundary temperature is immediately established from equation (3–98) and graphically determined as shown in figure 3–35, where $\Delta T_A + \Delta T_B =$ initial temperature difference between the two solids. The analysis then proceeds as two separate solids with the same constant boundary temperature. Other techniques and refinements of the graphical methods have been developed, and a good reference for the interested reader is Jakob [10].

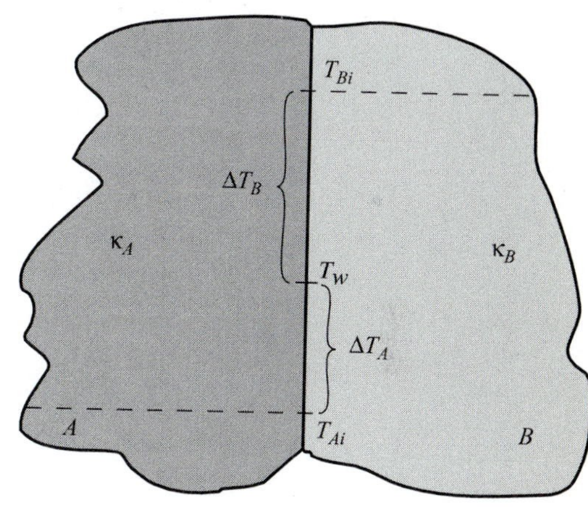

Figure 3–35 Graphical technique for determining the boundary temperature of two solids in contact.

3–8 SUMMARY

The conservation of energy applied to a system having transient heat transfer may be written

$$\sum_{\text{in}} \dot{Q} - \sum_{\text{out}} + \dot{E}_{\text{gen}} = \frac{d\text{Hn}}{dt} \tag{3–1}$$

where $\dot{E}_{\text{gen}}$ is the energy generated in the system and Hn is the enthalpy of the system. If the system is an incompressible solid, the heat transfers are by conduction, and the thermal conductivity is constant, this becomes,

$$\frac{\partial^2 T(x, y, z, t)}{\partial x^2} + \frac{\partial^2 T(x, y, z, t)}{\partial y^2} + \frac{\partial^2 T(x, y, z, t)}{\partial z^2} + \dot{e}_{\text{gen}} = \frac{1}{\alpha}\frac{\partial T(x, y, z, t)}{\partial t} \tag{3–4}$$

where $\dot{e}_{\text{gen}}$ is the energy generation per unit volume and α is the thermal diffusivity

$$\alpha = \frac{\kappa}{\rho c_p} \tag{3–3}$$

For systems experiencing transient heat transfers, the system can be treated as one having the same temperature throughout if the Biot Number, Bi, is less than about 0.1. The Biot Number is defined as

$$\text{Bi} = \frac{h\Delta x}{\kappa} \tag{3–6}$$

where Δx is some characteristic length such as the system volume divided by the system's surface area. For these situations where $\text{Bi} \le 0.1$, the system is sometimes referred to as a lumped heat capacity system, and if the heat transfer at the surface is convection to the surroundings, the system temperature is given by

$$\theta = \theta_i e^{-t/t_c} \tag{3–10}$$

where $\theta = (T - T_\infty)$, $\theta_i = (T_i - T_\infty)$, and t_c is the time constant, given by

$$t_c = \frac{\rho V_s c_p}{h A_s} \tag{3–11}$$

The thermal energy (in kJ or Btu) transferred between the lumped system and its surroundings is given by

$$Q = Q_o(1 - e^{-t/t_c}) \tag{3–15}$$

where

$$Q_o = \rho V c_p (T_i - T_\infty)$$

If the Biot Number is greater than about 0.1, then the system temperature should be considered to vary with position as well as time. For a semi-infinite homogeneous solid having initial temperature T_i throughout and boundary temperature T_B beginning and continuing from $t = 0$, the solid's temperature is given by

$$T(x, t) = T_B + (T_i - T_B)\operatorname{erf}\left(\frac{x}{2\sqrt{\alpha t}}\right) \tag{3–17}$$

The conduction heat transfer through the semi-infinite medium at any given location is determined by

$$\dot{q}_A = -\kappa(T_i - T_B)\frac{1}{\sqrt{\pi \alpha t}} e^{-x^2/4\alpha t} \tag{3–20}$$

For a semi-infinite homogeneous solid having initial temperature T_i throughout with boundary heat transfer $\dot{q}_B$ beginning and continuing from $t = 0$, the temperature of the solid is given by

$$T(x, t) = T_i + \frac{1}{\kappa}\left\{2\dot{q}_B\sqrt{\frac{\alpha t}{\pi}} e^{-x^2/4\alpha t} - \dot{q}_B x\left[1 - \operatorname{erf}\left(\frac{x}{2\sqrt{\alpha t}}\right)\right]\right\} \tag{3–22}$$

The semi-infinite homogenous solid having convection heat transfer at its boundary has a temperature history given by the equation

$$T(x, t) = T_i - (T_\infty - T_i)\left\{1 - \operatorname{erf}\left(\frac{x}{2\sqrt{\alpha t}}\right) - e^{\left(\frac{hx}{\kappa} + \frac{h^2 \alpha t}{\kappa}\right)}\left[1 - \operatorname{erf}\left(\frac{x}{2\sqrt{\alpha t}}\right) + \frac{h\sqrt{\alpha t}}{\kappa}\right]\right\} \tag{3–24}$$

An infinite plate or slab of a homogenous solid with convection heat transfer at both surfaces and with initial temperature T_i throughout has a temperature history given by the equation

$$\theta(x, t) = \theta_i \sum_{n=1}^{\infty} C_n e^{-\xi_n^2 \text{ Fo}} \cos(\xi_n x/L) \tag{3–26}$$

where $\theta(x, t) = T(x, t) - T_\infty$, $\theta_i = T_i - T_\infty$, Fo $=$ Fourier number $= \alpha t/L^2$, and ξ is the eigenvalue of the transcendental equation

$$\xi_n \tan \xi_n = \text{Bi} = \frac{h_\infty L}{\kappa} \tag{3–28}$$

The constant C_n is given by

$$C_n = \frac{4 \sin \xi_n}{2\xi_n + \sin 2\xi_n} \tag{3–27}$$

Equation (3–26) can be replaced with

$$\theta(x, t) = \theta_i C_i e^{-\xi_1^2 \text{Fo}} \cos(\xi_1 x/L) \tag{3–29}$$

if Fo ≥ 0.2. Also, the temperature history of the infinite plate can be approximated from charts presented in figure 3–12.

The thermal energy transfer between an infinite plate and its surroundings is evaluated, for Fo ≥ 0.2, by

$$\frac{Q}{Q_0} = 1 - C_1 e^{-\xi_1^2 \text{Fo}} \frac{\sin \xi_1}{\xi_1} \tag{3–36}$$

or from a graphical chart presented in figure 3–12c.

For the infinite cylinder of a homogeneous solid at a uniform initial temperature T_i subjected to a convection heat transfer with a fluid at temperature T_∞, the temperature history is given by

$$\theta(r, t) = \theta_i \sum_{n=1}^{\infty} C_n e^{-\beta_n^2 \text{Fo}_c} J_0(\beta_n r/r_0) \tag{3–38}$$

where $\text{Fo}_c = \alpha t/r_0^2$, β_n is the eigenvalue of the transcendental equation

$$\beta_n \frac{J_1(\beta_n)}{J_0(\beta_n)} = \frac{h_\infty r_0}{\kappa} = \text{Bi}_c \tag{3–40}$$

and the constant C_n is evaluated with

$$C_n = \frac{2J_0(\beta_n)}{\beta_n[J_0^2(\beta_n) + J_1^2(\beta_n)]} \tag{3–39}$$

If the cylindrical Fourier number, Fo_c, is greater than about 0.2, equation (3–38) may be replaced with

$$\theta(r, t) = \theta_i C_i e^{-\beta_1^2 \text{Fo}_c} J_0(\beta_1 r/r_0) \tag{3–41}$$

and the thermal energy transfer history can be predicted from

$$Q = Q_0[1 - 2C_1 e^{-\beta_1^2 \text{Fo}_c} J_0(\beta_1)/\beta_1] \tag{3–44}$$

Approximate solutions for the temperature and thermal energy transfers history of the infinite cylinder may be obtained from charts presented in figure 3–15.

For the sphere of a homogeneous solid at a uniform temperature T_i subjected to convection heat transfer with a fluid at T_∞, the temperature history can be predicted from

$$\theta(r, t) = \theta_i \sum_{n=1}^{\infty} C_n e^{-\zeta_n^2 \text{Fo}_r} \frac{r_0}{\zeta_n r} \sin\left(\zeta_n \frac{r}{r_0}\right) \tag{3–46}$$

where $\text{Fo}_r = \alpha t/r_0^2$, ζ_n's are the positive roots of the equation

$$1 - \zeta_n \cot \zeta_n = \text{Bi} = \frac{h_\infty r_0}{\kappa} \tag{3–49}$$

and the constant C_n is evaluated by the equation

$$C_n = \frac{4(\sin \zeta_n - \zeta_n \cos \zeta_n)}{2\zeta_n - \sin 2\zeta_n} \tag{3–48}$$

If the radial Fourier Number, Fo_r, is greater than about 0.2, equation (3–46) may be replaced by

$$\theta(r, t) = \theta_i C_1 e^{-\zeta_1^2 \mathrm{Fo}_r} \frac{r_0}{\zeta_n r} \sin\left(\zeta_1 \frac{r}{r_0}\right) \tag{3–50}$$

and the thermal energy transfer history may be evaluated by equation

$$Q = Q_0\left(1 - 3C_1 e^{-\zeta_1^2 \mathrm{Fo}_r} \frac{\sin \zeta_1 - \zeta_1 \cos \zeta_1}{\zeta_1^3}\right) \tag{3–53}$$

Charts presented in figure 3–17 may also be used to obtain approximate solutions to the temperature and thermal energy transfer histories.

Conduction heat-transfer applications include determination of cooling and heating rates for metals and other materials. The cooling rate for a *Jominy bar* modeled as an infinite lumped heat capacity cylinder in conjunction with a semi-infinite solid is given by

$$\frac{\partial T(x, t)}{\partial t} = (T_i - T_\infty)e^{-t/t_c}\left[-\frac{1}{t_c}\mathrm{erf}\left(\frac{x}{2\sqrt{\alpha t}}\right) - \frac{x}{2t^{3/2}\sqrt{\alpha\pi}}e^{-x^2/4\alpha t}\right] \tag{3–65}$$

The analysis of a semi-infinite solid subjected to a periodic boundary temperature given by the relationship

$$T(0, t) = A \cos \omega t$$

results in a temperature distribution

$$T(x, t) = Ae^{x\sqrt{\omega 2/2\alpha}} \cos\left(\omega t - x\sqrt{\frac{\omega}{2\alpha}}\right) \tag{3–70}$$

The line source of heat in an infinite medium has applications in buried electrical power cables and in the experimental determination of thermal conductivity in materials. For suitable conditions, the temperature distribution is given by

$$T(r, t) = T_0 - \frac{\dot{e}_{\text{gen}}}{4\pi\kappa}\left(\ln\frac{r^2}{4\alpha t} - \gamma\right) \tag{3–78}$$

where $\dot{e}_{\text{gen}}$ is the line source of heat. The Euler–Mascheroni constant γ has a value of 0.5772157 . . . .

Numerical methods of analysis may be used to analyze any transient conduction heat-transfer problem. The method of finite differences may be extended from steady state conditions to transient conditions where the time rate of change of enthalpy of an incompressible heat conducting material is defined as

$$\dot{H}_n = \frac{dHn}{dt} \approx \rho V c_p \frac{T_n^{\Delta t} - T_n}{\Delta t} \tag{3–83}$$

The temperature $T_n^{\Delta t}$ is that for node n after an elapsed time period Δt. In the explicit method the new node temperatures are directly computed from equations of the form

$$T_n^{\Delta t} = T_n + \frac{\Delta t}{\rho V c_p}\left(\sum_{\text{in}} \dot{Q} - \sum_{\text{out}} \dot{Q}\right) \tag{3–85}$$

These equations are restricted to certain maximum time intervals, $\Delta t \leq \Delta t_c$. The critical time interval, Δt_c, must be determined for each node, and the minimum interval of all the nodes of the particular system under analysis must then be used for all the nodes. For an internal node in two dimensions

$$\Delta t_c = \frac{\Delta x^2}{4\alpha} \qquad \textbf{(3–89)}$$

but is different for boundary nodes or one-dimensional or three-dimensional internal nodes. In the implicit method the node equations may take the form

$$T_n = T_n^{\Delta t} + \frac{\Delta t}{\rho V c_p}\left(\sum_{\text{out}} \dot{Q} - \sum_{\text{in}} \dot{Q}\right) \qquad \textbf{(3–86)}$$

and this method may be used to determine initial node temperatures or new temperatures. This method requires that all temperatures be determined simultaneously by matrix methods.

Graphical methods may be used for one-dimensional transient cases with convection boundary conditions.

DISCUSSION QUESTIONS

Section 3–1

3–1 Explain what is meant by a transient heat-transfer process.

3–2 Explain what thermal diffusivity is, mathematically and physically.

3–3 What mathematical terms are needed to account for a transient heat-transfer process?

Section 3–2

3–4 What is the Biot Number?

3–5 Is a "lumped heat capacity system" realistic?

3–6 Explain what is meant by a time constant in transient heat transfer.

Section 3–3

3–7 Why is a negative sign attached to the heat term of equation (3–20)?

3–8 What criteria would you use to decide whether to use the semi-infinite model to describe conduction heat transfer through a material?

3–9 What is the error function?

3–10 When are the Heisler Charts a more accurate description of heat transfer than lumped heat capacity analysis?

3–11 Is the thermal energy exchange function Q/Q_o a good method for predicting the degree of thermal energy transfer of an infinite cylinder with its surroundings? Is there a more accurate method?

Section 3–4

3–12 What is the Bessel Function?

3–13 Why may the Bessel Function solution to the heat transfer in a finite cylinder be approximated by the first term in the infinite series if Fo > 0.2?

Section 3–5

3–14 What is meant by *cooling rate*?

3–15 What is the physical significance of a line source of heat?

Section 3–6

3–16 Can you describe three phenomena that may be accounted as Q_{in} in the node equation (3–86)?

3–17 Describe the difference between the explicit and the implicit methods for determining node temperatures.

3–18 Why is the Second Law of Thermodynamics violated if the coefficient of the node temperature $\dot{T}_{n,m}$ is not positive or zero in the explicit method?

3–19 Can you think of any possible limitation for the length of the time step in the implicit method?

Section 3–7

3–20 Can you think of a situation where a graphical technique may be the most appropriate method for analyzing transient heat transfer?

3–21 Why is the boundary temperature between two conducting materials estimated by equation (3–98)?

PRACTICE PROBLEMS

Problems marked with an asterisk (*) are more challenging.

Section 3–1

3–1 Estimate the value of the thermal diffusivity for glycerin at 300 K.

3–2 Estimate the value of the thermal diffusivity for oxygen at 300 K and 101 kPa.

3–3 Estimate the temperature increase of 1 lbm of steel at 100°F if it is heated at the rate of 30 Btu/s.

Section 3–2

3–4 Determine the Biot Number, Bi, for a 0.5-in. ice cube at −10°F if the thermal conductivity is 1.27 Btu/hr · ft · °F and the convective heat transfer coefficient is 2.5 Btu/hr · ft^2 · °F.

3–5 Show that equation (3–8a),

$$hA_s[T(t) - T_\infty] = -\rho V c_p \frac{dT(t)}{dt}$$

is mathematically correct for either heating or cooling of a lumped heat capacity system.

3–6 Determine the time constant for a thermistor made of steel and having $h = 10$ W/m^2 · °C. Assume the thermistor is a sphere having a diameter of 1.2 mm.

3–7 An egg is being boiled in water at 100°C. If the egg is assumed to be a 4-cm-diameter sphere, if it has the same properties as water, and if the convective heat transfer coefficient is 80 W/m^2 · °C, estimate the time constant for the boiling egg.

3–8 Ball bearings 10 mm in diameter made of 40% nickel steel are heat-treated at 380°C. If the balls are placed in a heat-treat oven at 40°C, determine the time required to heat the balls to 370°C. Use a value of 5 W/m^2 · °C for the convective heat transfer coefficient.

3–9 Spherical rain drops $\frac{1}{8}$ in. in diameter fall at 10 ft/s in 10°F air. The convective heat-transfer coefficient is 15 Btu/hr · ft^2 · °F, and the rain forms at 38°F. Predict the distance through which the rain drops must fall before they are frozen.

3–10 In outer space there is no atmosphere and therefore no convection heat transfer, but there is radiation heat transfer. Using the expression

$$Q = A\sigma T^4$$

for the heat loss from a body having surface area A to outer space at zero kelvin, derive an expression for the lumped heat capacity temperature T as a function of time t.

3–11 Thermocouples of 0.02-in.-diameter steel are used to monitor the temperature inside a furnace. The convective heat-transfer coefficient is 12 Btu/hr · ft^2 · °F for the thermocouple. Determine the time constant of the thermocouple in the furnace and the time for the thermocouple to sense, within 0.2°F, an abrupt 10°F temperature change inside of the furnace.

3–12 A gravel bed is used to store and release thermal energy. The thermal energy is exchanged with an air stream. If the gravel has a typical particle size of 5 mm in diameter, if it has the same thermal properties as silicon, and if the convective heat-transfer coefficient is 18 W/m^2 · °C between the gravel particles and the air stream, estimate the time before 95% of the thermal energy of the gravel is exchanged with the air stream. Assume that the gravel particles are spherical.

***3–13** Lumped heat capacity analysis may be used on problems such as that diagrammed in figure 3–36. Show that the solution to this problem can be written, for constant heat transfer $\dot{q}_A$, as

$$\theta(t) = \theta_i e^{-mt} + (1 - e^{-mt})\frac{\dot{q}_A}{h}$$

where $\theta(t) = T(t) - T_\infty$, $\theta_i = T_i - T_\infty$, and $m = h/\rho L c_p$. (*Hint:* Use an energy balance and solve for the nonhomogenous differential equation.)

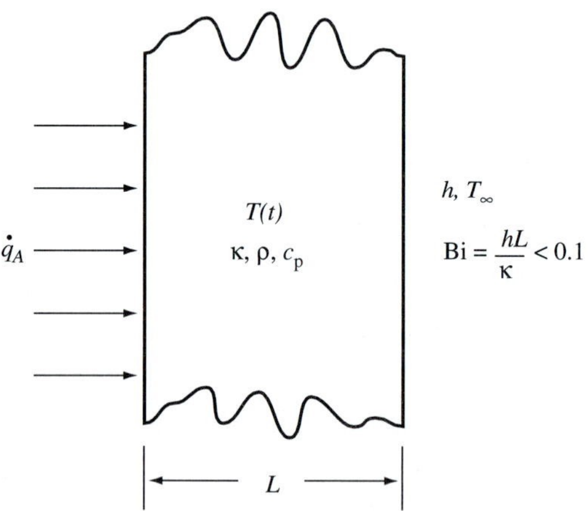

Figure 3–36 Lumped heat capacity for two boundary conditions.

3–14 A building wall is 0.5 m thick and has uniform properties $\rho = 500$ kg/m^3, $c_p = 473$ kJ/kg · °C, $\kappa = 43.0$ W/m · °C, and $T_0 = 15$°C. If one side of the wall is suddenly subjected to a heat loss of 200 W/m^2 and the other side to convection heat transfer with air at 20°C such that $h = 10$ w/m^2 · °C, determine the wall temperature after one hour. (*Hint:* See the result of problem 3–13.)

Section 3–3

3–15 Determine the value for the error function, erf, having an argument of 0.67.

3–16 Determine the value for the error function, erf(x), if its argument is 2.8.

3–17 A large slab of asphalt 0.5 m thick is at 40°C when it is suddenly covered with cold water and the surface temperature of the asphalt becomes 10°C. The properties of asphalt are $\kappa = 0.062$ W/m · °C, $c_p = 0.92$ kJ/kg · °C, and $\rho = 2115$ kg/m^3. Determine the surface heat transfer after the asphalt has been covered for 30 min. Also determine the maximum time period during which the semi-infinite solid may be used as a model to predict the thermal behavior of the asphalt.

3–18 Special heat shield material has the following properties $\kappa = 0.039$ Btu/hr · ft · °F, $c_p = 0.28$ Btu/lbm · °F, and $\rho = 45.0$ lbm/ft^3. It is used as a 2-in.-thick lining of a steel safe-deposit box. If the box is suddenly exposed to a constant high temperature during a fire, estimate the time before the inside of the box begins to change temperature.

3–19 Show that the conduction heat transfer through a semi-infinite solid having initial temperature T_i and subjected to a boundary heat transfer $\dot{q}_B$ may be predicted

$$\dot{q}_A = \dot{q}_B\left(1 - \text{erf}\frac{x}{2\sqrt{\alpha t}}\right) - \dot{q}_B\frac{x}{\sqrt{\pi}}e^{-\frac{x^2}{4\alpha t}}\left(2 - \frac{1}{\sqrt{\alpha t}}\right)$$

3–20 A steel plate is flame hardened by heating the surface with a constant heat flux of 80000 W/m^2 for 1 min. Determine the temperature of the steel plate at a depth of 1 cm in from the heated surface if the initial temperature of the plate is 20°C.

3–21 Determine the cooling rate of the surface of a large aluminum plate 2 min after it has been cooled from a uniform temperature of 250°F by cooling the surface at a rate of 1000 Btu/hr · ft^2. (*Hint:* The rate of change of temperature, dT/dt, needs to be determined.)

3–22 A large steel casting is to be case hardened over a portion of its area. This area is heated at 800°F for 1 min. If the casting is 80°F before heating and the surface is held at 800°F during the case hardening process, to what temperature will the casting be heated at a depth of 1 in. after 1 min?

3–23 A large pond of water is frozen with a layer of 1 ft of ice at 32°F. If cold air at −20°F and with a convective heat-transfer coefficient of 100 Btu/hr · ft^2 · °F blows over the ice surface, how long will it be until the water just below the ice begins to freeze?

3–24 Determine an equation for predicting the conduction heat transfer at any location, x, in a semi-infinite solid having an initial temperature T_i and subjected to convection heat transfer at the boundary surface. (*Hint:* Determine $\partial T/\partial x$.)

3–25 Sheets of plate glass 0.5 in. thick are cooled from 800°F by placing them in an air quench. The air is 100°F, and the convective heat transfer coefficient is 10 Btu/hr · ft^2 · °F. Determine the surface and the centerline temperature after 20 min of cooling. Also determine the thermal energy exchange after 20 min of cooling and the total heat exchanged per unit of area.

3–26 Determine the first five terms of the expansion equation (3–26), used to predict the temperature distribution through an infinite plate subjected to a convective boundary condition, if the Biot Number is 2.0. Assume that the thermal diffusivity is 1.0×10^{-6} m^2/s and L is 1.0. These terms will be functions of x and time, t. Then evaluate the truncated five-term series for x of 1 and at 20 s.

3–27 Determine the first five terms of the series expansion, equation (3–38), used to predict the temperature distribution through an infinite cylinder subjected to a convective boundary condition, if the Biot Number is 5.0. Assume that the thermal diffusivity is 2×10^{-6} ft^2/s and r_0 is 1 ft. Then evaluate the truncated five-term series for r of r_0, $r_0/4$, $r_0/2$, and $3r_0/4$, and at a time of 60 s.

3–28 Determine the first four terms of the series expansion, equation (3–46), used to predict the temperature distribution through a sphere subjected to a convective boundary condition, if the Biot Number is 0.5. Assume that the thermal diffusivity is 2.5×10^{-6} m^2/s and r_0 is 1.0 m. Then evaluate the truncated four-term series for r of r_0 and at a time of 120 s.

3–29 Steel ball bearings 50 mm in diameter are annealed by heating to 1150 K and then cooling to 400 K in air at 325 K. If the convective heat-transfer coefficient between the balls and air is 50 W/m^2 · °C and the properties of the ball bearings are $\rho = 7800$ kg/m^3, $c_p = 600$ J/kg · °C, and $\kappa = 10$ W/m · K, estimate the time required for cooling the ball bearings to 400 K.

3–30 Long 10-cm-diameter 1% carbon steel bars at 27°C are heated in an atmosphere at 1000°C with $h = 400$ W/m^2 · °C. Determine the time before the steel bars reach 400°C at a depth of 1 cm below the material surface.

3–31 Estimate the time required to cook a bratwurst in boiling water. Also determine the percentage of thermal energy exchange during this process. Assume the bratwurst is at 6°C when placed in the boiling water, that the convective heat transfer coefficient is 100 W/m^2 · °C, and that the bratwurst is cooked when its center temperature is 80°C. Use these properties for the bratwurst: $\rho = 880$ kg/m^3, $\kappa = 0.52$ W/m · °C, $c_p = 3350$ J/kg · °C, diameter = 20 mm, and $D \ll L$ (length of the bratwurst).

3–32 Steel ball bearings 12 mm in diameter are annealed by heating them to 1150 K and then cooling to 400 K in air at 50°C. If $h = 20$ W/m^2 · °C and for the steel balls $\rho = 7800$ kg/m^3, $c_p = 600$ J/kg · °C, and $\kappa = 40$ W/m · °C, estimate the time required for a ball's center to cool to 127°C.

3–33 Oranges 5 in. in diameter are placed in a cooler at 75°F. If the cooler temperature is 38°F, the convective heat transfer coefficient is 18 Btu/hr · ft^2 · °F, and the properties of oranges are the same as those of water, determine the time before the surface of the oranges are 40°F.

3–34 Some 10-cm ball bearings at 200°C are quenched in oil at 20°C, where h = 200 W/m^2 · °C. Using properties of 40% nickel steel for the ball bearings, determine

(a) Time for the surface to reach 80°C.

(b) Thermal energy lost to the oil by each ball in this time.

Section 3–4

3–35 Cubes of concrete are heated in a kiln. The cubes, shown in figure 3–37, are put in at 60°F, the kiln temperature is 400°F with h = 5 Btu/hr · ft^2 · °F, and the concrete's properties are κ = 0.79 Btu/hr · ft · °F, c_p = 0.21 Btu/lbm · °F, and $\alpha = 8.3 \times 10^{-6}$ ft^2/s. Determine the center temperature and the corner temperatures of the cubes after being heated in the kiln for 3 hr.

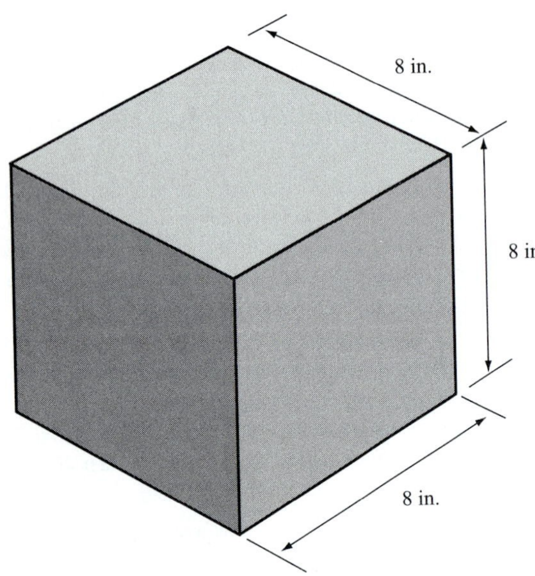

Figure 3–37 Concrete cubes.

3–36 Pyrex glass is molded into disks 4 cm thick by 10 cm in diameter. Upon leaving the mold the disks are at 1800°C are then air quenched at 30°C for 5 min before being placed in an oven to remove thermal stresses. It has been found that if the temperature difference in the Pyrex is more than about 200°C/cm irreversible cracks appear in the material, making the product unacceptable. Assuming that h = 10 W/m^2 · °C for the air quench and for the Pyrex, κ = 1.4 W/m · °C, c_p = 835 J/kg · °C, and ρ = 2225 kg/m^3, what recommendations do you have regarding the air quench time?

3–37 Long 2-in.-diameter oak dowels are heated at one end as shown in figure 3–38. Determine the time before the end reaches 100°F around the edge of the dowel, assuming an initial temperature of 50°F for the dowel.

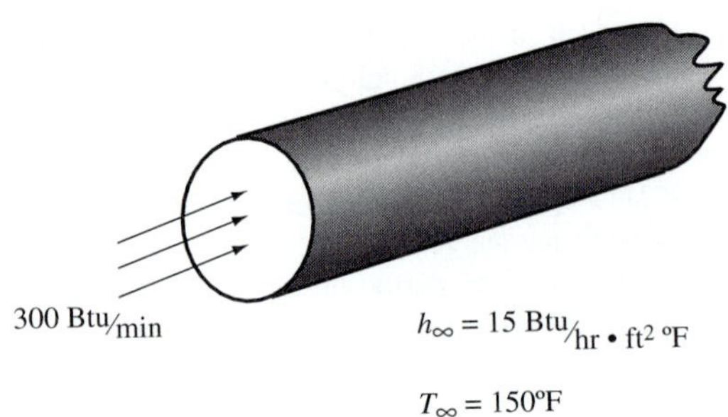

Figure 3–38

3–38 Long 20-cm × 20-cm square steel stock having κ = 50 W/m · °C and $\alpha = 1 \times 10^{-5}$ m^2/s is put in oil at 100°C when steel is at 300°C. If the convective heat transfer coefficient is 150 W/m^2 · °C determine the temperature of the center and the edge of the stock after being in the oil 2 min.

3–39 A 1 pound (lbm) square of butter, 2.5 in. × 2.5 in. × 5 in. at 50°F is placed in a freezer at 28°F as shown in figure 3–39. If h = 10 Btu/hr · ft^2 · °F, the thermal conductivity of butter is 0.114 Btu/hr · ft · °F, c_p = 0.52 Btu/lbm · °F, and ρ = 56 lbm/ft^3, determine the time for the center to reach 35°F.

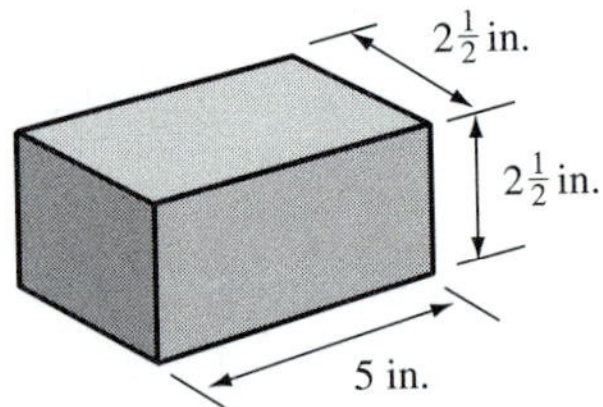

Figure 3–39

Section 3–5

3–40 Referring to figure 3–23, estimate the maximum cooling rate for a Jominy bar at a Jominy distance of 15 cm. At what time does this occur?

3–41 In figure 3–22 the Jominy distance, or distance from the quenched end of a Jominy bar, affects the metal hardness. If you wanted to obtain a surface hardness of R_C (Rockwell C) 55 instead of 59 as indicated in figure 3–22b, suggest a surface cooling rate.

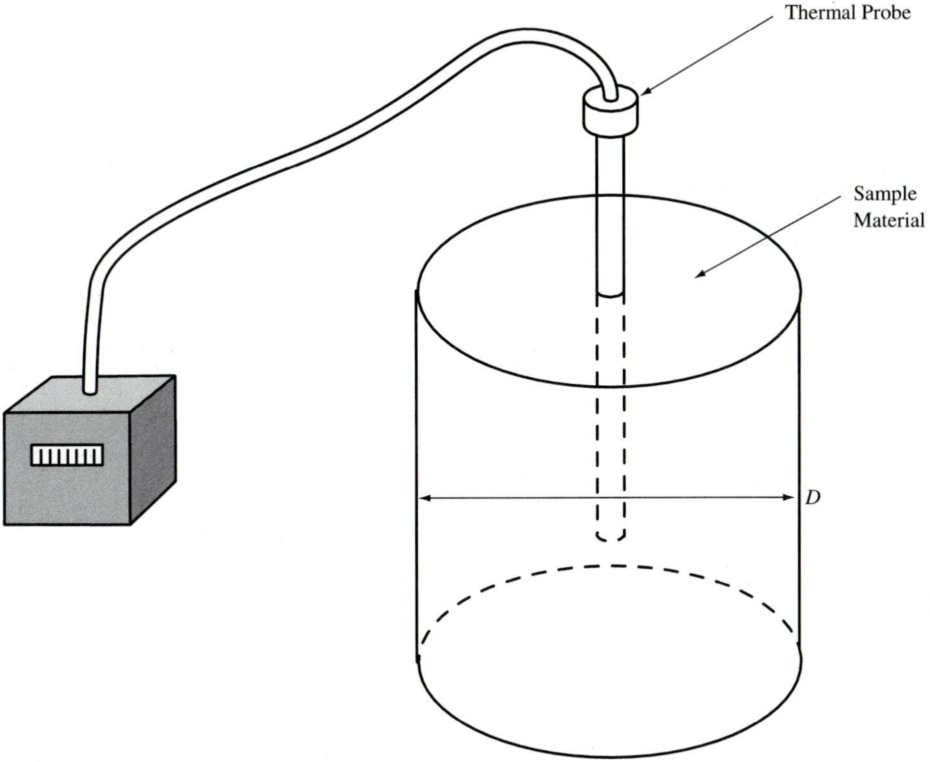

Figure 3–40 Thermal probe setup.

3–42 Predict the surface cooling rate of a Jominy bar at D_{qe} of 30 mm and 50 mm after 2 min.

3–43 Predict the coldest and hottest temperatures 1 ft below the earth's surface at a location where the surface temperature may be expected to vary seasonally from 100°F to −40°F in a periodic fashion. When do these temperatures occur if the location is in the northern hemisphere?

3–44 There has been some discussion about the long-range climatic cycling of the earth. Some people have suggested that every seven years the earth's seasonal temperatures are abnormally low. How would this seven-year cycle compare to yearly cycling of the earth's temperature, with regard to the changes in the earth's temperature?

3–45 If the earth's seasonal temperature at a particular location is described by

$$T(0, t) = (30°\text{C}) \cos \omega t$$

with $t = 0$ on July 24, estimate the maximum and minimum temperatures of the earth at a depth of 2 m below the surface.

3–46 Tests are to be conducted using a thermal probe device to determine the thermal conductivity of a sample of earth as shown in figure 3–40. Estimate the minimum diameter of sample D that can be used if the tests need to be run at least 3 min. Assume the thermal diffusivity of the material is no greater than 5×10^{-6} m²/s.

3–47 If the buried power cable of example 3–16 is subjected to a power transmission of 1200 amp at 440 volts, estimate the equilibrium temperature of the power cable.

Section 3–6

3–48 A large cast-iron plate at 300°C is suddenly subjected to a coolant at 15°C and where h = 100 W/m² · °C. Write the explicit node equation for nodes 1 and 2 as shown in figure 3–41. Then estimate the temperature of node 1 after 18 s.

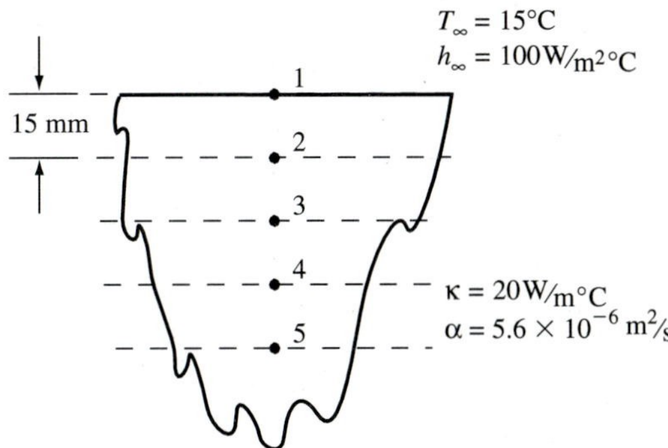

Figure 3–41 Large cast-iron plate cooled by convection.

3–49 Write the node equation for node number 12 in the composite medium shown in figure 3–42. For the explicit formulation of the node equations, determine the maximum time step permitted by this node if the following properties apply: $\rho_A = 2000\ \text{kg/m}^3$, $\rho_B = 8000\ \text{kg/m}^3$, $c_{pA} = 0.9\ \text{kJ/kg}\cdot{}^\circ\text{C}$, $c_{pB} = 0.4\ \text{kJ/kg}\cdot{}^\circ\text{C}$, $\kappa_A = 5\ \text{W/m}\cdot{}^\circ\text{C}$, and $\kappa_B = 80\ \text{W/m}\cdot{}^\circ\text{C}$.

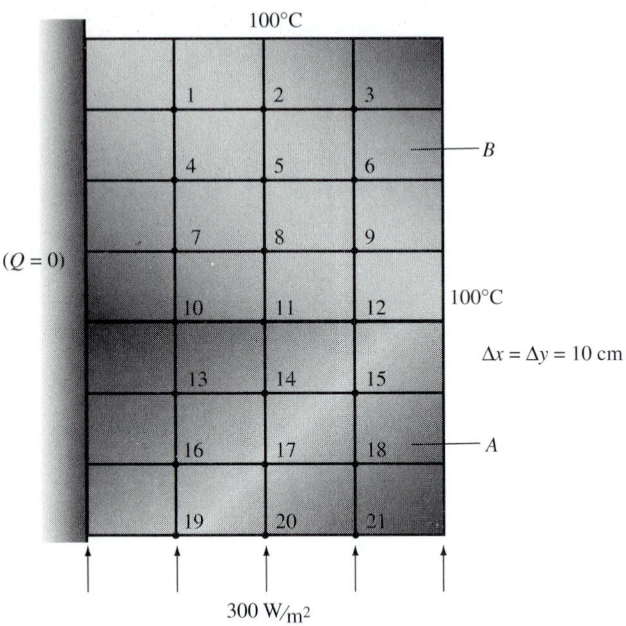

Figure 3–42

3–50 A model of a wall/ceiling of a building is shown in figure 3–43. Consider the corner indicated and write the node equations for nodes 4 and 5. Write these equations in both explicit and implicit form using $\kappa = 0.1\ \text{W/m}\cdot\text{K}$, $c_p = 1500\ \text{J/kg}\cdot\text{K}$, and $\rho = 100\ \text{kg/m}^3$. Then determine the maximum time step allowed for the explicit method by either of these two node equations.

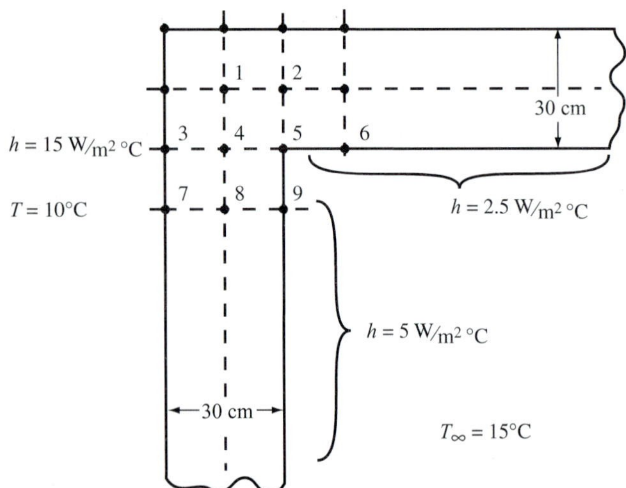

Figure 3–43

3–51 Consider a large steel and concrete boundary as shown in figure 3–44. Write the node temperature equation for transient conduction heat transfer at the boundary node m, n using explicit and implicit formulations. Also determine the maximum time step allowed for an explicit method of solution for the transient conduction heat transfer.

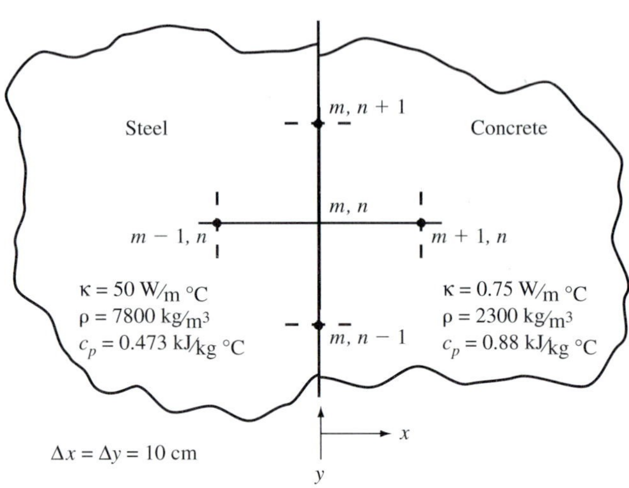

Figure 3–44

3–52 For the fin shown in figure 3–45 write the necessary node equations to solve for unknown nodal temperatures if the following data apply: $T_\infty = 10^\circ\text{C}$, T_i = initial temperature of fin = 250°C, T_0 = base temperature of fin = T_i, $h_\infty = 10\ \text{W/m}^2\cdot{}^\circ\text{C}$, and $\kappa = 200\ \text{W/m}\cdot{}^\circ\text{C}$, $\rho = 2000\ \text{kg/m}^3$, and $c_p = 900\ \text{J/kg}\cdot\text{K}$.

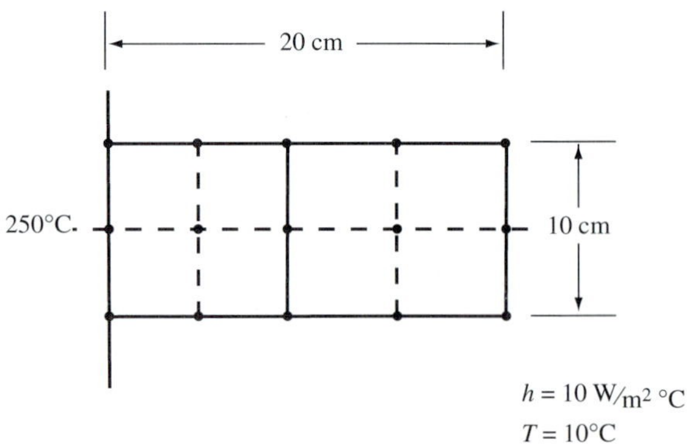

Figure 3–45

3–53 For the heat exchanger tube and fin arrangement shown in figure 3–46, write the node equations for nodes 2 and 5 to determine the temperatures during transient conduction heat transfer. Write the equations using both explicit and implicit formulations.

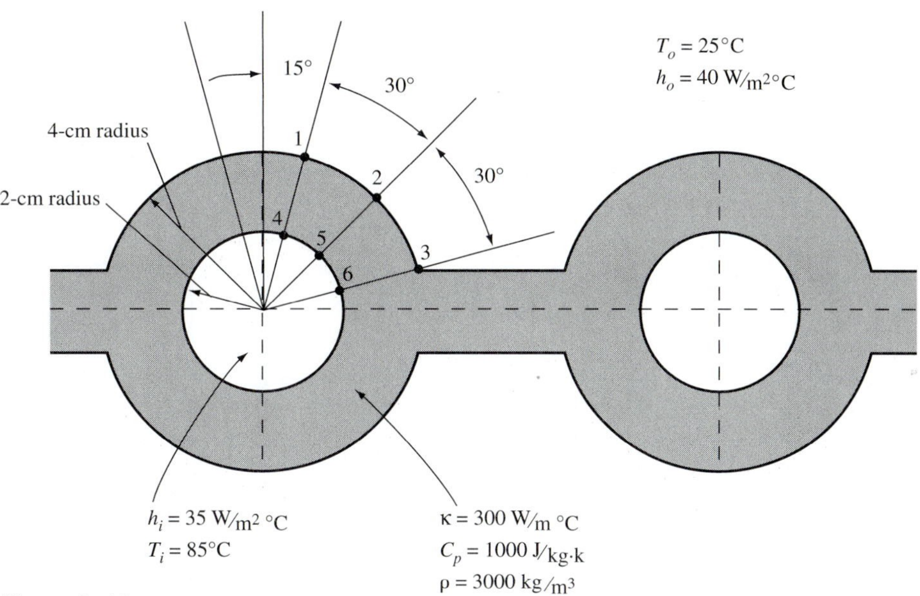

Figure 3–46

Section 3–7

3–54 A large spherical tank is filled with liquid nitrogen at −200°C. The tank wall is approximated by elements behaving as infinite walls with initial temperature of 10°C when the surroundings are at 10°C, as shown in figure 3–47. The tank is made of 1% carbon steel 2 cm thick with convective heat transfer coefficients of 40 W/m² · °C for air and 60 W/m² · °C for nitrogen. Predict the temperature profile through the wall after a minimum of 5 s using graphical methods. Also determine the time for the outside wall surface to drop to 5°C. Use $\kappa = 0.4$ W/m · K for nitrogen at −200°C.

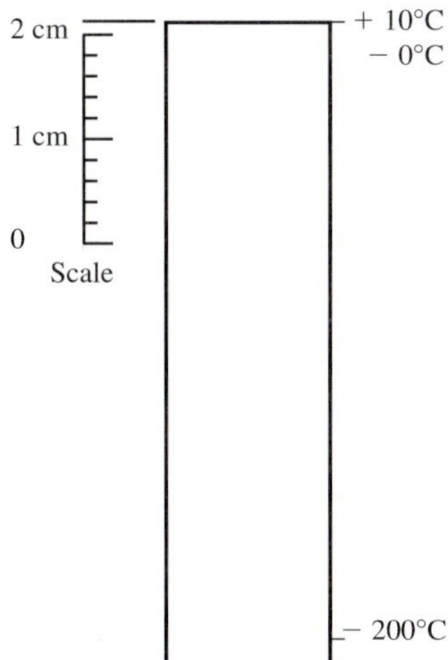

Figure 3–47

3–55 Using graphical methods as indicated in figures 3–33 and 3–34, determine the temperature distribution through the two materials after 60 min, using a grid of 1 in. as indicated in figure 3–48.

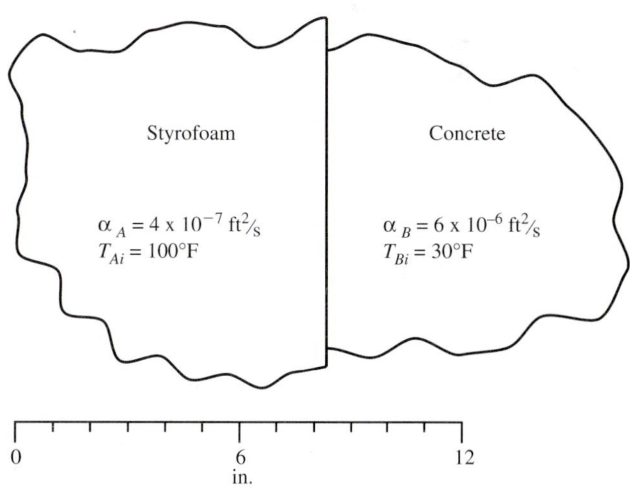

Figure 3–48

3–56 Using graphical methods, determine the temperature distribution through the two materials after 15 min, using a grid of 2 cm as indicated in figure 3–49.

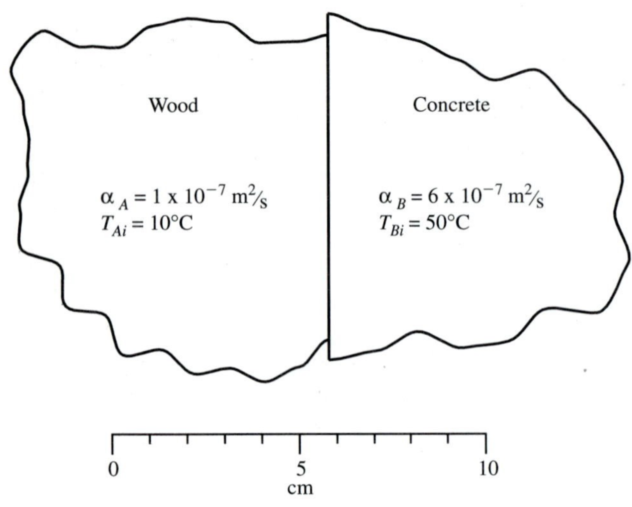

Figure 3–49

3–57 Using graphical methods, determine the temperature distribution through the material shown in figure 3–50 with the convective boundary condition, after 10 min. Use a grid of 1 cm.

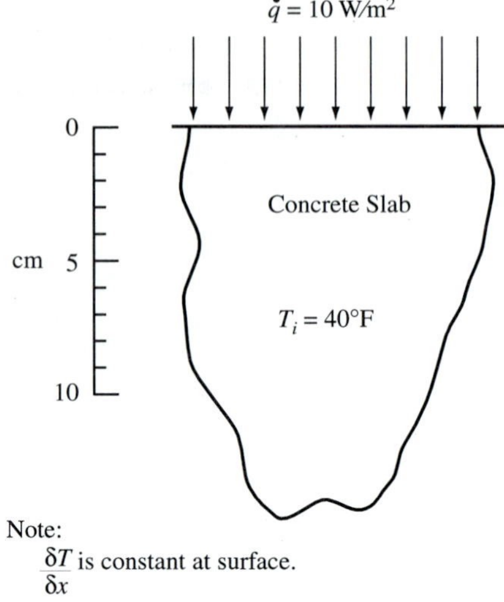

Figure 3–50

REFERENCES

1. Carslaw, H. S., and J. C. Jaeger, *Conduction of Heat in Solids,* 2nd edition, Oxford, UK, 1965.
2. Hill, J. M., and J. N. Dewynne, *Heat Conduction,* Blackwell Scientific, Oxford, UK, 1987.
3. Schneider, P. J., *Conduction Heat Transfer,* Addison-Wesley, Reading, MA, 1955.
4. Heisler, M. P., Temperature Charts for Induction and Constant Temperature Heating, *Trans. ASME,* 69, 227–236, 1947.
5. Grober, H., S. Erk, and U. Grigull, *Fundamentals of Heat Transfer,* 3rd edition, McGraw-Hill, New York, 1982.
6. Schneider, P. J., *Temperature Response Charts,* John Wiley & Sons, New York, 1963.
7. Langston, L. S., Heat Transfer from Multidimensional Objects Using One-Dimensional Solutions for Heat Loss, *Int. J. Heat Mass Transfer,* 25, 1, 149–150, 1982.
8. McGaw, R. W., A Full-Cycle Heating and Cooling Probe Method for Measuring Thermal Conductivity, *J. Heat Transfer,* ASME 84-WA/HT-109, 1984.
9. Rolle, K. C., and R. A. Wetzel, Experimental Determination of the Thermal Conductivity of Granular Materials as a Function of the Stress Level, *Proceedings of the 13th National Passive Solar Conference, Solar 88,* vol. 13, pp. 46–56, Cambridge, MA, June 1988.
10. Jakob, M., *Heat Transfer,* vol. 1, John Wiley & Sons, New York, 1949.

FORCED CONVECTION HEAT TRANSFER

In this chapter the general problem of convection heat transfer is set down with discussions of some of the important phenomena that contribute to the process. Fluid flow concepts are discussed, including Bernoulli's Equation applied to tube flow, to open channel flow, and to flow around objects; laminar flow is considered; and viscosity is defined. The Hagen–Poiseuille Equation is derived and discussed, and the causes for turbulent flow are then introduced. The Reynolds Number is defined, and the ideas of dimensional analysis are introduced. Then the boundary layer concept is developed and applied to thermal energy exchange as well as to momentum exchange. A derivation for the convective heat-transfer coefficient is sketched using the mass and energy balances over the boundary layers. Dimensional analysis is also appealed to as another method of obtaining analytic expressions for the convective heat-transfer coefficient.

Convection heat transfer along flat plates and around objects is analyzed, and empirical results are used whenever rigorous mathematical developments are too restrictive. The analysis of convection heat transfer of fluid flowing in closed channels is presented. Laminar and turbulent flow conditions are considered, and entrance effects are discussed. A few pertinent applications of forced convection heat transfer are presented to show how numerical techniques and the computer may be used to advantage: heat exchange around tube banks, heat exchange in packed beds, heat exchange in coulette flow as an approximation for predicting bearing lubrication cooling, and heat exchange in hermetically sealed chambers.

New Terms

C_D	drag coefficient	r_H	hydraulic radius
D_H	hydraulic diameter	δ delta	boundary layer thickness
Nu	Nusselt number	δ_T delta	thermal boundary layer thickness
Pr	Prandtl number	μ mu	dynamic viscosity
Re	Reynolds number	ν nu	kinematic viscosity
wp	wetted perimeter	τ tau	shear stress

4–1 THE GENERAL PROBLEMS OF CONVECTION HEAT TRANSFER

Convection heat transfer may be considered to be the exchange of thermal energy between a surface of a solid or liquid and a fluid that is immediately adjacent to the surface. If the surface is at a temperature T_s and the fluid has a temperature T_∞ that is lower than the surface temperature, then the equation used to describe such a heat transfer is equation (1–57)

$$\dot{Q} = hA_s(T_s - T_\infty) \qquad \textbf{(1 – 59, repeated)}$$

This equation is called Newton's Law of Cooling because the surface is being cooled by a fluid. If the fluid were hot and heated a cooler surface, then equation (1–59) could be written

$$\dot{Q} = hA_s(T_\infty - T_s)$$

and this equation would be Newton's Law of Heating. In a general form, Newton's Law of Convection may be written

$$\dot{Q} = hA_s\Delta T \tag{4–1}$$

where the ΔT is understood to be a positive temperature difference and the heat transfer is from the higher- to the lower-temperature region. If the convection coefficient, h, is known, then equation (4–1) is simple and straightforward, and in table 1–4 we saw some approximate values for h for five specific conditions.

Unfortunately, the engineer and technologist are faced with many more varied conditions than the five listed in table 1–4, and it is often necessary to analyze closely the mechanisms of convection to obtain a reliable value for the convection coefficient. In figure 4–1 is shown a situation where a fluid is cooling a surface. The fluid that is nearest the surface will become warm and conduct some of its newly acquired thermal energy to other regions of the fluid, regions further away from the surface. Also, the newly heated fluid may move up into the cooler regions of the fluid, carrying with it its thermal energy. As a result of the motion of the warm fluid away from the surface, some cool fluid moves in to occupy that region next to the surface. This mechanism is called *free* or *natural convection* to indicate that it occurs without any outside sources. The only conditions needed for free convection are a gravitational field and a difference in temperature between a fluid and a surface contacting that fluid. Free convection involves both the transfer of energy as heat and the transfer or motion of mass. We will study free convection more in Chapter 5.

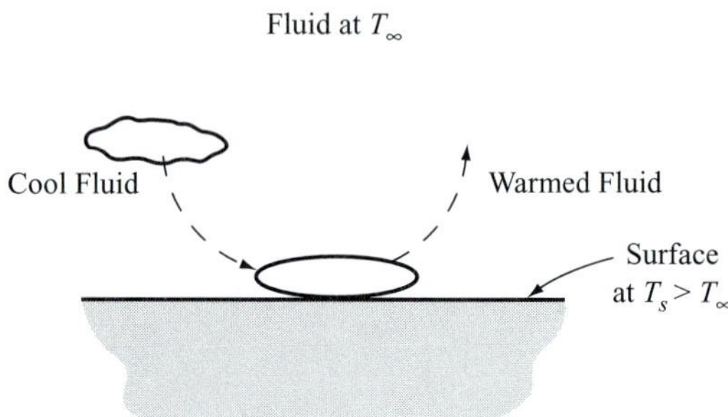

Figure 4–1 Free convection cooling mechanism.

If the fluid flows past the surface because of some external power source we call the heat transfer *forced convection* to distinguish it from *free convection*. In figure 4–2 is shown a situation of forced convection. Notice that the fluid receiving heat (for the case of surface cooling) flows downstream and fluid upstream is following along immediately to replace that warmed fluid. If the fluid particles are moving along parallel to each other and to the surface the flow is called *laminar flow*, however if the fluid particles tend to

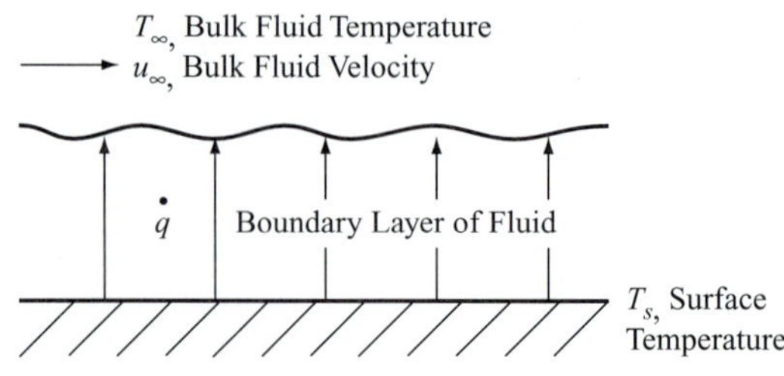

Figure 4–2 Forced convection with laminar flow of a fluid.

flow perpendicular as well as parallel they will quickly follow a random flow pattern, and such flow is called *turbulent flow*. In figure 4–3 is shown the difference between laminar flow, where the fluid seems to travel along well-defined paths or layers, and turbulent flow, where the fluid follows unpredictable paths. In forced convection where the fluid flow is turbulent, the process of heat transfer involves both energy and mass exchanges, but in laminar flow convection the energy exchange will be by conduction. As we will see later, there may also be momentum transfer between the fluid particles in laminar and turbulent flow, which can account for some convection heat transfer.

Figure 4–3 Sketch of the general character of flow for laminar and for turbulent flow.

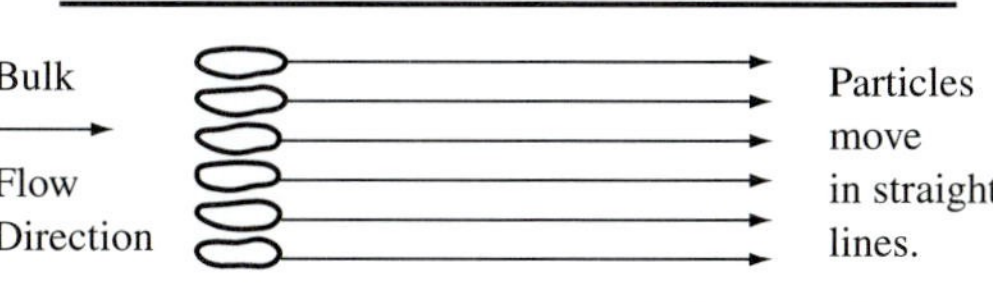

(a) General character for laminar flow of fluid in a tube or channel

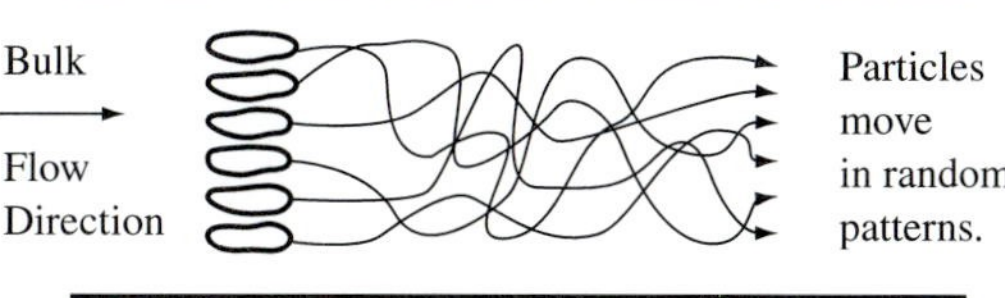

(b) Turbulent flow in a tube or channel

The power source in forced convection situations may be a pump, a fan, or a compressor that can push the fluid past the surface. For the case where the surface is moving in a stationary fluid, the power source may be a jet engine (if the surface is the outside of an aircraft for instance), an auto engine if the surface is the outside of an automobile, or an electric motor powering a rotating shaft if the surface is a bearing surface and the fluid is a lubricant in that bearing. There are many other physical situations for which the reader may identify different power sources. In any instance, some power is required to produce the motion of the fluid, independent of the heat transfer. In free convection we may interpret gravity to be the source of power that produces motion. All convection processes then involve some effective input power for motion, and that power may reveal itself as an energy generation in the fluid. It is not energy generation in the sense of some source of energy in the fluid but rather as internal frictional resistance that expends the power during fluid motion. This phenomenon is often called *viscous dissipation*, and it can be mathematically modeled as an energy generation.

In summary, the general problem of convection heat transfer is the application of Newton's Law of Convection

$$\dot{Q} = hA_s\Delta T \qquad \textbf{(4–1, repeated)}$$

with the stipulation that the convection heat-transfer coefficient h be a parameter used to describe the complicated mass, momentum, and energy exchanges in a fluid near a surface. Included in the coefficient h is the potential of significant energy generation in the fluid. The analysis of engineering problems involving the prediction of h values will necessitate the applications of at least the conservations of mass and energy, the associated Fourier's Law of Conduction, and the laws describing the fluid flow. We will consider the concepts of fluid flow in the next section.

4–2 CONCEPTS OF FLUID FLOW AND DIMENSIONAL ANALYSIS

Fluid flow often occurs with nearly constant temperature or isothermal conditions. If there is no heat transfer (or if the heat transfer is just enough to keep the fluid isothermal) and if there is no external or viscous work or power, fluid flow can be described by *Bernoulli's Equation*

$$\frac{dp}{\rho} + \mathbf{V}d\mathbf{V} + gdz = 0 \tag{4–2}$$

which results from a momentum balance of a fluid particle. A form of Bernoulli's Equation that can be applied to a macroscopic amount of fluid, rather than a particle, is obtained by integrating equation (4–2) between two locations, station 1 and station 2:

$$\frac{p_2}{\rho_2} - \frac{p_1}{\rho_1} + \frac{1}{2}(\mathbf{V}_2^2 - \mathbf{V}_1^2) + g(z_2 - z_1) \tag{4–3}$$

The application of Bernoulli's Equation to incompressible fluids where the density is constant gives

$$\frac{p_2 - p_1}{\rho} + \frac{1}{2}(\mathbf{V}_2^2 - \mathbf{V}_1^2) + g(z_2 - z_1) = 0 \tag{4–4}$$

which is more convenient than equation (4–3). Notice that these three equations are written such that each term has units of energy per unit mass, or specific energy. Often Bernoulli's Equation is written as pressure terms or length or "head" terms. For instance, if we consider an incompressible fluid, equation (4–4) may be written

$$(p_2 - p_1) + \frac{\rho}{2}(\mathbf{V}_2^2 - \mathbf{V}_1^2) + \gamma(z_2 - z_1) = 0 \qquad \text{(SI units)} \tag{4–5}$$

where γ is the specific weight, ρg. The terms of this equation all have units of pressure such as kPa. The English Engineering form of equation (4–5) would be written

$$(p_2 - p_1) + \frac{\rho}{2g_c}(\mathbf{V}_2^2 - \mathbf{V}_1^2) + \frac{g}{g_c}(z_2 - z_1) = 0 \qquad \text{(English units)} \tag{4–6}$$

where density, ρ, would be expected to be in units of lbm/ft^3.

Sometimes the mass unit of slug (1 slug = 1 lbf · s^2/ft) is used instead of lbm. Then Bernoulli's Equation in the form of (4–5) may be used. Also, equation (4–4) may be written in the form

$$\frac{p_2 - p_1}{\gamma} + \frac{1}{2g}(\mathbf{V}_2^2 - \mathbf{V}_1^2) + z_2 - z_1 = 0 \qquad \text{(SI or English units)} \tag{4–7}$$

where each term now has the units of length, such as meters or feet. Equation (4–7) is a common form of Bernoulli's Equation and may be used with either SI or English units. The specific weight $\gamma = \rho g$ (SI, or English if ρ is in slugs/ft^3) and $\gamma = \rho g/g_c$ in English units. Bernoulli's Equation has been used to analyze many systems involving fluid flow, a common and important problem being that of flow through a pipe, tube, or duct as shown in figure 4–4. Here fluid passes through tube section 1 and continues through a converging section to pass through section 2.

It is often convenient to ignore any fluid friction, particularly since Bernoulli's Equation in these forms does not account for any such phenomenon and also because fluids flowing at low velocities give the outward appearance of having no friction or internal resistance.

For the situation shown in figure 4–4 for frictionless fluids the fluid velocity is constant across a section and the velocity profile can be sketched as shown in the figure. The

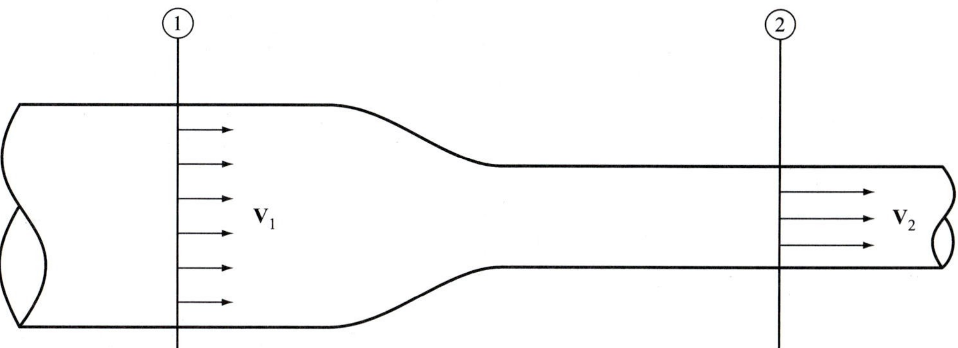

Figure 4–4 Tube flow of a fluid.

conservation of mass may be written for steady flow of the fluid, and it is equation (1–36),

$$\sum \dot{m}_{in} - \sum \dot{m}_{out} = 0$$

For the fluid flow the mass flow rates become

$$\sum \dot{m}_{in} = \rho_1 A_1 \mathbf{V}_1 \quad \text{and} \quad \sum \dot{m}_{out} = \rho_2 A_2 \mathbf{V}_2$$

but $\rho_1 = \rho_2$ for incompressible fluids so that we have for the conservation of mass of an incompressible fluid,

$$A_1 \mathbf{V}_1 = A_2 \mathbf{V}_2 \tag{4–8}$$

and it can be seen that if $A_2 < A_1$, as shown in figure 4–4, the velocity $\mathbf{V}_2 > \mathbf{V}_1$. The increase in velocity due to a decrease in area is indicated by the increased length of the velocity vectors in figure 4–4. The increase in velocity results in an increase in the kinetic energy term in Bernoulli's Equation and a reduction in either the pressure head or the elevation head.

EXAMPLE 4–1 Consider the flow of water at 27°C flowing through a pipe as shown in figure 4–5. There are 20 kg/s flowing and the local gravitational acceleration is 9.8 m/s^2. Estimate the water pressure at elevation point 2 if the pressure is 400 kPa at 1.

Solution Referring to figure 4–5, we can see that if the water is flowing through the pipe at 27°C in a steady flow manner, we may assume that the water is incompressible and apply Bernoulli's Equation in the form of (4–4), (4–5), or (4–7) to the pipe. Using equation (4–7), we have

$$\frac{p_2 - p_1}{\gamma} + \frac{\mathbf{V}_2^2 - \mathbf{V}_1^2}{2g} + z_2 - z_1 = 0$$

but the pipe has a constant cross-sectional area, which means that the velocity is constant. Bernoulli's Equation is then

$$\frac{p_2 - p_1}{\gamma} + z_2 - z_1 = 0$$

and the pressure at point 2 can be found from the rearrangement

$$p_2 = p_1 + \gamma(z_1 - z_2) = p_1 + \rho g(z_1 - z_2)$$

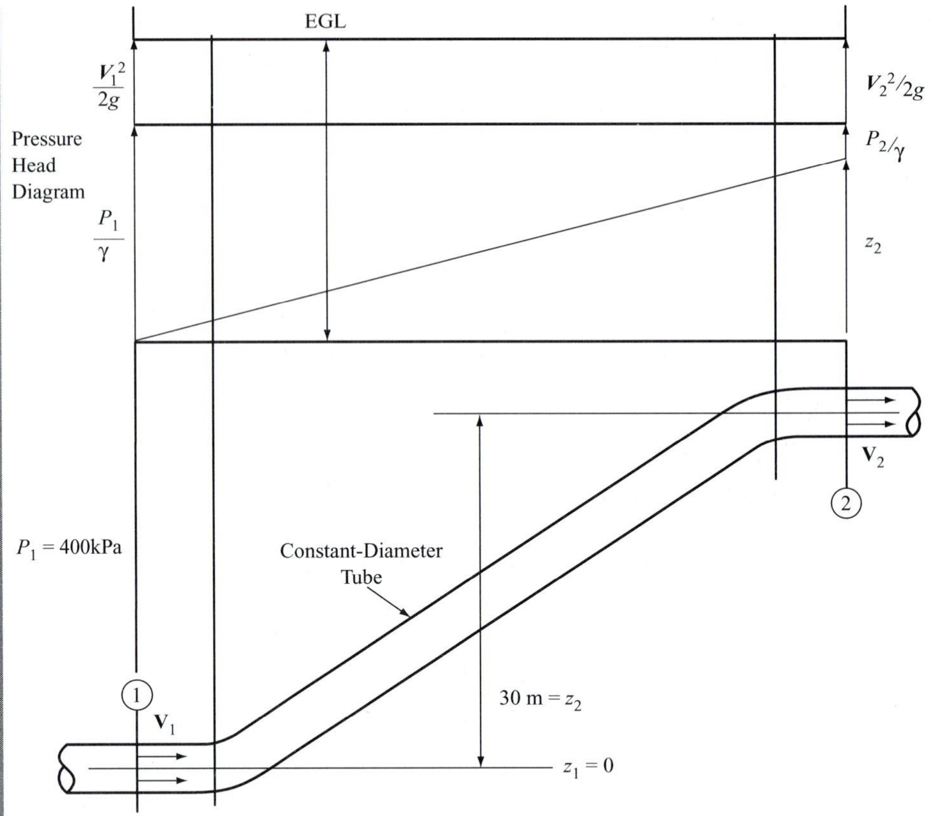

Figure 4–5 Fluid flow through a slanted pipe.

Using a value of 997 kg/m³ for water density from appendix table B–3 and substituting, we obtain

Answer

$$p_2 = 400 \text{ kPa} + \left(997 \frac{\text{kg}}{\text{m}^3}\right)\left(9.8 \frac{\text{m}}{\text{s}_2}\right)(-30 \text{ m}) = 106.881 \text{ kPa}$$

Notice in figure 4–5 that a pressure head diagram is included with the sketch of the slanted pipe. This diagram provides a graphic or visual means of picturing Bernoulli's Equation for many fluid flow conditions. In this example the line that indicates the sum of the kinetic energy head term, $\mathbf{V}^2/2g$, the pressure head term, p/γ, and the elevation head, z, is called the *energy grade line (EGL)*. It is a constant (in meters) for this particular situation of frictionless pipe flow.

EXAMPLE 4–2 A reservoir is filled to a depth of 36 ft with water at 60°F. A conduit 2 ft in diameter is opened at the bottom and water flows out as shown in figure 4–6. Estimate the mass flow of the water flowing through the pipe.

Solution The reservoir may be assumed to be large enough that its depth remains the same for a period of time, so we write Bernoulli's Equation from the water level to the conduit outlet as indicated in figure 4–6. The velocity at point 1 is then zero, $z_1 = 36$ feet, $z_2 = 0$, $p_1 =$ atmospheric pressure, and p_2 is also atmospheric pressure. Bernoulli's Equation, using equation (4–7), is

$$\frac{\mathbf{V}_2^2}{2g} - z_1 = 0$$

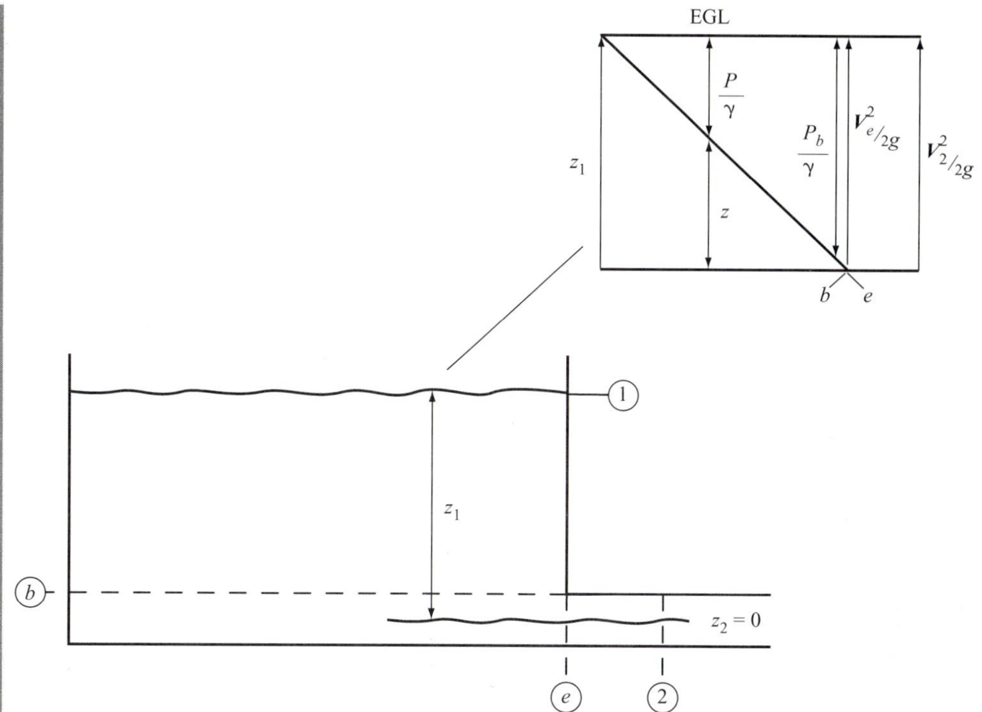

Figure 4–6 Water flowing from a reservoir.

and the velocity head at the conduit equals the elevation head, 36 ft. Solving for the velocity,

$$\mathbf{V}_2 = \sqrt{2gz_1} = \sqrt{2\left(32.2\,\frac{\text{ft}}{\text{s}^2}\right)(36\text{ ft})} = 48.15\,\frac{\text{ft}}{\text{s}}$$

and the mass flow is

Answer

$$\dot{m} = \rho A\mathbf{V} = \left(62.4\,\frac{\text{lbm}}{\text{ft}^3}\right)(\pi)(1\text{ ft}^2)\left(48.15\,\frac{\text{ft}}{\text{s}}\right) = 9439.1\,\frac{\text{lbm}}{\text{s}}$$

A pressure head diagram is included in figure 4–6. This diagram presents the energy grade line (EGL) for a unit mass of water as it flows from the top of the reservoir through the conduit at the bottom. Notice that the pressure head term increases as the elevation head decreases until the bottom is reached, where the velocity head includes all of the elevation head loss.

Many important problems involving fluid flow and heat transfer occur with open channel flow. This sort of flow can occur in a tube or pipe if a liquid is flowing but does not fill the tube. Such a situation is indicated in figure 4–7a. Other examples are the flow of a fluid in a trench, ditch, stream, or river. Condenser cooling water runoff channels at electrical power generating plants involves fluid flow with heat transfer, and the heat transfer plays a crucial role in the processes, particularly if the water is flowing into a cooling pond and is to be recirculated through the plant. For environmental concerns it is also important to understand the heat transfers of such a flow.

There are many other specific examples of open channel flow, and they all have the condition that one surface is exposed to air or some other vapor while the remaining portion of the flow cross section is contained by a pipe wall, earth bank, or some other solid surface. The exposed surface is sometimes called a *free surface* to indicate that it is not

Figure 4–7 Some open channel flow situations.

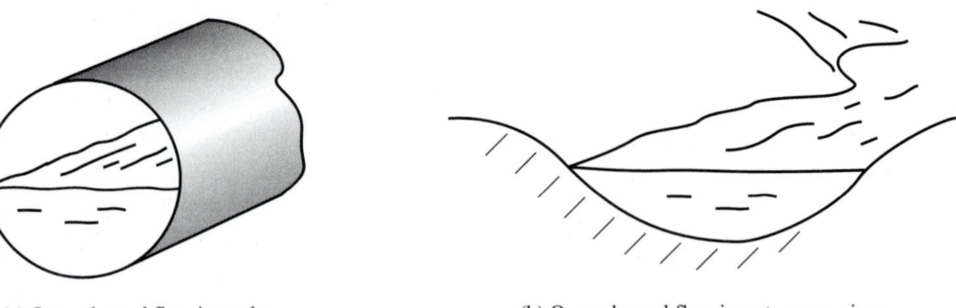

(a) Open channel flow in a tube

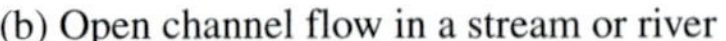

(b) Open channel flow in a stream or river

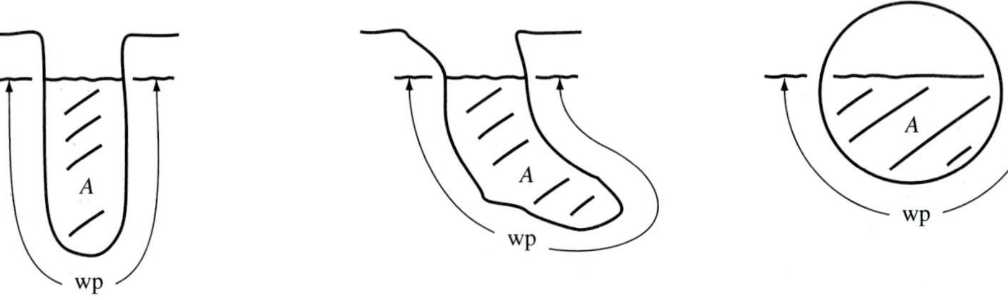

(b) Some other open channel flows with wetted perimeters

the same as the remaining surfaces. An important parameter used to analyze open channel flow and also flow through irregular channels is the hydraulic radius r_H defined as

$$r_H = \frac{A}{\text{wp}} \tag{4–9}$$

where A is the cross section of the fluid flow and wp is the wetted perimeter, or perimeter around the fluid cross section without the free area. Figure 4–7c shows some examples of the wetted perimeter and the hydraulic radius. The hydraulic diameter, D_H, is four times the hydraulic radius; that is,

$$D_H = 4r_H = \frac{4A}{\text{wp}} \tag{4–10}$$

and it is also a parameter that is used in fluid flow analysis. The purpose in defining the hydraulic diameter as four times the radius is to allow the hydraulic diameter to be equal to the actual diameter in the situation where the cross-sectional area is circular and the flow is tube flow. The hydraulic diameter is a convenient parameter for treating irregular cross sections.

EXAMPLE 4–3 Determine the hydraulic radius and hydraulic diameter for the trapezoid channel flow shown in figure 4–8.

Solution The open channel flow shown in figure 4–8 has a trapezoid cross section with an area given by

$$A = (2 \text{ m} \times 4 \text{ m}) + (2 \text{ m})(2 \text{ m})(\tan 30°) = 10.3 \text{ m}^2$$

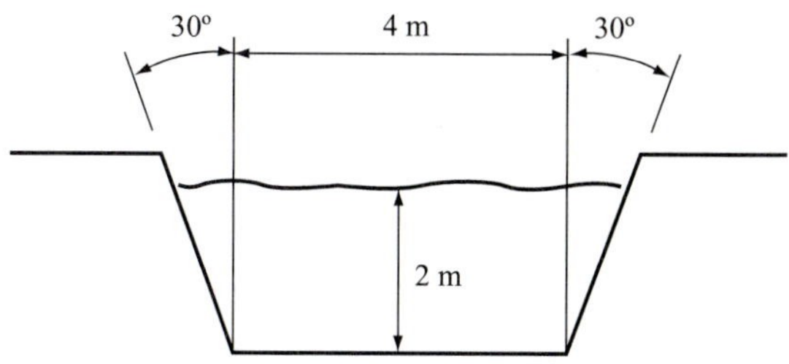

Figure 4–8 Trapezoidal cross section for a typical fluid flow.

The wetted perimeter, wp, is

$$wp = (4\ m) + 2\frac{(2\ m)}{\cos 30°} = 8.61\ m$$

From equation (4–9) we find the hydraulic radius,

Answer

$$r_H = \frac{A}{wp} = \frac{10.3}{8.61} = 1.196\ m$$

The hydraulic diameter is predicted from equation 4–10,

Answer

$$D_H = 4r_H = 4.874\ m$$

EXAMPLE 4–4 Water flows at 3 ft/s and at a depth of 3 ft in a wide open channel as shown in figure 4–9. If the channel is frictionless and has a slope of 2°, estimate the velocity of the water as it leaves the channel after traveling 100 ft. Then estimate the depth of the water at the exit.

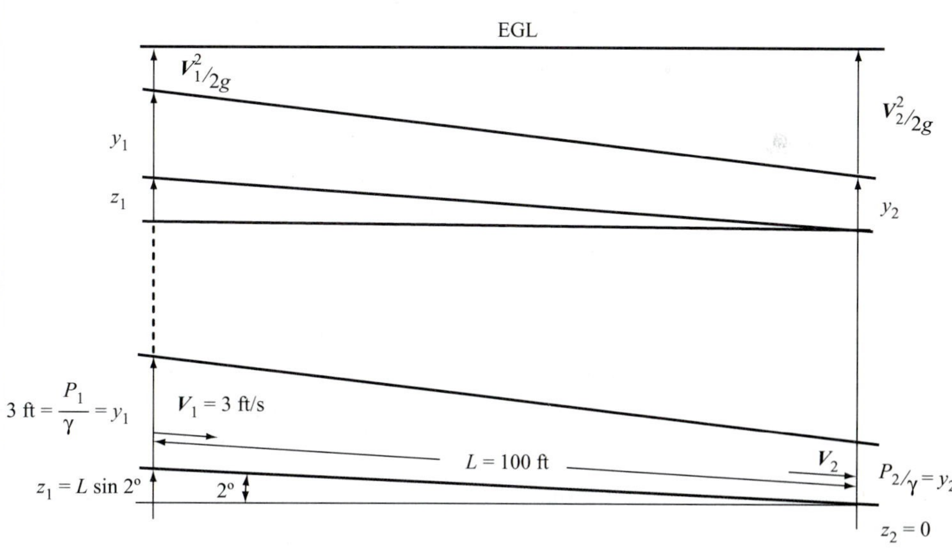

Figure 4–9 Water flowing through a sloped channel.

Solution

We may apply Bernoulli's Equation to the water as it flows through the channel from point 1 to point 2. For an open channel the pressure head is just the depth of the water, so Bernoulli's Equation is

$$L \sin 2° + y_1 + \frac{\mathbf{V}_1^2}{2g} = y_2 + \frac{\mathbf{V}_2^2}{2g}$$

Also, using the conservation of mass, assuming water to be incompressible, and recognizing that the channel is very wide, we have

$$y_1\mathbf{V}_1 = y_2\mathbf{V}_2$$

Combining these two equations gives

$$\mathbf{V}_2\left[L \sin 2^\circ + y_1 + \frac{\mathbf{V}_1^2}{2g}\right] = \mathbf{V}_2[\text{EGL}] = y_1\mathbf{V}_1 + \frac{\mathbf{V}_2^3}{2g}$$

and substituting values, we find the energy grade line (EGL) to be 6.6297 ft, $y_1 = 3$ ft, and $\mathbf{V}_1 = 3$ ft/s. Solving

$$\mathbf{V}_2[6.6297 \text{ ft}] = (3 \text{ ft})(3 \text{ ft/s}) + \frac{\mathbf{V}_2^3}{2(32.2 \text{ ft/s}^2)}$$

gives, after trial and error,

Answer
$$\mathbf{V}_2 = 19.94 \text{ ft}$$

The depth of the water is

Answer
$$y_2 = y_1\frac{\mathbf{V}_1}{\mathbf{V}_2} = (3 \text{ ft})\frac{3 \text{ ft/s}}{19.94 \text{ ft/s}} = 0.45 \text{ ft}$$

The subject of open channel flow is more complex than this example indicates, as we will see shortly. Flow around objects, or flow of an object that is submerged in a fluid, is another condition that often involves convection heat transfer. For frictionless flow, if the velocities are small enough, the fluid will always follow the contour of the object and the path of fluid particles can be traced along *path lines* as shown in figure 4–10. The term *streamline* is often used to describe the direction of the velocity of the fluid particles. If the flow is steady, then the path lines and the streamlines are the same. A good reference for visual examples of flow paths and streamlines is Van Dyke [1].

Figure 4–10 Fluid flow around objects.

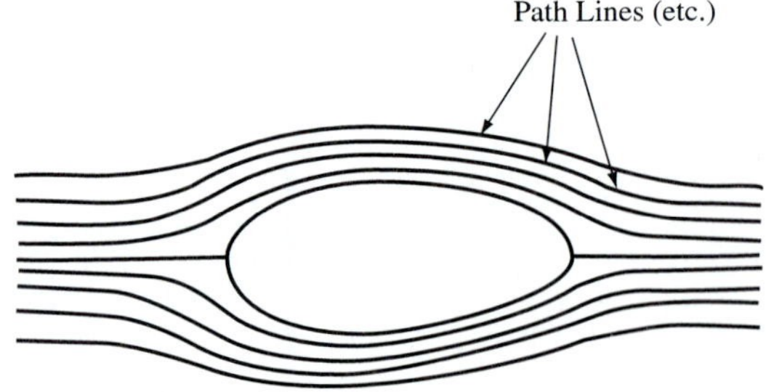

Fluid viscosity is defined by

$$\tau = \mu\frac{\partial \mathbf{v}}{\partial y} \tag{4–11}$$

where μ is called the dynamic viscosity of the fluid, τ is the shear stress applied to a fluid particle or layer of fluid, and $\partial\mathbf{v}/\partial y$ is the change in velocity $\mathbf{v}$ along an x direction in a perpendicular direction, y. The sketch shown in figure 4–11 will help to clarify the situation. Notice that the fluid layer shown in the figure is originally a rectangular shape and

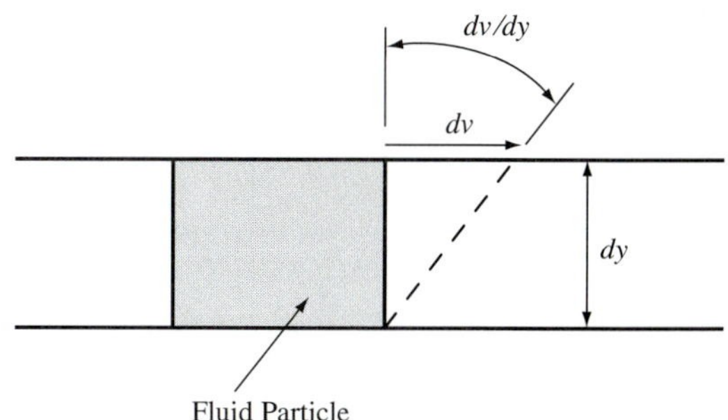

Figure 4–11 Shear of a viscous fluid layer.

is distorted because of the applied shear stress (force per unit area). The fluid remains fixed to the bottom surface relative to the top surface, and the velocity gradient is just the slope of the resulting parallelogram. The fluid layer is assumed to be very thin, of differential thickness, but the relationship of equation 4–11 can be applied to a finite thickness if the viscosity is a constant or able to be predicted for various shear stress conditions.

The units for viscosity may be taken in SI as $N \cdot s/m^2$ or $kg/s \cdot m$. The English units are $lbf \cdot s/ft^2$, or, if lbm units are to be used,

$$1\,\frac{lbf \cdot s}{ft^2} = 32.17\,\frac{lbm}{s \cdot ft}$$

A unit reported in the technical literature for viscosity is the poise (P), named after J. L. M. Poiseuille, a French physician. The definition of the poise is

$$10\ P = 1\,\frac{N \cdot s}{m^2} = 1\,\frac{kg}{s \cdot m}$$

Frequently the centipoise, cP, which is 10^{-2} P is used instead of the poise, because 1 cP happens to be the viscosity of liquid water at 20°C. The English conversion for viscosity is

$$4.788\ cP = 1\,\frac{lbf \cdot s}{ft^2} = 32.17\,\frac{lbm}{s \cdot ft}$$

Sometimes the kinematic viscosity is used, which is denoted as ν and defined by

$$\upsilon = \frac{\mu}{\rho} \tag{4–12}$$

and the units of the kinematic viscosity are m^2/s, ft^2/s, or other comparable units.

The viscosity of a fluid is a measure of the internal resistance of the fluid to flow or shear stresses. A fluid in a static state exhibits no viscosity, but if the fluid is subjected to shears or forces that tend to cause fluid flow, then the fluid will demonstrate its viscosity, or resistance to flow. All fluids have viscosity, but we will often use the term *viscous fluid* to mean a fluid that has a known or observed viscosity. Values for the viscosity of selected materials are given in table 4–1 and in appendix tables B–3 and B–4.

If we consider a differential element of a viscous fluid in a cylindrical tube where the flow is laminar, as shown in figure 4–12, and then sum the forces acting on the element, we obtain

$$\sum F_{\text{axial}} = 0 = p\pi\frac{D^2}{4} - (p - dp)\pi\frac{D^2}{4} - \tau\pi D dL$$

Table 4–1 Selected Values for Fluid Viscosities at 20°C (70°F)

Fluid	cP	N · s/m²	lbf · s/ft²
Honey	1500.0	1.500	3.13×10^{-2}
SAE 50 oil	800.0	0.800	1.67×10^{-2}
SAE 10 oil	70.0	0.070	1.46×10^{-3}
Mercury	1.50	0.0015	3.13×10^{-5}
Water	1.00	0.0010	2.09×10^{-5}
Air	0.018	18×10^{-6}	3.76×10^{-7}

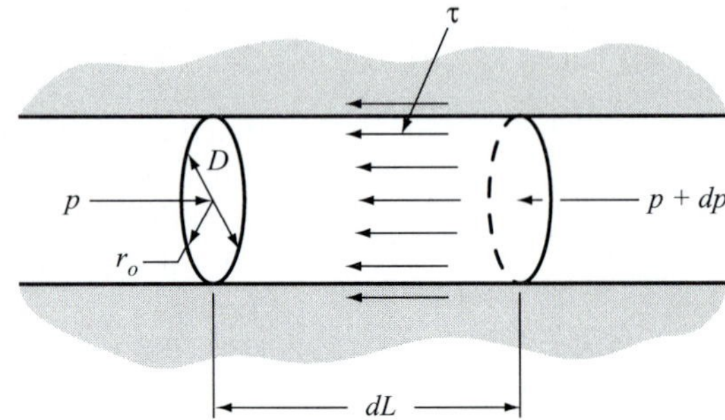

Figure 4–12 Differential element of fluid flowing in a uniform diameter tube.

Reducing this equation we obtain

$$dp = -\frac{4\tau}{D}dL$$

and upon integrating over a finite length L, we have

$$\Delta p = -\frac{4\tau}{D}L \tag{4–13}$$

Further, by using equation (4–11) for a viscous fluid, we have

$$\Delta p = -\frac{4L}{D}\mu\frac{\partial \mathbf{V}}{\partial r}$$

For steady laminar flow the velocity is uniform in the axial direction, so the velocity gradient is an ordinary differential. Then

$$\Delta p = -\frac{4L}{D}\mu\frac{d\mathbf{V}}{dr}$$

and by separating variables, assuming constant viscosity (called a *Newtonian fluid*), and integrating, we obtain

$$\mathbf{V}(r) = -\frac{\Delta p(r^2)}{4L\mu} + C$$

If we assume that the velocity $\mathbf{V}(r)$ is zero at the outer tube surface, where $r = r_o$, then

$$C = \frac{\Delta p(r_o^2)}{4L\mu}$$

and the velocity distribution is

$$\mathbf{V}(r) = \frac{\Delta p}{4L\mu}(r_o^2 - r^2) \tag{4–14}$$

Figure 4–13 Velocity profile in laminar flow of a fluid in a tube.

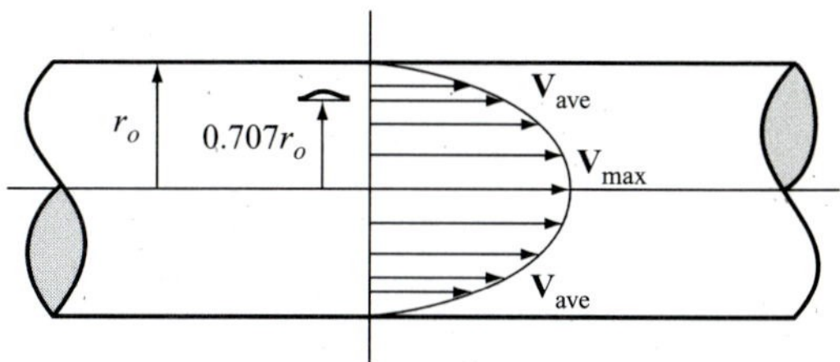

This result shows that the velocity distribution is a parabolic form with the maximum value at the center where $r = 0$. In figure 4–13 the velocity profile for this condition is shown.

Sometimes it is convenient to know the average velocity at a cross section. By using the definition for an average value

$$\mathbf{V}_{\text{average}} = \frac{1}{\int_0^{r_o} 2\pi r dr} \int_0^{r_o} \mathbf{V}(r) 2\pi r dr$$

it can be shown that the average value is

$$\mathbf{V}_{\text{average}} = \left(\frac{1}{2}\right)\frac{\Delta p}{4L\mu} r_o^2 = \left(\frac{1}{2}\right)\mathbf{V}_{\text{maximum}} \tag{4–15}$$

where the maximum velocity occurs at the center: $r = 0$. The average velocity occurs at $r = r_o/\sqrt{2}$, which is left as an exercise for the reader to verify (see practice problem 4–26). The volume flow rate $\dot{V}$ is the product of the average velocity and the cross-sectional area. Using equations (4–14) and (4–15) we have

$$\dot{V} = \mathbf{V}_{\text{average}} A = \frac{r_o^2 \Delta p}{8L\mu} \pi r_o^2 = \frac{\pi r_o^4 \Delta p}{8L\mu} \tag{4–16}$$

Notice that the volume flow rate is dependent on the pressure drop over the tube length L and is proportional to the 4th power of the tube radius, r_o. The result given by equation (4–16) is called the *Hagen–Poiseuille Equation* and was derived by G. H. L. Hagen, a German scientist, and J. L. M. Poiseuille. The Hagen–Poiseuille Equation is often used to predict the expected pressure drop, Δp, in a tube, but it should always be remembered that the result is restricted to laminar flow conditions. If the pressure drop is compared to the general form of Bernoulli's Equation (4–8) with the restriction of no elevation head loss we may write

$$\Delta p = \gamma \Delta(\text{KE}) = \frac{\gamma}{2g}(\mathbf{V}_1^2 - \mathbf{V}_2^2) \tag{4–17}$$

the purpose in writing equation (4–17) is to demonstrate that a pressure drop due to friction (viscosity) represents a potential drop in kinetic energy and it is convenient to set $\mathbf{V}_2$ equal to zero. Then equation (4–17) becomes

$$\Delta p = \frac{\gamma}{2g}\mathbf{V}^2 \tag{4–18}$$

and this equation represents a popular concept for defining tube and pipe frictional effects. The term $\Delta p/\gamma$ is referred to as the frictional head loss, h_L, and the friction factor, f, is defined from equation (4–18) by the relationship

$$h_L = f\frac{L}{D}\frac{1}{2g}\mathbf{V}^2 \tag{4–19}$$

The relationship of equation (4–19) is called the Darcy–Weisbach equation and is used to analyze fluid flow in tubes, whether the flow is laminar or turbulent. Also, the concept involved in equation (4–19) is used to predict head losses in valves, tube elbows and turns, and other restrictions.

Bernoulli's Equation in the form of equation (4–7) can be revised to include fluid friction by writing the modified Bernoulli Equation

$$\frac{p_1 - p_2}{\gamma} + \frac{1}{2g}\left[\mathbf{V}_1^2 - \mathbf{V}_1^2\right] + z_1 - z_2 = h_L \tag{4–20}$$

where h_L is predicted from equation (4–19) or a similar relationship and can represent a summation of various frictional head losses. The head loss, h_L, is an energy loss that is dissipated to the surroundings by heat transfer. Without making this observation it appears that energy may be destroyed in the viscous dissipation of the fluid flow, as the result of equation (4–20) seems to imply. Again, we should remember that heat transfer is necessary to prevent a temperature change of the fluid. We will return to this concept again in later sections.

EXAMPLE 4–5 Estimate the pressure drop in fluid flowing through a 10-m-long by 1-cm-diameter tube at 10 cm/s if the fluid viscosity is 1 cP and its density is 1000 kg/m^3. Also predict a friction factor applicable to this situation using the definition of equation (4–19).

Solution If the fluid flow is assumed to be laminar, then we may predict the pressure drop from equation (4–16). The average velocity is 0.1 m/s and the cross-sectional area is

$$\frac{\pi D^2}{4} = \frac{\pi(0.01\text{ m})^2}{4} = 7.85 \times 10^{-5}\text{ m}^2$$

The volume flow rate is then

$$\dot{V} = \mathbf{V}_{\text{average}} A = 7.85 \times 10^{-6}\,\frac{\text{m}^3}{\text{s}}$$

and from equation (4–16) we can estimate the pressure drop, Δp,

Answer
$$\Delta p = \frac{8L\mu}{\pi r_o^4}\dot{V} = \frac{8(10\text{ m})(0.001\text{ n}\cdot\text{s/m}^2)}{\pi(0.005\text{ m})^4}(7.85 \times 10^{-6}\text{ m}^3/\text{s}) = 320\,\frac{\text{N}}{\text{m}^2} = 0.32\text{ kPa}$$

The friction factor f can be predicted from equation (4–19). Using a gravitational acceleration of 9.8 m/s^2 and specific weight, γ, of 1000 kg/m^3 $\times$ 9.8 m/s^2 = 9800 N/m^3 gives

Answer
$$f = \frac{\Delta p}{\gamma}\left[\frac{D}{L}\right]\frac{1}{2g}\mathbf{V}^2 = \frac{320\text{ N/m}^2}{9800\text{ N/m}^3}\left(\frac{0.001\text{ m}}{10\text{ m}}\right)\frac{1}{2 \times 9.8\text{ m/s}^2}\left(0.1\,\frac{\text{m}}{\text{s}}\right)^2 = 0.00064$$

Another common fluid flow condition is that shown in figure 4–14, where two parallel surfaces or plates are separated by a viscous fluid. The top plate moves or slides relative to the bottom plate, and this results in a fluid flow between the surfaces. For a fluid obeying the relationship of equation (4–12) we have

$$\frac{F}{A} = \tau = \mu\frac{d\mathbf{V}}{dy}$$

or

$$d\mathbf{V} = \frac{\tau}{\mu}dy$$

and

$$\mathbf{V}(y) = \frac{\tau}{\mu}y + C$$

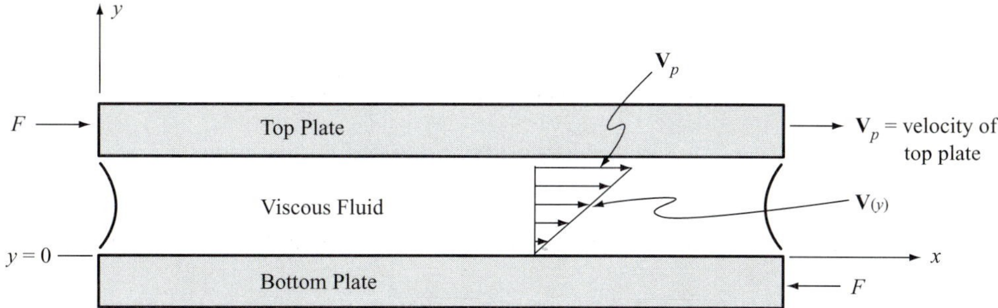

Figure 4–14 Condition of a fluid separating two parallel plates where one plate slides over the other.

If the bottom plate is stationary, as indicated in figure 4–14, then $\mathbf{V}(y) = 0$ at $y = 0$. This gives $C = 0$ and the velocity profile is linear with fluid thickness, y.

Notice again that the power required to move the top plate, $F\mathbf{V}$, is dissipated in the fluid friction and heat transfer occurs to the surroundings to prevent heating of the fluid. The action indicated in figure 4–14 is called *Couette Flow* and is an important starting point for studies of lubrication in rotating bearings. For instance, in figure 4–15 is shown a typical journal bearing and rotating shaft. The space between the shaft and bearing (called the *clearance*) is to be filled with a lubricant such as an oil or grease to reduce friction. The power required to rotate the shaft against the friction of the lubricant is $T\omega$, where ω is the angular velocity of the shaft in radians per unit time and T is the torque required to overcome the friction. The torque can be written

$$T = 2\pi r_i^2 L\tau \tag{4–21}$$

where L is the axial length of the bearing surface. The shear stress, τ, can be replaced with the lubricant viscosity and velocity gradient so that

$$T = 2\pi r_i^2 L\mu \left[\frac{d\mathbf{V}}{dr}\right]_{r=r_i} \tag{4–22}$$

Figure 4–15 Physical situation of a rotating shaft and journal bearing.

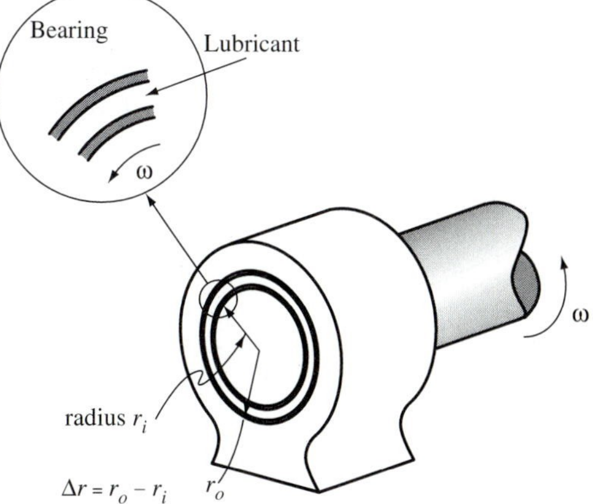

For small clearances the velocity gradient is approximately equal to $\omega r_i/(r_o - r_i)$, and the torque may then be written

$$T = 2\pi r_i^3 L\mu \frac{\omega}{r_o - r_i} \tag{4–23}$$

Equation (4–23) is called *Petroff's Equation*. The frictional power is

$$\dot{W}k = 2\mu\pi\omega^2 r_i^3 L \frac{1}{r_o - r_i} \tag{4–24}$$

and this must be dissipated as heat transfer from the bearing and the lubricant to prevent increase in temperature.

EXAMPLE 4–6 A 3.0-in.-diameter shaft is supported by two journal bearings each 4.25 in. long, having inside diameters of 3.001 in. If SAE 10 oil at 70°F is on the shaft and the shaft rotates at 600 rpm, estimate the power required to overcome the viscous friction of the oil.

Solution The power may be predicted from equation (4–24). The angular velocity is

$$\omega = \left(600\,\frac{\text{rev}}{\text{min}}\right)\left(2\pi\,\frac{\text{rad}}{\text{rev}}\right)\left(\frac{1\text{ min}}{60\text{ s}}\right) = 62.8\,\frac{\text{rad}}{\text{s}}$$

and the viscosity, from table 4–1, is 1.46×10^{-3} lbf · s/ft^2. The radius, r_i, is 1.5 in. or 0.125 ft, and the clearance, Δr, is 0.0005 in. or 4.167×10^{-5} ft. From equation (4–24) the power is, for two bearings,

Answer

$$\dot{W}k = 2\left(2\pi\mu\omega^2 L r_i^3 \frac{1}{\Delta r}\right) = 1202.46\,\frac{\text{ft}\cdot\text{lbf}}{\text{s}} = 2.186\text{ hp}$$

and this must be dissipated as heat from the oil.

Open channel flow of a viscous fluid is a more complex phenomenon than we considered in example 4–4. It has been determined that the velocity profile of fluid flowing in an open channel can be characterized as that shown in figure 4–16. The velocity is zero at the wetted perimeter and is maximum at a point furthest removed in the fluid stream from all of its outer surfaces. The velocity is reduced at the exposed surface because of viscous action of the two fluids there, but an analysis of a velocity profile is beyond the purposes of this text. If the average fluid velocity is considered, then the revised Bernoulli Equation may be applied to a sloping open channel flow where the friction head loss term represents the friction at the wetted perimeter. We neglect head losses between the two fluids at the open surface, but the heat transfer that is required to prevent temperature change of the fluid may occur at the open surface. In example 4–4 we applied Bernoulli's

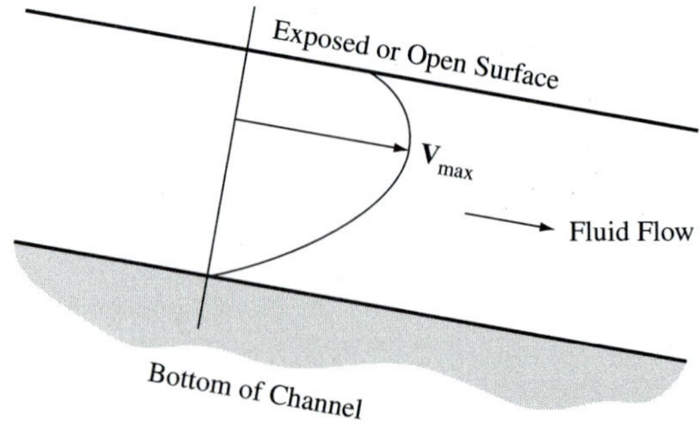

Figure 4–16 Typical velocity profile of viscous fluid in open channel flow.

Equation to an open channel flow problem. The equation can be modified to include frictional head losses by writing

$$z_1 + y_1 + \frac{1}{2g}\mathbf{V}_1^2 - z_2 - y_2 - \frac{1}{2g}\mathbf{V}_2^2 = h_L \tag{4–25}$$

where the terms on the left are the elevation head (z), the pressure head (y), and the kinetic or velocity head ($\mathbf{V}^2/2g$).

The frictional head loss, h_L, is more complex than it is for tube flow. For instance, the head loss requires some information about the velocity gradient, and this means knowing the velocity profile for the fluid. This is hard to determine in a direct manner. It is common to define the friction head loss rate, S, as

$$S = \frac{h_L}{L} \tag{4–26}$$

where L is the horizontal length of fluid flow. The rate is denoted with an S because it represents the slope of an energy grade line (EGL) over the horizontal flow length L. This is shown in figure 4–17. If the fluid flow in an open channel does not change its velocity, it is called uniform flow, the depth of the fluid is constant, and the slope of the bottom of the channel must be equal to the friction head loss rate.

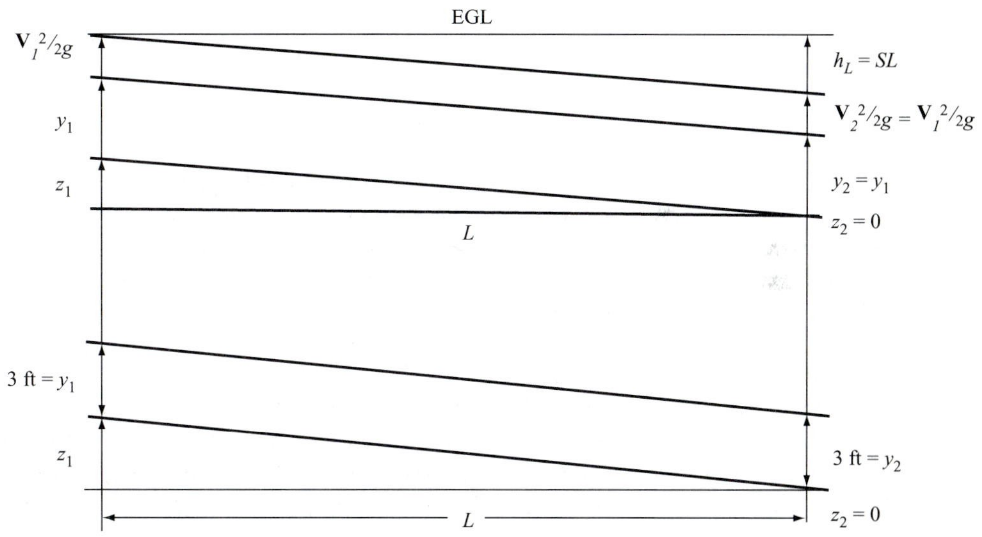

Figure 4–17 Open channel flow with friction head losses.

EXAMPLE 4–7 Consider a wide channel through which water at 70°F flows. If the water velocity is 3 ft/s, its depth is 3 ft, and the flow is uniform, determine the slope required at the bottom of the channel if the friction head loss is expected to be $\frac{1}{2}$ in. per foot of flow length. Also estimate the heat transfer required per width of the channel needed to satisfy isothermal conditions of the fluid.

Solution In figure 4–17 is shown the condition of the fluid flow. Applying equation (4–25) to this problem, we have $\mathbf{V}_1 = \mathbf{V}_2$ and $y_1 = y_2$. Equations (4–25) and (4–26) give

$$z_1 - z_2 = h_L = SL$$

but $z_1 - z_2)/L$ is the slope of the bottom of the channel, and it is also equal to the friction head loss rate, S. The friction head loss is $\frac{1}{2}$ in. per foot or $\frac{1}{24}$ ft per foot. The channel bottom must therefore have a slope of $\frac{1}{24}$ ft per foot. In figure 4–17 is shown the EGL sloping at this rate, and this energy loss must be dissipated as heat transfer to the surroundings. The energy loss is, from the form of the terms in equation (4–4), $h_L\gamma/\rho$ because $h_L\gamma$ is a pressure loss and pressure divided by density is energy per unit mass. The heat transfer necessary to dissipate the friction head loss is the energy per unit mass times the mass flow rate

$$\dot{Q} = \frac{h_L\gamma}{\rho}\dot{m}$$

in English units the term $\gamma/\rho = g/g_c = 1$ if g is numerically equal to $g_c(=32.17)$. The mass flow rate is

$$\dot{m} = \rho A\mathbf{V}_{\text{average}} = \left(62.8\,\frac{\text{lbm}}{\text{ft}^3}\right)(3\text{ ft} \times Z)(3\text{ ft/s})$$

where Z is the channel width. The heat transfer per unit length of channel width is then

Answer

$$\dot{q}_z = \frac{\dot{Q}}{Z} = \frac{h_L\gamma}{\rho Z}\dot{m} = \frac{(1\text{ ft/24 ft})\left(62.4\,\dfrac{\text{lbf}}{\text{ft}^3}\right)}{\left(62.4\,\dfrac{\text{lbm}}{\text{ft}^3}\right)Z}\left(561.6\,\frac{\text{lbm}}{\text{s}}\right)Z = 23.4\,\frac{\text{ft}\cdot\text{bf}}{\text{s}\cdot\text{ft}}$$

We have discussed the velocity profile of fluid flow for frictionless and for viscous fluids. For viscous fluids the shear stresses at the surfaces of the fluid stream are greatest because the velocity gradients are greatest at those surfaces. If the velocity gradient becomes too large, such as when the bulk fluid velocities become large, then the fluid will cease to follow laminar flow patterns of parallel paths but will follow agitated, random paths. This is the characteristic of turbulent flow, and a parameter that gives a quantitative measure of when laminar flow ceases and turbulent flow begins is the *Reynolds Number, Re*, defined as

$$\text{Re} = \frac{\rho\mathbf{V}L}{\mu} = \frac{\mathbf{V}L}{\nu} \tag{4–27}$$

where L is some characteristic length. The Reynolds Number also provides a measure for comparing two or more different fluids flowing in different conduits. As an example, SAE 50 oil (a viscous fluid, as shown from table 4–1) flowing at high velocity in a large pipe may have the same Reynolds Number as air (an inviscid fluid, from table 4–1) in a small tube traveling at low velocity. If the Reynolds Numbers are the same, then the character of the flow will be the same with the same velocity profile, or will have the same turbulent character. One procedure used to demonstrate that the Reynolds Number is important is that of *dimensional analysis.*

Dimensional Analysis

Dimensional analysis is useful in the study of processes or events that involve many separate variables. For instance, in scientific studies we seek to determine causes for certain effects. Simple effects such as the acceleration of a mass particle can be attributed to an external force acting on the particle. Newton's law of motion, $F = ma$, is familiar and useful. Similarly, the flow of fluids must obey this same law, but there are also other causes creating other effects and sometimes it is not so easy to predict what the causes and effects are. In the method of dimensional analysis the number of independent variables or "causes" is reduced to the least number without compromising the prediction of the "effects." The method is based on two propositions, or "laws":

1. Equations relating causes and effects or variables must be dimensionally homogenous. By this is meant that the fundamental dimensions must be the same in all terms of such equations. As an example, the equation $F = ma$ relates a force to mass times acceleration. In table 1–1 we saw that some basic dimensions are mass, length, time, force, and temperature. In dimensional analysis it is expected that mass (m), length (L), time (t), and temperature (T) are fundamental dimensions. For dimensional homogeneity force (F) must be equal to mLt^{-2}. Force therefore has a derived unit, as would energy or power. Some other common parameters and their derived units are given in table 4–2.

Table 4–2 Some Parameters Occurring in Dimensional Analysis and Their Units Derived from Mass (m), Length (L), Time (t), and Temperature (T).

Parameter	Derived Dimension
Acceleration	Lt^{-2}
Angle	0
Angular velocity	t^{-1}
Angular acceleration	t^{-2}
Area	L^2
Compressibility	$m^{-1}Lt^{-2}$
Density	mL^{-3}
Energy, work, heat	mL^2t^{-2}
Enthalpy	mL^2t^{-2}
Entropy	$mL^2t^{-2}T^{-1}$
Force	mLt^{-2}
Heat transfer, power	mL^2t^{-3}
Kinematic viscosity	L^2t^{-1}
Specific heat per unit mass	$L^2t^{-2}T^{-1}$
Strain	0
Stress, pressure	$mL^{-1}t^{-2}$
Temperature	T
Thermal conductivity	$mLt^{-3}T^{-1}$
Velocity	Lt^{-1}
Viscosity	$mL^{-1}t^{-1}$
Volume	L^3

2. The mathematical relationships describing physical interactions involving j parameters having n fundamental dimensions can be related by $j - n$ dimensionless quantities, called *Π-parameters* and formed from ratios of $j - n$ of the parameters with each of the remaining parameters.

An example of this second law, sometimes called *Buckingham's Pi-Theorem,* is the familiar equation $F = ma$. This equation describes the interaction of three parameters, F, m, and a, having three fundamental dimensions (m, L, and t). Since $j - n = 0$, we conclude that there are no Π-parameters for this phenomenon, called Newton's Law of Motion.

Another example is the case of a viscous fluid flowing from a tank, as shown in figure 4–18. In this example of fluid flow we assume the velocity of the fluid leaving the

Figure 4–18 Emptying a tank.

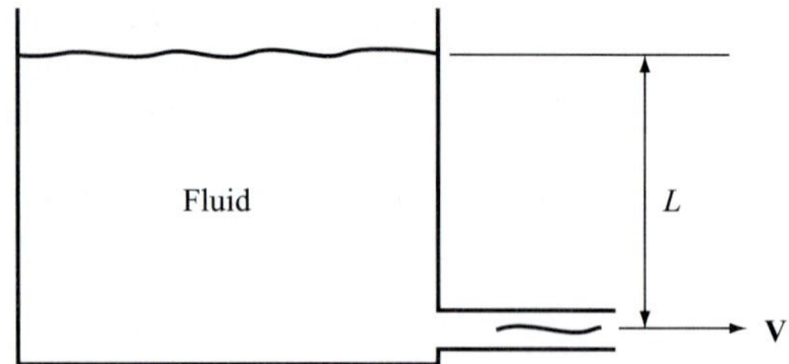

tank is affected by gravity and by the pressure head, L. For these three parameters there are two fundamental dimensions, L and t:

Velocity, $\mathbf{V} = Lt^{-1}$

Gravity, $g = Lt^{-2}$

Pressure head, $\dfrac{\Delta p}{\gamma} = L$

Therefore, $j = 3$, $n - 2$, and $j - n = 1$. We have one Π-parameter, which may be written

$$(\mathbf{V})^a(g)^b\left(\frac{\Delta p}{\gamma}\right)^c = C$$

or

$$(Lt^{-1})^a(Lt^{-2})^b(L)^c = C$$

Comparing the exponents we have for the L's, $a + b + c = 0$, and for the t's, $-a - 2b = 0$, from which we obtain $b = -\frac{1}{2}a$ and $c = -\frac{1}{2}a$. If we arbitrarily set $a = 1$ (we may set it to any value we wish), then $b = c = -\frac{1}{2}$ and the dimensionless Π-parameter is

$$(\mathbf{V})(g)^{-1/2}\left(\frac{\Delta p}{\gamma}\right)^{-1/2} = C = \frac{\mathbf{V}}{\sqrt{g\dfrac{\Delta p}{\gamma}}} = \mathbf{V}\sqrt{\frac{\gamma}{g\Delta p}} = \mathbf{V}\sqrt{\frac{\rho}{\Delta p}}$$

The result may also be written

$$\mathbf{V} = C\sqrt{\frac{\Delta p}{\rho}} = C\sqrt{L}$$

which we obtained from example 4–2. Dimensional analysis did not provide the information for the value of the constant C (which is $\sqrt{2}$ from example 4–2), but it does provide the information of how the various parameters were related. Let's consider now the example of viscous fluid flow in a tube.

EXAMPLE 4–8 Determine the Π-parameters that may be used to describe the flow of a viscous fluid in a tube of diameter D. We assume that the following parameters are needed to describe the fluid flow:

Tube length, L	(L)
Tube diameter, D	(L)
Fluid viscosity, μ	$(mL^{-1}t^{-1})$
Fluid density, ρ	(mL^{-3})
Fluid velocity, $\mathbf{V}$	(Lt^{-1})
Pressure drop, Δp	$(mL^{-1}t^{-2})$

Since there are six parameters with three fundamental dimensions, we have $j - n = 6 - 3 = 3$. The three Π-parameters that will describe the fluid flow are to be ratios of three parameters formed with the remaining parameters. If we choose D, μ, and $\mathbf{V}$ as the fundamental parameters then we need to form three Π-parameters or dimensional ratios with L, ρ, and Δp. The first parameter, Π_1, can be formed with L

$$\Pi_1 = D^a \mu^b \mathbf{V}^c L^d$$

For the second parameter, Π_2, we have a ratio formed with ρ,

$$\Pi_2 = D^e \mu^f \mathbf{V}^g \rho^h$$

and for the third, formed with Δp,

$$\Pi_3 = D^i \mu^j \mathbf{V}^k \Delta p^l$$

We could have chosen other combinations; however, it can be shown that the results will be similar for any combination. Writing the fundamental dimensions for these new Π-parameters we have

$$\begin{aligned} \Pi_1 &= (L)^a(mL^{-1}t^{-1})^b(Lt^{-1})^c(L)^d \\ \Pi_2 &= (L)^e(mL^{-1}t^{-1})^f(Lt^{-1})^g(mL^{-3})^h \\ \Pi_3 &= (L)^i(mL^{-1}t^{-1})^j(Lt^{-1})^k(mL^{-1}t^{-2})^m \end{aligned}$$

The exponents of L, m, and t in the first of these three equations gives, respectively,

$$\begin{aligned} a - b + c + d &= 0 \qquad (L \text{ exponents}) \\ b &= 0 \qquad (m \text{ exponents}) \\ -b - c &= 0 \qquad (t \text{ exponents}) \end{aligned}$$

The solution may be written $a = -d$ and so, if we set $a = 1$, then $d = -1$ and

$$\Pi_1 = \frac{D}{L} \left(\text{or } \frac{L}{D}\right)$$

The second Π-parameter can be determined similarly. Again, the exponent equations are

$$\begin{aligned} e - f + g - 3h &= 0 \qquad (L \text{ exponents}) \\ f + h &= 0 \qquad (m \text{ exponents}) \\ -f - g &= 0 \qquad (t \text{ exponents}) \end{aligned}$$

Then, the solution to these three equations can be written

$$e = g = h = -f$$

If we set $f = -1$, then

$$\Pi_2 = (D)(\mu)^{-1}(\mathbf{V})(\rho) = \frac{\rho \mathbf{V} D}{\mu} = \text{Re}_D \text{ (Reynolds Number based on } D)$$

Similarly, for the third Π-parameter,

$$\begin{aligned} i - j + k - m &= 0 \qquad (L \text{ exponents}) \\ j + m &= 0 \qquad (m \text{ exponents}) \\ -j - k - 2m &= 0 \qquad (t \text{ exponents}) \end{aligned}$$

from which we obtain $i = m = -j = -k$. Setting $i - 1$ gives

$$\Pi_3 = \frac{L \Delta p}{\mu \mathbf{V}}$$

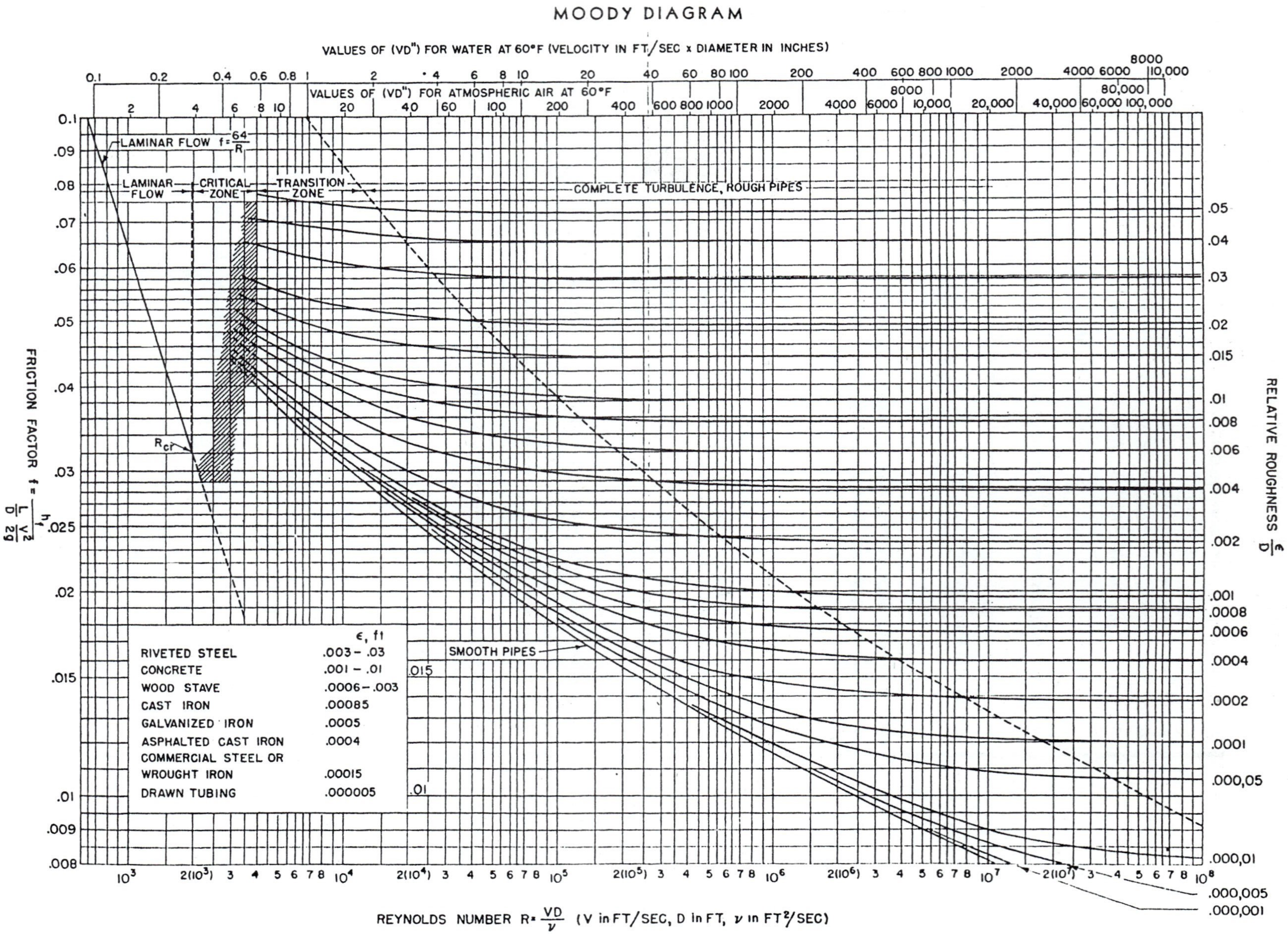

MOODY DIAGRAM

VALUES OF (VD'') FOR WATER AT 60°F (VELOCITY IN FT/SEC x DIAMETER IN INCHES)

VALUES OF (VD'') FOR ATMOSPHERIC AIR AT 60°F

LAMINAR FLOW $f = \frac{64}{R}$

LAMINAR FLOW — CRITICAL ZONE — TRANSITION ZONE — COMPLETE TURBULENCE, ROUGH PIPES

SMOOTH PIPES

R_{cr}

FRICTION FACTOR $f = \frac{h_f}{\frac{L}{D}\frac{V^2}{2g}}$

RELATIVE ROUGHNESS $\frac{\epsilon}{D}$

REYNOLDS NUMBER $R = \frac{VD}{\nu}$ (V in FT/SEC, D in FT, ν in FT²/SEC)

	ϵ, ft
RIVETED STEEL	.003 – .03
CONCRETE	.001 – .01
WOOD STAVE	.0006 – .003
CAST IRON	.00085
GALVANIZED IRON	.0005
ASPHALTED CAST IRON	.0004
COMMERCIAL STEEL OR WROUGHT IRON	.00015
DRAWN TUBING	.000005

Figure 4–19 Moody diagram for viscous fluid friction factor in tube or pipe flow.

The three parameters may be related as

$$\Pi_3 = f(\Pi_1)f(\Pi_2)$$

or

$$\frac{L\Delta p}{\mu \mathbf{V}} = \left(\frac{L}{D}\right)^n (\text{Re}_D)^m \tag{4–28}$$

Notice that equation (4–28) provides a relationship between a term involving a pressure drop, Δp, and the Reynolds Number and length/diameter ratio. Earlier, in equation (4–19), we saw another form of this sort of relationship called the Darcy–Weisbach Equation. In that equation,

$$h_L = f\frac{L}{D}\frac{1}{2g}\mathbf{V}^2 = \frac{\Delta p}{\gamma}$$

the head loss was the pressure head loss per unit of specific weight; that is, $h_L = \Delta_p/\gamma$. It can be shown (see practice problem 4–6) for laminar flow where the Hagen–Poiseuille Equation (4–16) applies, that the friction factor f in equation (4–19) can be written

$$f = \frac{64}{\text{Re}_D} \tag{4–29}$$

The friction factor is only a mathematical device to describe the viscous and tube wall resistance to flow, and it is encouraging to see that laminar flow can be as compactly described as equations (4–16), (4–19), and (4–29) indicate. The relationship of equation (4–29) applies only to laminar flow conditions in a tube and so is of limited use. From equation (4–28), however, we can see that pressure drops are affected by the length/diameter ratio and the parameter $\mathbf{V}\mu/L$ in addition to the Reynolds number. It has been experimentally verified by many investigators that the friction factor of equation (4–19) is affected by the tube or pipe roughness (which determines the effective flow area) for turbulent flow of fluid. The roughness also can be a factor contributing to the creation of turbulent flow from transition flow conditions. In figure 4–19 is shown a graph called the *Moody Diagram,* which presents the relationships of the friction factor to the Reynolds Number and tube or pipe roughness over a wide regime of flow conditions. The Moody Diagram, compiled in 1944 by L. F. Moody [2], is a convenient tool for predicting pressure drops in viscous flow of fluids in tubes, and with that information, the necessary heat transfer to provide for isothermal flow can be predicted.

EXAMPLE 4–9 Water at 70°F flows through a 4-in. ID wrought iron pipe at the rate of 1 lbm/s. Determine the pressure drop per foot of pipe and the heat transfer necessary so that the water does not change temperature.

Solution The pressure drop per foot of pipe length may be predicted from equation (4–19) after determining the friction factor. The necessary heat transfer to retain isothermal conditions is the power dissipated by the fluid friction, which is

$$\dot{Q} = \frac{\Delta p}{L}\mathbf{V}$$

The Reynolds Number can be determined with some rearranging,

$$\text{Re}_D = \frac{\rho \mathbf{V} D}{\mu} = \frac{4\dot{m}}{\mu D}$$

The viscosity of water is read as 2.09×10^{-5} lbf · s/ft from table 4–1, so

$$\text{Re}_D = \frac{4(1\ \text{lbm/s})}{(2.09 \times 10^{-5}\ \text{lbf} \cdot \text{s/ft})(4/12\ \text{ft})} = 5.74 \times 10^5$$

From figure 4–19 the roughness is 0.00015 ft, the relative roughness is (0.00015 ft)/($^4/_{12}$ ft) = 0.00045, and the friction factor is read as 0.0173. The velocity is

$$\mathbf{V} = \frac{\dot{m}}{\rho A} = \frac{1\ \text{lbm/s}}{(62.4\ \text{lbm/ft}^3)(\pi)(^2/_{12}\ \text{ft})^2} = 0.1836\ \frac{\text{ft}}{\text{s}}$$

and the head loss is

$$h_L = \frac{\Delta p}{\gamma} = f\frac{L}{D}\frac{1}{2g}\mathbf{V}^2 = (0.0173)\left(\frac{1\text{ft}}{4/12\ \text{ft}}\right)\left[\frac{1}{2(32.2\ \text{ft/s}^2)}\right]\left(0.1836\ \frac{\text{ft}}{\text{s}}\right)^2$$
$$= 2.72 \times 10^{-5}\ \text{ft}$$

The pressure drop is

Answer
$$\Delta p = \gamma h_L = \left(62.4\ \frac{\text{lbf}}{\text{ft}^3}\right)(2.72 \times 10^{-5}\ \text{ft}) = 170 \times 10^{-3}\ \frac{\text{lbf}}{\text{ft}^2}$$

The necessary volume rate is

$$\dot{V} = \frac{\dot{m}}{\rho} = \frac{1\ \text{lbm/s}}{62.4\ \text{lbm/ft}^3} = 0.016\ \frac{\text{ft}^3}{\text{s}}$$

and the necessary heat transfer is

Answer
$$\dot{Q}_L = \frac{\Delta p}{L} = \frac{0.0017\ \text{lbf/ft}^2}{1\ \text{ft}}(0.016\ \text{ft}^3/\text{s}) = 2.72 \times 10^{-5}\ \frac{\text{ft} \cdot \text{lbf}}{\text{s} \cdot \text{ft}}$$
$$= 3.5 \times 10^{-8}\ \frac{\text{Btu}}{\text{ft} \cdot \text{s}}$$

The Reynolds Number can be interpreted as a dimensionless ratio of the inertial forces to the viscous forces, which can be seen by multiplying top and bottom of the Reynolds Number by the velocity $\mathbf{V}$ to give

$$\text{Re} = \frac{\rho \mathbf{V}^2}{\mu \mathbf{V} D}$$

The $\rho\mathbf{V}^2$ term is proportional to kinetic energy, or an inertial force, and $\mu\mathbf{V}/D$ is proportional to a viscous force. We saw earlier that the Biot Number was interpreted as a ratio of the convection to the conduction heat transfers. In later sections we will make use of the dimensional analysis method and draw on the dimensionless ratio concept to better understand heat-transfer phenomena.

4–3 THE BOUNDARY LAYER CONCEPT

Fluid flow is resisted most at surfaces where the fluid contacts some solid. It is at such surfaces that the shear stresses are largest, and if the fluid temperature is different than that of the surface of the solid, the temperature gradient will also be largest. It is necessary, then, that we pay attention to the happenings at the surface for convection heat-transfer analysis. Consider a fluid at temperature T_∞ moving at a velocity $\mathbf{u}_\infty$ past a flat plate as shown in figure 4–20. As the fluid approaches the leading edge of the plate, it begins to be affected by the interaction of the fluid with the plate. We are considering a thin plate to avoid the effects of the fluid that impacts directly against the plate; that is, we assume that the plate is infinitely thin and the fluid flows above or below the plate. At the leading edge a certain portion of the fluid begins to become enveloped in the phenomena of the viscous and heat-transfer

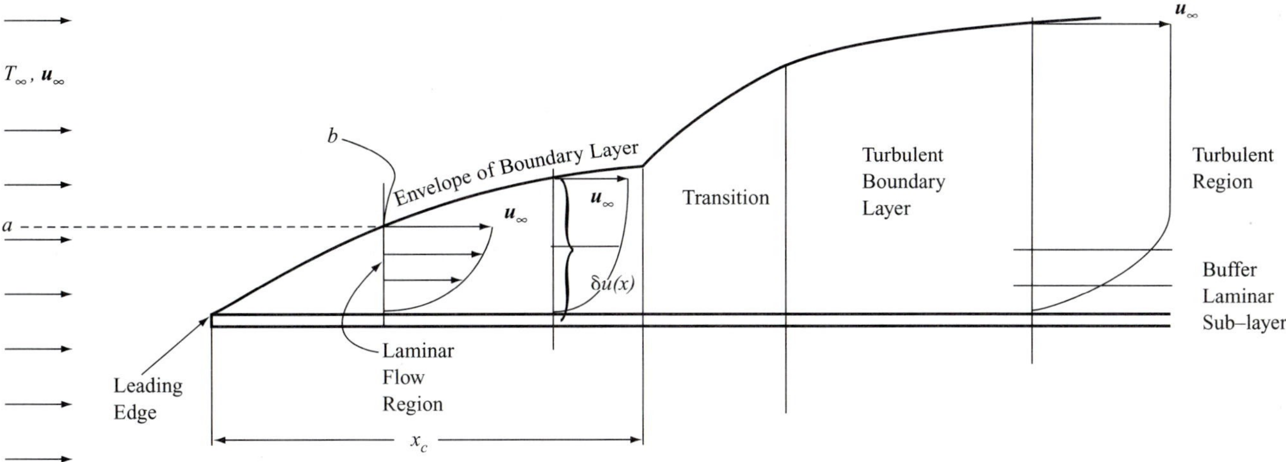

Figure 4–20 Boundary layer development along a flat thin plate.

effects. As the fluid moves further along the x direction past the plate, a greater amount of the fluid becomes affected, as indicated in figure 4–20 by the envelope of the boundary layer.

The boundary layer is the region of fluid between the solid surface and the envelope of the boundary layer. Its thickness δ continues to increase from the leading edge along the x direction, indicating that more and more of the fluid is becoming disturbed by the viscous and heat-transfer effects at the surface. Consider, for instance, a particle of fluid at point a in figure 4–20 flowing at $\mathbf{u}_\infty$. If the fluid is moving in a laminar fashion, the particle will reach the boundary layer at point b and have the same velocity as at point a. It then "enters" the boundary layer region and begins to decelerate as long as laminar flow conditions are present in the boundary layer. Notice that viscous effects at the surface are instrumental in creating this boundary layer of decelerating fluid particles and it can be shown that a Reynolds Number Re_x based on a distance from the leading edge, x, can be used to describe the qualitative behavior of the fluid flow. Thus,

$$\text{Re}_x = \frac{\rho \mathbf{u}_\infty x}{\mu} \tag{4–30}$$

where ρ and μ are properties of the fluid, is the appropriate parameter for comparing the inertial to the viscous forces for flow past a flat plate.

It has been experimentally determined that if a plate is long enough, eventually the boundary layer flow becomes turbulent. In figure 4–20 this sort of event is indicated as a transition flow leading into a turbulent boundary layer. Even then the turbulent boundary layer can be visualized as a turbulent region, a transition or buffer region, and a laminar sublayer. The boundary layer thickness tends to become proportionately thicker than the equivalent laminar boundary layer would tend to be. The length or location x_c where a laminar boundary layer converts to a turbulent boundary layer is such that Re_x is at least 1×10^5 and up to 3×10^6 or more, depending on the roughness of the surface and other effects. A value of 5×10^5 is sometimes cited as a critical Reynolds Number at which laminar flow in a boundary layer ceases to exist.

EXAMPLE 4–10 Compare the distance from the leading edge of a flat plate at which laminar flow may be expected to cease in a boundary layer of flow of air at 20°C, 101 kPa, and flow of water at 20°C. Assume that the bulk velocity is 2 m/s in both cases.

Solution The distance from the leading edge may be predicted from the relationship

$$x_c = \frac{\text{Re}_{x,c}\,\mu}{\rho u_\infty} = \frac{\text{Re}_{x,c}\,\nu}{u_\infty} \tag{4–31}$$

where the critical Reynolds Number, $\text{Re}_{x,c}$, is at the location where laminar flow ceases in the boundary layer. The viscosity values for air and water are obtained from table 4–1: $\mu_{\text{air}} = 18 \times 10^{-6}\,\text{N} \cdot \text{s/m}^2 = 18 \times 10^{-6}\,\text{kg/s} \cdot \text{m}$ and $\mu_{\text{water}} = 0.001$. The density for air may be taken from appendix table B–4 as 1.178 kg/m^3, and for water we use a value from appendix table B–3, 998 kg/m^3. The critical distances are, using a critical Reynolds number of 5×10^5 for air,

Answer

$$x_c = \frac{(5 \times 10^5)\left(18 \times 10^{-6}\,\frac{\text{kg}}{\text{s} \cdot \text{m}}\right)}{\left(1.178\frac{\text{kg}}{\text{m}^3}\right)\left(2\frac{\text{m}}{\text{s}}\right)} = 3.82\ \text{m}$$

and for water the critical distance is

Answer

$$x_c = \frac{5 \times 10^5\left(0.001\,\frac{\text{kg}}{\text{m} \cdot \text{s}}\right)}{\left(998\,\frac{\text{kg}}{\text{m}^3}\right)\left(2\,\frac{\text{m}}{\text{s}}\right)} = 0.25\ \text{m}$$

One conclusion to be drawn from these results is that water will have turbulent flow much sooner than will air, if both are moving with the same bulk fluid velocity.

The boundary layer concept has been extended to cover thermal energy transfers between a fluid and a solid surface. An analogy can be drawn between the velocity profile and the temperature profile, which results in a thermal boundary layer. If the fluid is at some elevated temperature, as shown in figure 4–21, then the thermal boundary layer is a device to visualize the temperature profile as it changes from the bulk fluid value, T_∞, to a surface value, T_s. Just as more fluid particles' velocities become affected by the interaction at the surface, so the temperature of more of the fluid particles also become affected.

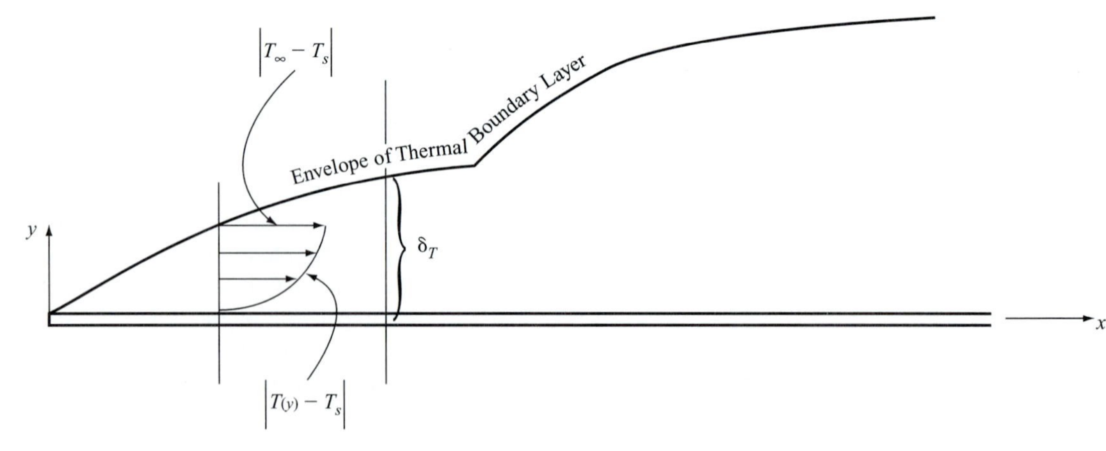

Figure 4–21 Thermal boundary layer at a thin flat plate.

The thermal boundary layer, δ_T, becomes thicker downstream from the leading edge in a manner similar to the behavior of the boundary layer thickness for velocity. In general the thermal boundary layer thickness is different from the boundary layer thickness, as we will see shortly. Notice in figure 4–21 that the thermal boundary layer concept creates the impression that the convection heat transfer mechanism, given by equation (1–59),

$$\dot{Q} = hA(T_\infty - T_s)$$

is an energy transfer facilitated through the thermal boundary layer. The convection heat-transfer coefficient h is then a parameter describing the conductance of heat through the thermal boundary layer. A corollary of this model is that the convection heat transfer is equal to the conduction heat transfer at the surface,

$$-\kappa_{\text{fluid}} A\left[\frac{\partial T(x, y)}{\partial y}\right]_{y=0} = hA(T_\infty - T_s) \tag{4–32}$$

and we will use this relationship to help predict the convection heat-transfer coefficient based on a knowledge of fluid properties. Methods based on mathematical analyses have been developed for these predictions, and we will outline some of those methods now.

Conservation of Mass and Momentum

Consider an element of the boundary layer, or a control volume, through which fluid and energy may pass as shown in figure 4–22. The element is shown as a differential square element for convenience. The third dimension (into the paper) is neglected, and the mechanisms are therefore considered to be two-dimensional in the x and the y directions. Also, steady state is assumed so that the mass within the control volume must be constant at all time. The mass flow rate entering from the left face in the x direction is

$$\rho(x, y) \cdot \mathbf{u}(x, y)dy$$

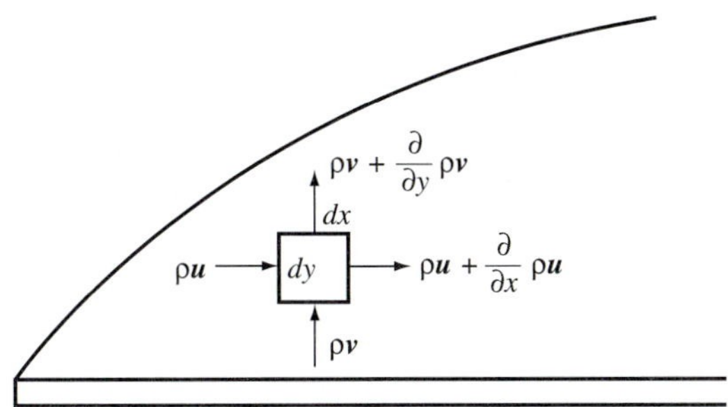

Figure 4–22 Differential control volume in a boundary layer and the mass balance for steady state.

where $\mathbf{u}(x, y)$ is the velocity component of the fluid in the x direction. Notice that the velocity is presumed to be a function of both x and y.

The mass entering from below in the v direction is

$$\rho(x, y) \cdot \mathbf{v}(x, y)$$

The mass flow rates are then shown leaving the opposite faces with possible changes in their values. From a summation of the mass flow rates in and the mass flow rates out we get

$$\rho\mathbf{u}dy + \rho\mathbf{v}dx - \rho\left(\mathbf{u} + \frac{\partial(\rho\mathbf{u})}{\partial x}dx\right)dy - \rho\left(\mathbf{v} + \frac{\partial(\rho\mathbf{v})}{\partial y}dy\right)dx = 0$$

The notation indicating the dependency of density and velocities on x and y has been dropped, but they are still functions of x and y. Canceling terms and recognizing that $dxdy = dydx$, this becomes

$$\frac{\partial(\rho\mathbf{u})}{\partial x} + \frac{\partial(\rho\mathbf{v})}{\partial y} = 0 \tag{4–33}$$

This equation is sometimes called the continuity equation as it applies to steady flow fluid flow conditions.

The control volume element must also obey the momentum balance, or Newton's Law of Motion. In figure 4–23 the various force and momentum terms acting on the element are indicated. The effects of gravitational attraction are assumed to be in the vertical or y direction. In the x direction we can write

$$\sum F_x = \frac{d(m\mathbf{u})}{dt} \tag{4–34}$$

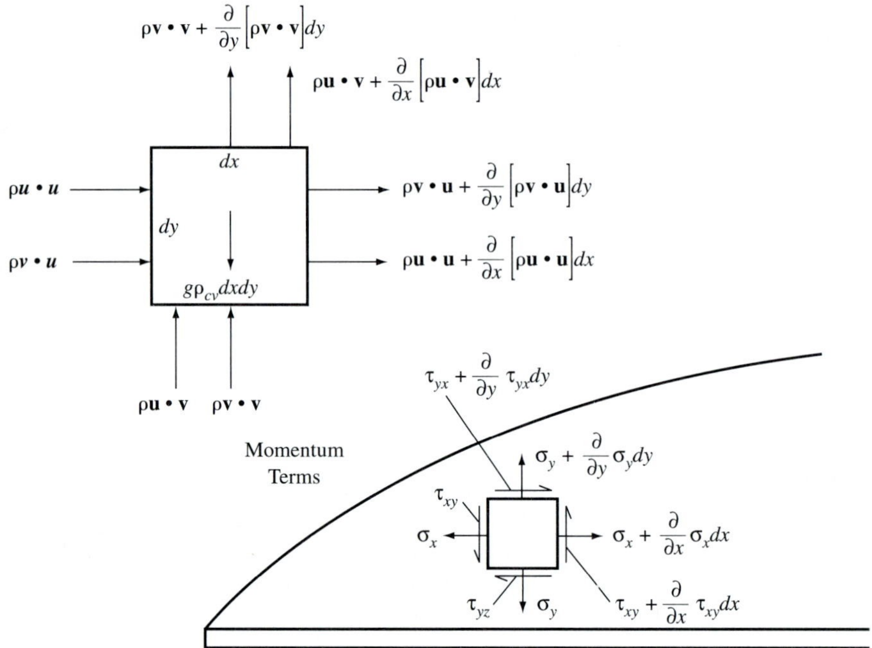

Figure 4–23 Differential control volume in a boundary layer and the momentum balance for steady state.

where the forces, F_x, are the shear stresses and normal stresses. Thus, taking positive forces to the right,

$$\sum F_x = -\sigma_x dy - \tau_{yx}dx + \left(\sigma_x + \frac{\partial\sigma_x}{\partial_x}dx\right)dy + \left(\tau_{yx} + \frac{\partial\tau_{yx}}{\partial y}dy\right)dx \tag{4–35}$$

The momentum change in the x direction, $d(m\mathbf{u})/dt$, is

$$\frac{d(m\mathbf{u})}{dt} = -(\rho\mathbf{u})\cdot\mathbf{u} - (\rho\mathbf{v})\cdot\mathbf{u} + (\rho\mathbf{u})\cdot\mathbf{u} + \frac{\partial[(\rho\mathbf{u})\cdot\mathbf{u}]}{\partial x}dx + (\rho\mathbf{v})\cdot\mathbf{u} + \frac{\partial[(\rho\mathbf{v})\cdot\mathbf{u}]}{\partial y}dy \tag{4–36}$$

where the · denotes a dot product of two vectors. Notice that the y-direction momentum can have momentum effects in the x direction as well. Equating equations (4–35) and (4–36) and canceling out gives

$$\frac{\partial \sigma_x}{\partial x}dxdy + \frac{\partial \tau_{yx}}{\partial y}dydx = -\frac{\partial[(\rho \mathbf{u}) \cdot \mathbf{u}]}{\partial x}dx - \frac{\partial[(\rho \mathbf{v}) \cdot \mathbf{u}]}{\partial y}dy \tag{4–37}$$

If the right side of this equation is expanded out and the continuity result of equation (4–33) is used, the steady state momentum balance in the x direction becomes

$$\frac{\partial \sigma_x}{\partial x}dxdy + \frac{\partial \tau_{yx}}{\partial y}dydx = -\rho \mathbf{u} \cdot \left(\frac{\partial \mathbf{u}}{\partial x}\right)dx - \rho \mathbf{v} \cdot \left(\frac{\partial \mathbf{u}}{\partial y}\right)dy \tag{4–38}$$

From the conservation of momentum in the y direction, $\Sigma F_y = dm\mathbf{v}/dt$, the steady state balance in the y direction is

$$\frac{\partial \tau_{xy}}{\partial x}dxdy + \left(\frac{\partial \sigma_y}{\partial y}\right)dydx + \rho_{cv}dxdy = -\frac{\partial[(\rho \mathbf{u}) \cdot \mathbf{v}]}{\partial x}dx - \frac{\partial[(\rho \mathbf{v}) \cdot \mathbf{v}]}{\partial y}dy \tag{4–39}$$

The third term on the left side is the effect of gravity or the weight in the differential control volume. It is often neglected because the stress terms frequently are an order of magnitude or more greater than the weight. Also, by expanding out the derivatives of the right side of this equation and using the continuity result of equation (4–33), the steady state momentum balance in the y direction of a boundary layer is

$$\frac{\partial \tau_{xy}}{\partial x}dxdy + \frac{\partial \sigma_y}{\partial y}dydx + \rho_{cv}dxdy = -\rho \mathbf{u} \cdot \frac{\partial \mathbf{v}}{\partial x}dy - \rho \mathbf{v} \cdot \frac{\partial \mathbf{v}}{\partial y}dy \tag{4–40}$$

For fluids it is useful to consider the normal stresses, σ_x and σ_y, to be made up of two independent components, pressure and viscous stress such that

$$\sigma_x = \sigma_{xp} + \sigma_{xv} \quad \text{and} \quad \sigma_y = \sigma_{yp} + \sigma_{yv} \tag{4–41}$$

The pressure components are

$$\sigma_{xp} = \sigma_{yp} = -p \qquad \text{(the hydrostatic pressure)} \tag{4–42}$$

and this pressure is the reaction of an element of a fluid to external pressures acting on it. The stress components due to viscosity, σ_{xv} and σ_{yv}, can be visualized as the reactions to volumetric distortions of the fluid element. For instance, a positive normal stress σ_x tends to elongate a rectangular element in the x direction and shorten the element in the y direction, as shown in figure 4–24. The shearing stresses, τ_{yx} and τ_{xy}, tend to distort the element in a rotational manner as indicated in figure 4–25.

Figure 4–24 Axial distortion due to normal stress on an element.

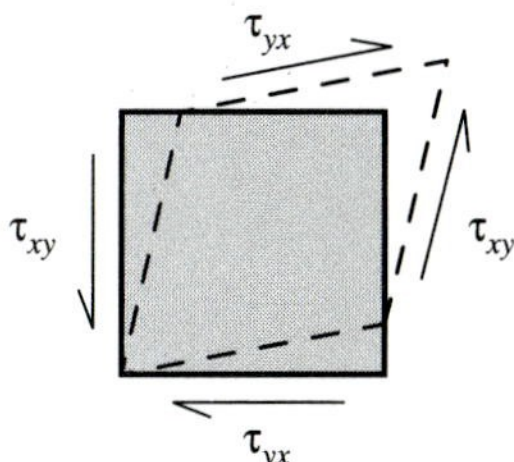

Figure 4–25 Angular distortion due to shear stress on an element.

For a Newtonian fluid (μ is constant) it has been shown [3] that the normal and the shear stresses can be written in terms of the velocities and the velocity gradients. The results are, using the concept of equation (4–41),

$$\sigma_x = -p + 2\mu\frac{\partial \mathbf{u}}{\partial x} - \frac{2}{3}\mu\left(\frac{\partial \mathbf{u}}{\partial x} + \frac{\partial \mathbf{v}}{\partial y}\right)$$

$$\sigma_y = -p + 2\mu\frac{\partial \mathbf{v}}{\partial y} - \frac{2}{3}\mu\left(\frac{\partial \mathbf{u}}{\partial x} + \frac{\partial \mathbf{v}}{\partial y}\right) \tag{4–43}$$

$$\tau_{xy} = \tau_{yx} = \mu\left(\frac{\partial \mathbf{u}}{\partial x} + \frac{\partial \mathbf{v}}{\partial y}\right)$$

When equations (4–43) are substituted into the momentum equations (4–38) and (4–40), the two-dimensional equations of motion for a Newtonian fluid at steady state are

$$-\frac{\partial p}{\partial x} + \frac{\partial}{\partial x}\left\{\mu\left[2\frac{\partial \mathbf{u}}{\partial x} - \frac{2}{3}\left(\frac{\partial \mathbf{u}}{\partial x} + \frac{\partial \mathbf{v}}{\partial y}\right)\right]\right\} + \frac{\partial}{\partial y}\mu\left(\frac{\partial \mathbf{u}}{\partial y} + \frac{\partial \mathbf{v}}{\partial x}\right) = \rho\left(\mathbf{u}\frac{\partial \mathbf{u}}{\partial x} + \mathbf{v}\frac{\partial \mathbf{u}}{\partial y}\right) \tag{4–44a}$$

$$\frac{\partial p}{\partial y} + \frac{\partial}{\partial y}\left\{\mu\left[2\frac{\partial \mathbf{v}}{\partial y} - \frac{2}{3}\left(\frac{\partial \mathbf{u}}{\partial x} + \frac{\partial \mathbf{v}}{\partial y}\right)\right]\right\} + \frac{\partial}{\partial x}\mu\left(\frac{\partial \mathbf{u}}{\partial y} + \frac{\partial \mathbf{v}}{\partial x}\right) + \rho_{cv}g\,dx\,dy = \rho\left(\mathbf{u}\frac{\partial \mathbf{v}}{\partial x} + \mathbf{v}\frac{\partial \mathbf{v}}{\partial y}\right) \tag{4–44b}$$

These two equations include all the terms necessary for a complete solution to steady state two-dimensional Newtonian fluid flow; however, they are too complicated for obtaining mathematical solutions to the velocity distributions, **u** and **v**. In the boundary layer it is customary to use the approximations that the density, specific heat, thermal conductivity and viscosity are constant. Also, using an order of magnitude analysis the x-direction velocity, **u**, is expected to be much greater than the y-direction velocity **v**; thus,

$$\mathbf{u} \gg \mathbf{v} \approx 0\ (?)$$

Further, the velocity gradients are expected to be such that

$$\frac{\partial \mathbf{u}}{\partial y} \gg \frac{\partial \mathbf{u}}{\partial x} \approx \frac{\partial \mathbf{v}}{\partial x} \approx \frac{\partial \mathbf{v}}{\partial y} \approx \frac{\partial}{\partial x}\frac{\partial \mathbf{u}}{\partial y} \approx 0(?)$$

From the approximation that density is constant the continuity equation (4–33) becomes

$$\frac{\partial \mathbf{u}}{\partial x} + \frac{\partial \mathbf{v}}{\partial y} = 0 \tag{4–45}$$

If the weight of the control volume is neglected, the momentum in the y direction, equation (4–44b), reduces to

$$\frac{\partial p}{\partial y} = 0 \tag{4–46}$$

which implies that $p = f(x)$ so that $\partial p/\partial x = dp/dx$. The momentum in the x direction, equation (4–44a), then becomes

$$-\frac{1}{\rho}\left(\frac{dp}{dx} - \mu\frac{\partial^2 \mathbf{u}}{\partial y^2}\right) = \mathbf{u}\frac{\partial \mathbf{u}}{\partial x} + \mathbf{v}\frac{\partial \mathbf{u}}{\partial y} \tag{4–47}$$

This equation can be normalized or set in a form such that it is dimensionless. This provides a form of the momentum equation that can be used to obtain general solutions of the velocity profiles independent of the material properties or flow conditions. That is, it will then indicate the parameters that affect all flow conditions. If we set

$$\mathbf{u}^* = \frac{\mathbf{u}}{\mathbf{V}_\infty}, \quad \mathbf{v}^* = \frac{\mathbf{v}}{\mathbf{V}_\infty}, \quad x^* = \frac{x}{L}, \quad y^* = \frac{y}{L} \quad \text{and} \quad p^* = \frac{p}{\rho \mathbf{V}_\infty^2}$$

then the derivatives become

$$\frac{\partial \mathbf{u}}{\partial x} = \frac{\mathbf{V}_\infty}{L}\frac{\partial \mathbf{u}^*}{\partial x}, \quad \frac{\partial \mathbf{v}}{\partial y} = \frac{\mathbf{V}_\infty}{L}\frac{\partial \mathbf{v}^*}{\partial y}, \quad \frac{dp}{dx} = \rho \mathbf{V}_\infty^2 \frac{dp^*}{dx}, \quad \frac{\partial^2 \mathbf{u}}{\partial y^2} = \frac{\mathbf{V}_\infty}{L}\frac{\partial^2 \mathbf{u}^*}{\partial y^2}$$

Equation (4–47) then becomes

$$\mathbf{u}^*\frac{\partial \mathbf{u}^*}{\partial x^*} + \mathbf{v}^*\frac{\partial \mathbf{u}^*}{\partial y^*} = -\frac{dp^*}{dxv} + \frac{\mu}{\rho \mathbf{V}_\infty L}\frac{\partial^2 \mathbf{u}^*}{\partial y^{*2}} \tag{4–48}$$

Notice that the parameter in the second term on the right, $\mu/\rho\mathbf{V}_\infty L$, is the inverse Reynolds Number for a plate of length L. Also, it is implied from equation (4–48) that the velocity $\mathbf{u}^*$ is a function of x^*, y^*, and the Reynolds Number. The velocity may also be a function of the pressure gradient dp^*/dx^* so that

$$\mathbf{u}^* = f\left(x^*, y^*, \frac{dp^*}{dx^*}, \mathrm{Re}_L\right) \tag{4–49}$$

We have used this result already when we expected turbulent flow to be predictable from the Reynolds Number. The result of equation (4–29) and figure 4–19, where the friction factor was predicted by the Reynolds Number, indicates that the velocity will also be dependent on the Reynolds Number since friction affects velocity.

Energy Balance of an Element in the Boundary Layer

If we consider the conservation of energy of the same element of the boundary layer on which we balanced the mass and momentum, and referring to figure 4–26, we obtain for steady state conditions

$$\begin{aligned}
&\rho\mathbf{u}\left(\mathrm{hn} + \frac{\mathbf{V}^2}{2}\right)dy + \rho\mathbf{v}\left(\mathrm{hn} + \frac{\mathbf{V}^2}{2}\right)dx - \kappa\frac{\partial T}{\partial x}dy - \kappa\frac{\partial T}{\partial y}dx + \dot{\mathrm{W}}\mathrm{k}_{\mathrm{net}} \\
&= \rho\mathbf{u}\left(\mathrm{hn} + \frac{\mathbf{V}^2}{2}\right)dy + \rho\mathbf{v}\left(\mathrm{hn} + \frac{\mathbf{V}^2}{2}\right)dx + \frac{\partial}{\partial \mathrm{x}}\left[\rho\mathbf{u}\left(\mathrm{hn} + \frac{\mathbf{V}^2}{2}\right)\right]dxdy \\
&+ \frac{\partial}{\partial y}\left[\rho\mathbf{v}\left(\mathrm{hn} + \frac{\mathbf{V}^2}{2}\right)\right]dxdy - \left(\kappa\frac{\partial T}{\partial x} + \frac{\partial}{\partial x}\kappa\frac{\partial T}{\partial x}dx\right)dy - \left(\kappa\frac{\partial T}{\partial y} + \frac{\partial}{\partial y}\kappa\frac{\partial T}{\partial y}dy\right)dx
\end{aligned} \tag{4–50}$$

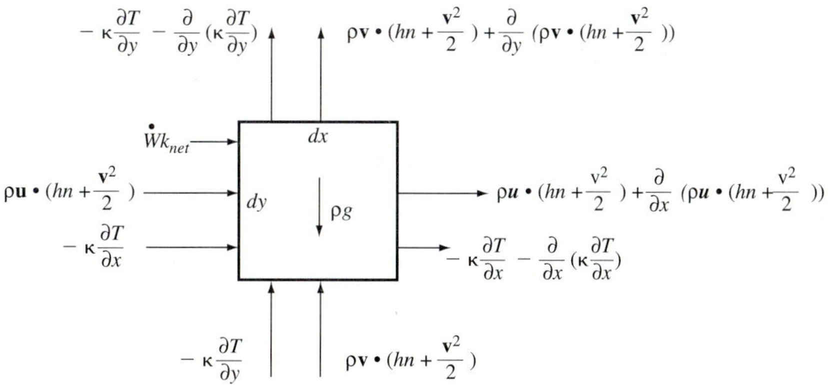

Figure 4–26 Energy balance on an element of the boundary layer.

In this energy balance and in figure (4–26) we have assumed that the heat transfers are conduction from or to adjacent elements, and there is assumed to be no energy generation. The energy accounted to the mass flows includes the enthalpy, hn $= u + pv$ from equation (1–17), and the kinetic energy only. These energies are carried into and out of the control volume, or *advected,* as described by Incropera and DeWitt [4]. The net power acting on the element, $\dot{\text{Wk}}_{\text{net}}$, is that done against the stresses,

$$\dot{\text{Wk}}_{\text{net}} = \frac{\partial \tau_{yx}\mathbf{u}}{\partial y}dydx + \frac{\partial \tau_{xy}\mathbf{v}}{\partial x}dxdy + \frac{\partial \sigma_{xv}\mathbf{u}}{\partial x}dxdy + \frac{\partial \sigma_{yv}\mathbf{v}}{\partial y}dydx \qquad \textbf{(4–51)}$$

where the stresses are defined in terms of the velocity gradients by equations (4–43).

In equations (4–41) and (4–43) the normal components due to viscosity are

$$\sigma_{xv} = \sigma_x + p = 2\mu\frac{\partial \mathbf{u}}{\partial x} - \frac{2}{3}\mu\left(\frac{\partial \mathbf{u}}{\partial x} + \frac{\partial \mathbf{v}}{\partial y}\right)$$

and

$$\sigma_{yv} = \sigma_y + p = 2\mu\frac{\partial \mathbf{v}}{\partial y} - \frac{2}{3}\mu\left(\frac{\partial \mathbf{u}}{\partial x} + \frac{\partial \mathbf{v}}{\partial y}\right) \qquad \textbf{(4–52)}$$

Substituting, rearranging, and reducing equation (4–50) gives for the boundary layer and the assumptions we made earlier (specifically that ρ, κ, and c_p are constant)

$$\mu\frac{\partial}{\partial y}\mathbf{u}\left(\frac{\partial \mathbf{u}}{\partial x} + \frac{\partial \mathbf{v}}{\partial y}\right)dydx + \mu\frac{\partial}{\partial x}\mathbf{v}\left(\frac{\partial \mathbf{u}}{\partial x} + \frac{\partial \mathbf{v}}{\partial y}\right)dxdy + 2\mu\frac{\partial^2 \mathbf{u}}{\partial x^2}dxdy + 2\mu\frac{\partial^2 \mathbf{v}}{\partial y^2}$$
$$- \frac{2}{3}\mu\left[\frac{\partial}{\partial x}\left(\frac{\partial \mathbf{u}}{\partial x} + \frac{\partial \mathbf{v}}{\partial y}\right)\right]dxdy - \frac{2}{3}\mu\left[\frac{\partial}{\partial y}\left(\frac{\partial \mathbf{u}}{\partial x} + \frac{\partial \mathbf{v}}{\partial y}\right)\right]dydx + \kappa\left(\frac{\partial^2 T}{\partial x^2} + \frac{\partial^2 T}{\partial y^2}\right)$$
$$= \frac{\partial}{\partial x}\left(\text{hn} + \frac{\mathbf{V}^2}{2}\right) + \frac{\partial}{\partial y}\left(\text{hn} + \frac{\mathbf{V}^2}{2}\right) \qquad \textbf{(4–53)}$$

Equation (4–53) may be reduced further to a thermal energy balance by removing the kinetic energy. This is accomplished by using the equations of motion, (4–44), multiplying the first or x-direction motion equation by $\mathbf{u}$ and multiplying the second or y-direction motion equation by $\mathbf{v}$. Manipulating these two equations with equation (4–53) can give the result

$$\rho\mathbf{u}\frac{\partial \text{hn}}{\partial x} + \rho\mathbf{v}\frac{\partial \text{hn}}{\partial y} = \kappa\left(\frac{\partial^2 T}{\partial x^2} + \frac{\partial^2 T}{\partial y^2}\right) + \mathbf{u}\frac{dp}{dx} + \mathbf{v}\frac{dp}{dy} + \mu\Phi \qquad \textbf{(4–54)}$$

where the term

$$\mu\Phi = \mu\left\{\left[\frac{\partial \mathbf{u}}{\partial x} + \frac{\partial \mathbf{v}}{\partial y}\right]^2 + 2\left[\left(\frac{\partial \mathbf{u}}{\partial x}\right)^2 + \left(\frac{\partial \mathbf{v}}{\partial y}\right)^2\right] - \frac{2}{3}\left[\frac{\partial \mathbf{u}}{\partial x} + \frac{\partial \mathbf{v}}{\partial y}\right]^2\right\} \qquad \textbf{(4–55)}$$

is the power dissipated because of the fluid viscosity. It is sometimes called the *viscous dissipation,* and it can be counted as an energy rate that must be dispersed through the convection processes. Using the boundary layer approximations that we observed to reduce the equations of motion to those of equation (4–44) the viscous dissipation reduces to

$$\mu\Phi = \mu\left(\frac{\partial \mathbf{u}}{\partial x}\right)^2$$

and the thermal energy equation (4–54) then becomes

$$\rho\mathbf{u}\frac{\partial \text{hn}}{\partial x} + \rho\mathbf{v}\frac{\partial \text{hn}}{\partial y} = \kappa\frac{\partial^2 T}{\partial y^2} + \mu\left(\frac{\partial \mathbf{u}}{\partial x}\right)^2 \qquad \textbf{(4–56)}$$

In many engineering problems the viscous dissipation is an order of magnitude or more less than the flow energy term (the two terms on the left side of equation 4–56) or the conduction heat transfer (the first term on the right side), and then the energy equation is

$$\rho \mathbf{u}\frac{\partial \text{hn}}{\partial x} + \rho \mathbf{v}\frac{\partial \text{hn}}{\partial y} = \kappa \frac{\partial^2 T}{\partial y^2} \tag{4–57}$$

For fluids that behave as ideal gases the enthalpy hn is a function of the temperature and the specific heat, c_p. The thermal energy equation then becomes, if c_p is assumed as a constant,

$$\rho c_p \mathbf{u}\frac{\partial T}{\partial x} + \rho c_p \mathbf{v}\frac{\partial T}{\partial y} = \kappa \frac{\partial^2 T}{\partial y^2}$$

or, upon rearranging,

$$\mathbf{u}\frac{\partial T}{\partial x} + \mathbf{v}\frac{\partial T}{\partial y} = \alpha \frac{\partial^2 T}{\partial y^2} \tag{4–58}$$

where α is the thermal diffusivity defined in Chapter 3 as

$$\alpha = \frac{\kappa}{\rho c_p}$$

If this equation (4–58) is normalized in the same way as the motion equations, setting

$$\mathbf{u}^* = \frac{\mathbf{u}}{\mathbf{V}_\infty}, \quad \mathbf{v}^* = \frac{\mathbf{v}}{\mathbf{V}_\infty}, \quad x^* = \frac{x}{L}, \quad y^* = \frac{y}{L}, \quad T^* = \frac{T - T_s}{T_\infty - T_s}$$

then equation (4–58) becomes

$$\mathbf{u}^*\frac{\partial T^*}{\partial x^*} + \mathbf{v}^*\frac{\partial T^*}{\partial y^*} = \left(\frac{\alpha}{\mathbf{V}_\infty L}\right)\frac{\partial^2 T^*}{\partial y^{*2}} \tag{4–59}$$

The parameter $\alpha/\mathbf{V}_\infty L$ may be rearranged by writing it as

$$\frac{\alpha}{\mathbf{V}_\infty L} = \frac{\mu\rho\alpha}{\mu\rho\mathbf{V}_\infty L} = \left(\frac{\mu}{\rho\mathbf{V}_\infty L}\right)\left(\frac{\alpha\rho}{\mu}\right)$$

The quantity $\rho\mathbf{V}_\infty L/\mu$ is our Reynolds Number, Re_L, and the quantity $\mu/\alpha\rho = \nu/\alpha$ is called the Prandtl Number, Pr. The Prandtl Number may be written in various forms,

$$\text{Pr} = \frac{\mu}{\alpha\rho} = \frac{\nu}{\alpha} = \frac{\mu c_p}{\kappa} \tag{4–60}$$

The Prandtl Number is often described as the ratio of the viscous or momentum diffusion to the thermal diffusion in a fluid. Equation (4–59) may be revised to read

$$\mathbf{u}^*\frac{\partial T^*}{\partial x^*} + \mathbf{v}^*\frac{\partial T^*}{\partial y^*} = \left(\frac{1}{\text{Re}_L}\right)\left(\frac{1}{\text{Pr}}\right)\frac{\partial^2 T^*}{\partial y^{*2}} \tag{4–61}$$

and from this equation we can infer that the temperature distribution T^* is affected by x^*, y^*, Re_L, and Pr; that is,

$$T^* = f(x^*, y^*, \text{Re}_L, \text{Pr}) \tag{4–62}$$

Earlier we discussed the idea that convection heat transfer may be compared to conduction heat transfer through a fluid at the surface of the solid. This was expressed as an equality in equation (4–32). Using equation (4–32) and the definitions for the normalized temperature T and displacement y*, we may write that

$$hA(T_\infty - T_s) = -\kappa_{\text{fluid}} A\left[\frac{\partial T}{\partial y}\right]_{y=0} = -\kappa_{\text{fluid}} A\left(\frac{T_\infty - T_s}{L}\right)\frac{\partial T^*}{\partial y^*}$$

or

$$h = -\kappa_{\text{fluid}}\left(\frac{1}{L}\right)\frac{\partial T^*}{\partial y^*} \tag{4–63}$$

Then, referring to equation (4–62), we can write that

$$h = \left(\frac{\kappa_{\text{fluid}}}{L}\right) f(x^*, y^*, \text{Re}_L, \text{Pr})$$

or

$$\frac{hL}{\kappa_{\text{fluid}}} = f(x^*, y^*, \text{Re}_L, \text{Pr}) \tag{4–64}$$

It is customary to define the parameter hL/κ_{fluid} as the *Nusselt Number*, Nu_L, and it may be visualized as the ratio of the convection heat transfer through a fluid to the conduction heat transfer through the same fluid. Notice that earlier, in Chapter 3, we defined the Biot Number, Bi, as the ratio of the convection heat transfer through a fluid surrounding a solid object to the conduction through the solid. The crucial difference between the Nusselt Number and the Biot Number is the use of the thermal conductivity of the fluid for the Nusselt Number and the thermal conductivity of the solid for the Biot Number.

Equation (4–64) may be restated in an algebraic form as

$$\text{Nu}_L = C(\text{Re}_L)^n(\text{Pr})^m \tag{4–65}$$

where C, n, and m are constants that need to be determined by experiment or measurements. This equation will be used to provide methods for predicting convective heat transfer coefficients from empirical results for many different forced convection situations. Notice also that the Nusselt Number here is defined by a characteristic length, L, like the Reynolds Number. We will see that the Nusselt Number can be defined with a characteristic diameter, written Nu_D, as we have done with the Reynolds Number. Equation (4–65) then can be written

$$\text{Nu}_D = C(\text{Re}_D)^n(\text{Pr})^m \tag{4–65}$$

If the reader is unclear in reading through to this point how the result of equation (4–65) comes about, the same result follows from dimensional analysis, and we will now show that development to give you a further perspective and perhaps better understand the rationale for the analysis.

Dimensional Analysis for Forced Convection Heat Transfer

Using the methods of dimensional analysis, we want to relate the convection heat transfer in a flowing fluid to those parameters that affect the phenomena. We first assume that convection heat transfer is affected by fluid velocity, density, thermal conductivity, viscosity, temperature difference in the fluid, and size. Thus

$$\dot{Q}_{\text{convection}} = f(\mathbf{V}, \rho, \kappa_{\text{fluid}}, \mu, c_p, L, \Delta T) \tag{4–66}$$

We know that convection heat transfer can also be described as $\dot{Q}_{\text{convection}} = hA\Delta T$, so that by applying the separation of variables the convection heat transfer coefficient, h, is a function of all of the parameters of equation (4–66) except ΔT. Thus

$$h = f(\mathbf{V}, \rho, \kappa_{\text{fluid}}, \mu, c_p, L) \tag{4–67}$$

There are seven parameters that have four fundamental dimensions. Based on Buckingham's Pi-Theorem there will be three ($7 - 4 = 3$) Π-parameters that relate the seven parameters. We choose arbitrarily h, c_p, and μ as the three fundamental parameters from which we will form the three Π-parameters. This selection of parameters is arbitrary, and if others are chosen there will likely be different combinations for the Π-parameters. Continuing, we have

$$\Pi_1 = h^a \rho^b \kappa^c \mathbf{V}^d L^e$$

The dimensional form of this equation is

$$\Pi_1 = (mt^{-3}T^{-1})^a (mL^{-3})^b (mLt^{-3}T^{-1})^c (Lt^{-1})^d (L)^e$$

and setting each of the dimensions to zero:

$$\begin{aligned} &\text{for } m: && a + b + c = 0 \\ &\text{for } L: && -3b + c + d + e = 0 \\ &\text{for } t: && -3a - 3c - d = 0 \\ &\text{for } T: && -a - c = 0 \end{aligned}$$

Since we have five exponents and only four equations, we shall set $a = 1$. Then $c = -1$, $b = 0$, $d = 0$, and $e = 1$. This gives

$$\Pi_1 = \frac{hL}{\kappa_{\text{fluid}}} = \frac{hL}{\kappa} \qquad \text{(the Nusselt Number)}$$

The second Π-parameter is formed from c_p, ρ, κ, $\mathbf{V}$, and L;

$$\Pi_2 = c_p^a \rho^b \kappa^c \mathbf{V}^d L^e$$

Dimensionally,

$$\Pi_2 = (L^2t^{-2}T^{-1})^a (mL^{-3})^b (mLt^{-3}T^{-1})^c (Lt^{-1})^d (L)^e$$

$$\begin{aligned} &\text{for } m: && b + c = 0 \\ &\text{for } L: && 2a - 3b + c + d + e = 0 \\ &\text{for } t: && -2a - 3c - d = 0 \\ &\text{for } T: && -a - c = 0 \end{aligned}$$

Again we have four equations and five unknowns so if we set $a = 1$ then $c = -1$, $d = 1$, $b = 1$, and $e = 1$. The second Π-parameter is

$$\Pi_2 = \frac{c_p \rho \mathbf{V} L}{\mu}$$

(sometimes called the Peclet Number, Pe). By a similar procedure, using the remaining parameter μ with ρ, κ, $\mathbf{V}$, and L, gives that the third Π-parameter is the Reynolds Number, Re_L,

$$\Pi_3 = \frac{\rho \mathbf{V} L}{\mu} = \text{Re}_L$$

Then we can write

$$\Pi_1 = Cf(\Pi_2, \Pi_3) \tag{4–68}$$

or

$$\text{Nu}_L = C\text{Re}^j \text{Pe}^k$$

The Peclet Number can be rearranged to read

$$\text{Pe} = \frac{c_p \rho \mathbf{V} L \mu}{\mu \kappa} = \left(\frac{c_p \mu}{\kappa}\right)\left(\frac{\rho \mathbf{V} L}{\mu}\right) = \text{Pr} \cdot \text{Re}_L$$

and equation (4–68) then corresponds to equation (4–65):

$$\mathrm{Nu}_L = C\,\mathrm{Re}_L^m\,\mathrm{Pr}^n \tag{4–69}$$

The Nusselt Number may be related to the Reynolds and Prandtl Numbers in a manner other than the power relationship indicated in equations (4–65) and (4–69); however, the important idea is that the complicated happenings of convection heat transfer may be expressed by such functional relationships. It has been customary to treat the relationships in the form of equations (4–65) and (4–69).

Reynolds–Colburn Analogy

The shear stress in a flowing fluid has been identified from equation (4–11), and the shear stress acting on the solid surface is

$$\tau_s = \mu\left[\frac{\partial \mathbf{u}}{\partial y}\right]_{y=0} \tag{4–70}$$

where y is zero at the surface, as indicated in figure 4–20. The shear stress may be visualized as a friction drag on the solid plate, so we may define a friction drag coefficient, C_f, as

$$C_f\left(\rho\frac{\mathbf{V}^2}{2}\right) = \mu\left[\frac{\partial \mathbf{u}}{\partial y}\right]_{y=0} \tag{4–71}$$

and this can be rewritten, using the dimensionless velocity and displacement, as

$$C_f\left(\rho\frac{\mathbf{V}_\infty^2}{2}\right) = \left(\frac{\mu\mathbf{V}_\infty}{L}\right)\left[\frac{\partial \mathbf{u}^*}{\partial y^*}\right]_{y=0} \tag{4–72}$$

This result can be rearranged to be

$$C_f = \left(\frac{2}{\mathrm{Re}_L}\right)\left[\frac{\partial \mathbf{u}^*}{\partial y^*}\right]_{y=0} \tag{4–73}$$

From equation (4–49) we may conclude that $\dfrac{\partial \mathbf{u}^*}{\partial y^*}$ is a function of x^*, y^*, Re_L, and a pressure gradient, dp^*/dy^*. From equation (4–32) and (4–63) we have

$$\frac{hL}{\kappa_{\text{fluid}}} = \left[\frac{\partial T^*}{\partial y^*}\right]_{y=0} \tag{4–74}$$

Then, from equation (4–62) we may conclude that $\partial T^*/\partial y^*$ is a function x^*, y^*, Re_L, and Pr for a given pressure gradient. Reynolds suggested that convection heat transfer is related to fluid friction at the containing surface [5], which implies that equations (4–73) and (4–74) may be compared. Other investigators have extended Reynolds' suggestion by assuming that the forms of the functions indicated in equations (4–49) and (4–62) are identical for the same physical conditions; that is, that $\mathbf{u}^*$ and T^* and their gradients have similar functional relationships. This gives

$$\frac{\partial T^*}{\partial y^*} = \frac{\partial \mathbf{u}^*}{\partial y} f(\mathrm{Pr}) \tag{4–75}$$

and by using the results from equations (4–73) and (4–74), we have

$$\mathrm{Nu}_L = \frac{C_f}{2}\mathrm{Re}_L f(\mathrm{Pr}) \tag{4–76}$$

It was suggested by Colburn [6] that the Prandtl function $f(\mathrm{Pr})$ is

$$f(\mathrm{Pr}) = \mathrm{Pr}^{1/3} \tag{4–77}$$

so that equation (4–76) becomes

$$\frac{\mathrm{Nu}_L}{\mathrm{Re}_L\,\mathrm{Pr}^{1/3}} = \mathrm{St}\cdot\mathrm{Pr}^{2/3} = \frac{1}{2}C_f \tag{4–78}$$

where $\mathrm{St} = \mathrm{Nu}_L/(\mathrm{Re}_L \cdot \mathrm{Pr})$ is called the Stanton Number. Equation (4–78) is sometimes called the *Colburn Equation* or the *Reynolds–Colburn Equation*. The result expressed by equation (4–78) is appealing because it allows for the experimental determination of the convective heat-transfer coefficient without resorting to measurements of the boundary layer temperatures. The friction drag coefficient must be determined, but it is obtained through mechanical measurements of the forces acting on the plate, and these are usually routine. The relationship given in equation (4–78) is usually restricted to fluids having Prandtl Numbers between 0.6 and 50.

4–4 CONVECTION HEAT TRANSFER AT A FLAT PLATE

The flat plate is the physical configuration that is the least complicated of all surfaces where convection heat transfer needs to be analyzed. The flat plate is also a reasonable approximation for many important engineering applications: heat transfer at a building or structural surface, heat transfer at surfaces of a heat exchanger, heat transfer at outside surfaces of moving vehicles, heat transfer at the earth's surface, heat transfer within engine components and nozzles, and others. In figure 4–20 is shown the situation of fluid flow parallel to a flat plate with a developing boundary layer beginning at the leading edge of a flat plate and continuing through a location where the boundary layer contains turbulent fluid flow. For the flow in the boundary layer we have the equations developed in section 4–3:

Conservation of Mass

$$\frac{\partial \mathbf{u}}{\partial x} + \frac{\partial \mathbf{v}}{\partial y} = 0 \tag{4–45, repeated}$$

Conservation of Momentum in the *x* Direction

$$\mathbf{u}\frac{\partial \mathbf{u}}{\partial x} + \mathbf{v}\frac{\partial \mathbf{u}}{\partial y} = \frac{\mu}{\rho}\frac{\partial^2 \mathbf{u}}{\partial y^2} - \frac{1}{\rho}\frac{dp}{dx} \tag{4–47, repeated}$$

Conservation of Energy

$$\mathbf{u}\frac{\partial T}{\partial x} + \mathbf{v}\frac{\partial T}{\partial y} = \alpha\frac{\partial^2 T}{\partial y^2} \tag{4–58, repeated}$$

These equations are applicable for the following conditions:

1. Steady state.
2. Incompressible substance with constant specific heats.
3. No viscous dissipation.
4. Homogeneous and isotropic properties.
5. Subjected to order-of-magnitude analysis.

For laminar, parallel flow over the flat plate, the pressure gradient in the x direction can be assumed to be zero, or nearly so. That means that $dp/dx = 0$ and equation (4–47) becomes

$$\mathbf{u}\frac{\partial \mathbf{u}}{\partial x} + \mathbf{v}\frac{\partial \mathbf{u}}{\partial y} = \frac{\mu}{\rho}\frac{\partial^2 \mathbf{u}}{\partial x^2} \tag{4–79}$$

The three equations, (4–45), (4–79), and (4–58) represent the set of equations describing heat transfer and fluid flow so that a temperature and velocity distribution can be predicted in the boundary layer. A solution can be obtained with the boundary conditions:

$$\begin{aligned} &1.\quad T' = \frac{T - T_s}{T_\infty - T_s} && \text{at } x \geq 0,\ y = 0 \\ &2.\quad T' = 1 && \text{at } x \geq 0,\ y \to \infty \\ &3.\quad \mathbf{u} = 0 && \text{at } x \geq 0,\ y = 0 \\ &4.\quad \mathbf{u} = \mathbf{u}_\infty && \text{at } x \geq 0,\ y \to \infty \end{aligned} \tag{4–80}$$

The analysis for this problem is somewhat simplified because the physical properties are constant, and this implies that the equations of motion, (4–45) and (4–79), can be solved independently of the energy equation (4–58). Thus, by solving for the velocity profiles first, the solution to the temperature profile through the energy equation can be obtained as a separate problem.

The solution to the motion equations is called the hydrodynamic solution and a particular solution was first given by Blasius [7]. Others have considered the same problem and similar solutions have been obtained. The method of obtaining a solution, as outlined by Schlichting [3] and following Blasius, begins by assuming a *stream function*, $\psi(x, y)$, that can satisfy the conditions

$$\mathbf{u} = \frac{\partial \psi(x, y)}{\partial y} \quad \text{and} \quad \mathbf{v} = -\frac{\partial \psi(x, y)}{\partial x} \tag{4–81}$$

and these relationships mean that equation (4–45) is automatically satisfied (see problem 4–19). Then, through judicious selection, two new variables are defined,

$$\eta = y\sqrt{\frac{\mathbf{u}_\infty}{\nu x}} \tag{4–82}$$

$$f(\eta) = \frac{\psi(x, y)}{\sqrt{\mathbf{u}_\infty \nu x}} \tag{4–83}$$

Also, for flow in the laminar region the velocity profile is assumed to be of the same character at any location along the plate. This means that $\mathbf{u}/\mathbf{u}_\infty$ is some function of y/δ

$$\frac{\mathbf{u}}{\mathbf{u}_\infty} = \phi\left(\frac{y}{\delta}\right) \tag{4–84}$$

If the boundary layer thickness is assumed to be described by the relationship

$$\delta = \sqrt{\frac{\nu x}{\mathbf{u}_\infty}} \tag{4–85}$$

then equation (4–84) becomes

$$\frac{\mathbf{u}}{\mathbf{u}_\infty} = \phi\left(y\sqrt{\frac{\mathbf{u}_\infty}{\nu x}}\right) = \phi(\eta) \tag{4–86}$$

From equations (4–81), (4–82), and (4–83) we have

$$\mathbf{u} = \frac{\partial \psi}{\partial y} = \frac{\partial \psi}{\partial \eta}\frac{\partial \eta}{\partial y} = \sqrt{\mathbf{u}_\infty \nu x}\frac{\partial f(\eta)}{\partial \eta} \cdot \sqrt{\frac{\mathbf{u}_\infty}{\nu x}} = \mathbf{u}_\infty \frac{\partial f(\eta)}{\partial \eta} = \mathbf{u}_\infty \frac{df(\eta)}{d\eta} \tag{4–87}$$

Also,

$$\mathbf{v} = -\frac{\partial\psi}{\partial x} = \frac{1}{2}\sqrt{\frac{\mathbf{u}_\infty \nu}{x}}\left(\eta\frac{df(\eta)}{d\eta} - f(\eta)\right) \tag{4–88}$$

The differentials of these velocity terms are

$$\frac{\partial \mathbf{u}}{\partial x} = -\left(\frac{\mathbf{u}_\infty}{2x}\right)\eta\frac{d^2f(\eta)}{d\eta^2} \tag{4–89}$$

$$\frac{\partial \mathbf{u}}{\partial y} = \mathbf{u}_\infty\sqrt{\frac{\mathbf{u}_\infty}{\nu x}}\frac{d^2f(\eta)}{d\eta^2} \tag{4–90}$$

and

$$\frac{\partial^2 \mathbf{u}}{\partial y^2} = \frac{\mathbf{u}_\infty^2}{\nu x}\frac{d^3f(\eta)}{d\eta^3} \tag{4–91}$$

Substituting the relationships of (4–87) through (4–91) into equation (4–79), the x-direction momentum balance, gives

$$2\frac{d^3f(\eta)}{d\eta^3} + f(\eta)\frac{d^2f(\eta)}{d\eta^2} = 0 \tag{4–92}$$

The purpose of the manipulations to this point are to reduce the partial differential equation of equation (4–79) into an ordinary differential equation, equation (4–92). This is a nonlinear, third-order ordinary differential equation, and the boundary conditions that can be associated with it for our problem are

$$\begin{aligned} &\text{B.C. 1} \quad \frac{df(\eta)}{d\eta} = 0 \quad &&\text{at } \eta = 0 \\ &\text{B.C. 2} \quad f(\eta) = 0 \quad &&\text{at } \eta = 0 \\ &\text{B.C. 3} \quad \frac{df(\eta)}{d\eta} = 1 \quad &&\text{as } \eta \to \infty \end{aligned} \tag{4–93}$$

A closed-form solution to this problem has not been developed, but solutions have been obtained by means of series expansions. Howarth [8] used numerical methods to compute solutions that have been quoted often and which seem to be sufficiently accurate for engineering purposes. These results are presented in table 4–3 and shown in the graph of figure 4–27 in a condensed form. From the table it can be seen that at $\eta = 5.0$ the term $df(\eta)/d\eta = \mathbf{u}/\mathbf{u}_\infty = 0.9916$.

Many investigators have chosen to define the envelope of the boundary layer such that

$$y = \delta \qquad \text{at } \frac{\mathbf{u}}{\mathbf{u}_\infty} = 0.9916 \tag{4–94}$$

For this condition and using equations (4–82) and (4–85), we have

$$5.0 = y\sqrt{\frac{\mathbf{u}_\infty}{\nu x}}$$

Since $\text{Re}_x = \mathbf{u}_\infty x/\nu$, this can be rearranged to give

$$\delta = 5.0\, x/\sqrt{\text{Re}_x} \tag{4–95}$$

Table 4–3 Results of the Function $f(\eta)$ for Parallel Boundary Layer Flow along a Flat Surface as Determined by Howarth [8]

$\eta = y\sqrt{\frac{u_\infty}{\nu x}}$	$f(\eta)$	$\frac{df(\eta)}{d\eta} = \frac{\mathbf{u}}{\mathbf{u}_\infty}$	$\frac{d^2f(\eta)}{d\eta^2}$
0	0	0	0.3321
0.2	0.0066	0.0664	0.3320
0.4	0.0266	0.1328	0.3315
0.6	0.0597	0.1989	0.3301
0.8	0.1061	0.2647	0.3274
1.0	0.1656	0.3298	0.3230
1.4	0.3230	0.4563	0.3079
1.8	0.5295	0.5748	0.2829
2.2	0.7812	0.6813	0.2484
2.6	1.0725	0.7725	0.2065
3.0	1.3968	0.8461	0.1614
3.4	1.7470	0.9018	0.1179
3.8	2.1161	0.9411	0.0801
4.2	2.4981	0.9670	0.0505
4.6	2.8883	0.9827	0.0295
5.0	3.2833	0.9916	0.0159
5.4	3.6809	0.9962	0.0079
5.8	4.0799	0.9984	0.0037
6.2	4.4795	0.9994	0.0016
6.6	4.8793	0.9998	0.0006
7.0	5.2793	0.9999	0.0002
7.4	5.6792	1.0000	0.0001
7.8	6.0792	1.0000	0.0000
8.2	6.4792	1.0000	0.0000

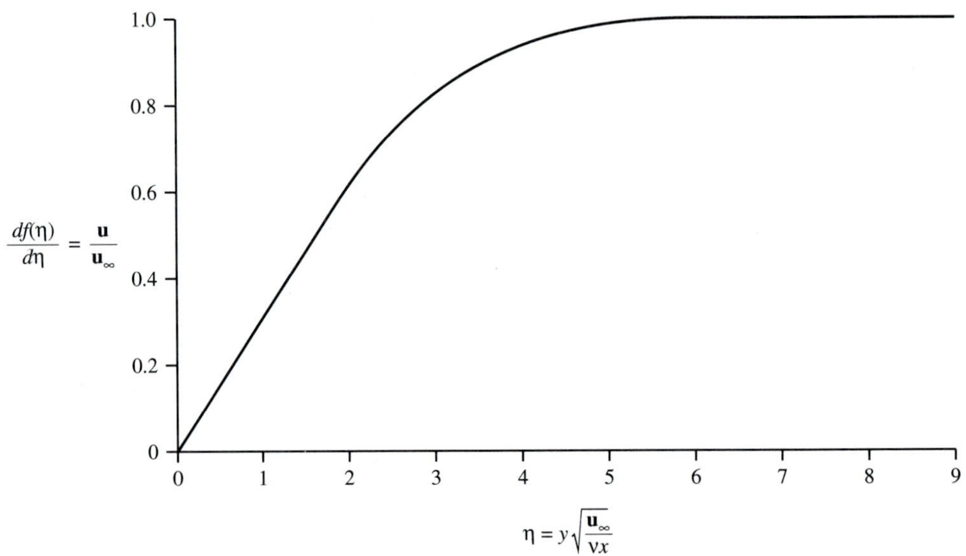

Figure 4–27 Laminar boundary layer flow profile along a flat surface.

Also, recalling that the shear stress acting on the fluid is given by equation (4–70), $\tau_s = \mu[\partial \mathbf{u}/\partial y]_{y=0}$, we may use the result of equation (4–90) to give

$$\tau_s = \mu \mathbf{u}_\infty \sqrt{\frac{\mathbf{u}_\infty}{\nu x}} \left[\frac{d^2 f(\eta)}{d\eta^2}\right]_{\eta=0}$$

From table 4–3 we have

$$\frac{d^2 f(\eta)}{d\eta^2} = 0.332 \qquad \text{at } \eta = 0$$

so that we have

$$\tau_s(x) = 0.332 \mu \mathbf{u}_\infty \sqrt{\frac{\mathbf{u}_\infty}{\nu x}} \qquad \textbf{(4–96)}$$

Notice that the shear stress is a function of the distance from the leading edge, x. The friction drag coefficient, C_f, defined earlier by equation (4–71), is

$$C_f = \frac{2\tau_s}{\rho \mathbf{u}_\infty^2}$$

From equation (4–96), the friction drag coefficient can then be written as

$$C_f = 0.664 \mu \frac{1}{\rho \mathbf{u}_\infty} \sqrt{\frac{\mathbf{u}_\infty}{\nu x}}$$

Recalling the definition for the Reynolds Number gives the result

$$C_f = \frac{0.664}{\sqrt{\text{Re}_x}} \qquad \textbf{(4–97)}$$

The friction drag coefficient given by equation (4–97) is the local value, and the average value over a full surface would be

$$C_{f,\text{ave}} = \frac{1}{L} \int_0^L C_f dx$$

For the case of laminar flow over an entire flat surface, equation (4–97) may be used to obtain an explicit equation for the average drag coefficient (see practice problem 4–14),

$$C_{f,\text{ave}} = \frac{1.328}{\sqrt{\text{Re}_L}} \qquad \textbf{(4–98)}$$

EXAMPLE 4–11 Air at 27°C flows over a flat roof at 5 m/s. Estimate the location from the leading edge of the roof where the air flow boundary layer becomes transitional from laminar to turbulent, then estimate the boundary layer thickness and the friction drag coefficient at this location.

Solution The transition from laminar to turbulent flow along a flat surface can be estimated by assuming a critical Reynolds Number, $\text{Re}_L = 4 \times 10^5$. Thus

Answer

$$x = \frac{\text{Re}_{xc}\nu}{\mathbf{u}_\infty} = \frac{(4 \times 10^5)(15.86 \times 10^{-6}\ \text{m}^2/\text{s})}{5\ \text{m/s}} = 1.271\ \text{m}$$

The boundary layer thickness may be estimated by using equation (4–95),

Answer

$$\delta = \frac{5.0x}{\sqrt{\text{Re}_L}} = \frac{(5.0)(1.271\ \text{m})}{\sqrt{4 \times 10^5}} = 1.0\ \text{cm}$$

The friction drag coefficient C_f can be determined from equation (4–97),

Answer

$$C_f = \frac{0.664}{\sqrt{\text{Re}_x}} = \frac{0.664}{\sqrt{4 \times 10^5}} = 0.00105$$

EXAMPLE 4–12 Assume that the roof over which the air flows for the conditions of example 4–11 is 40 m wide, as indicated in figure 4–28. Then estimate the shear force acting on the roof due only to the first 1.271 m of roof. That is, estimate the shear force caused by the laminar boundary layer shear stress.

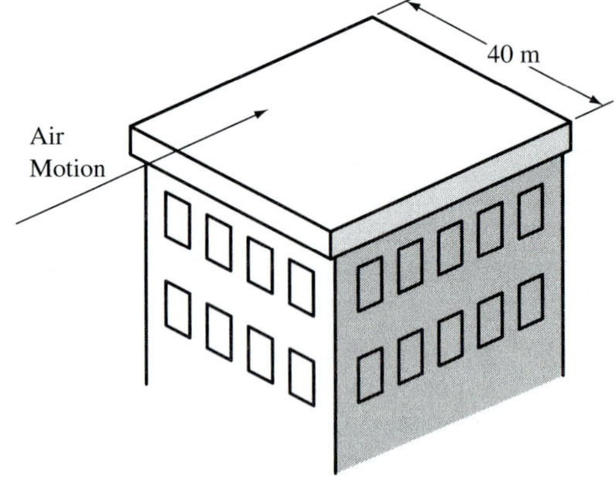

Figure 4–28 Air flowing over a flat roof.

Solution The shear force may be estimated by first estimating the shearing stress, $\tau_{x,\text{ave}}$, and then using the definition of the shear stress to determine the force, $F_s = \tau_{x,\text{ave}} \cdot A$, where A is the surface area on which the shear stress acts. The shear stress must be considered to be an average over the entire surface, and it is estimated from the equation

$$\tau_{s,\text{ave}} = C_{f,\text{ave}} \rho \frac{\mathbf{u}_\infty^2}{2}$$

where the average friction drag coefficient is defined by equation 4–98,

$$C_{f,\text{ave}} = \frac{1.328}{\sqrt{\text{Re}_L}} = \frac{1.328}{\sqrt{4 \times 10^5}} = 2.1 \times 10^{-3}$$

The average shear stress is then

$$\tau_{s,\text{ave}} = (2.1 \times 10^{-3})\left(1.1614\ \frac{\text{kg}}{\text{m}^3}\right)\frac{(5\ \text{m/s})^2}{2} = 0.030\ \text{Pa}$$

and the shear force is

Answer

$$F_s = \left(0.030\ \frac{\text{N}}{\text{m}^2}\right)(40\ \text{m} \times 1.271\ \text{m}) = 1.525\ \text{N}$$

Using the results for the velocity distributions, the energy equation (4–58) can be solved for a temperature distribution. The accepted approach to obtaining a solution has been to use the dimensionless temperature T' where this parameter is defined in the first boundary condition of equation (4–80). This dimensionless temperature profile is assumed to have the same character at any location along the surface, and the temperature is describable as a function of η, namely $T' = T'(\eta)$. The energy equation is then written

$$\frac{d^2f(\eta)}{d\eta^2} + \frac{\text{Pr}}{2}f(\eta)\frac{dT'(\eta)}{d\eta} = 0 \tag{4–99}$$

where $f(\eta)$ is the same function used to solve for the momentum equation and given in table 4–3. The boundary conditions for the situation of parallel flow in a boundary layer along a flat surface can be written

$$\begin{array}{lll} \text{B.C. 1} & T'(\eta) = 0 & \text{at } \eta = 0 \\ \text{B.C. 2} & T'(\eta) = 1 & \text{as } \eta \to \infty \end{array} \tag{4–100}$$

A solution to equation (4–99) with these associated boundary conditions has been obtained by numerical methods, and Pohlhausen [9] shows that for $50 \geq \text{Pr} \geq 0.6$,

$$\left[\frac{dT'(\eta)}{d\eta}\right]_{\eta=0} = 0.332\,\text{Pr}^{1/3} \tag{4–101}$$

Then, using the result of equation (4–32), we write for the local convection heat transfer coefficient, h_x,

$$h_x = \frac{\kappa_{\text{fluid}}}{T_\infty - T_s}\left[\frac{\partial T}{\partial y}\right]_{y=0}$$

Since $T' = (T - T_s)/(T_\infty - T_s)$, we have $\partial T/\partial y = (T_\infty - T_s)\partial T'/\partial y$ and

$$h_x = \kappa_{\text{fluid}}\left[\frac{\partial T'}{\partial y}\right]_{y=0}$$

Using the previous dimensional parameter η we have

$$\frac{\partial T'}{\partial y} = \sqrt{\frac{\mathbf{u}_\infty}{\nu x}}\frac{\partial T'(\eta)}{\partial \eta}$$

Then

$$h_x = \sqrt{\frac{\mathbf{u}_\infty}{\nu x}}\kappa_{\text{fluid}}\left[\frac{\partial T'(\eta)}{\partial \eta}\right]_{\eta=0} = 0.332\sqrt{\frac{\mathbf{u}_\infty}{\nu x}}\kappa_{\text{fluid}}\,\text{Pr}^{1/3}$$

or

$$h_x = 0.332\kappa_{\text{fluid}}\frac{1}{x}\sqrt{\frac{\mathbf{u}_\infty x}{\nu}}\,\text{Pr}^{1/3} = 0.332\frac{\kappa_{\text{fluid}}}{x}\text{Re}_x^{1/2}\,\text{Pr}^{1/3} \tag{4–102}$$

This equation is more often put in a nondimensional form so that

$$\text{Nu}_x = 0.332\text{Re}^{1/2} \cdot \text{Pr}^{1/3} \tag{4–103}$$

Notice how this equation is in agreement with the general equation (4–69) that was obtained by dimensional analysis. The reader should remember that equations (4–102) and (4–103) are applicable only if the flow is laminar, $\text{Re} \leq 4 \times 10^5$, and $50 \geq \text{Pr} \geq 0.6$.

The thermal boundary layer δ_T can be shown to be related to the velocity boundary layer by

$$\delta_T = \delta \text{Pr}^{-1/3} = 05.0 \times \text{Re}^{-1/2} \cdot \text{Pr}^{-1/3} \quad \textbf{(4–104)}$$

Again, this result is restricted to laminar boundary layer flow and for Prandtl numbers less than 50 but greater than 0.6. The reader should also notice that the boundary layer increases in thickness from the leading edge while the convective heat-transfer coefficient decreases. The average convective heat-transfer coefficient, h_{ave}, can be defined as

$$h_{\text{ave}} = \frac{1}{L}\int_0^L h_x dx \quad \textbf{(4–105)}$$

where L is the full length of the surface described by the average convective heat-transfer coefficient. Using equation (4–102) or (4–103) and assuming laminar flow throughout a flat surface, the average convective heat-transfer coefficient is, from equation (4–105),

$$h_{\text{ave}} = \frac{\kappa_{\text{fluid}}}{L}(0.664\ \text{Re}_L^{1/2} \cdot \text{Pr}^{1/3}) = 2h_L \quad \textbf{(4–106)}$$

where h_L is the local value for the convective heat-transfer coefficient at $x = L$. For fluids with small Prandtl Numbers, the previous development must be modified and the results of equations (4–102) through (4–106) do not apply. Fluids having small Prandtl Numbers ($\text{Pr} = \mu/\rho\alpha = \mu c_p/\kappa$) tend to have thermal boundary layers much thicker than the velocity boundary layer; that is, $\delta_T \gg \delta$. As a consequence of this, the velocity in the thermal boundary layer may be assumed to be a constant, $\mathbf{u} = \mathbf{u}_\infty$. It can be shown that the following relationship is appropriate

$$\text{Nu}_x = 0.565(\text{Re}_x\text{Pr})^{1/2} \quad \textbf{(4–107)}$$

for $\text{Pr} \leq 0.05$ and $\text{Re} \geq 2000$. Liquid metals, such as sodium and mercury, are important fluids having Prandtl Numbers less than 0.05. An equation recommended by Churchill and Ozoe [10] that can be used for fluids having any Prandtl Number is

$$\text{Nu}_x = 0.3387\frac{\text{Re}_x^{1/2} \cdot \text{Pr}^{1/3}}{\left[1 + \left(\dfrac{0.0468}{\text{Pr}}\right)^{2/3}\right]^{1/4}} \quad \textbf{(4–108)}$$

provided that $\text{Re} \cdot \text{Pr} \geq 100$. Results gotten from equation (4–108) must be considered somewhat less reliable than results obtained from the previous equations, except within the range of Prandtl numbers between 0.05 and 0.6 equation (4–108) is probably more accurate than the other forms.

Turbulent Flow Many convection heat-transfer situations involve fluid flows that can be characterized as turbulent. Turbulent flow has been well studied, and there are still no well-developed analytic tools that allow for reliable flow predictions. Due to the chaotic nature of turbulent flow, there is reason to believe that reliable predictions of the velocities of a fluid in a turbulent boundary layer may not be possible. As a result of this barrier, the heat flows and analytic predictions for convective heat-transfer coefficients in the turbulent boundary layers are not yet tractable. Experimental data has shown that a turbulent boundary layer velocity profile as shown in figure 4–20 can be expected.

A rather uniform velocity of $\mathbf{u} \approx \mathbf{u}_\infty$ is prevalent throughout the region except near the solid surface, where a thin laminar sublayer occurs immediately adjacent to the surface. The laminar sublayer is presumed to be separated from the turbulent layer by a mixed buffer region or transition between laminar and turbulent flow characters. Experimental results have been presented by Schlichting [3] and others indicating that, for $10^8 \geq \mathrm{Re}_x \geq 5 \times 10^5$,

$$C_f = \frac{0.0592}{\mathrm{Re}_x^{1/5}} \tag{4–109}$$

within an experimental accuracy of 15%. Further, the velocity boundary layer may be approximated by

$$\delta = 0.37x/\mathrm{Re}_x^{1/5} \tag{4–110}$$

and from these two relationships it can be seen that the Prandtl Number does not appear to be a significant variable for the prediction of drag coefficients or boundary layer thicknesses.

Extending this observation to the thermal phenomena, we may assume that the thermal boundary layer and the velocity boundary layers are identical; that is,

$$\delta_T = \delta \qquad \text{(for the turbulent boundary layers)} \tag{4–111}$$

Using the Reynolds–Colburn analogy of equation (4–78) with equation (4–109) gives

$$\mathrm{Nu}_x = 0.0296\ \mathrm{Re}_x^{4/5} \cdot \mathrm{Pr}^{1/3} \tag{4–112}$$

The average convective heat-transfer coefficient for a flat surface where both laminar and turbulent flow patterns occur in the boundary layer may be predicted by assuming that laminar flow ceases at a unique location, x_c, and then

$$h_{\mathrm{ave}} = \frac{1}{L}\left(\int_0^{x_c} h_{x,\mathrm{laminar}}\,dx + \int_{x_c}^{L} h_{x,\mathrm{turb}}\,dx\right) \tag{4–113}$$

where $h_{x,\mathrm{laminar}}$ is identified with equation (4–102) and $h_{x,\mathrm{turb}}$ is identified with equation (4–112), noting that $h_{x,\mathrm{turb}} = \mathrm{Nu}_x \kappa/x$.

Another form of the result given in equation (4–113) can be written

$$\mathrm{Nu}_{L,\mathrm{ave}} = \frac{1}{L}\left(\int_0^{x_c} \mathrm{Nu}_{x,\mathrm{laminar}}\,dx \int_{x_c}^{L} \mathrm{Nu}_{x,\mathrm{turb}}\,dx\right) \tag{4–114}$$

where $\mathrm{Nu}_{x,\mathrm{laminar}}$ is that of equation (4–103) and $\mathrm{Nu}_{x,\mathrm{turb}}$ is that of equation (4–112). If the critical Reynolds Number is taken as 4×10^5 and the integration of equation (4–114) is carried out, we find

$$\mathrm{Nu}_{L,\mathrm{ave}} = (0.037\ \mathrm{Re}_L^{4/5} - 702)\mathrm{Pr}^{1/3} \tag{4–115}$$

Some investigators have used a critical Reynolds Number of 5×10^5 for the instantaneous transition from laminar to turbulent flow, and for this value the Nusselt Number is

$$\mathrm{Nu}_{L,\mathrm{ave}} = (0.037\ \mathrm{Re}_L^{4/5} - 871)\mathrm{Pr}^{1/3} \tag{4–116}$$

Again, if equation (4–115) or (4–116) is used to obtain a Nusselt Number and subsequently predict a convective heat transfer coefficient, the fluid Prandtl Number must be between 0.6 and 60. For flow conditions over very long flat surfaces or where the flat surface is rough so that turbulence occurs over nearly all of the surface, equations (4–115) and (4–116) reduce to the form

$$\mathrm{Nu}_{L,\mathrm{ave}} = 0.037\ \mathrm{Re}_L^{4/5} \cdot \mathrm{Pr}^{1/3} \qquad \text{(turbulent flow exclusively)} \tag{4–117}$$

For the special case where flow in a boundary layer is exclusively turbulent and there is no identifiable leading edge or total length it has been recommended [11] that $L \approx 0.6$ m or 2 ft. We will apply this parameter in example 4–13.

There are many published results giving equations that relate the Nusselt Number to the Reynolds and Prandtl Numbers for laminar, turbulent, or mixed boundary layer flow over a flat surface. The interested reader may want to pursue this study further, and a good reference for such a study is Churchill [10, 12]. In all cases where a convective heat transfer coefficient is desired, the student should expect discrepancies of 25% or more between values obtained from different equations. Because of the variation in property values such as the Prandtl Number, the density, the specific heats, or other crucial terms; because of the fluctuations in actual flow conditions compared to those assumed in deriving or developing certain equations; or because of unexpected phenomena the agreement between theory and practice is not always as good as one would like it to be. Also, since property values for a fluid vary with the local temperatures it is usually expected that the fluid properties are evaluated at the *film temperature,* T_f, defined as

$$T_f = \frac{1}{2}(T_\infty + T_s) \qquad \textbf{(4–118)}$$

where T_s is the surface temperature. Occasionally an equation will be presented in which the properties need to be identified at some temperature other than the film temperature. When such an equation is presented, the proper temperature or temperatures will be defined.

EXAMPLE 4–13 Atmospheric air at 60°F is moving across a 1000-ft square pond of water that is used as a cooling pond for a power plant. The air is moving at 15 mph (22 ft/s), and the surface temperature of the pond water is 100°F. Estimate the heat loss from the pond per square foot of surface area.

Solution The heat loss will be estimated by predicting a convective heat transfer coefficient and then applying Newton's Law of Cooling. Because the air can be assumed to be moving over the earth around the pond as well, we will assume that the air flow in the boundary layer is completely turbulent. Also, since the character of the flow will likely change somewhat over the pond as compared to flow over the surrounding terrain, we will assume that the characteristic length used to describe the flow is 2 ft, based on an assumption that the particular turbulent flow patterns over the pond commence within 2 ft of the shoreline and that beyond that point turbulent "cells" of 2-ft length occur. Also, Brown and Marco [11] have suggested this same value for the characteristic length. Then, using a film temperature of 80°F [= (100°F + 60°F)/2] for evaluation of the air properties,

$$\mathrm{Re}_L = \frac{\mathbf{u}_\infty L}{\nu} = \frac{(22\ \mathrm{ft/s})(2\ \mathrm{ft})}{(17.1 \times 10^{-5}\ \mathrm{ft^2/s})} = 2.57 \times 10^5$$

From equation (4–117) we have

$$\mathrm{Nu}_{L,\mathrm{ave}} = 0.037\ \mathrm{Re}_L^{4/5}\ \mathrm{Pr}^{1/3} = 0.037(2.57 \times 10^5)^{4/5}(0.71)^{1/3} = 702.4$$

The average convective heat transfer coefficient for heat transfer on the pond is

$$h_{\mathrm{ave}} = \frac{\mathrm{Nu}_{L,\mathrm{ave}}\,\kappa}{L} = \frac{(702.4)\left(0.0139\ \dfrac{\mathrm{Btu}}{\mathrm{hr \cdot ft \cdot {}^\circ F}}\right)}{(2\ \mathrm{ft})} = 4.88\ \frac{\mathrm{Btu}}{\mathrm{hr \cdot ft^2 \cdot {}^\circ F}}$$

and the heat loss per unit area is

Answer

$$\dot{q}_A = h_{\mathrm{ave}}(T_s - T_\infty) = \left(4.88\ \frac{\mathrm{Btu}}{\mathrm{hr \cdot ft^2 \cdot {}^\circ F}}\right)(100^\circ\mathrm{F} - 60^\circ\mathrm{F}) = 195.2\ \frac{\mathrm{Btu}}{\mathrm{hr \cdot ft^2}}$$

In this example the Reynolds Number is less than 4×10^5 and would indicate that flow should be laminar. Experimental evaluations have shown that turbulent flow can occur much below the critical value of 4×10^5 if the surface is rough or if the fluid stream is irregular in direction or magnitude. Notice that a crucial assumption was the characteristic length used to predict the Reynolds Number. If a leading edge can be positively identified and a total length can also be defined, then the value for L should be that of the total length.

Unheated/Heated Flat Surfaces

Occasionally a convection heat transfer will occur along a flat surface when only a portion of the surface is to be heated or cooled. Such a situation is indicated schematically in figure 4–29, where fluid passes over a flat surface without convection heat transfer occurring. Such a condition can be approximated by the one where the fluid and the flat surface are at the same temperature and viscous dissipation is negligible; that is, the flow velocity is low. If the flow velocity approaches the velocity of sound, then viscous dissipation must be included, but here we exclude that possibility. As the fluid passes over the flat surface, there is a sudden step temperature increase or decrease at the surface and at a location $x = x_H$ along the surface from the leading edge. In figure 4–29 the schematic indicates a heater embedded in the plate; however, a cooling element could provide the opposite condition of fluid cooling instead of fluid heating. The boundary layers are indicated by their envelopes, and it can be shown by methods used previously for the isothermal surface that

$$\mathrm{Nu}_x = 0 \qquad \text{for } x \le x_H$$

$$\mathrm{Nu}_x = 0.332\frac{\mathrm{Re}^{1/2} \cdot \mathrm{Pr}^{1/3}}{\left[1 - (x_H/x)^{3/4}\right]^{1/3}} \qquad \text{for } x > x_H \tag{4–119}$$

and for $50 \ge \mathrm{Pr} \ge 0.6$, $\mathrm{Re} \le 4 \times 10^5$ (laminar flow).

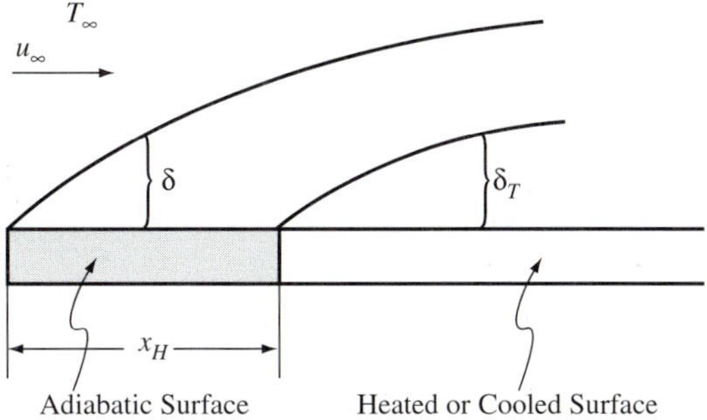

Figure 4–29 Convective heat transfer along a flat surface with adiabatic conditions over initial portion.

For turbulent boundary layer flow conditions of this same problem

$$\mathrm{Nu}_x = 0.0296\frac{\mathrm{Re}^{4/5} \cdot \mathrm{Pr}^{1/3}}{\left[1 - (x_H/x)^{9/10}\right]^{1/9}} \tag{4–120}$$

for $50 \ge \mathrm{Pr} \ge 0.6$ and $\mathrm{Re} > 4 \times 10^5$.

Uniform Heat Transfer

In the discussions up to this point the flat surfaces have been assumed to be isothermal; that is, the surface temperature was assumed to be constant. Since the local Nusselt Number and convective heat transfer coefficient tend to vary along the flow surface, the local heat transfer can also be expected to vary. If the convective heat transfer per unit area is assumed to be constant over a flat surface, then the local surface temperatures can be expected to vary, and the following equation has been derived for such a condition, assuming laminar flow in the boundary layer,

$$\mathrm{Nu}_x = 0.453\ \mathrm{Re}_x^{1/2} \cdot \mathrm{Pr}^{1/3} \tag{4–121}$$

for $\mathrm{Re}_x \leq 4 \times 10^5$.

The equation relating the Nusselt Number to the Reynolds and Prandtl Numbers for turbulent boundary layer flow of fluid across a flat surface is

$$\mathrm{Nu}_x = 0.308\ \mathrm{Re}_x^{4/5} \cdot \mathrm{Pr}^{1/3} \tag{4–122}$$

If the heat transfer per unit area is known, then the local surface temperature $T_s(x)$ can be predicted from Newton's Law of Heating/Cooling in the form

$$T_s(x) = T_\infty + \frac{\dot{q}_A}{h_x} \tag{4–123}$$

where $\dot{q}_A$ is the heat transfer per unit area in power per unit area units (W/m², Btu/hr · ft², etc.).

EXAMPLE 4–14 An electrical hot plate 20 cm × 20 cm uses 200 W of power when operating at steady state conditions. The surface of the plate is subjected to a stream of 27°C air moving at 10 m/s parallel to the plate, as shown in figure 4–30. Estimate the minimum and maximum plate temperatures at equilibrium and steady state.

Figure 4–30

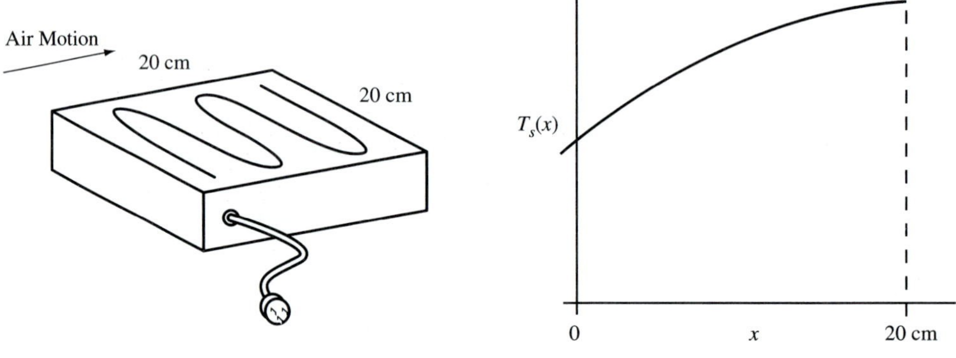

Solution The plate is assumed to dissipate the 200 W of power as heat in a uniform manner over its full area. Thus,

$$\dot{q}_A = \frac{200\ \mathrm{W}}{400\ \mathrm{cm}^2} = 5000\ \frac{\mathrm{W}}{\mathrm{m}^2}$$

The flow characteristics are predicted by determining the Reynolds Number,

$$\mathrm{Re}_L = \frac{\mathbf{u}_\infty L}{\nu} = \frac{(10\ \mathrm{m/s})(0.2\ \mathrm{m})}{(15.86 \times 10^{-6}\ \mathrm{m^2/s})} = 1.25 \times 10^5$$

We assume laminar flow throughout the plate surface. Then, from equation (4–121),

$$\mathrm{Nu}_x = \frac{h_x x}{\kappa} = 0.453\mathrm{Re}^{1/2} \cdot \mathrm{Pr}^{1/3} = 320\sqrt{x}$$

The local convective heat-transfer coefficient is

$$h_x = 320\frac{\kappa}{\sqrt{x}} = \frac{8.42}{\sqrt{x}}\frac{\text{W}}{\text{m}^2\cdot{}^\circ\text{C}}$$

and the local surface temperature $T_s(x)$ is, from equation (4–123),

$$T_s(x) = T_\infty + \frac{\dot{q}_A}{h} = T_\infty + \frac{594}{\sqrt{x}}$$

The minimum temperature occurs at $x = 0$ so that $T_{s,\text{min}}(0) = 27°\text{C}$, and the maximum temperature occurs at the trailing edge where $x = 0.2$ m. Thus

Answer

$$T_{s,\text{max}} = 27°\text{C} + \frac{594}{\sqrt{0.2}} = 293°\text{C}$$

4–5 CONVECTION HEAT TRANSFER AROUND OBJECTS

In section 4–4 we discussed convection heat transfer at a flat surface. Here we will extend that exposition to convection heat transfer between an object and a fluid passing around that object. The configuration we will first consider is shown in figure 4–31, where a fluid has a free stream velocity $\mathbf{V}_\infty$ and flows over the cylinder in a direction that is normal to the cylinder's axis, called cross-flow. The cylinder has a diameter *D,* and in figure 4–31 the stagnation point is indicated. This point represents a location where the fluid velocity is zero: It flows to the cylindrical surface and stops. In the schematic of figure 4–31 the path lines of fluid are indicated, and if the fluid were frictionless and inviscid the lines would by symmetrical in both the x and the y directions. This behavior has been discussed earlier and demonstrated in figure 4–10. Here we will consider a viscid and frictional fluid where the fluid path lines are symmetrical in the y direction but asymmetric in the x direction. Because of the viscous effects, there will be a boundary layer at the circular outer surface of the cylinder and the boundary layer equations we developed in section 4–4 may be used to analyze the fluid flow. Since the flow is around a circular surface, the coordinates are r (radius) and θ (angular distance from the stagnation point), and the first condition we consider is the development of a laminar boundary layer along the

Figure 4–31 General configuration of cylinder in a fluid flowing across the cylinder.

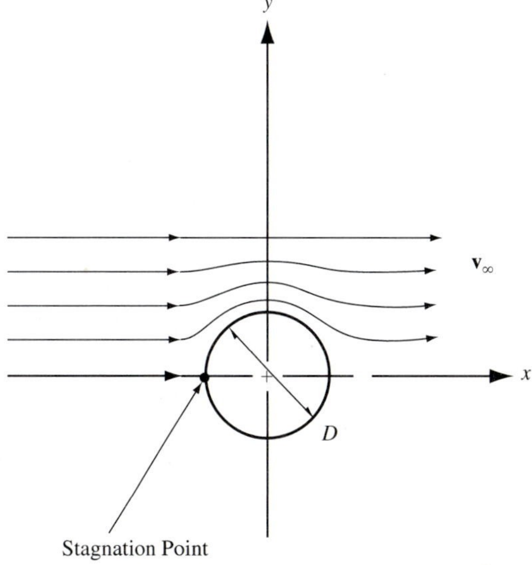

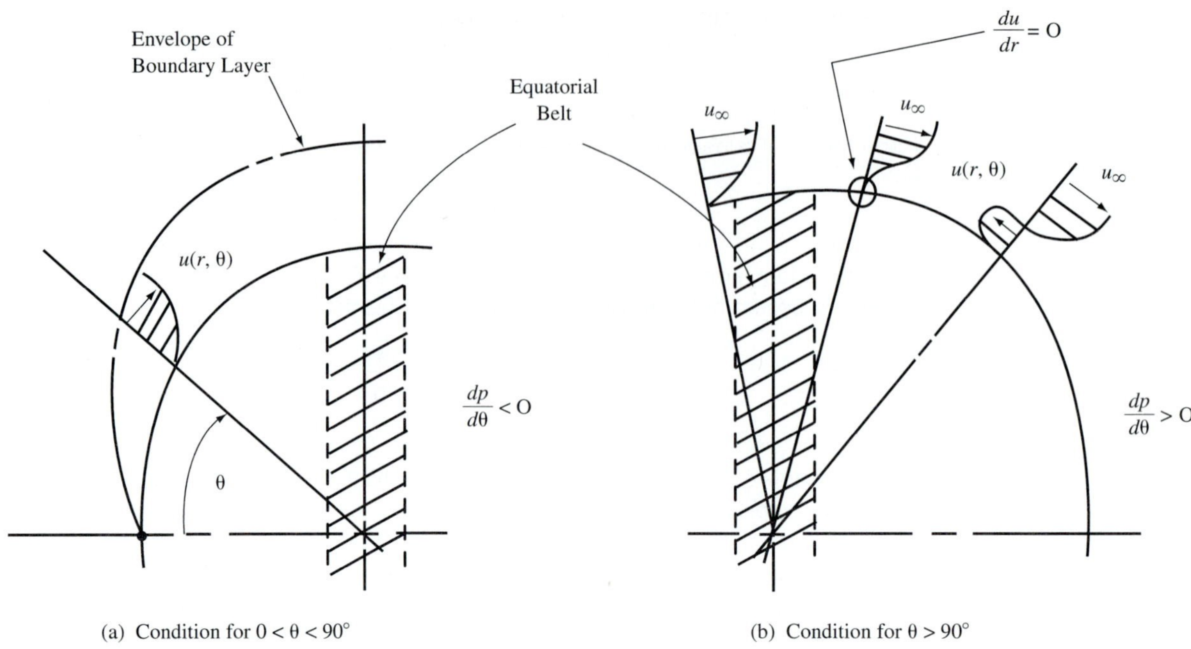

Figure 4–32 Velocity profiles in the boundary layer at a circular cylinder.

cylinder. In figure 4–32a is shown the development of a boundary layer along a circular surface from the stagnation point. Notice that the fluid moving with a free stream velocity $\mathbf{V}_\infty$ is diverted at the cylinder because of viscous and momentum actions so that the stream velocities change along the outer envelope $\mathbf{u}_\infty$ of the boundary layer. Remember that $\mathbf{u}_\infty$ is zero at the stagnation point and increases around the circular surface, reaching a maximum at about $\theta \approx 90°$, while the boundary layer thickness increases.

The action of increasing free stream velocity $\mathbf{u}_\infty$ can be accounted as a decrease in the fluid pressure along the circular surface. This can be seen by considering that, because of the geometric shape of the cylinder and the fluid flow pattern around the cylinder, the fluid will tend to decrease in magntiude as θ increases. Then, according to Bernoulli's Equation applied to a particle,

$$-\frac{1}{\rho}\frac{dp}{d\theta} = \frac{1}{2g_c}\frac{d(\mathbf{u}^2)}{d\theta} \tag{4–124}$$

The fluid pressure is maximum at the stagnation point and decreases at greater θ; that is, $dp/d\theta < 0$ from equation (4–124). This decrease in pressure and increase in velocity, along with an increase in the boundary layer thickness, continues to the equatorial belt. Because of the diverging geometry of the fluid flow region beyond the hemispheric belt, the fluid free stream velocity will then tend to decrease and the pressure to increase again. The velocity profile will begin to develop a point of inflection, as indicated in figure 4–32b. Further downstream at greater θ a velocity profile occurs such that the velocity gradient $d\mathbf{u}(r)/dr = 0$ at the surface. This is called the *separation point,*

$$\frac{d\mathbf{u}(r)}{dr} = 0$$

and from this point downstream the fluid stream separates from the surface of the cylinder. This action creates a partial vacuum at the surface, and it becomes occupied with fluid flowing in the reversed or upstream direction, as indicated in figure 4–32b at the furthest downstream velocity profile.

Such reversal of flow leads to a turbulent action, and the region downstream from the cylinder, sometimes called the wake, is characterized as a turbulent flow region. For the case of low velocity the wake is small and little or no turbulence is present in the wake, but for conditions of increasing velocity the wake increases in size as indicated in the diagrams of figure 4–33. Defining a Reynolds Number for this flow situation by

$$\text{Re}_D = \frac{\mathbf{V}_\infty D}{\nu} \tag{4–125}$$

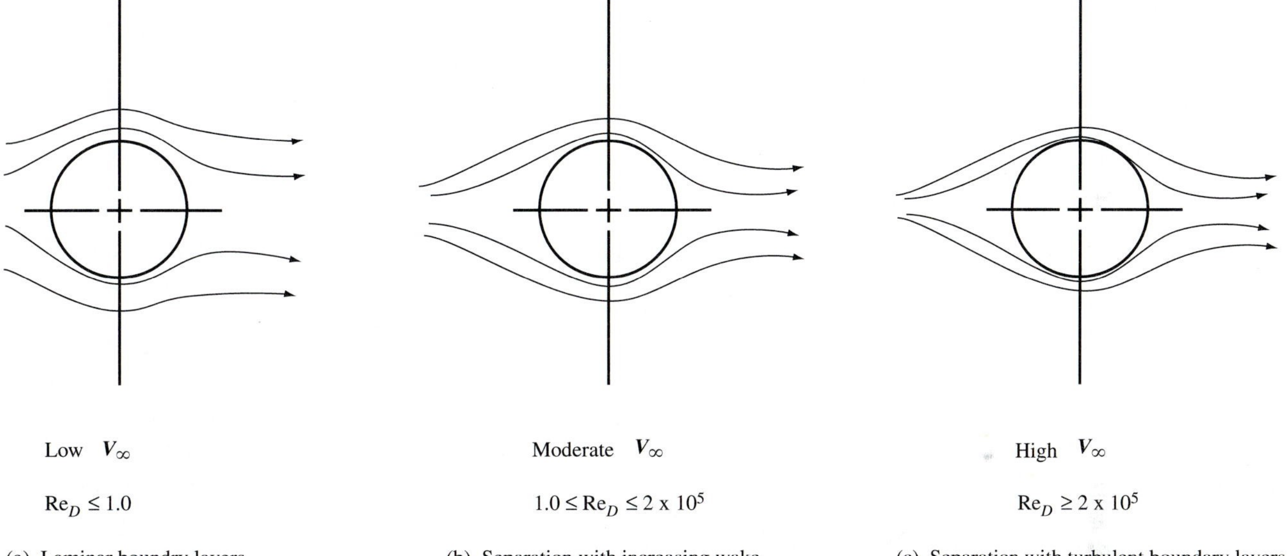

Figure 4–33 Some typical flow lines around a cylinder of a fluid stream moving at $\mathbf{V}_\infty$.

where D is the diameter of the cylinder, allows for a quantification of the flow. For a Reynolds Number of about 1.0 or less there is no separation and the boundary layer contains only laminar flow. As the Reynolds Number increases, the laminar boundary layer occurs only on the upstream region; separation occurs at the equatorial belt as we discussed. For Reynolds Numbers of greater than about 2×10^5 the flow in the boundary layer at the cylinder may be classified as turbulent, although a small portion of the boundary layer near the stagnation point may be laminar. Turbulent boundary layers seem to retard separation, and this results in a smaller wake. This phenomenon was observed in the study of the frictional drag around spheres by Eiffel [13] and is discussed by Schlichting [3].

The drag coefficient C_D may be defined as

$$C_D = F_D \frac{2g_c}{A_f \rho \mathbf{V}_\infty^2} \tag{4–126}$$

where A_f is the frontal area. For a cylinder A_f would be DL (L = cylinder length). A relationship between C_D and the Reynolds Number is presented by Schlichting [3], and this

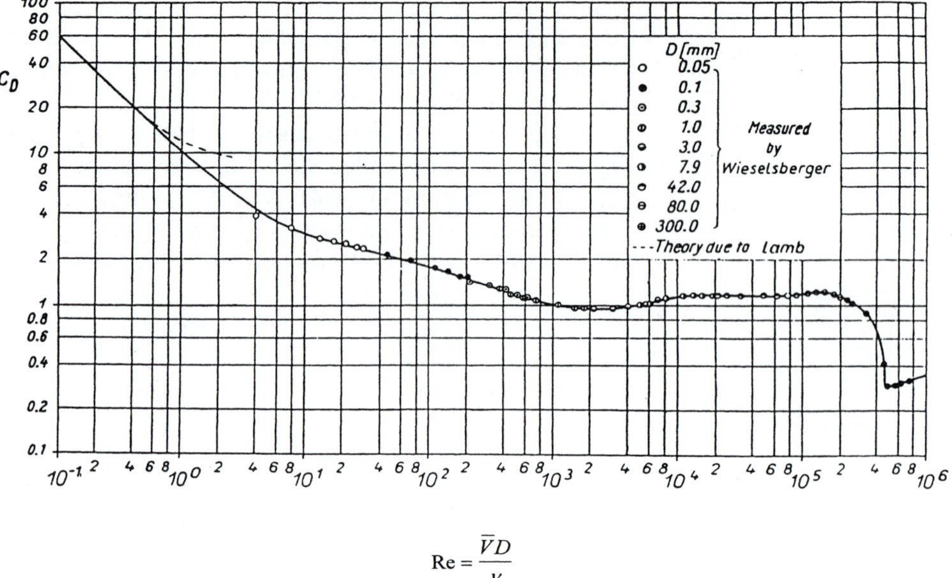

Figure 4–34 Drag coefficient for cross flow over a cylinder as a function of the Reynolds Number. (After Schlichting [3])

$$\mathrm{Re} = \frac{\bar{V}D}{\nu}$$

is given in figure 4–34. Notice in the graph that the value of C_D is relatively constant between Reynolds Numbers of 400 to 1×10^5. This is the region of laminar boundary layer flow with increasing wake. From equation (4–126) we can see that for a given fluid having constant viscosity ν the drag force F_D increases with increasing Reynolds Number. At a condition where $\mathrm{Re}_D \approx 2 \times 10^5$ the drag coefficient drops significantly, and at $\mathrm{Re}_D \approx 5 \times 10^5$ it reaches a minimum. This minimum occurs at turbulent boundary layers and with a minimum wake. Beyond Re_D of 5×10^5 the value of C_D increases and approaches a value of 1.0. The results in figure 4–34 were obtained with the implication that the fluid was incompressible; however, as the fluid velocity increases to values approximating the sonic velocity, the fluid will become significantly compressible and the drag coefficient will become less and less dependent on the Reynolds Number. At compressible flow conditions the drag coefficient is affected by the mach number

$$M = \frac{\mathbf{V}}{\mathbf{V}_{sc}} \tag{4–127}$$

where $\mathbf{V}_{sc}$ is the sonic velocity. At Mach numbers greater than about 0.8 the drag coefficient is nearly independent of the Reynolds Number. We will not concern ourselves with compressible flow.

The reader can see that the complexities of describing the fluid flow around a cylinder can become great and that any attempts to derive the heat transfer and predict the convective heat transfer coefficients will be more difficult. Investigators have resorted to experimental determinations of the convective heat transfer coefficients for heat transfer between a cylinder and a cross-flowing fluid. In particular, local values for the convective heat transfer coefficient $h(\theta)$ have been reported by Schmidt and Wenner [14], and these are given in figure 4–35. Notice that the minimum $h(\theta)$ occurs around the equatorial region of the cylinder, or $\theta \approx 90°$. Engineers usually require information of the average convective heat transfer coefficient, and particular results for air flow around a cylinder are presented in figure 4–36. These results have been verified by various investigators and may be extended to other gases, provided the Prandtl Number is greater than about 0.7 and that the temperatures are not excessively high.

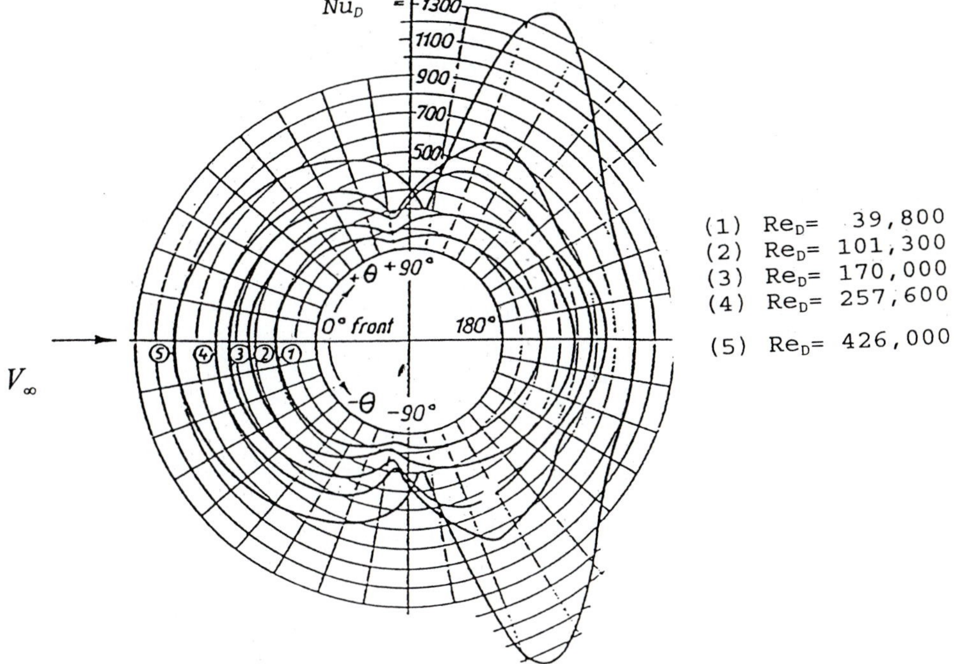

Figure 4–35 Local convective heat transfer coefficient as a function of the Reynolds Number for cross-flow of a fluid over a circular cylinder. (After Schmidt and Wenner [14])

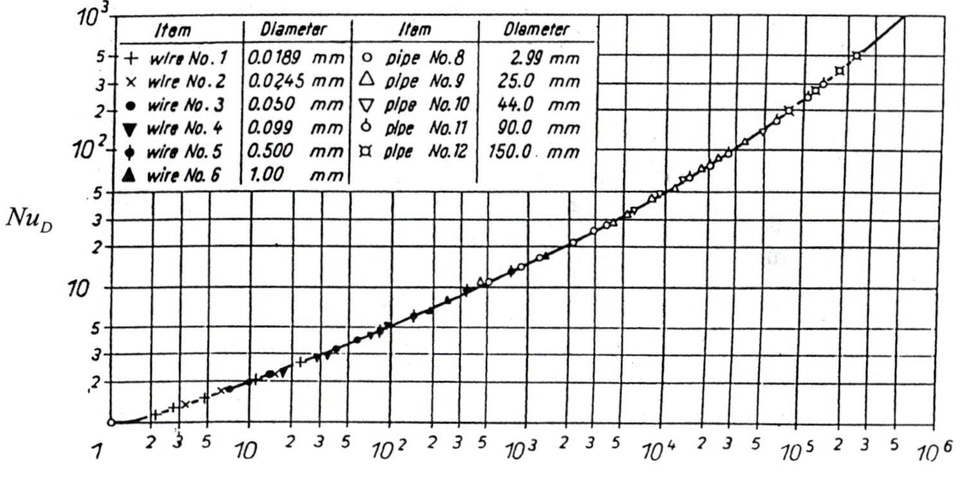

Figure 4–36 Nusselt Number as a function of the Reynolds Number for cross-flow of air over a circular cylinder.

$$\mathrm{Re}_D = \frac{V_\infty D}{\nu}$$

An empirical equation that was suggested by Hilpert [15] is

$$\mathrm{Nu}_{D,\mathrm{ave}} = \frac{h_{\mathrm{ave}} D}{\kappa_f} = C\,\mathrm{Re}_D^m \mathrm{Pr}^{1/3} \tag{4–128}$$

where the parameters C and m are functions of the Reynolds Number and are given in table 4–4. Equation (4–128) may be used for gases and liquids provided that the Prandtl Number is greater than about 0.7. All properties of the fluid are to be evaluated at the film temperature of the fluid. Also, equation (4–128) has been used for fluids with cross-flow

Table 4–4 *C* and *m* for Equation (4–128) after Hilpert [15] and Knudsen and Katz [16]

Re_D	C	m
0.4–4	0.989	0.330
4–40	0.911	0.385
40–4000	0.683	0.466
4000–40,000	0.193	0.618
40,000–400,000	0.027	0.805

Table 4–5 *C* and *m* for Equation (4–128) after Jakob [17] for Noncircular Cross-Sectional Objects in Gas Cross-Flow

Cross Section	$Re_{D,ef}$	C	m
Diamond ($\mathbf{u}_\infty$, D_{eff})	5×10^3–10^5	0.246	0.588
Square (D_{eff})	5×10^3–10^5	0.102	0.675
Hexagon (D_{eff})	5×10^3–1.95×10^4	0.160	0.638
	1.95×10^4–10^5	0.0385	0.782
Hexagon (D_{eff})	5×10^3–10^5	0.153	0.638
Rectangle ($\mathbf{u}_\infty$, D_{eff})	4×10^3–1.5×10^4	0.228	0.731

Note: $Re_{D,ef} = \rho \mathbf{u}_\infty D_{eff}/\mu$ where D_{eff} is defined in the figures.

of noncircular objects. For the situation of cross-flow of fluids over diamonds, squares, hexagons, or rectangular objects, table 4–5 provides the values of C and m to be used with equation (4–128).

Another method for predicting the average Nusselt Number for cross-flow over a circular cylinder is with an equation proposed by Churchill and Bernstein [18],

$$Nu_{D,ave} = 0.3 + \frac{(0.62\, Re_D^{1/2} \cdot Pr^{1/3})\left[1 + (Re_D/282{,}000)^{5/8}\right]^{4/5}}{\left[1 + (0.4/Pr)^{2/3}\right]^{1/4}} \quad \textbf{(4–129)}$$

where the fluid properties are all to be evaluated at the film temperature. It is suggested that equation (4–129) is applicable for all Reynolds Numbers and $Re_D \cdot Pr \geq 0.2$. Other equations have been suggested for the application of convection heat transfer of objects in a cross-flow of a fluid and the interested reader may want to refer to Morgan [19] for a discussion of correlations for the smooth circular cylinder. Also, the reader can see that the form of the equations relating the Nusselt Numbers to the Reynolds and Prandtl

Numbers are in agreement with the results obtained by dimensional analysis in equation (4–69). Also, the student should be aware that discrepancies in using any of the various equations can probably be 25% or greater.

EXAMPLE 4–15 Predict the convective heat transfer coefficient for a power line 6 cm in diameter when air at 0°C and 10 m/s (36 km/h) flows across the wire.

Solution We assume steady state conditions and also that the wire temperature is 10°C. Then we may predict the convective heat transfer coefficient by three different methods:

1. Use the graph of figure 4–36 for predicting Nu_D and calculate h_{ave}.
2. Use equation (4–128) for predicting Nu_D and calculate h_{ave}.
3. Use equation (4–129) for predicting Nu_D and calculate h_{ave}.

We use 5°C for the film temperature and obtain the properties of air:

$$\kappa_f = 24.54 \times 10^{-3}\,\text{W/m}\cdot{}^\circ\text{C}$$
$$\nu = 13.93 \times 10^{-6}\,\text{m}^2/\text{s}$$
$$\text{Pr} = 0.713$$

The Reynolds Number is

$$\text{Re}_D = \frac{\mathbf{V}_\infty D}{\nu} = \frac{(10\text{ m/s})(0.06\text{ m})}{(13.93 \times 10^{-6}\text{ m}^2/\text{s})} = 43{,}073$$

From figure 4–34 we read $\text{Nu}_{D,ave} \approx 140$, and the average convective heat transfer coefficient is

Answer

$$h_{ave} = \text{Nu}_{ave}\frac{\kappa_f}{D} = 140\,\frac{24.54 \times 10^{-3}\text{ W/m}\cdot{}^\circ\text{C}}{0.06\text{ m}} = 57.26\,\frac{\text{W}}{\text{m}^2\cdot{}^\circ\text{C}}$$

Using equation (4–128) to predict a Nusselt Number,

$$\text{Nu}_{D,ave} = C\text{Re}_D^m\,\text{Pr}^{1/3}$$

From table 4–4 we read $C = 0.027$ and $m = 0.805$ for $\text{Re}_D = 43{,}070$, so

$$\text{Nu}_{D,ave} = 0.027(43{,}073)^{0.805}(0.713)^{1/3} = 129.7$$

Answer and

$$h_{ave} = 53.0\,\frac{\text{W}}{\text{m}^2\cdot{}^\circ\text{C}}$$

Finally, using equation (4–129) gives

Answer

$$h_{ave} = 53.4\,\frac{\text{W}}{\text{m}^2\cdot{}^\circ\text{C}}$$

Notice that the predicted values for the convective heat transfer coefficient are in good agreement using the three methods presented in this text. For a hand calculation the use of the Nusselt Number from the graph or the use of equation (4–128) is probably quickest, but for an extensive study of the effects of changing weather patterns on heat transfer of a power line the use of equation (4–129) with a computer is probably most convenient.

EXAMPLE 4–16 Long nylon strips $\frac{1}{2}$ in. wide by $\frac{1}{16}$ in. thick are hung vertically and preheated with 200°F air, shown in the schematic of figure 4–37. If the strips are 70°F when they are hung and they should be 160°F after 30 s of preheating, estimate the air velocity required to accomplish this preheating.

Figure 4–37 Example 4–16.

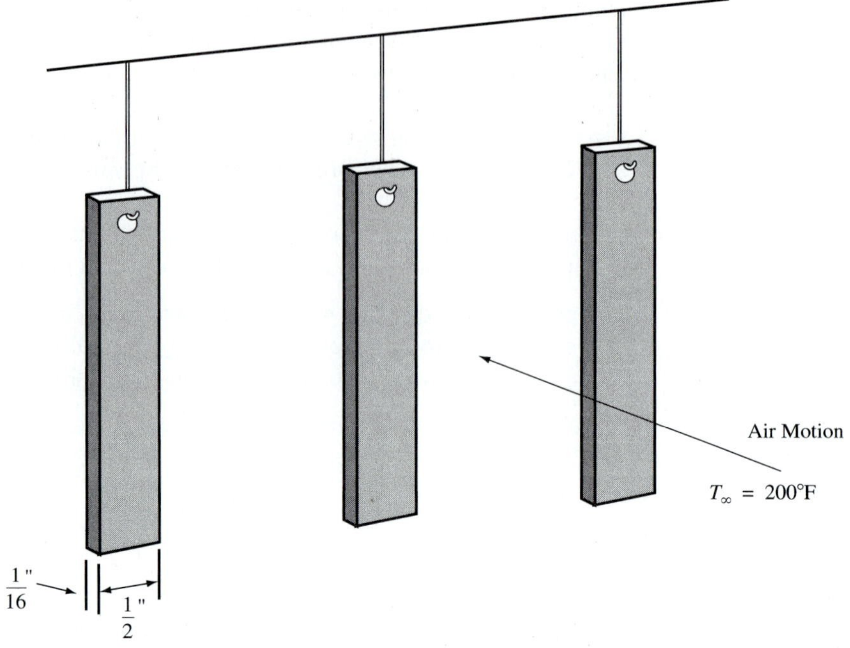

Solution

This problem may be analyzed by estimating the convective heat-transfer coefficient from a transient conduction condition, determining the Nusselt Number, using equation (4–128) and table 4–5 to predict a Reynolds Number, and then obtaining an air velocity from the Reynolds Number. If we assume lumped heat capacity (see section 3–2) then

$$\frac{T - T_\infty}{T_i - T_\infty} = e^{-t/t_c} = \frac{160 - 200}{70 - 200} = 0.3077$$

and

$$t_c = \frac{t}{\ln 0.3077} = 25.45 \text{ s}$$

The time constant t_c is also given by

$$t_c = \frac{\rho V c_p}{h_{\text{ave}} A_s} = \frac{\rho A_c c_p}{h_{\text{ave}} P}$$

where A_c is the cross-sectional area of the strip and P is the perimeter. Using properties the same as Teflon for the nylon strips, then

$$t_c = \frac{\left(2200 \frac{\text{kg}}{\text{m}^3}\right)\left(\frac{1}{2} \text{ in.} \times \frac{1}{16} \text{ in.}\right)\left(0.0254 \frac{\text{m}}{\text{in.}}\right)^2\left(1.0 \frac{\text{kJ}}{\text{kg} \cdot {}^\circ\text{C}}\right)}{h_{\text{ave}}(1.125 \text{ in.})(0.0254 \text{ m/in.})} = 25.45 \text{ s}$$

Solving for h,

$$h_{\text{ave}} = 0.06099 \frac{\text{kW}}{\text{m}^2 \cdot {}^\circ\text{C}} = 10.55 \frac{\text{Btu}}{\text{hr} \cdot \text{ft}^2 \cdot {}^\circ\text{F}}$$

We should check the Biot Number to be sure that the lumped heat capacity criteria is satisfied. The Biot Number is

$$\text{Bi} = \frac{h_{\text{ave}} L}{\kappa_{\text{nylon}}} = \frac{(60.99 \text{ W/m}^2 \cdot {}^\circ\text{C})(0.0023 \text{ ft})(0.3048 \text{ m/ft})}{0.35 \text{ W/m} \cdot {}^\circ\text{C}} = 0.122$$

Since the Biot Number is nearly 0.10, we will assume that the lumped heat capacity criterion is satisfied. The characteristic length L in this calculation is A_c/P and used in the time constant calculation. The Nusselt Number for this flow condition is

$$\text{Nu}_{\text{ave}} = \frac{h_{\text{ave}}L}{\kappa_{\text{fluid}}} = \frac{(10.55\ \text{Btu/hr}\cdot\text{ft}^2\cdot{}^\circ\text{F})(0.0023\ \text{ft})}{(0.026\ \text{W/m}\cdot{}^\circ\text{C})\left(0.5779\ \dfrac{\text{Btu/hr}\cdot\text{ft}\cdot{}^\circ\text{F}}{\text{W/m}\cdot{}^\circ\text{C}}\right)} = 1.615$$

where the air thermal conductivity is evaluated at 158°F, the average of the film temperatures during the preheating. Using equation (4–128) and the values $C = 0.228$ and $m = 0.731$ from table 4–5 gives

$$\text{Re}_{\text{eff}} = \left[\frac{\text{Nu}_{\text{ave}}}{0.228}\left(\frac{1}{\text{Pr}^{1/3}}\right)\right]^{1/0.731} = 16.98$$

Answer and

$$\mathbf{V}_\infty = \frac{\text{Re}_L\nu}{L} = \frac{16.98(245\times10^{-6}\ \text{ft}^2/\text{s})}{0.0023\ \text{ft}} = 1.81\ \frac{\text{ft}}{\text{s}}$$

This velocity is very small, and air flows of this magnitude occur in natural convection without use of fans. If the air is to be moved horizontally, however, as indicated in figure 4–37, then a small fan may be used to achieve the preheating.

The Sphere

Convective heat transfer between a spherical object and a surrounding fluid has a number of important applications and has been well studied. The flow of a fluid around a sphere has also been studied, and Schlichting [3] presents a graph of the frictional drag coefficient as experimentally determined by various investigators. This graph is given in figure 4–38, and a comparison between these results and the drag coefficient for a circular cylinder as given in figure 4–34 demonstrates the close similarity of the results. The drag coefficient decreases rapidly as the Reynolds Number increases at laminar flow conditions, remains nearly constant at moderately turbulent conditions, drops precipitously to

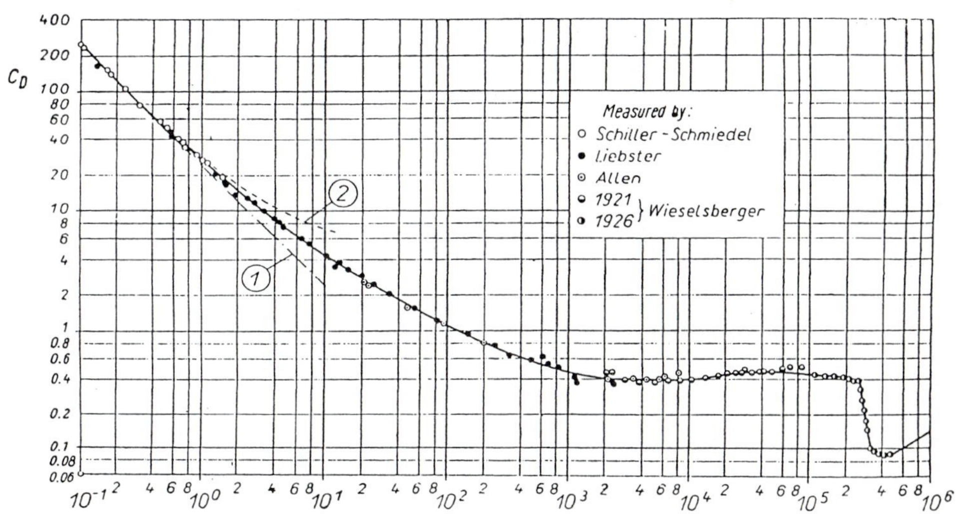

Figure 4–38 Drag coefficient for cross-flow over a sphere as a function of the Reynolds Number. (After Schlichting [3])

a minimum value near a Reynolds Number of 4×10^5, and then increases at higher Reynolds Numbers. The frictional drag coefficient for the sphere is defined by equation (4–126), where the frontal area A_f is the circular area $\pi D^2/4$. The convective heat transfer coefficient may be predicted from equations used to determine the Nusselt Number. For the flow of a gas around spheres, McAdams [20] suggests the equation

$$\text{Nu}_D = 0.37\,\text{Re}_D^{0.6} \tag{4–130}$$

for $70{,}000 > \text{Re}_D > 17$. The properties are to be evaluated at the film temperature. An equation that has been suggested by Whitaker [21] for gases and liquids flowing around spheres is

$$\text{Nu}_D = 2.0 + (0.4\,\text{Re}_D^{1/2} + 0.06\,\text{Re}_D^{2/3})(\mu_\infty/\mu_w)^{1/4}\,\text{Pr}^{0.4} \tag{4–131}$$

for $8 \times 10^4 > \text{Re}_D > 3.5$ and $380 > \text{Pr} > 0.7$. The properties are to be evaluated at the free stream temperature rather than the film temperature, except for the viscosity μ_w, which is evaluated at the surface temperature of the sphere.

EXAMPLE 4–17

Small spheres of lead are formed by dropping molten lead down a high tower. The frictional resistance of the air as the lead falls down the tower helps to form the spherically shaped "shot" and to cool the lead. If 2-mm-diameter lead shot is dropped in a tower when the air is 17°C, estimate the terminal velocity or steady state velocity of the shot as it drops through the tower and the heat transfer from each shot. Assume the lead is at 227°C when terminal velocity is reached.

Solution

The shot will drop and increase in velocity until the frictional drag force is equal to the weight of the shot. In figure 4–39 is shown a free body diagram to indicate the steady state condition reached at terminal velocity. From the free body diagram we have that

$$W_{\text{shot}} = C\rho A_f \frac{1}{2g_c}\mathbf{V}_\infty^2$$

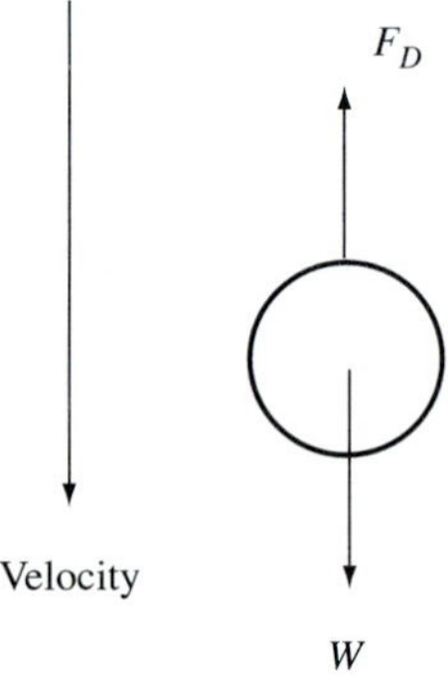

Figure 4–39 Lead shot dropping at terminal velocity with forces acting on a shot.

In SI units $g_c = 1\ \text{kg}\cdot\text{m/s}^2\cdot\text{N}$ and the frontal area of the shot is

$$A_f = \pi r^2 = \pi(0.001\ \text{m})^2 = 3.1415 \times 10^{-6}\,\text{m}^2$$

The air density may be taken at the free stream of the air, $T = 17°\text{C}$ so that $\rho = 1.208\ \text{kg/m}^3$. The weight of a lead shot is

$$W_{\text{shot}} = (\rho g V)_{\text{lead}} = \left(11{,}340\,\frac{\text{kg}}{\text{m}^3}\right)\left(9.81\,\frac{\text{m}}{\text{s}^2}\right)\left(\frac{4}{3}\pi\right)(0.001\ \text{m})^3 = 0.000466\ \text{N}$$

Then we have

$$C_d\mathbf{V}_\infty^2 = W_{\text{shot}}\frac{2}{\rho A_f} = (0.000466\ \text{N})\frac{2}{(1.208\ \text{kg/m}^3)(3.1415\times10^{-6}\ \text{m}^2)} = 246\ \frac{\text{m}^2}{\text{s}^2}$$

Reynolds Number is related to the velocity by

$$\text{Re}_D = \frac{\mathbf{V}_\infty D}{\nu} = \mathbf{V}_\infty\frac{(0.002\ \text{m})}{(15\times10^{-6}\ \text{m}^2/\text{s})} = 133.3\ \mathbf{V}_\infty$$

By iteration we may obtain the terminal velocity, $\mathbf{V}_\infty$, from the data of figure 4–38. Assuming $\mathbf{V}_\infty = 20$ m/s gives $\text{Re}_D = 2667$, and from figure 4–38 $C_D = 0.4$. From the relationship $C_D\mathbf{V}_\infty^2 = 246$ we get $C_D = 0.615$, which indicates that the velocity of 20 m/s is incorrect. Noting on the graph of figure 4–38 that the drag coefficient is nearly constant at 0.4 over a range of Reynolds Numbers we assume $C_D = 0.4$ and then

Answer

$$\mathbf{V}_\infty = \sqrt{246/C_D} = 24.8\ \text{m/s}$$

The Reynolds Number is

$$\text{Re}_D = 24.8(133.3) = 3306$$

and the Nusselt Number can be predicted from equation 4–131.

$$\text{Nu}_D = 2.0 + (0.4\ \text{Re}_D^{1/2} + 0.06\ \text{Re}_D^{2/3})\left(\frac{\mu_\infty}{\mu_w}\right)^{1/4}\text{Pr}^{0.4}$$

$$= 2.0 + (0.4(3306)^{1/2} + 0.06(3306)^{2/3})\left(\frac{17.96\times10^{-6}}{27.01\times10^{-6}}\right)^{1/4}(0.71)^{0.4} = 28.6$$

and

$$h_{\text{ave}} = \frac{\text{Nu}_D\kappa_{\text{fluid}}}{D} = \frac{28.6(0.0252\ \text{W/m}\cdot{}^\circ\text{C})}{0.002\ \text{m}} = 360.2\ \text{W/m}^2\cdot{}^\circ\text{C}$$

The heat transfer is then

Answer

$$\dot{Q} = h_{\text{ave}}A\Delta T = (360.2\ \text{W/m}^2\cdot{}^\circ\text{C})(4\pi)(0.001\ \text{m})^2(227 - 17^\circ\text{C}) = 0.95\ \text{W}$$

This example treated a shot or sphere of lead as a separate object that was unaffected by surrounding objects. In later sections we will see how to analyze packed beds, or large numbers of spheres in close contact experiencing convection heat transfer.

4–6 CONVECTION HEAT TRANSFER IN CLOSED CHANNELS

In this section we will discuss convection heat transfer of fluids flowing through closed channels, where the term *closed channel* means a duct, pipe, tube, or other conduit that completely encloses the fluid flowing inside the channel. In figure 4–40 is shown a typical circular closed channel or pipe, which is a configuration of many applications of convection heat transfer where the fluid flowing through the channel occupies all of the

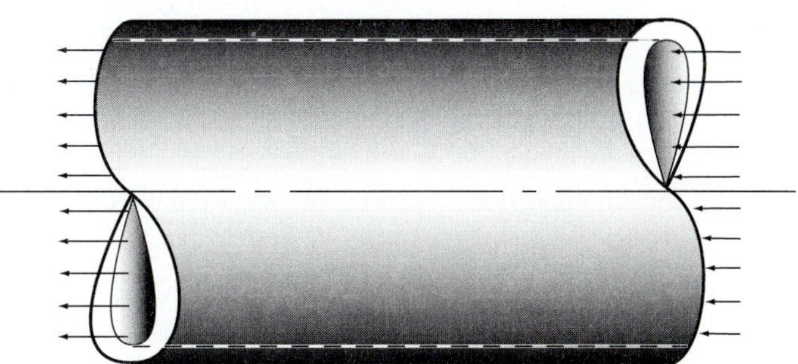

Figure 4–40 Closed channel flow in a pipe or tube.

channel volume. Earlier, in section 4–2, we discussed fluid flow in circular channels, and the velocity profile of a fully developed laminar flow was described by equation (4–14),

$$\mathbf{V}(r) = \frac{\Delta p}{4\mu L}(r_o^2 - r^2) \qquad \textbf{(4–14, repeated)}$$

Further, the Hagen–Poiseuille Equation, equation (4–16), relates the pressure drop, the volume flow rates, and the channel size for laminar flow. A relationship between the pressure drop and the kinetic energy was given by equation (4–18)

$$\Delta p = \gamma\frac{\mathbf{V}^2}{2g} \qquad \textbf{(4–18, repeated)}$$

and this equation is extended to turbulent flow in channels by the Darcy–Weisbach Equation (4–19)

$$h_L = \frac{\Delta p}{\gamma} = f\frac{L}{D}\frac{\mathbf{V}^2}{2g} \qquad \textbf{(4–19, repeated)}$$

where the friction factor f can be related to the Reynolds Number for channel flow, and in figure 4–19 this data is presented.

In convection heat-transfer phenomena involving fluid flow in channels, the convection heat transfer often occurs over the full length of a channel. The analysis that gives equations (4–16), (4–18), and (4–19) assumes that the fluid flow is fully developed, but if we consider a finite length channel having a specific entrance as shown in figure 4–41, it can be seen that the fluid flow characteristics change, beginning at the entrance to the channel and continuing until a fully developed flow condition is reached.

At the entrance to the channel a boundary layer forms at the channel walls, and the thickness of this layer increases in a manner similar to that for fluid flow past a flat surface as we considered earlier. For a circular channel or pipe, as indicated in figure 4–41, the fluid flow becomes fully developed at a location downstream where the boundary layer thickness δ ($= \delta_F$) is equal to the pipe radius, r_o. A number of possible flow patterns may occur in the entrance and fully developed regions of a closed channel, with the two clearest being (1) laminar boundary layers at the entrance with developed laminar flow and (2) laminar/turbulent boundary layers at the entrance with developed turbulent flow.

The situation where laminar/turbulent boundary layers with developed turbulent flow occurs is common in engineering problems involving fluid flow. As we saw earlier, turbulent flow can be expected if the Reynolds Number is greater than about 2300 and where the Reynolds Number is $\mathbf{u}_{\text{ave}}D/\nu$. For laminar flow conditions the analysis is perhaps easier to visualize, and in figure 4–42 is shown the condition of fluid flow between two parallel plates and with an entrance region with developing flow patterns. For the entrance region the boundary layer thickness of laminar flow fluid may be approximated by equation (4–95),

$$\delta = 5.0\frac{x}{\sqrt{\text{Re}_x}} \qquad \textbf{(4–95, repeated)}$$

and for closed channel flow between parallel plates the entrance region length x_F can be predicted by noting that $\delta = y_0/2$ when the flow is just developed. Thus, $\delta_F = y_0/2$ at this location and from equation (4–95) we have that $x_F = y_0\frac{\sqrt{\text{Re}_{xF}}}{10}$ where $\text{Re}_{xF} = \frac{\mathbf{u}_\infty x_F}{\nu}$. Then we have

$$x_F = \frac{y_0}{10}\sqrt{\frac{\mathbf{u}_\infty x_F}{\nu}} = y_0^2\frac{\mathbf{u}_\infty}{100\nu} \qquad \textbf{(4–132)}$$

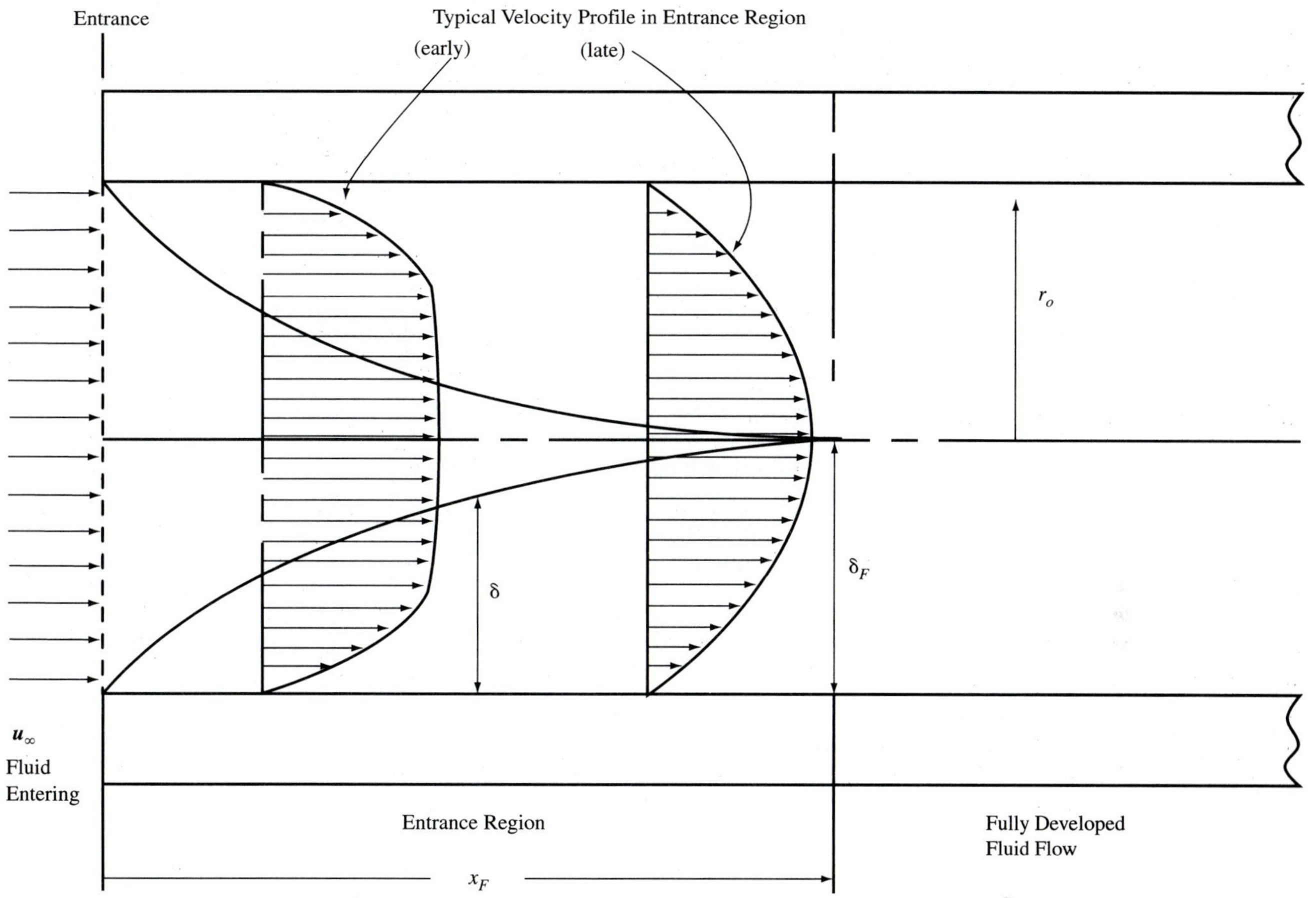

Entrance
Typical Velocity Profile in Entrance Region
(early)
(late)
r_o
δ
δ_F
u_∞
Fluid
Entering
Entrance Region
Fully Developed
Fluid Flow
x_F

Figure 4–41 Fluid flow development in a closed circular channel.

Figure 4–42 Entrance condition of closed channel flow between two parallel plates.

y_o
δ_F
Boundary Layer
δ
x_F
x
Fluid
Flow Between
Two Parallel
Plates

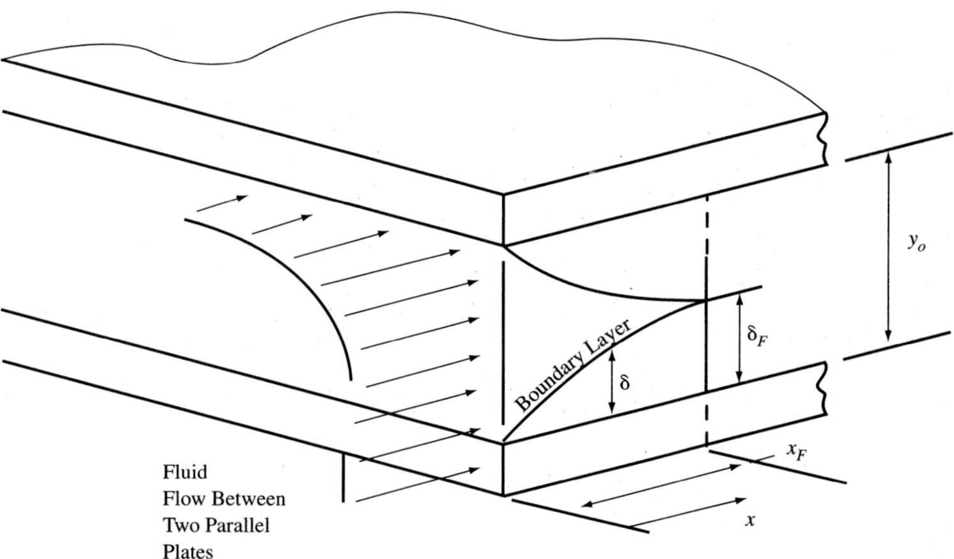

Equation (4–132) can be used to predict the length of the entrance region for flow of a fluid between two parallel plates. Langhaar [22] has shown that a similar expression can be obtained to approximate the entrance region length x_F for flow of a fluid in a circular channel as in figure 4–41. This expression is

$$x_F = \frac{\text{Re}_D D}{20} = \mathbf{u}_{\text{ave}}\frac{r_o^2}{5\nu} \tag{4–133}$$

where the flow must be laminar throughout and so that $\text{Re}_D \leq 2300$. For turbulent flow conditions in a circular channel Kays and Crawford [23] suggest that

$$120\, r_o \geq x_F \geq 20\, r_o \tag{4–134}$$

EXAMPLE 4–18 Water flows from a large condenser reservoir through a 10-cm ID pipe at 12 m^3/min. Determine the distance downstream from the entrance where the water flow may first be expected to be fully developed. Assume the water density is 1000 kg/m^3.

Solution The schematic of the system is shown in figure 4–43, and we see that the distance x_F represents the location downstream where the water flow may first be expected to be fully developed. We first check to see if the flow will be laminar or turbulent by determining the Reynolds Number;

$$\text{Re}_D = \frac{\mathbf{u}_{\text{ave}} D}{\nu} = \frac{4\dot{V}}{\pi \nu D}$$

Figure 4–43 Water flow from a condenser through a pipe.

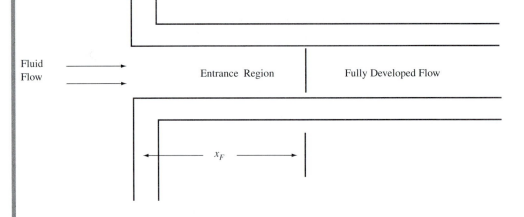

Using for the kinematic viscosity $\nu = 1.25 \times 10^{-6}$ m^2/s we have

$$\text{Re}_D = \frac{4(12\ \text{m}^3/\text{min})}{\pi(1.25 \times 10^{-6}\ \text{m}^2/\text{s})(60\ \text{s/min})(0.1\ \text{m})} = 2.037 \times 10^6$$

and so the flow will be turbulent since $\text{Re}_D > 2300$. The entrance length may be predicted, from equation (4–134), to be at least

Answer

$$x_F = 20 r_o = 20(0.05\ \text{m}) = 1\ \text{m}$$

Heat transfer in a closed channel may occur whenever the fluid and the channel walls have different temperatures. For the circular channel shown in figure 4–44a the convection heat transfer begins at the entrance to the channel and is denoted as a combined condition. In the schematic of figure 4–44a the channel wall temperature is assumed to be higher than the fluid temperature and the profile of the fluid temperature as the fluid flows through the channel is depicted at three locations: early entrance, late entrance, and fully developed. Except for one special case, which we will mention later, the development of the fluid flow and the development of the thermal profile do not proceed at the same rate, so that the fluid will have developed flow either before or after its temperature profile becomes fully developed. For the illustrated case of figure 4–44a the fluid flow develops more rapidly than do the thermal conditions, which usually occurs with laminar flow of fluids having Prandtl Numbers greater than about 1.0. For the case of turbulent flow the thermal conditions develop quite rapidly and usually entrance effects are then ignored.

In figure 4–44b is shown the situation where a heater (or a cooler) is placed along the channel some distance from the entrance. Here it can be seen again that the thermal development, as indicated by the temperature profile, proceeds through the thermal entrance length before fully developed thermal conditions are attained. The situation sketched in figure 4–44a for simultaneous development of fluid flow and thermal conditions and that sketched in figure 4–44b for separate development are the two extreme types of phenomena, and the reader should recognize that there are many combinations in between these two cases. For our purposes, however, we will only consider these two. The temperature profiles sketched in figure 4–44 indicate that the fluid temperature varies in some manner across the channel. It is often desirable to identify an average or bulk or mixed temperature of the fluid at a particular location. We define this temperature as

$$T_m = \frac{1}{\dot{m}c_p}\int_{\text{Area}} \rho\mathbf{u}(r)c_pT(r)dA \tag{4–135}$$

and it may be described as the temperature that the fluid would have if it were mixed and brought to rest. It is the average temperature across a particular cross section of the channel as indicated in figure 4–45. The temperature distribution must be known if the mixed temperature is to be evaluated from equation (4–135). An arithmetic average of between an initial and a final temperature is frequently used if an average fluid temperature over a length of a channel is to be determined. That is, the average fluid temperature within a channel as sketched in figure 4–45 may be taken as

$$T_{\text{ave}} = \frac{1}{2}(T_{m1} + T_{m2}) \tag{4–136}$$

It is sometimes important to be able to predict the thermal entrance distance x_T, and we will discuss that topic. However, first we will consider the steady state fully developed flow and fully developed thermal condition.

Steady State Fully Developed Convection Heat Transfer

Consider a closed circular channel having laminar fluid flow and axisymmetric conditions so that convection heat transfer can occur radially and axially but not angularly about the cross section of the channel. A differential element of the fluid is shown in figure 4–46, and the various heat flow terms are identified as follows: The radial heat flow into the element by conduction is

$$\dot{q}_r = -2\pi\kappa r\frac{\partial T}{\partial r}dz$$

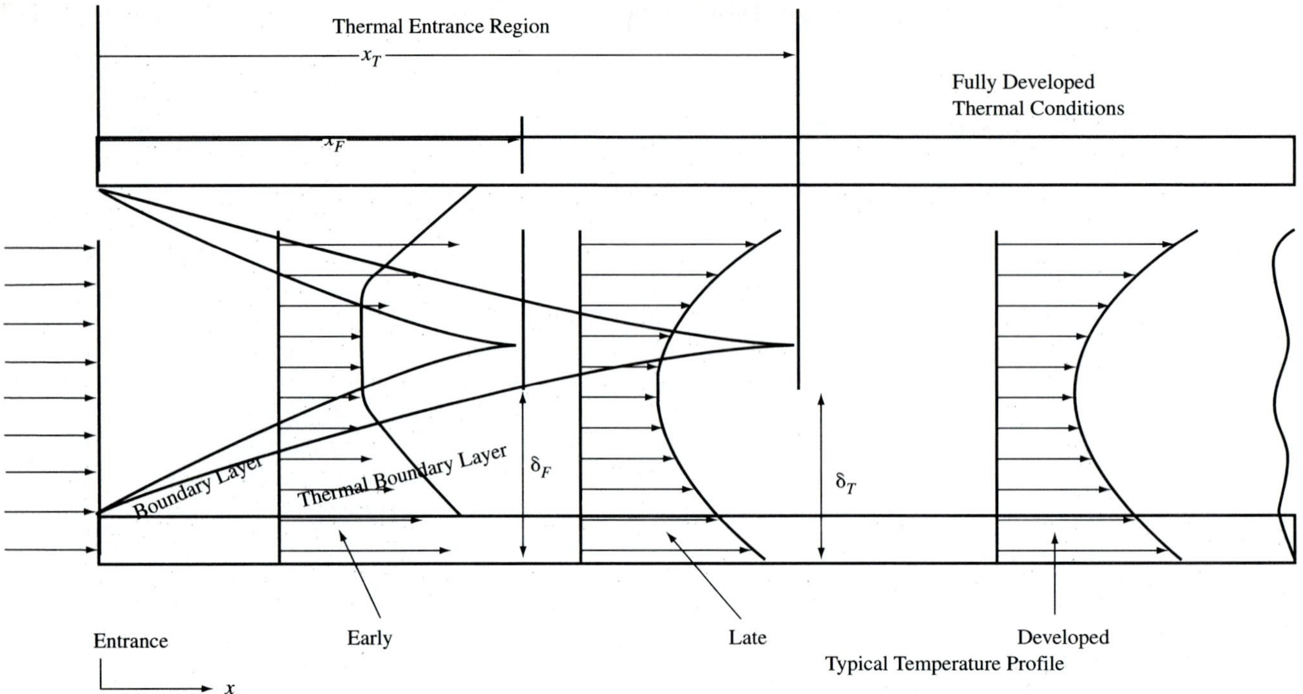

(a) Simultaneous development of fluid flow and thermal conditions

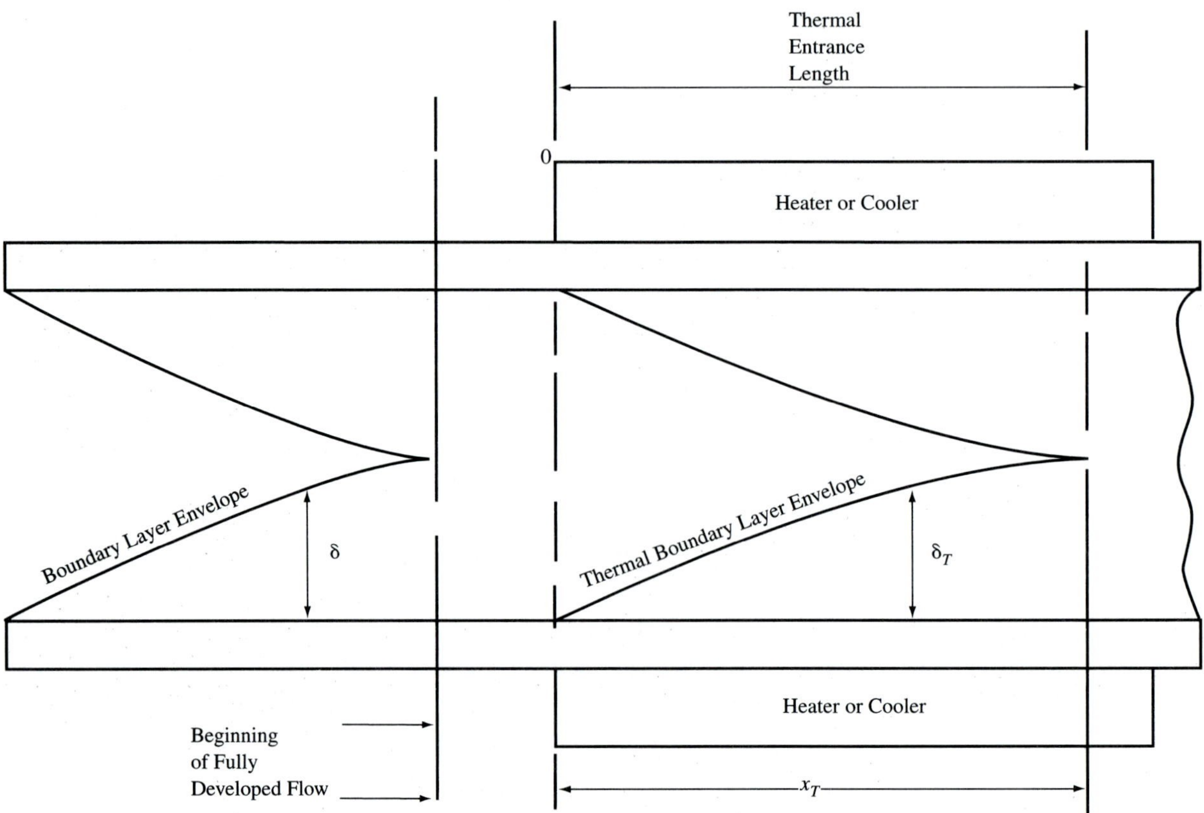

(b) Development of thermal conditions after fluid flow development

Figure 4–44

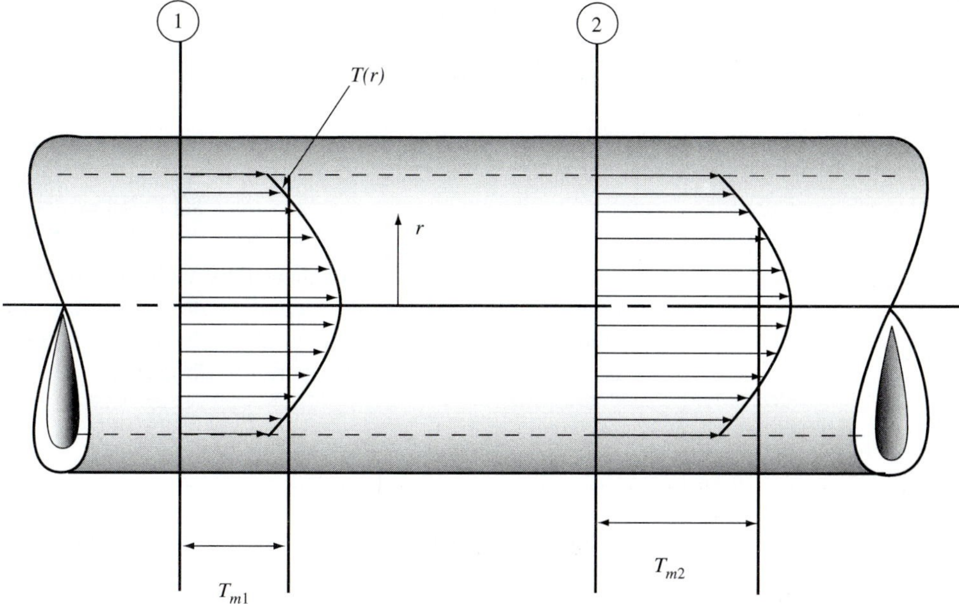

Figure 4–45 Illustration of the mixed temperature and the average temperature of a fluid flowing in a closed channel.

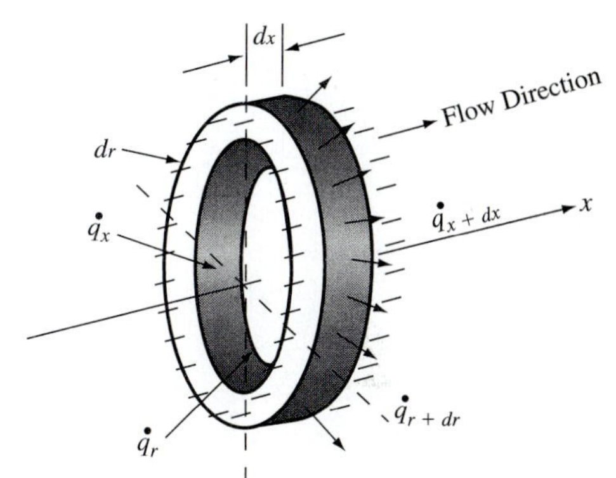

Figure 4–46 Differential element of fluid element flowing in a circular channel with convection heat transfer.

The radial heat flow out of the element by conduction is

$$\dot{q}_{r+dr} = -2\pi\kappa(r + dr)\left(\frac{\partial T}{\partial r} + \frac{\partial^2 T}{\partial r^2}dr\right)dz$$

The axial heat flow into the element due to fluid flow is

$$\dot{q}_z = 2\pi\rho c_p r\mathbf{u}(r)T_z dr$$

and the axial heat flow out of the element due to fluid flow is

$$\dot{q}_{z+dz} = 2\pi\rho c_p r\mathbf{u}(r)\left(T_z + \frac{\partial T_z}{\partial z}dz\right)dr$$

The energy balance may be written as

$$\dot{q}_r + \dot{q}_z = \dot{q}_{r+dr} + \dot{q}_{z+dz} \tag{4–137}$$

and after substituting we get

$$-2\pi\kappa r dz\frac{\partial T}{\partial r} + 2\pi\rho c_p r\mathbf{u}(r)T_z dz = -2\pi\kappa(r + dr)\left(\frac{\partial T}{\partial r} + \frac{\partial^2 T}{\partial r^2}dr\right)dz + 2\pi\rho c_p r\mathbf{u}(r)\left(T_z + \frac{\partial T_z}{\partial z}dz\right)dz$$

Expanding the terms of this equation and neglecting second-order terms gives

$$\frac{1}{r\mathbf{u}(r)}\frac{\partial}{\partial r}\left(r\frac{\partial T}{\partial r}\right) = \frac{1}{\alpha}\frac{\partial T}{\partial z} \tag{4–138}$$

For the case of uniform heat flow with the flowing fluid, as shown in the sketch of figure 4–47, the temperature gradient along the z axis, $\partial T/\partial z$, is constant and the temperature profile across the channel at any location is similar to that at any other location. It is convenient to define a temperature difference

$$\theta(r) = T(r, z) - T_w \tag{4–139}$$

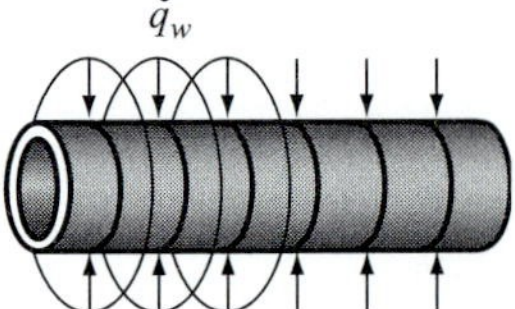

Figure 4–47 Uniform heat transfer per unit area at a circular channel with a fluid in laminar flow.

where T_w is the channel wall temperature. Notice that, because of similarity of the temperature profiles, θ is a function of r only and also

$$\frac{\partial T}{\partial r} = \frac{d\theta}{dr} \quad \text{and} \quad \frac{\partial^2 T}{\partial r^2} = \frac{d^2 T}{dr^2}$$

so that equation (4–138) becomes

$$\frac{1}{u\mathbf{u}(r)}\frac{d}{dr}\left(r\frac{d\theta}{dr}\right) = \frac{1}{\alpha}\frac{dT}{dz} \tag{4–140}$$

with the boundary conditions

$$\text{B.C. 1} \qquad \frac{d\theta}{dr} = 0 \qquad \text{at } r = 0$$

$$\text{B.C. 2} \qquad \theta = 0 \qquad \text{at } r = r_o$$

Equation (4–140) with its associated boundary conditions can be solved to obtain the temperature distribution. It can also be used to lead to a relationship for the convection heat transfer coefficient. First substituting the laminar velocity distribution from equation (4–14) into equation (4–140) gives

$$d\left(r\frac{d\theta}{dr}\right) = \frac{1}{\alpha}\left(\frac{\partial T}{\partial z}\right)\frac{\Delta p}{4\pi\mu L}(r_o^2 r - r^3)dr$$

and integrating once, recognizing that $\partial T/\partial z$ = constant

$$r\frac{d\theta}{dr} = \frac{\Delta p}{4\pi\alpha\mu L}\left(\frac{\partial T}{\partial z}\right)\left(\frac{1}{2}r_o^2 r^2 - \frac{1}{3}r^4\right) + C_1$$

Separating variables again and integrating gives

$$\theta(r) = \frac{\Delta p}{4\pi\mu\alpha L}\left(\frac{\partial T}{\partial z}\right)\left(\frac{1}{4}r_o^2 r^2 - \frac{1}{16}r^4\right) + C_1 \ln r + C_2 \tag{4–141}$$

From B.C. 1 we must have that $C_1 = 0$ or $\partial\theta/\partial r \rightarrow \infty$ as $r \rightarrow 0$. Applying B.C. 2 gives

$$C_2 = -\frac{\Delta p}{4\pi\mu\alpha L}\left(\frac{\partial T}{\partial z}\right)\frac{3r_o^4}{16}$$

and the temperature distribution is

$$\theta(r) = \frac{\Delta p}{4\pi\mu\alpha L}\left(\frac{\partial T}{\partial z}\right)\left(\frac{1}{4}r_o^2 r^2 - \frac{1}{16}r^4 - \frac{3}{16}r_o^2\right) \tag{4–142}$$

or

$$T(r, z) = T_w(x) + \frac{\Delta p}{4\pi\mu\alpha L}\left(\frac{\partial T}{\partial z}\right)\left(\frac{1}{4}r_o^2 r^2 - \frac{1}{16}r^2 - \frac{3}{16}r_o^2\right) \tag{4–143}$$

We may determine the mixed fluid temperature by substituting the temperature distribution of equation (4–143) and the velocity distribution of equation (4–14) into equation (4–135) and performing the integration to obtain

$$T_m = T_c + \frac{7}{16}\left(\frac{\mathbf{u}_c r_o^2}{\alpha}\right)\frac{\partial T}{\partial z} \tag{4–144}$$

where $\mathbf{u}_c$ is the maximum velocity at the channel center, or where $r = 0$, and T_c is the temperature at the center; that is, from equation (4–143) we have

$$T_c = T(0, z) = T_w(z) - \frac{\Delta p}{4\pi\mu\alpha L}\left(\frac{\partial T}{\partial z}\right)\frac{3}{16}r_o^2 \tag{4–145}$$

The heat transfer between the fluid and the channel may be described by equation (4–32) as it relates to channel flow; that is,

$$h(T_m - T_w) = -\kappa_{\text{fluid}}\left[\frac{\partial T}{\partial r}\right]_{r=r_o} \tag{4–146}$$

Combining the relationships of equations (4–144) and (4–145) into equation (4–146) gives

$$h\frac{\Delta p}{4\pi\mu\alpha L}\left(\frac{\partial T}{\partial z}\right)\frac{3}{16}r_o^2 = \kappa_{\text{fluid}}\left[\frac{\partial T}{\partial r}\right]_{r=r_o}$$

and if the temperature gradient $\partial T/\partial r$ is obtained from equation (4–143) then we have, at $r = r_o$,

$$h\frac{\Delta p}{4\pi\mu\alpha L}\left(\frac{\partial T}{\partial z}\right)\frac{3}{16}r_o^2 = \kappa_{\text{fluid}}\frac{\mathbf{u}_o r_o}{4\alpha}\frac{\partial T}{\partial x}$$

which reduces to

$$h = \kappa_{\text{fluid}}\frac{24}{11r_o} = \kappa_{\text{fluid}}\frac{48}{11D_o} \tag{4–147}$$

The Nusselt Number, Nu_D, may be defined as $hD_o/\kappa_{\text{fluid}}$ and is, from equation (4–147),

Uniform Convection Heat Transfer in a Laminar Fluid Flow Circular Channel

$$\mathrm{Nu}_D = \frac{48}{11} = 4.364 \tag{4–148}$$

which shows that the heat transfer is independent of the velocity of the fluid for laminar flow in a circular channel with uniform heat transfer per unit area of the channel. For the situation of uniform channel wall temperature, or where T_w is a constant, it can be shown that

Uniform Circular Channel Wall Temperature with Laminar Fluid Flow

$$\text{Nu}_D = 3.657 \qquad \textbf{(4–149)}$$

Fully Developed Turbulent Flow

The analysis of heat transfer with turbulent flow in a circular channel uses empirical information and dimensional analysis techniques. Rigorous mathematical methods become too complicated and require simplifying assumptions, which return to empirical information. The result obtained in section 4–3, equation (4–69), in the form

$$\text{Nu}_D = \frac{hD}{\kappa_{\text{fluid}}} C(\text{Re}_D)^m(\text{Pr})^n \qquad \textbf{(4–150)}$$

has been applied to turbulent fluid flow in a circular channel. There have been many experimental evaluations made of the constant C and the exponents m and n in equation (4–150), and there have also been numerical revisions made to the form of that equation to better fit experimental data to the dimensional premise that the Nusselt Number depends on the Reynolds and Prandtl Numbers. One of the most useful equations is the *Dittus–Boelter Equation*

$$\text{Nu}_D = 0.023\ \text{Re}_D^{0.8} \cdot \text{Pr}^n \qquad \textbf{(4–151)}$$

where $n = 0.4$ for fluid heating and $n = 0.3$ for fluid cooling. This equation gives reliable results for $160 \geq \text{Pr} \geq 0.7$, for $\text{Re}_D \geq 10{,}000$, and for channel lengths L greater than about 10 times the diameter. The fluid properties in the Dittus–Boelter Equation should all be evaluated at the mixed temperature of the fluid. Convective heat-transfer coefficients are generally within 25% of experimentally verified values.

Another equation that has been suggested by Petukhov [24] to provide agreement within 10% or less between the predicted and experimentally verified values is

$$\text{Nu}_D = \frac{f}{8} \frac{\text{Re}_D \cdot \text{Pr}}{1.07 + 12.7(f/8)^{1/2}(\text{Pr}^{2/3} - 1)} \qquad \textbf{(4–152)}$$

where f is the friction factor obtained from figure 4–19 (the Moody Diagram), $2000 \geq \text{Pr} \geq 0.5$, and $5 \times 10^6 \geq \text{Re}_D \geq 10{,}000$. The fluid properties are to be evaluated at the mixed fluid temperature T_m. Equation (4–152) may be used for turbulent flow down to the transition conditions (where $\text{Re}_D \rightarrow 2300$), but errors may be greater than 10% when predicting the convective heat-transfer coefficient.

EXAMPLE 4–19 Ethyl glycol flows at an average velocity of 20 m/s through a 10-cm-diameter drawn tube. The mixed or average temperature of the glycol is 10°C, and the tube wall surface is 70°C. Compare the value for the convective heat transfer coefficient predicted by the Dittus–Boelter equation to that predicted by the Petukov–Kirillov–Popov (PKP) equation.

Solution Using properties of ethyl glycol from the appendix table, the Reynolds Number is

$$\text{Re}_D = \frac{\rho \mathbf{u}_{\text{ave}} D}{\mu} = \frac{(1116\ \text{kg/m}^3)(20\ \text{m/s})(0.1\ \text{m})}{(0.0214\ \text{kg/m} \cdot \text{s})} = 104{,}299$$

and the Prandtl Number is

$$\text{Pr} = \frac{\mu c_p}{\kappa} = \frac{(0.0214 \text{ kg/m}\cdot\text{s})(2385 \text{ J/kg}\cdot\text{K})}{(0.257 \text{ J/s}\cdot\text{m}\cdot\text{k})} = 198.6$$

Using the Dittus–Boelter equation (4–151), setting $n = 0.4$ because the fluid is being heated (note that this equation is not necessarily reliable if Pr > 160), we obtain

$$h = \frac{\kappa}{D} 0.023\text{Re}_D^{0.8}\text{Pr}^{0.4} = \frac{(0.257 \text{ W/m}\cdot\text{K})}{0.1 \text{ m}}(0.023)(104{,}299)^{0.8}(198.6)^{0.4}$$

Answer $= 5075.5 \text{ W/m}^2\cdot\text{K}$

Using the PKP equation (4–152), we need the friction factor. Using a roughness value, ε, of 0.000005 for drawn tubing and a relative roughness, ε/D, of 0.000005/0.328 ft = 0.000015 with a Reynolds Number of 104299 gives a value of 0.0175 for the friction factor from figure 4–19. Then

Answer $$h = \frac{\kappa}{D}\frac{f}{8}\frac{\text{Re}_D\text{Pr}}{1.07 + 12(f/8)^{1/2}(\text{Pr}^{2/3} - 1)} = 5626.92 \text{ W/m}^2\cdot\text{K}$$

The convective heat-transfer coefficient values are in reasonably good agreement, even though the Dittus–Boelter equation is not necessarily reliable for a Prandtl Number over 160.

Thermal Entry Conditions

It is sometimes important to be able to identify a thermal entry region or region where the heat transfer between a flowing fluid and a channel has begun but a stable steady state thermal condition has not yet been achieved. Because of the developing nature of the thermal condition at the entrance, the convection heat transfer may be different from that with steady state conditions. The thermal entry length δ_T is often used to describe the length of the thermal entry region, and if the flow is turbulent the thermal entry length is short and the use of steady state conditions may be assumed for the full length of the heat transfer region. Thus, using relationships such as equations (4–151) or (4–152) is appropriate for all of the closed circular channel of turbulent flow where convection heat transfer occurs.

For laminar fluid flow a mathematical analysis can be carried out for predicting the entrance length during which thermal conditions are developing. It has been shown that the Nusselt Number is a function of the Reynolds and Prandtl Numbers. A convenient parameter, the *Graetz Number,* Gz_D, defined as

$$\text{Gz}_D = \frac{D}{x}\text{Re}_D \cdot \text{Pr} \tag{4–153}$$

where x is the distance downstream from the beginning of the thermal entry region, is related to the local Nusselt Number, Nu_{Dx}, and the average Nusselt Number, Nu_D. The local Nusselt Number is

$$\text{Nu}_{Dx} = \frac{h_x D_o}{\kappa_{\text{fluid}}} \tag{4–154}$$

where h_x is the convection heat transfer coefficient at a particular location, x. The average Nusselt Number is

$$\text{Nu}_D = \frac{h_{\text{ave}} D_o}{\kappa_{\text{fluid}}} \tag{4–155}$$

where h_{ave} is a convection heat transfer coefficient used to describe the heat transfer from the beginning of the thermal region to the particular location x. The relationship between the Nusselt Numbers and the Graetz Number is shown graphically in Figure 4–48a for the situation of thermal development in fully developed fluid flow. The reader can see that the Nusselt Numbers are functions of the inverse Graetz Number. Also, notice that the Nusselt Numbers approach constant values beyond Gz^{-1} of about 0.4, corresponding to a Graetz Number of 2.5, and that the values for the average and local Nusselt Numbers are 4.364 for uniform heat transfer (which agrees with the result given in equation 4–148) and 3.657 for uniform channel wall temperature.

Figure 4–48

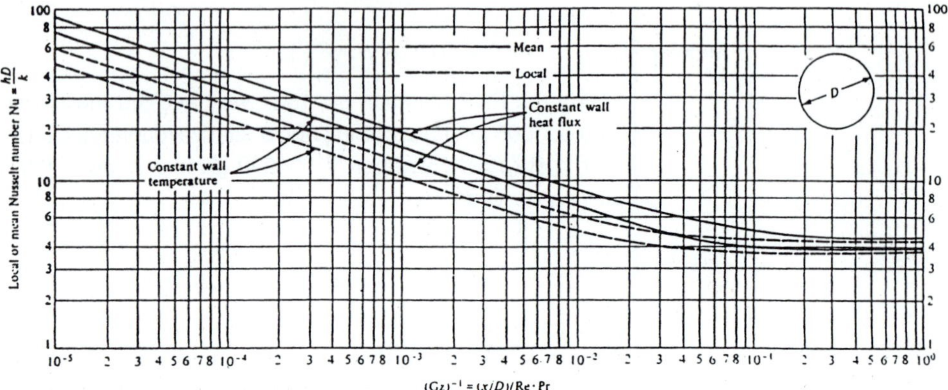

(a) Variation of the Nusselt number with entrance distance in thermally developing/developed fluid flow

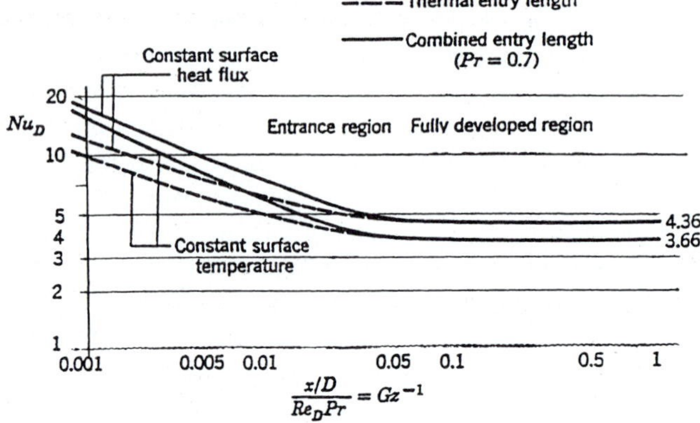

(b) Variation in Nusselt numbers for simultaneously developing thermal and fluid flow conditions

For situations of simultaneously developing thermal and fluid flow conditions the relationship between the Nusselt Numbers and the inverse Graetz Number is more complex, and figure 4–48b shows those relationships for a Prandtl Number of 0.7 (air or a similar substance). The Nusselt Numbers again approach the steady state values of 4.364 for uniform heat transfer and 3.657 for uniform channel wall temperature. The thermal entry length for simultaneously developing thermal and fluid flow occurs at a Graetz Number of about 20, or an inverse Graetz Number of 0.05.

Convection Heat Transfer in Noncircular Channels

Convection heat transfer and the methods of predicting the convection heat transfer coefficients may be extended to the situation of noncircular-cross-section channels by using the concept of the hydraulic diameter, defined by equation (4–9),

$$D_H = \frac{4A_c}{\text{wp}} \qquad \textbf{(4–9, repeated)}$$

Thus, the Nusselt and Reynolds Numbers may be determined by using the hydraulic diameter in place of the diameter and using the correlations between the Nusselt Number and the Reynolds, Prandtl, and Graetz Numbers accordingly. For fully developed laminar flow conditions the Nusselt number may be obtained for various cross sections by using table 4–6.

Table 4–6 Nusselt Number for Fully Developed Laminar Flow in Noncircular-Cross-Section Channels

Cross Section	b/a	Nusselt Number, Nu_D: Uniform Channel Temperature	Nusselt Number, Nu_D: Uniform Heat Transfer
Circle		3.657	4.364
Rectangle (a × b)	1.0	2.98	3.61
Rectangle (a × b)	0.7	3.08	3.73
Rectangle (a × b)	0.5	3.39	4.12
Rectangle (a × b)	1/3	3.96	4.79
Rectangle (a × b)	1/4	4.44	5.33
Rectangle (a × b)	1/8	5.60	6.49
Parallel lines	0 (parallel plates)	7.54	8.23
Triangle		2.47	3.11

Ussed with permission of Kays and Crawford [23].

EXAMPLE 4–20 A rectangular intake air duct is 36 in. wide and 24 in. high. It is inside a warehouse and is at 50°F, but one end of the duct opens to the outside where outside air can enter the duct as shown in figure 4–49. If 3600 cfm (cubic feet per minute) of outside air at 12°F enters the duct, estimate the air temperature 20 ft inside the duct.

Solution The analysis for estimating the air temperature involves applying an energy balance of the air flow over the 20-ft section and including Newton's Law of Heating. First the character of the air flow is determined, whether laminar or turbulent. The cross section is rectangular, so the hydraulic diameter must be determined:

Figure 4–49 Intake air duct, example 4–20.

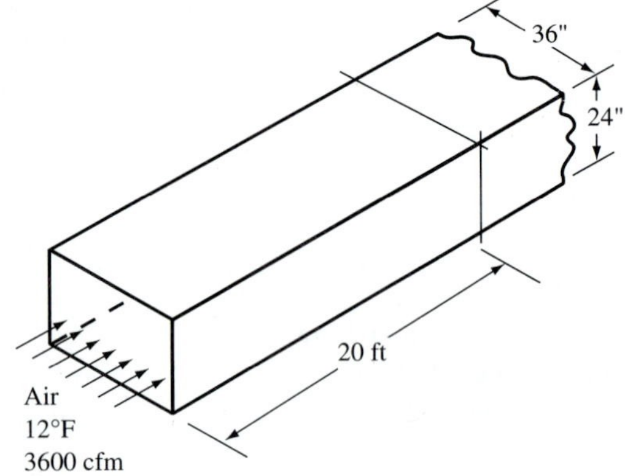

$$D_H = \frac{4A_c}{wp} = \frac{4(24 \times 36\ \text{in}^2)}{(120\ \text{in.})} = 28.8\ \text{in.} = 2.4\ \text{ft}$$

The air properties are obtained from appendix table B–4, and an approximate average air temperature of 30°F is assumed:

$$\rho = \frac{p}{RT} = \frac{(14.7\ \text{lbf/in}^2)(144\ \text{in}^2/\text{ft}^2)}{(53.3\ \text{ft} \cdot \text{lbf/lbm} \cdot {}^\circ\text{R})(490^\circ\text{R})} = 0.08\ \text{lbm/ft}^3$$

$$c_p = 0.24\ \text{btu/lbm} \cdot {}^\circ\text{R},$$

$$\kappa = 0.0139\ \text{btu/hr} \cdot \text{ft}^2 \cdot {}^\circ\text{F}$$

$$\nu = 15.5 \times 10^{-4}\ \text{ft}^2/\text{s}$$

$$\text{Pr} = 0.715$$

The air velocity at the entrance is

$$\mathbf{V} = \frac{\dot{V}}{A} = \frac{3600\ \text{ft}^3/\text{min}}{6\ \text{ft}^2} = 600\ \text{ft/min} = 10\ \text{ft/s}$$

and the Reynolds Number is

$$\text{Re}_D = \frac{\mathbf{V}D}{\nu} = \frac{(10\ \text{ft/s})(2.4\ \text{ft})}{(15.5 \times 10^{-4}\ \text{ft}^2/\text{s})} = 15{,}484$$

The air flow is turbulent because $\text{Re}_D > 2300$, and the entrance conditions can be ignored for heat-transfer purposes. The energy balance for the air stream can be written as

$$\dot{Q} = \dot{m}c_p(T_{m2} - T_{m1}) = hA(T_w - T_{\text{average}})$$

where the heat transfer to the air increases the enthalpy of the air and is the same heat transfer identified as convection heat transfer. The mass flow rate is

$$\dot{m} = \dot{V}\rho = (60\ \text{ft}^3/\text{s})(0.08\ \text{lbm/ft}^3) = 4.8\ \text{lbm/s}$$

the mixed air temperature T_{m1} is 12°F, the convection surface area A is PL (perimeter × length) = 10 ft × 20 ft = 200 ft², the duct wall temperature T_w is 50°F, and the average air temperature, T_{average}, is $T_{\text{average}} = \frac{1}{2}(T_{m1} + T_{m2})$.

The convection heat transfer coefficient h can be predicted from the Dittus–Boelter equation (4–151) or from the more elaborate PKP equation (4–152). Using equation (4–151), with $n = 0.4$ because the air is being heated, we have

$$\text{Nu}_D = 0.023\ \text{Re}_D^{0.8}\ \text{Pr}^{0.4} = 0.023(15{,}484)^{0.8}(0.715)^{0.4} = 45.2$$

The convective heat-transfer coefficient is

$$h = \frac{\text{Nu}_D \kappa}{D} = \frac{45.2(0.0139 \text{ Btu/hr}\cdot\text{ft}\cdot{}^\circ\text{F})}{2.4 \text{ ft}} = 0.262 \frac{\text{Btu}}{\text{hr}\cdot\text{ft}^2\cdot{}^\circ\text{F}}$$

and the energy balance can then be written

$$\dot{m}c_p(T_{m2} - T_{m1}) = hPL[T_w - 1/2(T_{m1} + T_{m2})]$$

and

$$T_{m2} = \frac{1}{1 + hPL/\dot{m}c_p}\left[\left(\frac{hPL}{\dot{m}c_p}\right)\left(T_w - \frac{1}{2}T_{m1}\right) + T_{m1}\right]$$

which gives, after substituting values,

Answer

$$T_{m2} = 43.3^\circ\text{F}$$

Another solution can be obtained by using equation (4–152) instead of (4–151), and that gives

$$\text{Nu}_D = (f/8)\frac{\text{Re}_D\text{Pr}}{1.07 + 12.7(f/8)^{1/2}(\text{Pr}^{2/3} - 1)}$$

or

$$h = \frac{(\kappa/D_H)(f/8)\text{Re}_D\text{Pr}}{1.07 + 12.7(f/8)^{1/2}(\text{Pr}^{2/3} - 1)}$$

From figure 4–19 we have at $\text{Re}_D = 15{,}484$ that $f = 0.0275$ for a smooth duct. Substituting values into this equation we get

$$h = 0.259 \text{ Btu/hr}\cdot\text{ft}^2\cdot{}^\circ\text{F}$$

and for the mixed air temperature at station 2,

Answer

$$T_{m2} = 43.3^\circ\text{F}$$

Notice that the predicted temperature is the same for both methods of estimating the convective heat-transfer coefficient. The analysis may continue by noting that the average air temperature is (12 + 43.3)/2 = 27.65°F. Since the air properties were selected at 30°F we could revise those values slightly and complete the computational procedure again, but such iterations are unnecessary because the predicted values for the convective heat transfer coefficient are not closer than 10%.

EXAMPLE 4–21 Engine oil passes through a 1-cm-diameter tube from a crankcase. Under certain conditions when the temperature is low the oil needs to be heated to increase its ability to flow. It is suggested to install a heating element around the tube to affect this requirement, as shown in the sketch of figure 4–50. If engine oil at 0°C enters the tube at 10 cm/s and a

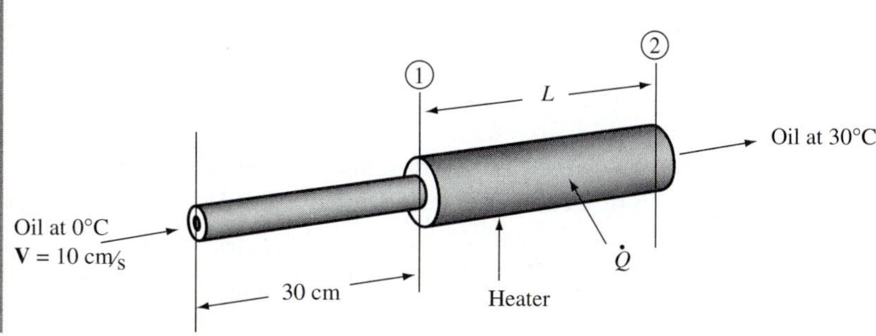

Figure 4–50 Engine oil heater, example 4–21.

heater is placed 30 cm downstream, determine the necessary length of the heater and how much heat it must supply if the oil leaving the heating section is to be 30°C and the specifications of the heat indicate that the inside tube wall temperature will be 60°C when the heater is operating.

Solution The analysis for this problem can proceed by applying conservation of energy to the oil,

$$\dot{Q} = \dot{m}c_p(T_{m2} - T_{m1}) = hA(T_w - T_{\text{average}})$$

where $T_{\text{average}} = (T_{m1} + T_{m2})/2 = (0°\text{C} + 30°\text{C})/2 = 15°\text{F}$. The character of the oil flow needs to be determined, and the Reynolds Number will provide that information. We obtain the properties of engine oil from appendix table B–3, using for an oil temperature of 0°C for the unheated section, $\kappa = 0.147$ W/m · °C, $\nu = 0.00428$ m^2/s, and Pr = 47,100.

The Reynolds Number is

$$\text{Re}_D = \frac{\mathbf{V}D}{\nu} = \frac{(0.1 \text{ m/s})(0.01 \text{ m})}{(0.00428 \text{ m}^2/\text{s})} = 0.2336$$

This indicates that the flow is laminar, since $\text{Re}_D \ll 2300$. We must determine if the fluid flow is developed at a point 30 cm from the entrance. From equation (4.133) we have

$$x_F = \frac{r_o^2}{5\nu}\mathbf{V}_{\text{average}} = \frac{(0.005 \text{ m})^2}{5(0.00428 \text{ m}^2/\text{s})}(0.1 \text{ m/s}) = 0.117 \text{ mm}$$

which indicates that the fluid flow develops rapidly and the heater will be located in a region of fully developed fluid flow. The amount of energy or the heat required to heat up the oil to 30°C is, from the energy balance, using the averaged mixed oil temperature of 15°C to be determined by the properties $\rho = 890.63$ kg/m^3, $c_p = 1859$ J/kg° · C, $\kappa = 0.145$ W/m · °C, $\nu = 0.00243$ m^2/s, and Pr = 27,000.
The energy balance is

$$\dot{Q} = \dot{m}c_p(T_{m2} - T_{m1}) = \rho\mathbf{V}_{\text{average}}c_p(T_{m2} - T_{m1})$$
$$= (890.63 \text{ kg/m}^3)(0.1 \text{ m/s})\pi(0.005 \text{ m})^2\left(1859 \frac{\text{kJ}}{\text{kg} \cdot °\text{C}}\right)(30°\text{C}) = 390 \text{ W}$$

Then we may write that

$$390 \text{ W} = hA(T_w - T_{\text{average}}) = h\pi DL(T_w - T_{\text{average}})$$

This equation can be solved for L, the heater length, if the convective heat-transfer coefficient is known. Since the flow is laminar, we may assume that the Nusselt Number, Nu_D, is 3.657, which is the value for thermally developed flow. The convective heat transfer coeffficient is then

$$h = \text{Nu}_D\kappa/D = (3.657)(0.145 \text{ W/m} \cdot °\text{C})/(0.01 \text{ m}) = 53 \text{ W/m}^2 \cdot °\text{C}$$

This gives

$$L = 390 \text{ W}\frac{1}{h\pi D}(T_w - T_{\text{average}}) = 52 \text{ m}$$

We check the thermal development with the Graetz Number, $\text{Gz} = (D/x)(\text{Re}_D \cdot \text{Pr})$. If Gz is 2.5 or less, then our assumption that $\text{Nu}_D = 3.657$ is correct. This means that

$$x = D\frac{\text{Re}_D\text{Pr}}{\text{Gz}} = (0.01 \text{ m})\frac{\text{Re} \cdot \text{Pr}}{2.5}$$

For the oil at the averaged mixed temperature,

$$\text{Re}_D = \mathbf{V}D/\nu = (0.1 \text{ m/s})(0.01 \text{ m})/(0.00243 \text{ m}^2/\text{s}) = 0.412$$

and

$$x = (0.01 \text{ m})(0.412)(27000)/(2.5) = 44.5 \text{ m}$$

Comparing this distance to that predicted by the energy balance indicates that we must iterate to obtain a heater length less than 44.5 m and greater than 5.2 m. Using $\text{Re}_D = 0.412$, $\text{Pr} = 27{,}000$, and the graphical relationship in figure 4–48a we iterate to obtain the expected heater length. If we assume $\text{Nu}_D = 7.0$, then the inverse Graetz Number is about 0.012 and $\text{Gz} = 83.33$. The heater length is then predicted from the relationship

$$x = L = \frac{D\text{Re}_D \cdot \text{Pr}}{\text{Gz}} = \frac{(0.01 \text{ m})(0.412)(27{,}000)}{83.33} = 1.33 \text{ m}$$

and also from the energy balance,

$$L = \frac{390 \text{ W}}{h\pi D(T_w - T_{\text{average}})}$$

where $h = \text{Nu}_D\kappa/D = (7.0)(0.145 \text{ W/m} \cdot {}^\circ\text{C})(0.01 \text{ m}) = 101.5 \text{ W/m}^2 \cdot {}^\circ\text{C}$ and $L = (390 \text{ W})/[(101.5 \text{ W/m}^2 \cdot {}^\circ\text{C})(\pi)(0.01 \text{ m})(45^\circ\text{C})] = 2.7 \text{ m}$.

These two results still do not agree, so another iteration needs to be made. If Nu_D is taken as 5.3 we have, from figure 4–48a, that the inverse Graetz Number is 0.032 and $\text{Gz} = 31.25$. Again we compute new values for the length:

$$x = L = \frac{(0.01 \text{ m})(0.412)(27{,}000)}{31.25} = 3.56 \text{ m}$$

and

$$L = \frac{390 \text{ W}}{(h)(\pi)(D)(T_w - T_{\text{average}})}$$

where $h = \text{Nu}_D\kappa/D = (5.3)(0.145 \text{ W/m} \cdot {}^\circ\text{C})/(0.01\text{m}) = 76.85 \text{ W/m}^2 \cdot {}^\circ\text{C}$. Then $L = 3.59$ m, and this is nearly the same as x. Thus, we may assume the heater to be, as an optimum,

Answer

$$\text{Heater length} = 3.6 \text{ m}$$

4–7 APPLICATIONS OF CONVECTION HEAT TRANSFER

In this section we will consider some engineering problems that are impacted by results obtained from convection heat-transfer analysis. Each of these may be studied as a separate item.

Convection Heat Transfer in Tube Banks

Tube banks are a group of many tubes arranged in rows and columns as shown in figure 4–51. Fluid flows through the tube bank to transfer heat by convection. In figure 4–51a is shown an arrangement of tubes that are *in line*. It is often convenient to use an in-line arrangement for construction purposes and to provide less resistance for the fluid to flow through the tube bank. The *staggered* arrangement of tubes, as shown in figure 4–51b, provides a better arrangement for heat transfer because the fluid has more stagnation points on the tubes, turbulence is more likely, and the mixing temperatures of the fluid are better distributed. Tube banks are used in many devices and systems that require heat transfers between the fluids, one fluid passing through the tubes and one passing around the tubes. Economizers and superheaters in steam power generating stations function effectively because they are tube banks extracting thermal energy from combustion gases. Cooling and heating coils are tube banks used to cool or heat air for buildings, and the condenser of a refrigerator or air conditioner is a row of tubes for cooling a refrigerant

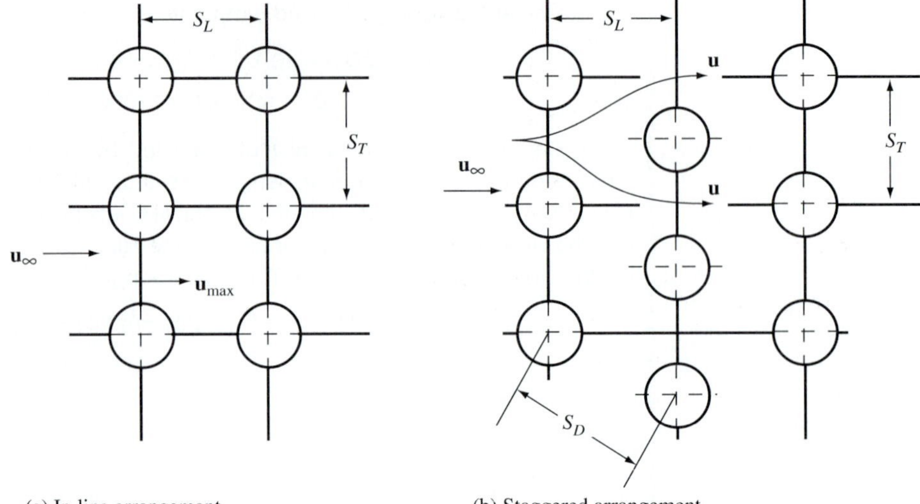

Figure 4–51 Tube bank arrangements.

by passing air over the tubes. There are many more situations where tube banks are an integral part of a heat exchanger or represent a model for analyzing heat transfer.

The analysis of tube banks is made by assuming steady state heat transfer. The heat transfer in the tube bank may then be written as

$$\dot{Q} = h_e A \Delta T \tag{4–156}$$

where ΔT is the temperature difference between the tube and the fluid. Since these temperatures are varying from one location to another, even in steady state, the fluid temperature T_f can be approximated by an arithmetic mean between the inlet T_i and the outlet T_o; that is,

$$T_f = 1/2(T_i + T_o) \tag{4–157}$$

The wall temperatures may also be approximated as arithmetic means between the inlet and the outlet. The tube wall temperature is often assumed to be constant, particularly in those applications where the fluid flowing through the tubes is evaporating or condensing. The effective area A_e is usually assumed to be the total surface area of the tubes,

$$A_e = \pi DL(i \times \mathrm{j}) \tag{4–158}$$

where L is the length of the tube, D is the tube diameter, i is the number of rows, and j is the number of columns. Frequently the heat transfer is determined for a unit length of tube banks in equation (4–156)

$$\dot{q}_L = \frac{\dot{Q}}{L} = h_e \frac{A_e}{L} \Delta T = h_e a_e \Delta T \tag{4–159}$$

where

$$a_e = \pi D(i \times j) \tag{4–160}$$

In section 4–5 we studied fluid flow around tubes and objects and the resulting heat transfer. There we obtained relationships for predicting the convection heat-transfer coefficient h_e. In particular, equation (4–128) gives the relationship between the Nusselt Number and the Reynolds and Prandtl Numbers. It has been useful to extend this result to tube banks and write

$$\mathrm{Nu}_e = \frac{h_e D}{\kappa} = C\mathrm{Re}_{D,\max}^m \mathrm{Pr}^{1/3} \tag{4–161}$$

where the Reynolds Number is determined at the maximum fluid velocity condition. Referring to figure 4–51a, the maximum fluid velocity will occur between the tubes. Assuming incompressible fluid flow this gives

$$\mathbf{u}_{\max} = (S_T - D) = \mathbf{u}_\infty S_T \tag{4–162}$$

or

$$\mathbf{u}_{\max} = \mathbf{u}_\infty\left[\frac{S_T}{S_T - D}\right] \tag{4–163}$$

For staggered tube arrangements the maximum fluid velocity may occur between tubes of the same row or between tubes diagonally separated in adjacent rows. For the case where the maximum fluid velocity occurs in diagonal tubes, we have

$$\mathbf{u}_{\max}(2A_D) = \mathbf{u}_\infty S_T$$

where

$$A_D = S_D - D$$

so that

$$2\mathbf{u}_{\max}(S_D - D) = \mathbf{u}_\infty S_T$$

and

$$\mathbf{u}_{\max} = \mathbf{u}_\infty \frac{S_T}{2(S_D - D)} \tag{4–164}$$

Thus, for staggered tube arrangements we have the maximum fluid velocity determined by the *maximum* of either equation (4–163) or (4–164). The fluid properties are to be determined at the temperature T_f. The Prandtl Number for air is normally equal to 0.7 so that equation (4–161) becomes

$$\mathrm{Nu}_e = C\,\mathrm{Re}^m_{D,\max}(0.7)^{1/3} = 0.888C\,\mathrm{Re}^m_{D,\max} = C_1\,\mathrm{Re}^m_{D,\max} \tag{4–165}$$

where $C_1 = 0.888C$. The limitations for using equation (4–165) are N (number of columns) ≥ 10, $2{,}000 \leq \mathrm{Re}_D \leq 40{,}000$, and $\mathrm{Pr} = 0.7$

Grimison [25] has reported values of m and C, and these are given in table 4–7. For tube banks containing less than 10 columns, Kays and Lo [26] have suggested applying

Table 4–7 Constants for Equation 4–165, Forced Convection Heat Transfer over Tube Banks of Ten or More Columns of Tubes

	S_T/D							
	1.25		1.5		2.0		3.0	
S_L/D	C_1	m	C_1	m	C_1	m	C_1	m
Aligned								
1.25	0.348	0.592	0.275	0.608	0.100	0.704	0.0633	0.752
1.50	0.367	0.586	0.250	0.620	0.101	0.702	0.0678	0.744
2.00	0.418	0.570	0.299	0.602	0.229	0.632	0.198	0.648
3.00	0.290	0.601	0.357	0.584	0.374	0.581	0.286	0.608
Staggered								
0.600	—	—	—	—	—	—	0.213	0.636
0.900	—	—	—	—	0.446	0.571	0.401	0.581
1.000	—	—	0.497	0.558	—	—	—	—
1.125	—	—	—	—	0.478	0.565	0.518	0.560
1.250	0.518	0.556	0.505	0.554	0.519	0.556	0.522	0.562
1.500	0.451	0.568	0.460	0.562	0.452	0.568	0.488	0.568
2.000	0.404	0.572	0.416	0.568	0.482	0.556	0.449	0.570
3.000	0.310	0.592	0.356	0.580	0.440	0.562	0.428	0.574

Abstracted from [21].

a correction factor, C_F, to the Nusselt Number obtained from equation (4–161) or (4–165), so that

$$\mathrm{Nu}_{e(<10\,\mathrm{col})} = C_F \mathrm{Nu}_e \tag{4–166}$$

where C_F is tabulated in table 4–8 for tube banks having 9 columns or less.

Table 4–8 Correction Factor C_F for Tube Banks for Less Than Ten Columns, Equation 4–166

N_L	1	2	3	4	5	6	7	8	9
Aligned	0.64	0.80	0.87	0.90	0.92	0.94	0.96	0.98	0.99
Staggered	0.68	0.75	0.83	0.89	0.92	0.95	0.97	0.98	0.99

Abstracted from [21].

In tube bank design and analysis it is frequently important to predict the pressure drops experienced by the fluid flow through the tube bank. Such information is needed to predict the power required to sustain fluid flow. It has been suggested by Zhukauskas [27] that the tube bank pressure drop be written, for j columns, as

$$\Delta p = j\chi f\left(\frac{1}{2}\rho \mathbf{u}_{\max}^2\right) \quad \text{(SI units)} \tag{4–167}$$

$$\Delta p = j\chi f\left(\frac{1}{2g_c}\rho \mathbf{u}_{\max}^2\right) \quad \text{(English units)}$$

where f is a friction factor dependent on the Reynolds Number $\mathrm{Re}_{D,\max}$. The factor χ is a correction factor to use for in-line or staggered arrangements that are not uniform; that is, if $S_L = S_T$ for uniform array χ is 1.0, and if $S_D = S_T$ for staggered arrangements χ is 1.0. Corrections for other geometric patterns are given in figures 4–52 and 4–53.

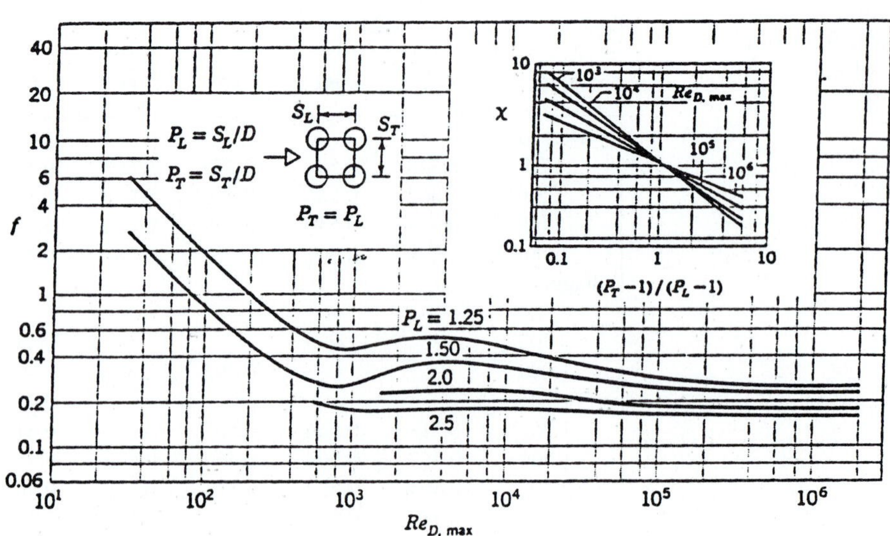

Figure 4–52 Friction factors in in-line tube banks [27].

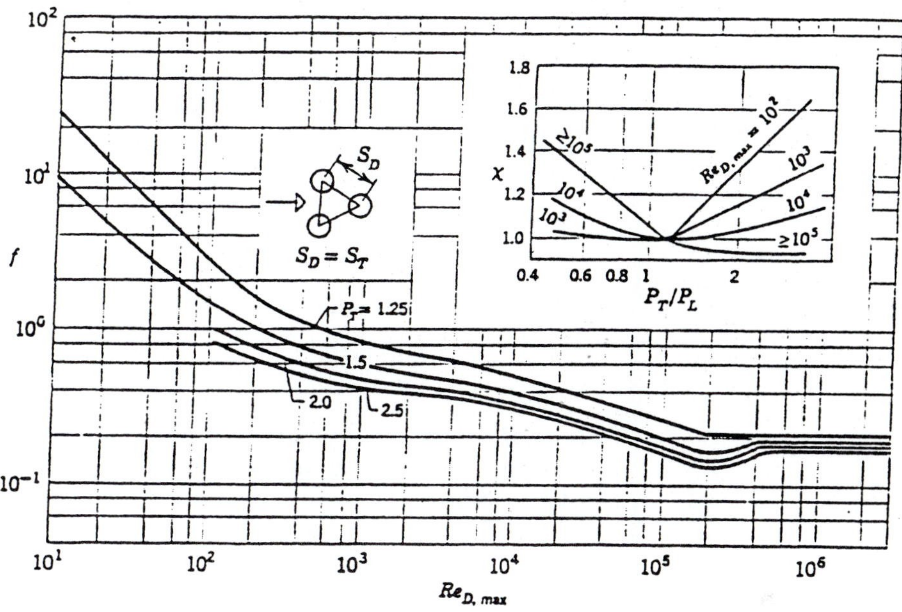

Figure 4–53 Friction factors in staggered tube banks [27].

EXAMPLE 4–22 Thirty m^3/min of air at 35°C are cooled in a chilled water coil. The coil is a set of 8 in-line columns of 2.5-cm-diameter by 50-cm-long tubes as shown in Figure 4–54. There are 10 tubes per column, and the tube wall temperature is kept at a constant 5°C. Predict the air temperature leaving the coil and the pressure drop through the coil.

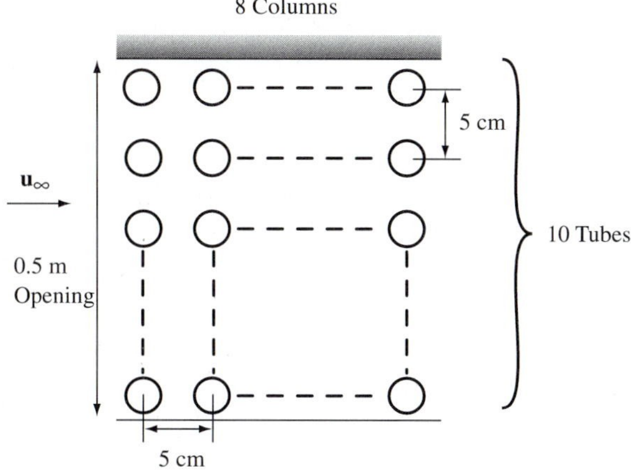

Figure 4–54 Chilled water coil, example 4–22.

Solution

The free stream air velocity entering the coil is found from

$$\dot{V} = A\mathbf{u}_\infty = 30 \text{ m}^3/\text{min}$$

The cross-sectional area is $A = 0.5 \text{ m} \times 0.5 \text{ m} = 0.25 \text{ m}^2$, and the free stream velocity is

$$\mathbf{u}_\infty = \frac{\dot{V}}{A} = \frac{30 \text{ m}^3/\text{min}}{0.25 \text{ m}^2} = 120 \text{ m/min} = 2 \text{ m/s}$$

The maximum velocity throughout the tube bank is found from equation (4–163):

$$\mathbf{u}_{max} = \mathbf{u}_\infty \frac{S_T}{S_T - D} = \left(\mathbf{2}\frac{\text{m}}{\text{s}}\right)\left(\frac{5}{5 - 2.5}\right) = \mathbf{4}\ \text{m/s}$$

The Reynolds Number may be determined if the film temperature of the air is known. If we estimate the film temperature to be 300 K, then the Reynolds Number is

$$\text{Re}_{D,\text{max}} = \frac{\rho \mathbf{u}_{max} D}{\mu} = \frac{(1.178\ \text{kg/m}^3)(4\ \text{m/s})(0.025\ \text{m})}{(0.0181\ \text{g/m}\cdot\text{s})} = 6508$$

The Nusselt Number is determined from equation (4–165) (if Pr = 0.7):

$$\text{Nu}_e = \frac{h_e D}{\kappa} = C_1 \text{Re}_D^m$$

Since $S_T/D = 2.0$ and $S_L = 2.0$, from table 4–7 we have $C_1 = 0.229$ and $m = 0.632$. Also, for air at 300 K the Prandtl Number is 0.701. Then we may compute the convective heat transfer coefficient:

$$h_e = \frac{\kappa}{D} C_1 Re_{D,\text{max}}^m = \frac{0.026\ \text{W/m}\cdot\text{K}}{0.025\ \text{m}}(0.229)(6508)^{0.632} = 61.23\ \frac{\text{W}}{\text{m}^2\cdot\text{K}}$$

Since there are only eight columns we have, from table 4–8,

$$h_{(e,10\ \text{col})} = C_F h_e = 0.98(61.23\ \text{W/m}^2\cdot\text{K}) = 60.0\ \text{W/m}^2\cdot\text{K}$$

The effective area is the surface area of the tubes, given by

$$A_e = \pi D(i \times j)L = \pi(0.025\ \text{m})(8 \times 10)(0.5\ \text{m}) = 3.14\ \text{m}^2$$

The heat transfer is then set equal to the decrease in the thermal energy of the air:

$$\dot{Q} = h_e A_e \Delta T = \dot{m} c_p (T_1 - T_2)$$

For ΔT we have that

$$\Delta T = 1/2(T_1 + T_2) - 5°\text{C}$$

For the mass flow rate

$$\dot{m} = \dot{V}\rho = (0.5\ \text{m}^3/\text{s})(1.178\ \text{kg/m}^3) = 0.589\ \frac{\text{kg}}{\text{s}}$$

Then we may solve for the outlet air temperature from the energy equation written above to give that

Answer $$T_2 = 19.85°\text{C}$$

The pressure drop can be predicted from equation (4–167)

$$\Delta p = j\chi f\left(\frac{1}{2}\rho \mathbf{u}_{max}^2\right)$$

From figure 4–49 we read $f = 0.28$, $\chi = 1.0$ and then

Answer $$\Delta p = (8\ \text{columns})(1)(0.28)[1/2(1.178\ \text{kg/m}^3)(4\ \text{m/s})^2] = 21.1\ \text{Pa}$$

Packed Beds A packed bed is a volume occupied by loosely fitting particles. The particles may be rocks or pebbles, carbon particles, a desiccant (used to dry gases), sands, salt, or metallic particles. When a gas or vapor flows through the bed an exchange of heat may occur, and this mechanism may be visualized as a flow around tube banks where the tubes are touching one another, as shown in figure 4–55. The heat transfer may then be assumed to be describable with the equation

$$\dot{Q} = hA_s\Delta T \tag{4–168}$$

where A_s is the total surface are of the packed bed particles. A_s can be estimated by using the concept of the specific surface, S_m, defined as the surface area of the particles per unit of volume. If the particles are assumed to be uniform-diameter spheres and are arranged in a cubic pattern as shown in figure 4–56, then the specific surface is

$$S_m = \pi\frac{D^2}{D^3} = \frac{\pi}{D}\left(\frac{\text{cm}^2}{\text{cm}^3}, \frac{\text{m}^2}{\text{m}^3}, \frac{\text{ft}^2}{\text{ft}^3}\right)$$

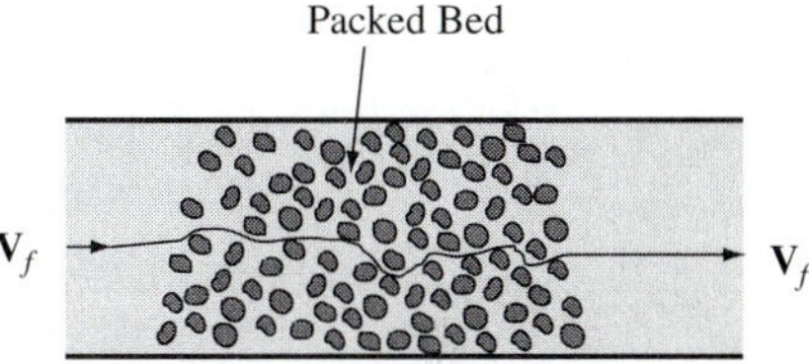

Figure 4–55 Flow of a vapor through a packed bed.

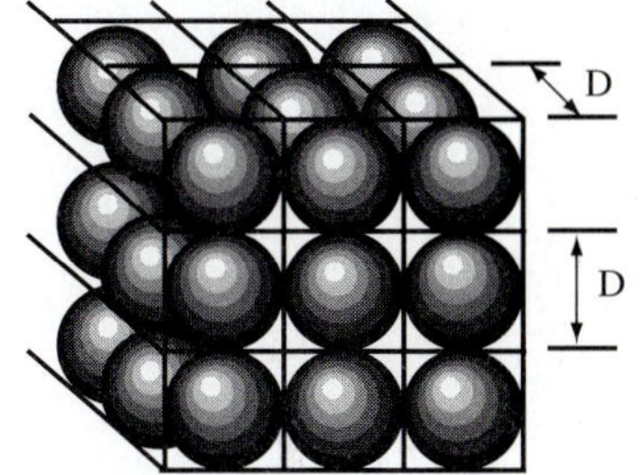

Figure 4–56 Face-centered spherical particles in a packed bed.

The total surface for a packed bed of volume V is

$$A_s = S_m V \tag{4–169}$$

The convection heat transfer coefficient may be approximated from the Reynolds–Colburn analogy in the form

$$\frac{C_f}{2} = \text{St} \cdot \text{Pr}^{2/3} \tag{4–170}$$

if the Reynolds Number is defined as

$$\text{Re}_D = \frac{\mathbf{V}_f D_p}{\nu_{\text{fluid}}} \tag{4–171}$$

where $\mathbf{V}_f$ is the free stream or face velocity, D_p is the diameter of a typical particle, and ν_f is the kinematic viscosity of the vapor at the average film temperature,

$$T_{\text{fluid}} = \frac{1}{2}(T_i + T_e) \tag{4–172}$$

Correlations recommended by Gamson, et al. [28] using a form of the Colburn analogy are

$$\left(\frac{h_\infty}{\rho c_p \mathbf{V}_f}\right) \cdot \mathrm{Pr}^{2/3} = 1.064\ \mathrm{Re}_D^{-0.41} \tag{4–173}$$

$$\left(\frac{h_\infty}{\rho c_p \mathbf{V}_f}\right) \cdot \mathrm{Pr}^{2/3} = 18.1\left(\frac{1}{\mathrm{Re}_D}\right) \qquad \text{for } \mathrm{Re}_D < 350$$

A special kind of packed bed is a *rock bed,* which is often used to store thermal energy from solar energy or from industrial processes. By passing air or other vapor through a rock bed, or packed bed, one can exchange heat between the rocks and the air. A typical arrangement is shown in figure 4–57, where the fluid or vapor flow enters horizontally into a plenum, is redirected downward through the rock bed, and is then redirected again to flow horizontally out. The fluid flow can be reversed without affecting the performance of the bed. A rock bed performance map provides graphic information of the bed pressure drop as a function of the vapor velocity and the rock diameter, D_p. In figure 4–58 is shown a typical rock bed performance map suggested by Cole, et al. [29]. From the information we may find the pressure gradient $\Delta p/L$ within the rock bed and then predict the power required to have proper fluid flow through the bed.

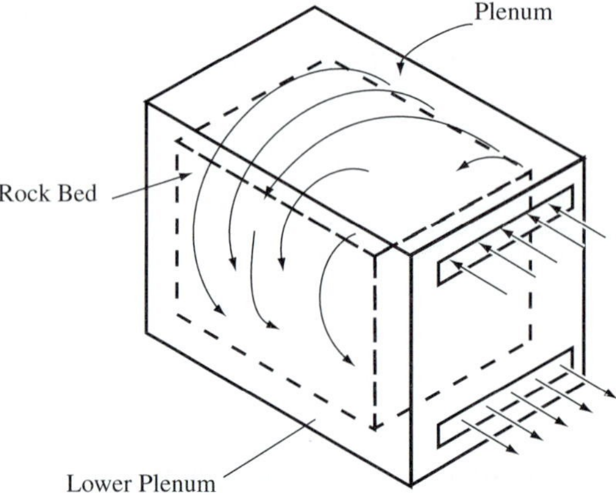

Figure 4–57 Rock bed.

EXAMPLE 4–23 A rock bed is 2 m high, 1.5 m wide, and 1.5 m long as shown in figure 4–57. It has 3.8-cm-nominal-diameter rocks in the bed. If 25 m^3/min of air is to pass through the bed, estimate the convective heat-transfer coefficient for heat transfer between the rocks and the air.

Solution The face velocity or free stream velocity is

$$\mathbf{V}_f = \frac{\dot{V}}{A_f} = \frac{25\ \text{m}^3/\text{min}}{2.25\ \text{m}^2} = 0.185\ \text{m/s}$$

Using the correlation of Gamson, et al., equation (4–173), where the Reynolds Number Re_D is

$$\mathrm{Re}_D = \frac{\rho \mathbf{V}_f D_p}{\mu} = \frac{(1.15\ \text{kg/m}^3)(0.185\ \text{m/s})(0.038\ \text{m})}{(0.0181\ \text{g/m} \cdot \text{s})} = 446.7$$

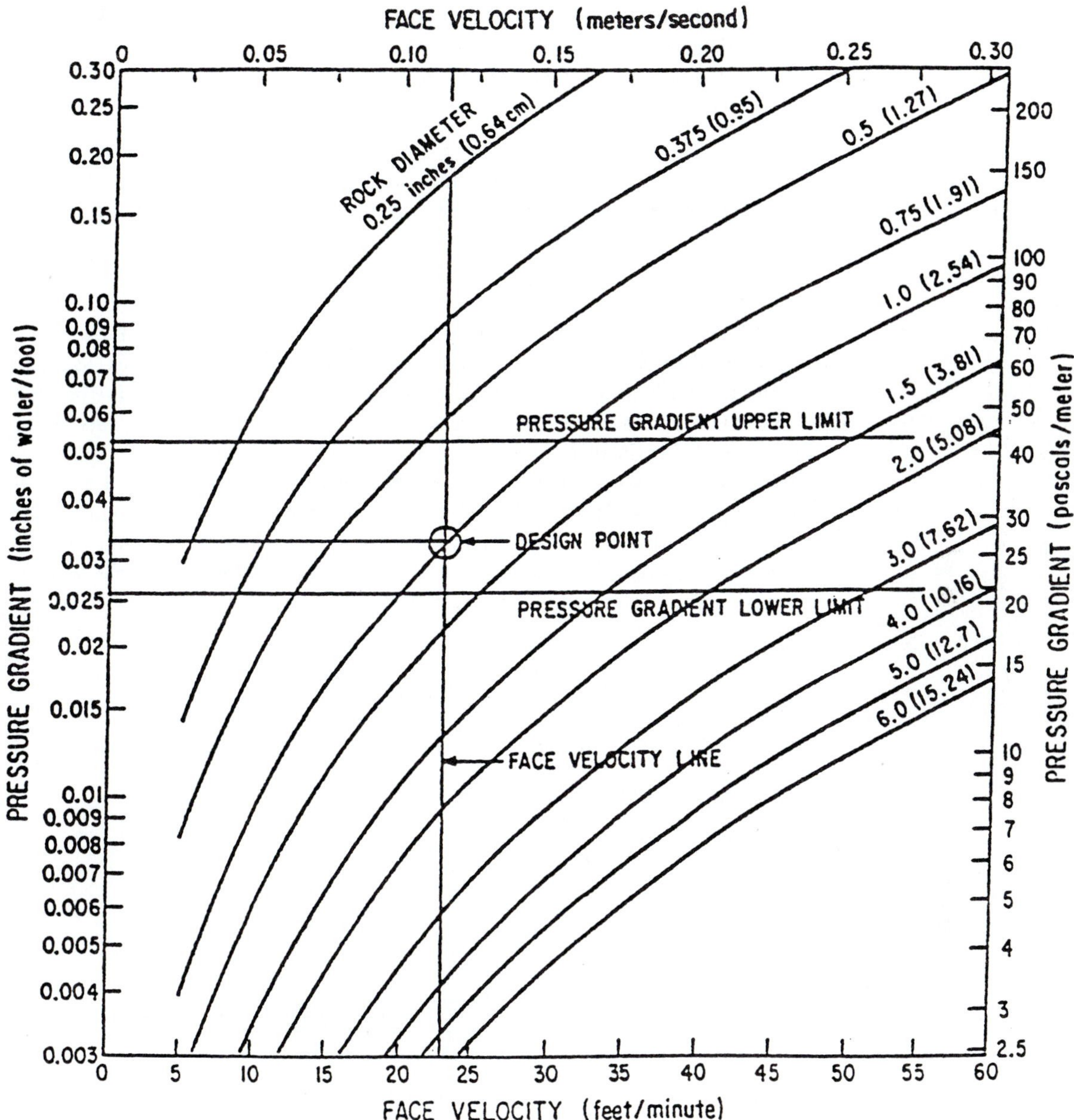

Figure 4–58 Rock bed performance map. (After Ref. [29])

gives

$$h = 1.064 \frac{\rho c_p \mathbf{V}_f}{\mathrm{Re}^{0.41} \cdot \mathrm{Pr}^{2/3}}$$

Answer

$$= 1.064 \frac{(1.15\ \mathrm{kg/m^3})(1007\ \mathrm{kJ/kg \cdot K})(0.185\ \mathrm{m/s})}{(446.7)^{0.41}(0.701)^{2/3}} = 23.6 \frac{\mathrm{W}}{\mathrm{m \cdot K}}$$

This result indicates that the rock bed can be an effective heat exchanger. There has been much study and discussion on using the Reynolds–Colburn analogy, which seems to be good for flow through closed channels, and extending it to include flow through packed beds as suggested by Gamson to predict the convective heat transfer coefficient from the frictional effects in packed beds. The interested reader may want to refer to references [28], [30], and [31] for further discussions of this matter.

Couette Flow

We considered in section 4–2 the flow of a fluid between two plates sliding relative to one another, as shown in figure 4–59. The fluid velocity between the plates is given by

$$\mathbf{u} = \mathbf{u}_\infty \frac{y}{L} \tag{4–174}$$

for the bottom plate stationary. This physical state is called Couette Flow and represents the model often used to analyze bearing lubrication for thrust bearings, journal bearings, or ball or roller bearings. In such analyses the plates would represent the bearing surfaces and the fluid would be the oil or grease between the surfaces. Lubricants are used to reduce frictional resistance and also to cool the bearings. Thus, heat transfer needs to be considered with Couette Flow. The heat transfer flow may be analyzed by predicting the fluid temperatures. Using the energy equation (4–56) and assuming (1) the fluid temperature T varies only in the y direction, that is, $T = T(y)$, and (2) the fluid enthalpy or thermal energy is proportional to temperature, $\text{hn} = \text{hn}(y)$, flow is laminar so that the fluid velocity in the y direction is zero, $\mathbf{v} = 0$. We then have

$$\frac{d^2T(y)}{dy^2} = -\frac{\mu}{\kappa}\left(\frac{d\mathbf{u}}{dy}\right)^2 \tag{4–175}$$

where the viscous dissipation, $\frac{\mu}{\kappa}\left(\frac{d\mathbf{u}}{dy}\right)^2$, can be a crucial part of the energy analysis. For the velocity function of equation (4–174) we have

$$\frac{d^2T(y)}{dy^2} = -\frac{\mu \mathbf{u}_\infty^2}{\kappa L^2} \tag{4–176}$$

for $0 < y < L$, $T(y) = T_o$ at $y = 0$, and $T(y) = T_1$ at $y = L$.

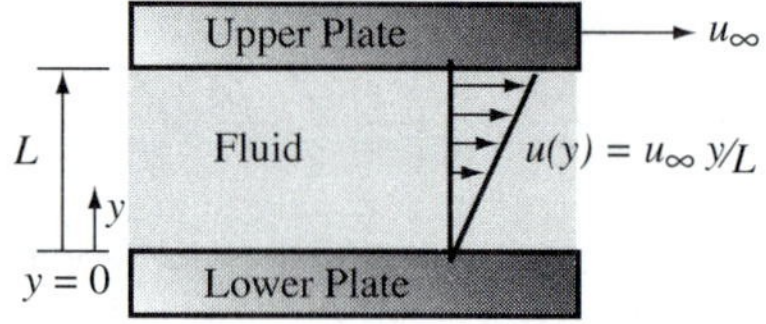

Figure 4–59 Couette Flow.

The solution to this equation and its boundary conditions is

$$T(y) = T_0 + \frac{y}{L}\left[(T_1 - T_0) + \frac{\mu \mathbf{u}_\infty^2}{2\kappa}\left(1 - \frac{y}{L}\right)\right] \tag{4–177}$$

For the situation where $T_1 = T_0$ or where both plates at the same temperature, equation (4–177) becomes

$$T(y) = T_0 + \frac{y}{L}\left[\frac{\mu \mathbf{u}_\infty^2}{2L\kappa}\left(1 - \frac{y}{L}\right)\right] \tag{4–178}$$

This gives the fluids temperatures for the situation where the viscous dissipation is the only term contributing to the thermal energy of the fluid. The distribution of the temperature through the fluid is parabolic in y and the maximum temperature occurs at the center where $y = L/2$. Then,

$$T_{max} = T_0 + \frac{\mu \mathbf{u}_\infty^2}{8\kappa} \tag{4–179}$$

For situations where the lubricant or fluid acts specifically as a coolant, the fluid bulk temperature is below that of the plates or surfaces. In this instance it is convenient to define a temperature function

$$\theta(y) = T(y) - T_f \tag{4–180}$$

and the fluid temperature can then be written

$$\theta(y) = \theta_0 + \frac{y}{L}\left[(\theta_1 - \theta_0) + \frac{\mu \mathbf{u}_\infty^2}{2\kappa}\left(1 - \frac{y}{L}\right)\right] \tag{4–181}$$

where $\theta_0 = T_0 - T_f$

and $\theta_1 = T_1 - T_f$

If the plates are at the same temperatures, then $\theta_1 = \theta_0$ and

$$\theta(y) = \theta_0 + \frac{y}{L}\left[\frac{\mu \mathbf{u}_\infty^2}{2\kappa}\left(1 - \frac{y}{L}\right)\right] \tag{4–182}$$

The heat transfer may be written

$$\begin{aligned} \dot{Q} &= h_{eff} A \theta_0 \\ &= -\kappa A \left[\frac{\partial \theta}{\partial y}\right]_{y=0} \\ &= -\kappa A\left(\frac{\mu \mathbf{u}_\infty^2}{2\kappa} - \frac{\mu \mathbf{u}_\infty^2}{\kappa}\right) \end{aligned} \tag{4–183}$$

The effective convection heat transfer coefficient, h_{eff}, is then

$$h_{eff} = \frac{\mu \mathbf{u}_\infty^2}{2\theta_0 L} \tag{4–184}$$

EXAMPLE 4–24 A hermetically sealed 90-hp electric motor needs more effective cooling than would be achieved with air flow around the housing. This is accomplished by circulating cold refrigerant inside the housing and around the rotor. The motor is expected to have a power loss of 6 hp when it develops 90 hp. This loss is due to the frictional electrical resistance in the motor wire windings. The motor is designed to have a maximum operating temperature of 95°F, and it is proposed to pump R-22 (Freon) through the motor housing because this is an electrically nonconducting material. Assume that the heat is to be dissipated to the R-22 in the annular volume between the rotor and the housing, as shown in the cross-sectional view of figure 4–60. The annular volume is 2 ft long axially through the motor housing, and the motor runs at 1875 rpm. The viscosity of R-22 at 95°F may be taken as 1.98×10^{-4} kg/m · s. Estimate the appropriate gap between the rotor and the housing, L, as shown in figure 4–57, for adequate cooling of the motor.

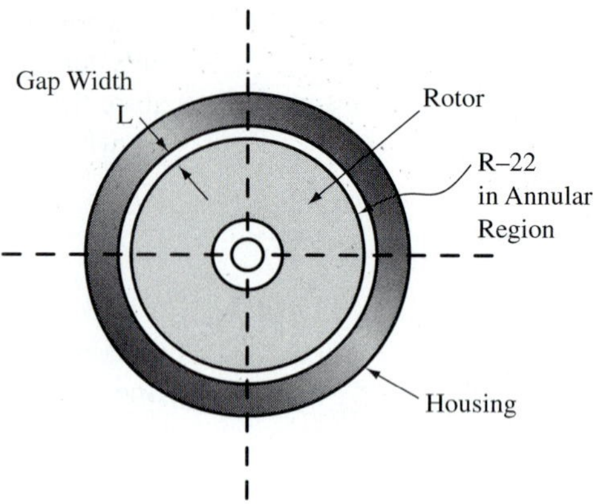

Figure 4–60 Cross-sectional view of a hermetically sealed electrical motor.

Solution

Since 6 hp needs to be dissipated as heat through the refrigerant, we may predict the heat transfer from equations (4–183) and (4–184):

$$\dot{Q} = h_{\text{eff}} A \Delta T = \frac{\mu \mathbf{u}_\infty^2}{2L} A$$

The velocity $\mathbf{u}_\infty$ is, approximately,

$$\mathbf{u}_\infty = (1875 \text{ rpm})(2\pi \text{ rad/rev})(1 \text{ ft})\left(\frac{1}{60} \text{ s/min}\right) = 196.3 \text{ ft/s}$$

The viscosity is

$$\begin{aligned} \mu &= 19.8 \times 10^{-4} \text{ Pa} \cdot \text{s} \qquad \text{(at 300 K)} \\ &= 4.135 \times 10^{-6} \text{ lbf} \cdot \text{s/ft}^2 \qquad \text{(as a saturated liquid)} \end{aligned}$$

Also, the area is

$$A = \pi(2 \text{ ft})(2 \text{ ft}) = 12.57 \text{ ft}^2$$

the heat transfer is

$$\dot{Q} = 6 \text{ hp} = 3300 \frac{\text{ft} \cdot \text{lbf}}{\text{s}} = \frac{\mu \mathbf{u}_\infty^2}{2L} A$$

and

Answer

$$L = 0.00364 \text{ in.}$$

The gap should not be more than 0.004 in. Also, it should be noted that the R-22 was assumed to be a saturated liquid throughout the process, and this is probably not fully representative of the exact state in the motor housing.

4–8 SUMMARY

Convection heat transfer is caused by fluid flowing near a solid surface having a temperature different from that of the ambient fluid temperature. Heat transfer occurs through the region of the fluid nearest the surface, called the boundary layer.

Forced convection describes heat transfer when fluid flow is caused by an external force.

Free convection describes heat transfer when fluid flow is caused by gravitational effects.

Laminar flow is fluid flow in a parallel fashion, occurring at low fluid velocities.

Turbulent flow is fluid flow in an irregular pattern, occurring at high fluid velocities.

Fluid flow without heat or power is described by *Bernoulli's Equation*

$$\frac{p_2}{\rho_2} - \frac{p_1}{\rho_1} + \frac{1}{2}[\mathbf{V}_2^2 - \mathbf{V}_1^2] + g(z_2 - z_1) = 0 \tag{4–3}$$

Conservation of mass can be expressed as

Mass flow rate in − Mass flow rate out = Mass accumulated

For *steady flow, steady state* the mass accumulated is zero.

Mass flow rate is $\rho A\mathbf{V}$.

For steady flow, steady state of an incompressible fluid

$$A_1\mathbf{V}_1 = A_2\mathbf{V}_2 \tag{4–8}$$

Hydraulic diameter is used to describe the size of cross sections of noncircular sections, defined as

$$D_H = \frac{4A}{\text{wp}} \tag{4–10}$$

where wp is the wetted perimeter.

Viscosity is the fluid resistance to motion, or fluid friction.

The Hagen–Poiseuille Equation is used to describe laminar fluid flow in a circular tube,

$$\dot{V} = A\mathbf{V} = \frac{\pi r_o^2 \Delta p}{8L\mu} \tag{4–16}$$

The *Reynolds Number*, Re, is used to describe the character of fluid flow

$$\text{Re}_L = \frac{\rho \mathbf{V} L}{\mu} = \frac{\mathbf{V}L}{\nu} \tag{4–27}$$

Dimensional analysis is used to aid in developing methods for describing complex interactions involving many variables. It requires the identification of parameters having *fundamental dimensions (mass, length, temperature,* and *time)* or *derived dimensions*. There can be $j - n$ Π-parameters written to describe this interaction, where j is the number of parameters and n is the number of fundamental dimensions involved.

A general equation relating convective heat transfer to fluid flow is the Reynolds–Colburn Analogy,

$$\frac{1}{2}C_f = \frac{\text{Nu}_L}{\text{Re}_L \cdot \text{Pr}^{1/3}} = \text{St} \cdot \text{Pr}^{2/3} \tag{4–78}$$

For fluid flow along a flat surface, the boundary layer flow will be laminar until the Reynolds Number reaches a value of about 5×10^5, called the critical Reynolds Number,

$$\text{Re}_{\text{critical}} = 5 \times 10^5$$

For laminar boundary layer flow, the boundary layer thickness is

$$\delta = 5.0\frac{x}{\sqrt{\text{Re}_x}} \tag{4–95}$$

The local friction drag coefficient C_f is

$$C_f = 0.664\frac{1}{\sqrt{\mathrm{Re}_x}} \tag{4–97}$$

and the average from the leading edge,

$$C_{f,\mathrm{ave}} = 1.328\frac{1}{\sqrt{\mathrm{Re}_L}} \tag{4–98}$$

The *thermal boundary layer*,

$$\delta_T = \delta\frac{1}{\mathrm{Pr}^{1/3}} \tag{4–104}$$

where the *Prandtl Number* is

$$\mathrm{Pr} = \frac{\mu c_p}{\kappa} = \frac{\mu}{\alpha\rho} = \frac{\nu}{\alpha} \tag{4–60}$$

The convection heat transfer can be predicted through the *Nusselt Number*

$$\mathrm{Nu} = \frac{hL}{\kappa_{\mathrm{fluid}}} \tag{4–64}$$

which is, for laminar boundary layer fluid flow along a flat surface

$$\mathrm{Nu}_{\mathrm{ave}} = 0.664\ \mathrm{Re}_L^{1/2} \cdot \mathrm{Pr}^{1/3} \tag{4–106}$$

For small Prandtl Numbers, less than about 0.05,

$$\mathrm{Nu}_{\mathrm{ave}} = 0.565(\mathrm{Re}_L \cdot \mathrm{Pr})^{1/2} \tag{4–107}$$

For *turbulent* fluid flow in a boundary layer along a flat surface,

$$\delta = 0.37\ x/\mathrm{Re}_x^{1/5} \tag{4–110}$$

$$\delta_T = \delta \tag{4–111}$$

and

$$\mathrm{Nu}_x = 0.0296\ \mathrm{Re}_x^{4/5} \cdot \mathrm{Pr}^{1/3} \tag{4–112}$$

The average Nusselt Number is, using $\mathrm{Re}_{\mathrm{crit}} = 5 \times 10^5$,

$$\mathrm{Nu}_{\mathrm{ave}} = (0.037\ \mathrm{Re}_L^{4/5} - 871)\mathrm{Pr}^{1/3} \tag{4–116}$$

For very long surfaces, where turbulent boundary layer flow prevails

$$\mathrm{Nu}_{\mathrm{ave}} = 0.037\ \mathrm{Re}_L^{4/5} \cdot \mathrm{Pr}^{1/3} \tag{4–117}$$

If there is no identifiable leading edge use L of 0.6 m or 2 ft. In all cases, unless otherwise specified, use fluid properties at a *film temperature* of

$$T_f = \frac{1}{2}(T_\infty + T_s) \tag{4–118}$$

For the case where a portion of the leading surface is at the same temperature ($T_\infty = T_s$), and heat transfer occurs a distance x_H from the leading edge,

$$\text{(laminar)} \qquad \mathrm{Nu} = 0.332\frac{\mathrm{Re}_x^{1/2} \cdot \mathrm{Pr}^{1/3}}{\left[1 - \left(\frac{x_H}{x}\right)^{3/4}\right]^{1/3}} \tag{4–119}$$

$$\text{(turbulent)} \qquad \mathrm{Nu}_x = 0.0296\frac{\mathrm{Re}_x^{4/5}\cdot \mathrm{Pr}^{1/3}}{\left[1-\left(\frac{x_H}{x}\right)^{9/10}\right]^{1/9}} \qquad \textbf{(4–120)}$$

Convection heat transfer with fluid flow around a circular cylinder can be described by

$$\mathrm{Nu}_D = 0.3 + (0.62\,\mathrm{Re}_D^{1/2}\cdot \mathrm{Pr}^{1/3})\frac{[1+(\mathrm{Re}_D/282{,}000)^{5/8}]^{4/5}}{[1+(0.4/\mathrm{Pr})^{2/3}]^{1/4}} \qquad \textbf{(4–129)}$$

or

$$\mathrm{Nu}_D = C\,\mathrm{Re}_D^m\,\mathrm{Pr}^{1/3} \qquad \textbf{(4–128)}$$

Equation (4–128) can also be used for flow around noncircular objects if D is replaced by an effective D and c and m are empirically determined.

Convection heat transfer with fluid flow through ducts and tubes can be described for laminar flow ($\mathrm{Re}_D < 2000$) by

$$\mathrm{Nu}_D = 4.364 \qquad \text{for } \dot{q} \text{ uniform} \qquad \textbf{(4–148)}$$

$$\mathrm{Nu}_D = 3.657 \qquad \text{for } T_s \text{ uniform} \qquad \textbf{(4–149)}$$

For turbulent flow ($\mathrm{Re}_D > 10{,}000$, $160 > \mathrm{Pr} > 0.7$) the Dittus–Boelter Equation is often used,

$$\mathrm{Nu}_D = 0.023\,\mathrm{Re}_D^{0.8}\cdot \mathrm{Pr}^n \qquad \textbf{(4–151)}$$

Entrance conditions are described by use of the *Graetz Number*

$$\mathrm{Gz}_D = \frac{D}{x}\mathrm{Re}_D\cdot \mathrm{Pr} \qquad \textbf{(4–153)}$$

and the relationship between the Graetz Number the Nusselt Number needs to be given, as in figure 4–48a and b. The fluid properties need to be determined at the bulk fluid temperature

$$T_m = \frac{1}{\dot{m}c_p}\int_{\text{Area}} \rho\mathbf{u}\, c_p T_d A = \text{average temperature at a cross section} \qquad \textbf{(4–135)}$$

Flow through noncircular channels can be analyzed with the hydraulic diameter.

Heat transfer at tube banks can be analyzed with the equation

$$\mathrm{Nu}_e = C\,\mathrm{Re}_{D,\max}^m\cdot \mathrm{Pr}^{1/3} \qquad \textbf{(4–161)}$$

where C and m are affected by the tube arrangement and the maximum fluid velocity around the tube bank, $\mathbf{u}_{\max}$. For in-line tube arrangements the maximum velocity is given by

$$\mathbf{u}_{\max} = \mathbf{u}_\infty\frac{S_T}{S_T - D} \qquad \textbf{(4–163)}$$

and for staggered tube arrangements

$$\mathbf{u}_{\max} = \mathbf{u}_\infty\frac{S_t}{2(S_d - D)} \qquad \textbf{(4–164)}$$

or equation (4–163), whichever is greater. The pressure drop across a tube bank can be predicted by

$$\Delta p = j\chi f\left(\frac{\rho}{2g_c}(\mathbf{u}_{\max})^2\right) \qquad \text{(English units)} \qquad \textbf{(4–167)}$$

For packed beds the convection heat transfer is approximated in a manner similar to flow around tube banks. An equation suggested for determining the convection heat transfer coefficient of the fluid in the packed beds is

$$\frac{h_\infty}{\rho c_p \mathbf{V}_f}\mathrm{Pr}^{2/3} = 1.064\,\mathrm{Re}_D^{-0.41} \qquad (\mathrm{Re}_D > 350) \tag{4–173}$$

$$\frac{h_\infty}{\rho c_p \mathbf{V}_f}\mathrm{Pr}^{2/3} = 18.1\frac{1}{\mathrm{Re}_D} \qquad (\mathrm{Re}_D < 350)$$

Couette Flow is usually assumed to be laminar and the temperature distribution through the fluid is, if the two plates are at temperatures T_0 and T_1,

$$T(y) = T_0 + \frac{y}{L}\left[(T_1 - T_0) + \frac{\mu \mathbf{u}_\infty^2}{2\kappa}\left(1 - \frac{y}{L}\right)\right] \tag{4–177}$$

For the situation where the plates have the same temperatures,

$$T(y) = T_0 + \frac{y}{L}\frac{\mu \mathbf{u}_\infty^2 y}{2\kappa L}\left(1 - \frac{y}{L}\right) \tag{4–178}$$

or

$$\theta(y) = \theta_0 + \frac{y}{L}\left[\frac{\mu \mathbf{u}_\infty^2}{2\kappa}\left(1 - \frac{y}{L}\right)\right] \tag{4–182}$$

where $\theta(y) = T(y) - T_f$

DISCUSSION QUESTIONS

Section 4–1

4–1 What is h?

4–2 What is viscous dissipation?

4–3 Describe the difference between forced and free convection.

Section 4–2

4–4 What is meant by a free surface of a fluid?

4–5 What is viscosity?

4–6 What is meant by the term *dimensionally homogeneous*?

4–7 What is the difference between fundamental dimensions and derived dimensions?

Section 4–3

4–8 What is meant by a boundary layer?

4–9 What is a thermal boundary layer?

4–10 What is meant by a critical Reynolds Number?

4–11 What is hydrostatic pressure?

Section 4–4

4–12 The Blasius Solution to the laminar flow of fluid along a flat surface gives $\mathbf{u}/\mathbf{u}_\infty$ of 0.9916 at η of 5.0. Why is this usually given as the boundary layer thickness?

4–13 What is the Prandtl Number?

4–14 What is the film temperature?

4–15 What is the purpose of the Reynolds–Colburn Analogy?

Section 4–5

4–16 What is meant by a separation point?

4–17 What is meant by stagnation point?

4–18 What is the drag coefficient, C_D?

4–19 In figure 4–34 the drag coefficient C_D drops significantly at a Reynolds Number of about 4×10^5. What is the explanation for this?

Section 4–6

4–20 What is meant by an entrance region?

4–21 What is the bulk fluid temperature?

4–22 What is the Dittius–Boelter equation used for?

Section 4–7

4–23 What is meant by a packed bed?

4–24 Describe Couette Flow.

PRACTICE PROBLEMS

Section 4–2

4–1 Determine the wetted perimeter, wp, and the hydraulic diameter, D_H, for a circular tube filled with R-134 liquid as shown in figure 4–61.

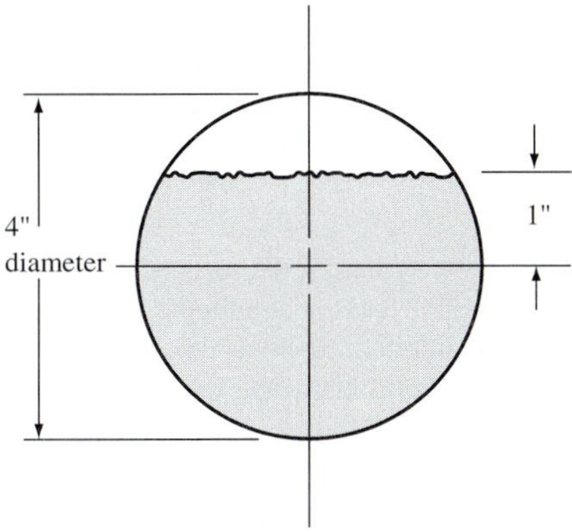

Figure 4–61

4–2 Determine the viscosity and kinematic viscosity for air at 300 K and 350 K. Express your answer in cP units, lbf · s/ft^2, and lbm/s · ft.

4–3 Determine the viscosity and kinematic viscosity of water at 300 K and 350 K.

4–4 SAE 50 oil at 70°F flows through a 2-ft-long copper tube of $^1/_8$-in. inside diameter. What must the pressure drop be through the tube if 10 lbm/min of oil must flow? Use $g = 32.2$ ft/s^2 and an oil density of 58 lbm/ft^3.

4–5 Water at 20°C flows at 5 m/s through a 2-cm-diameter tube. If the tube has roughness of 2×10^{-5} m, estimate the friction factor defined by equation 4–19.

4–6 Show that the friction factor f in the Darcy Equation (4–19) can be written as $f = 64/\text{Re}_D$ for laminar flow through a circular tube having diameter D.

Section 4–3

4–7 Syrup ($\rho = 1000$ kg/m^3) having viscosity of 1250 cP flows through a deep trough at 10 cm/s. Determine the Reynolds Number, Re_L, at a distance 2 m downstream from the leading edge. Is the flow in the boundary layer laminar or turbulent?

4–8 Determine the Prandtl Number, Pr, for the syrup of problem 4–7. The syrup has thermal conductivity of 1 W/mK and specific heat, c_p, of 4 kJ/kgK.

4–9 Determine the Prandtl Number of liquid potassium at 600 K.

4–10 Determine the Prandtl Number for helium at 200 K.

Section 4–4

4–11 Predict the boundary layer thickness as a function of the distance from the leading edge for steam at 573 K flowing at 10 m/s over a flat boiler surface.

4–12 The energy equation for flow between two parallel plates, neglecting flow in the y direction, is

$$\rho c_p \mathbf{u}(y)\frac{\partial T(x, y)}{\partial x} = \kappa \frac{\partial^2 T(x, y)}{\partial y^2}$$

Setting $\mathbf{U} = \mathbf{u}/\mathbf{u}_0$, $Y = y/L$, $X = \alpha x/\mathbf{u}_0 L^2$, and $\theta = (T(x, y) - T_0)/\Delta T$, where $\mathbf{u}_0$ and T_0 are the bulk velocity and temperature of the fluid between the two plates that are spaced L distance apart and having temperature difference ΔT, express the energy equation in dimensionless form; that is, in terms of θ, Y, X, and $\mathbf{U}$.

4–13 For laminar flow of a fluid over a flat plate the boundary layer thickness may be described by

$$\delta = 4.64\sqrt{\frac{\nu x}{\mathbf{u}_\infty}}$$

and the velocity profile by

$$\frac{\mathbf{u}}{\mathbf{u}_\infty} = \frac{3}{2}\frac{y}{\delta} - \frac{1}{2}\left(\frac{y}{\delta}\right)^2$$

For a fluid having viscosity μ write an expression for the shear stress on the plate as a function of x, the distance from the leading edge.

4–14 Show that the average drag coefficient for flow of a fluid over a flat surface, defined in section 4–4 as

$$C_{f,\text{ave}} = \frac{1}{L}\int_0^L C_f dx$$

may be written

$$C_{f,\text{ave}} = 1.328\frac{1}{\sqrt{\text{Re}_L}}$$

for the case of a laminar boundary layer.

4–15 A fluid at $T_\infty = 40$°C flows at $\mathbf{u}_\infty = 8$ m/s along a flat plate that is 3 m long. The plate is maintained at 100°C. Determine the local convective heat-transfer coefficient h_x and the average h_m, using 4×10^5 for the critical Reynolds Number, for
(a) Air at 100 kPa.
(b) Water at 100 kPa.

4–16 A humidifier water pan shown in figure 4–62 has a surface water temperature of 95°F when the surrounding air temperature is 80°F. If the pan is 2 ft square and the air moves horizontally over the water surface at 2 ft/s, estimate the average convective heat-transfer coefficient.

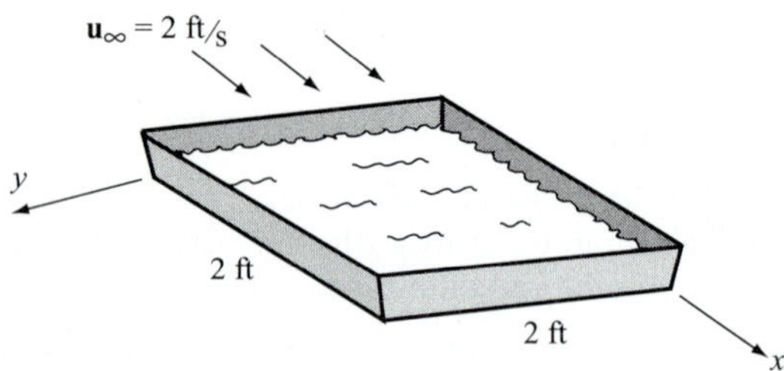

Figure 4–62

4–17 A wing of a particular airplane, shown in figure 4–63, is 10 m wide. During flight conditions air at 80 kPa and −10°C passes around the wing at 400 m/s. If the wing surface temperature is taken as 80°C, determine
(a) How much of the boundary layer flow is turbulent. Use $Re_{critical} = 5 \times 10^5$.
(b) $h_{average}$.
(c) Heat transfer per unit of wing length.

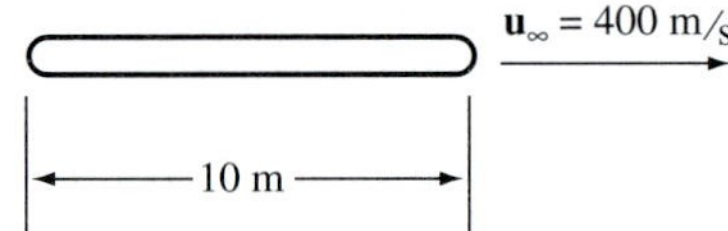

Figure 4–63

4–18 For the wing of problem 4–17, if the heat loss from the wing is expected to be 400 W/m^2 uniformly on both sides of the wing (upper and lower surfaces) and the wing leading edge is at 80°C, determine the temperature distribution along the wing.

4–19 Show that conservation of mass in two-dimensional flow (x and y) is satisfied by the stream function ψ (x, y), which satisfies the relationships

$$\mathbf{u} = \frac{\partial \psi}{\partial y} \quad \text{and} \quad \mathbf{v} = -\frac{\partial \psi}{\partial x}$$

Section 4–5

4–20 Determine the drag force acting per unit length in a 1-in.-diameter steel cable when air at 60°F, 14.7 psia moves across the cable at 30 mph.

4–21 A 6-cm-diameter cooling coil tube has a surface temperature of 5°C when air at 25°C and 100 kPa flows around it at 10 m/s. Estimate the location on the tube, referring to figure 4–64, where the maximum convection heat transfer occurs and determine its amount in W/m^2.

4–22 A cylindrical rod of 3 cm diameter has a surface temperature of 300°C when in a free stream of engine oil at 80°C moving at 12 m/s. Determine the heat transfer per unit length of the rod.

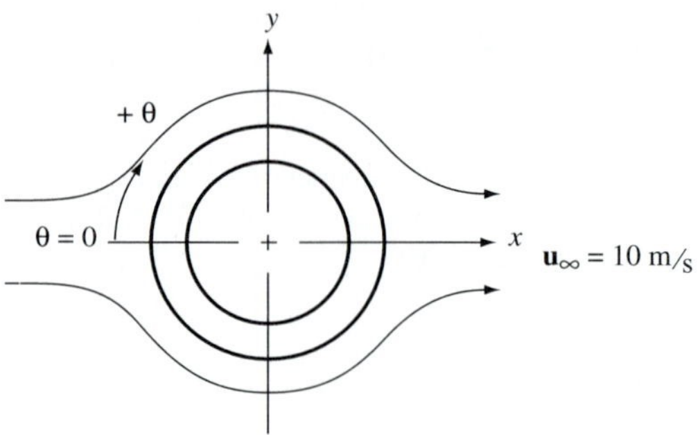

Figure 4–64

4–23 Estimate the heat loss due to convection from the vertical walls of a square apartment building if air at −10°C moves 30 km/h as shown in figure 4–65 across the building. The building has a uniform outside surface temperature of 10°C.

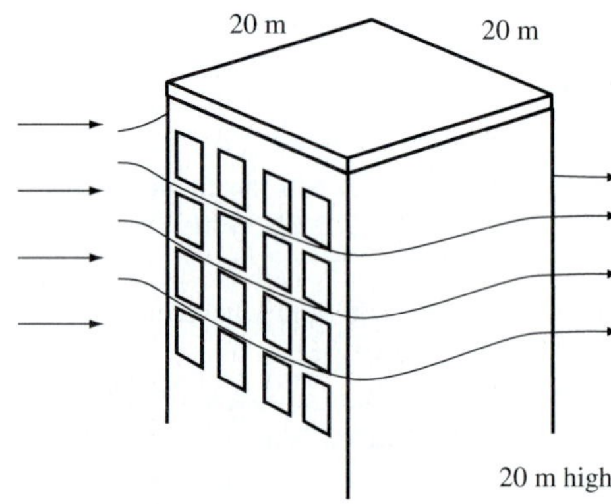

Figure 4–65

4–24 Air at 0°C blows across a 3-cm-diameter power cable at 20 m/s. Assuming that the outer surface of the cable is at 50°C, estimate
(a) Heat transfer by convection per unit of cable length.
(b) Drag force in newtons per meter of cable length.

4–25 Estimate the convective heat transfer coefficient for air at 101 kPa, 30°C passing around the square fin shown in figure 4–66. The fin is used to cool microprocessor chips in a computer and have a surface temperature of 90°C.

Section 4–6

4–26 For laminar flow in a circular tube the velocity is given by

$$\mathbf{v}(r) = \frac{\Delta p}{4\mu L}(r_0^2 - r^2)$$

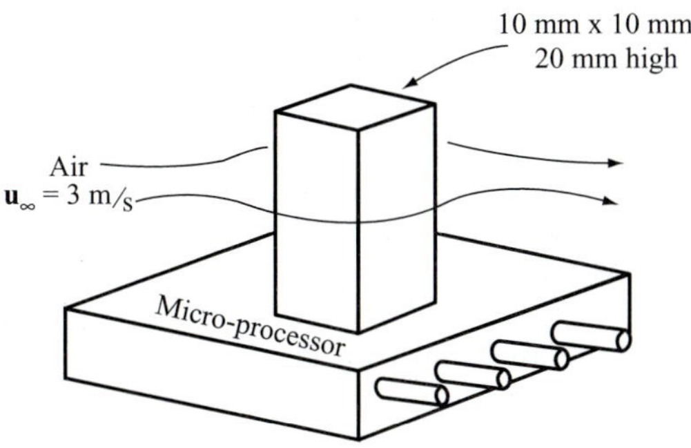

Figure 4–66

where r is the only variable in the equation. Show that the average velocity, defined as

$$\mathbf{v}_{\text{ave}} = \frac{1}{\int 2\pi r dr}\int \mathbf{v}(r) 2\pi r dr$$

becomes

$$\mathbf{v}_{\text{ave}} = \frac{r_0^2 \Delta p}{8L\mu}$$

4–27 Consider laminar flow of a fluid through a circular duct of radius R, diameter D, and constant temperature on the wall of T_w. The mean fluid temperature is T_∞ and the fluid temperature is a function of the radius, $T(r)$, expressed as

$$\theta(r) = \frac{T(r) - T_w}{T_\infty - T_w} = \frac{40}{7}\left[\frac{4}{25} + \frac{1}{25}\left(\frac{r}{R}\right)^2 - \frac{1}{5}\left(\frac{r}{R}\right)^3\right]$$

Using the concept of the convective heat-transfer coefficient $h(r)$ given by the equation

$$h(r) = -\kappa\left[\frac{\partial\theta}{\partial r}\right]_{r=R}$$

determine the Nusselt Number for this condition.

4–28 Water at 70°C flows through a 1-cm-ID tube at 6 kg/min. If the tube is used to cool the water, what is the fluid flow and thermal entrance length?

4–29 Engine oil at 100°F passes through a $^1/_8$-in.-ID tube at a rate of 50 ft/min. Estimate the heat transfer to the oil per unit area of the tube if the surface temperature is 150°F.

4–30 Air flows through a 1-cm-diameter tube at 8 m/s. The air is at 40°C and the tube is 100°C. If the tube is 3.5 m long,
(a) Determine the entrance length for the flow.
(b) Estimate the heat transfer to the air.

4–31 Air at 40°C passes through an 80-cm × 60-cm rectangular duct 80 m long at 1 m/s.
(a) Estimate the outlet temperature when the duct wall temperature is 15°C.
(b) Estimate the hydrodynamic and thermal entrance length for the duct.

4–32 The economizer stage of a steam generating unit is fabricated of stainless steel tubes having an ID of 2 cm. Saturated water at 80°C flows through the tubes at 200 g/s and the economizer is surrounded by exhaust gases at 300°C. Assume that the inside surfaces of the tubes are at 200°C.
(a) Determine the heat transfer to the water per unit length.
(b) After 10 years of operation the inside of the tubes become coated with a film so that the effective ID is 1.9 cm and the rms roughness is 0.2 mm. Predict the heat transfer per unit of length.

4–33 An office building is constructed with a 6-in. space between the floor and the ceiling of the next lower floor, as shown in figure 4–67. If air at 65°F passes through the space at 10 ft/s, determine the heat transfer per unit area between the air and the upper and lower surfaces at 75°F.

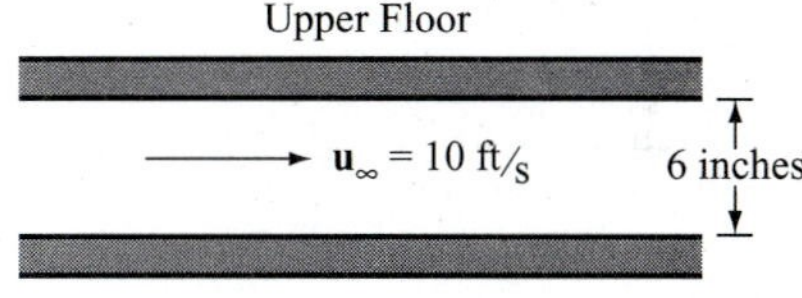

Figure 4–67

4–34 A flat plate solar collector heats water by means of a 1-in.-diameter copper tube. At a certain time the inside surface temperature of the tube is 115°F and the water enters at 50°F and is required to leave at 95°F with a flow rate of 140 ft^3/hr.
(a) Determine the Reynolds Number for the flow.
(b) Determine the length of tube needed.

4–35 Ethyl glycol at 90°C flows at 10 m/s through 0.2-cm-diameter heat exchanger tubes. If the tubes are arranged as shown in figure 4–68 and the temperature of the tube surfaces is 30°C at the entrance, predict the temperatures of the tube surfaces and the ethyl glycol at the outlet.

4–36 Air at 14.7 psia and 85°F enters a square 1-ft × 1-ft heating duct at 20 ft/s. If the duct is 80 ft long and has an inside surface temperature of 65°F:
(a) Determine the hydraulic diameter of the duct.
(b) Determine the convective heat-transfer coefficient.
(c) Estimate the air temperature leaving the duct.

Section 4–7

4–37 A tube bank consists of 30 columns of 3-cm-diameter tubes spaced 4.5 cm apart (center-to-center) in-line both ways. Ethylene glycol at 90°C is to be cooled to 60°C after passing over the bank at 2 m/s. Estimate the tube temperature.

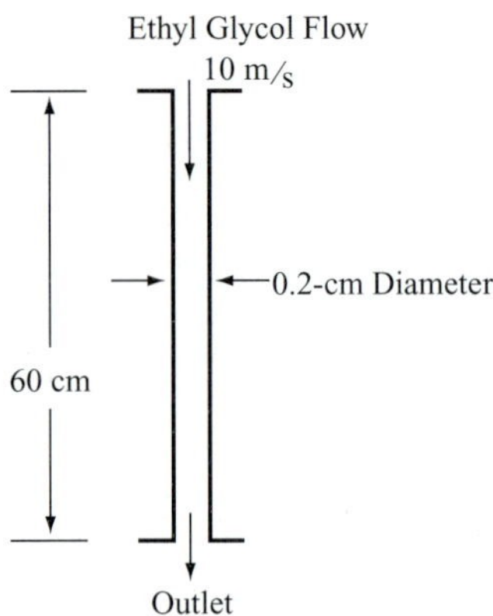

Figure 4–68

4–38 Hot flue gases at 700°C pass over a bank of superheater tubes at 6 m/s. The tube bank is composed of 20 columns of 2-cm-diameter tubes with 40 tubes per column. The arrangement of the tubes is staggered, as shown in figure 4–69, with $S_T = S_D = 4.0$ cm. If the flue gases have the same properties as air, determine the pressure drop across the tube bundles, assuming that the tubes have an average surface temperature of 500°C.

4–39 A heat exchanger uses a staggered tube arrangement of 1-in.-diameter tubes and with $S_D = S_T = 3$ in. There are 4 columns of 10 tubes per column, and all of the tube surfaces are at 190°F. Atmospheric air at 80°F and 14.5 psia is forced across the tube bank at 15 ft/s.

(a) Determine the air temperature downstream from the heat exchanger.

(b) Determine the downstream pressure.

4–40 A rock bed used for thermal storage has washed rocks $^1/_2$ in. in nominal diameter in a bed 5 ft high, 5 ft wide, and 8 ft long, and 1000 cubic feet per minute (cfm) of air pass vertically through the bed. Estimate the convective heat-transfer coefficient for the rock bed under this flow condition.

4–41 A journal bearing and rotating shaft shown in figure 4–70 is lubricated with oil having viscosity of 800 cP. If the shaft is rotating at 2000 rpm, determine the maximum oil temperature when the shaft and bearing are at 60°C.

4–42 A lapping disk travels at 30 m/s over a surface to be lapped. The lapping compounds acts as water with the same properties as water. Determine the maximum fluid temperature. *Hint:* A maximum occurs when $\partial T/\partial y = 0$. See figure 4–71.

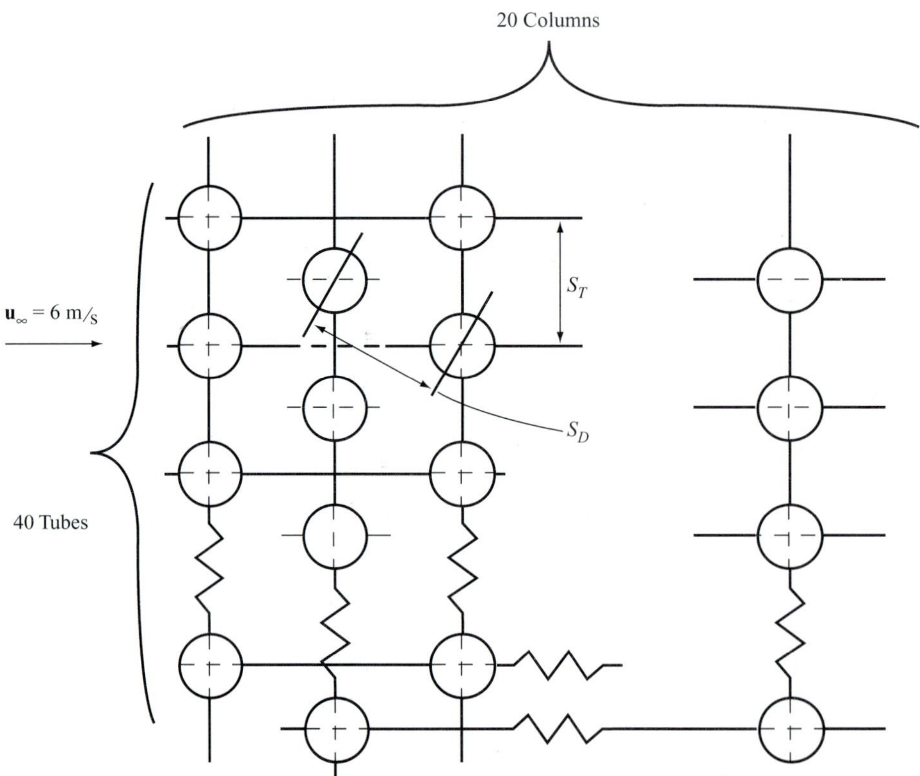

Figure 4–69 Staggered tube bundle for superheater.

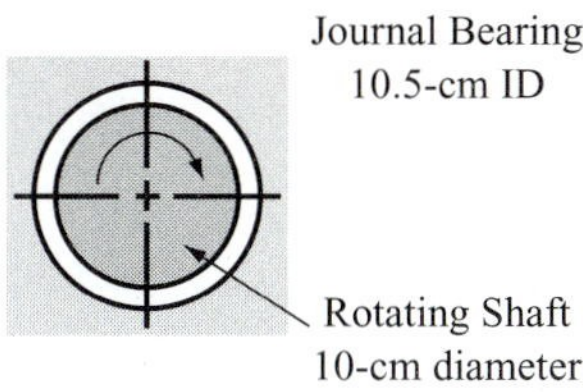

Figure 4–70

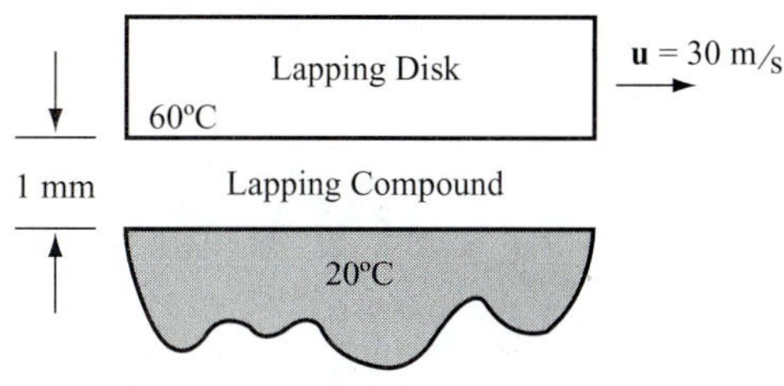

Figure 4–71 Lapping process, problem 4–42.

REFERENCES

1. Van Dyke, M., *An Album of Fluid Flow,* Parabolic Press, Stanford, CA, 1982.
2. Moody, L. F., Friction Factors for Pipe Flow, *Trans. ASME,* 66, 671–677, 1944.
3. Schlichting, H., *Boundary Layer Theory,* 6th edition (transl. by J. Kestin), McGraw-Hill, New York, 1968.
4. Incropera, F. P., and D. P. DeWitt, *Introduction to Heat Transfer,* 2nd edition, John Wiley & Sons, New York, 1990.
5. Reynolds, O., On the Extent and Action of the Heating Surfaces for Steam Boilers, *Proc. Manchester Lit. Phil. Soc.,* 144, 7, 1874.
6. Colburn, A. P., A Method of Correlating Forced Convection Heat Transfer Data and a Comparison with Fluid Friction, *Trans. AIChE,* 29, 174–210, 1933.
7. Blasius, H., Grenzschichten in Flüssigkeiten mit kleiner Reibung. *Z. Math. u. Phys.* 56, 1–37, 1908. (English transl. in NACA TM 1256.)
8. Howarth, L., On the Solution of the Laminar Boundary Layer Equations, *Proc. Roy. Soc. London,* A164, 547–579, 1938.
9. Pohlhausen, E., Der Wärmeaustausch zwischen festen Körpern und Flüssigkeiten mit kleiner Reibung und kleiner Wärmeleitung, *Z. Angew. Math. Mech.* 1, 115, 1921.
10. Churchill, S. W., and H. Ozoe, Correlations for Laminar Forced Convection with Uniform Heating in Flow over a Plate and in Developing and Fully Developed Flow in a Tube, *J. Heat Transfer,* 95, 78, 1973.
11. Brown, A. I., and S. M. Marco, *Introduction to Heat Transfer,* 3rd edition, McGraw-Hill, New York, 1958, p. 155.
12. Churchill, S. W., A Comprehensive Correlating Equation for Forced Convection from Flat Plates, *AIChEJ,* 22, 2, 264–268, 1976.
13. Eiffel, G., Sur la résistance des sphères dans l'air en mouvement. *Comptes Rendus,* 155, 1597, 1912.
14. Schmidt, E., and K. Wenner, Wärmeabgabe über den Umfang eines angeblasenen geheizten Zylinders, *Forsch. Ingenieurwes.,* 12, 65–73, 1941.
15. Hilpert, R., Wärmeabgabe von geheizten Drähten und Rohren, *Forsch. Geb. Ingenieurwes.,* 4, 215, 1933.
16. Knudsen, J. D., and D. L. Katz, *Fluid Dynamics and Heat Transfer,* McGraw-Hill, New York, 1958.
17. Jakob, M., *Heat Transfer,* vol. 1, John Wiley & Sons, New York, 1949.
18. Churchill, S. W., and M. Bernstein, A Correlating Equation for Force Convection from Gases and Liquids to a Cylinder in Crossflow, *J. Heat Transfer,* 99, 300, 1977.
19. Morgan, V. T., The Overall Convection Heat Transfer from Smooth Circular Cylinders, in T. F. Irvine, Jr. and J. P. Hartnett, eds., *Advances in Heat Transfer,* vol. 11, Academic Press, New York, 1975.
20. McAdams, W. H., *Heat Transmission,* 3rd edition, McGraw-Hill, New York, 1954.
21. Whitaker, S., Forced Convection Heat Transfer Correlations for Flow in Pipes, Past Flat Plates, Single Cylinders, Single Spheres, and for Flow in Packed Beds and Tube Bundles, *AIChEJ,* 18, 361–371, 1972.
22. Langhaar, H. L., Steady Flow in Transition Length of Straight Tube, *J. Appl. Mech., ASME,* 9, 2, A55–A58, 1942.
23. Kays, W. M., and M. E. Crawford, *Convection Heat and Mass Transfer,* McGraw-Hill, New York, 1980.
24. Petukhov, B. S., Heat Transfer and Friction in Turbulent Pipe Flow with Variable Physical Properties, in J. P. Hartnett and T. F. Irvine, Jr., eds., *Advances in Heat Transfer,* vol. 6, Academic Press, New York, 1970.
25. Grimison, E. D., Correlation and Utilization of New Data on Flow Resistance and Heat Transfer for Cross Flow of Gases over Tube Banks, *Trans. ASME,* 59, 583–594, 1937.
26. Kays, W. M., and R. K. Lo, Basic Heat Transfer and Flow Friction Design Data for Gas Flow Normal to Banks of Staggered Tubes—Use of a Transient Technique, Stanford University Technical Report no. 15, 1952.
27. Zhukauskas, A., Heat Transfer from Tubes in Cross Flow, in J. P. Hartnett and T. F. Irvine, Jr., eds., *Advances in Heat Transfer,* vol. 8, Academic Press, New York, 1972.
28. Gamson, B. W., G. Thodos, and O. A. Hougens, Heat, Mass, and Momentum Transfer in the Flow of Gases Through Granular Solids, *Trans. AIChE,* 39, 1943.
29. Cole, R. L., and K. J. Nield, R. R. Rohde, and R. M. Wobsewicz, *Design and Installation Manual for Thermal Energy Storage,* 2nd edition, U.S.D.O.E. ANL-79-15, January, 1980.
30. Green, L., Jr., Heat, Mass, and Momentum Transfer in Flow Through Porous Media, *Trans. ASME,* 57-HT-19, 1957.
31. Kays, W. M., and A. L. London, *Compact Heat Exchangers,* 3rd edition, McGraw-Hill, New York, 1984.

5 FREE CONVECTION HEAT TRANSFER

In this chapter the concepts of free or natural convection are considered. First a review or reminder of the mechanisms of bouyancy and the resulting chimney effects is presented. Then convection heat transfer along a flat vertical surface is discussed and analyzed. This presentation includes the introduction of the *Grashof Number*. The analysis is then extended to free convection at horizontal and inclined surfaces. Free convection around submerged objects such as horizontal and vertical cylinders and tube banks is considered. Then natural convection within enclosed spaces is presented, followed by a presentation of combined forced and free convection. Finally, the chapter closes with a compilation of simplified free convection heat-transfer equations for air as the fluid.

New Terms

β_P isobaric compressibility

Gr Grashof Number

5–1 THE GENERAL CONCEPTS OF FREE CONVECTION

Free convection is the phenomenon where fluid motion is caused by a difference in specific weight through the fluid volume. Free convection can only occur if there is a gravitational field, such as on the earth. A fluid having a lower specific weight (or density) than that of a surrounding fluid will tend to move upward or rise in the gravitational field, and a fluid with greater specific weight will tend to descend. The mechanism whereby the sun is able to affect high and low atmospheric pressures on the earth and cause wind currents is a direct consequence of the phenomenon of free convection.

The mechanism of free convection can be explained through the concept of the buoyant force, which is defined as the weight of a submersed body (or fluid volume) if it had the specific weight of the surrounding fluid. Thus, as shown in figure 5–1, the body that

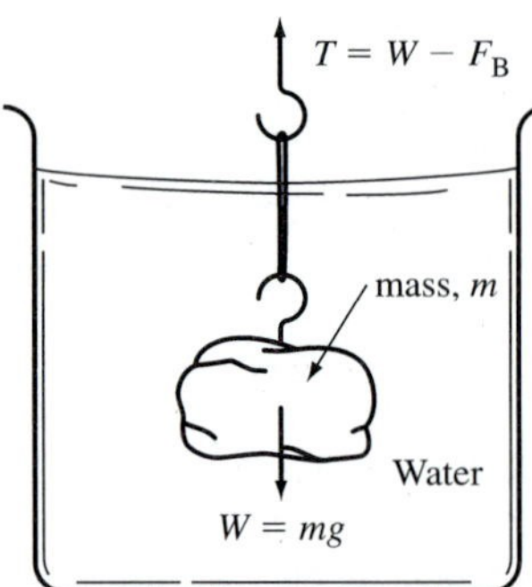

Figure 5–1 Example of buoyancy.

is hanging from a cord in a tank of a fluid will exert a force on the cord equal to the body weight, W, minus the buoyant force that is acting upward on the body. The buoyant force is equal to the specific weight of the fluid times the volume occupied by the body. In this example it is assumed that the weight of the body is more than the buoyant force acting on the body; otherwise the body would float in the fluid. An important and instructive engineering application of the action of buoyancy is the chimney effect. Consider a vertical chimney, or stack, as shown in figure 5–2 filled with air or a gas at some temperature T_s; a temperature that is greater than the surrounding air temperature at T_∞. The weight of the gas in the stack is

$$W_s = \gamma_s V = \gamma_s A_s L \qquad \textbf{(5–1)}$$

Figure 5–2 The chimney effect.

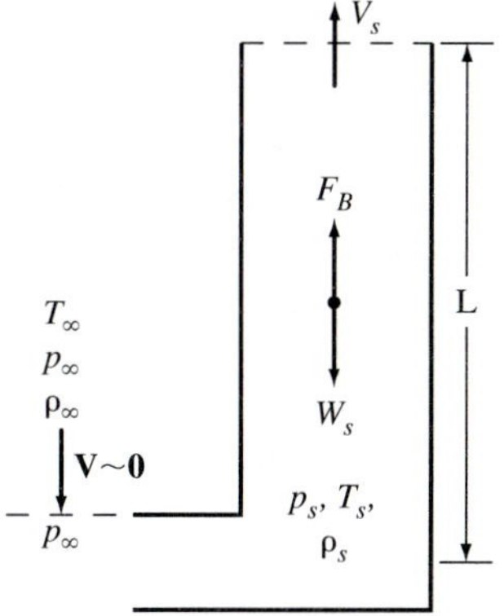

The gas is assumed to be an ideal gas so that the specific weight is given by the equation

$$\gamma_s = g\rho_s = \frac{gp_s}{R_s T_s} \qquad \textbf{(5–2)}$$

For many situations where T_s does not vary too much through the stack the weight of the gases in the stack is

$$W_s = \frac{gp_s}{R_s T_s} A_s L \qquad \textbf{(5–3)}$$

The buoyant force acting on the column of stack gas is

$$F_B = \gamma_\infty V = \frac{gp_\infty}{R_\infty T_\infty} A_s L \qquad \textbf{(5–4)}$$

Then the sum of the vertical forces acting on the gases for steady flow of gases through the stack, discounting any friction on the sides of the stack, is

$$\sum F_{\text{vertical}} = 0 = p_s A_s + F_B - W_s - p_\infty A_s \qquad \textbf{(5–5)}$$

and substituting the results from equations (5–3) and (5–4) gives

$$0 = p_s A_s - p_\infty A_s + \frac{gp_\infty}{R_\infty T_\infty} A_s L - \frac{gp_s}{R_s T_s} A_s L$$

which can then be rearranged to give

$$p_\infty - p_s = gL\left(\frac{p_\infty}{R_\infty T_\infty} - \frac{p_s}{R_s T_s}\right)$$

If the stack gases and the surrounding air have the same gas constant (i.e., $R_\infty = R_s = R$) and the pressures are not too greatly different, then this equation is

$$p_\infty - p_s = \frac{g p_\infty L}{R}\left(\frac{1}{T_\infty} - \frac{1}{T_s}\right) \qquad \text{(SI units)}$$

$$p_\infty - p_s = \frac{g p_\infty L}{R g_c}\left(\frac{1}{T_\infty} - \frac{1}{T_s}\right) \qquad \text{(English units)} \tag{5–6}$$

We may also apply Bernoulli's Equation to the gas flow through the stack and obtain

$$p_\infty = p_s + \rho_s \frac{\mathbf{V}_s^2}{2} \tag{5–7}$$

where $\mathbf{V}_s$ is the velocity of the stack gases. Rearranging equation (5–7) and comparing it to equation (5–6) we get

$$\frac{1}{2}\mathbf{V}_s^2 = \frac{g p_\infty L}{R \rho_s}\left(\frac{1}{T_\infty} - \frac{1}{T_s}\right)$$

and solving for the stack gas velocity, with $\rho_s = p_\infty / RT_s$, we have

$$\mathbf{V}_s = \sqrt{2gLT_s\left(\frac{1}{T_\infty} - \frac{1}{T_s}\right)} \tag{5–8}$$

EXAMPLE 5–1 A chimney effect can occur in any open vertical passageway. Estimate the velocity of air that may be induced in an open stairwell of a 20-story building (12 ft per story) during a fire when the air temperature is 300°F. Assume that surrounding air is at 60°F and 14.7 psia.

Solution From equation (5–8) we have

Answer

$$V_s = \sqrt{2(32.17\ \text{ft/s}^2)(20 \times 12\,\text{ft})(760°\text{R})\left[\frac{1}{520°\text{R}} - \frac{1}{760°\text{R}}\right]} = 84.4\ \text{ft/s} = 57.6\ \text{mph}$$

It can be seen that the buoyancy and velocity will increase if the vertical height L is larger. This is what makes open stairwells of tall buildings dangerous as fire escapes, and it is also one reason chimneys and stacks are made tall: to have an effective free convection of gases up the stack. We also see that the velocity of free convection may in general be written, after Gebhart et. al. [1],

$$\mathbf{V} = f\left(\sqrt{g}, \sqrt{L}, \sqrt{\frac{d\rho}{\rho}}\right) \tag{5–9}$$

Consider a differential element of a fluid in a field; as shown in figure 5–3, the gradient of the pressure in the fluid, dp/dy, is greater in the stack than in the surroundings. That is,

$$\frac{dp_s}{dy} > \frac{dp_\infty}{dy} \qquad \text{for } p_s > p_\infty$$

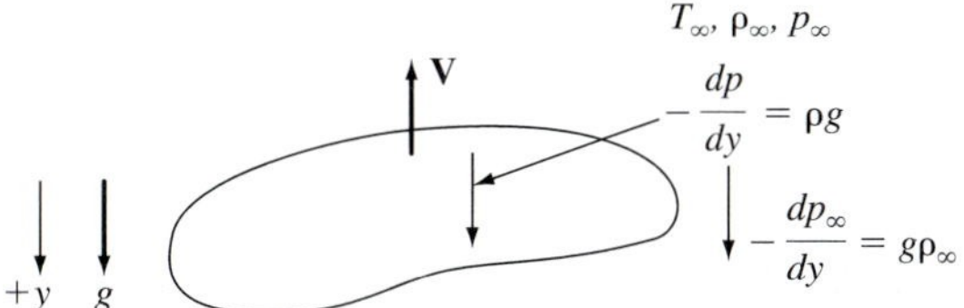

Figure 5–3 Action of the pressure gradient in a stack.

It can be seen that the surrounding pressure will eventually become greater than the fluid pressure and thereby induce the free convection demonstrated in example 5–1. It is crucial to recognize that the surrounding conditions for quantifying free convection need to be those of the fluid that will eventually be able to replace the buoyancy fluid. Heat transfer is necessary to keep free convection in a steady state condition. The fluid occupying the defined control volume of figure 5–3 needs to be kept at a temperature T_c, and this requires thermal energy. Typically this is heat transfer from an immediately adjacent surface at temperature T_s, and this affects the convection heat transfer to (or from) the fluid. The free convection and heat transfer can be in either direction, as we will see in the next section.

5–2 ANALYSIS OF FREE CONVECTION ALONG VERTICAL SURFACES

We have seen in Chapter 4 that the Reynolds Number provides a quantitative method of describing the character of fluid flow. The character of free convection likewise may be described by a free convection Reynolds Number, defined as

$$\mathrm{Re}_{\mathrm{fc}} = \frac{\mathbf{V}L}{\nu} \qquad \textbf{(5–10)}$$

Using equation (5–9) we have

$$\mathrm{Re}_{\mathrm{fc}} = f\left(\sqrt{gL\frac{\Delta\rho}{\rho}}, \frac{L}{\nu}\right)$$

or

$$\mathrm{Re}_{\mathrm{fc}} = f\left(\sqrt{gL^3\frac{\Delta\rho}{\rho\nu^2}}\right) \qquad \textbf{(5–11)}$$

The Grashof Number, Gr, is

$$\mathrm{Gr} = gL^3\frac{d\rho}{\rho\nu^2} = \mathrm{Re}_{\mathrm{fc}}^{\;2} \qquad \textbf{(5–12)}$$

and this is used to describe fluid flow in free convection. The Grashof Number is often defined as the ratio of the buoyant force ($gL\Delta\rho \cdot A_s$) to the viscous force ($\mu \cdot L \cdot A_s$). Through the methods of divisional analysis it can also be shown that the Grashof Number, a dimensionless number, will arise, and it is sometimes useful to write the Grashof Number as

$$\mathrm{Gr} = g\frac{L^3}{\nu^2}\frac{\Delta\rho}{\rho} \qquad \textbf{(5–13)}$$

which includes two dimensionless numbers, gL^3/ν^2 and $\Delta\rho/\rho$. The isobaric compressibility (constant pressure) of a substance is defined as

$$\beta_P = -\frac{1}{\rho}\left[\frac{\partial\rho}{\partial T}\right]_P \qquad \textbf{(5–14)}$$

which is, as a first approximation,

$$\beta_P \approx \frac{\Delta\rho}{\rho\Delta T} \tag{5–15}$$

Therefore, equation (5–15) gives

$$\frac{\Delta\rho}{\rho} = \beta_P \Delta T$$

and the Grashof Number can then be written

$$\text{Gr} = \frac{gL^3\Delta T}{\nu^2}\beta_P \tag{5–16}$$

For free convection heat transfer the temperature difference, ΔT, is defined as

$$\Delta T = |T_s - T_\infty| \tag{5–17}$$

where T_s is the surface temperature. The properties of the fluid or gas in free convection are usually evaluated at the film temperature of the fluid, T_f. The film temperature is the arithmetic average of T_s and T_∞. Further, from the results of Forced convection analysis of Chapter 4 we have that the Nusselt Number in free convection is a function of the Grashof and Prandtl Numbers.

$$\text{Nu} = f(\text{Gr}, \text{Pr}) \tag{5–18}$$

and the Gr · Pr product is called the Rayleigh Number, Ra:

$$\text{Ra} = \text{Gr} \cdot \text{Pr} \tag{5–19}$$

The characteristic length L is the vertical height of the stack or chimney. For horizontal or inclined surfaces the characteristic length is defined as

$$L = \frac{A_s}{P} \tag{5–20}$$

where P is the perimeter of the area A_s.

EXAMPLE 5–2 Determine the Grashof Number for water at 300 K at the bottom and along the sides in an open boiler shown in figure 5–4. Assume the water is saturated liquid and the boiler surfaces are at 450 K. Assume atmospheric pressure of 101 kPa.

Figure 5–4 Cylindrical boiler.

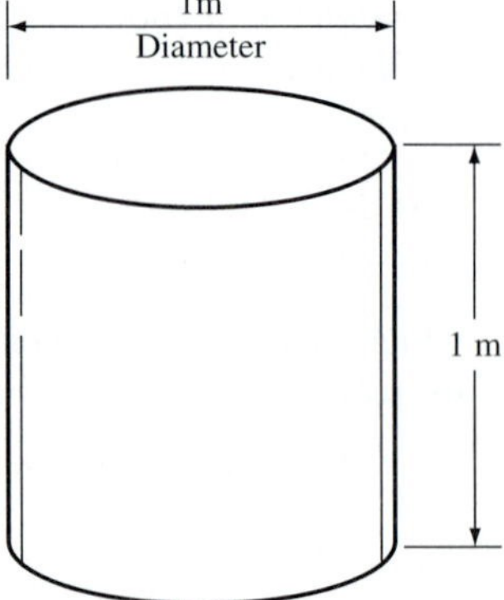

Solution

The film temperature of the water is

$$T_f = \frac{1}{2}(T_s + T_\infty) = \frac{1}{2}(300\text{K} + 450\text{K}) = 375\text{K}$$

From appendix table B–6 we interpolate

$$\beta_P = 761 \times 10^{-6}\ 1/\text{K at } 375\ \text{K}$$

$$\nu = \frac{\mu}{\rho} = \frac{0.274\ \text{g/m}\cdot\text{s}}{957\ \text{kg/m}^3} = 2.86 \times 10^{-7}\text{m}^2/\text{s}$$

The characteristic length of the bottom surface is

$$L = \frac{A_s}{P} = \frac{\pi D^2}{4\pi D} = \frac{D}{4} = 0.25\ \text{m}$$

For the sides,

$$L = 1\ \text{m}$$

Then the Grashof Numbers are

$$\text{Gr}_{\text{bottom}} = \frac{gL^3\Delta T}{\nu^2}\beta_P = \frac{(9.8\ \text{m/s}^2)(0.25\ \text{m})^3(450 - 300\ \text{K})}{(2.86 \times 10^{-7}\ \text{m}^2/\text{s})^2}(761 \times 10^{-6}\ 1/\text{K})$$

Answer

$$= 21.37 \times 10^{10}$$

For the sides,

Answer

$$\text{Gr}_{\text{sides}} = \frac{(9.8)(1^3)(150)}{(2.86 \times 10^{-7})^2}(761 \times 10^{-6}) = 1.37 \times 10^{13}$$

The isothermal compressibility β_P for ideal gases can be shown to be (see practice problem 5–8)

$$\beta_P = \frac{1}{T} \qquad \text{(Ideal gas)} \tag{5–21}$$

and the Grashof number then reduces to, for ideal gases,

$$\text{Gr}_{\text{Ideal gas}} = \frac{gL^3\Delta T}{\nu^2 T_f} \tag{5–22}$$

The flow of fluids in free convection conditions can be classified as laminar or turbulent, and the heat transfer will be significantly affected by the flow conditions. It is usually assumed that, for vertical surfaces, the following criteria hold:

$$\begin{aligned} &\text{Laminar flow at Gr} \cdot \text{Pr} = \text{Ra} < 10^9 \\ &\text{Turbulent flow at Gr} \cdot \text{Pr} = \text{Ra} \geq 10^9 \end{aligned} \tag{5–23}$$

It is useful to analyze the boundary layer flow in free convection heat transfer by considering a vertical surface as shown in figure 5–5. The y direction (vertical) momentum equation (4–40) can be reduced to the form

$$\rho\left(\mathbf{v}\frac{\partial \mathbf{v}}{\partial y} + \mathbf{u}\frac{\partial \mathbf{v}}{\partial x}\right) = -\rho g - \frac{\partial p}{\partial y} + \mu\frac{\partial^2 \mathbf{v}}{\partial y^2} \tag{5–24}$$

For the boundary layer we may also write that

$$\frac{\partial p}{\partial y} = -\rho_\infty g \tag{5–25}$$

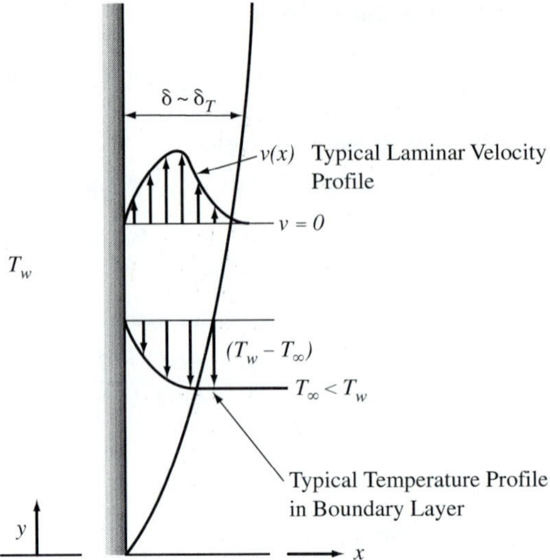

Figure 5–5 Boundary layer in free convection of a fluid along a vertical surface.

where the density is that of the free stream fluid. The momentum in the y direction in the boundary layer is then

$$\rho\left(\mathbf{v}\frac{\partial \mathbf{v}}{\partial y} + \mathbf{u}\frac{\partial \mathbf{v}}{\partial x}\right) = -(\rho - \rho_\infty)g + \mu\frac{\partial^2 \mathbf{v}}{\partial y^2} \tag{5–26}$$

From equation (5–13) we have, for an ideal gas,

$$(\rho - \rho_\infty) = -\beta_P(T - T_\infty)\rho$$

and substituting this into equation (5–26) gives

$$\rho\left(\mathbf{v}\frac{\partial \mathbf{v}}{\partial y} + \mathbf{u}\frac{\partial \mathbf{v}}{\partial x}\right) = \beta_P\rho g(T - T_\infty) + \mu\frac{\partial^2 \mathbf{v}}{\partial y^2} \tag{5–27}$$

or

$$\mathbf{v}\frac{\partial \mathbf{v}}{\partial y} + \mathbf{u}\frac{\partial \mathbf{v}}{\partial x} = g\beta_P(T - T_\infty) + \nu\frac{\partial^2 \mathbf{v}}{\partial y^2} \tag{5–28}$$

Using equation (5–28) along with the conservation of mass, equation (4–33),

$$\frac{\partial \mathbf{u}}{\partial x} + \frac{\partial \mathbf{v}}{\partial y} = 0$$

the conservation of energy,

$$\mathbf{v}\frac{\partial T}{\partial y} + \mathbf{u}\frac{\partial T}{\partial x} = \alpha\frac{\partial^2 T}{\partial y^2}$$

and associated boundary conditions,

B.C. 1	$\mathbf{u} = 0, \mathbf{v} = 0$	for $T = T_w$	at $y = 0$
B.C. 2	$T = T_\infty$		as $y \to \infty$

a solution can be obtained by exact methods or approximation methods.

For an ideal gas and where the Prandtl Number is approximately equal to 1.0, it can be shown that the thermal boundary layer and the momentum boundary layer have nearly the same thickness, referring to figure 5–5, $\delta = \delta_T$. The boundary layer thickness is a function of the height, y, and the temperature profile inside the boundary layer is

$$T(x, y) = T_\infty + (T_s - T_\infty)\left(1 - \frac{x}{\delta(y)}\right)^2 \tag{5–29}$$

The velocity profile inside the boundary layer is approximated by the equation

$$\mathbf{v} = \mathbf{v}_0\left(\frac{x}{\delta(y)}\right)\left(1 - \frac{x}{\delta(y)}\right)^2 \tag{5–30}$$

where $\mathbf{v}_0$ is the maximum velocity in the boundary layer and is proportional to the square root of the vertical height, y; that is, $\mathbf{v}_0 = f(\sqrt{y})$. This can then be written as

$$\mathbf{v}_0 = C_0\sqrt{y} \tag{5–31}$$

where C_0 is a constant to be determined. It can be shown (see Holman [2]) that this expression becomes

$$\mathbf{v}_0 = \sqrt{\mathrm{Gr}_y}\; 5.17\nu/y\sqrt{\left(\frac{20}{21} + \frac{\nu}{\alpha}\right)} \tag{5–32}$$

where the Grashof Number is evaluated locally, or

$$\mathrm{Gr}_y = \frac{gy^3\Delta T}{\nu^2}\beta_P = \left(\frac{gy^3\Delta T}{\mu^2}\right)\rho^2\beta_P$$

The boundary layer thickness δ will be affected by whether the flow is laminar or turbulent in the boundary layer. For laminar flow the Rayleigh number $\mathrm{Ra} = \mathrm{Gr} \cdot \mathrm{Pr} < 10^9$ and the boundary layer may be approximated by the equation

$$\delta(y) = y\left(\frac{3.93(0.952 + \mathrm{Pr})^{1/4}}{\mathrm{Pr}^{1/2} \cdot \mathrm{Gr}_y^{1/4}}\right) \tag{5–33}$$

The local Nusselt Number, Nu_y, is defined as

$$\mathrm{Nu}_y = \frac{h_y y}{\kappa} \tag{5–34}$$

The heat transfer at the surface, where $x = 0$, may be written

$$\dot{q}_A = -\kappa\left[\frac{\partial T}{\partial x}\right]_{x=0} = h_y(T_s - T_\infty) \tag{5–35}$$

and if we combine equations (5–34) and (5–35) we obtain for the Nusselt Number,

$$\mathrm{Nu}_y = -\frac{y}{T_s - T_\infty}\left[\frac{\partial T}{\partial x}\right]_{x=0} \tag{5–36}$$

From the temperature distribution equation (5–29) we have

$$\left[\frac{\partial T}{\partial x}\right]_{x=0} = -2\frac{T_s - T_\infty}{\delta} \tag{5–37}$$

and the Nusselt Number then is

$$\mathrm{Nu}_y = \frac{2y}{\delta} \tag{5–38}$$

From equation (5–33) we find, for laminar flow along a vertical surface,

$$\mathrm{Nu}_y = 0.508\sqrt{\mathrm{Pr}}(\mathrm{Gr}_y^{1/4})\frac{1}{(0.952 + \mathrm{Pr})^{1/4}} \tag{5–39}$$

This result agrees with the previous result for convective heat transfer obtained from dimensional analysis. The average convective heat-transfer coefficient over the full height, L, of a vertical surface is obtained by the usual defining equation for averaging,

$$h_{\mathrm{ave}} = \frac{1}{L}\int_0^L h_y dy \tag{5–40}$$

From equation (5–39) we have

$$h_{\mathrm{ave}} = \frac{4}{3}h_L \tag{5–41}$$

where h_L is the local value from equation (5–39) obtained at $y = L$. Thus, the average value may be written

$$h_{\mathrm{ave}} = \frac{4\kappa}{3L}(0.508)\sqrt{\mathrm{Pr}}\left(\frac{\mathrm{Gr}_L}{0.952 + \mathrm{Pr}}\right)^{1/4} \tag{5–42}$$

for $\mathrm{Gr}_L \cdot \mathrm{Pr} < 10^9$. Other equations based on empirical results have been suggested. One of these is the general equation suggested by McAdams [3]

$$h_{\mathrm{ave}} = C\frac{\kappa}{L}(\mathrm{Gr}_L \cdot \mathrm{Pr})^n \tag{5–43}$$

where n and C are to be determined from table 5–1.

Table 5–1 Constants C and n for Equation (5–43)

Flow condition	Gr · Pr	C	n	Reference
Laminar	10^4 to 10^9	0.59	1/4	[3]
Turbulent	10^9 to 10^{13}	0.10	1/3	[4, 5]

Other correlations suggested by Churchill and Chu [6] are, for laminar flow

$$h_{\mathrm{ave}} = \frac{\kappa}{L}\left(\frac{0.68 + 0.67(\mathrm{Gr} \cdot \mathrm{Pr})^{1/4}}{[1 + (0.492/\mathrm{Pr})^{9/16}]^{4/9}}\right) \tag{5–44}$$

for all Prandtl Numbers and for the condition $10^{-1} < \mathrm{Gr}_L \cdot \mathrm{Pr} < 10^9$. A more general equation including laminar and turbulent flow and for all Prandtl Numbers is

$$h_{\mathrm{ave}} = \frac{\kappa}{L}\left(\frac{0.825 + 0.387(\mathrm{Gr}_L \cdot \mathrm{Pr})^{1/6}}{[1 + (0.492/\mathrm{Pr})^{9/16}]^{8/27}}\right)^2 \tag{5–45}$$

EXAMPLE 5–3

A south-facing 40-m-wide wall 15 m high is heated during a still day (no wind) by sunlight, and its surface temperature reaches 36°C at 3 o'clock in the afternoon. Estimate the heat transfer by free convection from the wall if the air temperature is 10°C.

Solution

We first determine the film temperature of the air:

$$T_f = \frac{1}{2}(T_s + T_\infty) = \frac{1}{2}(36°\mathrm{C} + 10°\mathrm{C}) = 23°\mathrm{C} = 296\ \mathrm{K}$$

The Grashof Number is

$$\mathrm{Gr}_L = \frac{gL^3\Delta T}{\mu^2}\rho^2\beta_P = \frac{(9.8\ \mathrm{m/s^2})(15\ \mathrm{m})^3(26\ \mathrm{K})}{(0.0181\ \mathrm{g/m \cdot s})^2}(1178\ \mathrm{kg/m^3})^2\left(\frac{1}{296\ \mathrm{K}}\right)$$

$$= 1.23 \times 10^{13}$$

For air the Prandtl Number is about 0.707, so

$$\mathrm{Gr}_L \cdot \mathrm{Pr} = 1.23 \times 10^{13}(0.707) = 8.7 \times 10^{12}$$

which indicates that the flow is turbulent in the boundary layer. If we use equation 5–43 to predict the convective heat transfer coefficient, then, from table 5–1, C = 0.10 and n = 1/3, and

$$h_{\mathrm{ave}} = C\frac{\kappa}{L}(\mathrm{Gr}_L \cdot \mathrm{Pr})^n = 0.10\frac{0.026\ \mathrm{W/m \cdot K}}{15\ \mathrm{m}}(8.7 \times 10^{12})^{1/3} = 3.57\ \mathrm{W/m^2 \cdot K}$$

The free convection heat transfer is then

Answer $$\dot{Q} = h_{\mathrm{ave}}A\Delta T = (3.57\ \mathrm{W/m^2 \cdot K})(15 \times 40\ \mathrm{m^2})(26\ \mathrm{K}) = 55.614\ \mathrm{kW}$$

5–3 FREE CONVECTION ALONG HORIZONTAL AND INCLINED SURFACES

For horizontal surfaces there are four configurations, which need to be treated separately. Consider a hot plate or surface such that $T_s > T_\infty$ as shown in figure 5–6a. The fluid will naturally move up and be replaced by cooler fluid. Free convection can occur without stagnation of the fluid flow. In figure 5–6b the fluid will tend to flow around the plate if it has the opportunity. If flow is restricted and the fluid cannot pass around the plate, then the fluid flow will stagnate, the fluid will reach thermal equilibrium with the surface temperature, and the fluid will be cooler further below the plate surface. In the situation depicted in figure 5–6c, the fluid will tend to stagnate at the surface because of the tendency for cold fluid to have a greater density and specific weight. The cold surface with a fluid below it, as shown in figure 5–6d, will tend to behave as the unrestricted free convection of the condition of figure 5–6a. Thus, we have two distinct cases to consider:

I. Hot surface/cold fluid above and cold surface/hot fluid below
II. Hot surface/cold fluid below and cold surface/hot fluid above

Figure 5–6 Four conditions of free convection at a horizontal surface.

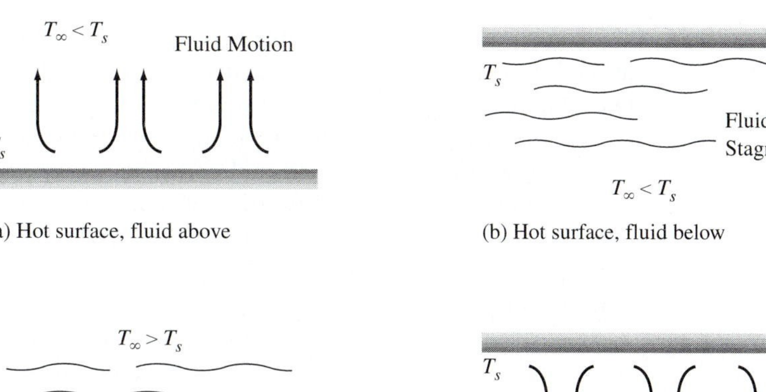

(a) Hot surface, fluid above

(b) Hot surface, fluid below

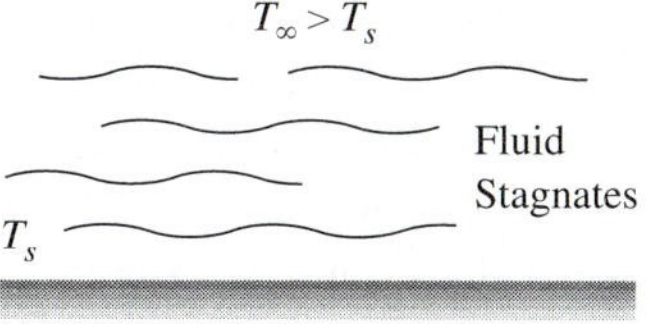

(c) Cold surface, fluid above

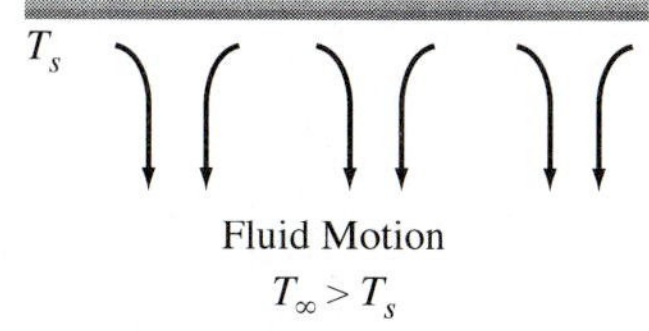

(d) Cold surface, fluid below

Analytical developments to describe the free convection heat transfer have not been completely successful because of the complicated nature of the flow, and empirical results have been used to predict the heat transfer. The average Nusselt Number for these cases may be predicted from the following empirically determined expressions using the Rayleigh Number Ra_L, where $\mathrm{Ra}_L = \mathrm{Gr}_L \cdot \mathrm{Pr}$:

Case I ($L \leq 0.6$ m)

$$\mathrm{Nu}_{\mathrm{ave}} = 0.54\ \mathrm{Ra}_L^{1/4} \qquad 10^4 \leq \mathrm{Ra}_L \leq 10^7 \tag{5–46}$$

$$\mathrm{Nu}_{\mathrm{ave}} = 0.15\ \mathrm{Ra}_L^{1/3} \qquad 10^7 < \mathrm{Ra}_L \leq 10^{11} \tag{5–47}$$

Case II ($L \leq 0.6$ m)

$$\mathrm{Nu}_{\mathrm{ave}} = 0.27\ \mathrm{Ra}_L^{1/4} \qquad 10^5 \leq \mathrm{Ra}_L \leq 10^{10} \tag{5–48}$$

The characteristic length L is defined as the ratio of the surface area to the perimeter. It has been observed that convective heat transfer remains unaffected if the value of L exceeds about 2 ft, or 0.6 m [7]. As a first approximation, an L of 0.6 m may be used to evaluate h in the expressions (5–46) through (5–48). These equations for free convection on horizontal surfaces may be extended to include the analysis of free convection along inclined surfaces, as shown in figure 5–7, if the Grashof Number is replaced by the "inclined plane Grashof Number"

$$\mathrm{Gr}_{\mathrm{IL}} = \frac{gL^3\Delta T}{\nu^2}\beta_P \cos\theta_H \tag{5–49}$$

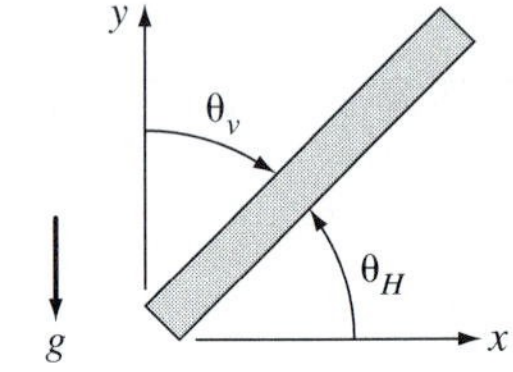

Figure 5–7 The inclined surface with free convection heat transfer.

This model is only applicable for angles of θ_H up to about 45°. For inclined surfaces at angles greater than 45°, the results from the analyses of vertical surfaces can be extended to inclined surfaces at angles greater than 45°, if the angle θ_v, shown in figure 5–7, is used in equation (5–49) for the inclined plane Grashof Number.

For the situation of a cold inclined surface with a warmer fluid below it, the convective flow is complicated by the periodic separation of the boundary layer as indicated in figure 5–8a. A similar phenomenon of periodic separation can occur if a hot inclined surface has a cold fluid above it, as shown in figure 5–8b. A more stable flow pattern usually occurs if the inclined surface is hot with a cold fluid below it (figure 5–8c) or the surface is cold and the warm fluid is above it (figure 5–8d). If the azimuth angle of the inclined surface, θ_v, becomes small such that the surface has a steep incline, then the free convection becomes more like that the free convection along a vertical surface. The analysis of free convection heat transfer along an inclined surface requires judgment on the part of the engineer to obtain reasonable results and an adequate design.

Figure 5–8 Inclined surface with free convection heat transfer.

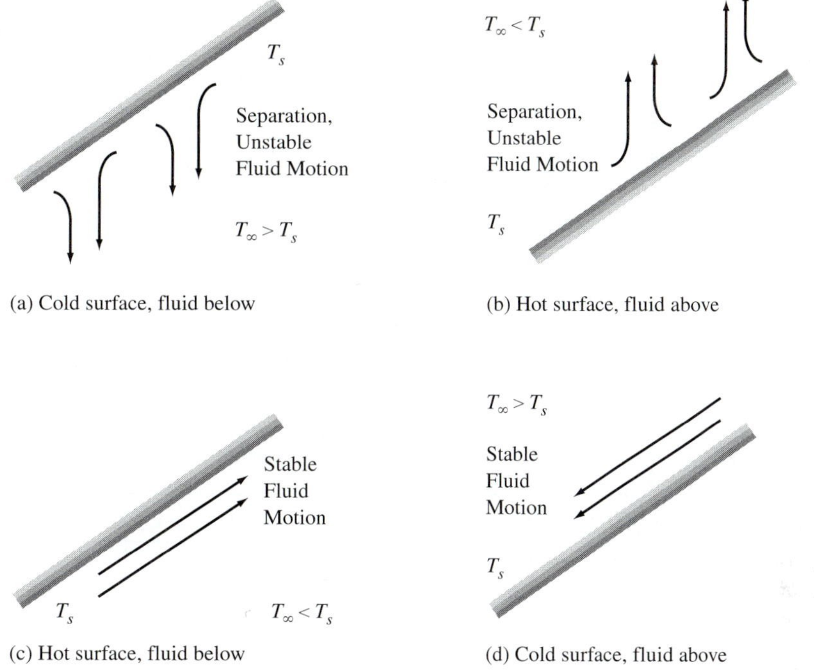

EXAMPLE 5–4 The flight runway at a particular airport is 2 miles long and 60 yards wide. On a hot, still day, when the air temperature is 90°F the runway surface temperature is 110°F. Estimate the heat transfer per unit area on the runway surface due to free convection.

Solution We first will determine the Rayleigh Number (Ra) from the relationship

$$\text{Ra}_L = \text{Gr}_L \cdot \text{Pr} = \frac{gL^3\Delta T}{\nu^2}\beta_P \cdot \text{Pr}$$

We will need to evaluate the air properties at the film temperature, $T_f = (110°\text{F} + 90°\text{F})/2 = 100°\text{F}$. From appendix table B–4 we approximate the properties at 300 K Pr $\approx$ 0.707, $\mu = 0.0181$ g/m · s = 0.0000181 kg/m · s, and $\rho = 1.178$ kg/m^3 so that $\nu = 0.0000181/1.178 = 1.54 \times 10^{-5}$m^2/s = 1.66×10^{-4}ft^2/s. The characteristic length, L, may be estimated by first assuming

$$L = \frac{\text{Surface area}}{\text{Perimeter}} = \frac{(2\text{ miles})(5280\text{ ft/mile})(180\text{ ft})}{21{,}120 + 360\text{ ft}} = 88.5\text{ ft}$$

Since this exceeds 2 ft, we will use the value of 2 ft for L. The Rayleigh Number is then found to be

$$\text{Ra}_L = 2.358 \times 10^8$$

Using equation (5–47), we may approximate a solution by predicting a value for the convective heat transfer coefficient:

$$\text{Nu}_{\text{ave}} = \frac{h_{\text{ave}}L}{\kappa} = 0.15\ \text{Ra}_L^{1/3}$$

or $$h_{ave} = 0.15\frac{\kappa}{L}\text{Ra}_L^{1/3} = (0.15)\frac{(0.026 \times 0.5779\ \text{Btu/hr}\cdot\text{ft}^2\cdot{}^\circ\text{F})}{2\ \text{ft}}(2.358 \times 10^8)^{1/3}$$

$$= 0.696\frac{\text{Btu}}{\text{hr}\cdot\text{ft}^2\cdot{}^\circ\text{F}}$$

Answer Then $$\dot{q}_A = h_{ave}\Delta T = \left(0.696\frac{\text{Btu}}{\text{hr}\cdot\text{ft}^2\cdot{}^\circ\text{F}}\right)(20^\circ\text{F}) = 13.92\frac{\text{Btu}}{\text{hr}\cdot\text{ft}^2}$$

EXAMPLE 5–5 A small room, 10 m × 10 m square, is heated from steam flowing through tubes embedded in the ceiling, as shown in figure 5–9. The ceiling temperature is 80°C at maximum heating load conditions when the room air temperature is 30°C. Estimate the convective heat transfer coefficient for this condition.

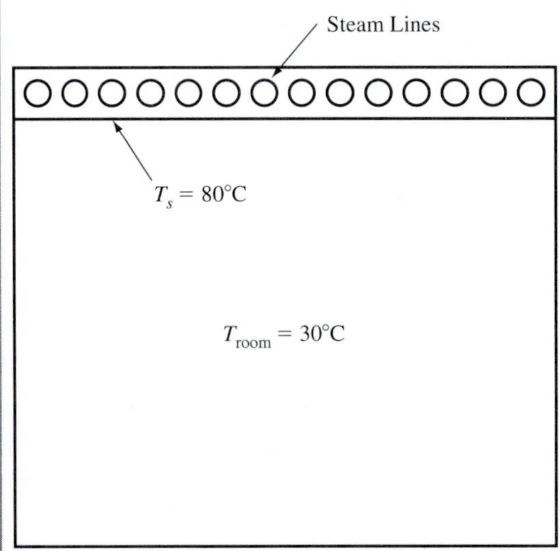

Figure 5–9 Steam heat in a ceiling.

Solution We first determine the Rayleigh Number:

$$\text{Ra}_L = \text{Gr}_L \cdot \text{Pr} = \frac{gL^3\Delta T}{\nu^2}\beta_P \cdot \text{Pr}$$

Using 55°C = 328 K as the film temperature for the air, we interpolate on appendix table B–4 to get Pr = 0.704, β_P = 1/328 K, and $\nu = 15.4 \times 10^{-6}$ m²/s, and for the characteristic length, $L = A$/perimeter = 2.5 m. The Rayleigh Number is, using L = 0.6 m,

$$\text{Ra}_L = 9.62 \times 10^8$$

We use case II and equation (5–48) so that

Answer $$h_{ave} = 0.27\frac{\kappa}{\text{L}}\text{Ra}_L^{1/4} = 0.27\frac{0.028\ \text{W/m}\cdot\text{K}}{0.6\ \text{m}}(9.62 \times 10^8)^{1/4} = 2.219\ \frac{\text{W}}{\text{m}^2\cdot\text{K}}$$

EXAMPLE 5–6 Water is in a large cauldron and is heated from below, as shown in figure 5–10. If the surface temperature of the cauldron is 90°C when the water is nominally 30°C, determine the convective heat transfer per unit area along the inclined portion of the cauldron. Assume the cauldron is much longer than the 1-m length of the inclined surface.

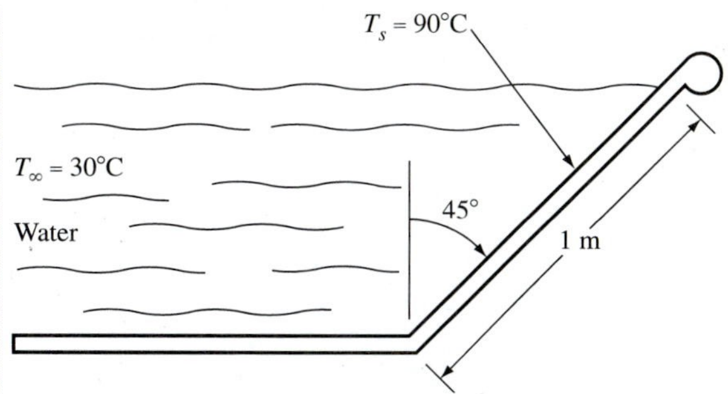

Figure 5–10 Water cauldron with inclined sides.

Solution

The inclined plane Grashof Number is, from equation (5–49),

$$\mathrm{Gr}_{\mathrm{IL}} = \frac{gL^3\Delta T}{\nu^2}\beta_P \cos\theta_H$$

Since the length is presumed to be much greater than the 1-m incline we may assume that the characteristic length L is

$$L = \frac{\text{Area}}{\text{Perimeter}} = \frac{(1\text{ m})(W)}{2W + 2(1\text{ m})} \approx \frac{(1\text{ m})W}{2W} = 0.5\text{ m}$$

For water at (90°C + 30°C)/2 = 60°C = 333 K, Pr ≈ 3.01, $\beta_P = 5.23 \times 10^{-4}\text{ K}^{-1}$ (from appendix table B–6), and $\nu = (4.71 \times 10^{-4}\text{ kg/m}\cdot\text{s})/(983.3\text{ kg/m}^3) = 4.79 \times 10^{-7}\text{ m}^2/\text{s}$. Then,

$$\mathrm{Ra}_{\mathrm{IL}} = \mathrm{Gr}_{\mathrm{IL}}\cdot\mathrm{Pr} = \frac{gL^3\Delta T}{\nu^2}\beta_P\cos\theta\cdot\mathrm{Pr}$$

$$= \frac{(9.8\text{ m/s}^2)(0.5\text{ m})^3(60\text{ K})}{(4.79\times 10^{-7}\text{ m}^2/\text{s})^2}(5.93\times 10^{-4}\text{K}^{-1})(\cos 45°)(3.01) = 4.04\times 10^{11}$$

We use equation (5–47):

$$\mathrm{Nu}_{\mathrm{ave}} = \frac{h_{\mathrm{ave}}L}{\kappa} = 0.15\,\mathrm{Ra}_{\mathrm{IL}}^{1/3}$$

and

Answer

$$h_{\mathrm{ave}} = 0.15\frac{\kappa}{L}\,Ra_L^{1/3} = 0.15\frac{0.658\text{ W/m}\cdot\text{K}}{0.5\text{ m}}(4.04\times 10^{11})^{1/3} = 1460\,\frac{\text{W}}{\text{m}^2\cdot\text{K}}$$

5–4 FREE CONVECTION ALONG HORIZONTAL AND VERTICAL CYLINDERS

Horizontal Cylinders

Free convection heat transfer between a horizontal cylinder and a surrounding fluid has been well studied. In figure 5–11 is shown a specific situation of a cylinder whose surface is at a higher temperature than the surrounding fluid, which results in the free convection of the fluid upward. It has been most useful to resort to empirical results in conjunction with results from the analysis of flat surfaces to describe the heat transfer. Using an equation of the form

$$\mathrm{Nu}_{\mathrm{ave}} = \frac{h_{\mathrm{ave}}D}{\kappa k} = \mathrm{CRa}_D^n \tag{5–50}$$

where

$$\mathrm{Ra}_D = \mathrm{Gr}_D\cdot\mathrm{Pr} = \frac{gD^3\Delta T}{\nu^2}\beta_{\mathrm{P}}\cdot\mathrm{Pr} \tag{5–51}$$

we can predict the convective heat-transfer coefficient. The constants C and n may be determined from table 5–2 and for various Rayleigh Number values.

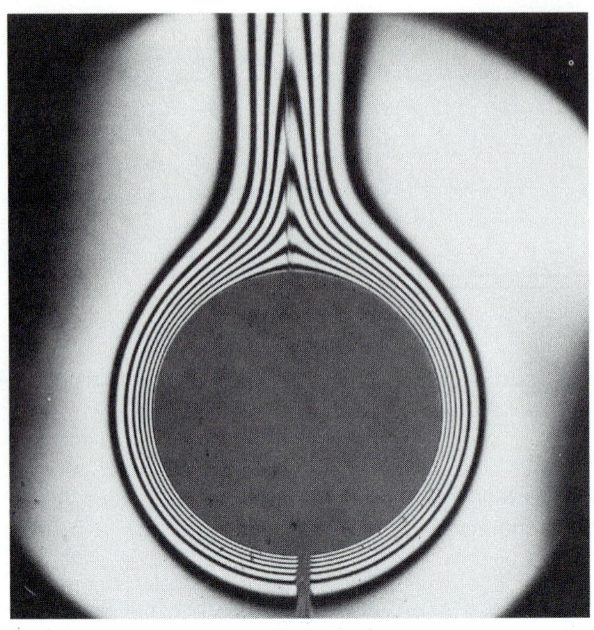

Figure 5–11 Free convection at a horizontal cylinder. A horizontal 6-cm-diameter cylinder, 60 cm long, is heated 9 Celsius degrees above the surrounding air temperature to give a Grashof Number of 30,000. (Photograph by U. Grigull and W. Hauf. Method described in [8].)

Table 5–2 Constants C and n for Use in Equation (5–50)

$\mathbf{Ra}_D$	C	n
10^{-10} to 10^{-2}	0.675	0.058
10^{-2} to 10^{2}	1.02	0.148
10^{2} to 10^{4}	0.850	0.188
10^{4} to 10^{7}	0.480	0.25
10^{7} to 10^{12}	0.125	0.333

Churchill and Chu [6] have provided a relationship that can be used to predict the Nusselt Number and convective heat transfer coefficient over the full range of Rayleigh Numbers from 10^{-5} to 10^{12}, which is

$$\mathrm{Nu}_{\mathrm{ave}} = \left(0.60 + \frac{0.387\ \mathrm{Ra}_D^{1/6}}{[1 + (0.559/\mathrm{Pr})^{9/16}]^{8/27}}\right)^2 \qquad \text{for } 10^{-5} < \mathrm{Ra}_D < 10^{12} \qquad \textbf{(5–52)}$$

EXAMPLE 5–7 A certain electric hot water heater element may be approximated as a horizontal cylinder of 1-in. diameter and 12-in. length, as shown in figure 5–12. Estimate the convective heat transfer from the element to the water when the element surface temperature is 200°F and the water is 80°F.

Solution The water film temperature is 140°F $[=(200°\text{F} + 80°\text{F})/2]$ and the water properties, read from appendix table B–6 for saturated liquid, at 60°C ≈ 140°F are $\rho_f = 983.28\ \text{kg/m}^3$,

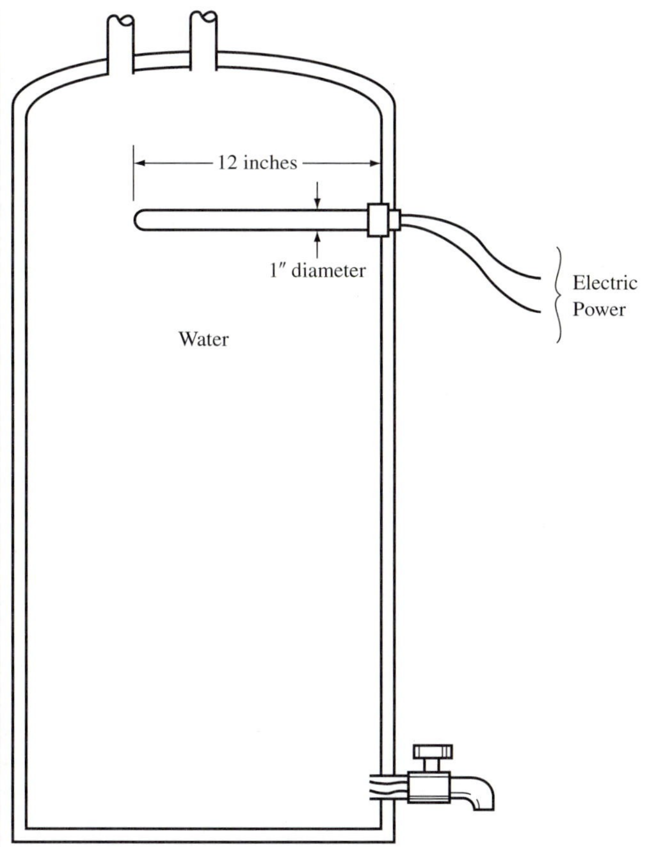

Figure 5–12 Electric hot water heater element.

$c_{p,f} = 4186 \dfrac{\text{J}}{\text{kg}\cdot\text{K}}$, $\kappa_f = 0.658 \dfrac{\text{W}}{\text{m}\cdot\text{K}} = 0.36966 \dfrac{\text{Btu}}{\text{hr}\cdot\text{ft}\cdot{}^\circ\text{R}}$, $\mu_f = 0.478 \times 10^{-3} \dfrac{\text{g}}{\text{m}\cdot\text{s}}$, and $\beta_P = 5.23 \times 10^{-3}\ \text{K}^{-1} = 2.91 \times 10^{-6}\ {}^\circ\text{R}^{-1}$. Then

$$\nu_f = \frac{\mu_f}{\rho_f} = 4.86 \times 10^{-7}\ \frac{\text{m}}{\text{s}} = 5.23 \times 10^{-6}\ \frac{\text{ft}^2}{\text{s}}$$

$$\alpha_f = \frac{\kappa_f}{\rho_f c_{p,f}} = 1.6 \times 10^{-7}\ \frac{\text{m}^2}{\text{s}}$$

$$\text{Pr} = \frac{\nu_f}{\alpha_f} = 3.04$$

Then

$$\text{Ra}_D = \text{Gr}_D \cdot \text{Pr} = \frac{gD^3\Delta t}{\nu^2}\beta_P \cdot \text{Pr}$$

$$= \frac{(32.2\ \text{ft/s}^2)(1/12\ \text{ft})^3(120^\circ\text{F})}{(5.23 \times 10^{-6}\ \text{ft}^2/\text{s})^2}(2.91 \times 10^{-4}\ {}^\circ\text{R}^{-1})(3.04) = 7.23 \times 10^7$$

If we use equation (5–52),

$$\text{Nu}_D = \frac{h_{\text{ave}}D}{\kappa} = \left(0.60 + \frac{0.387\ \text{Ra}_D^{1/6}}{[1 + (0.559/\text{Pr})^{9/16}]^{8/27}}\right)^2$$

$$= \left(0.6 + \frac{0.387(7.23 \times 10^7)^{1/6}}{[1 + (0.559/3.04)^{9/16}]^{8/27}}\right)^2 = 15.43$$

Then

$$h_{ave} = \frac{Nu_{ave}\kappa}{D} = 68.45\,\frac{Btu}{hr\cdot ft^2\cdot {}^\circ F}$$

and

Answer

$$\dot{Q} = h_{ave}\,A\Delta T = \left(68.45\,\frac{Btu}{hr\cdot ft^2\cdot {}^\circ F}\right)(\pi)\left(\frac{1}{12}\,ft\right)(1\,ft)(120^\circ F) = 2150\,\frac{Btu}{hr}$$

Vertical Cylinders

Vertical cylinders in a quiescent fluid may be treated, as shown in figure 5–13, as vertical flat surfaces provided that

$$\frac{L}{D}\frac{1}{Gr_L^{1/4}} < 0.025 \tag{5–53}$$

and $Pr \approx 1.0$.

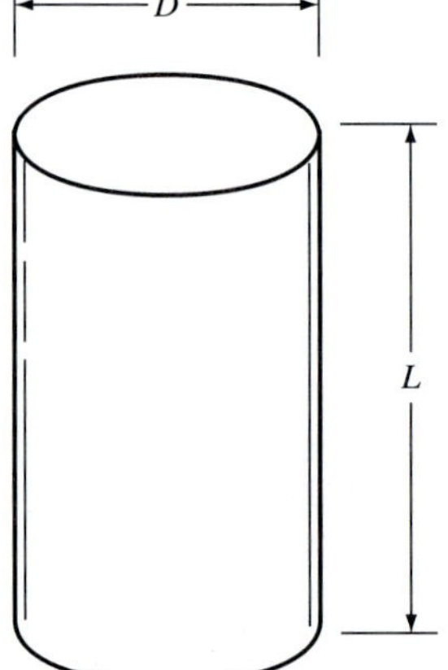

Figure 5–13 Vertical cylinder.

If equation 5–53 is not satisfied, then it has been suggested by Cebeci [9] that the information of figure 5–14 be used to account for the variance of results from using the vertical plate. In the figure the ratio of the average Nusselt Number of a vertical cylinder, using L as the characteristic length,

$$(Nu_{ave})_{cylinder} = \frac{h_{ave}L}{\kappa}$$

to the average Nusselt Number for a flat plate,

$$(Nu_{ave})_{flat\ plate} = \frac{hL}{\kappa}$$

is given as a function of the cylinder radius, L/R, and the Grashof Number based on length, Gr_L.

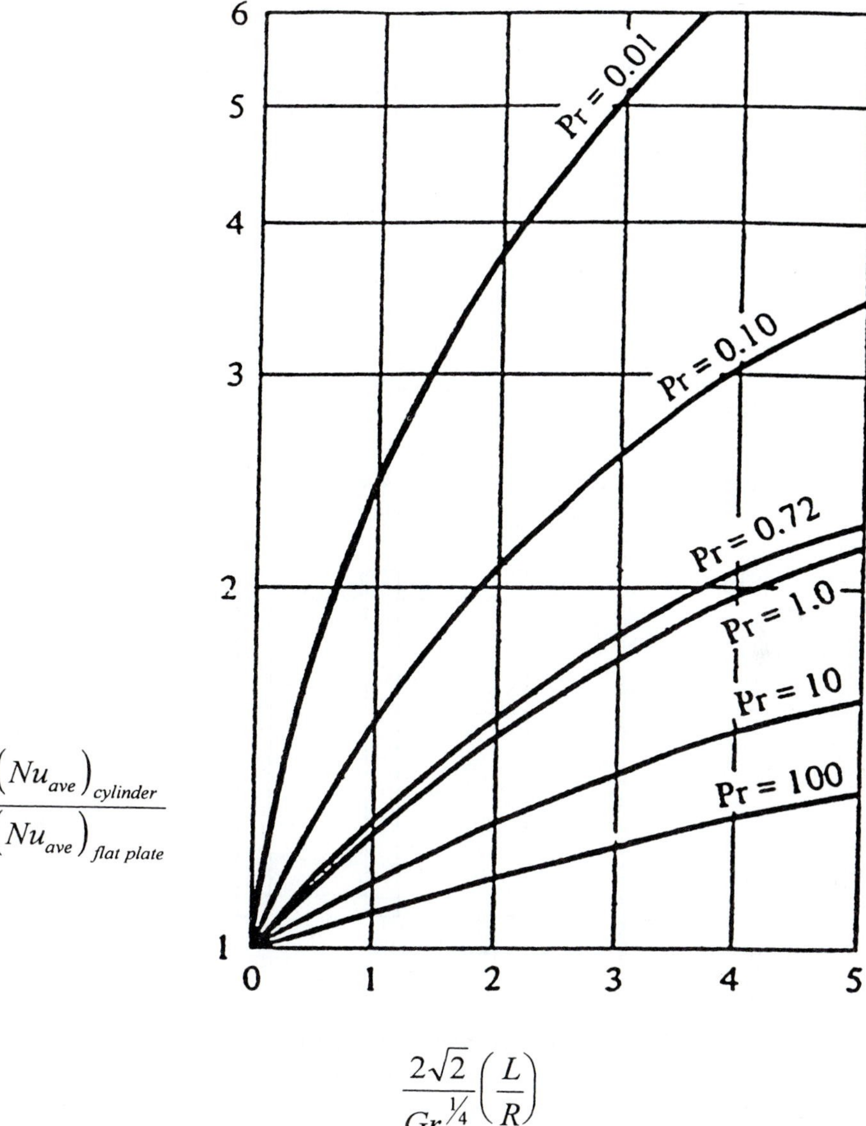

Figure 5–14 Nusselt Number for convection heat transfer at a vertical cylinder for various Prandtl numbers and slenderness ratios. (From T. Cebeci [9])

EXAMPLE 5–8 A flue gas stack at a foundry is 20 m high and 4 m in diameter. The outer surface temperature of the stack is 80°C, and the surrounding air temperature is 20°C. Estimate the heat loss from the stack surface by free convection, assuming there is no wind.

Solution We first need to predict the Grashof Number,

$$\mathrm{Gr}_L = \frac{gL^3\Delta T}{\nu^2}\beta_P$$

For air at a film temperature of 50°C [=(80°C + 20°C)/2] we have Pr ≈ 0.70, $\nu \approx 18 \times 10^{-6}$ m²/s, and

$$\mathrm{Gr}_L = \frac{(9.8\ \mathrm{m/s^2})(20\ \mathrm{m})^3(60\ \mathrm{K})}{(18 \times 10^{-6}\ \mathrm{m^2/s})^2}\left(\frac{1}{323\ \mathrm{K}}\right) = 44.9 \times 10^{12}$$

Checking the criterion of equation (5–53):

$$\frac{L}{D}\frac{1}{\mathrm{Gr}_L^{1/4}} = \frac{20\ \mathrm{m}}{4\ \mathrm{m}}\frac{1}{(44.9 \times 10^{12})^{1/4}} = 0.00193 < 0.025$$

Thus, we may assume that the vertical stack convection heat transfer behaves like a vertical flat surface or plate. Using equation (5–43) with $\mathrm{Gr}_L \cdot \mathrm{Pr} = (44.9 \times 10^{12})(0.7) = 3.14 \times 10^{13}$, from table 5–1 we read $C = 0.10$ and $n = 1/3$. Then

$$h_{\mathrm{ave}} = \frac{0.031\ \mathrm{W/m \cdot k}}{20\ \mathrm{m}}(0.10)(3.14 \times 10^{13})^{1/3} = 4.89\ \mathrm{W/m^2 \cdot K} = 4.89\ \mathrm{W/m^2 \cdot {}^\circ C}$$

The heat loss can be predicted from the calculation

Answer

$$\dot{Q} = h_{ave} A \Delta T = (4.89\ \mathrm{W/m^2 \cdot {}^\circ C})(\pi)(4\ \mathrm{m})(20\ \mathrm{m})(60^\circ\mathrm{C}) = 73.7\ \mathrm{kW}$$

EXAMPLE 5–9 Nuclear fuel rods are submersed in liquid sodium as shown schematically in figure 5–15. If the fuel rods have a surface temperature of 1100 K when the liquid sodium is 800 K, estimate the heat transfer from one rod.

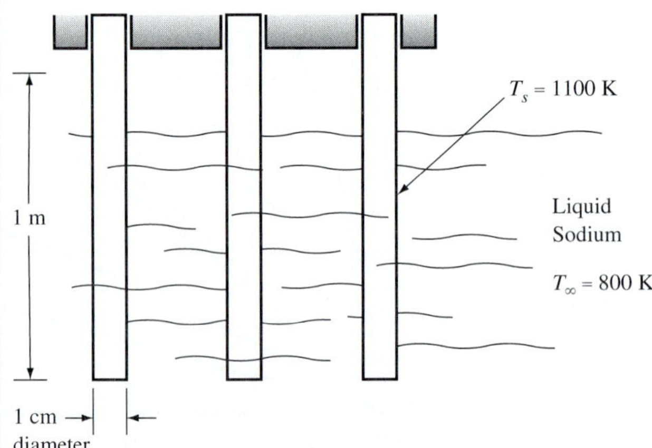

Figure 5–15 Nuclear fuel rods.

Solution

We approximate by interpolation the properties of liquid sodium at the film temperature of 950 K from appendix table B–3, $\rho = 782.6\ \mathrm{kg/m^3}$, $c_p = 1262\ \mathrm{J/kg \cdot K}$, $\kappa = 60.25\ \mathrm{W/m \cdot K}$, $\mu = 0.1225 \times 10^{-3}\ \mathrm{kg/m \cdot s}$, and $\beta_P = 0.376 \times 10^{-3}\ \mathrm{K^{-1}}$. Then

$$\nu = \frac{\mu}{\rho} = 1.57 \times 10^{-7}\frac{\mathrm{m^2}}{\mathrm{s}} \quad \text{and} \quad \mathrm{Pr} = 0.00256$$

The Grashof Number is then approximately,

$$\mathrm{Gr}_L = \frac{(9.8\ \mathrm{m/s^2})(1\ \mathrm{m})^3(300\ \mathrm{K})}{(1.57 \times 10^{-7}\mathrm{m^2/s})^2}(3.76 \times 10^{-4}\ \mathrm{K^{-1}}) = 44.85 \times 10^{12}$$

Using equation (5–52) to see if we may use equation (5–43) to estimate the Nusselt Number,

$$\frac{1\ \mathrm{m}}{0.01\ \mathrm{m}}\frac{1}{(44.85 \times 10^{12})^{1/4}} = 0.03864 > 0.025$$

so we must use the corrections from figure 5–14. The parameter ξ is

$$\xi = \frac{2\sqrt{2}}{\mathrm{Gr}_L^{1/4}}\left(\frac{L}{R}\right) = 0.218$$

From figure 5–14 we interpolate to read for a Prandtl Number of 0.00256 that

$$\frac{(\text{Nu})_{\text{cylinder}}}{(\text{Nu})_{\text{flat plate}}} \approx 1.8$$

The Grashof–Prandtl Number product is

$$\text{Gr}_L \cdot \text{Pr} = 11.48 \times 10^{10}$$

and from equation (5–43), with $C = 0.10$ and $n = 1/3$,

$$(h_{\text{ave}})_{\text{cylinder}} = \frac{59.7\ \text{W/m} \cdot \text{K}}{1\ \text{m}}(0.10)(11.48 \times 10^{10})^{1/3}(1.8) = 52{,}229\ \text{W/m}^2 \cdot \text{K}$$

and the heat transfer for one rod is

Answer

$$\dot{Q} = (h_{\text{ave}})_{\text{cylinder}} A\Delta T = (52{,}229\ \text{W/m}^2 \cdot \text{K})(\pi)(0.01\ \text{m})(1\ \text{m})(300\ \text{K}) = 492\ \text{kW/rod}$$

Multiple Cylinders

Free convection around a horizontal cylinder in a series of tubes can be a situation of interest. The flow pattern can be altered due to adjacent cylinders as shown in the photograph of figure 5–16. The mechanisms are similar to those of forced convection through

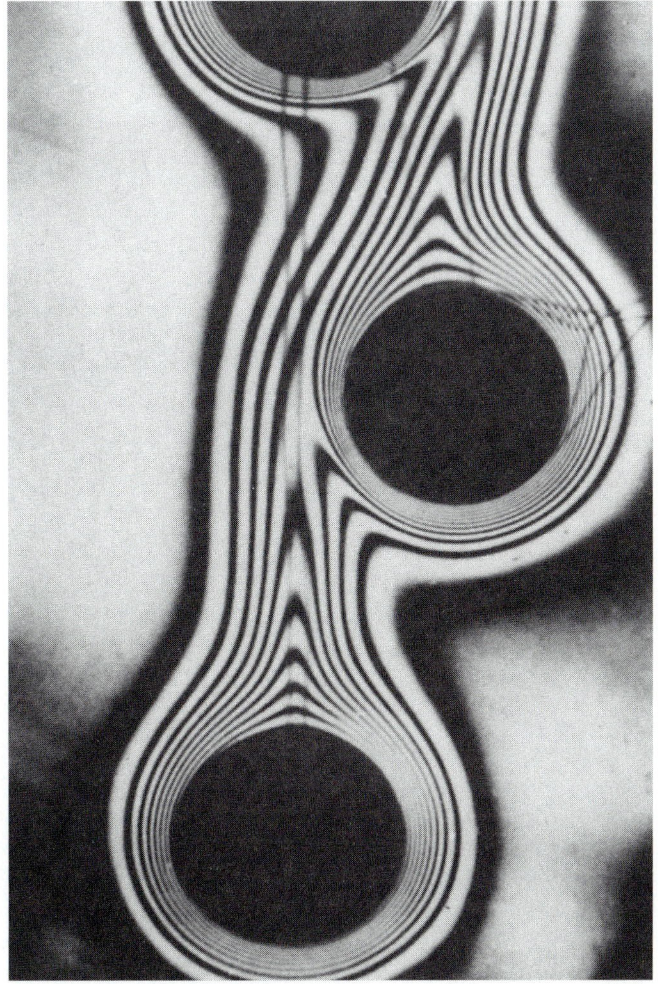

Figure 5–16 Free convection from three horizontal heated cylinders, showing the laminar plume in air as it interacts with each of the cylinders. The plume is rising from bottom to top. (From E. Eckert and E. Soehngen [10])

a tube bank. However, there has not been extensive work done on predicting these effects on free convection. If the spacing between tubes is greater than twice the diameter, one may use the results from single horizontal cylinders to obtain approximations.

5–5 FREE CONVECTION IN ENCLOSED SPACES

An enclosed space is a volume that is contained on all sides by solid surfaces. Free convection will occur in the enclosed space if the fluid and surfaces are each at different temperatures. In figure 5–17 is shown a rectangular enclosure. It could be a room, an oven, or a furnace, and there are other physical situations that can be modeled by the rectangular enclosure. In figure 5–17, the enclosure is oriented so that one direction coincides with the gravitational field g. There are three situations, or cases, that we will consider:

1. Heated enclosure from above, sides adiabatic.
2. Heated from below, sides adiabatic.
3. Heated from one side to the other side, other sides, top, and bottom are adiabatic surfaces.

For case 1, the heat transfer will be predominantly conduction because any free convection is discouraged by the configuration of a hot surface adjacent to a fluid below and a cold surface adjacent to a fluid above. The heat transfer for steady state conduction would be written

$$\dot{Q} = \kappa_{\text{eff}}\frac{A}{L}\Delta T \qquad \textbf{(5–54)}$$

where κ_{eff} is the average or effective thermal conductivity of the gas or fluid in the enclosure and A is the surface area, XY, as indicated in figure 5–17.

Figure 5–17 Enclosed space.

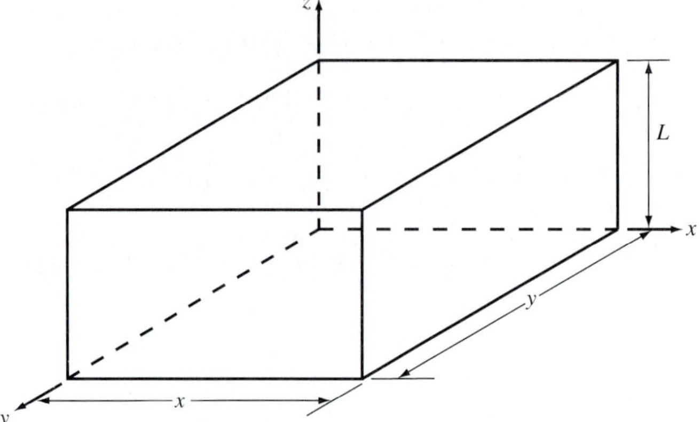

Case 2 is that of the typical free convection with hot horizontal surface below and cold surface above the enclosed fluid. For a $\text{Ra}_L = (gL^3\Delta T/\alpha\nu)\beta_P < 1708$ the viscous forces will dominate and the enclosed fluid will not move vertically or horizontally but will remain relatively quiescent. For Ra_L below about 1108, the heat transfer for case 2 can best be described by the pure conduction of equation (5–54). Thus, for Ra_L between 1108 and 1708, the heat transfer is in transition between conduction and free convection; however, approximate heat transfer can be predicted from equation (5–54). For the situation where $\text{Ra}_L > 1708$, free convection of the fluid will begin, and because there must

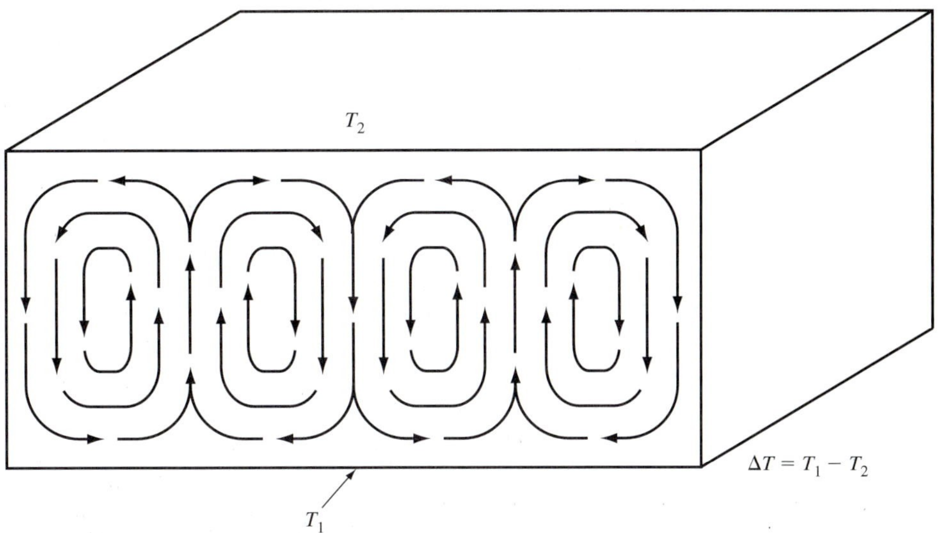

Figure 5–18 Bénard cells in free convection of enclosed space.

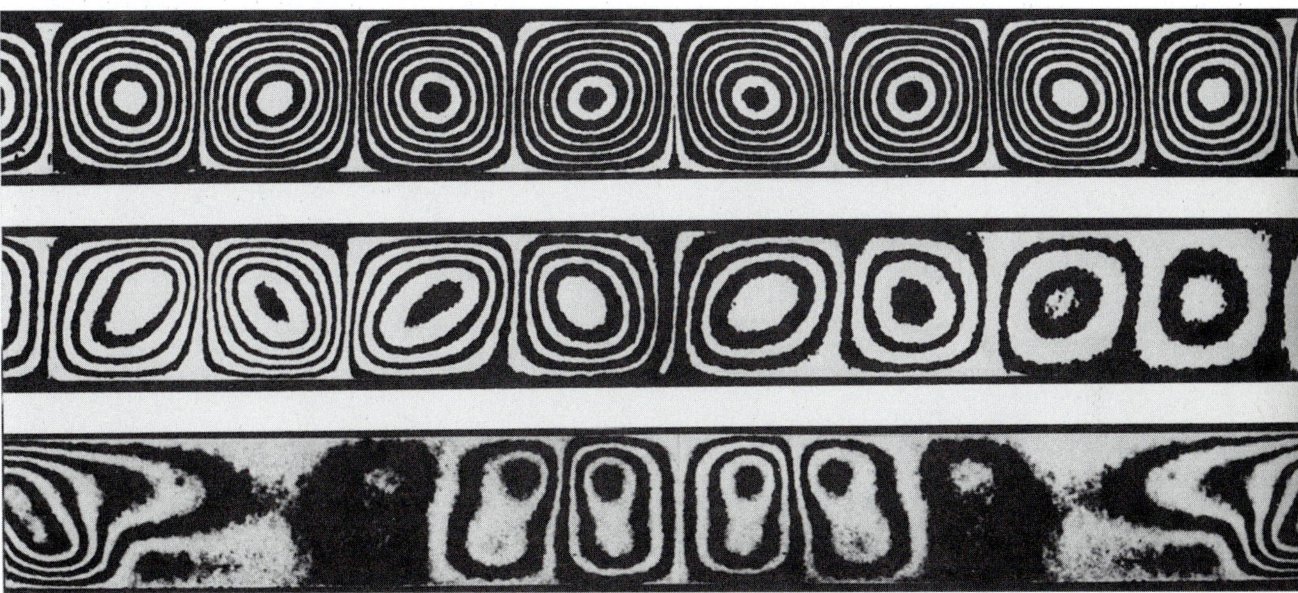

Figure 5–19 Bouyancy-driven convection rolls. Differential interforograms show side views of convection instabilities of silicone oil in a rectangular box of relative dimensions 10:4:1 (width:depth:height) heated from below. At the top is the classical Bénard cell situation: Uniform heating produces rolls parallel to the shorter side. In the middle photograph the temperature difference and hence the amplitude of motion increases from right to left. At the bottom the box is rotating about a vertical axis. (H. Oertel, Jr. and K. Kirchartz [13])

be a replacement of fluid along the hot and cold surfaces there may be a pattern of "cells" formed, called Bénard cells. This phenomenon is sketched in figures 5–18 and 5–19. For $\text{Ra}_L > 5 \times 10^4$, turbulent motion dominates and the Bénard cells may disappear. A more complete discussion of Bénard cells and free convection can be found in Jaluria [11]. The heat transfer under steady state conditions in the enclosed region is

$$\dot{Q} = h_{\text{ave}} A \Delta T$$

where the average convective heat transfer coefficient may be predicted from an empirical equation by Globe and Dropkin [12],

$$h_{ave} = \frac{\kappa}{L}(0.069)\,\mathrm{Ra}_L^{1/3}\,\mathrm{Pr}^{0.074} \qquad \textbf{(5–55)}$$

where $3 \times 10^5 < \mathrm{Ra}_L < 7 \times 10^9$. All properties are evaluated at the average enclosure temperature, $(T_1 + T_2)/2$.

For case 3, when two opposite vertical surfaces are subject to uniform heat transfer, the fundamental flow will be that shown in figure 5–20. Notice that L represents a horizontal length for this case whereas L represented a vertical height in the other cases of enclosures.

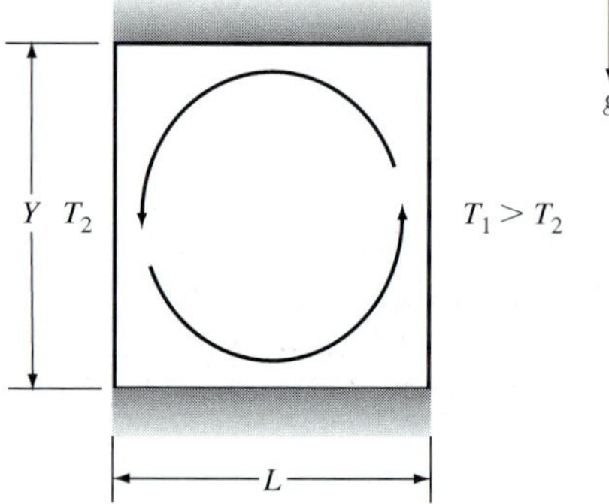

Figure 5–20 Case 3, free convection heat transfer in enclosures.

The average Nusselt Number may be predicted from

$$\mathrm{Nu}_{ave} = \frac{h_{ave}L}{\kappa} = \left[0.22\left(\frac{\mathrm{Ra}_L \cdot \mathrm{Pr}}{0.2 + \mathrm{Pr}}\right)^{0.28}\right]\left(\frac{L}{Y}\right)^{1/4} \qquad \textbf{(5–56)}$$

where $10 > Y/L > 1$, $10^3 < \mathrm{Ra}_L < 10^{10}$, and $1 < \mathrm{Pr} < 20$. For a wider range of slenderness ratios Y/L, MacGregor and Emery [14] suggest the equation

$$\mathrm{Nu}_{ave} = \frac{h_{ave}L}{\kappa} = 0.046\,\mathrm{Ra}_L^{1/3} \qquad \textbf{(5–57)}$$

for $40 > Y/L > 1$, $10^6 < \mathrm{Ra} < 10^9$, and $1 < \mathrm{Pr} < 20$. For other relationships the reader is referred to Catton [15].

The situation of convection heat transfer in enclosures tilted at an angle to the horizontal has been analyzed and the reader is referred to, for instance, the work by Bejan [16].

EXAMPLE 5–10 Consider a 10-m × 10-m room that is 2.5 m high. If the room is completely closed off when its floor temperature is 30°C and its ceiling temperature is 25°C, estimate the heat transfer between the floor and ceiling if the walls are assumed to be well insulated.

Solution We determine the Rayleigh Number:

$$\mathrm{Ra}_L = \frac{gL^3\Delta T}{\alpha\nu}\beta_P$$

using $\Delta T = 5°C$ and 300 K for the air temperature. Then, using appendix table B–4, $\rho = 1.178$ kg/m³, $c_p = 1007$ J/kg · K, $\kappa = 0.026$ W/m · K, $\mu = 0.0181 \times 10^{-3}$ kg/m · s, and then $\nu = 0.154 \times 10^{-4}$ m²/s.

$$\text{Ra}_L = \frac{(9.8\text{ m/s}^2)(2.5\text{ m})^3(5°\text{C})}{(15.365 \times 10^{-6}\text{ m}^2/\text{s})(0.219 \times 10^{-4}\text{ m}^2/\text{s})}\left(\frac{1}{300\text{ K}}\right) = 7.58 \times 10^9$$

If we use equation (5–55) to predict the convective heat-transfer coefficient, then

$$h_{\text{ave}} = \frac{0.026\text{ W/m} \cdot \text{K}}{2.5\text{ m}}(0.069)(7.58 \times 10^9)^{1/3}(0.708)^{0.074} = 1.374\text{ W/m}^2 \cdot °\text{C}$$

The heat transfer is then

Answer
$$\dot{Q} = h_{\text{ave}}A\Delta T = (1.374\text{ W/m}^2 \cdot °\text{C})(100\text{ m}^2)(5°\text{C}) = 687\text{ W}$$

5–6 COMBINED FREE AND FORCED CONVECTION HEAT TRANSFER

Convection heat transfer seldom involves only free or forced mechanisms. That is, convection heat transfer can usually be described as a combination of the two mechanisms. In section 5–1 we introduced the Grashof Number as a dimensionless parameter for describing the character of free convection flow. In that discussion it was found that the effective Reynolds Number was $\sqrt{\text{Gr}}$ so that if we are to compare forced convection with free convection we should compare the Grashof Number to the square of the Reynolds Number. Therefore, the ratio

$$\frac{\text{Gr}}{\text{Re}^2} = \frac{gL\Delta T}{\mathbf{V}_0^2}\beta_P \tag{5–58}$$

where $\mathbf{V}_0$ may be taken as the free stream forced convection velocity of the fluid, can be used to determine in a qualitative manner whether free convection dominates. We may suggest that

$\text{Gr/Re}^2 \gg 1$ (say 4) implies that free convection dominates.

$\text{Gr/Re}^2 \ll 1$ (say 0.25) implies that forced convection dominates.

$\text{Gr/Re}^2 \approx 1$ (say 0.25 to 4) implies that the convection is a combination of forced and free mechanisms.

One example of empirical information that has been used for analysis of a vertical surface where it is not clear whether forced or free convection is dominant is shown in figure 5–21. In this figure the coordinate system is designated such that x represents the vertical and y the horizontal direction. The typical dependent variable here is the Nusselt Number, which deviates from the relationships developed for only forced or free convection. An expression that may be used to combine the results from the separate analyses is

$$\text{Nu}_x^n = \text{Nu}_{x,\text{free}}^n + \text{Nu}_{x,\text{forced}}^n \tag{5–59}$$

The exponent n is an empirical parameter, and for vertical surfaces a value of $n = 3$ has been suggested. For horizontal flat surfaces a value of $n = 3.5$ has been suggested, and for horizontal cylinders a value of $n = 4$ is recommended. There is much work that can be done to clarify the phenomena of combined forced and free convection, and further discussion is beyond our purposes in this text.

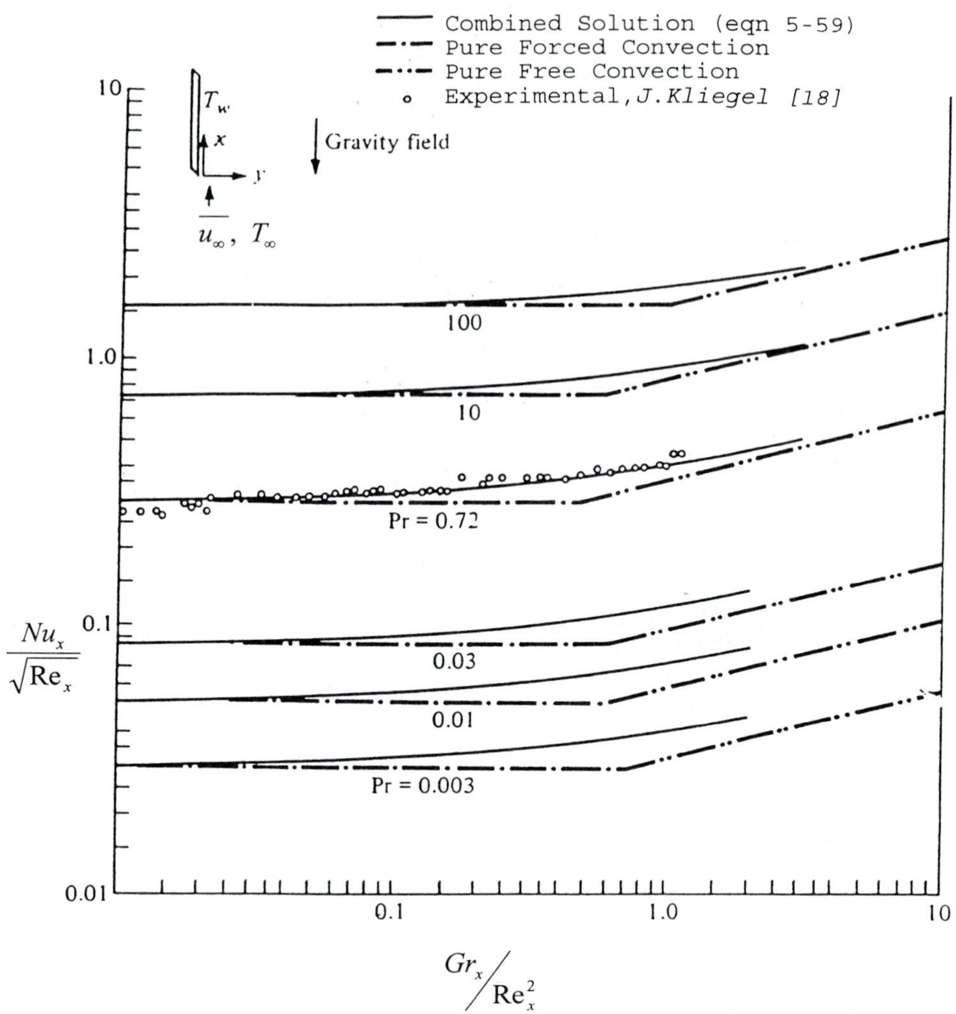

Figure 5–21 Nusselt Numbers for combined forced and free convection heat transfer on a vertical surface. Forced convection would be due to fluid moving upward with a velocity $\mathbf{u}_\infty$. (From J. Lloyd and E. M. Sparrow, [17])

EXAMPLE 5–11

A large cooling pond has water at 110°F when the air temperature is 50°F on a day when the prevailing winds are 2 mph. Estimate the convective heat-transfer coefficient at the pond surface, considering both free and forced convection.

Solution

The film temperature of the air may be taken as 80° F = 555°R. Using properties of air from appendix table B–4, we find Pr ≈ 0.708, $\kappa = 0.026 \times 0.5779 = 0.015$ Btu/hr · ft · °F, $\nu = 16.84 \times 10^{-6}$ m²/s $= 1.81 \times 10^{-4}$ ft²/s. Since the pond is large we may use for the characteristic length a value of 2 ft. Then

$$\text{Gr}_L = \frac{gL^3\Delta T}{\nu^2}\beta_P = \frac{gL^3\Delta T}{\nu^2 T_f} = \frac{(32.2\text{ ft/s}^2)(2\text{ ft})^3(110 - 50°\text{R})}{(1.81 \times 10^{-4}\text{ ft}^2/\text{s})^2(540°\text{R})} = 8.79 \times 10^8$$

The Reynolds Number is

$$\text{Re}_L = \frac{\mathbf{u}_\infty L}{\nu} = \frac{(2.94\text{ ft/s})(2\text{ ft})}{1.81 \times 10^{-4}\text{ ft}^2/\text{s}} = 3.25 \times 10^4$$

Then, from equation (5–58),

$$\frac{\mathrm{Gr}_L}{\mathrm{Re}_L^2} = 0.832$$

which indicates that combined free and forced convection applies. For free convection we use equation (5–48)

$$\mathrm{Nu}_{\text{ave,free}} = 0.15\ \mathrm{Ra}_L^{1/3} = 0.15[(8.79 \times 10^8)(0.708)]^{1/3} = 128$$

For forced convection we use the relationship for turbulence at a flat surface,

$$\mathrm{Nu}_{\text{ave,forced}} = 0.037\ \mathrm{Re}_L^{4/5}\mathrm{Pr}^{1/3} = 0.037(3.25 \times 10^4)^{4/5}(0.708)^{1/3} = 134.2$$

From equation (5–58) we have

$$\mathrm{Nu}_{\text{ave}} = [\mathrm{Nu}_{\text{ave,forced}}^{3.5} + \mathrm{Nu}_{\text{ave,free}}^{3.5}]^{1/3.5} = 160$$

The convective heat-transfer coefficient is

Answer
$$h_{\text{ave}} = \frac{\mathrm{Nu}_{\text{ave}}\kappa}{L} = 1.2\frac{\text{Btu}}{\text{hr}\cdot\text{ft}^2\cdot{}^\circ\text{F}}$$

5–7 APPROXIMATE EQUATIONS FOR FREE CONVECTION OF AIR

Many engineering applications of free convection involve air as the convecting medium. If the standard properties for air at 20°C and 1 atm pressure are inserted into the equations we have developed in the preceding sections, these equations reduce to a simplified form. Using for the Grashof Number the form

$$\mathrm{Gr}_L = \frac{gL^3\Delta T}{\nu^2}\beta_P = \frac{gL^3\Delta T}{\nu^2 T_f}$$

where

$$\Delta T = |T_s - T_\infty| \quad \text{and} \quad T_f = \frac{1}{2}(T_s + T_\infty)$$

these relationships reduce to those given in table 5–3, and they may also be used for exhaust gases, flue gases, nitrogen gas, oxygen, carbon dioxide, or carbon monoxide.

Table 5–3 Approximate Equations for Free Convection of Air

Configuration	Characteristic Length, L	Type of Flow	Range of $\mathrm{Gr}_L \cdot \mathrm{Pr}$	Convective Heat-Transfer Coefficient, h_∞ (W/m² · °C)
Vertical plates and cylinders	Height	Laminar	10^4 to 10^9	$1.42\left(\frac{\Delta T}{L}\right)^{1/4}$
		Turbulent	10^9 to 10^{13}	$1.31(\Delta T)^{1/3}$
Horizontal plates:				
Hot upper or cold lower surface	Area/perimeter	Laminar	10^5 to 2×10^7	$1.32\left(\frac{\Delta T}{L}\right)^{1/4}$
		Turbulent	2×10^7 to 3×10^{10}	$1.52(\Delta T)^{1/3}$
Hot lower or cold upper surface	Area/perimeter	Laminar	3×10^5 to 3×10^{10}	$0.59\left(\frac{\Delta T}{L}\right)^{1/4}$
Horizontal cylinders	Diameter	Laminar	10^4 to 10^9	$1.32\left(\frac{\Delta T}{D}\right)^{1/4}$
		Turbulent	10^9 to 10^{12}	$1.24(\Delta T)^{1/3}$

This data may be extended to situations where the gas pressure p is higher if the convective heat-transfer coefficient h_∞ is multiplied by the correction factor

$$\sqrt{\frac{p}{p_0}} \qquad \text{if flow is laminar}$$

$$\left(\frac{p}{p_0}\right)^{2/3} \qquad \text{if flow is turbulent}$$

The pressure p_0 is atmospheric pressure of 1 atmosphere, 101 kPa, or 14.7 psia.

5–8 SUMMARY

Buoyancy and *gravity* are the causes of free or natural convection heat transfer. The *chimney effect* is the upward motion of fluid due to its buoyancy or reduced density from that of the surrounding media, usually the atmosphere. The velocity of the fluid can be approximated by

$$\mathbf{V}_s = \sqrt{2gLT_s\left(\frac{1}{T_\infty} - \frac{1}{T_s}\right)} \tag{5–8}$$

The *Grashof Number,* Gr, is the dimensionless parameter used to describe the character of free convection fluid flow, given by

$$\mathrm{Gr} = gL^3\frac{d\rho}{\rho\nu^2} \tag{5–12}$$

or

$$\mathrm{Gr} = \frac{gL^3\Delta T}{\nu^2}\beta_P \tag{5–16}$$

where β_P is the isobaric compressibility of the fluid,

$$\beta_P = -\frac{1}{\rho}\left[\frac{\partial\rho}{\partial T}\right]_P \tag{5–14}$$

For *ideal gases*

$$\beta_P = \frac{1}{T} \tag{5–21}$$

Laminar flow of fluid for free convection along a vertical flat surface is characterized by

$$\mathrm{Gr}_L \cdot \mathrm{Pr} = \mathrm{Ra}_L < 10^9$$

and turbulent flow by

$$\mathrm{Ra}_L > 10^9$$

where Ra is the *Rayleigh Number.*

For laminar free convection along a vertical surface the temperature distribution in the boundary layer is

$$T(x,y) = T_\infty + (T_s - T_\infty)\left(1 - \frac{x}{\delta(y)}\right)^2 \tag{5–29}$$

where δ is the boundary layer thickness of the fluid,

$$\delta = y\left(\frac{3.93(0.952 + \mathrm{Pr})^{1/4}}{\mathrm{Pr}^{1/2} \cdot \mathrm{Gr}_L^{1/4}}\right) \tag{5–33}$$

The Nusselt Number for laminar flow along a vertical surface is

$$\mathrm{Nu}_y = 0.508\sqrt{\mathrm{Pr}}(\mathrm{Gr}_y^{1/4})\frac{1}{(0.952 + \mathrm{Pr})^{1/4}} \tag{5–39}$$

and, for Ra $< 10^9$,

$$h_{\mathrm{ave}} = \frac{4\kappa}{3L}(0.508)\sqrt{\mathrm{Pr}}\frac{\mathrm{Gr}_L^{1/4}}{(0.952 + \mathrm{Pr})^{1/4}} \tag{5–42}$$

For turbulent flow, as well as laminar flow, the relation

$$h_{\mathrm{ave}} = C\frac{\kappa}{L}(\mathrm{Gr}_L \cdot \mathrm{Pr})^n$$

can be used, where C and n are empirically determined. For all Pr and for laminar or turbulent flow, another equation that can be used is

$$h_{\mathrm{ave}} = \frac{\kappa}{L}\left(\frac{0.825 + 0.387(\mathrm{Gr}_L \cdot \mathrm{Pr})^{1/6}}{[1 + (0.492/\mathrm{Pr})^{9/16}]^{8/27}}\right)^2 \tag{5–45}$$

For free convection on horizontal or inclined surfaces the equation

$$\mathrm{Nu} = C(\mathrm{Ra})^n$$

where C and n are determined by one of two conditions: *case I,* where the surface is hot and the fluid is above the surface or the surface is cold and fluid is below the surface, and *case II,* where the surface is hot with fluid below, or the surface is cold with fluid above. For horizontal surfaces L is the surface area/perimeter or 0.7 m as a maximum. For inclined surfaces, up to 45°, a modified Grashof Number

$$\mathrm{Gr}_{\mathrm{IL}} = \cos\theta_H \frac{gL^3\Delta T}{\nu^2}\beta_P \tag{5–49}$$

is used in conjunction with the horizontal surface relationships. For surfaces inclined greater than 45° the use of the vertical surface relationships is recommended.

For convection around horizontal cylinders

$$\mathrm{Nu}_{\mathrm{ave}} = C(\mathrm{Ra}_D)^n$$

is recommended, or for $10^{-5} < \mathrm{Ra}_D < 10^{12}$

$$\mathrm{Nu}_{\mathrm{ave}} = \left(0.60 + 0.387\frac{\mathrm{Ra}_D^{1/6}}{[1(0.559/\mathrm{Pr})^{9/16}]^{8/27}}\right)^2 \tag{5–52}$$

For free convection along vertical cylinders the parameter

$$\frac{2\sqrt{2}}{\mathrm{Gr}_L^{1/4}}\left(\frac{L}{R}\right)$$

determines how the *Nusselt Number* is related to the Nusselt Number for a vertical flat surface. For *enclosed spaces* there are three cases to consider

Case 1: heated from above with sides adiabatic,

$$\dot{Q} = \kappa_{\mathrm{eff}}\frac{A}{L}\Delta\mathrm{T} \tag{5–54}$$

Case 2: heated from below, sides adiabatic,

For $\mathrm{Ra}_L < 1708$, use equation (5–54)

For $\mathrm{Ra}_L > 1708$, use

$$h_{\mathrm{ave}} = \frac{\kappa}{L}(0.069)\mathrm{Ra}_L^{1/3}\mathrm{Pr}^{0.074} \tag{5–55}$$

Case 3: heated from one side to the opposite side, other surfaces adiabatic, use

$$h_{\mathrm{ave}} = 0.22\frac{\kappa}{L}\left(\frac{\mathrm{Pr}\cdot\mathrm{Ra}_L}{0.2+\mathrm{Pr}}\right)^{0.28}\left(\frac{L}{Y}\right)^{1/4} \tag{5–56}$$

For combined forced and free convection,

If $\mathrm{Gr/Re}^2$ is 0.25 or less use forced convection relationships.

If $\mathrm{Gr/Re}^2$ is 4 or more, use free convection.

If $0.25 < \mathrm{Gr/Re}^2 < 4$, then use

$$\mathrm{Nu}_x^n = \mathrm{Nu}_{x,\mathrm{free}}^n + \mathrm{Nu}_{x,\mathrm{forced}}^n \tag{5–59}$$

DISCUSSION QUESTIONS

Section 5–1

5–1 What is buoyancy?

5–2 What is the chimney effect?

5–3 What is meant by draft?

Section 5–2

5–4 What does the Grashof Number mean?

5–5 What is meant by compressibility?

5–6 How is laminar or turbulent free convection determined?

Section 5–3

5–7 Why do you suspect that for $L > 0.6$ m the free convection is unaffected by size?

5–8 How is L then determined?

5–9 Why modify Gr for inclined surfaces?

Section 5–4

5–10 State what a plume is.

5–11 Why are vertical cylinders approximated by vertical flat surfaces?

5–12 How would you analyze multiple horizontal cylinders?

Section 5–5

5–13 What is a Bénard cell?

5–14 What would prevent free convection at very low Grashof Numbers in an enclosed space?

5–15 What are some examples of enclosed spaces?

Section 5–6

5–16 What is meant by combined forced and free convection?

PRACTICE PROBLEMS

Section 5–1

5–1 An 85-lbm cast-iron bracket is hung from a wire and a spring weigh scale as shown in figure 5–22. What would be the reading of the weigh scale and what is the apparent density of the bracket. Assume that the density of iron is 535 lbm/ft^3 and the density of water is 62.4 lbm/ft^3.

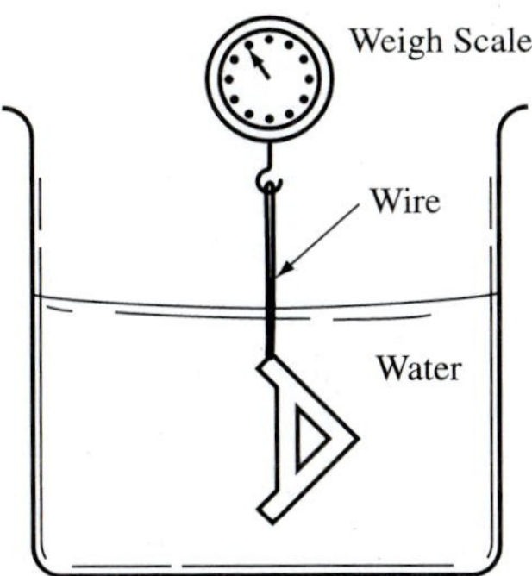

Figure 5–22

5–2 A steel stack 4 m in diameter is 60 m high and is used to exhaust gases that are at an average temperature of 600°C. The stack wall temperature is 100°C when the surrounding air temperature is 20°C. Estimate the stack gas velocity, using air properties for the stack gases.

Section 5–2

5–3 A picture window in a house is 3 ft high by 6 ft long. If the window is 50°F and the outside air temperature is −10°F, estimate the heat transfer from the window by natural convection.

5–4 A cooling tower, shown schematically in figure 5–23, has an outside surface temperature of 70°C when quiescent outside air is at 10°C. If the nominal diameter of the tower is 20 m, estimate

(a) The vertical distance up from the base where turbulent flow ensues.

(b) The total heat transfer (in kW) due to natural convection.

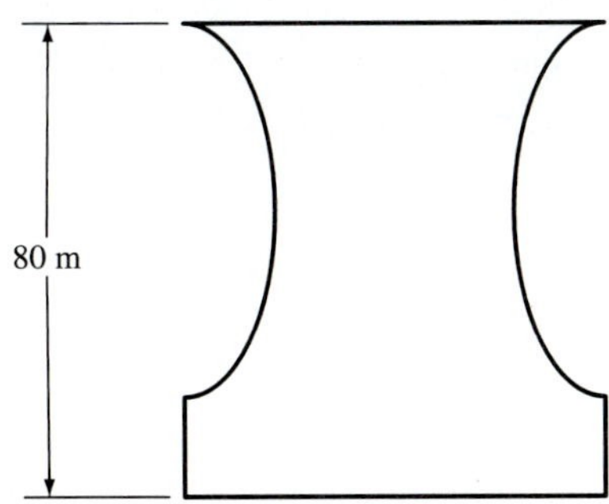

Figure 5–23 Cooling tower.

5–5 A 3-m-diameter by 2-m-high tank contains sulfuric acid at 100°C. If the sides of the tank are suddenly heated to 150°C, determine the boundary layer thickness of the acid as it moves up the sides by natural convection. Use for sulfuric acid that $\beta_P = 5.0 \times 10^{-3}\ \text{K}^{-1}$, $\nu = 1.2 \times 10^{-6}\ \text{m}^2/\text{s}$, and $\text{Pr} = 1.5$.

5–6 For the sulfuric acid in problem 5–5 estimate the velocity profile along the tank vertical wall in terms of the vertical height y and the distance from the wall x.

5–7 Ethylene glycol at 100°C is cooled along a vertical flat plate. The plate is 0.5 m wide by 1 m high and is at 20°C. Estimate the heat transfer from the ethylene glycol.

5–8 Show that the equation for isobaric compressibility,

$$\beta_P = -[(1/\rho)\partial\rho/\partial T]_P$$

reduces to

$$\beta_P = 1/T$$

for ideal gases.

5–9 A vertical outside wall of a building receives 400 W/m^2 of direct solar radiation and transfers 50 W/m^2 heat uniformly by convection to the surrounding air at 10°C.

(a) Determine the vertical distance up the wall where turbulent air flow begins, assuming an average wall temperature of 20°C.

(b) Determine the wall surface temperature at the location where turbulence begins.

5–10 A vertical side of a cooling tower for an air conditioning system is approximately 90°C when the surrounding air temperature is 25°C. If the cooling tower is 30 m high, estimate the maximum vertical local air velocity due to natural convection.

Section 5–3

5–11 Predict the heat transfer per unit area for warm water at 70°C placed in a large pan at 10°C.

5–12 In August the pavement of highways can become excessively hot on some days. Estimate the heat loss per unit area from a blacktop highway at 110°C if the air is still and 28°C.

5–13 Estimate the heat transfer from a house 15 m × 10 m × 4 m high to the eaves, as shown in figure 5–24, with a 45° pitch on the gabled roof. The air surrounding the house is still and 0°C, and the surfaces of the house are all 15°C.

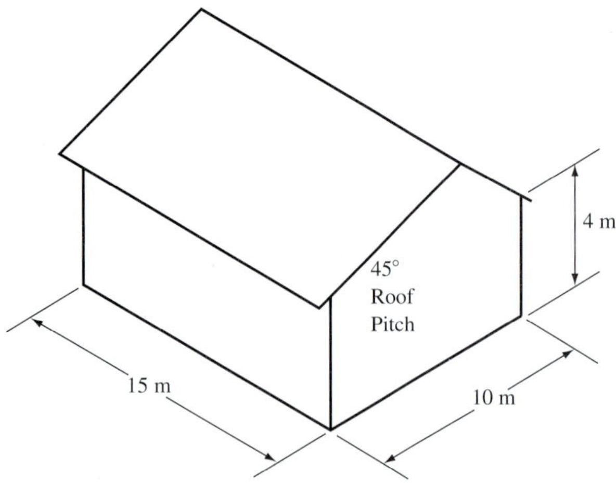

Figure 5–24

Section 5–4

5–14 A 2-in.-diameter fluorescent light is mounted horizontally as shown in figure 5–25. Estimate the heat transfer per unit area if the light surface is 85°F and the surrounding air is 65°F.

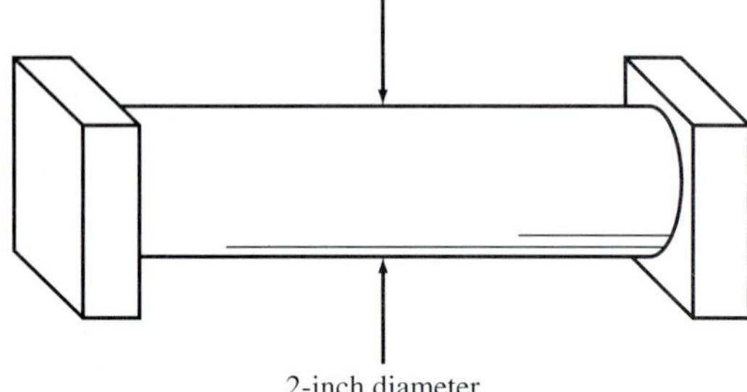

Figure 5–25

5–15 Consider the horizontal cylinder shown in figure 5–11.
(a) Estimate the heat transfer per unit length.
(b) Discuss the dependency of the fluid temperature surrounding the cylinder with position, radially (r) and angularly (θ).

5–16 Hot air flows through a square 60-cm × 60-cm steel heating duct, and the outside surfaces are at 45°C when the surrounding quiescent air is at 15°C. Estimate the heat loss from the duct per unit of duct length.

5–17 A stack 60 m high by 5 m in diameter has an average temperature of 300°C on its outside surfaces when exhaust gases pass through it. Estimate the heat loss from the stack into still air at 25°C.

5–18 The vertical plate shown in figure 5–26 is 10 cm tall and is heated in still air at 20°C.
(a) Estimate the plate surface temperature.
(b) Determine the heat transfer per unit area of plate in W/m^2.

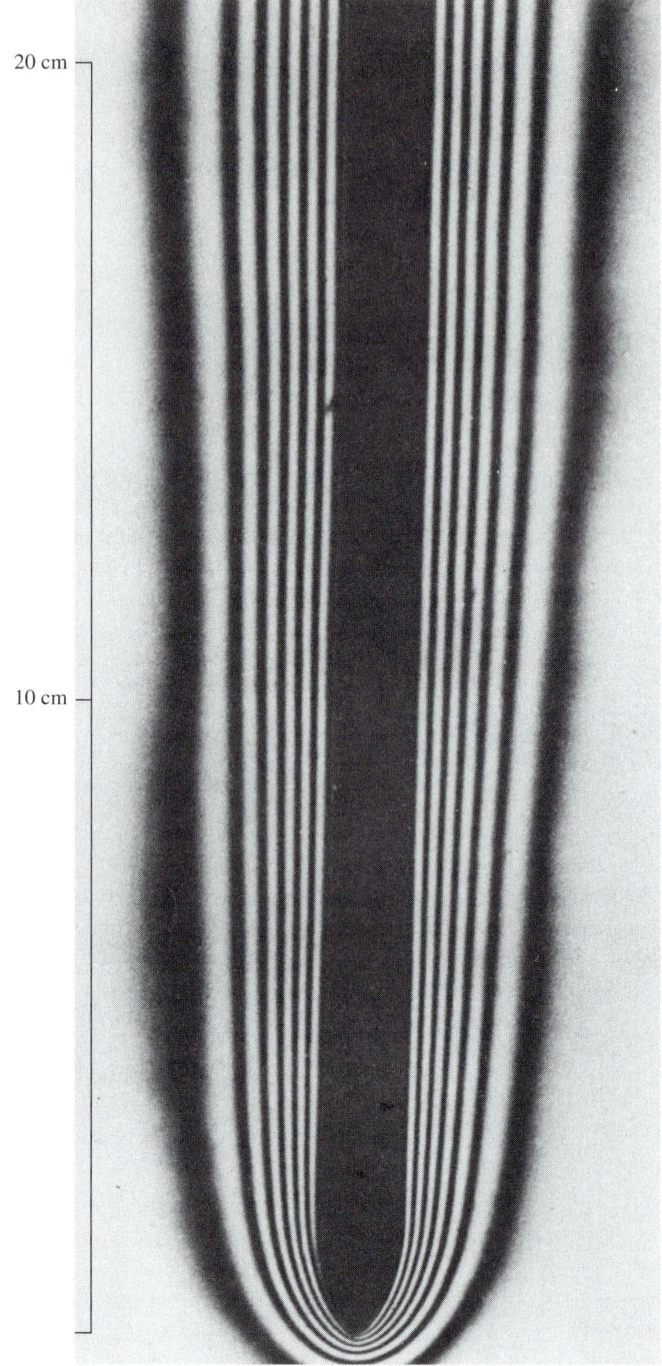

Figure 5–26 Free convection from a uniformly heated vertical plate in air so that the Grashof Number is about five million at 10 cm up from the lower end of the plate. (Photograph by E. Eckert and E Soehngen [10])

5–19 If 3-in.-diameter bars of steel are to be air dried in still air, recommend a horizontal arrangement of the bar placement so that the adjacent bars do not interfere with the cooling. (*Hint:* See figure 5–16.)

Section 5–5

5–20 A room is heated with electrical resistance wires embedded in the 2.2-m-high ceiling. Estimate the heat transfer per unit of floor area when the ceiling is at 80°C and the floor is 10°C. Neglect any radiation and heat transfer between the walls of the room.

5–21 A 30-ft-high vertical wall is separated by a stagnant air mass from a glass wall parallel to it, as shown in figure 5–27. When the wall surface temperature is 85°F and the glass temperature is 25°F, estimate the heat transfer between the two surfaces.

5–22 A rectangular oven 1 m × 1 m × 1 m is heated from the bottom surface at 200°C. If the top surface is 80°C at a given time and the sides are assumed to be well insulated, determine the heat transfer through the oven.

Section 5–6/5–7

5–23 A large stratified brine/water pond has a surface temperature of 140°F, and air at 70°F moves across the pond at 15 mph. Estimate the convection heat transfer at the pond surface per unit area.

5–24 A 50-m-long by 5-m-diameter rocket at launch is moving up at 45 km/hr when its sides are at −45°C and the surrounding air is at 25°C. Estimate the heat transfer from the surrounding air.

5–25 Air at 20°C moves over a horizontal roof at 20 cm/s. The roof is at 30°C and is 40 m × 40 m. Estimate the heat transfer by convection.

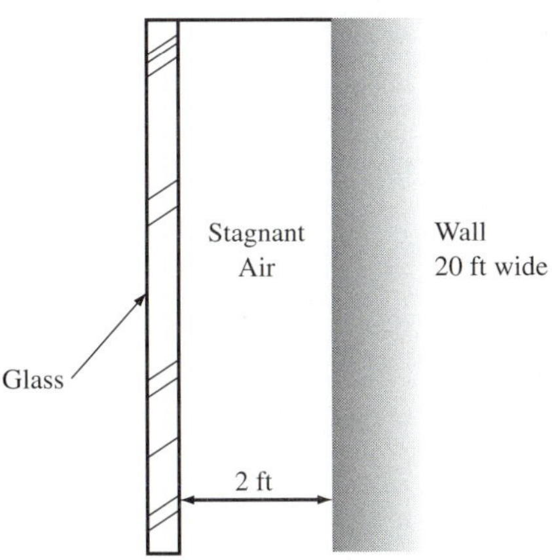

Figure 5–27

REFERENCES

1. Gebhart, B., Y. Jaluria, R. L. Mahajan, and B. Sammakia, *Boundary Induced Flows and Transport,* Hemisphere Publishing, 1988.
2. Holman, J. P., *Heat Transfer,* 7th edition, McGraw-Hill, New York, 1990.
3. McAdams, W. H., *Heat Transmission,* 3rd edition, McGraw-Hill, New York, 1954.
4. Bayley, F. J., An Analysis of Turbulent Free Convection Heat Transfer, *Proc. Inst. Mech. Engr.,* London, UK, 169, 361, 1955.
5. Warner, C. Y., and V. S. Arpaci, An Investigation of Turbulent Natural Convection in Air at Low Pressure Along a Vertical Heated Flat Plate, *Int. J. Heat Mass Transfer,* 11, 397, 1968.
6. Churchill, S. W., and H. H. S. Chu, Correlating Equations for Laminar and Turbulent Free Convection from a Vertical Plate, *Int. J. Heat Mass Transfer,* 18, 1323, 1975.
7. Brown, A. I., and S. M. Marco, *Introduction to Heat Transfer,* 3rd edition, McGraw-Hill, New York, 1958.
8. Grigull, U., and W. Hauf, *Proc. 3rd Int. Heat Transfer Conf.* 2, 182–195, 1966.
9. Cebeci, T., Laminar-Free-Convective Heat Transfer from the Outer Surface of a Vertical Slender Circular Cylinder, *Proc. 5th Int. Heat Transfer Conf.,* 3, NC 1.4, 15–19, 1974.
10. Eckert, E. R. G, and E. Soehngen, *U.S. Air Force Tech. Rept. 5747,* 1948.
11. Jaluria, Y., *Natural Convection Heat and Mass Transfer,* Pergamon Press, Oxford, UK, 1980.
12. Globe, S., and D. Dropkin, Natural Convection Heat Transfer in Liquids Confined by Two Horizontal Plates and Heated from below, *J. Heat Transfer,* 81, 24–28, 1959.

13. Oertel, H. Jr., and K. R. Kirchartz, Influence of Initial and Boundary Conditions on Bénard Convection, in U. Müller, K. G. Roesner, and B. Schmidt, eds., *Recent Developments in Theoretical and Experimental Fluid Mechanics,* Berlin, Springer-Verlag, 1979, pp. 355–366.

14. MacGregor, R. K., and A. P. Emery, Free Convection through Vertical Plane Layers: Moderate and High Prandtl Number Fluids, *J. Heat Transfer,* 91, 391, 1969.

15. Catton, I., Natural Convection in Enclosures, *Proc. 6th Int. Heat Transfer Conference, Toronto,* 6, 13–31, 1978.

16. Bejan, A., *Convection Heat Transfer,* Wiley Interscience, New York, 1984.

17. Lloyd, J. R., and E. M. Sparrow, Combined Forced and Free Convection on Vertical Surfaces, *Int. J. Heat Mass Transfer,* 13, 434–438, 1970.

18. Kliegel, J. R., Laminar Free and Forced Convection Heat Transfer from a Vertical Flat Plate, Ph.D. Thesis, University of California, Berkeley, CA, 1959.

6 THE NATURE OF RADIATION HEAT TRANSFER

In this chapter we will discuss thermal radiation as a mode of heat transfer. The spectral and directional characteristics of thermal radiation will be discussed, and the concepts of black and gray bodies will be introduced. Diffuse and directional radiation will be discussed, and absorptivity, emissivity, reflectivity, and transmissivity will be defined. Kirchhoff's Law relating emissivity to absorptivity and opaque and transparent materials will be discussed.

The view factors or shape factors are considered for discussing the solid geometry of the interaction of two separate surfaces. Some applications of spectral and geometric responses of material to thermal radiation will be considered such as window glass, absorbing materials, and concentrating solar collection.

New Terms

α_r	absorptivity	ν	frequency (hertz, cps)
ε_r	emissivity	λ	wavelength
τ_r	transmissivity	θ	plane angle
ρ_r	reflectivity	F_{i-j}	radiation shape factor or view factor
ϕ	lateral angle	$\dot{I}$	spectral intensity
Θ	azimuth angle	$\hbar$	Planck's Constant
ω	solid angle		

6–1 ELECTROMAGNETIC RADIATION

Thermal radiation is the transfer of energy from one surface to another through an intervening space that cannot be explained as conduction or convection, shown schematically in figure 6–1. In section 1–4 we were introduced to thermal radiation as the model used to describe, for instance, how solar energy is able to travel from the sun to the earth through a vast distance of essentially no heat conducting material. Thermal radiation seems to operate best when the intervening space is a vacuum and is degraded or is not as effective when a gas, liquid, or solid is placed between the two surfaces. Thermal radiation is visualized as that energy associated with photons emitted from, or absorbed by, a material.

Conduction heat transfer may be described as the energy flow causing molecular vibration, rotation, and other kinetic or potential energy. These molecular energies are direct functions of the material temperature. Consequently, we saw that conduction is driven by a temperature gradient. Thermal radiation or radiant heat transfer, however, is the flow of photons, which are the pockets of energy emitted or absorbed by an electron changing energy levels in an atom or molecule. Yet thermal radiation behaves as a wave and is a form of what is called EM radiation, so it may be described as a wave traveling

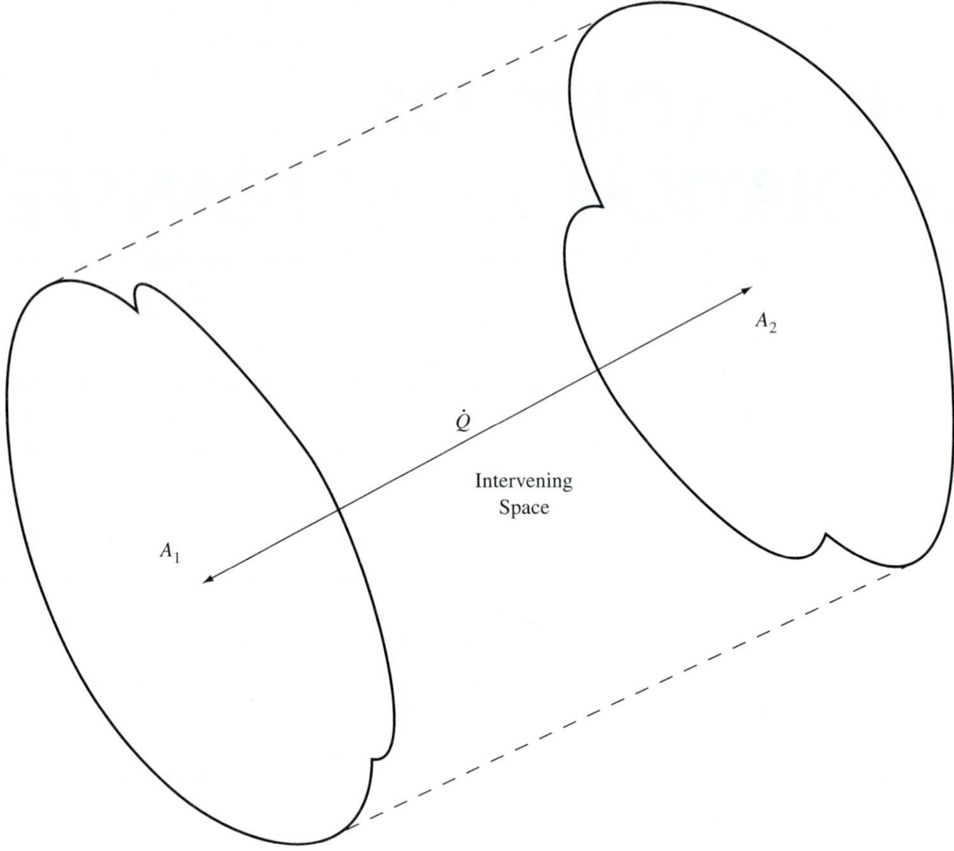

Figure 6–1 Schematic of thermal radiation from surface A_1 to A_2.

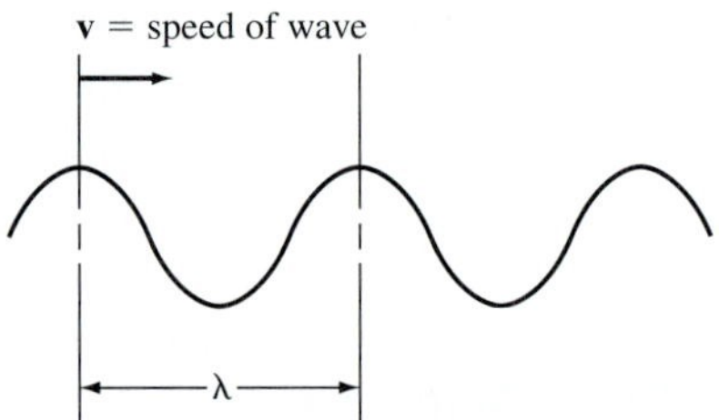

Figure 6–2 Schematic of typical wave motion.

at a speed and having a wavelength and a frequency. In figure 6–2 is shown such a wave motion. The frequency of the wave, ν is given by equation (6–1) and it represents a parameter for describing a wave form or electromagnetic (EM) radiation.

$$\nu = \frac{\mathbf{v}}{\lambda} \tag{6–1}$$

Here **v** is the speed or velocity of the wave, λ is the wavelength, and ν is the frequency of the wave. Thermal radiation travels at the speed of light, and in a vacuum the speed of light is approximately 2.998×10^8 m/s. Electromagnetic radiation may be described by its wavelength, and thermal radiation and the radiation heat-transfer portion of the EM spectrum are found to occur between wavelengths of 10^2 μm ($=10^{-4}$ m) and 10^{-1} μm ($=10^{-7}$ m). We say that the spectrum of thermal radiation is from 10^{-1} μm to 10^2 μm.

In figure 6–3 is shown the electromagnetic spectrum based on wavelength. Notice that the visible light portion of the spectrum, or "light," including the color spectrum, is part of the thermal radiation spectrum. There are other forms of electromagnetic radiation that also are conveyors of energy; however, thermal radiation is the form that accounts for thermal equilibrium. Consider, for instance, a closed container being perfectly evacuated (a vacuum), having inner surface temperature T_1, and having a mass at temperature T_2 not equal to T_1 inside it, as shown in figure 6–4. Only thermal radiation can bring these two surfaces to thermal equilibrium, where finally $T_2 = T_1$. It can be seen from the EM spectrum of figure 6–3 that at a surface, emitted or absorbed thermal radiation will be affected by wavelength (or frequency). Radiation heat transfer analysis requires understanding this spectral behavior. In the next section we will discuss this concept further.

Figure 6–3 Thermal radiation portion of the electromagnetic spectrum.

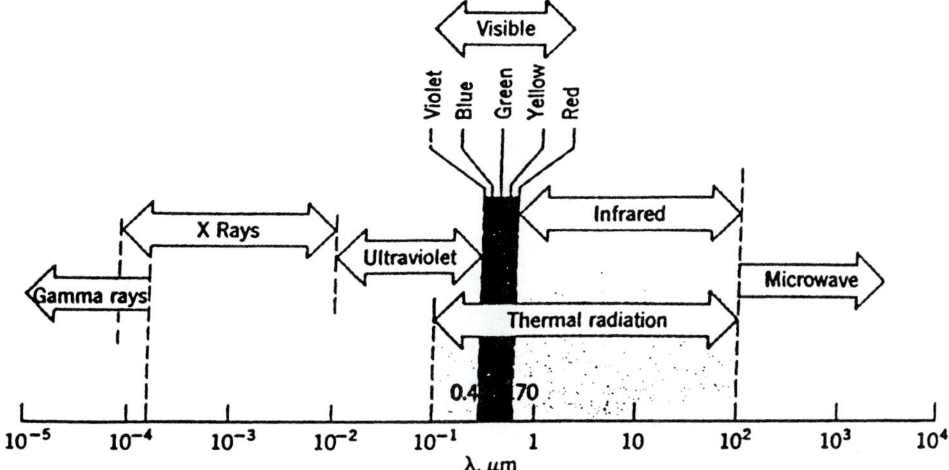

Figure 6–4 Evacuated chamber with mass.

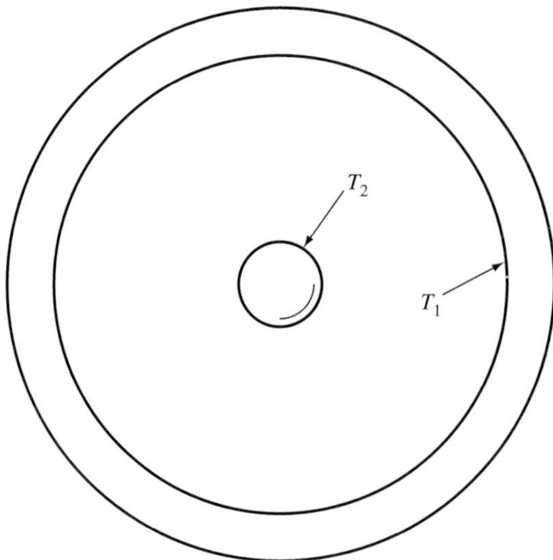

Thermal radiation emitted from a surface can be affected by direction; for instance, consider a surface A_1 shown in figure 6–5, emitting radiation. At one specific point of the surface, the differential area dA_1, thermal radiation can be emitted in all directions in the hemisphere above the surface. Due to irregularities in the surface smoothness, molecular or atomic makeup, or other reasons, the emission can likely be greater or less in different directions, so radiation heat-transfer analysis requires an understanding of the directional nature of thermal radiation. The reader should also recognize that a complicating feature of the radiation from surface A_1 is that other portions of A_1 can emit radiation, so the thermal radiation emanating from surface dA_1 is extremely difficult to experimentally determine.

As a concept for better understanding radiation we consider the differential amount of thermal energy emitted from surface dA_1 in the direction shown in figure 6–6, denoted as $d\dot{E}_{dA_1}$. The angle Θ is the azimuth angle to the surface dA_1 and ϕ is the lateral angle

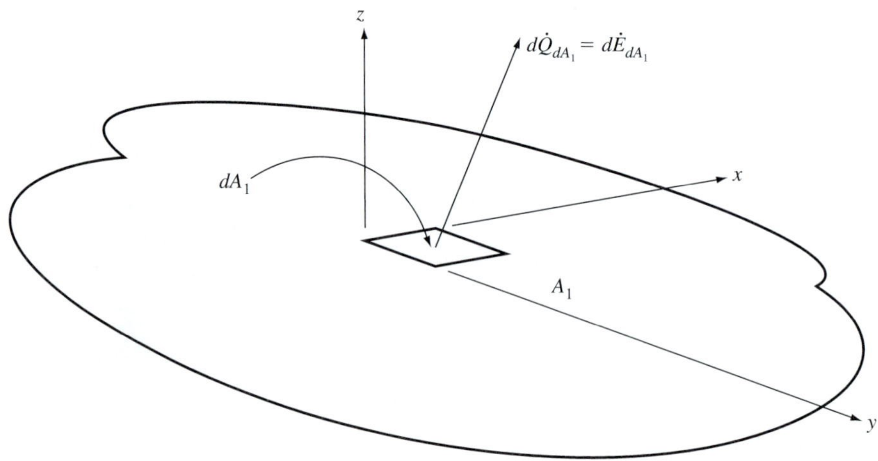

Figure 6–5 Thermal radiation from a portion of a surface.

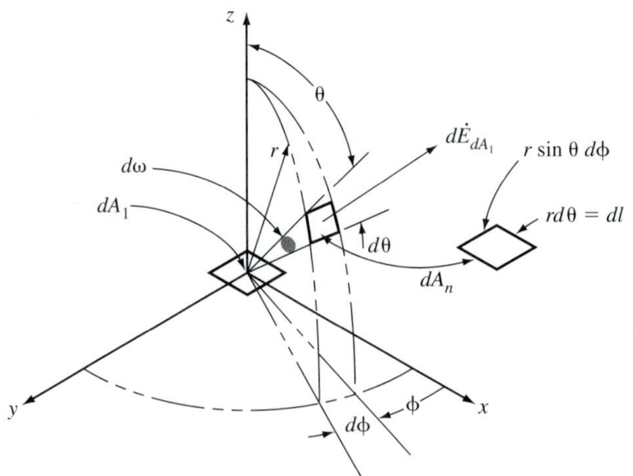

Figure 6–6 Geometry of thermal radiation.

on surface A_1. The surface dA_n is a surface normal to $d\dot{E}_{dA_1}$, and the solid angle $d\omega$ is defined as

$$d\omega = \frac{dA_n}{r^2} \qquad [\text{Units of steradians, str (ab)}] \tag{6–2}$$

As a point of reference, the reader may remember that an angle in the two-dimensional plane can be described in radians, with 2π radians equalling 360 degrees. In three-dimensional space the solid angle is described in steradians, and there are 4π steradians in a sphere of rotation or 2π steradians in a hemisphere of rotation. The differential angle, $d\Theta$, illustrated in figure 6–6 may be written

$$d\Theta = \frac{dl}{r} \qquad (\text{radians}) \tag{6–3}$$

and then

$$dl = r d\Theta$$

From figure 6–6 we write

$$dA_n = r \sin\Theta \, d\Theta d\phi$$

and the solid angle $d\omega$ is then, from figure 6–6,

$$d\omega = \sin\Theta \, d\Theta d\phi \tag{6–4}$$

The construction of the solid angle $d\omega$ through which the thermal energy $d\dot{E}_{dA_1}$ is directed is indicated as a point at the origin.

It is important to recognize that the area dA is not a point but rather a differential area; however, it is useful to approximate the origin of the solid angle $d\omega$ as a point. It is also useful to define the spectral intensity as

$$\dot{I}_{dA_1}(\lambda, T, \Theta, \phi) = \frac{1}{dA_n d\omega d\lambda} d\dot{E}_{dA_1} = \frac{1}{dA_1 \cos\Theta \, d\omega d\lambda} d\dot{E}_{dA_1} \,(\text{W/m}^2 \cdot \text{str} \cdot \mu\text{m}) \tag{6–5}$$

where $d\lambda$ is a differential bandwidth in the EM spectrum. Notice that the spectral intensity is based on the area from which the energy eminates, dA_1, projected in the azimuthal direction ϕ. Thus, the projected area along direction ϕ is $dA_n = dA_1 \cos\phi$. The spectral intensity is a function of temperature because the emitted energy $d\dot{E}_{dA_1}$, is a function of temperature. Notice that the spectral intensity is not a function of the distance from the emitting surface, r. Thus, $\dot{I}_{dA_1}$ is a constant in the radial direction. On the other hand, as we discussed earlier, the radiation may change with azimuth angle or lateral angle.

It is convenient to define the concept of diffuse radiation as radiation that has spectral intensity that is invariant in space. Thus, a diffuse surface emits diffuse radiation, where the spectral intensity is only a function of wavelength and temperature,

$$\dot{I}_{dA_1}(\lambda, T) \qquad (\text{for a diffuse radiator})$$

The opposite of diffuse radiation is direct radiation, which is defined as radiation traveling in a straight path without dispersing. That is, direct radiation emitted from a normal area dA_n will follow a path such that the area dA_n through which it passes is always the same:

$$dA_n = dA_i \qquad \text{for all } i \tag{6–6}$$

as shown in figure 6–7. Sunlight on a clear day and a laser beam of light are considered to be direct radiation, although radiation is neither completely diffusion nor direct (sunlight has characteristics of both). In this text we will usually assume radiation to be diffuse except when it is noted otherwise.

Figure 6–7 Direct radiation.

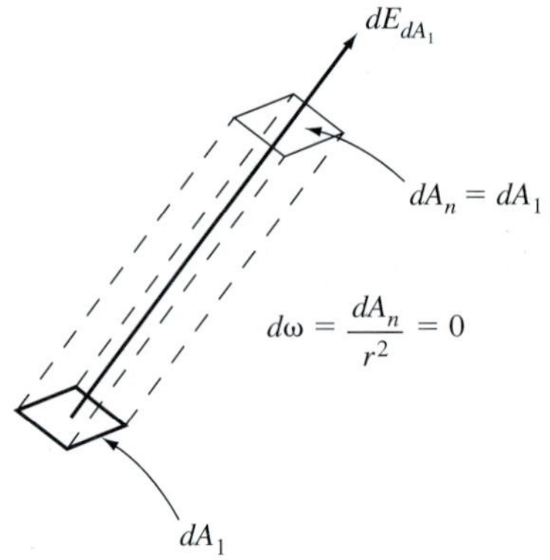

From equation (6–5) we may write the spectral differential energy per unit area, $\dot{E}_{dA}(\lambda, T, \Theta, \phi)$, as

$$\frac{d\dot{E}_{dA_1}}{dA_1} = \dot{I}(\lambda, T, \Theta, \phi) \cos\Theta \, d\omega d\lambda \tag{6–7}$$

For diffuse radiation the spectral differential energy per unit area is

$$\frac{d\dot{E}_{dA_1}}{dA_1} = \frac{d\dot{E}_{dA_1}(\lambda, T)}{dA_1} = \dot{I}(\lambda, T) \cos\Theta \, d\omega d\lambda = \dot{I}(\lambda, T) \cos\Theta \sin\Theta \, d\Theta d\phi d\omega d\lambda$$

The total spectral energy emitted from dA_1 in all directions (in the hemispherical area outward from the surface) and for the full thermal spectrum is

$$\dot{E}_{dA_1} = \int_0^{\lambda=\infty} \int_0^{\phi=2\pi} \int_0^{\Theta=\pi/2} \dot{I}(\lambda, T) \cos\Theta \sin\Theta \, d\Theta d\phi d\lambda \tag{6–8}$$

For diffuse radiation this reduces to

$$\dot{E}_{dA_1} = \pi \int_0^{\infty} \dot{I}(\lambda, T) d\lambda \tag{6–9}$$

EXAMPLE 6–1 Diffuse radiation is emitted from a very small differential surface area. The radiation has intensity of 2800 W/m^2 · str, and a partial shield is placed over the source as shown in figure 6–8. Determine the radiation that is not intercepted by the shield.

Solution We may use a form of equation (6–8) with the following adjustments:

1. Θ integrated from 0 to $\pi/6$ (30°)
2. $d\omega = \sin\Theta \, d\Theta \, d\phi$

Then

$$\dot{E}(\lambda, T) = \dot{I}(\lambda, T) \int_0^{2\pi} \int_0^{\pi/6} \cos\Theta \sin\Theta \, d\Theta \, d\phi$$

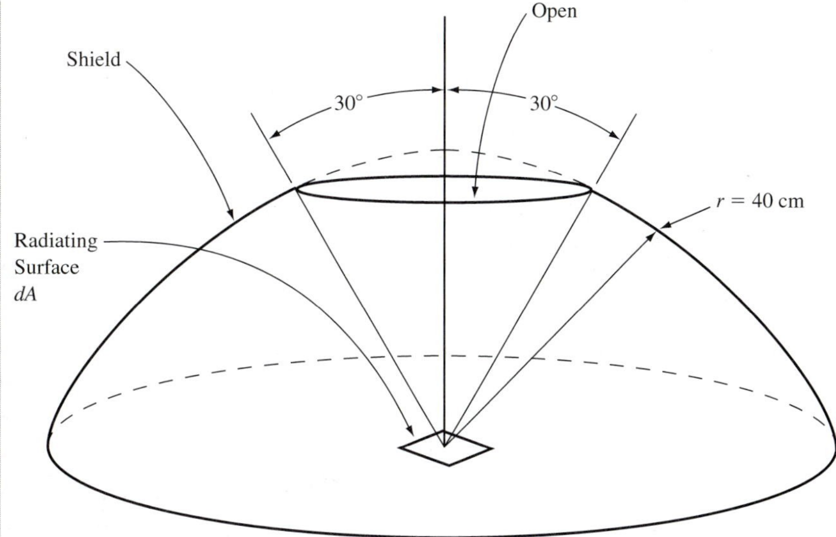

Figure 6–8 Shield placed over a differential source of thermal radiation.

and since $\cos\Theta\sin\Theta = 1/2\sin 2\Theta$, we get

$$\dot{E}(\lambda, T) = \dot{I}(\lambda, T)\int_0^{2\pi}\int_0^{\pi/6}\frac{1}{2}\sin 2\Theta\, d\Theta\, d\phi = 1800\frac{\mathrm{W}}{\mathrm{m}^2\cdot\mathrm{str}}(2\pi)\left[-\frac{1}{2}\cos 2\Theta\right]_0^{\pi/6}$$

Answer

$$= 1800\frac{\mathrm{W}}{\mathrm{m}^2\cdot\mathrm{str}}\left(\frac{\pi}{4}\mathrm{str}\right) = 1413.6\frac{\mathrm{W}}{\mathrm{m}^2}$$

Notice that the full hemispherical emission when the shield is removed or ignored and Θ is integrated from 0 to $\pi/2$, the total radiation is

$$\dot{E}(\lambda, T) = \pi[\dot{I}(\lambda, T)] = 5654.8\frac{\mathrm{W}}{\mathrm{m}^2}$$

EXAMPLE 6–2 Estimate the solid angle between the two surfaces A_1 and A_2 shown in figure 6–9.

Figure 6–9 Arrangement of two surfaces A_1 and A_2.

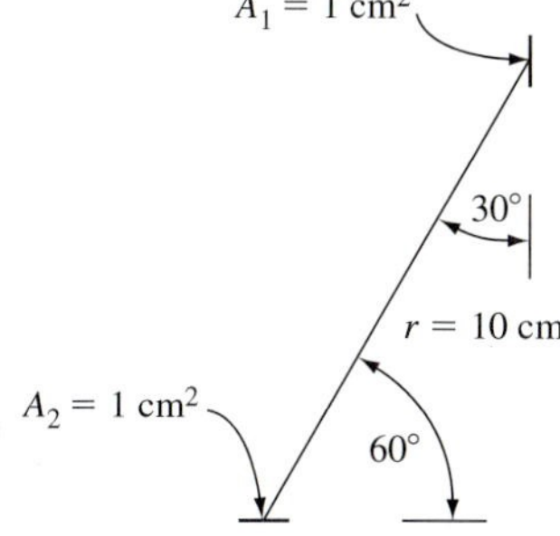

Solution

Since the distance between the two surfaces, r, is much greater than the square root of either of the surface areas, we may write, from equation (6–2) that

$$\omega_{12} = (\text{solid angle from } A_1 \text{ to } A_2) = \frac{A_{2n}}{r^2} = \frac{A_2\cos\Theta_2}{r^2}$$

$$\omega_{21} = (\text{solid angle from } A_2 \text{ to } A_1) = \frac{A_{1n}}{r^2} = \frac{A_1\cos\Theta_1}{r^2}$$

Then

Answer
$$\omega_{12} = \frac{(1 \times 10^{-4}\ \text{m}^2)\cos 30^\circ}{(10\ \text{m})^2} = 0.866 \times 10^{-6}\ \text{steradians (str)}$$

Answer
$$\omega_{21} = \frac{(1 \times 10^{-4}\ \text{m}^2)\cos 60^\circ}{(10\ \text{m})^2} = 0.5 \times 10^{-6}\ \text{str}$$

The solid angle from A_1 to A_2 is larger than that from A_2 to A_1.

The absorption or receiving of thermal radiation from a surface as shown in figure 6–10 (for instance, surface dA_2 receiving radiation from dA_1) can be handled geometrically the same as emission.

Figure 6–10

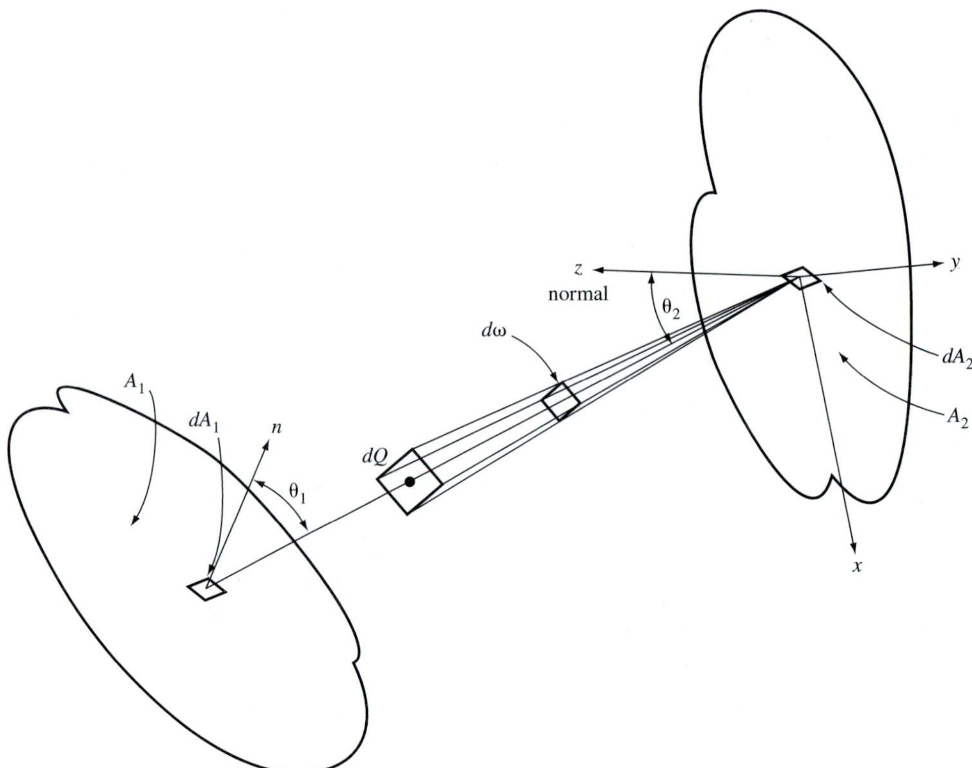

In figure 6–10 is shown the situation where the surface dA_2 is the surface of interest. The solid angle $d\omega$, the azimuthal and lateral angles, Θ and ϕ, the projected area $dA_2 \cos\Theta$ and the normal area $dA_n = r^2 \cos\Theta\, d\Theta d\phi$ are all comparable to that for the analysis of emission, treated in figure 6–6.

We call the energy directed to a surface dA_2 *irradiation* as opposed to *emission,* where energy is emitted. The irradiated energy on a diffuse surface dA_2 is written

$$\dot{E}_{dA_2} = \pi \int_0^\infty \dot{I}(\lambda, T)\, d\lambda \qquad \textbf{(6–10)}$$

where

$$\dot{I}(\lambda, T) = \frac{1}{dA_2 \cos \Theta_2 d\omega_{21} d\lambda} d\dot{E}_{dA_2} \tag{6–11}$$

we will make use of these equations in the discussions of radiation heat transfer between two surfaces in the next sections.

6–2 BLACK BODY RADIATION

The concept of the black body has been useful for analyzing and understanding thermal radiation. The black body has the following characteristics:

1. Diffuse radiator and absorber.
2. Absorbs all irradiation.
3. Emits radiation at the maximum amount at which photons arrive at the surface from inside the material.

The black body is an ideal thermal radiator or absorber, and although there is no real surface that is a black body, some surfaces can be approximated as black bodies. The term *black body* does not mean that all black surfaces are black bodies, nor does it mean that only black surfaces are black bodies. Some surfaces that are colored black do show some characteristics of black bodies; however, surface color does not determine whether a surface can be approximated as a black body. The inside surface of the configuration shown in figure 6–4 will absorb all radiation on it, and if a pinhole were put in the container, then the hole would act as a black body from the outside. It has been shown by Planck [1] that the spectral radiation for a black body radiator is

$$\dot{I}_B(\lambda, T) = \frac{2\hbar c_0^2}{\lambda^5 (e^{\hbar c_0/\lambda T} - 1)} \tag{6–12}$$

where $\hbar$ is Planck's Constant, 6.62554×10^{-34} J · s, and c_0 is the velocity of light in a vacuum. Further, the black body has a diffuse surface so that the total spectral power emitted from its surface is

$$\dot{E}_B = \pi \int_0^{\infty} \dot{I}_B(\lambda, T) d\lambda \qquad (\text{W/m}^2 \text{ or Btu/hr} \cdot \text{ft}^2) \tag{6–13}$$

In figure 6–11 is shown the spectral intensity emitted by a black body as a function of wavelength and temperature. In the graph of figure 6–11 it can be seen that the spectral emission is strongly affected by the wavelength for a constant temperature. Also, the temperature affects the magnitude of the emission at any given wavelength and also affects the particular wavelength where the spectral emission is maximum. The maximum emission can be determined by differentiating equation (6–13) with respect to the wavelength and setting this to zero,

$$\frac{\partial \dot{E}_B}{\partial \lambda} = 0$$

From this operation it can be found that the maximum spectral energy emitted by a black body occurs at the condition that

$$\lambda T = 2897.8 \mu\text{m} \cdot \text{K} \tag{6–14}$$

which is referred to as *Wein's Displacement Law.* The locus of these maxima, or Wein's Displacement Law, is plotted on the graph of figure 6–11 as a phantom line. Notice from the results of the graph or from equation (6–14) that the maximum energy emitted from

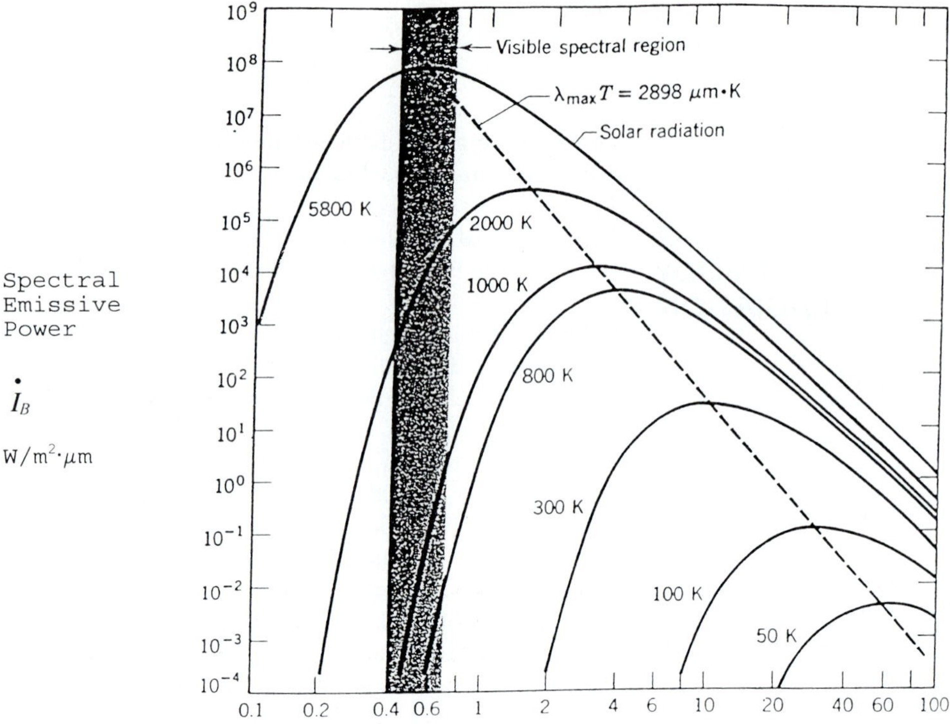

Figure 6–11 Spectral intensity of a black body.

a black body occurs at increasingly shorter wavelengths as the temperature is increased. Also, the integral of the spectral energy over the full spectrum of EM radiation gives the total energy emitted by a black body, and it is

$$\dot{E}_B = \pi \int_0^\infty \dot{I}(\lambda, T)d\lambda = \sigma T^4 \,(\text{W/m}^2 \text{ or Btu/hr} \cdot \text{ft}^2) \tag{6–15}$$

where σ is called the *Stefan–Boltzmann Constant*. The value of the Stefan–Boltzmann Constant is usually taken as

$$\sigma = 0.174 \times 10^{-8} \frac{\text{Btu}}{\text{hr} \cdot \text{ft}^2 \cdot {}^\circ\text{R}^4} = 5.669 \times 10^{-8} \frac{\text{W}}{\text{m}^2 \cdot \text{K}}$$

By definition, the total power $\dot{E}_B$ is the area under the spectral energy curve of figure 6–11 at a constant temperature T.

EXAMPLE 6–3 Compare the power emitted by a black body at room temperature of 70°F and the same black body when it is at 600°F.

Solution At room temperature we have

Answer

$$\dot{E}_B = \sigma T^4 = \left(0.174 \times 10^{-8} \frac{\text{Btu}}{\text{hr} \cdot \text{ft}^2 \cdot {}^\circ\text{R}^4}\right)(530^\circ\text{R})^4 = 137.3 \frac{\text{Btu}}{\text{hr} \cdot \text{ft}^2}$$

At 600°F we have

Answer

$$\dot{E}_B = (0.174 \times 10^{-8})(1060)^4 = 2196.7 \frac{\text{Btu}}{\text{hr} \cdot \text{ft}^2}$$

Notice that there is more than an order of magnitude difference between the emitted power of the black body at room temperature and that when it is at a high temperature. Also, when the black body is at 70°F the spectral power is greatest at a wavelength of

$$\lambda = \frac{2897.8\ \mu\text{m} \cdot \text{K}}{T} = \frac{2897.8\ \mu\text{m} \cdot \text{K}}{530°\text{R}(5\ \text{K}/9°\text{R})} = 5.47\,\mu\text{m}$$

while the maximum occurs at

$$\lambda = \frac{2897.8}{1060 \times 5/9} = 2.7\,\mu\text{m}$$

for 600°F.

It is sometimes important to know the black body radiation over a portion of the EM spectrum. This information can be obtained by integrating equation (6–15) between the limits of the wavelengths bounding the particular spectrum desired. A more general method of handling this problem is to determine the energy as a function of the product λT. Then the fraction of the total power or rate of energy emission of a black body is

$$\frac{\dot{E}_{B,0-\lambda T}}{\sigma T^4} = \frac{\pi}{\sigma T^4} \int_0^{\lambda T} \dot{I}(\lambda, T)\, d\lambda \qquad \textbf{(6–16)}$$

as is presented in table 6–1. The same information is given in graphical form in figure 6–12.

The function $\dot{E}_{B,0-\lambda T}/\sigma T^4$ approaches 1.0 (100%) as the product λT approaches infinity, or becomes large. Notice that the bandwidth from $\lambda T = 0$ to

Figure 6–12 Accumulated fraction of emissive power from a black body radiator.

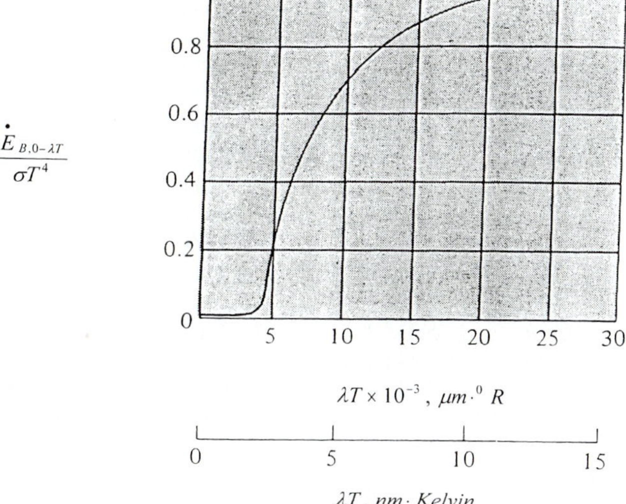

Table 6–1 Radiation Functions

λT		$\frac{\dot{E}_{B,0-\lambda T}}{\sigma T^4}$	λT		$\frac{\dot{E}_{B,0-\lambda T}}{\sigma T^4}$
μm · °R	μm · K		μm · °R	μm · K	
1,000	555.6	0.170×10^{-7}	10,200	5,666.7	0.70754
1,200	666.7	0.756×10^{-6}	10,400	5,777.8	0.71806
1,400	777.8	0.106×10^{-4}	10,600	5,888.9	0.72813
1,600	888.9	0.738×10^{-4}	10,800	6,000.0	0.73777
1,800	1,000.0	0.321×10^{-3}	11,000	6,111.1	0.74700
2,000	1,111.1	0.00101	11,200	6,222.2	0.75583
2,200	1,222.2	0.00252	11,400	6,333.3	0.76429
2,400	1,333.3	0.00531	11,600	6,444.4	0.77238
2,600	1,444.4	0.00983	11,800	6,555.6	0.78014
2,800	1,555.6	0.01643	12,000	6,666.7	0.78757
3,000	1,666.7	0.02537	12,200	6,777.8	0.79469
3,200	1,777.8	0.03677	12,400	6,888.9	0.80152
3,400	1,888.9	0.05059	12,600	7,000.0	0.80806
3,600	2,000.0	0.06672	12,800	7,111.1	0.81433
3,800	2,111.1	0.08496	13,000	7,222.2	0.82035
4,000	2,222.2	0.10503	13,200	7,333.3	0.82612
4,200	2,333.3	0.12665	13,400	7,444.4	0.83166
4,400	2,444.4	0.14953	13,600	7,555.6	0.83698
4,600	2,555.6	0.17337	13,800	7,666.7	0.84209
4,800	2,666.7	0.19789	14,000	7,777.8	0.84699
5,000	2,777.8	0.22285	14,200	7,888.9	0.85171
5,200	2,888.9	0.24803	14,400	8,000.0	0.85624
5,400	3,000.0	0.27322	14,600	8,111.1	0.86059
5,600	3,111.1	0.29825	14,800	8,222.2	0.86477
5,800	3,222.2	0.32300	15,000	8,333.3	0.86880
6,000	3,333.3	0.34734	16,000	8,888.9	0.88677
6,200	3,444.4	0.37118	17,000	9,444.4	0.90168
6,400	3,555.6	0.39445	18,000	10,000.0	0.91414
6,600	3,666.7	0.41708	19,000	10,555.6	0.92462
6,800	3,777.8	0.43905	20,000	11,111.1	0.93349
7,000	3,888.9	0.46031	21,000	11,666.7	0.94104
7,200	4,000.0	0.48085	22,000	12,222.2	0.94751
7,400	4,111.1	0.50066	23,000	12,777.8	0.95307
7,600	4,222.2	0.51974	24,000	13,333.3	0.95788
7,800	4,333.3	0.53809	25,000	13,888.9	0.96207
8,000	4,444.4	0.55573	26,000	14,444.4	0.96572
8,200	4,555.6	0.57267	27,000	15,000.0	0.96892
8,400	4,666.7	0.58891	28,000	15,555.6	0.97174
8,600	4,777.8	0.60449	29,000	16,111.1	0.97423
8,800	4,888.9	0.61941	30,000	16,666.7	0.97644
9,000	5,000.0	0.63371	40,000	22,222.2	0.98915
9,200	5,111.1	0.64740	50,000	27,777.8	0.99414
9,400	5,222.2	0.66051	60,000	33,333.3	0.99649
9,600	5,333.3	0.67305	70,000	38,888.9	0.99773
9,800	5,444.4	0.68506	80,000	44,444.4	0.99845
10,000	5,555.6	0.69655	90,000	50,000.0	0.99889
			100,000	55,555.6	0.99918

$\lambda T = 30 \times 10^3 \mu$in. · °R (≈16 nm · K) includes nearly 99.6% of the total emitted energy from a black body. Emitted energy over a band of the spectrum from λ_1 to λ_2 at temperature T can be obtained from the relationship

$$\dot{E}_{B,\lambda_1 T-\lambda_2 T} = \pi \int_{\lambda_1 T}^{\lambda_2 T} \dot{I}(\lambda, T)\,d\lambda = \sigma T^4\left(\frac{\dot{E}_{B,0-\lambda_2 T}}{\sigma T^4} - \frac{\dot{E}_{B,0-\lambda_1 T}}{\sigma T^4}\right) \tag{6–17}$$

where the two terms inside the bracket on the right side of the equation are obtained from table 6–1 or figure 6–11.

EXAMPLE 6–4 Determine the energy emitted in the visible range of the EM spectrum from a black body at 1000°C.

Solution The visible range as indicated in the spectrum of figure 6–3 is between $\lambda_1 = 0.4\ \mu$m and $\lambda_2 = 0.7\ \mu$m. Then

$$\lambda_1 T = (0.4\ \mu\text{m})(1273\ \text{K}) = 509.2\ \mu\text{m}\cdot\text{K} \quad \text{and} \quad \lambda_2 T = (0.7\ \mu\text{m})(1273\ \text{K}) = 891.1\ \mu\text{m}\cdot\text{K}$$

From table 6–1 we read that

$$\frac{\dot{E}_{B,0-\lambda_1 T}}{\sigma T^4} < 0.170 \times 10^{-7} \approx 0 \quad \text{and} \quad \frac{\dot{E}_{B,0-\lambda_2 T}}{\sigma T^4} = 0.74 \times 10^{-4}$$

From equation (6 – 17) we have

Answer

$$\dot{E}_{B,\lambda_1 T-\lambda_2 T} = \sigma T^4(0.74 \times 10^{-4} - 0) = 11.02\ \text{W/m}^2$$

The black body at 1000°C is not at a high enough temperature to emit a significant amount of energy in the visible range. That is, of the energy emitted from the black body only 0.0074% is in the visible range.

6–3 GRAY BODY MECHANISMS

Most surfaces do not behave as black bodies, and to analyze radiation heat transfer for actual surfaces we need to consider what happens to irradiation, or thermal radiation, directed toward such a surface. The irradiation, $\dot{E}_{irr}$, as shown in figure 6–13, is either

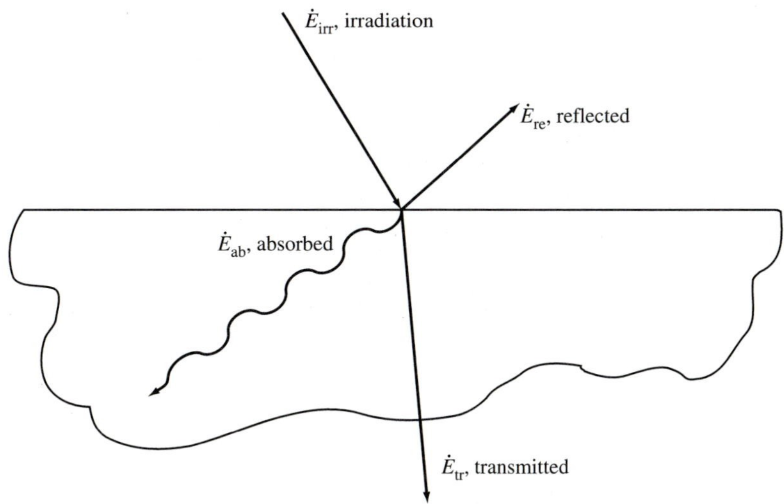

Figure 6–13 Schematic of the paths taken by irradiation onto a surface.

absorbed into the surface as $\dot{E}_{\text{ab}}$, reflected off the surface as $\dot{E}_{\text{re}}$, or transmitted through the surface material as $\dot{E}_{\text{tr}}$. We can then write that

$$\dot{E}_{\text{irr}} = \dot{E}_{\text{ab}} + \dot{E}_{\text{re}} + \dot{E}_{\text{tr}} \tag{6–18}$$

and if we express these terms as fractions,

$$\frac{\dot{E}_{\text{ab}}}{\dot{E}_{\text{irr}}} + \frac{\dot{E}_{\text{re}}}{\dot{E}_{\text{irr}}} + \frac{\dot{E}_{\text{tr}}}{\dot{E}_{\text{irr}}} = 1.0 \tag{6–19}$$

These fractions are defined as

$$\frac{\dot{E}_{\text{ab}}}{\dot{E}_{\text{irr}}} = \alpha_r = \text{absorptivity} \tag{6–20}$$

$$\frac{\dot{E}_{\text{re}}}{\dot{E}_{\text{irr}}} = \rho_r = \text{reflectivity} \tag{6–21}$$

$$\frac{\dot{E}_{\text{tr}}}{\dot{E}_{\text{irr}}} = \tau_r = \text{transmissivity} \tag{6–22}$$

and equation (6–19) can then be written

$$\alpha_r + \rho_r + \tau_r = 1.0 \tag{6–23}$$

The reflectivity, absorptivity, and transmissivity are radiation response properties of the material of the surface, and they may be functions of the material temperature and the wavelength of the spectral radiation. The spectral radiation is dependent on the source of that radiation, so we have an intriguing situation in that the radiation response of a particular surface is dependent not only on the material properties but also on the properties of an external surface, far removed from the point of interest.

We saw in section 6–2 that a black body emits radiation diffusely and with a magnitude of σT^4 for a unit area of the surface. By definition, the black body emits the maximum possible amount of thermal radiation at a given temperature and wavelength, so an actual surface will be expected to emit some fraction of that black body radiation. We define *emissivity,* ε_r for a surface of a body as

$$\varepsilon_r = \frac{1}{\sigma T^4} \int_0^{\infty} \int_0^{2\pi} \int_0^{\pi/2} \varepsilon_{r,\lambda}(T, \Theta, \phi, \lambda) \cdot \dot{I}_B(T, \Theta, \phi, \lambda) \cos\Theta \sin\Theta \, d\Theta d\phi d\lambda \tag{6–24}$$

which is the expression that includes the hemispherical radiation emitted from a surface, accounting for any directional dependency of the radiation and for the temperature. Notice that the emissivity is the hemispherical radiation as a fraction of that of a black body. In equation (6–24), the spectral emissivity, or emissivity at a particular wavelength, is

$$\varepsilon_{r,\lambda}(T, \Theta, \phi, \lambda) = \frac{1}{\dot{I}_B(\lambda, T)} \dot{I}_{dA}(T, \Theta, \phi, \lambda) \tag{6–25}$$

For a diffuse radiator surface the spectral emissivity and the intensity are only functions of the wavelength and the temperature. Then equation (6–24) reduces to, for a diffuse radiator,

$$\varepsilon_r = \frac{1}{\sigma T^4} \int_0^{\infty} \varepsilon_{r,\lambda}(\lambda, T) \cdot \dot{E}_B(\lambda, T) d\lambda \tag{6–26}$$

EXAMPLE 6–5 A diffuse surface has spectral emissivity given by the graph of figure 6–14. Determine the emissivity of the surface at a temperature of 35°C.

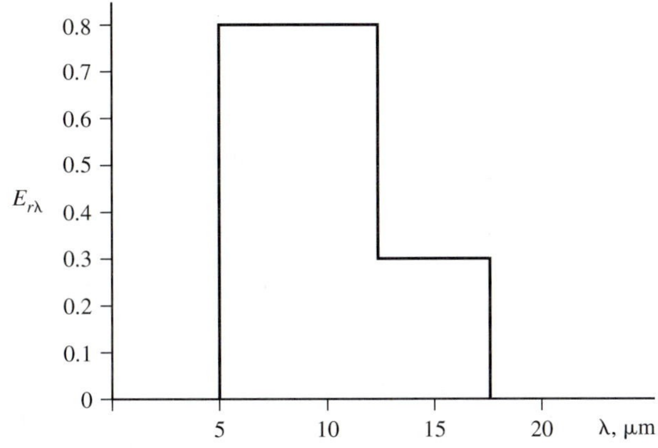

Figure 6–14 Spectral emissivity.

Solution We use equation (6–26) to obtain the emissivity. The integration can be done in two parts:

$$\varepsilon_r = \int_{\lambda_1 = 5\mu m}^{\lambda_2 = 12\mu m} \varepsilon_{r,\lambda}(\lambda, T)_{\lambda_1}^{\lambda_2} \cdot \frac{\dot{E}_B(\lambda, T)_{\lambda_1}^{\lambda_2}}{\sigma T^4} d\lambda + \int_{\lambda_2 = 12\mu m}^{\lambda_3 = 18\mu m} \varepsilon_{r,\lambda}(\lambda, T)_{\lambda_2}^{\lambda_3} \cdot \frac{\dot{E}_B(\lambda, T)_{\lambda_2}^{\lambda_3}}{\sigma T^4} d\lambda$$

The emissivity is zero outside these integral domains of the wavelength spectrum. Then, from figure 6–14,

$$\varepsilon_{r,\lambda}(\lambda, T)_{\lambda_1}^{\lambda_2} = 0.8$$

$$\varepsilon_{r,\lambda}(\lambda, T)_{\lambda_2}^{\lambda_3} = 0.3$$

$$\varepsilon_r = 0.8 \int_{\lambda_1}^{\lambda_2} \frac{\dot{E}_B(\lambda, T)_{\lambda_1}^{\lambda_2}}{\sigma T^4} d\lambda + 0.3 \int_{\lambda_2}^{\lambda_3} \frac{\dot{E}_B(\lambda, T)_{\lambda_2}^{\lambda_3}}{\sigma T^4} d\lambda$$

From equation (6–17) we have

$$\int_{\lambda_1}^{\lambda_2} \frac{\dot{E}_B(\lambda, T)_{\lambda_1}^{\lambda_2}}{\sigma T^4} d\lambda = \frac{\dot{E}_{B,0-\lambda_2 T}}{\sigma T^4} - \frac{\dot{E}_{B,0-\lambda_1 T}}{\sigma T^4}$$

$$\int_{\lambda_2}^{\lambda_3} \frac{\dot{E}_B(\lambda, T)_{\lambda_2}^{\lambda_3}}{\sigma T^4} d\lambda = \frac{\dot{E}_{B,0-\lambda_3 T}}{\sigma T^4} - \frac{\dot{E}_{B,0-\lambda_2 T}}{\sigma T^4}$$

Thus, we may determine the emissivity from the spectral information provided in table 6–1. We have at $T = 35°C = 308$ K the following data read from table 6–1:

	λT, μm · K	**$\dfrac{\dot{E}_{B,0-\lambda T}}{\sigma T^4}$**
$\lambda_1 = 5\ \mu m$	1540	0.0162
$\lambda_2 = 12\ \mu m$	3696	0.418
$\lambda_3 = 18\ \mu m$	5544	0.69

Then we have

Answer

$$\varepsilon_r = 0.8(0.418 - 0.0162) + 0.3(0.69 - 0.418) = 0.403$$

The surface of example 6–5 is called a *selective surface* because its emissivity varies with wavelength. Selective surfaces may also have radiation response properties which vary with wavelength. For instance, the transmissivity, defined by equation (6–22), can be written for a diffuse selective surface as

$$\tau_r = \frac{1}{\sigma T^4}\int_0^\infty \tau_{r,\lambda}(\lambda, T)\cdot \dot{E}_B(\lambda, T)d\lambda \tag{6–27}$$

where $\tau_{r,\lambda}(\lambda, T)$ is called the spectral transmissivity for a diffuse surface and can be a function of the wavelength of the irradiation as well as of the surface temperature.

In a similar manner we may define the reflectivity as

$$\rho_r = \int_0^\infty \rho_{r,\lambda}(\lambda, T)\cdot \frac{\dot{E}_B(\lambda, T)}{\sigma T^4}d\lambda \tag{6–28}$$

and the absorptivity as

$$\alpha_r = \int_0^\infty \alpha_{r,\lambda}(\lambda, T)\cdot \frac{\dot{E}_B(\lambda, T)}{\sigma T^4}d\lambda \tag{6–29}$$

where $\rho_{r,\lambda}(\lambda, T)$ and $\alpha_{r,\lambda}(\lambda, T)$ are the spectral reflectivity and spectral absorptivity respectively. The spectral reflectivity and the spectral absorptivity can be functions of the radiation wavelength and the surface temperature.

We define a *gray body* or *gray surface* as a diffuse surface whose emissivity is a constant, or does not depend on the wavelength. Therefore, for a gray body

$$\varepsilon_r(\lambda, T) = \varepsilon_r(T) \tag{6–30}$$

which reduces the result of equation (6–26) to a constant. It is appropriate to call a surface whose emissivity is a fraction of 1.0 a gray body.

Kirchhoff's Law of Radiation

Consider again the vacuum chamber with the enclosed mass of figure 6–4. For simplicity we assume mass m_2 to be a gray body. The emitted energy of mass m_2 is

$$\dot{E}_{2,\text{emitted}} = \varepsilon_{r2}A_2T_2^4 \tag{6–31}$$

and the irradiation on the mass is

$$\dot{E}_{\text{irr}} = \alpha_{r2}A_2T_1^4 \tag{6–32}$$

The net heat transfer to the mass m_2 is the irradiation to it minus its emitted radiation, or

$$\dot{Q}_{\text{net}} = \dot{\text{E}}_{\text{irr}} - \dot{E}_{2,\text{emitted}} = \alpha_{r2}A_2T_1^4 - \varepsilon_{r2}A_2T_2^4 \tag{6–33}$$

At thermal equilibrium the net heat transfer must be zero, and thermal equilibrium implies that the temperatures are same, $T_1 = T_2$. Then, for thermal equilibium, equation (6–33) reduces to

$$\alpha_{r2} = \varepsilon_{r2} \tag{6–34}$$

We may extend this argument to include selective surfaces so that in general

$$\alpha_{r,\lambda}(\lambda, T) = \varepsilon_{r,\lambda}(\lambda, T) \tag{6–35}$$

These two equations, (6–34) and (6–35), are called *Kirchhoff's Law of Radiation*. It should be noted that the crucial assumption is one of thermal equilibrium between the two interacting surfaces and that, unless there is no net heat transfer with surroundings that have an identical temperature, the emissivity may not be equal to the absorptivity. We may now modify equation (6–23), based on the result of Kirchhoff's Law, to

$$\varepsilon_r + \rho_r + \tau_r = 1.0 \tag{6–36}$$

The form of this equation allows for better experimental determinations of the radiation response properties because absorptivity is difficult, if not impossible, to measure.

Two specific models are used to describe most of the materials involving radiation heat transfer:

1. Opaque gray body.
2. Transparent gray body.

The opaque gray body allows no transmission of irradiation, so its transmissivity is zero. Then equation (6–36) becomes

$$\varepsilon_r + \rho_r = 1.0 \qquad \text{(Opaque gray body)} \tag{6–37}$$

Emissivity values are given for selected surfaces in appendix table B–5, and more extensive information about the radiation response properties is given by Touloukian [2]. The experimental determination of emissivity was made by assuming test materials to be opaque, and reflectivity was then measured in controlled thermal tests. This experimental procedure is discussed further in section 6–5. The majority of engineering analysis involving radiation heat will be based on the opaque gray body model.

The transparent body is defined as one having no absorptivity. Therefore, $\alpha_r = \varepsilon_r = 0$ and equation (3–36) becomes

$$\rho_r + \tau_r = 1.0 \qquad \text{(Transparent body)} \tag{6–38}$$

Glass is one of the most obvious examples of a transparent body, although it is usually a selective material, giving it some important characteristics, as we shall see in section 6–5.

EXAMPLE 6–6 A material is found to have the spectral reflectivity and transmissivity at all temperatures as shown in figure 6–15. Determine the spectral emissivity.

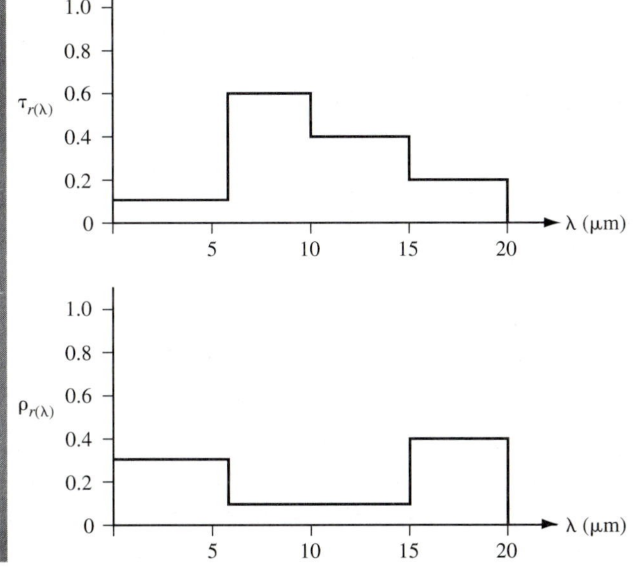

Figure 6–15 Example of spectral reflectivity and transmissivity.

Solution

We assume the validity of Kirchhoff's Law and use equation (6–35) to determine

$$\varepsilon_{r,\lambda}(\lambda) = 1 - \tau_{r,\lambda}(\lambda) - \rho_{r,\lambda}(\lambda)$$

The result is given in figure 6–16.

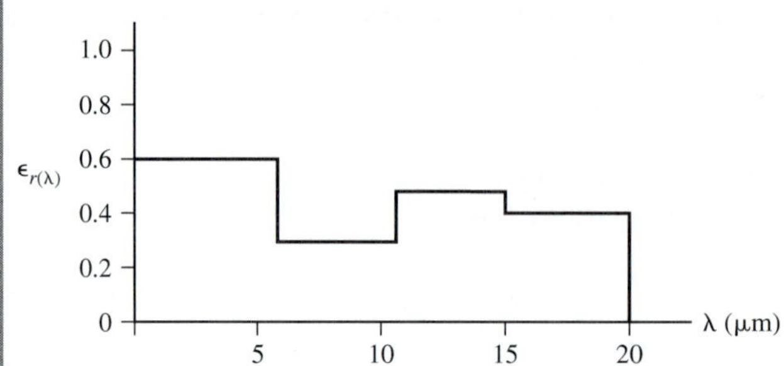

Figure 6–16 Spectral emissivity for surface of example 6–6.

6–4 GEOMETRY OF RADIATION

Radiation heat transfer is strongly affected by the geometric configuration or direction of the radiation. Even if the surfaces radiating or receiving radiation are diffuse surfaces, one must still account for the directional geometric effects. Consider two surfaces A_1 and A_2 as shown in figure 6–17, with radiation emitted from surface dA_1 and irradiating surface dA_2. As we saw in figure 6–5, the radiant rate of energy flow from surface dA_1 to dA_2, $\dot{E}_{dA_1-dA_2}$ can be identified as radiation heat transfer $\dot{Q}_{dA_1-dA_2}$ and given by

$$d\dot{Q}_{dA_1-dA_2} = \dot{I}_{dA_1}(T_1, \Theta_1, \phi_1) dA_1 \cos\Theta_1 \, d\omega_{12} \tag{6–39}$$

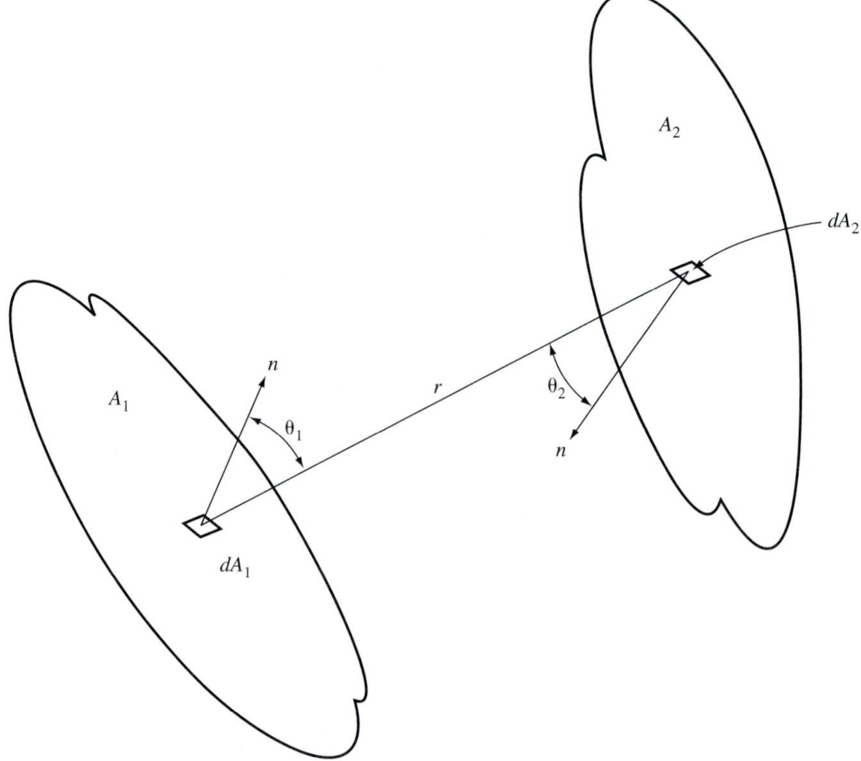

Figure 6–17 Geometric view factor.

where we are treating the full spectrum of wavelengths so that

$$\dot{I}_{dA_1}(T,\Theta,\phi) = \int_0^{\infty} \dot{I}_{dA_1}(T,\Theta,\phi,\lambda)d\lambda$$

The solid angle $d\omega_{12} = \sin\Theta_1 d\Theta_1 d\phi_1$ from equation (6–4) and also the normal area, dA_{n2}, at surface dA_2 is $dA_2\cos\Theta_2$ so that the solid angle may also be written as

$$d\omega_{12} = \cos\Theta_2 \frac{dA_{n2}}{r^2} \tag{6–40}$$

Upon substituting these terms into equation (6–39), we have

$$d\dot{Q}_{dA_1-dA_2} = \dot{I}_{dA_1}(T_1,\Theta_1,\phi_1)\cos\Theta_1\cos\Theta_2\frac{dA_2}{r^2}dA_1 \tag{6–41}$$

For a diffuse surface dA_1 this equation becomes

$$d\dot{Q}_{dA_1-dA_2} = \dot{I}_{dA_1}(T_1)\cos\Theta_1\cos\Theta_2\frac{dA_2}{r^2}dA_1 \tag{6–42}$$

The total hemispherical radiation emitted from a diffuse surface dA_1 is, using equation (6–8),

$$\dot{Q}_{dA_1} = \int_0^{2\pi}\int_0^{\pi/2} \dot{I}_{dA_1}(T_1)\cos\Theta\sin\Theta\, d\Theta d\phi dA_1 = \pi\dot{I}_{dA_1}(T_1)dA_1$$

and the fraction of this that reaches surface dA_2 is just

$$F_{dA_1-dA_2} = \frac{d\dot{Q}_{dA_1-dA_2}}{\dot{Q}_{dA_1}} = \cos\Theta_1\cos\Theta_2\frac{dA_2}{\pi r^2} \tag{6–43}$$

This fraction is called the *radiation shape factor* or *view factor*. The view factor from surface dA_1 to a finite surface A_2 is

$$F_{dA_1-A_2} = \frac{1}{\pi}\int_{A_2} \cos\Theta_1\cos\Theta_2\frac{dA_2}{r^2} \tag{6–44}$$

Notice that the determination of the view factor is strictly an analytic geometry problem, even though the term represents a fraction of diffuse radiation that leaves a surface and irradiates another surface.

The view factor from a finite surface A_1 to another finite surface A_2 can be obtained by revising equation (6–44) to read

$$F_{A_1-A_2}A_1 = \int_{A_1} F_{dA_1-A_2}dA_1 = \frac{1}{\pi}\int_{A_1}\int_{A_2} \cos\Theta_1\cos\Theta_2\frac{dA_2}{r^2}dA_1 \tag{6–45}$$

Using a shorter notation, $F_{A_1-A_2} = F_{1-2}$, equation (6–45) can be written

$$F_{1-2} = \frac{1}{\pi A_1}\int_{A_1}\int_{A_2} \cos\Theta_1\cos\Theta_2\frac{dA_2}{r^2}dA_1 \tag{6–46}$$

The view factor from surface A_2 to A_1 can be developed in the same manner to give

$$F_{2-1} = \frac{1}{\pi A_2}\int_{A_2}\int_{A_1} \cos\Theta_2\cos\Theta_1\frac{dA_1}{r^2}dA_2 \tag{6–47}$$

In figures 6–18 through 6–24 are presented some of the most often used view factors and in table 6–2 analytic expressions for some of these terms are given. The relationships of equations (6–44), (6–46), and (6–47) are the forms most used to determine or derive the expressions for these various view factors. View factors for other geometries may be found in Hottel and Sarofim [3] and Hamilton and Morgan [4].

Figure 6–18 View factor for differential area dA_1 to a parallel flat rectangular area A_2 having one corner directly opposed to the differential area.

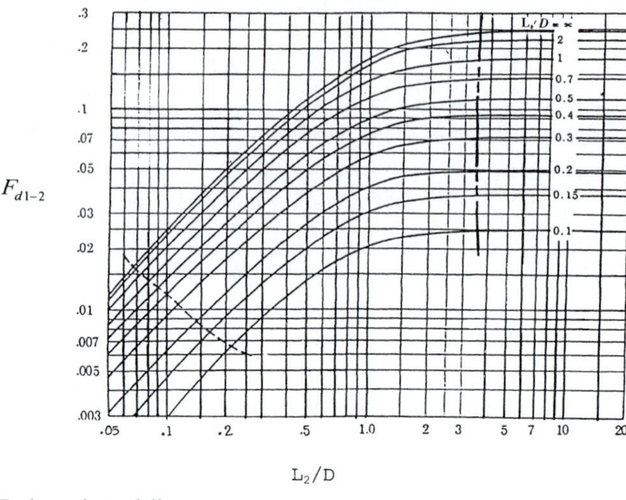

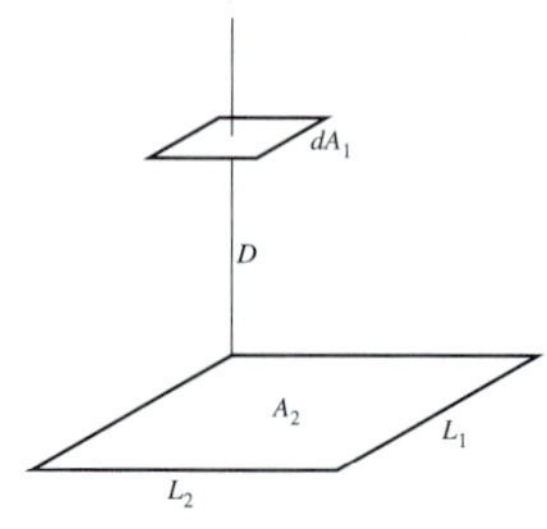

Below dotted line:

$$F_{d1-2} = \frac{L_1 L_2}{\pi\left[D^2 + \frac{1}{8}\left(L_1^2 + L_2^2\right)\right]}$$

To right of phantom line:

$$F_{d1-2} = \frac{L_1}{4\sqrt{L_1^2 + D^2}}$$

See table 6–2(1) for other analytic expressions.

Figure 6–19 View factor for a differential area dA_1 to a rectangular flat area A_2 in a plane normal to dA_1, having one edge in plane of dA_1, where dA_1 is on a normal from one of the corners of A_2, as shown in the figure.

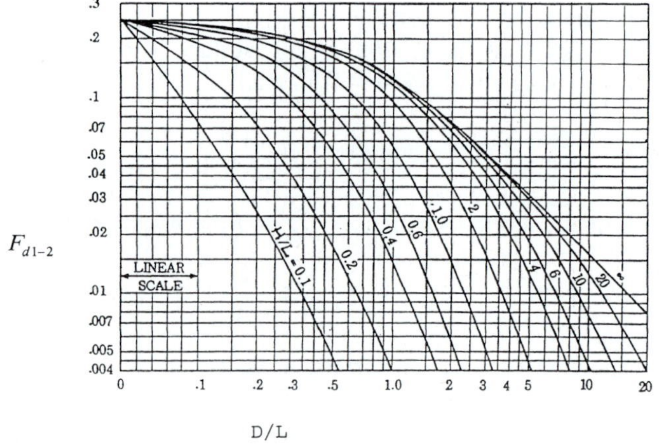

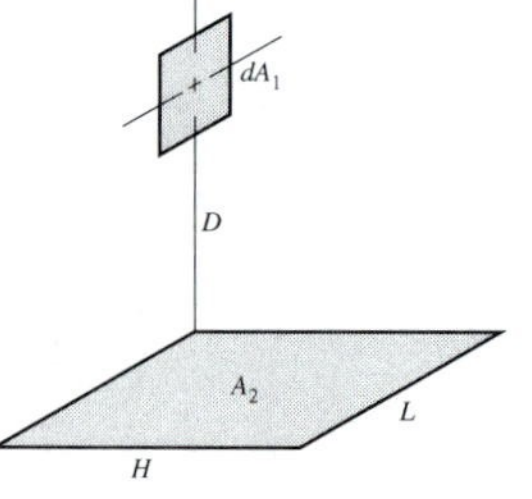

See table 6–2(2) for the analytic expression.

Figure 6–20 The view factor for differential planes and spheres to a large sphere.

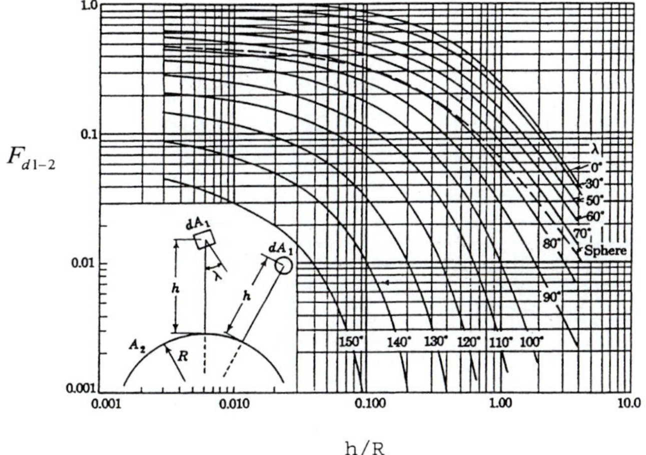

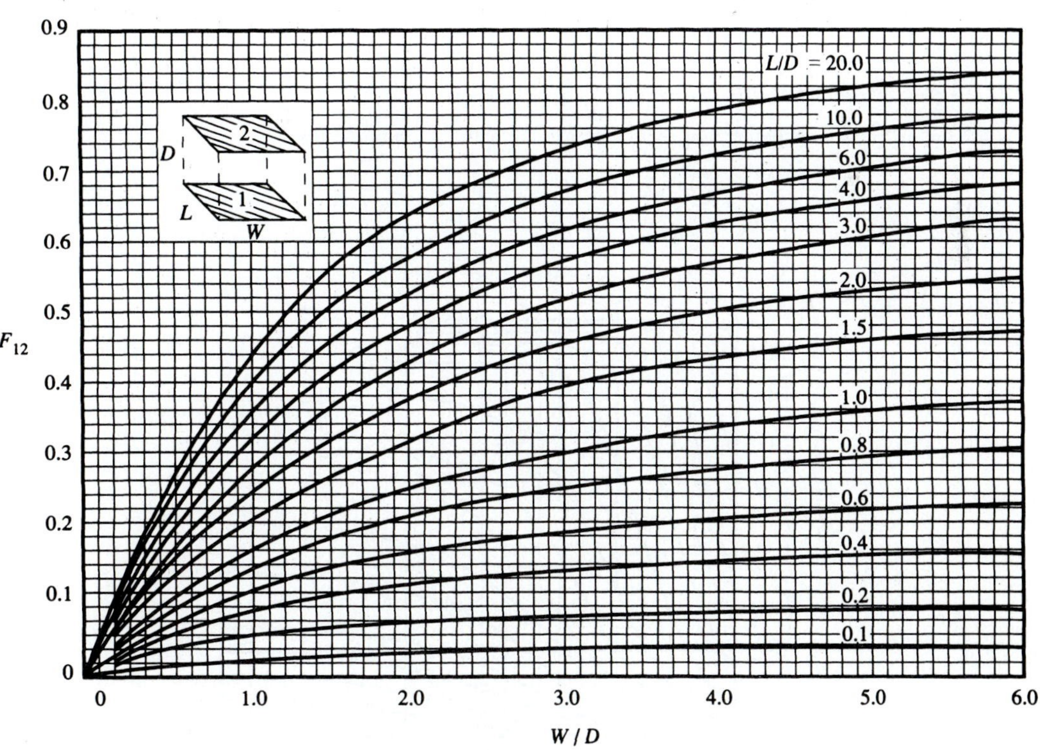

See table 6–2(3) for the analytic expression.

Figure 6–21 View factor for two parallel, directly opposed rectangular areas.

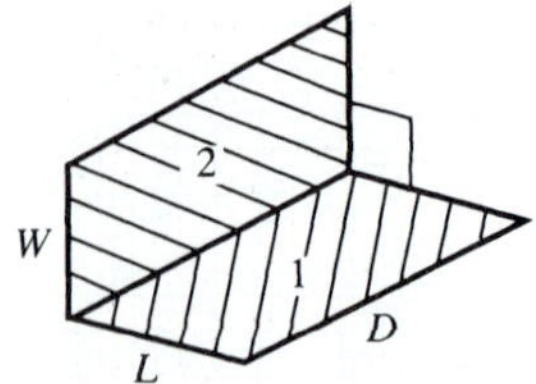

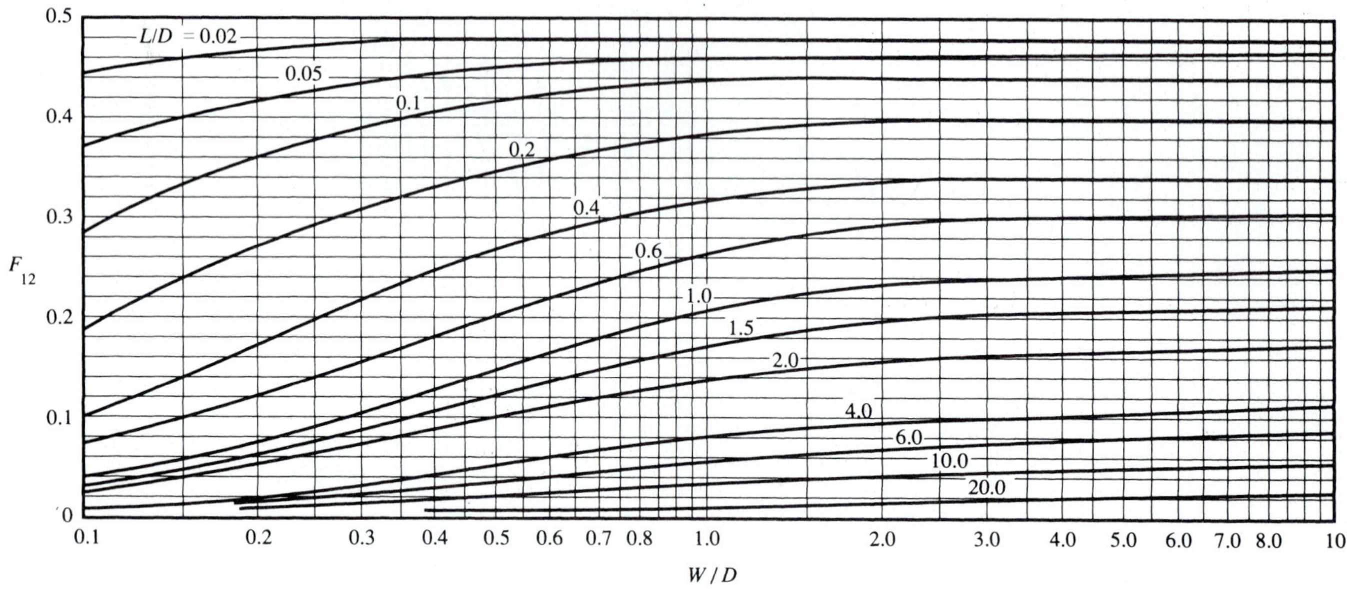

See Table 6–2(4) for analytic expression.

Figure 6–22 View factor for two perpendicular rectangular areas with a common edge.

Figure 6–23 View factor for parallel, concentric disks.

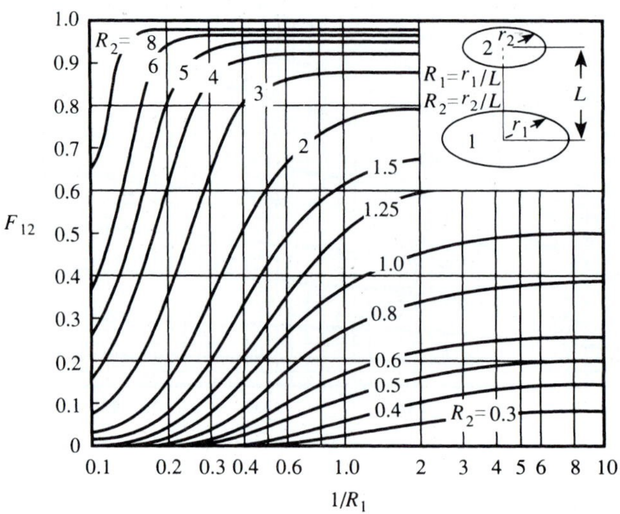

See Table 6–2(5) for analytic expression.

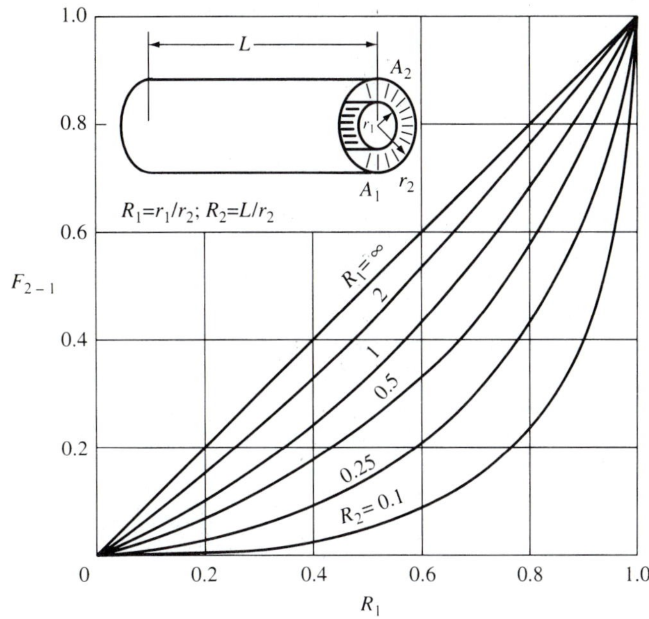

Figure 6–24 The radiation shape factor for concentric cylinders of finite length.

EXAMPLE 6–7 A ventilator grill is located in the corner of the ceiling of a room 20 ft × 30 ft × 10 ft high, as shown in figure 6–25. Determine the view factor from the grill to the floor and the opposite wall indicated in figure 6–25.

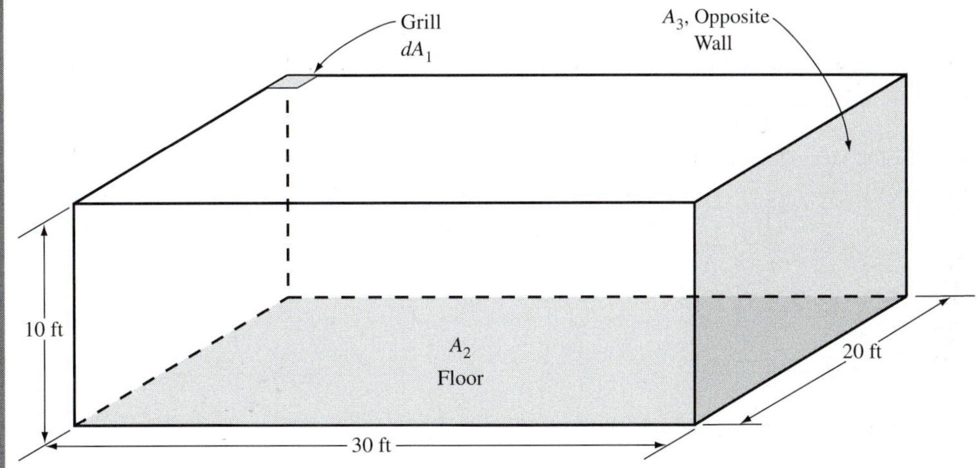

Figure 6–25

Solution We will assume that the grill is very small in relation to the size of the room so that the grill's area may be approximated by the area dA_1. The view factor of the grill to the floor is written more concisely as

$$F_{dA_1-A_2} = F_{d1-2}$$

and using figure 6–18 for reference, $D = 10$ ft, $L_1 = 20$ ft, and $L_2 = 30$ ft. Then $L_2/D = 30 \text{ ft}/10 \text{ ft} = 3$ and $L_1/D = 2$. From figure 6–18 we read

Table 6–2 View Factors for Selected Geometric Configurations

Geometric Configuration	View Factor Expression	Figure
1. Differential area to finite parallel area	$F_{dA_1-A_2} = \frac{1}{2\pi}\left[\frac{X}{\sqrt{1+Y^2}}\tan^{-1}\left(\frac{Y}{\sqrt{1+X^2}}\right) + \frac{Y}{\sqrt{1+Y^2}}\tan^{-1}\left(\frac{X}{\sqrt{1+Y^2}}\right)\right]$ where $X = \frac{L_1}{D}$ and $Y = \frac{L_2}{D}$	6–18
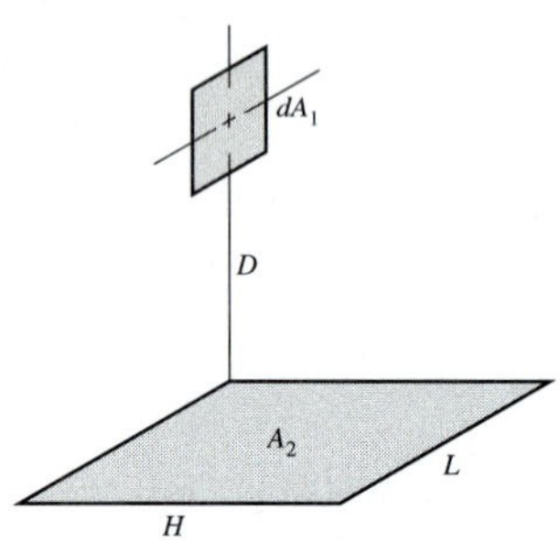 2. Differential area to finite perpendicular area	$F_{dA_1-A_2} = \frac{1}{2\pi}\left[\tan^{-1}\left(\frac{1}{X}\right) - \frac{1}{\sqrt{1-(Y/X)^2}}\tan^{-1}\left(\frac{1}{\sqrt{X^2+Y^2}}\right)\right]$ where $X = \frac{D}{L}$ and $Y = \frac{H}{L}$	6–19
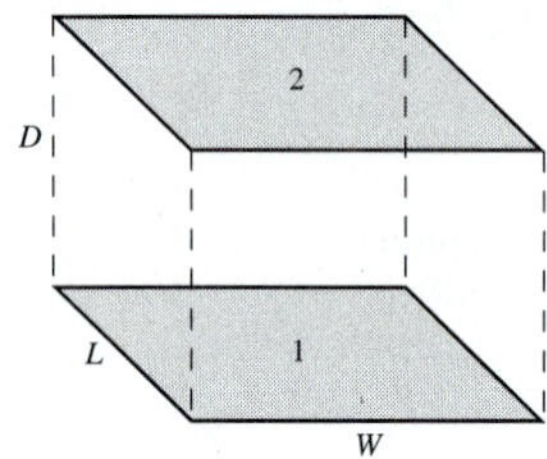 3. Parallel rectangular areas	$F_{1-2} = \frac{2}{\pi XY}\left[\ln\sqrt{\frac{(1+X^2)(1-Y^2)}{1+X^2+Y^2}} + X\sqrt{1+Y^2}\tan^{-1}\left(\frac{X}{\sqrt{1+Y^2}}\right)\right]$ $+\frac{2}{\pi XY}\left[Y\sqrt{1+X^2}\tan^{-1}\left(\frac{Y}{\sqrt{1+X^2}}\right) - X\tan^{-1}(X) - Y\tan^{-1}(Y)\right]$ where $X = \frac{W}{D}$ and $Y = \frac{L}{D}$	6–21
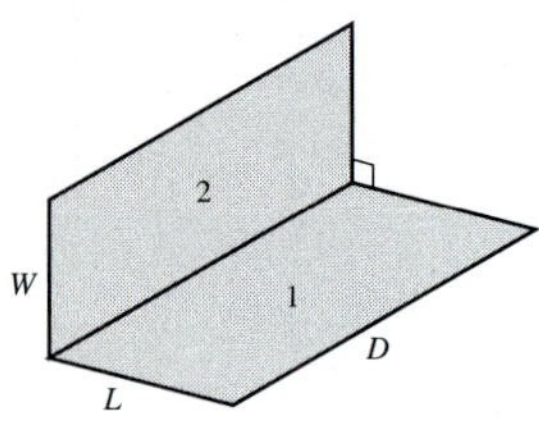 4. Perpendicular rectangular areas	$F_{1-2} = \frac{1}{\pi R_1}\left[R_1\tan^{-1}\left(\frac{1}{R_1}\right) + R_2\tan^{-1}\left(\frac{1}{R_2}\right) - \sqrt{R_1^2+R_2^2}\tan^{-1}\left(\frac{1}{\sqrt{R_1^2+R_2^2}}\right)\right]$ $+\frac{1}{4\pi R_1}\ln\left[\frac{(1+R_1^2)(1+R_2^2)}{1+R_1^2+R_2^2}\left(\frac{R_1^2(1+R_1^2+R_2^2)}{(1+R_1^2)(R_1^2+R_2^2)}\right)^{R_1^2}\left(\frac{R_2^2(1+R_1^2+R_2^2)}{(1+R_2^2)(R_1^2+R_2^2)}\right)^{R_2^2}\right]$ where $R_2 = \frac{W}{D}$ and $R_1 = \frac{L}{D}$	6–22
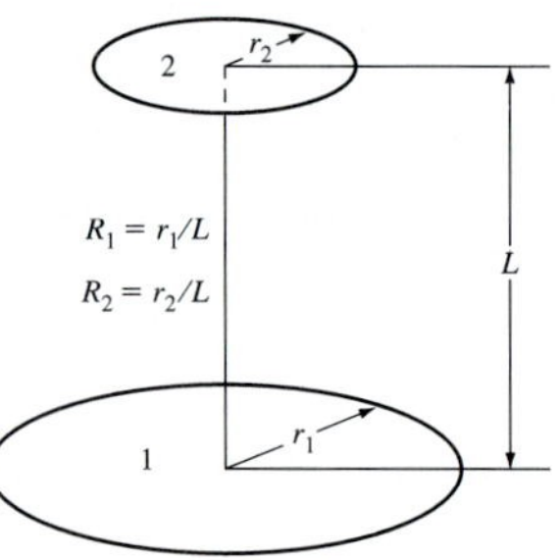 5. Coaxial parallel disks	$F_{1-2} = \frac{1}{2}\left(S^2 - \sqrt{S^2 - 4(r_2/r_1)^2}\right)$ where $R_1 = \frac{r_1}{L}$, $R_2 = \frac{r_2}{L}$, and $S = 1 + \frac{1+R_2^2}{R_1^2}$	6–23

Answer

$$F_{d1-2} = 0.22$$

The view factor for the grill with the opposite wall we refer to figure 6–19, where $D = 30$ ft, $L = 20$ ft, and $H = 10$ ft. Then $D/L = 1.5$ and $H/L = 0.5$, and reading from figure 6–19,

Answer

$$F_{dA_1-A_3} = F_{d1-3} = 0.0085$$

Cross-String Method

A method suggested by Hottel [5] to determine the view factors for two-dimensional situations is the cross-string method as sketched in figure 6–26. The method involves extending straight lines or "strings" between the two ends of each of the two surfaces. The view factor is then defined as

$$F_{1-2} = \frac{1}{2L_1}[(L_3 + L_4) - (L_5 + L_6)] \qquad \textbf{(6–48)}$$

Often the cross-string method is used with a graphical technique, but it must always be remembered that it is restricted to two-dimensional situations. By the principle of reciprocity (see equation 6–50) we can determine the view factor from 2 to 1,

$$F_{2-1} = \frac{1}{2L_2}[(L_3 + L_4) - (L_5 + L_6)] = F_{1-2}\left(\frac{L_1}{L_2}\right)$$

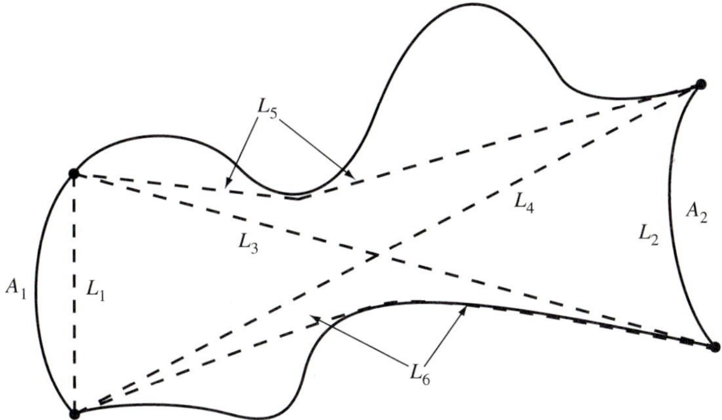

Figure 6–26 Cross-string method for shape factor determination.

EXAMPLE 6–8

A long reflector and a long light source, parallel to the reflector, are placed as shown in figure 6–27a. An obstruction or shield is located between the reflector and the light source. Determine the view factors between the reflector (area 1) and the light source (area 2).

Solution

By using some analytic geometry the lengths indicated in figure 6–27 and corresponding to the designation given in figure 6–26, are found to be

$L_1 = 2$ ft and $L_2 = \pi$(1-ft diameter) $= 3.14159$ ft (see figure 6–27b).
$L_3 = 3$ ft $+ 0.785$ ft $+ 0.165$ ft $= 3.95$ ft (see figure 6–27c).
$L_4 = 3.414$ ft $+ 0.785$ ft $+ 0.245$ ft $= 4.444$ ft (see figure 6–27c).
$L_5 = 3.785$ ft and $L_6 = 3.414$ ft $+ 0.785$ ft $= 4.199$ ft (see figure 6–27d).

Figure 6–27 Example 6–8.

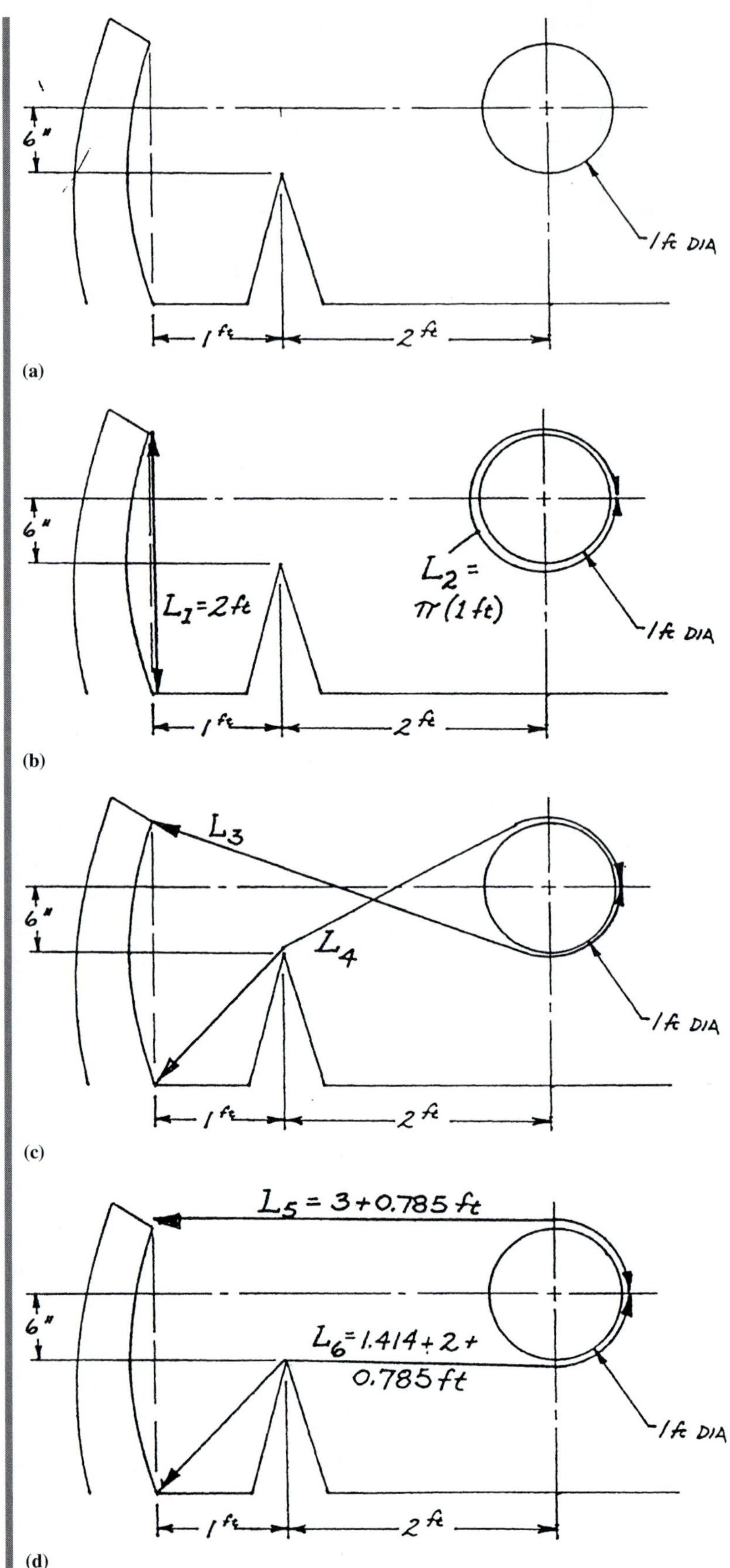

and the view factors can be determined from equation (6–48):

Answer
$$F_{1-2} = \frac{1}{2L_1}[(L_3 + L_4) - (L_5 + L_6)] = 0.1025$$

By reciprocity,

Answer
$$F_{2-1} = F_{1-2}\left(\frac{L_1}{L_2}\right) = 0.06525$$

Shape Factor Algebra It can be seen from close scrutiny of the various relationships given in equations (6–44), (6–46), and (6–47) that the view factors are fractions of the hemispherical viewing region of the surface. Therefore the sum of all the view factors of a particular area or surface must be identically 1.0. That is,

$$\sum_{j=1}^{n} F_{i-j} = 1.0 \tag{6–49}$$

where F_{i-j} is the view factor of surface i to another surface j and where there are n areas or surfaces. If surface i, or A_i, is flat or convex then $F_{i-i} = 0$ (which means that surface A_i cannot "see" itself). If the surface is concave then $F_{i-i} > 0.0$. For instance, in figure 6–24 the inside surface of the outer concentric cylinder is concave, so the shape factor for A_2 to A_1 is less than 1.0, except for the limiting case where $r_1 = r_2$. Thus, for a very long cylinder such that $1 \rightarrow \infty$ and $R_2 \rightarrow \infty$, suppose that $r_1 = r_2/2$ (for illustrative purposes). Then $R_1 = 0.5$, and from figure 6–24 we read $F_{2-1} = 0.5$. Thus, from equation (6–49) we have

$$F_{2-1} + F_{2-2} = 1.0 \quad \text{and} \quad F_{2-2} = 1.0 - 0.5 = 0.5$$

Equation (6–49) is an important result in the determination of view factors for complicated configurations. Further, if we compare the results of equations (6–46) and (6–47) we have $A_1F_{1-2} = A_2F_{2-1}$. In general for surfaces i and j,

$$A_iF_{i-j} = A_jF_{j-i} \tag{6–50}$$

which is called the *reciprocity relationship*. This is another important analytic tool for the determination of view factors that may otherwise be difficult to obtain.

It is important to note that an emitting or a receiving surface, A_j, can be visualized as a sum of n smaller surface portions such that

$$A_j = \sum_{l=1}^{n} A_l \tag{6–51}$$

Further, the view factor from a surface A_i to a surface A_j that is composed of smaller portions can be written as a summation, or corollary of the view factor summation equation (6–49),

$$F_{i-j} = \sum_{l=1}^{n} F_{i-l} \tag{6–52}$$

For an area A_i that is a sum of m smaller portions, the view factor with another area A_j is

$$A_iF_{i-j} = \sum_{k=1}^{m} A_kF_{k-j} \tag{6–53}$$

EXAMPLE 6–9 Determine an expression for predicting the view factor F_{A-B} from surface A to surface B as shown in figure 6–28.

Figure 6–28

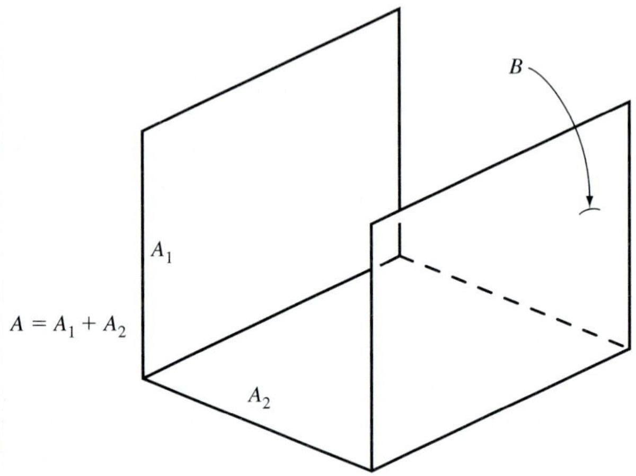

Solution Surface A is composed of two flat surfaces, area 1 and area 2. Using equation (6–53) we obtain

$$A_AF_{A-B} = A_1F_{1-B} + A_2F_{2-B}$$

or

$$F_{A-B} = \frac{1}{A_A}(A_1F_{1-B} + A_2F_{2-B})$$

The view factors F_{1-B} and F_{2-B} can be determined using the information from figures 6–21 and 6–22.

Using the reciprocity relationship, equation (6–50), we can revise equation (6–53) to read

$$A_iF_{i-j} = \sum_{l=1}^{n} A_jF_{j-l} \tag{6–54}$$

For example, suppose an area A_1 is composed of two areas, A_2 and A_3. The view factor of this area A_1 to another separate area A_4 can be written, using equation (6–53), as

$$F_{1-4} = \frac{1}{A_1}(A_2F_{2-4} + A_3F_{3-4})$$

Using equation (6–54), we can also write this same view factor as

$$F_{1-4} = \frac{1}{A_1}(A_2F_{4-2} + A_3F_{4-3})$$

EXAMPLE 6–10 Determine the view factor F_{1-2} for the annular disk to the solid disk shown in figure 6–29.

Figure 6–29

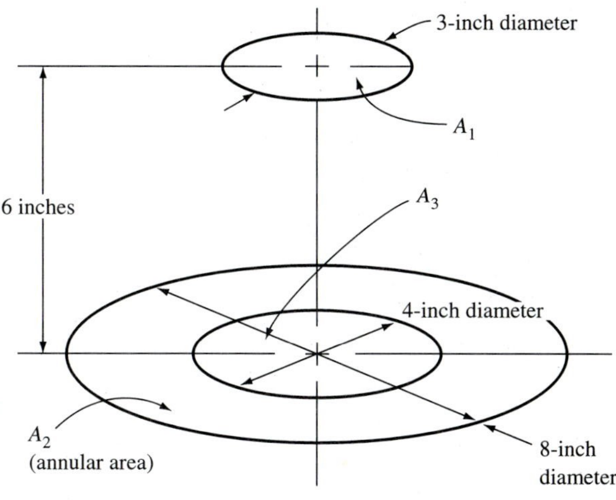

Solution The view factor F_{1-2} is given by equation (6–52),

$$F_{1-2,3} = F_{1-2} + F_{1-3}$$

and

$$F_{1-2} = F_{1-2,3} - F_{1-3}$$

From figure 6–23, for determining $F_{1-2,3}$, we have $L = 6$ in., $r_1 = 1.5$ in., and $r_2 = 4$ in. so that $R_1 = 1.5/6 = 0.25$, $1/R_1 = 4$, and $R_2 = 4/6 = 0.667$. We read from figure 6–23 that

$$F_{1-2,3} \approx 0.3$$

For determining F_{1-3} we have $L = 6$ in., $r_1 = 1.5$ in., and $r_2 = 2$ in. so that $1/R_1 = 4$ and $R_2 = 2/6 = 0.333$. We read from figure 6–23 that $F_{1-3} \approx 0.1$. The view factor then is

Answer

$$F_{1-2} = 0.3 - 0.1 = 0.2$$

EXAMPLE 6–11 Determine the view factor F_{1-8} for surface A_1 to surface A_8 as shown in figure 6–30.

Solution The analysis for determining the view factor F_{1-8} requires manipulation of the relationships we have discussed. We have $A_{1,2,3,4} = A_1 + A_2 + A_3 + A_4$ and $A_{5,6,7,8} = A_5 + A_6 + A_7 + A_8$. Then the view factors, $F_{1,2,3,4-5,6,7,8}$ and $F_{1,2,3,4-5,6}$ can be determined from figure 6–22. Using equation (6–52) we have

$$F_{1,2,3,4-7,8} = F_{1,2,3,4-5,6,7,8} - F_{1,2,3,4-5,6}$$

and from equation (6–53)

$$A_{1,2,3,4}F_{1,2,3,4-7,8} = A_1F_{1-7,8} + A_2F_{2-7,8} + A_{3,4}F_{3,4-7,8}$$

where, from equation (6–53)

$$A_{3,4}F_{3,4-7,8} = A_3F_{3-7,8} + A_4F_{4-7,8}$$

Figure 6–30

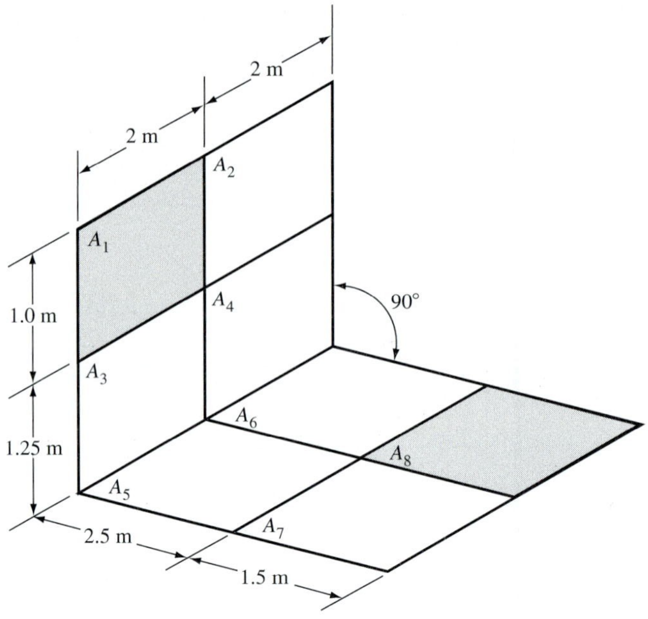

Also,

$$F_{3,4-7,8} = F_{3,4-5,6,7,8} - F_{3,4-5,6}$$

where the view factors $F_{3,4-5,6,7,8}$ and $F_{3,4-5,6}$ can be determined from figure 6–22.

From the third preceding equation we can write that

$$A_1F_{1-7,8} + A_2F_{2-7,8} = A_{1,2,3,4}F_{1,2,3,4-5,6,7,8} - A_{3,4}F_{3,4-7,8}$$

and, since $A_1 = A_2$, $F_{1,7-8} = F_{2,7-8}$ by symmetry. This reduces the equation to the relationship

$$F_{1-7,8} = (1/2A_1)(A_{1,2,3,4}F_{1,2,3,4-5,6,7,8} - A_{3,4}F_{3,4-7,8})$$

where all of the terms on the right side may be determined. From equation (6–52) we have $F_{1-7,8} = F_{1-7} + F_{1-8}$ or $F_{1-8} = F_{1-7,8} - F_{1,7}$. The view factor F_{1-7} can be determined from the development

$$A_1F_{1-5,7} + A_3F_{3-5,7} = A_{1,3}F_{1,3-5,7}$$

or $$F_{1-5,7} = (1/A_1)(A_{1,3}F_{1,3-5,7} - A_3F_{3-5,7}) = F_{1-5} + F_{1-7}$$

and $$F_{1-7} = F_{1-5,7} - F_{1-5}$$

By the reciprocity relationship,

$$F_{1-5} = (A_5/A_1)F_{5-1}$$

with $F_{5-1} = F_{5-1,3} - F_{5-3}$.

We now evaluate the various terms:

$$A_1 = A_2 = 2\text{ m}^2 \qquad A_3 = A_4 = 2.5\text{ m}^2 \qquad A_5 = A_6 = 5\text{ m}^2 \qquad A_7 = A_8 = 3\text{ m}^2$$

For $F_{5-1,3}$,

$$L = 2.5\text{ m},\ W = 2.25\text{ m, and } D = 2\text{ m}$$

Then $L/D = 2.5/2 = 1.25$ and $W/D = 2.25/2 = 1.125$

From figure 6–22 we read that $F_{5-1,3} \approx 0.18$. For F_{5-3},

$$L/D = 1.25 \quad \text{and} \quad W/D = 0.625$$

From figure 6–22 we read that $F_{5-3} \approx 0.145$, so

$$F_{5-1} = 0.18 - 0.145 = 0.035$$

By reciprocity,

$$F_{1-5} = (5/2)(0.035) = 0.0875$$

For determining $F_{1,3-5,7}$,

$$L/D = 2.25/2 = 1.125 \quad \text{and} \quad W/D = 4/2 = 2.0$$

We read from figure 6–22 that $F_{1,3-5,7} \approx 0.22$. For determining $F_{3-5,7}$,

$$L/D = 1.25/2 = 0.625 \quad \text{and} \quad W/D = 4/2 = 2.0$$

From figure 6–22 we read again that $F_{3-5,7} \approx 0.288$ so that we can now determine $F_{1-5,7}$:

$$F_{1-5,7} = (\tfrac{1}{2}\,\text{m}^2)[(4.5\,\text{m}^2)(0.22) - (2.5\,\text{m}^2)(0.288)] = 0.135$$

Then $F_{1-7} = F_{1-5,7} - F_{1,5} = 0.135 - 0.0875 = 0.0475$.

We now determine $F_{1-7,8}$ by first considering $F_{1,2,3,4-5,6,7,8}$, where

$$L/D = 2.25/4 = 0.5625 \quad \text{and} \quad W/D = 4/4 = 1.0$$

From figure 6–22 we read $F_{1,2,3,4-5,6,7,8} \approx 0.279$. For determining $F_{1,2,3,4-5,6}$,

$$L/D = 2.25/4 = 0.5625 \quad \text{and} \quad W/D = 2.5/4 = 0.625$$

Then we read from figure 6–22 that

$$F_{1,2,3,4-5,6} \approx 0.25$$

By summation,

$$F_{1,2,3,4-7,8} = F_{1,2,3,4-5,6,7,8} - F_{1,2,3,4-5,6} = 0.279 - 0.25 = 0.029$$

For $F_{3,4-5,6,7,8}$,

$$L/D = 1.25/4 = 0.3125 \quad \text{and} \quad W/D = 4/4 = 1.0$$

Then from figure 6–22 we read that $F_{3,4-5,6,7,8} \approx 0.348$. Then

$$F_{1-7,8} = \frac{1}{4\,\text{m}^2}[(4 \times 2.25\,\text{m}^2)(0.03) - (5\,\text{m}^2)(F_{3,4-7,8})]$$

where $$F_{3,4-7,8} = F_{3,4-5,6,7,8} - F_{3,4-5,6}$$

For determining $F_{3,4-5,6}$ we have $L/D = 1.25/4 = 0.3125$ and $W/D = 2.5/4 = 0.625$. From figure 6–22 we read $F_{3,4-5,6} \approx 0.33$, so $F_{3,4-7,8} = 0.34 - 0.33 = 0.01$ and

$$F_{1-7,8} = \frac{1}{4\,\text{m}^2}[(9\,\text{m}^2)(0.03) - (5\,\text{m}^2)(0.01)] = 0.055$$

Finally,

Answer $$F_{1-8} = F_{1,7,8} - F_{1,7} = 0.055 - 0.0475 = 0.0075$$

6-5 APPLICATIONS OF RADIATION

In this section we will consider four examples of how the nature of radiation is used to advantage in engineering applications. We consider in order (1) the selective properties of window glass, (2) the combined properties of transmissivity and absorptivity to utilize solar energy, (3) the reflective properties of a material and a hohlraum to experimentally determine radiation response properties, and (4) the geometric configurations appropriate to concentrate solar or other thermal radiation for generating high temperatures. These four examples are given only to provide the student with some background and encouragement for further applications.

Figure 6-31 Spectral transmissivity of glass for varying amounts of Fe_2O_3: (a) 0.02, (b) 0.10, (c) 0.15, (d) 0.50. (From Dietz, A. G. H., Diathermanous Materials and Properties of Surfaces, in R. W. Hamilton, ed., *Space Heating with Solar Energy,* MIT Press, 1954 [6])

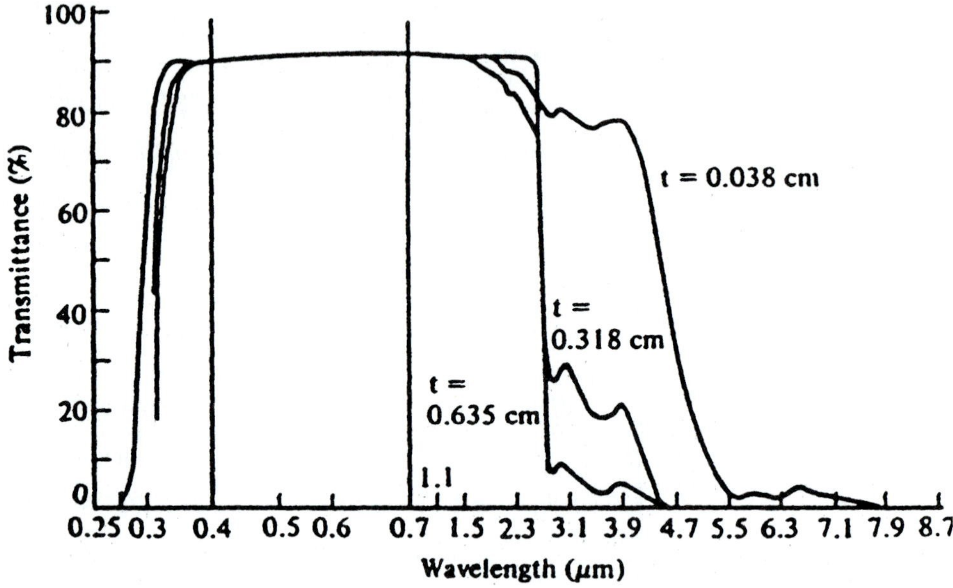

(a)

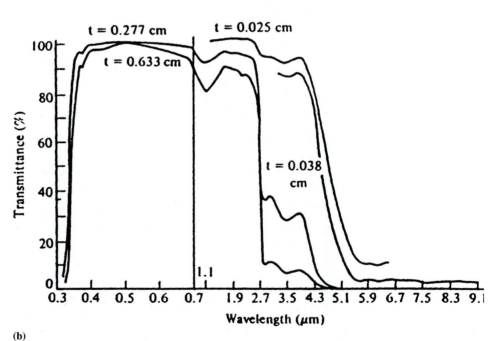

(b)

Selective Properties of Glass

An example of a selective radiation material that can be used to advantage is window glass. Glass is a nearly transparent material, so the important radiation response properties are transmissivity and reflectivity. In figure 6–31 are shown spectral transmissivities for four different glass composites, each containing small percentages of ferrous oxide, Fe_2O_3. Notice that the spectral transmissivities vary significantly over the thermal radiation spectrum, and it is this characteristic that makes window glass so advantageous.

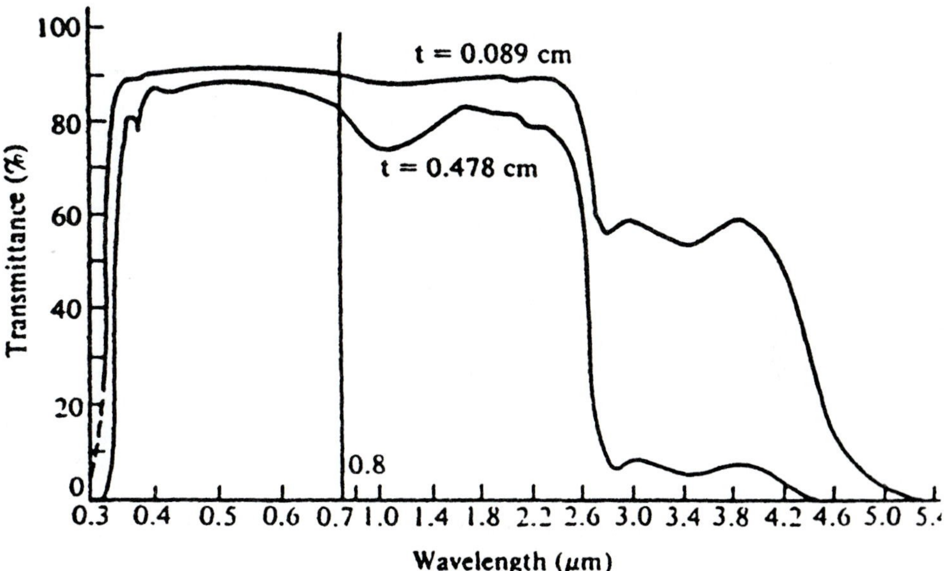

(c)

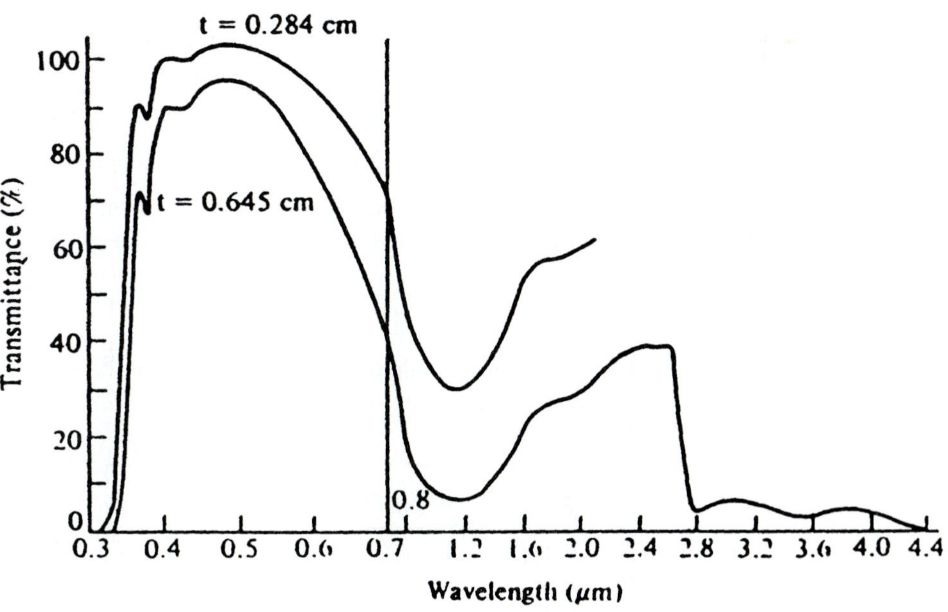

(d)

EXAMPLE 6–12 Consider a pane of 0.15% Fe_2O_3 glass 0.478 cm thick to be used in an atrium. As shown in figure 6–32, assume solar energy $\dot{Q}_{sol} = 600\ \text{W/m}^2$ arrives as black body radiation from a source at 6000 K and is transmitted through the pane of glass in an amount $\dot{Q}_{tr,sol}$. There is also radiation from inside the atrium, which is assumed to be black body radiation at 300 K, transmitted through in the amount $\dot{Q}_{tr,in}$. Determine the net thermal radiation transmitted into the atrium.

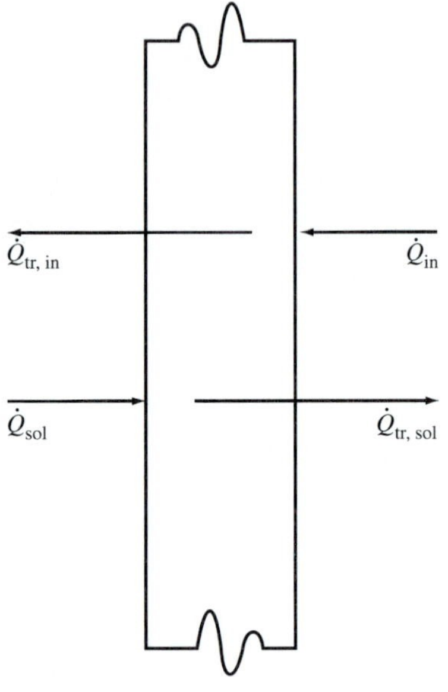

Figure 6–32 Schematic of radiation transmission through a pane of glass.

Solution The net radiation through the pane of glass is just

$$\dot{Q}_{tr,net} = \dot{Q}_{tr,sol} - \dot{Q}_{tr,in}$$

where

$$\dot{Q}_{tr,sol} = \tau_{r,6000\,\text{K}}\dot{Q}_{sol} = \tau_{r,6000\,\text{K}}\left(600\frac{\text{W}}{\text{m}^2}\right)$$

and

$$\dot{Q}_{tr,in} = \tau_{r,300\,\text{K}}\dot{Q}_{in} = \tau_{r,300\,\text{K}}\sigma T^4 = \tau_{r,300\,\text{K}}\sigma(300\ \text{K})^4$$

From figure 6–31 we will approximate the transmissivities as sketched in figure 6–33. Then the transmissivities may be estimated from the sum

$$\tau_r \approx \tau_{r,0.3-2.6\,\mu\text{m}}\frac{\dot{E}_{0.3-2.6\,\mu\text{m}}}{\sigma T^4} + \tau_{r,2.6-4.2\,\mu\text{m}}\frac{\dot{E}_{2.6-4.2\,\mu\text{m}}}{\sigma T^4}$$

For the transmissivity of the solar energy we have, at $T = 6000$ K, the following results read from table 6–1:

$\lambda(\mu)$m	$\lambda T\ (\mu\text{m}\cdot\text{K})$	$\dot{E}_{B,0-\lambda T}/\sigma T_4$
0.3	1,800	0.05
2.6	15,600	0.972
4.2	25,200	0.991

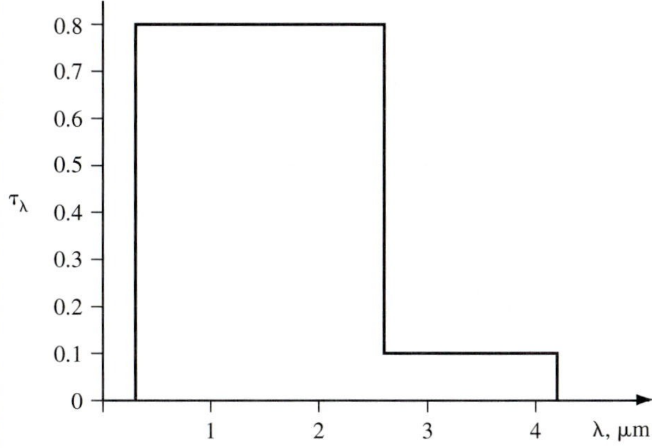

Figure 6–33 Approximate spectral transmissivity of 0.15% Fe_2O_3 glass.

Then

$$\tau_{r,6000\text{ K}} = (0.8)(0.972 - 0.050) + (0.06)(0.991 - 0.972) = 0.73874$$

and the transmitted solar energy is

$$\dot{Q}_{\text{tr,sol}} = (0.73874)\left(600\frac{\text{W}}{\text{m}^2}\right) = 443.2\frac{\text{W}}{\text{m}^2}$$

For the inside radiation we have $T = 300$ K and the following results from Table 6–1:

$\lambda(\mu)$m	$\lambda T(\mu\text{m}\cdot\text{K})$	$\dot{E}_{B,0-\lambda T}/\sigma T_4$
0.3	90	≈0
2.6	780	≈0
4.2	1260	≈0.0026

Then

$$\tau_{r,300\text{ K}} = (0.8)(0 - 0) + (0.06)(0.0026 - 0) = 0.000156$$

and

$$\dot{Q}_{\text{tr,in}} = (0.000156)\sigma(300\text{ K})^4 = 0.0716\frac{\text{W}}{\text{m}^2}$$

The net radiation transmitted into the atrium is

Answer

$$\dot{Q}_{\text{tr,net}} = 443.2 - 0.0716 = 443.13\frac{\text{W}}{\text{m}^2}$$

This net input can only occur if the solar energy is directed toward the glass surface. Thus, we must assume that the 600 W/m^2 solar radiation is only an approximation of a portion of the day. The implication of this result is clear; however, selective materials can provide passive benefits by virtue of their response characteristics.

Collection of Solar Energy

The radiation heat transfer from the sun is a complex subject involving not only engineering, mathematics, and physics, but also geography, chemistry, astronomy, and others. The sun is approximately 150 million km (93 million miles) from the earth, and it is convenient to assume that the radiation is direct after traveling so far from its source. Recent measurements indicate that the total intensity I_{sc}, called the solar constant, is 1353 W/m^2 at the earth's distance from the sun but still out of the earth's atmosphere. It is not known if the solar radiation is diffuse, but in all likelihood it is not. In figure 6–34 is shown the

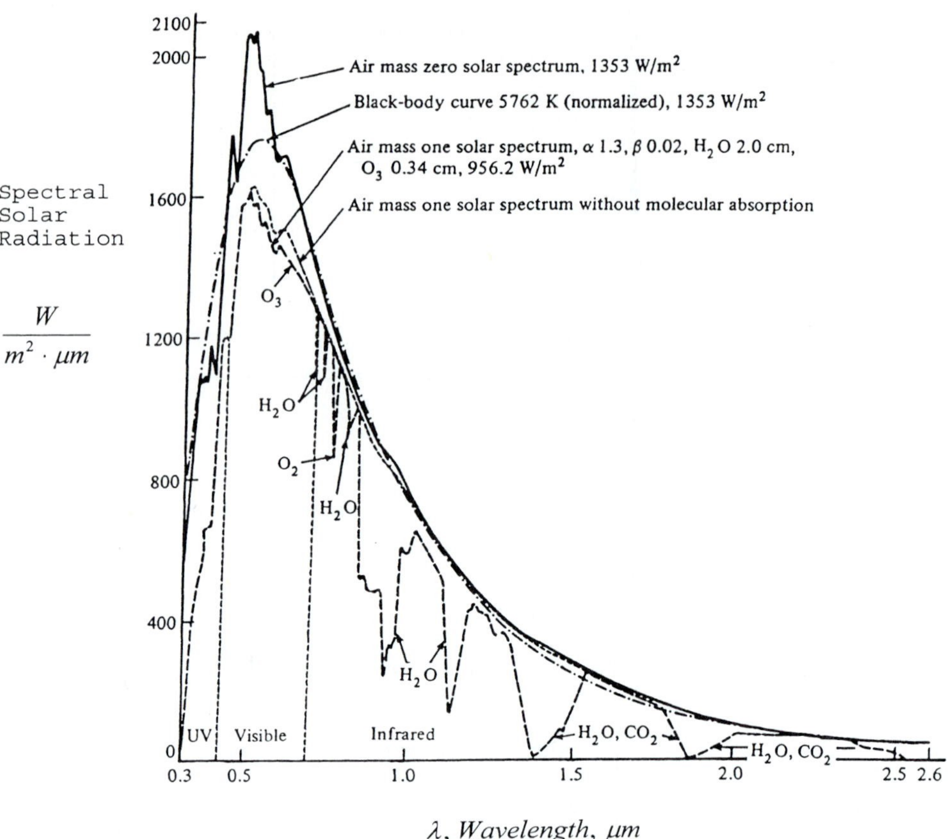

Figure 6–34 Solar spectral intensity at the earth's surface.

spectral intensity $I(\lambda)$ of solar energy as measured outside the earth's atmosphere. The area under the curve is the solar constant; that is,

$$\dot{I}_{sc} = \int_0^\infty \dot{I}(\lambda)d\lambda = 1353\frac{\text{W}}{\text{m}^2} \tag{6–55}$$

As the solar energy is transmitted through the earth's atmosphere and irradiates the earth, it is reduced in intensity, or attenuated. This attenuation is strongly dependent on the local earth weather conditions and the time of day. At noon the sun is directly overhead when viewed from the earth, and the solar energy is said to pass through 1 *air mass.* As shown in figure 6–35, this is the distance through which the radiation must pass from outer space to the earth's surface when the sun is directly overhead. We will discuss the attenuation of radiation in gases further in Chapter 7.

The number of air masses for any time of the day may be approximated from

$$m_a = \frac{1}{\cos\zeta_s} \tag{6–56}$$

where m_a is the number of air masses and ζ_s is the zenith angle of the sun as measured from the earth, as defined in figure 6–36. At solar noon (when the sun is directly overhead) the zenith angle is zero (0), and when the zenith angle is, for instance, 75.5° the air mass is approximately 4. For a zenith angle of 84.3° the air mass is 10, or the solar energy

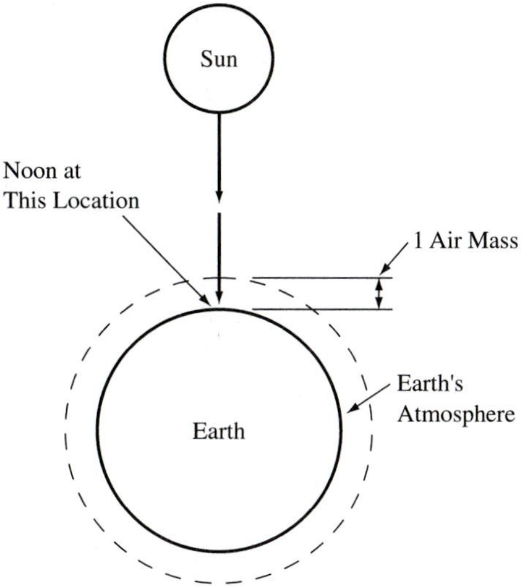

Figure 6–35 Illustration of atmospheric attenuation of solar energy.

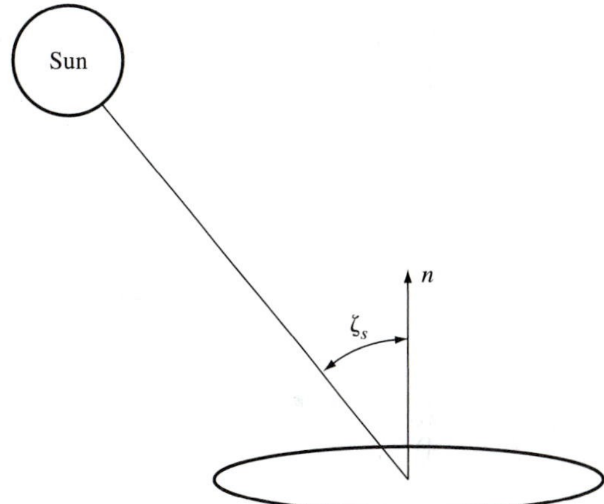

Figure 6–36 Zenith angle of the sun.

must pass through 10 times more of the atmosphere than at solar noon. Figure 6–37 indicates the attenuation of solar energy at the earth's surface for air masses of 4, 7, and 10. It can be seen that there is a significant decrease in the intensity from that at an air mass of 1.

Since the solar energy is assumed to be direct, the intensity of radiation $\dot{I}$ is the same as the radiation,

$$\dot{q}_A = \dot{I} \qquad (\text{W/m}^2)$$

As mentioned before, the daily and seasonal motion of the earth with respect to the sun as well as the local weather patterns strongly affect the intensity of solar radiation at the earth's surface, and it is beyond our purposes here to consider this matter at further length. The reader is referred to, for instance, the text by Duffie and Beckman [7] for further discussions of predicting solar intensity.

The collection of solar energy involves the transmitting and absorbing characteristics of materials. A typical flat-plate solar collector is shown in figure 6–38. The cover is a glass plate through which solar energy is transmitted to the absorber, where the energy is

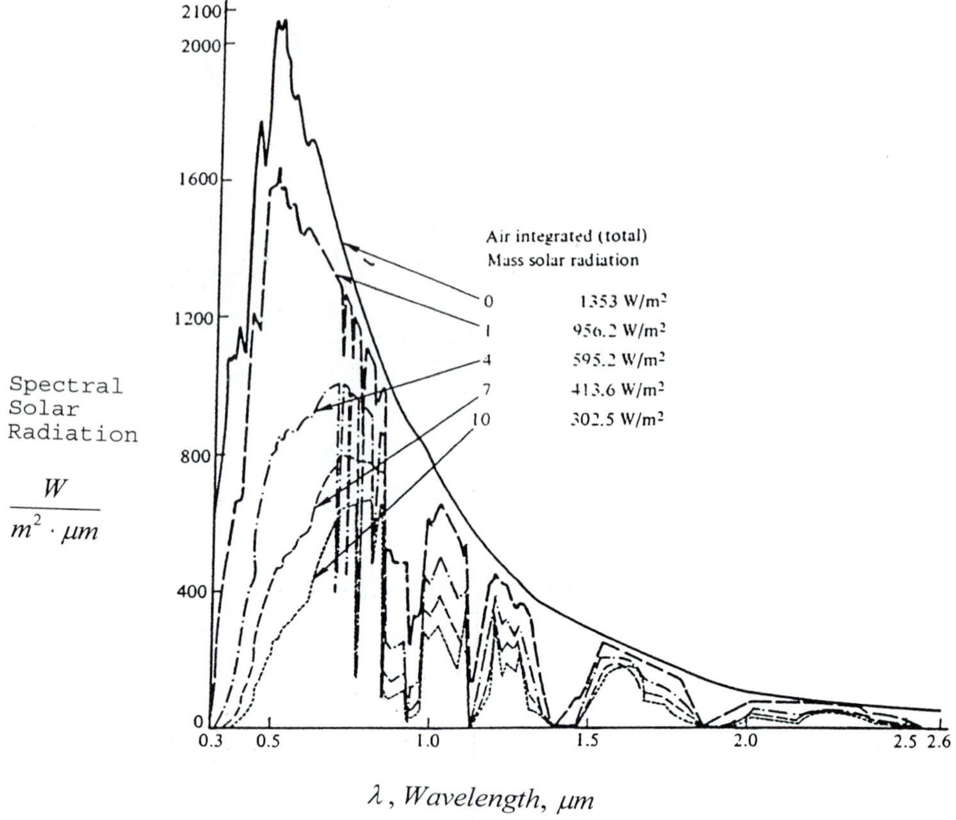

Figure 6–37 Spectral solar radiation for different air masses and for U.S. standard atmosphere, 20 mm precipitable water vapor, 3.4 mm ozone, and very clear air.

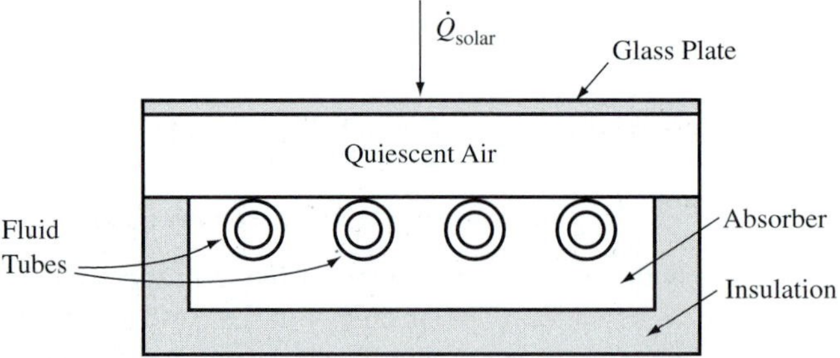

Figure 6–38 Cross section of a typical flat-plate solar collector.

then used to increase the thermal energy of a fluid, such as air or water, which is passed through the collector. In the configuration sketched in figure 6–38 the fluid flows through the tubes. The maximum solar energy that may be collected by the fluid is given by

$$\dot{Q}_{max} = \tau_{r,glass}\alpha_{r,abs}\dot{Q}_{solar} \tag{6–57}$$

It is tempting to remove the glass cover and thereby have a transmissivity of 1.0; however, there will be convection heat transfer losses and radiation heat transfer losses, as we saw in example 6–12, which usually more than offset the increase in transmission. Insulation at the bottom and the sides of the collector is recommended to reduce the thermal losses to a minimum. In the solar energy literature the value of $\tau_{r,glass}\alpha_{r,abs}$ is called $(\tau\alpha)_{effective}$, and Duffie and Beckman [7] suggest its value is typically between 0.65 and 0.75 for flat-plate collectors.

Experimental Determination of Radiant Response Properties

The measurement of the radiant response properties for opaque surfaces can be accomplished by using a method and equipment suggested by Gier and Dunkle [8] that is sketched in figure 6–39. The apparatus is a closed container much like that of figure 6–4, except that there is a provision to insert a cylindrical test sample at one surface and a pinhole opposite the sample. The radiation, spectral or total ($\dot{A}_\lambda$), may be measured from the surface of the test sample and the spectral or total radiation from the inside surface of the container ($\dot{B}_\lambda$) can simultaneously be measured. The term *hohlraum* (literally meaning "empty space" or "cavity" in *German*) is used to describe the container, which is also nearly a black body. The reflectivity of the surface of the test sample is

$$\rho_r = \frac{\dot{A}_\lambda}{\dot{B}_\lambda} \tag{6–58}$$

and the emissivity and absorptivity follow directly by using equation (6–37).

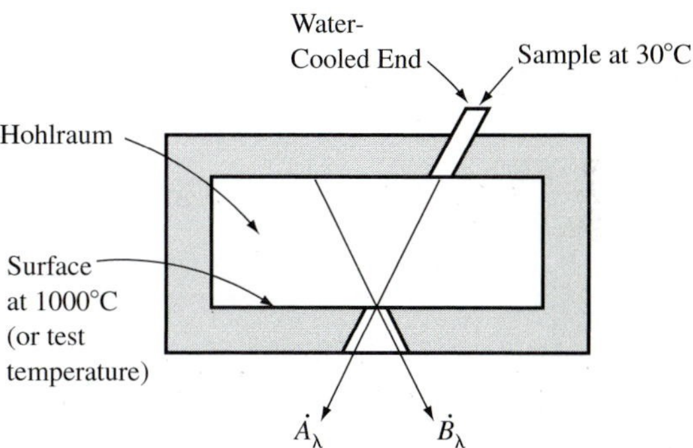

Figure 6–39 Sketch of an apparatus to measure spectral reflectivity, ρ_λ.

Geometric Configuration for Concentrating Radiation

Solar radiation or direct radiation may be concentrated into a smaller area, thereby increasing its intensity, by means of optical lenses or reflectors. The concentrator uses parabolic forms for reflecting direct radiation, and for approximate concentration circular forms may be used. In figure 6–40 is shown a typical concentrator. The focal point represents the point or line at which the direct radiation over the area A_1 is concentrated and the concentration factor is A_a/A_F. For a flat-plate solar collector as discussed earlier, the concentration factor is 1.0; however, typical concentrating collectors have concentration

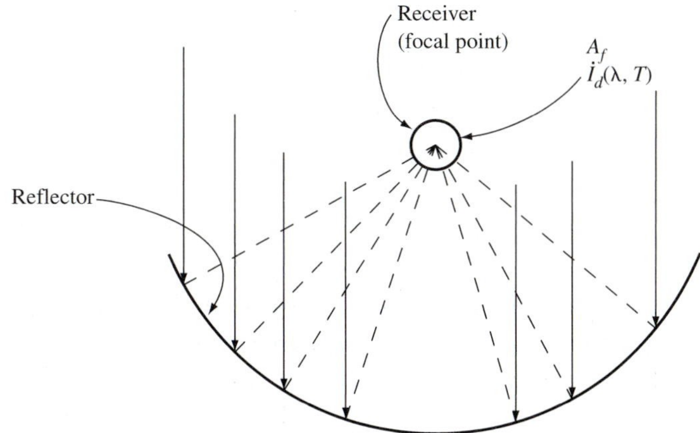

Figure 6–40 Concentrating collector.

factors or 4 to 10 or more. The radiation intensity at the receiver, if the reflector has reflectivity of 1.0, is

$$\dot{I} = \frac{A_1}{A_F}\dot{I}_d(\lambda, T) \tag{6–59}$$

If the emitted radiation from the reflecting surface of the concentrator is ignored, then the radiation intensity at the receiver is

$$\dot{I} = \rho_r \frac{A_1}{A_F}\dot{I}_d(\lambda, T) \tag{6–60}$$

In the following chapter we will discuss how radiation heat transfer between two or more surfaces is analyzed.

6–6 SUMMARY

$d\dot{E}_{dA_1}$ = Heat transfer from area dA_1 (W or Btu/hr)

Solid angle,

$$d\omega = \frac{dA_n}{r^2} \tag{6–2}$$

Spectral intensity,

$$\dot{I}_{dA_1} = \frac{1}{dA_n d\omega d\lambda} d\dot{E}_{dA_1} = \frac{1}{dA_1 \cos\Theta_1 d\omega d\lambda} d\dot{E}_{dA_1} \qquad (\text{W/m}^2 \cdot \text{str} \cdot \mu\text{m}) \tag{6–5}$$

Heat transfer per unit area,

$$\dot{E}_{dA_1} = \int_0^{\lambda=\infty}\int_0^{\phi=2\pi}\int_0^{\Theta=\pi/2} \dot{I}(\Theta, \phi, T, \lambda)\cos\Theta_1 \sin\Theta_1 d\Theta d\phi d\lambda$$

$$(\text{W/m}^2 \text{ or Btu/hr} \cdot \text{ft}^2) \tag{6–8}$$

Heat transfer per unit area for diffuse surface,

$$\dot{E}_{dA_1} = \pi \int_0^{\lambda=\infty} \dot{I}(\lambda, T) d\lambda \tag{6–9}$$

Spectral intensity for a black body,

$$\dot{I}_B(\lambda, T) = \frac{2\hbar c_0^2}{\lambda^5 (e^{\hbar c_0/\lambda T} - 1)} \qquad (W/m^2 \cdot \mu m) \tag{6–12}$$

Black body spectral power, or heat transfer per unit area per unitwavelength,

$$\dot{E}_{B,dA} = \pi \int_0^{\lambda=\infty} \dot{I}_B(\lambda, T) d\lambda \qquad (\text{W/m}^2) \tag{6–13}$$

reduces to

$$\dot{E}_B = \sigma T^4 \qquad (\text{W/m}^2) \tag{6–15}$$

The fraction of the total power emitted by a black body radiator for a bandwidth of radiation from 0 to some wavelength λ is

$$\frac{\dot{E}_{B,0-\lambda T}}{\sigma T^4} = \frac{\pi}{\sigma T^4}\int_0^{\lambda T} \dot{I}_B(\lambda, T) d\lambda \tag{6–16}$$

Wein's Displacement Law gives the wavelength/temperature for maximum spectral intensity of a black body

$$\lambda T = 2897.8 \mu\text{m} \cdot \text{K} \tag{6–14}$$

For radiation response characteristics of a surface,

$$\alpha_r = \frac{\dot{E}_{\text{ab}}}{\dot{E}_{\text{irr}}} = \text{absorptivity} \tag{6–20}$$

$$\rho_r = \frac{\dot{E}_{\text{re}}}{\dot{E}_{\text{irr}}} = \text{reflectivity} \tag{6–21}$$

$$\tau_r = \frac{\dot{E}_{\text{tr}}}{\dot{E}_{\text{irr}}} = \text{transmissivity} \tag{6–22}$$

and for a surface

$$\alpha_r + \rho_r + \tau_r = 1.0 \tag{6–23}$$

The spectral emissivity from a surface is

$$\varepsilon_{r,\lambda}(T, \Theta, \phi, \lambda) = \frac{1}{\dot{I}_B(\lambda, T)} \dot{I}_{dA}(T, \Theta, \phi, \lambda) \tag{6–25}$$

and for the emissivity,

$$\varepsilon_r = \frac{1}{\sigma T^4} \int_0^{\infty} \int_0^{2\pi} \int_0^{\pi/2} \varepsilon_{r,\lambda}(T, \Theta, \phi, \lambda) \cdot \dot{I}_B(T, \Theta, \phi, \lambda) \cos\Theta \sin\Theta \, d\Theta d\phi d\lambda \tag{6–24}$$

For a gray body surface, or a diffuse surface,

$$\varepsilon_r = \frac{1}{\sigma T^4} \int_0^{\infty} \varepsilon_{r,\lambda}(\lambda, T) \cdot \dot{E}_B(\lambda, T) d\lambda \tag{6–26}$$

Kirchhoff's Law of Radiation is

$$\varepsilon_r = \alpha_r \tag{6–34}$$

For gray, opaque surfaces,

$$\alpha_r + \rho_r = \varepsilon_r + \rho_r = 1.0 \tag{6–37}$$

and for transparent bodies,

$$\rho_r + \tau_r = 1.0 \tag{6–38}$$

The radiation shape factor or view factor for area i to area j is

$$F_{i-j} = \frac{1}{\pi A_i} \int_{A_i} \int_{A_j} \cos\Theta_i \cos\Theta_j \frac{dA_j}{r^2} dA_i \tag{6–46}$$

The summation of view factors from i interacting with n areas is

$$\sum_{j=1}^{n} F_{i-j} = 1.0 \tag{6–49}$$

The reciprocity relationship is

$$A_i F_{i-j} = A_j F_{j-i} \tag{6–50}$$

For an area i that is a sum of m smaller portions, the view factor with another area j is

$$F_{i-j} = \frac{1}{A_i} \sum_{k=1}^{m} A_k F_{k-j} \tag{6–53}$$

DISCUSSION QUESTIONS

Section 6–1

6–1 Does all electromagnetic radiation travel at the speed of light?

6–2 What is diffuse radiation?

6–3 Why is spectral intensity not diminished by the distance from the source of radiation?

6–4 Define the solid angle.

Section 6–2

6–5 What is a black body?

6–6 What is Wein's Displacement Law?

6–7 At what temperatures is radiation emitted from a black body?

Section 6–3

6–8 What is a gray body?

6–9 What is Kirchhoff's Law?

6–10 What is a selective surface?

6–11 What is an opaque surface?

6–12 What is a transparent surface?

Section 6–4

6–13 What is shape factor?

6–14 State the reciprocity law.

6–15 What is the cross-string method?

Section 6–5

6–16 Can diffuse radiation be concentrated?

6–17 What is a flat plate solar collector?

6–18 What is the hohlraum?

PRACTICE PROBLEMS

Section 6–1

6–1 If thermal radiation passes through an atmosphere such that its speed is decreased to 2.1×10^8 m/s, determine the wavelength bandwidth of the visible portion of the thermal radiation. Assume that the frequency bandwidth remains unchanged from that of radiation passing through a vacuum.

6–2 A small area of $1\,\text{cm}^2$ is viewed from three locations: I, II, and III, as shown in figure 6–41. Determine the solid angles from each of these three points subtended by the 1-cm^2 area.

6–3 Estimate the solid angle from area A_1, subtended by area A_2 as shown in figure 6–42, as area A_1 moves from position i to position ii to position iii. That is, determine ω_{12i}, ω_{12ii}, and ω_{12iii}.

Section 6–2

6–4 Determine the temperature associated with the radiation emitted from a black body radiator such that its color appears to be
(a) Red (b) Blue (c) Green (d) Yellow
(e) Purple
(*Hint:* Use Wein's Displacement Law and refer to a physics textbook for the wavelength associated with a color.)

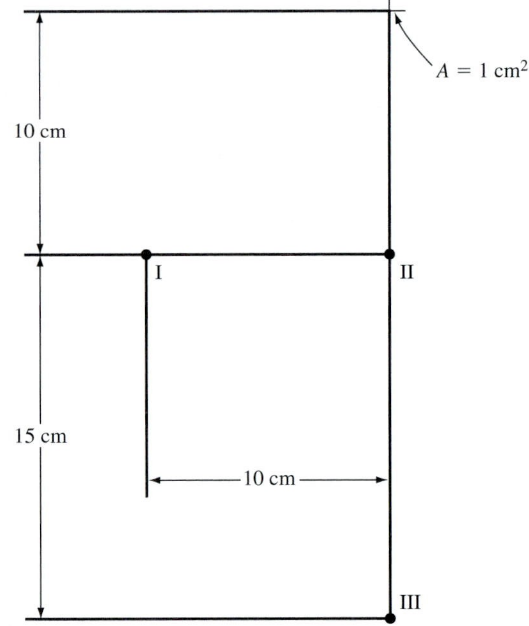

Figure 6–41

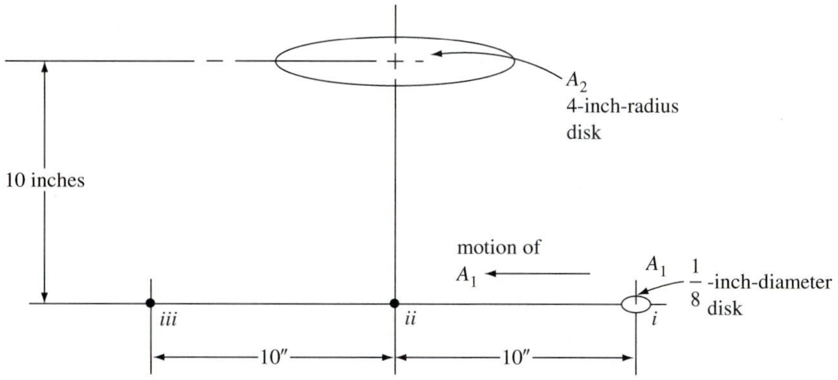

Figure 6–42

6–5 Determine the black body radiation emitted in the bandwidth $\lambda_1 = 0.2\,\mu\text{m}$ to $\lambda_2 = 5.0\,\mu\text{m}$ from a surface at a temperature of
(a) 100°C
(b) 1000°C
(c) 3000°C

6–6 A certain window glass has spectral transmissivity as shown in figure 6–43. Determine the effective transmissivity of thermal energy from a source at 1000°R.

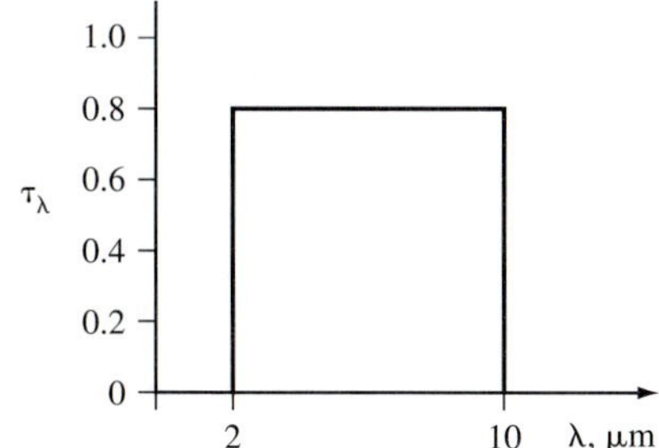

Figure 6–43

Section 6–3

6–7 A Plexiglas wall has reflectivity of 0.1. Estimate the transmissivity if its emissivity is 0.3.

6–8 If a gray opaque body has emissivity of 0.6, what is its reflectivity?

6–9 A transparent surface has transmissivity of 0.8. What would be its reflectivity.

6–10 If a dull metal surface is assumed to be nonreflecting and its absorptivity is 0.65, determine its transmissivity.

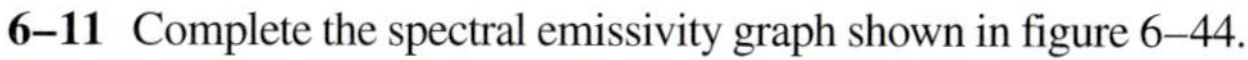

6–11 Complete the spectral emissivity graph shown in figure 6–44.

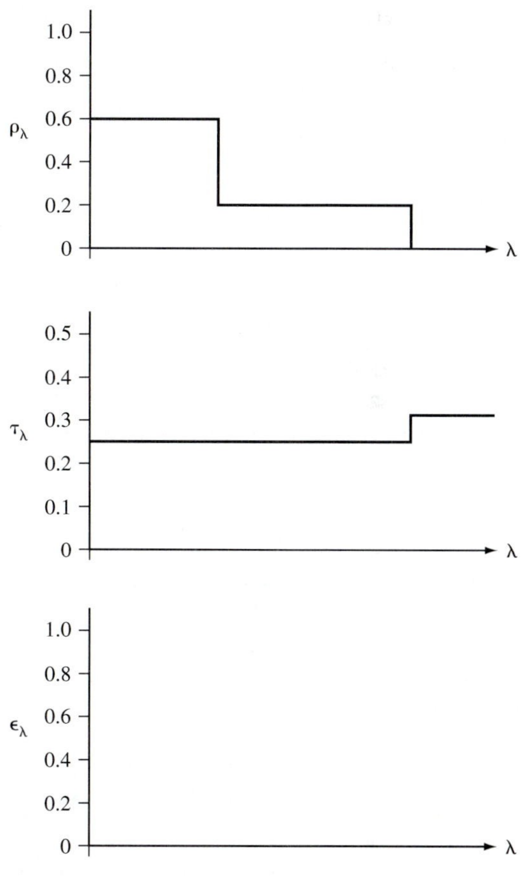

Figure 6–44

Section 6–4

6–12 Three surfaces are interacting with each other through thermal radiation. If the following data apply:

$$A_1 = 2 \text{ ft}^2 \text{ (convex surface)} \qquad F_{1-2} = 0.2$$
$$A_2 = 3 \text{ ft}^2 \text{ (convex surface)}$$
$$A_3 = 4 \text{ ft}^2 \text{ (concave surface)}$$

determine F_{1-3} and F_{2-3}.

6–13 A polished aluminum plate, 0.3 cm thick, is used to partially shield against thermal radiation. Determine the shape factor, F_{1-2}, between one side, A_1, and the other side, A_2, if there are 1.2-cm-diameter holes punched in the plate on 3-cm centers as shown in figure 6–45.

6–14 A large solar furnace reflects solar radiation from a parabolic dish to a collector. The dish and collector are very long and have a configuration shown in figure 6–46. Determine the shape factor from the dish to the collector.

6–15 A 1-cm-diameter thick shield is placed around a furnace. What diameter hole should be put in the shield so that 50% of the incident radiation on the hole passes through unaffected by the hole?

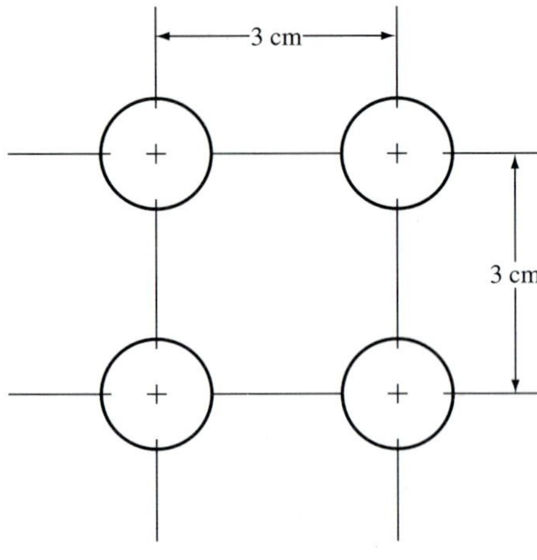

Figure 6–45

6–16 Determine the shape factors F_{1-2} and F_{2-1} for the surfaces shown in figure 6–47. Surfaces A_1 and A_2 are flat and perpendicular to each other.

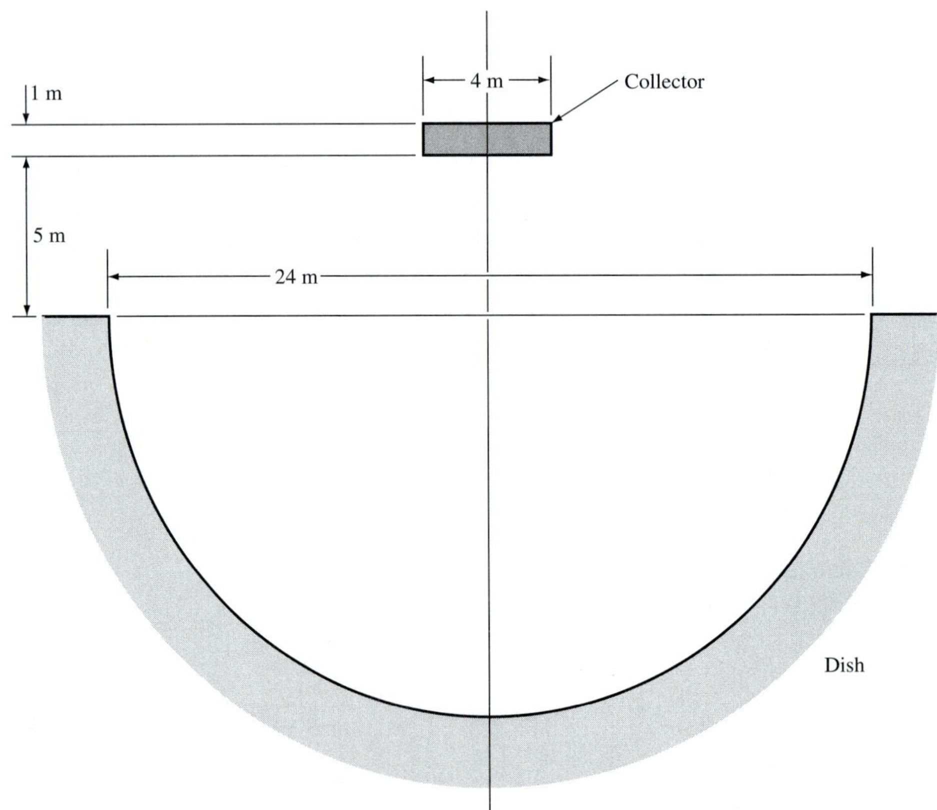

Figure 6–46

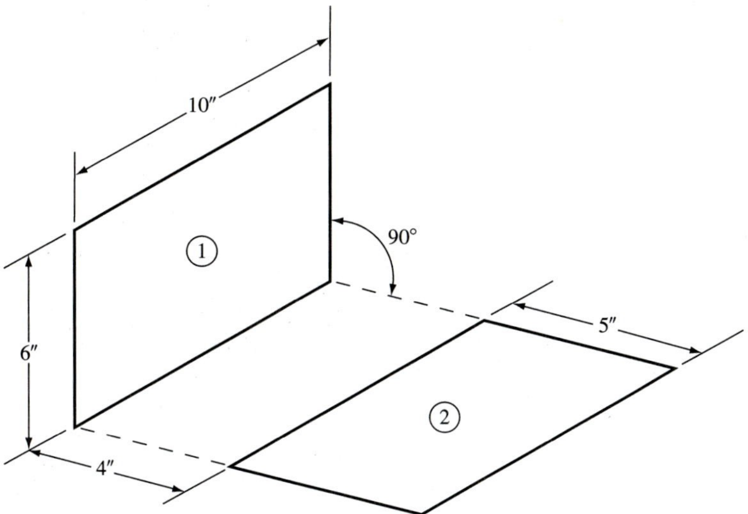

Figure 6–47

6–17 Determine the shape factors F_{1-2}, F_{2-1}, F_{1-3}, F_{2-3}, F_{3-1}, and F_{3-2} for the configurations of surfaces A_1, A_2, and A_3 shown in figure 6–48.

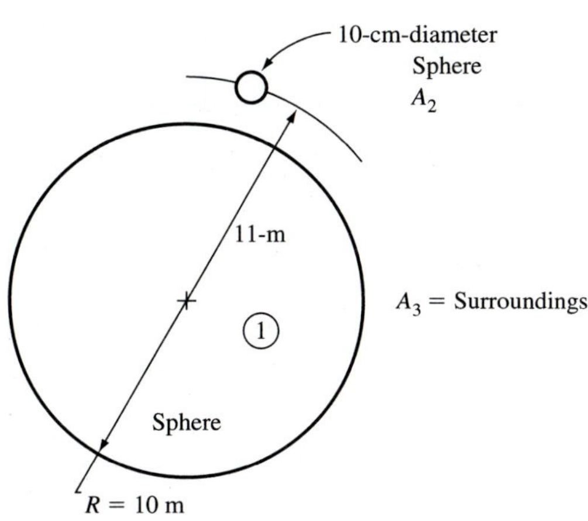

Figure 6–48

6–18 Two flat rings, each having outer diameter of 6 in. and inner diameter of 3 in. are parallel and 4 in. apart as shown in figure 6–49. Determine the shape factor $F_{1-2} = F_{2-1}$.

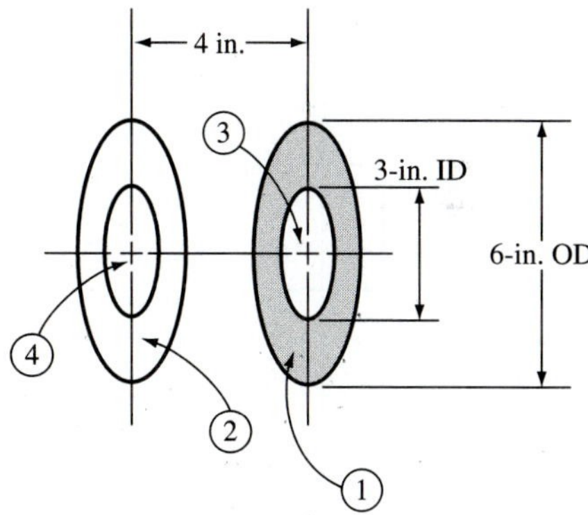

Figure 6–49

6–19 Determine the shape factors F_{1-2}, F_{1-3}, F_{2-1}, and F_{3-1} between the surfaces shown in figure 6–50.

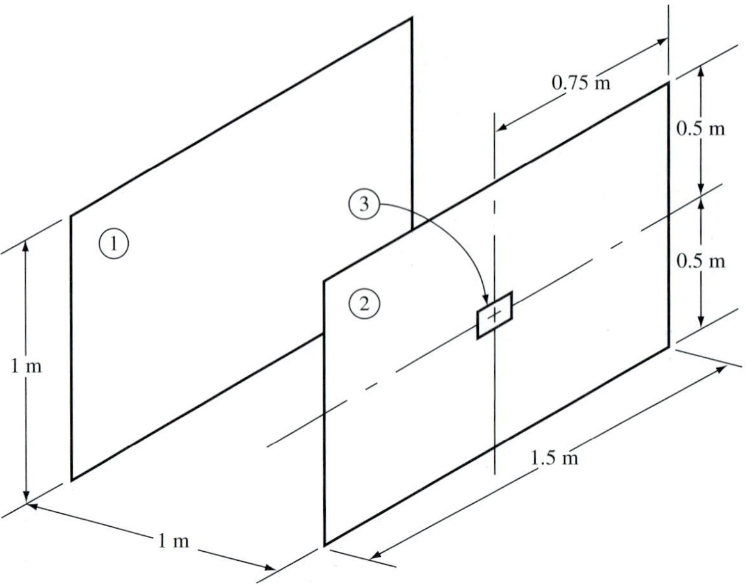

Figure 6–50

6–20 Determine the shape factors between the 4 × 4-in. square long rod and the 4-in.-diameter long rod shown in figure 6–51. That is, find the shape factors F_{1-2} and F_{2-1}. (*Hint:* Use of the cross-string method is advantageous.)

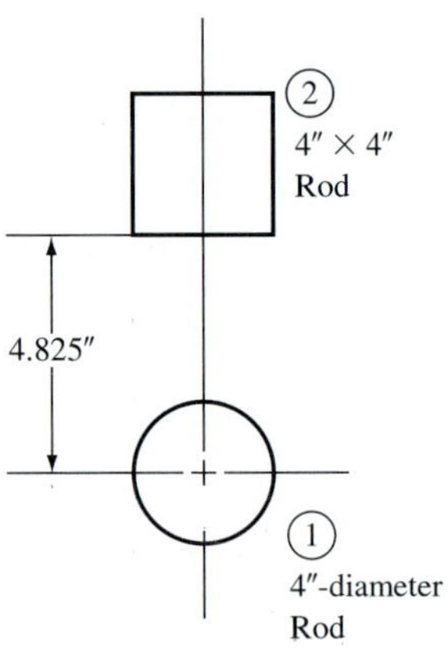

Figure 6–51

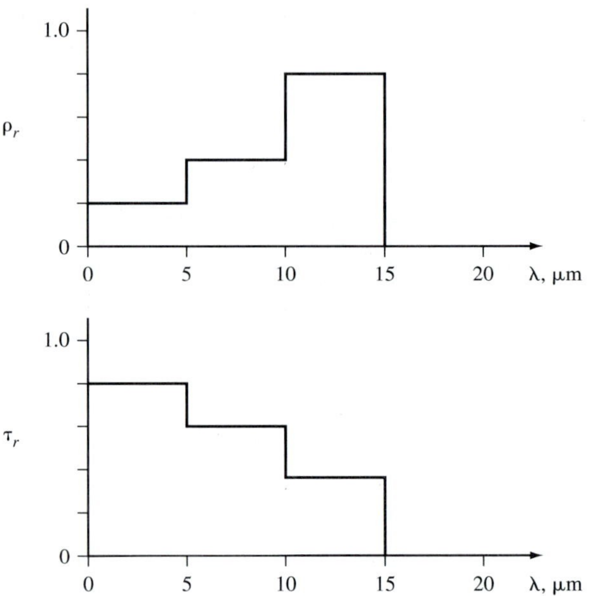

Figure 6–52

Section 6–5

6–21 A gray transparent surface has the spectral reflectivity shown in figure 6–52. Determine the average reflectivity and transmissivity at 500 K and at 850 K.

6–22 A pane of glass having transmissivity

$$\tau_\lambda = 0.8 \qquad 0 < \lambda \leq 4\ \mu\text{m}$$
$$\tau_\lambda = 0.0 \qquad \lambda > 4\ \mu\text{m}$$

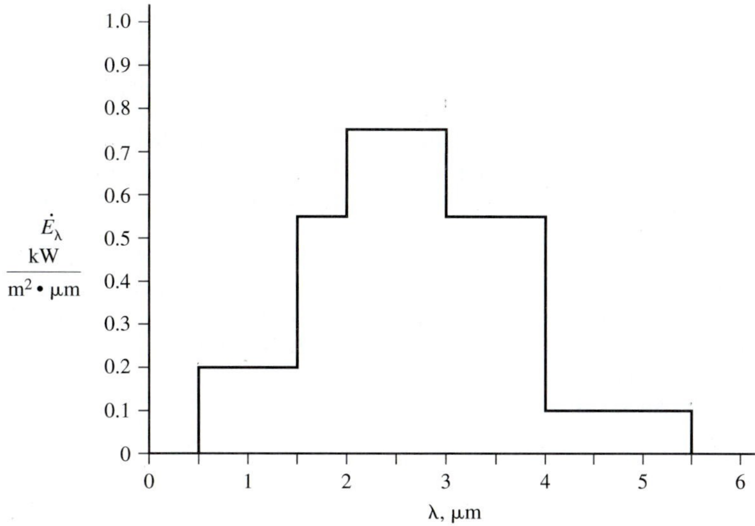

Figure 6–53

is subjected to direct thermal radiation from the sun on one side and from a room on the other side. Assume the sun is a black body at 6000 K and the room is a black body at 300 K. Determine the net thermal radiation through the glass.

6–23 For the monochromatic emissive power distribution shown in figure 6–53 determine the radiant energy emitted for
(a) The full spectrum of wavelengths.
(b) The wavelength band of 1 μm to 2 μm.

6–24 For the gray opaque surface having the spectral emissivity given in figure 6–54, determine the spectral reflectivity.

6–25 Determine the number of air masses through which the solar energy from the sun must pass to reach the earth at the equator for each hour on the equinox (the day of the year when the sun is in the sky for 12 hr).

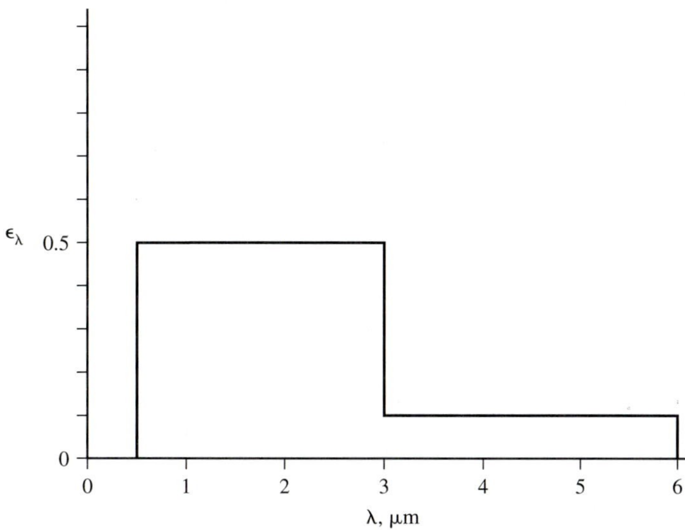

Figure 6–54

REFERENCES

1. Planck, M., *The Theory of Heat Radiation,* Dover Publications, New York, 1959.
2. Touloukian, Y. S. et al., *Thermophysical Properties of Matter,* vol. 7 (1970), vol. 8 (1972), vol. 9 (1972), Plenum Data Corporation, New York.
3. Hottel, H. C., and A. F. Sarofim, *Radiative Heat Transfer,* McGraw-Hill, New York, 1967.
4. Hamilton, D. C., and W. R. Morgan, *Radiant Interchange Configuration Factors,* National Advisory Committee for Aeronautics (NACA) Technical Note 2836, 1952.
5. Hottel, H. C., Radiant Heat Transmission, in W. H. McAdams, ed., *Heat Transmission,* 3rd edition, McGraw-Hill, New York, 1954.
6. Dietz, A. G. H., Diathermanous Materials and Properties of Surfaces, in R. W. Hamilton, ed., *Space Heating with Solar Energy,* MIT Press, Cambridge, MA, 1954.
7. Duffie, J. A., and W. A. Beckman, *Solar Engineering of Thermal Processes,* John Wiley & Sons, New York, 1980.
8. Gier, J. T., and R. V. Dunkle, *Selective Spectral Characteristics as an Important Factor in the Efficiency of Solar Collectors,* Transactions of the Conference of the Use of Solar Energy, vol. 2, part I, University of Arizona Press, Tucson, AZ, 1958, p. 41.

ANALYSIS OF RADIATION HEAT TRANSFER

In this chapter the concepts introduced and discussed in Chapter 6 will be used to analyze radiation heat transfer between two or more bodies. Radiosity will be introduced, and then the two-body radiation situation will be considered. The thermal resistance concept will be used to develop a consistent procedure for analyzing radiation heat transfer. Radiation heat transfer between three or more surfaces and gas radiation will be considered. A section will be devoted to applications of radiation involving radiation, conduction, and convection heat transfer. Some examples will be developed that illustrate how engineering design can require an understanding of radiation heat transfer.

New Terms

K_λ extinction coefficient

J radiosity

7–1 RADIOSITY

We considered irradiation of a surface and sketched that phenomenon in figure 6–13. Part of the irradiation is absorbed, part is transmitted, and part is reflected. The surface also emits radiation and if we want to account for all of the radiation arriving and leaving the surface we have the situation sketched in figure 7–1. Notice in the schematic that the total radiation leaving the surface is

$$\dot{E}_{\text{emitted}} + \dot{E}_{\text{reflected}} = J \tag{7–1}$$

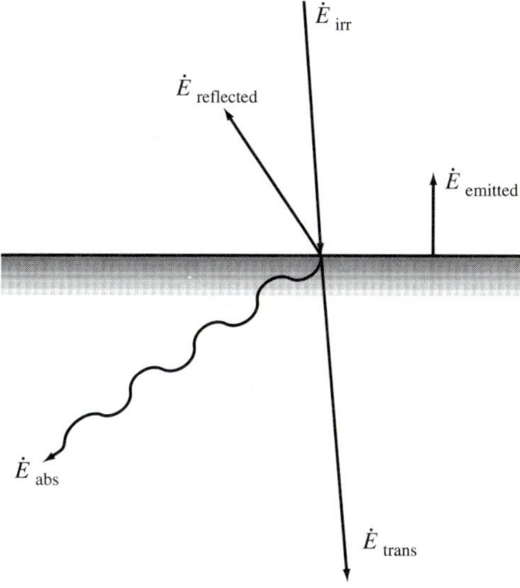

Figure 7–1 Radiation at a surface.

and we call this the radiosity, J. Radiosity has units of power/area, such as W/m^2, Btu/hr · ft^2, and so on. For a gray opaque surface the radiosity is

$$J = \varepsilon_r \sigma T^4 + \rho_r \dot{E}_{\text{irr}} \tag{7–2}$$

Since $\rho_r = 1 - \varepsilon$, this becomes

$$J = \varepsilon_r \sigma T^4 + (1 - \varepsilon_r)\dot{E}_{\text{irr}} \tag{7–3}$$

The net radiation leaving the surface per unit area is

$$\dot{q}_{A,\text{net}} = \frac{\dot{Q}_{\text{net}}}{A} = J - \dot{E}_{\text{irr}} \tag{7–4}$$

or, using equation (7–3),

$$\dot{q}_{A,\text{net}} = \varepsilon_r \sigma T^4 + (1 - \varepsilon_r)\dot{E}_{\text{irr}} - \dot{E}_{\text{irr}} \tag{7–5}$$

and

$$\dot{q}_{A,\text{net}} = \varepsilon_r(\sigma T^4 - \dot{E}_{\text{irr}}) \tag{7–6}$$

It is more useful to solve for $\dot{E}_{\text{irr}}$ in equation (7–3):

$$\dot{E}_{\text{irr}} = \frac{1}{1 - \varepsilon_r}(J - \sigma T^4) \tag{7–7}$$

and then substitute equation (7–7) into (7–6) to obtain

$$\dot{q}_{A,\text{net}} = \frac{\varepsilon_r}{1 - \varepsilon_r}(\sigma T^4 - J)$$

or

$$\dot{Q}_{\text{net}} = \frac{\varepsilon_r A}{1 - \varepsilon_r}(\sigma T^4 - J) \tag{7–8}$$

where $\dot{Q}_{\text{net}}$ is the total net radiation leaving surface A. The form of equation (7–8) suggests a thermal resistance analogy in which the flux, $\dot{Q}_{\text{net}}$, is driven by the thermal potential, $(\sigma T^4 - J)$, and resisted by the thermal resistance, $(1 - \varepsilon_r)/\varepsilon_r A$. In figure 7–2 is shown the electrical analogy to the surface radiations.

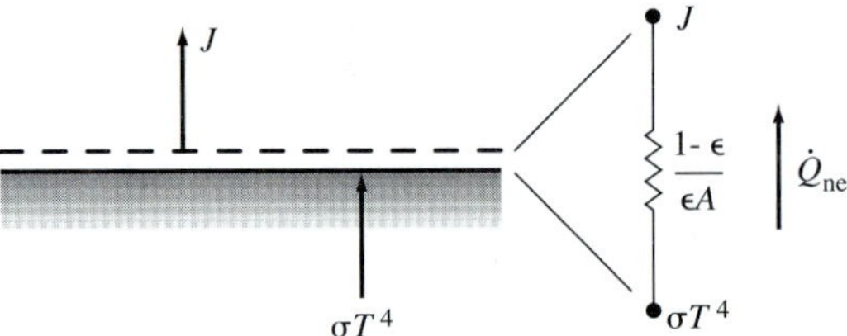

Figure 7–2 Thermal resistance at a radiating surface.

The analogy tends to be more abstract than those in earlier discussions of the analogy between conduction heat transfer and Ohm's Law; however, the comparison is useful. The net heat transfer, $\dot{Q}_{\text{net}}$ has units of flux or power, W; however, the thermal potential terms σT^4 and J have units of W/m^2, or power per unit area, and the thermal resistance has units of m^{-2}, or inverse area. From the surface sketched in figure 7–2 we see that the model or analogy implies that the gray opaque surface is driven by a black body source (σT^4) manifested as a radiosity (J). The thermal resistance decreases as the surface area increases or as the emissivity increases. In the limit as $\varepsilon_r \to 1$, the surface then acts as a black body and the thermal resistance is zero, $(1 - \varepsilon_r)/\varepsilon_r A = 0$. For zero thermal resistance we have

$$J = \sigma T^4 \tag{7–9}$$

which is a result we came across earlier; black bodies do not reflect radiation but emit perfectly. At the other extreme, as $\varepsilon_r \to 0$ the radiosity is $J = \dot{E}_{\text{irr}}$ (perfect reflector).

EXAMPLE 7–1 The emissivity of a certain surface is temperature dependent by the equation $\varepsilon_r(T) = 0.6 - 0.0002T$ $(T \le 2600\ \text{K})$. Estimate the temperature at which the net heat is maximized for a constant irradiation $\dot{E}_{\text{irr}} = 400\ \text{W/m}^2$.

Solution The net heat transfer is

$$\dot{Q}_{\text{net}} = A(J - \dot{E}_{\text{irr}}) = \varepsilon_r A(\sigma T^4 - \dot{E}_{\text{irr}})$$

and this is maximized when the radiosity is maximized.

We have $$J = \varepsilon_r \sigma T^4 + (1 - \varepsilon_r)\dot{E}_{\text{irr}}$$
and since $$\varepsilon_r(T) = 0.6 - 0.0002T$$
then $$J(T) = (0.6 - 0.0002T)\sigma T^4 + (0.4 + 0.0002T)\dot{E}_{\text{irr}}$$
or $$J(T) = 0.6\sigma T^4 - 0.0002\sigma T^5 + 0.0002\dot{E}_{\text{irr}} + 0.4\dot{E}_{\text{irr}}$$

For $\dot{E}_{\text{irr}} = 400\ \text{W/m}^2$ this becomes

$$J(T) = 0.6\sigma T^4 - 0.0002\sigma T^5 + 0.08T + 160$$

Determination of the maximum and minimum for $J(T)$ is made by the Newton–Raphson method, by trial and error substitution, or by the maximum and minimum principle of calculus. Using the calculus we have

$$\frac{dJ(T)}{dT} = J' = 2.4\sigma T^3 - 0.001\sigma T^4 + 0.08$$

and $$\frac{d^2J(T)}{dT^2} = J'' = 7.2\sigma T^2 - 0.004\sigma T^3$$

Points of inflection occur when $J'' = 0$. This gives $T = 0$ or $T = 1800$ at points of inflection. Maxima or minima occur at $J' = 0$, and this gives $T \approx 2400$ K with $J = 227{,}287\ \text{W/m}^2$. The emissivity at 2400 K is $\varepsilon_r = 0.6 - 0.0002(2400) = 0.12$, and

Answer $$\frac{\dot{Q}_{\text{net}}}{A} = \dot{q}_{A,\text{net}} = J - \dot{E}_{\text{irr}} = 226{,}887\ \text{W/m}^2$$

In figure 7–3 is shown the plot of the net radiation at a surface for constant $\dot{E}_{\text{irr}} = 400\ \text{W/m}^2$.

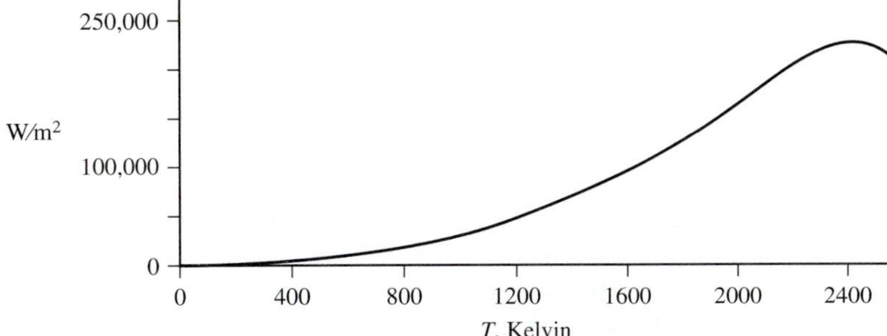

Figure 7–3 Net radiation from a surface at constant $\dot{E}_{\text{irr}} = 400\ \text{W/m}^2$ for example 7–1.

7–2 ANALYSIS OF TWO-SURFACE RADIATION HEAT TRANSFER

Two-surface radiation heat transfer is the phenomenon in which radiation occurs only between two surfaces or, in some cases, in which one of the two surfaces interacts with the other surface exclusively.

Many physical situations involving radiation heat transfer can be described as an interaction between two surfaces, and in figure 7–4 are shown some common two-surface configurations. Case (a) represents the situation where the two surfaces can only see each other. In case (b) surface A_1 can only see A_2 while A_2 can see part of itself because it is concave. In case (c) the surface A_1, which is flat or convex, is so small that it can only see A_2, which is much larger. The radiation heat transfer from surface A_1 to A_2 is

$$\dot{Q}_{1-2} = A_1F_{1-2}J_1 \tag{7–10}$$

and the heat transfer from A_2 to A_1 is

$$\dot{Q}_{2-1} = A_2F_{2-1}J_2$$

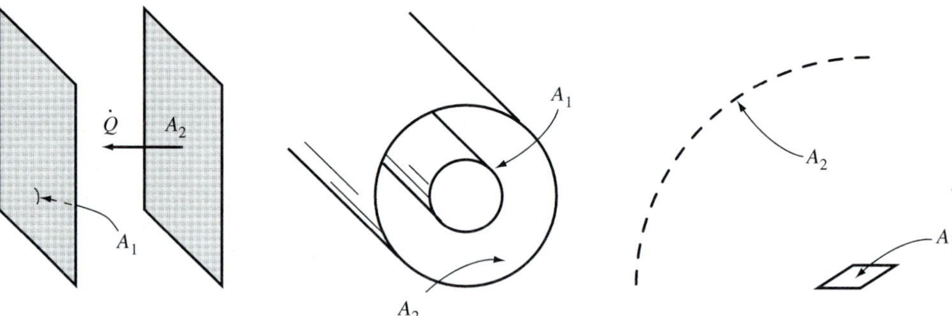

Figure 7–4 Two-surface radiation heat-transfer configuration.

By the reciprocity relationship, equation (6–50). $A_1F_{1-2} = A_2F_{2-1}$ so that we could write

$$\dot{Q}_{2-1} = A_1F_{1-2}J_2$$

The net radiation from surface A_1 to A_2 is

$$\dot{Q}_{\text{net},1-2} = \dot{Q}_{1-2} - \dot{Q}_{2-1} = A_1F_{1-2}(J_1 - J_2) \tag{7–11}$$

or, in general, between any two surfaces A_i and A_j,

$$\dot{Q}_{\text{net},i-j} = A_iF_{i-j}(J_i - J_j) \tag{7–12}$$

The form of equations (7–11) and (7–12) suggests thermal resistance concepts, and the thermal resistance will be

$$\frac{1}{A_1F_{1-2}} = \frac{1}{A_2F_{2-1}}$$

or, in general

$$\frac{1}{A_iF_{i-j}} = \frac{1}{A_jF_{j-i}} \tag{7–13}$$

The thermal potential is $J_i - J_j$ and the flux is $\dot{Q}_{i-j}$. Thus we have a model of net radiation between two surfaces as sketched in figure 7–5. The complete analogy of radiation heat transfer between two surfaces where temperatures are the thermal potentials is constructed by combining the thermal resistance of the surfaces A_1 and A_2 from equations (7–8) and (7–11). The schematic of the circuit is shown in figure 7–6.

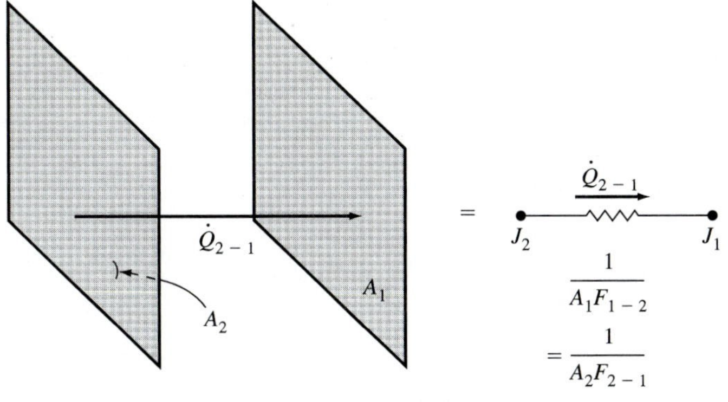

Figure 7–5 Thermal resistance analogy of net radiation between two surfaces.

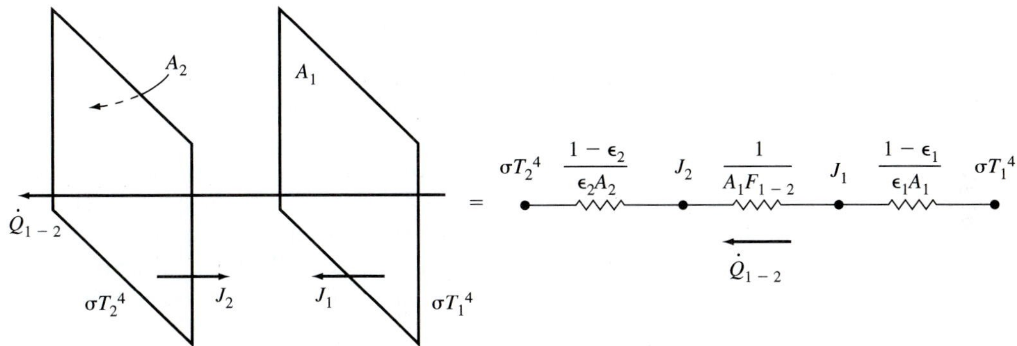

Figure 7–6 Schematic of radiation exchange between two surfaces.

The net heat transfer can be written in at least one of the following forms:

$$\dot{Q}_{\text{net},1-2} = \frac{\varepsilon_{r,1}A_1}{1-\varepsilon_{r,1}}(\sigma T^4 - J_1) \tag{7–14}$$

$$= \frac{\varepsilon_{r,2}A_2}{1-\varepsilon_{r,2}}(J_2 - \sigma T_2^4) \tag{7–15}$$

$$= A_1F_{1-2}(J_1 - J_2) \tag{7–16}$$

With some algebra manipulations, the radiosity terms can be eliminated and the net heat transfer written

$$\dot{Q}_{\text{net},1-2} = \frac{\sigma(T_1^4 - T_2^4)}{\dfrac{1-\varepsilon_{r,1}}{\varepsilon_{r,1}A_1} + \dfrac{1}{A_1F_{1-2}} + \dfrac{1-\varepsilon_{r,2}}{\varepsilon_{r,2}A_2}} \tag{7–17}$$

Equation 7–17 expresses the summation of thermal resistances in series between two abstract (or imaginary) black bodies as indicated in the circuit diagram of figure 7–6. This is similar to the results of conduction heat transfer where the heat transfer was driven by a temperature difference and resisted by a thermal resistance. In equation (7–17) and in figure 7–6 the heat transfer is driven by a difference in temperature to the 4th power and

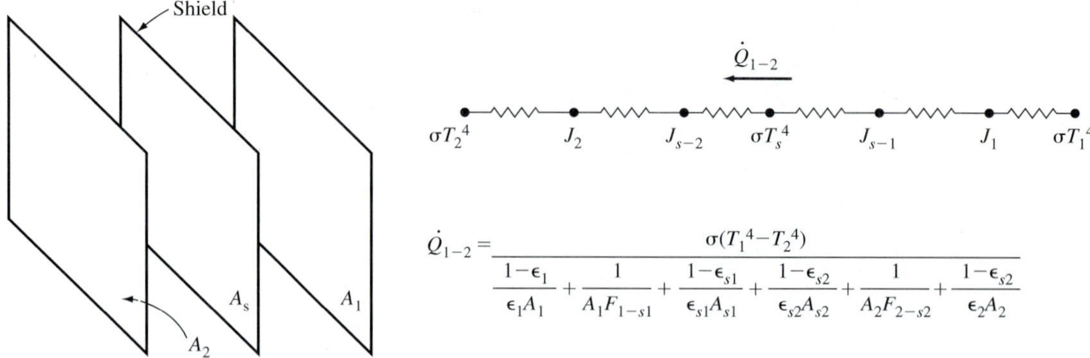

Figure 7–7

resisted by thermal resistances representing gray body behavior and by the geometric restraints of the shape factors. The net radiation heat transfer $\dot{Q}_{net,1-2}$ can be reduced between two surfaces by increasing the thermal resistance. Such an increase in thermal resistance occurs if a thermal shield is placed between the two surfaces, as shown in figure 7–7. In the schematic of an analogous electrical circuit it can be seen that the total thermal resistance includes three additional terms representing the thermal resistance of the shield. This increases the total thermal resistance between A_1 and A_2. For special situations the result of equation 7–17 simplifies. We consider some of them:

1. If $A_2 \gg A_1$ then

$$\dot{Q}_{net,1-2} = \frac{A_1 \sigma (T_1^4 - T_2^4)}{\dfrac{1 - \varepsilon_{r,1}}{\varepsilon_{r,1}} + \dfrac{1}{F_{1-2}}} \tag{7–18}$$

2. If $A_2 \gg A_1$ and $F_{1-2} = 1.0$, then

$$\dot{Q}_{net,1-2} = A_1 \varepsilon_{r,1} \sigma (T_1^4 - T_2^4) \tag{7–19}$$

3. If $A_2 \gg A_1$ and A_1 is a black body (A_2 may be a gray body)

$$\dot{Q}_{net,1-2} = A_1 F_{1-2} \sigma (T_1^4 - T_2^4) \tag{7–20}$$

4. If $A_2 \approx A_1$ and both surfaces are black bodies

$$\dot{Q}_{net,1-2} = A_1 F_{1-2} \sigma (T_1^4 - T_2^4) \tag{7–21}$$

5. If A_1 is convex for either case 3 or 4, then

$$F_{1-2} = 1.0 \qquad \text{(two-surface problem)}$$

and

$$\dot{Q}_{net,1-2} = A_1 \sigma (T_1^4 - T_2^4) \tag{7–22}$$

If the radiation heat transfers between two surfaces in a steady state situation, then energy must be supplied and rejected. For instance, for the schematic shown in figure 7–6, there must be an energy source to surface A, such as convection heat transfer, conducted heat, or some other external radiation if the steady state condition is to be satisfied. Also, there must be a heat rejection at surface A_2. If surface A_1 and/or surface A_2 is that of a material experiencing a phase change at constant temperature, then to a limited extent the analysis of the two-surface radiation can proceed as if steady state conditions prevailed.

EXAMPLE 7–2 Cylindrical plutonium nuclear rods of 2-in. diameter at 1000°F with emissivity of 0.85 are surrounded by a steel tubing of 7-in. inside diameter. Consider two separate conditions:

1. When $\varepsilon_r = 0.75$ steel tubing is at 100°F, determine the net heat transfer from the rods per unit length,
2. If a black body shield is placed concentrically between the rod and the tubing, determine the shield temperature and the net heat transfer from the rod.

Solution For the unshielded arrangement of a nuclear rod inside a steel tubing we can use equation (7–17) and, as sketched in figure 7–8a,

$$A_1 = \pi(7/12 \text{ ft}) = 1.8326 \text{ ft}^2/\text{ft}$$
$$A_2 = \pi(2/12 \text{ ft}) = 0.5236 \text{ ft}^2/\text{ft}$$
$$F_{1-2} = \frac{A_2}{A_1}F_{2-1} = \frac{0.5236}{1.8326}(1.0) = 0.286$$

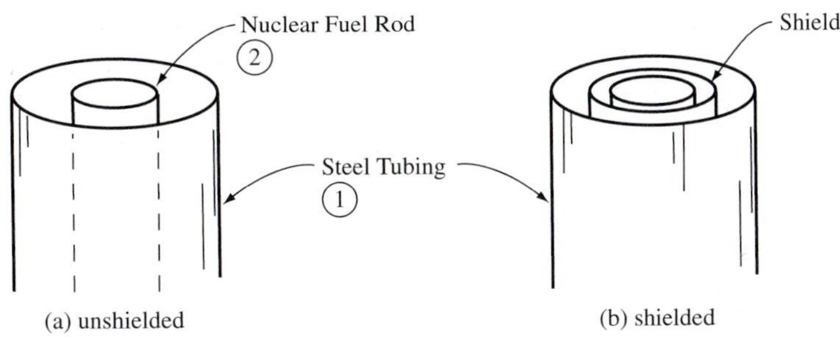

Figure 7–8 Nuclear rods in steel tubing.

Then, for equation (7–17),

$$\dot{Q}_{\text{net},1-2} = \frac{(0.174 \times 10^{-8} \text{ Btu/hr} \cdot \text{ft}^2 \cdot {}^\circ\text{R}^4)[(1460^\circ\text{R})^4 - (560^\circ\text{R})^4]}{\dfrac{1-0.85}{0.85(0.5236 \text{ ft}^2/\text{ft})} + \dfrac{1}{0.286(1.8326 \text{ ft}^2/\text{ft})} + \dfrac{1-0.75}{0.75(1.8326 \text{ ft}^2/\text{ft})}}$$

Then,

Answer
$$\dot{Q}_{\text{net},1-2} = 3184.7 \text{ Btu/hr} \cdot \text{ft}$$

For the shielded arrangement of figure 7–8b we have

$$A_s = \pi(5/12 \text{ ft}) = 1.309 \text{ ft}^2/\text{ft}$$

and, again using equation (7–17), with the added thermal resistance caused by the shield,

$$\dot{Q}_{\text{net},1-2} = \frac{(0.174 \times 10^{-8})(1460^4 - 560^4)}{\dfrac{1-0.85}{0.85(0.5236)} + \dfrac{1}{0.5236} + \dfrac{1}{1.309} + \dfrac{1-0.75}{0.75(1.8326)}}$$

Answer
$$= 2422.7 \text{ Btu/hr} \cdot \text{ft}$$

Thus, the heat transfer has been reduced by placing a shield between the two surfaces. To determine the shield temperature we need to consider the heat transfer between the shield and the tubing:

$$\dot{Q}_{net,1-s} = \frac{\sigma(1460^4 - T_s^4)}{\dfrac{1-\varepsilon_{r,1}}{\varepsilon_{r,1}A_1} + \dfrac{1}{A_1F_{1-s}} + \dfrac{1-\varepsilon_{r,s}}{\varepsilon_{r,s}A_s}} = 2422.7 \text{ Btu/hr} \cdot \text{ft}$$

Solving for the shield temperature T_s, gives

Answer

$$T_s = 1352.4°\text{R}$$

EXAMPLE 7–3 An oven has been turned off at 1000°C and its door is opened, as shown in figure 7–9. The oven inside walls have an emissivity of 0.95, and the surroundings may be taken as a black body at 300 K. Estimate the heat loss from the oven just after opening the door.

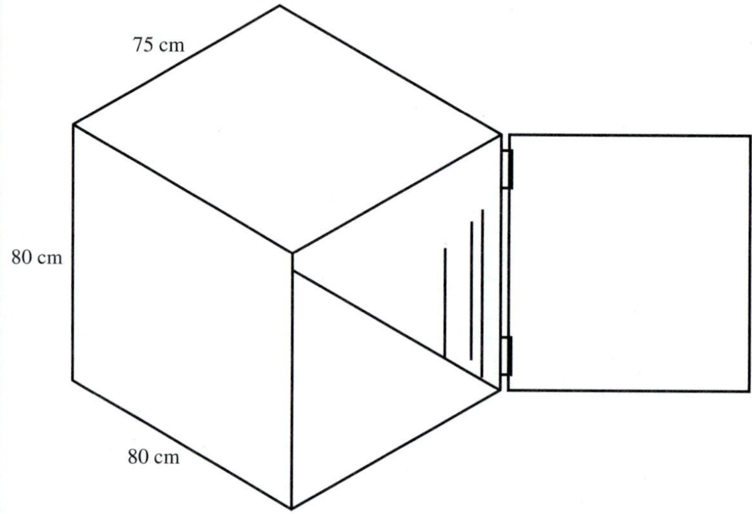

Figure 7–9 Cooling an oven.

Solution

This problem can be approximated as a two-surface problem with one black body so that we may use equation (7–17) with the revision that $\varepsilon_{r,2} = 1.0$:

$$\dot{Q}_{net,1-2} = \frac{\sigma(T_1^4 - T_2^4)}{\dfrac{1-\varepsilon_{r,1}}{\varepsilon_{r,1}A_1} + \dfrac{1}{A_1F_{1-2}}}$$

The shape factor F_{1-2} can be determined from the reciprocity relationship

$$F_{1-2} = \frac{A_2}{A_1}F_{2-1}$$

where $F_{2-1} = 1$, $A_1 = 3.08 \text{ m}^2$, and $A_2 = 0.6 \text{ m}^2$. Then

$$F_{1-2} = \frac{0.6 \text{ m}^2}{3.08 \text{ m}^2}(1.0) = 0.195$$

Then

Answer

$$\dot{Q}_{net,1-2} = \frac{\sigma(1273 \text{ K}^4 - 300 \text{ K}^4)}{\dfrac{1-0.95}{0.95(3.08 \text{ m}^2)} + \dfrac{1}{3.08 \text{ m}^2(0.195)}} = 88.7 \text{ kW}$$

EXAMPLE 7–4 A proposal is made to heat special 3-cm-diameter cylindrical tubing by a radiant heating element placed concentrically around the tubing. The production method would involve feeding continuous tubing at 50 cm/s and not allow the tubing to touch the heater, as sketched in figure 7–10. The radiant heater is 10 cm in diameter and has an emissivity of 0.9. It can be furnished in any axial length. Estimate the length needed to heat the tubing from 15°C to 85°C if the specific heat per unit length of tubing is 20 J/mK.

Figure 7–10 Example 7–4.

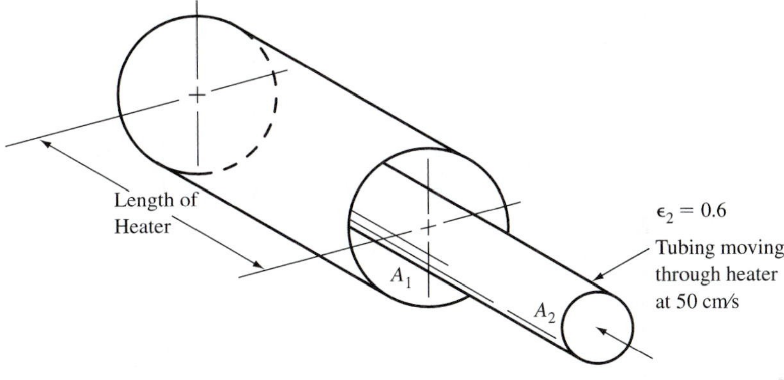

Solution This is an example of an concentric cylinder and the heat transfer may be written

$$\dot{Q}_{\text{net},1-2} = \frac{\sigma(T_1^4 - T_2^4)}{\dfrac{1-\varepsilon_{r,1}}{\varepsilon_{r,1}A_1} + \dfrac{1}{A_1F_{1-2}} + \dfrac{1-\varepsilon_{r,2}}{\varepsilon_{r,2}A_2}}$$

If we assume $T_2 \approx 50°\text{C}$ as an average tubing temperature, then

$$\dot{Q}_{\text{net},1-2} = \frac{(5.7 \times 10^{-8}\ \text{W/m}^2 \cdot \text{K}^4)[(500\ \text{K})^4 - (323\ \text{K})^4]}{\dfrac{1-0.9}{0.9(0.1\pi\ \text{m}^2/\text{m})} + \dfrac{1}{0.03(1.0)} + \dfrac{1-0.6}{0.6(0.03\pi\ \text{m}^2/\text{m})}} = 163.1\ \frac{\text{W}}{\text{m}}$$

For the tubing, sketched in figure 7–11, we apply the control volume concept and write that

$$\dot{Q}_{\text{tubing}} = \dot{m}c_p(T_o - T_i) = \rho \mathbf{V} A c_p(T_o - T_i) = \mathbf{V} c_{p,l}(T_o - T_i)$$

where $c_{p,l}$ is the specific heat per unit length of tubing and the temperatures are the tubing temperature at the inlet (T_i) and at the outlet (T_o). Then,

$$\dot{Q}_{\text{tubing}} = (0.5\ \text{m/s})(20\ \text{J/m} \cdot \text{K})(70\ \text{K}) = 700\ \text{W}$$

which is the net heat transfer required of the heater. Thus

$$\dot{Q}_{\text{tubing}} = 700\ \text{W} = 163.1\ \text{W/m (heater length)}$$

Answer

$$\text{Heater length} = \frac{700\ \text{W}}{163.1\ \text{W/m}} = 4.29\ \text{m}$$

Figure 7–11 Tubing of example 7–4.

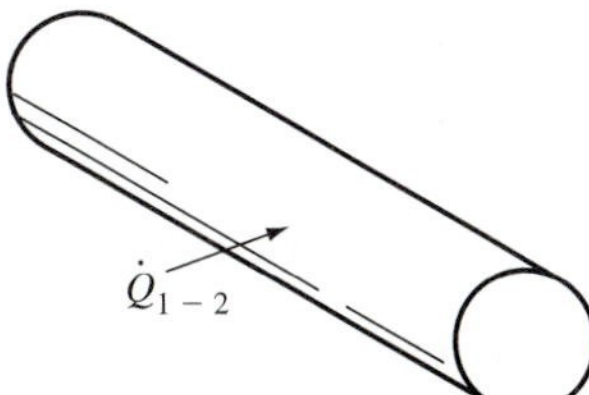

EXAMPLE 7–5 A radiant broiler at 800 K is used to toast 1.5-cm-thick bread slices (see figure 7–12). A bread slice has the same properties as water and its surface area is much smaller than the broiler, so its shape factor to this broiler is 1.0. If the bread is 15° when placed under the broiler, determine the time required for it to reach 120°C. Assume the emissivity of bread is 0.9.

Figure 7–12

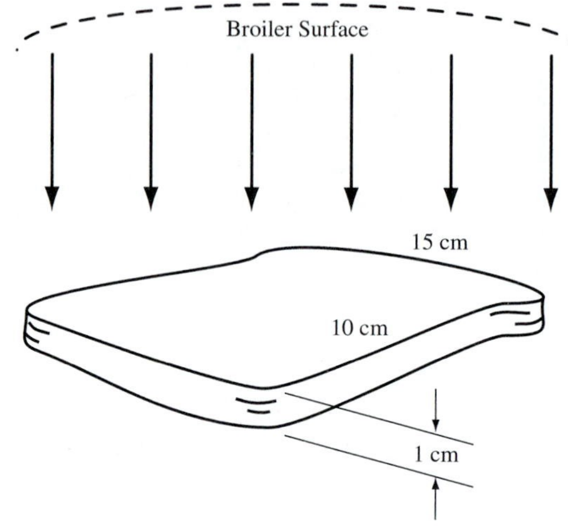

Solution Assuming that the bread does not lose heat by convection or conduction on its surface, we may write A_1 = bread surface and A_2 = heater in surface,

$$\dot{Q}_{2-1} = \frac{dE}{dt} = \varepsilon_{r,1} A_1 \sigma (T_2^4 - T_1^4)$$

If we now assume lumped heat capacity for the bread, we have that

$$\rho V c_p \frac{dT}{dt} = \rho A_1 (\text{thickness}) c_p \frac{dT}{dt} = \varepsilon_{r,1} A_1 \sigma (T_2^4 - T_1^4)$$

or, separating variables,

$$dt = \rho c_p \frac{(\text{thickness})}{\varepsilon_{r,1}\sigma} \frac{dT}{(T_2^4 - T_1^4)}$$

or

$$\int_0^t dt = t = \frac{\rho c_p (\text{thickness})}{\varepsilon_{r,1}\sigma} \int_{288\text{ K}}^{393\text{ K}} \frac{1}{T^4 - 800\text{ K}^4} dT$$

From the table of integrals, appendix table A–4, we have for the right side integral

$$\int \frac{1}{800^4 - T^4} dT = \frac{1}{2 \times 800^3} \left(\frac{\ln\left(\frac{800 - T}{800 + T}\right)}{2} - \tan^{-1}\left(\frac{800}{T}\right) \right) = -\int \frac{1}{T^4 - 800^4} dT$$

and evaluating the integral between 288 K and 393 K gives

Answer

$$t = 180 \text{ s}$$

7–3 ANALYSIS OF THREE-SURFACE RADIATION HEAT TRANSFER

Many radiation heat transfer problems can be analyzed and reasonable solutions obtained by using the two-surface analysis discussed in the previous section. However, in general, radiant heat transfer will involve at least three surfaces. A typical arrangement of two surfaces is shown in figure 7–13, where the surroundings represent a third surface. The reader can extend this further and notice that there may be four, five, six, or more surfaces exchanging radiation heat transfer. For now, we will limit the discussion to three-surface situations and then generalize the analysis to multiple surfaces. In figure 7–13 it can be seen that the two surfaces A_1 and A_2 exchange radiation with the surroundings, called A_3. Using the electrical analogy to describe the heat transfer of the three-surface configuration of figure 7–13 gives the network shown in figure 7–14. This analogy was presented by Oppenheim [1] and is a popular and convenient method for analyzing many radiation heat-transfer problems involving gray or black body opaque surfaces. Notice in the network of figure 7–14 that it is the radiosities J that represent the nodes of interaction between the surfaces. Continuing with the network analogy and applying a conservation of heat flux at each of the three radiosity nodes, J_1, J_2, and J_3, gives for node J_1:

$$\frac{\varepsilon_{r,1}A_1}{1-\varepsilon_{r,1}}(\sigma T_1^4 - J_1) + A_1F_{1-2}(J_2 - J_1) + A_1F_{1-3}(J_3 - J_1) = 0 \quad \textbf{(7–23)}$$

for node J_2,

$$\frac{\varepsilon_{r,2}A_2}{1-\varepsilon_{r,2}}(\sigma T_2^4 - J_2) + A_2F_{2-1}(J_1 - J_2) + A_2F_{2-3}(J_3 - J_2) = 0 \quad \textbf{(7–24)}$$

and for node J_3,

$$\frac{\varepsilon_{r,3}A_3}{1-\varepsilon_{r,3}}(\sigma T_3^4 - J_3) + A_3F_{3-1}(J_1 - J_3) + A_3F_{3-2}(J_2 - J_3) = 0 \quad \textbf{(7–25)}$$

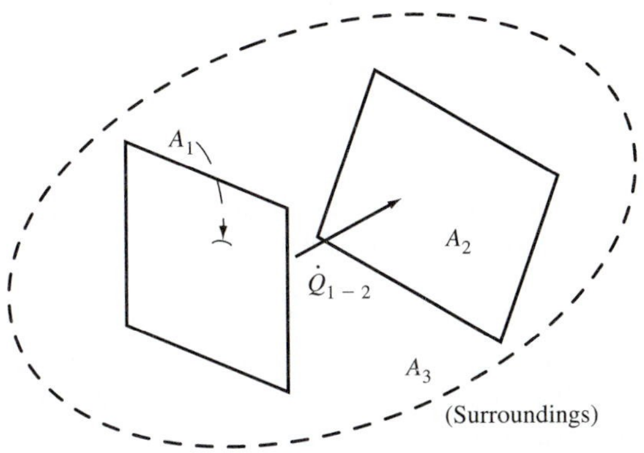

Figure 7–13 Three-surface radiation heat transfer.

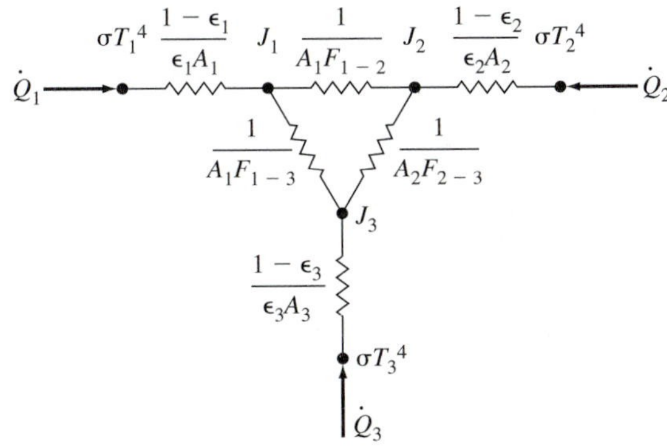

Figure 7–14 Electrical analogy for radiation heat transfer between three gray opaque surfaces.

In equation (7–23) the first term, $\varepsilon_1 A_1(T_1^4 - J_1)/(1 - \varepsilon_1)$, represents the net radiation leaving surface A_1 and is the rate of energy that must be supplied to the surface if its temperature is to remain constant, or that its thermal energy remain constant. This is shown schematically in figure 7–15, where

$$\dot{Q}_1 = \frac{\varepsilon_{r,1} A_1}{1 - \varepsilon_{r,1}}(T_1^4 - J_1) \tag{7–26}$$

is an amount of the rate of energy from some other external source such as convection or conduction heat transfer, or from an internal energy source. The second term, $A_1 F_{12}(J_2 - J_1)$, is the net heat transfer from surface A_2 to surface A_1, which we may write as

$$\dot{Q}_{\text{net},2-1} = A_1 F_{1-2}(J_2 - J_1) = A_2 F_{2-1}(J_2 - J_1) \tag{7–27}$$

The third term, $A_1 F_{1-3}(J_3 - J_1)$, is the net heat transfer from A_3 to A_1,

$$\dot{Q}_{\text{net},3-1} = A_3 F_{3-1}(J_3 - J_1) = A_1 F_{1-3}(J_3 - J_1) \tag{7–28}$$

Figure 7–15

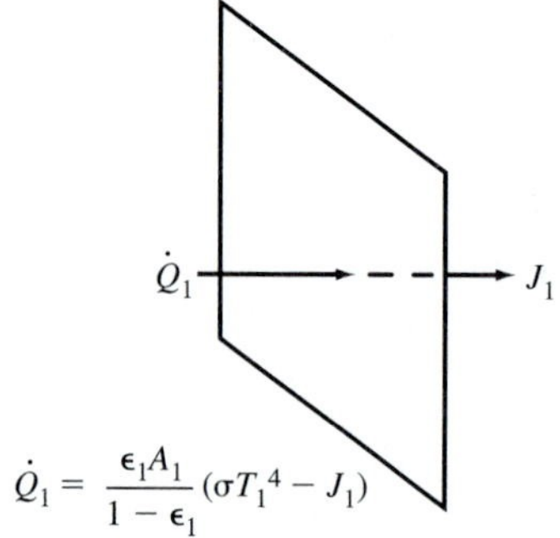

Similar physical interpretations can be made to the terms in equations (7–24) and (7–25), and an energy balance over the full system of three surfaces gives, for steady state,

$$\dot{Q}_1 + \dot{Q}_2 + \dot{Q}_3 = 0 \tag{7–29}$$

The set of equations (7–23) through (7–25) can be solved for the radiosities if the emissivities, shape factors, and surface temperatures are known. Thus, these three equations can be written as a matrix, after rearranging the equations to read

$$\left(A_1F_{1-2} + A_1F_{1-3} + \frac{\varepsilon_{r,1}A_1}{1 - \varepsilon_{r,1}}\right)J_1 + (-A_1F_{1-2})J_2 + (-A_1F_{1-3})J_3 = \frac{\varepsilon_{r,1}A_1\sigma}{1 - \varepsilon_{r,1}}T_1^4$$

$$(-A_1F_{1-2})J_1 + \left(A_2F_{2-1} + A_2F_{2-3} + \frac{\varepsilon_{r,2}A_2}{1 - \varepsilon_{r,2}}\right)J_2 + (-A_2F_{2-3})J_3 = \frac{\varepsilon_{r,2}A_2}{1 - \varepsilon_{r,2}}\sigma T_2^4$$

$$(-A_1F_{1-3})J_1 + (-A_3F_{3-2})J_2 + \left(A_3F_{3-1} + A_3F_{3-2} + \frac{\varepsilon_{r,3}A_3}{1 - \varepsilon_{r,3}}\right)J_3 = \frac{\varepsilon_{r,3}A_3}{1 - \varepsilon_{r,3}}\sigma T_3^4$$

or

$$[M]\{J_i\} = \left\{\frac{\varepsilon_{r,i}A_i}{1 - \varepsilon_{r,i}}\sigma T_i^4\right\} \tag{7–30}$$

where $[M]$ is a matrix of the coefficients of the radiosities, $\{J_i\}$ is the column vector of the radiosities,

$$J_i = \begin{Bmatrix} J_1 \\ J_2 \\ J_3 \end{Bmatrix} \tag{7–31}$$

and $\left\{\dfrac{\varepsilon_{r,i}A_i}{1-\varepsilon_{r,1}}\sigma T_i^4\right\}$ are the column vectors of the black body radiators.

Occasionally radiation problems require solutions to the surface temperatures when a heat flux is known. Then equation (7–26) can be written, in general, for surface A_i,

$$\dot{Q}_i = \frac{\varepsilon_{r,i}A_i}{1-\varepsilon_{r,i}}(\sigma T_i^4 - J_i) \tag{7–32}$$

and

$$J = \sigma T_i^4 - \frac{1-\varepsilon_{r,i}}{\varepsilon_{r,i}A_i}\dot{Q}_i \tag{7–33}$$

The matrix of equation (7–30) can then be revised to be written with T_i^4 as one of the column vector components of $\{J\}$ instead of J_i. While this manipulation is straightforward, the result is a nonlinear matrix in temperature, the unknown parameter. This can lead to some complications in computations because linear matrices are most often considered.

EXAMPLE 7–6 The oven of example 7–3 is turned on when the oven temperature is 250°C and with the door open. The heater is located in the top surface of the inside of the oven, has an emissivity of 0.9, and quickly reaches a steady state temperature of 1100°C. Assume the oven is well insulated and determine the equilibrium temperature of the inside walls of the oven when the door is left open.

Solution The oven considered in example 7–3 was analyzed by using a two-surface radiation model. However, this condition requires that three surfaces be considered: top of the oven (heater), walls and bottom surface, and the open door. Since the oven is well insulated we may assume there is no heat transfer through the walls. We call this a *reradiating surface* or *adiabatic surface.* In figure 7–16 is shown the network for the oven, where A_1 is the heater surface, A_2 is the oven wall surface, and A_3 is the area of the open door. From the network of figure 7–16 it can be seen that the inside walls of the oven provide a parallel path for heat flow from the heater to the open door. The equivalent resistance between J_1 and J_3 is

$$R_{eq} = \frac{\left(\dfrac{1}{A_1F_{1-3}}\right)\left(\dfrac{1}{A_1F_{1-2}} + \dfrac{1}{A_2F_{2-3}}\right)}{\dfrac{1}{A_1F_{1-2}} + \dfrac{1}{A_1F_{1-3}} + \dfrac{1}{A_2F_{2-3}}}$$

as shown in figure 7–16. The heat loss is

$$\dot{Q}_1 = \dot{Q}_3 = \frac{\sigma(T_1^4 - T_3^4)}{R_1 + R_{eq} + R_3}$$

We have from figure 7–9 or figure 7–16 that

$$A_1 = 0.8 \times 0.75 = 0.6\ \text{m}^2$$
$$A_2 = (0.8 \times 0.8)(2) + (0.75)(0.8)(2) = 2.48\ \text{m}^2$$

and

$$A_3 = 0.6\ \text{m}^2$$

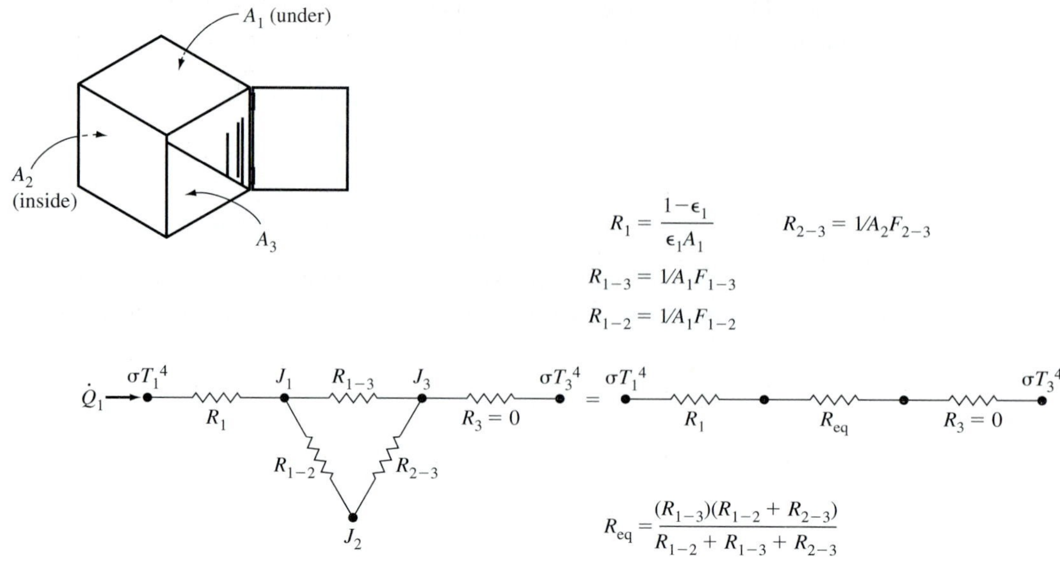

Figure 7–16 Schematic of oven with door open and heater on.

For determining F_{1-3} we have $W = 0.8$ m, $L = 0.8$ m, and $D = 0.75$ m, so $W/D = 0.8/0.75 = 1.067$ and $L/D = 1.067$. Then, from figure 6–21 we read that $F_{1-3} \approx 0.2$. By reciprocity, $F_{3-1} = 0.2$ and from a summation of shape factors, $F_{3-1} + F_{3-2} = 1$ so that

$$F_{3-2} = 1 - F_{3-1} = 1 - 0.2 = 0.8 = F_{1-2}$$

The equivalent resistance is $R_{eq} = 2.777\ \text{m}^{-2}$ so that

Answer
$$\dot{Q}_1 = \dot{Q}_3 = 68.2\ \text{kW}$$

The equilibrium oven wall temperature may be determined by first determining the radiosity J_1, then J_2, and subsequently calculating the temperature. We have

$$\dot{Q}_1 = 68{,}200\ \text{W} = \frac{\varepsilon_{r,1}A_1}{1 - \varepsilon_{r,1}}(\sigma T_1^4 - J_1)$$

or, solving for J_1,

$$J_1 = 189{,}931.9\ \text{W/m}^2$$

Also $\quad J_3 = \sigma T_3^4 = (5.7 \times 10^{-8}\ \text{W/m}^2 \cdot \text{K}^4)(300\ \text{K})^4 = 461.7\ \text{W/m}^2$

At node J_1 we have

$$\dot{Q}_1 + (J_3 - J_1)A_1F_{1-3} + (J_2 - J_1)A_1F_{1-2} = 0$$

so that

$$J_2 = J_1 - \frac{\dot{Q}_1}{A_1F_{1-2}} + \frac{A_1F_{1-3}}{A_1F_{1-2}}(J_1 - J_3) = 95{,}216.1\ \text{W/m}^2 = \sigma T_2^4$$

giving

Answer
$$T_2 = 1136.9\ \text{K} = 633.9°\text{C}$$

Notice in this example that we treated the walls of the oven as adiabatic surfaces and therefore there was no heat transfer through them. Notice also that the adiabatic (or reradiating) surfaces reduced the three-surface problem to an equivalent two-surface problem, with the equivalent resistance accounting for the heat reflecting from the walls.

EXAMPLE 7–7 A portable room space heater is designed to radiate heat into a room, as shown in figure 7–17. The heater has a face area of 0.375 m^2, and the heater element has an emissivity of 0.8. It is rated at 1200 W of heating power. A person stands 1 m from the heater surface when the room has an ambient temperature of 22°C. If the person has an effective emissivity of 0.6 and a frontal area facing the heater of 0.8 m^2, determine the equilibrium temperature of the heating element and the person. Assume that the room acts as a black body.

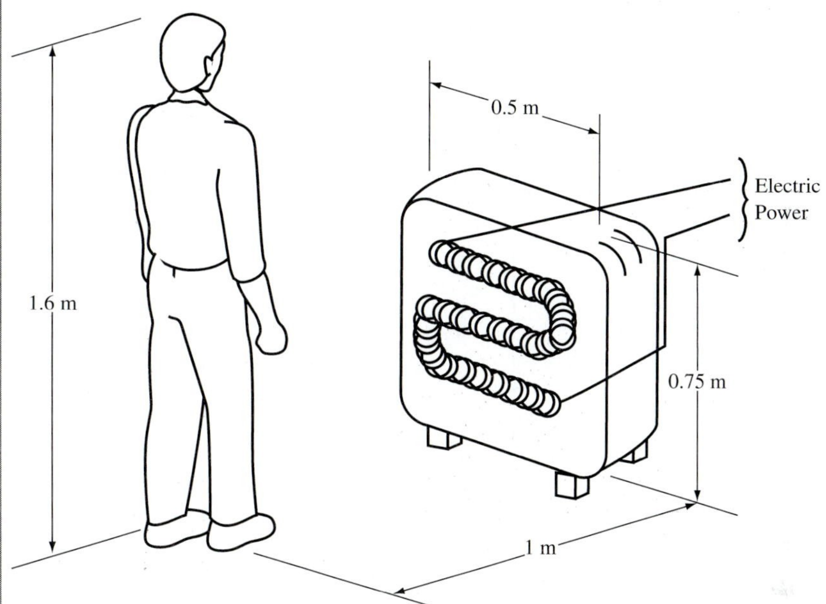

Figure 7–17 Using a space heater.

Solution

The network diagram for the situation, two gray surfaces and one black body surface, is shown in figure 7–18. The heat transfer arrows $\dot{Q}_1$, $\dot{Q}_2$, and $\dot{Q}_3$ are given directions that are intuitively correct, although the analysis may show them to be incorrect. We have for the analysis

$$A_1 = 0.375\ \text{m}^2 \quad \text{(heater)} \qquad A_2 = 0.8\ \text{m}^2 \quad \text{(person)}$$
$$\varepsilon_{r,1} = 0.8 \qquad \varepsilon_{r,2} = 0.6$$

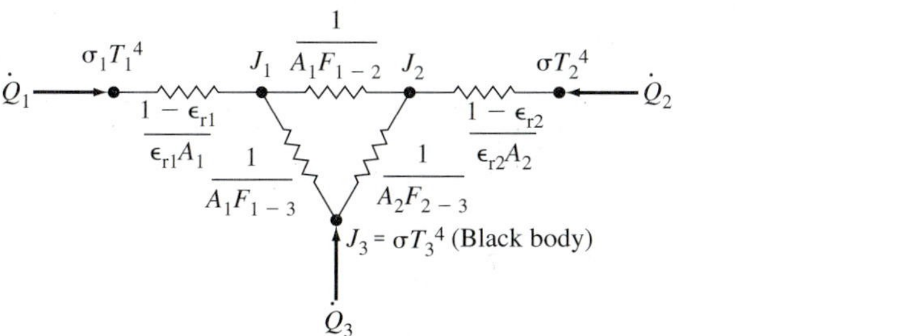

Figure 7–18 Network for space heater.

If we assume the surfaces A_1 (heater) and A_2 (person) are parallel rectangles, we may estimate the shape factor F_{1-2} by first visualizing two parallel rectangles, $A_{14} = A_1 + A_4$, and $A_2 = A_5 + A_6$, as shown in figure 7–19. We then write

$$A_{14}F_{14-2} = A_1F_{1-5} + A_1F_{1-6} + A_4F_{4-5} + A_4F_{4-6} = A_2F_{2-14}$$

By reciprocity $A_4F_{4-5} = A_5F_{5-4}$, and by symmetry, $F_{5-4} = F_{1-6}$. Then $F_{4-5} = (A_5/A_4)F_{1-6}$. Substituting these relationships into the preceding equation gives

$$A_2F_{2-14} = A_1F_{1-5} + A_1F_{1-6} + A_4(A_5/A_4)F_{1-6} + A_4F_{4-6}$$

and then
$$F_{1-6} = \frac{1}{A_5 + A_1}(A_2F_{2-14} - A_1F_{1-5} - A_4F_{4-6})$$

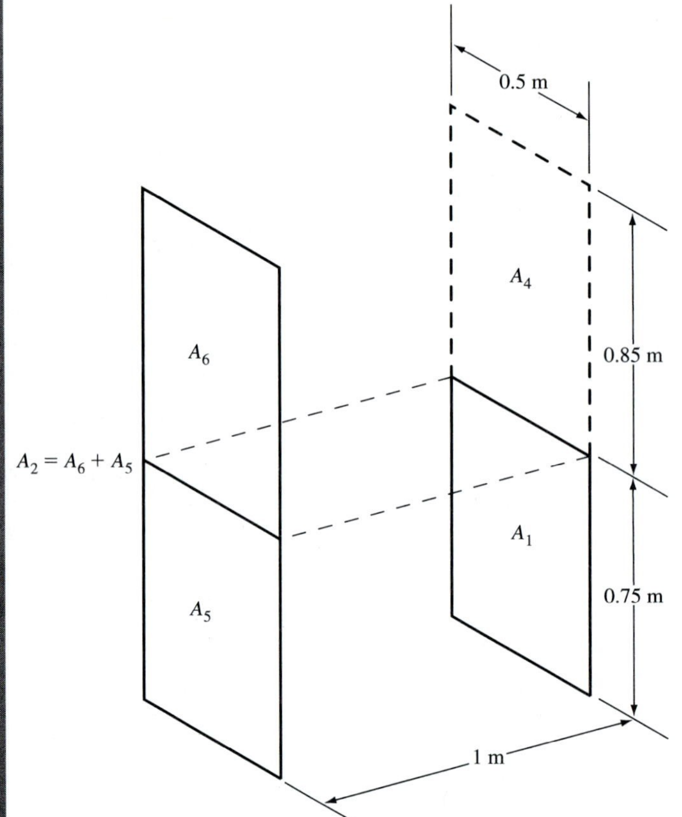

Figure 7–19 Shape factor determination for heater–person.

We can determine in a straightforward manner all of the terms in this equality. The shape factor F_{1-2} (the heater to the person) is then $F_{1-2} = F_{1-6} + F_{1-5}$. We have $A_4 = A_6 = 0.425\ \text{m}^2$ and $A_5 = A_1 = 0.375\ \text{m}^2$. Then, for determining F_{2-14}, $D = 1$ m, $L = 1.6$ m, and $W = 0.5$ m, so that $F_{2-14} \approx 0.145$ from figure 6–20.

For determining F_{1-5} we have $D = 1$ m, $W = 0.5$ m, and $L = 0.75$ m so that $F_{1-5} \approx 0.09$ from figure 6–20.

For determining F_{4-6} we have $D = 1$ m, $W = 0.5$ m, and $L = 0.85$ m so that $F_{4-2} \approx 0.10$ from figure 6–20 and then

$$F_{1-6} = \frac{0.8(0.145) - 0.425(0.10) - 0.375(0.09)}{2(0.8)} = 0.0248$$

Also, we have

$$F_{1-2} = F_{1-5} + F_{1-6} = 0.09 + 0.0248 = 0.1148$$

By summation of shape factors, equation (6–45),

$$F_{1-3} = 1.0 - F_{1-2} = 0.8852$$

and by reciprocity,

$$F_{2-1} = (A_1/A_2)F_{1-2} = (0.375/0.8)(0.1148) = 0.0538$$

Again by summation of shape factors,

$$F_{2-3} = 1.0 - F_{2-1} = 0.9462$$

For node J_1 we have

$$\dot{Q}_1 + A_1F_{1-3}(\sigma T_3^4 - J_1) + A_1F_{1-2}(J_2 - J_2) = 0$$

Making substitutions and reducing this equation we get

$$0.6J_1 - 0.06888J_2 = 1429.3 \qquad (J \text{ in W/m}^2)$$

For node J_2 we obtain

$$-0.04305J_1 + 1.178J_2 - Q_2 = 490.15$$

and for node J_3 we obtain

$$0.332J_1 + 1.135J_2 + Q_2 = 1833.45$$

These three equations can be solved by matrix methods, Gauss elimination methods, or algebra manipulations. The solution is

$$\dot{Q}_2 = -14.3 \text{ W}$$
$$J_1 = 3182.7 \text{ W/m}^2$$
$$J_2 = 697 \text{ W/m}^2$$

from which we may obtain the equilibrium temperatures of the heater surface and the person:

$$1200 \text{ W} = (\sigma T_1^4 - J_1)\varepsilon_{r,1}A_1/(1 - \varepsilon_{r,1})$$

giving

$$T_1^4 = \frac{(1200 \text{ W})(1 - 0.8)}{0.8(0.375 \text{ m}^2)(5.7 \times 10^{-8} \text{ W/m}^2 \cdot \text{K}^4)} + \frac{3182.7 \text{ W/m}^2}{5.7 \times 10^{-8} \text{ W/m}^2 \cdot \text{K}^4}$$

Answer or $T_1 = 514$ K

The equilibrium temperature for the person is found from

$$\dot{Q}_2 = \frac{\varepsilon_{r,2}A_2}{1 - \varepsilon_{r,2}}(J_2 - \sigma T_2^4)$$

or

$$T_2^4 = \frac{-(-14.3 \text{ W})(0.4)}{0.6(1.2 \text{ m}^2)(5.7 \times 10^{-8} \text{ W/m}^2 \cdot \text{K}^4)} + \frac{697 \text{ W/m}^2}{5.7 \times 10^{-8} \text{ W/m}^2 \cdot \text{K}^4}$$

Answer $T_2 = 333 \text{ K} = 60°\text{C}$

The network method for solving three-surface radiation problems can be extended to radiation between n surfaces, and it can be seen from the preceding example that the equations written to account for heat fluxes at the surfaces have the general form, for the mth node:

$$\sum_{i=1}^{n}\left[J_iA_iF_{i-m} - J_m\left(A_iF_{i-m} + \frac{\varepsilon_{r,m}A_m}{1-\varepsilon_{r,m}}\right) + \frac{\varepsilon_{r,m}A_m}{1-\varepsilon_{r,m}}\sigma T_m^4\right] = 0 \qquad \textbf{(7–34)}$$

The analysis is restricted to opaque gray or black surfaces, and it is expected that the various emissivities, surface temperatures, and shape factors are able to be determined. As we discussed before, and demonstrated in example 7–7, if a surface temperature is not known, then the heat flux must be known.

7–4 GAS RADIATION

We have discussed thermal radiation as an "action at a distance," with the surfaces of objects receiving, reflecting, absorbing, transmitting, or emitting radiation. Sometimes it is necessary to consider the gas or substance in the intervening region between the surfaces in the analysis of radiation heat transfer to obtain reasonably accurate results, that is, to consider the gas as a three-dimensional radiation heat-transfer system. For instance, exhaust gases from a power plant or steel furnace may emit radiation to the smokestack or chimney through which they flow. Hot gases in a furnace or incinerator exchange thermal radiation with the walls of the device. If we consider direct radiation of intensity $\dot{I}_0$ impacting on a mass of a gas, as shown in figure 7–20, then it is reasonable to assume that the intensity of radiation is attenuated or reduced as the thermal radiation progresses through the gas. Assuming that K_λ is an *extinction coefficient* such that

$$K_\lambda = -\frac{1}{\dot{I}(x)}\frac{d\dot{I}(x)}{dx} \qquad \textbf{(7–35)}$$

this parameter may be used to predict the attenuation of radiation intensity in a gas. If the extinction coefficient is taken as constant, then from equation 7–35 we have

$$\int_{\dot{I}_0}^{\dot{I}} \frac{1}{\dot{I}(x)}d\dot{I}(x) = -K_\lambda \int_0^L dx$$

which gives

$$\ln\left(\frac{\dot{I}(x)}{\dot{I}_0}\right) = -K_\lambda L$$

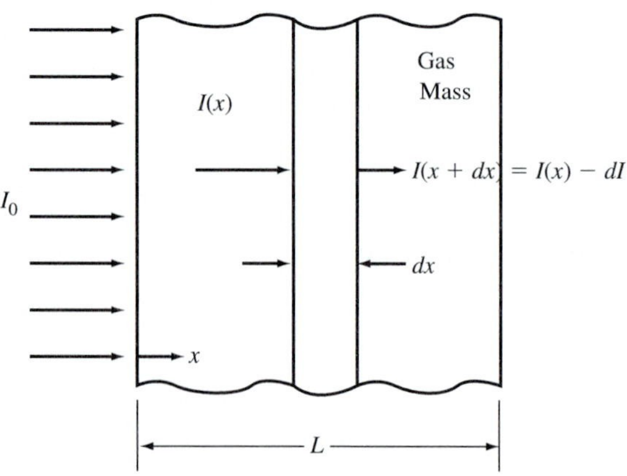

Figure 7–20 Schematic of thermal radiation attenuation through a gas.

Using the definition of transmissivity, this can be written

$$\tau_{r,g} = \frac{\dot{I}(x)}{\dot{I}_0} = e^{-K_\lambda L} \quad \text{(7–36)}$$

which describes an exponential decay function of thermal radiation.

The extinction coefficient is usually assumed to be a function of the gas temperature and the concentration of the components of the gas. The concentration of an ideal gas may be determined from the partial pressure of the gas so that, as a first approximation,

$$K_\lambda = \sum_{i=1}^{n} K_{\lambda i} p_i \quad \text{(7–37)}$$

where $K_{\lambda i}$ is the extinction coefficient for the ith component of the gas mixture and p_i is its partial pressure. Thus, equation (7–36) may be written

$$\tau_{r,g} = e^{-L\Sigma K_{\lambda i} p_i} \quad \text{(7–38)}$$

where the summation is understood to proceed from $i = 1$ to n. Many gases, particularly oxygen (O_2), nitrogen (N_2), and argon, are nearly transparent to thermal radiation so that their extinction coefficients are nearly zero. Also, air is principally composed of these three gases so that air may be assumed to be transparent to thermal radiation. Water vapor, carbon dioxide, carbon monoxide, sulfur dioxide, and soot absorb significant amounts of thermal radiation, and humid air containing carbon dioxide gas can highly attenuate thermal radiation.

If we assume a nonreflecting gas, then from equation (6–23) we have,

$$\tau_{r,g} + \alpha_{r,g} = 1.0 \quad \text{(7–39)}$$

If we apply Kirchhoff's Law, then equation (7–39) becomes

$$\varepsilon_{r,g} = 1.0 - \tau_{r,g} \quad \text{(7–40)}$$

and from equation (7–38),

$$\varepsilon_{r,g} = 1.0 - e^{-\Sigma K_{\lambda i} p_i L} \quad \text{(7–41)}$$

The extinction coefficient is expected to be a function of temperature, and experimental data has been available since about 1930 to verify this expectation. In figure 7–21 are presented effective gas emissivities for water vapor and dry air, ε_w, at various temperatures and partial pressures. The partial pressure of the water vapor is indicated as p_w. Thus, an isotherm (constant temperature curve) in the graph of figure 7–21 would provide emissivity data equivalent to that from equation (7–41), provided that the extinction coefficients were known. Similar information is presented in figure 7–22 for a mixture of carbon dioxide and air. The partial pressure of the carbon dioxide is denoted as p_c and the total pressure is p_T.

From Dalton's Law of Partial Pressures we have for a mixture of water vapor in air,

$$p_T = p_{\text{air}} + p_w$$

and for carbon dioxide in air

$$p_T = p_{\text{air}} + p_c$$

For a mixture of water vapor and carbon dioxide in air this is

$$p_T = p_{\text{air}} + p_w + p_c \quad \text{(7–42)}$$

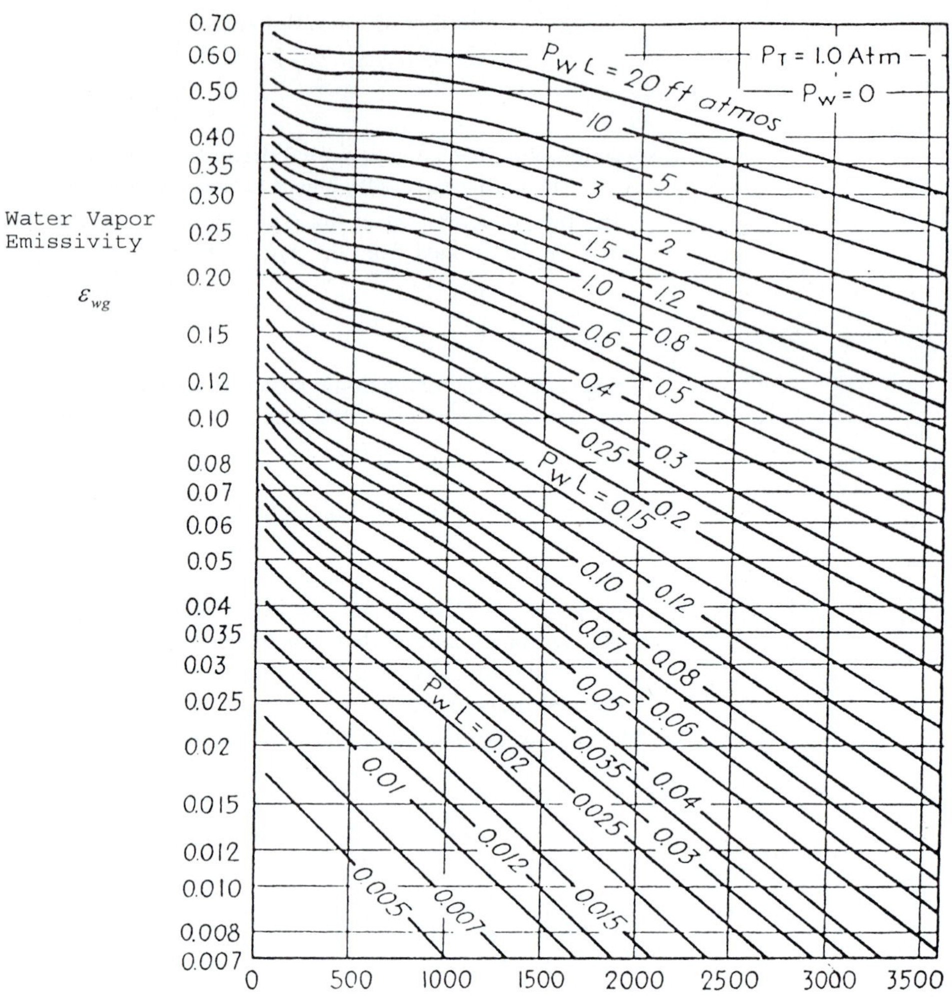

Figure 7–21 Emissivity of water vapor in air. (From H. Hottel and R. Egbert [2])

Since water vapor in air has such a broad range of partial pressures, figure 7–21 represents the emissivity as the partial pressure p_w approaches zero; that is, as $p_w \to 0$. The corrected emissivity for water vapor in air at 1 atm total pressure is,

$$\varepsilon_{r,wg} = \varepsilon_{r,w} C_w \tag{7–43}$$

where C_w is the pressure broadening correction factor determined by Hottel and Egbert [2] and shown in the graph of figure 7–23.

EXAMPLE 7–8 Determine the emissivity, ε_{wg}, for a 100-ft-thick air bank at 90°F, 80% relative humidity.

Solution We assume the total pressure p_T is 1 atm. From appendix table B–6 we interpolate to determine the saturation pressure of water vapor at 90°F (33°C):

$$p_g = 3.49 \text{ kPa} = 0.506 \text{ psia}$$

Figure 7–22 Emissivity of carbon dioxide in air. (From H. Hottel and R. Egbert [2])

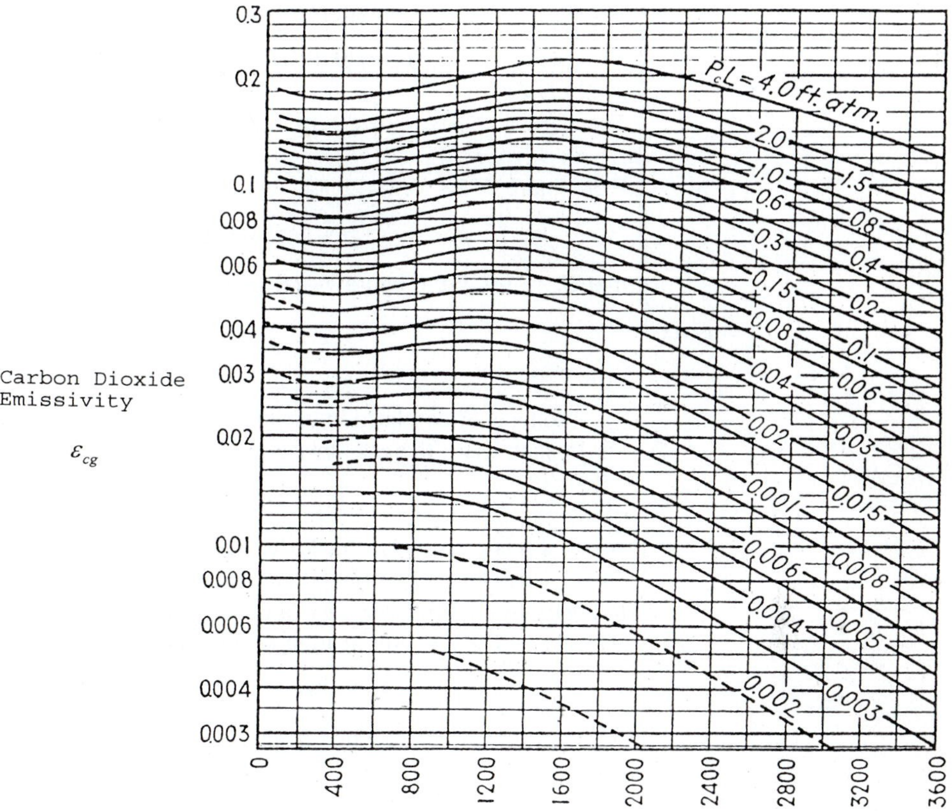

Figure 7–23 Correction factor to account for air–water vapor mixture emissivities at pressures other than 1 atm. (After Hottel and Egbert [2])

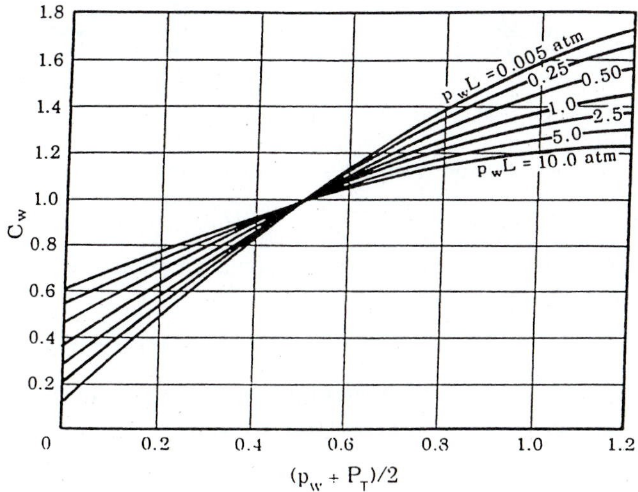

and the water vapor partial pressure is

$$p_w = (0.8)(0.506 \text{ psia}) = 0.4048 \text{ psia} = 0.027 \text{ atm}$$

Then

$$\frac{1}{2}(p_w + p_T) = \frac{1}{2}(0.027 + 1.0 \text{ atm}) = 0.5135 \text{ atm}$$

From figure 7–23 we read, for

$$p_wL = (0.027 \text{ atm})(100 \text{ ft}) = 2.7 \text{ atm} \cdot \text{ft}$$

that $C_w \approx 1.0$. From figure 7–21 we read

Answer $$\varepsilon_{r,wg} \approx 0.42$$

Products of combustion usually contain mixtures of carbon dioxide, water vapor, and air. In such cases the gas emissivity ε_g is given by the relationship

$$\varepsilon_{r,g} = \varepsilon_{r,wg} + \varepsilon_{r,cg} - \Delta\varepsilon \quad \textbf{(7–44)}$$

where $\Delta\varepsilon$ is a correction factor to account for the spectral overlap of the CO_2 and water vapor. This correction may be predicted from the graphs of figure 7–24. Some rough interpolations may be necessary to obtain a correction factor for gases that are not at 260°F, 1000°F, or 1700°F. Application of Kirchhoff's law gives

$$\alpha_{r,g} = \varepsilon_{r,g} = \alpha_{r,g} + \alpha_{r,g} - \Delta\alpha \quad \textbf{(7–45)}$$

where $\Delta\alpha = \Delta\varepsilon$ is obtained from figure 7–24.

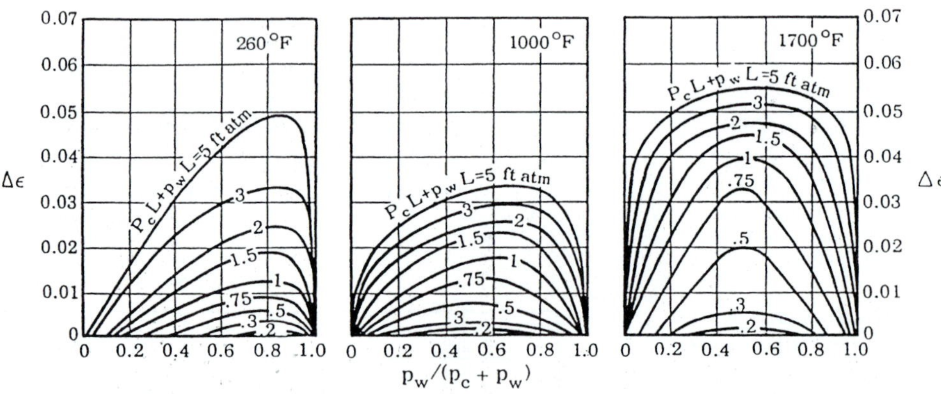

Figure 7–24 Correction factors to account for spectral overlap of water and carbon dioxide in air in the determination of emissivity. (From H. Hottel and R. Egbert [2])

The radiation from the gas is determined from

$$\dot{Q}_{\text{emit}} = \varepsilon_g A_s \sigma T_g^4 \quad \textbf{(7–46)}$$

where A_s is the surface area of a container vessel for the gas. The net heat transfer to the container would be

$$\dot{Q}_{\text{net}} = A_s(\varepsilon_{r,g}\sigma T_g^4 - \alpha_{r,g}\sigma T_s^4) \quad \textbf{(7–47)}$$

where α_s is the absorptivity of the containing vessel surface and T_s is its temperature. For heat transfer to a gas in a vessel, equation (7–47) may be written

$$\dot{Q}_{\text{net}} = A_s\sigma(\varepsilon_{r,g}T_s^4 - \alpha_{r,g}T_g^4) \quad \textbf{(7–48)}$$

In those situations where a uniform gas thickness L cannot be identified, a mean beam length of radiation L_m is used instead of L in equations for the determination of $\tau_{r,g}$, $\varepsilon_{r,g}$, and $\alpha_{r,g}$. Hottel and Sarofim [3] have published mean beam lengths, and some of these are given in table 7–1.

Table 7–1 Mean Beam Length for Gas Surface Radiation

Shape of Vessel Containing Gas	L_m	Characteristic Dimension
Sphere	$0.63D$	Diameter, D
Cube	$0.60L$	Edge, L
Infinite cylinder, radiation to sides	$0.94D$	Diameter, D
Right circular cylinder ($D = L$), radiation to full surface	$0.60D = 0.6L$	Diameter, D, or length, L
Infinite parallel plates	$1.76L$	Distance between plates, L
Gas mass of any shape	$3.5V/A$	Volume to surface area, V/A

After Hottel and Sarofim [3].

7–5 APPLICATIONS OF RADIATION HEAT TRANSFER

In this section we consider some examples where radiation is a significant or dominating mode of heat transfer. First we consider the effect of radiation on temperature measurements made with thermometers and thermocouples. We discuss the concept of the mean radiation temperature, used in human thermal comfort analysis, and then consider radiation heat transfer through a gas.

Radiation Effects on Temperature Measurements

In temperature measurements made with mercury-in-glass thermometers, with thermocouples, or with thermistors, it is usually assumed that thermal equilibrium is reached through convection or conduction heat transfer. In particular, to determine the air temperature the temperature sensor usually is analyzed as shown in figure 7–25a and with equation (7–49),

$$\dot{Q} = h_{\infty}A_s(T_{\infty} - T_s) \tag{7–49}$$

At thermal equilibrium $\dot{Q}$ is zero and then $T_{\infty} = T_s$. The instrument is then expected to be sensing the gas or liquid temperature. Consider now the situation where thermal radiation is also contributing heat transfer to the process as indicated in Figure 7–25b. Then,

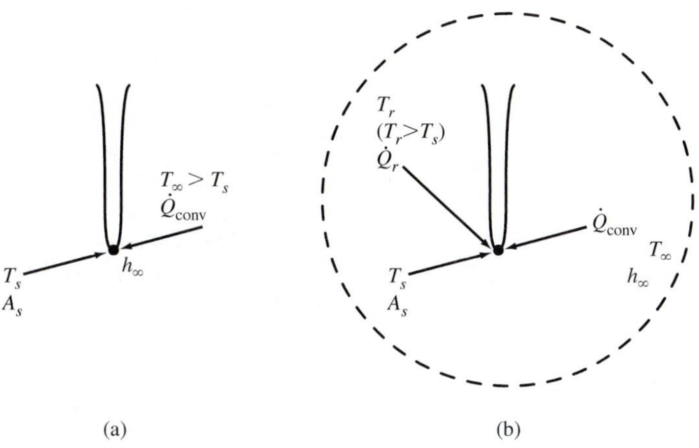

Figure 7–25 Temperature measurement in a gas or liquid.

for the situation where the surrounding radiation temperature, T_r, is greater than the ambient fluid temperature, T_∞, thermal equilibrium will be obtained when

$$\alpha_{r,s} A_s \sigma (T_r^4 - T_s^4) = \dot{Q}_r = -\dot{Q} = h_\infty A_s (T_s - T_\infty) \tag{7–50}$$

If the surrounding radiation temperature is less than the ambient fluid temperature, then equation (7–50) is

$$\varepsilon_{r,s} A_s \sigma (T_s^4 - T_r^4) = h_\infty A_s (T_\infty - T_s) \tag{7–51}$$

and in either case there will be a discrepancy between the sensor temperature and the gas temperature. This discrepancy or error can be reduced by placing a shield around the sensor, by using a sensor that has an extremely low emissivity, or by increasing the convection heat transfer through agitation of the fluid.

EXAMPLE 7–9 An unshielded thermocouple is placed midstream inside of a wind tunnel to monitor the air temperature, as shown in figure 7–26. During a particular test the thermocouple/readout device indicates that the air temperature inside the wind tunnel is 80°F, and it is known that the thermocouple emissivity is 0.85, that the convection heat transfer coefficient is 12 Btu/hr · ft² · °F, and that the surfaces of the wind tunnel were at 68°F. What is the actual air temperature?

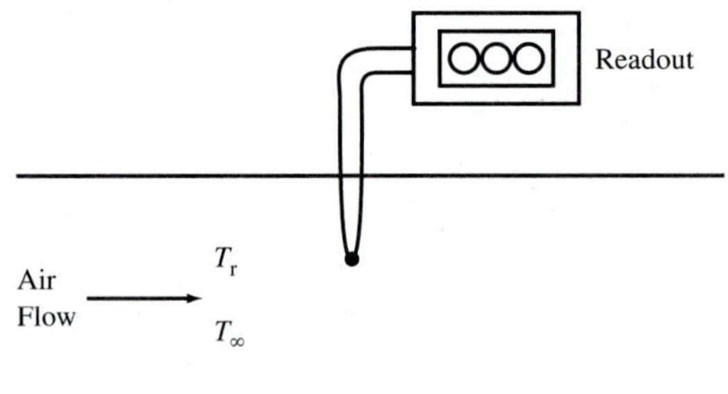

Figure 7–26 Wind tunnel temperature monitor.

Solution Since the air temperature is indicated as 80°F and the wind tunnel surfaces are 68°F, it is reasonable to expect that the thermocouple has reached thermal equilibrium by convection heat transfer from the air and by radiation with the tunnel surfaces. Thus, we may use equation (7–51), where T_r = 68°F = 528°R and T_s = 80°F = 540°R. Then, solving for the air temperature in equation (7–51) gives

$$T_\infty = \frac{\varepsilon_{r,s}\sigma}{h_\infty}(T_s^4 - T_r^4) + T_s$$

Answer

$$= \frac{(0.85)(0.174 \times 10^{-8}\ \text{Btu/hr} \cdot \text{ft}^2 \cdot {}^\circ\text{R}^4)}{(12\ \text{Btu/hr} \cdot \text{ft}^2 \cdot {}^\circ\text{R})}[(540^\circ\text{R})^4 - (528^\circ\text{R})^4] + 80^\circ\text{F} = 80.9^\circ\text{F}$$

The thermocouple reads approximately 0.9°F low.

Mean Radiation Temperature

The surface temperature of surrounding surfaces, T_r, which we considered in figures 7–25 and 7–26, can sometimes be difficult to determine. An approximation to this temperature is the *mean radiation temperature,* T_{rm}, defined as [4]

$$T_{\text{rm}} = \frac{1}{360^\circ}\int_0^{360^\circ} T(\theta)d\theta \tag{7–52}$$

where the angle θ is measured from some point of observation as shown in figure 7–27. The surface temperature $T(\theta)$ or effective surface temperature for the case of a gas

Figure 7–27

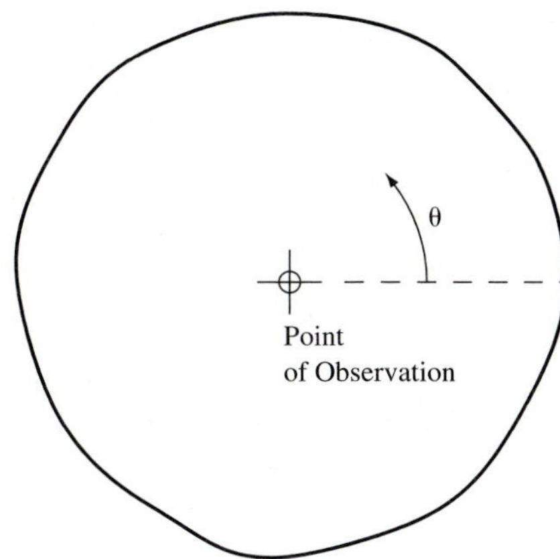

surrounding the point of observation may change with angular position θ. Notice that the mean radiant temperature only accounts for variations in temperature in two dimensions. A more elaborate definition could include the third dimension by using solid angle variations of temperature; however, that is beyond the purposes of this text. The mean radiant temperature may be used to replace the parameter T_r in those analyses such as in example 7–9.

EXAMPLE 7–10 An office plan view is shown in figure 7–28, and the expected extreme surface temperature distribution $T(\theta)$ is indicated in figure 7–28. Determine T_{rm} at points of observation *A* and *B*.

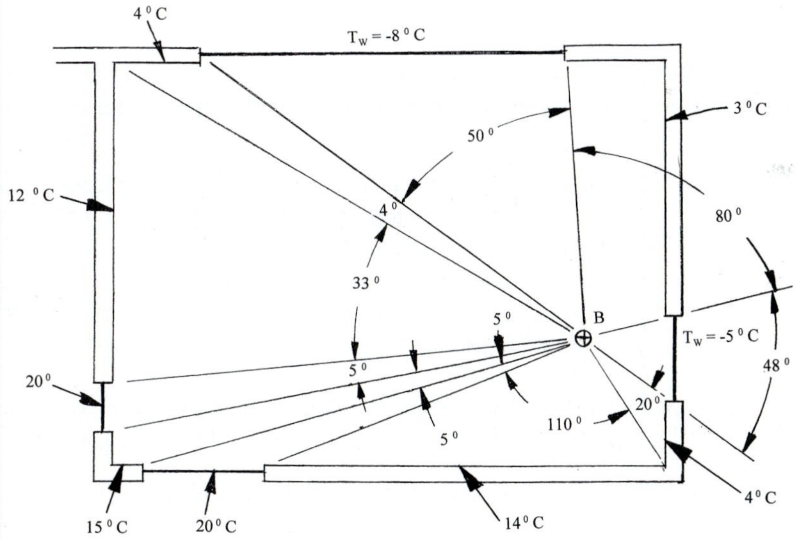

Figure 7–28 Plan view of office with point of observation *B*.

Solution The mean radiant temperature can be determined by a piecewise integration of equation (7–52). For point of observation *B* we have

$$T_{rm} = \frac{1}{360^\circ}[(-5^\circ\text{C})(48^\circ) + (3^\circ\text{C})(80^\circ) + (-8^\circ\text{C})(50^\circ) + (4^\circ\text{C})(4^\circ)$$

$$+ (12^\circ\text{C})(33^\circ) + (20)(5) + (15)(5) + (20)(5) + (14)(110) + (4)(20)]$$

Answer

$$= 3.45^\circ\text{C}$$

At point A, from figure 7–29, we have

Answer

$$T_{rm} = \frac{1}{360^\circ}[(-5)(14) + (3)(42) + (-8)(60) + (4)(15) + (12)(70) + (20)(15) + (20)(35) + (15)(14) + (20)(15) + (14)(85) + (4)(10)] = 8.1^\circ\text{C}$$

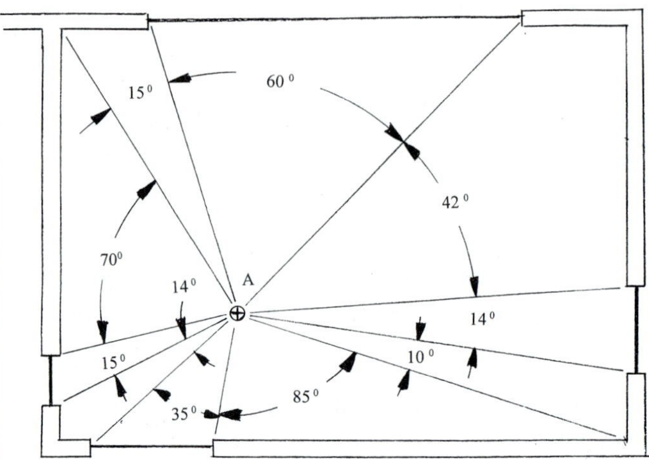

Figure 7–29 $T(\theta)$ from point of observation A.

It can be seen that the mean radiant temperature depends on the overall temperature distribution and also the location of the point of observation. This is a rough approximation to the actual radiant temperature, and the reader should be able to see that the concepts of the shape factor are used to approximate the temperature distribution.

Radiation Heat Transfer Through a Gas

Radiation heat transfer can be effective in many instances when a liquid or gas is the interchanging medium. As an example we consider a cooling pond of water.

EXAMPLE 7–11

A 10-hectare (10,000 m^2) cooling pond is used to cool condenser water from a power plant. On a particular night the pond water is at 33°C, the surface emissivity of the water is 0.85, the air temperature is 20°C, the emissivity of the air is 0.7, and the free convective heat transfer coefficient is 5 W/m^2 · °C. The night sky is assumed to be an effective black body at −40°C outside of the atmosphere, and the atmosphere is 10 km thick and of uniform temperature. Estimate the heat loss from the pond water in megawatts.

Solution

In figure 7–30 is shown a diagram of the situation for the cooling pond. The heat loss may be determined from an energy balance of the cooling pond as

$$\dot{Q}_{\text{loss}} = \dot{Q}_{\text{convection}} + \dot{Q}_{\text{radiation}}$$

where

$$\dot{Q}_{\text{convection}} = h_\infty A_s(T_w - T_\infty) = (5\ \text{W/m}^2 \cdot \text{K})(10{,}000\ \text{m}^2)[33^\circ\text{C} - 20^\circ\text{C}] = 6.5\ \text{MW}$$

The radiation heat transfer loss is that from the pond surface to the effective night sky with the 20°C, 10-km-thick air mass between. For this situation we notice that the thermal radiation leaving the pond, J_w, that is transmitted through the air mass is

$$\dot{Q}_{w-\text{sky}} = J_w A_w F_{w-\text{sky}} \tau_m$$

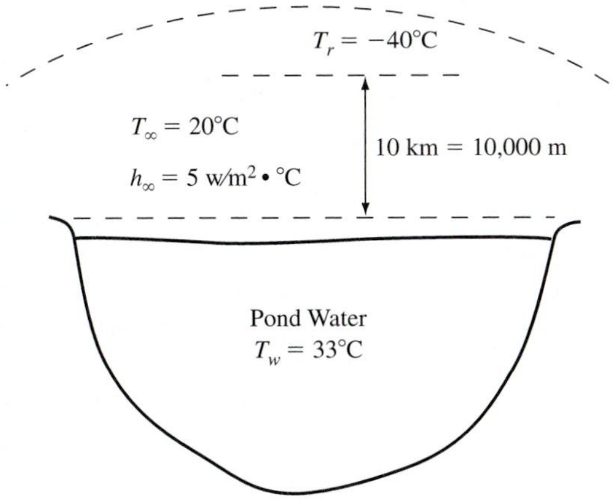

Figure 7–30 Cooling pond water at night.

where τ_m is the transmissivity of the air mass. The thermal radiation leaving the sky that reaches the pond is

$$\dot{Q}_{\text{sky}-w} = J_{\text{sky}} A_{\text{sky}} F_{\text{sky}-w} \tau_m$$

Using reciprocity we have

$$A_{\text{sky}} F_{\text{sky}-w} = A_w F_{w-\text{sky}}$$

The net heat loss from the pond to the sky is

$$\dot{Q}_{\text{net}} = \dot{Q}_{w-\text{sky}} - \dot{Q}_{\text{sky}-w} = A_w F_{w-\text{sky}} \tau_m (J_w - J_{\text{sky}})$$

Since $\tau_m = 1.0 - \varepsilon_m$,

$$\dot{Q}_{\text{net}} = A_w F_{w-\text{sky}} (1 - \varepsilon_{r,m})(J_w - J_{\text{sky}})$$

Notice that the thermal resistance term between the radiosities is $1/A_w F_{w\text{-sky}}(1 - \varepsilon_{\text{r,m}})$ for radiation through the transmitting gas, or air. The network analogy is given in figure 7–31 for this example. Then we have

$$\dot{Q}_{\text{net}} = \frac{\sigma A_w [T_w^4 - T_{\text{sky}}^4]}{\dfrac{1 - \varepsilon_{r,m}}{\varepsilon_{r,m}} + \dfrac{1}{1 - \varepsilon_{r,m}}}$$

since $F_{w-\text{sky}} = 1.0$. Then

$$\dot{Q}_{\text{net}} = \dot{Q}_{\text{radloss}} = \frac{\sigma(10{,}000\ \text{m}^2)[(306\ \text{K})^4 - (293\ \text{K})^4]}{\dfrac{1 - 0.7}{0.7} + \dfrac{1}{1 - 0.7}} = 2.1\ \text{MW}$$

The total heat loss is then

Answer $$\dot{Q}_{\text{loss}} = \dot{Q}_{\text{convection}} + \dot{Q}_{\text{radloss}} = 6.5\ \text{MW} + 2.1\ \text{MW} = 8.6\ \text{MW}$$

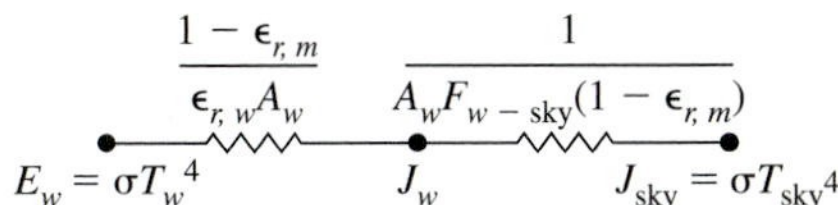

Figure 7–31 Network of cooling pond to night sky radiation.

Notice in this last example that the thermal resistance term between the radiosities of two surfaces when a transmitting gas occupies the region between those surfaces includes the term $(1 - \varepsilon_w)$ rather than only the area and shape factor.

EXAMPLE 7–12 A steam generating unit using fossil fuels is a large combustion chamber filled with combustion gases, which radiate thermal energy to the vertical walls of the chamber. The vertical walls are composed of water tubes, which receive the energy and subsequently boil the water in the tubes. Assume that a steam generator is to burn coal and have combustion gases at 1700°F throughout the chamber, and is constructed as a cube of sides 30 ft in length. Estimate the amount of radiation heat transfer to the vertical walls. The emissivity of the vertical walls may be taken as 0.86, and the surface temperature of the walls is 900°F.

Solution We assume steady state conduction and neglect convection heat transfer between the hot gases and the vertical walls. Then

$$\dot{Q}_{\text{add}} = A_s\sigma[\varepsilon_{r,m}T_m^4 - \alpha_{r,s}T_s^4]$$

and the area of the vertical surfaces is

$$A = 30 \text{ ft} \times 30 \text{ ft} \times 4 = 3600 \text{ ft}^2$$

The emissivity of the gases may be predicted by using table 7–1, with $L_m = 0.6L = 18$ ft, $p_c = 0.25$ atm, and $p_cL = 4.5$ atm · ft. Then $\varepsilon_{r,m} \approx 0.23 = \varepsilon_{r,c}$ from figure 7–22 and

$$\dot{Q}_{\text{add}} = (3600 \text{ ft}^2)(0.174 \times 10^{-8} \text{ Btu/hr} \cdot \text{ft}^2 \cdot {}^\circ\text{R}^4)[(0.23)(2160)^4 - (0.86)(1360)^4]$$

Answer

$$= 12{,}932{,}166 \text{ Btu/hr}$$

Notice that the temperature of the surface of the steam generator is crucial in determining the radiation to the walls. If this temperature becomes too great, the heat transfer will be reduced, and safety problems, with weakening of the tubes, will also be created. This last example is a simplified analysis of what an actual generator will do. For instance, the assumption that the chamber has the same temperature throughout is not realistic. Hottel and Sarofim [3] describe how a cubic container may be sectioned into smaller isothermal compartments and with the aid of the computer a temperature field may be predicted in the combustion gas volume. Other discussions of radiation heat transfer may be found, for instance, in Siegel and Howell [5], Sparrow and Cess [6], Tien [7], and Dunkle [8].

7–6 SUMMARY

Radiosity, J, is the total radiation leaving a surface. For a *gray opaque* surface

$$J = \varepsilon_r\sigma T^4 + \rho_r\dot{E}_{\text{irr}} \tag{7–2}$$

where $\rho_r = 1 - \varepsilon_r$.

The radiosity from a black body is

$$J = \sigma T^4 \tag{7–9}$$

The *net radiation* leaving a gray opaque surface is

$$\dot{Q}_{\text{net}} = \frac{\varepsilon_r A}{1 - \varepsilon_r}(\sigma T^4 - J) \tag{7–8}$$

The net radiation heat transfer between *two-surface* problems, for gray opaque surfaces, is

$$\dot{Q}_{\text{net},1-2} = \frac{\sigma(T_1^4 - T_2^4)}{\dfrac{1-\varepsilon_{r,1}}{\varepsilon_{r,1}A_1} + \dfrac{1}{A_1F_{1-2}} + \dfrac{1-\varepsilon_{r,2}}{\varepsilon_{r,2}A_2}} \tag{7–17}$$

For the case where $A_2 \gg A_1$

$$\dot{Q}_{\text{net},1-2} = \frac{\sigma A_1(T_1^4 - T_2^4)}{\dfrac{1-\varepsilon_{r,1}}{\varepsilon_{r,1}} + \dfrac{1}{F_{1-2}}} \tag{7–18}$$

For the case where A_1 is a black body and $A_2 \gg A_1$

$$\dot{Q}_{\text{net},1-2} = \sigma A_1F_{1-2}(T_1^4 - T_2^4) \tag{7–20}$$

For *three-surface* or more radiation heat transfer problems, the analysis can be made by identifying a circuit and writing a square matrix

$$[M]\{J_i\} = \left\{\frac{\varepsilon_{r,i}A_i}{1-\varepsilon_{r,i}}\sigma T_i^4\right\} \tag{7–30}$$

where $\{J_i\}$ is the radiosity vector, $\left\{\frac{\varepsilon_{r,i}A_i}{1-\varepsilon_{r,i}}\sigma T_i^4\right\}$ is the black body vector, and $[M]$ is the matrix of coefficients of the radiosities. When a gas occupies the region between two or more radiating surfaces, the gas can reduce the radiation as it passes through. The extinction coefficient, K_λ, given by the definition

$$K_\lambda = -\frac{1}{\dot{I}(x)}\frac{d\dot{I}(x)}{dx} \tag{7–35}$$

can be used to quantify the radiation reduction. Assuming an extinction coefficient that is constant

$$\tau_{r,g} = e^{-K_\lambda L} = \frac{\dot{I}(x)}{\dot{I}_0} \tag{7–36}$$

and the gas emissivity can be approximated as

$$\varepsilon_{r,g} = 1 - \tau_{r,g} \tag{7–40}$$

From Kirchhoff's Law

$$\varepsilon_{r,g} = \alpha_{r,g}$$

The radiation from a surrounding surface can be analyzed through the concept of the *mean radiation temperature,* T_{rm}, defined as

$$T_{\text{rm}} = \frac{1}{360°}\int_0^{360°} T(\theta)\,d\theta \tag{7–52}$$

DISCUSSION QUESTIONS

Section 7–1

7–1 What is radiosity?

7–2 What is meant by thermal resistance at a surface?

7–3 What are the units of radiosity?

Section 7–2

7–4 Can you identify any two-surface radiation configurations other than those shown in figure 7–4?

7–5 How does a shield affect radiation?

Section 7–3

7–6 What is the difference between reradiating surfaces and black body surfaces?

7–7 What is meant by an adiabatic surface?

Section 7–4

7–8 What does the *extinction coefficient* describe?

7–9 Why would the pressure of a gas affect the extinction coefficient?

7–10 What is meant by the mean beam length?

Section 7–5

7–11 Does radiation affect the measurement of temperature with a thermocouple? How?

7–12 What does the mean radiant temperature represent?

PRACTICE PROBLEMS

Section 7–1

7–1 A gray opaque surface has an emissivity of 0.8 at 400°F. Determine the radiosity if it is irradiated at 1000 Btu/hr · ft^2.

7–2 For the surface of example 7–1, where $\varepsilon = 0.6 - 0.0002T$ and $T \leq 2600\,\text{K}$, if the irradiation is decreased to 200 W/m^2, determine the temperature at which maximum net emitted heat transfer occurs.

Section 7–2

7–3 A 5-cm-diameter exhaust pipe at 700 K has an opaque outer surface and emissivity of 0.9. A galvanized gray sheet iron tube 15 cm in diameter is placed concentrically around the pipe. If the tube is to have a temperature of no more than 350 K, determine the maximum heat transfer that can be dissipated per unit length of the pipes.

7–4 An oven has inside walls with emissivity of 0.6. When the oven is at 200°C the door is opened and the oven is turned off, as shown in figure 7–32. Determine the net radiation from the oven to the surrounding room. Assume that the room acts as a black body at 20°C.

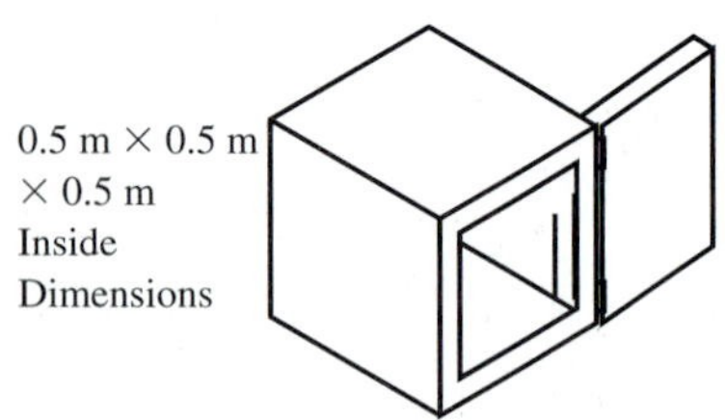

Figure 7–32

7–5 A solar collector is at 25°C and is facing the night sky at −50°C. If the collector emissivity is 0.9 and its absorptivity is 0.85, determine the net radiation of the solar collector.

7–6 A 7-cm-diameter heat exchanger rod at 400°C is surrounded by a 15-cm-diameter heat shield. If the surrounds are a black body at 20°C, determine the equilibrium temperature of the shield if

(a) Both rod and shield are black bodies.

(b) The emissivity of both is 0.7.

(c) The rod is a black body, and the shield has an emissivity of 0.7.

Section 7–3

7–7 A solarium is approximated by the rectangular parallelopiped shown in figure 7–33. The floor is well insulated, and all walls and ceiling have emissivity of 0.2. The floor temperature is 40°F, temperature of the walls and ceiling is 67°F, and the window is at 12°F.

(a) Construct an equivalent circuit to analyze the heat transfers.

(b) Determine the heat loss by radiation from the solarium. (*Hint:* for the window, $\tau_{out} = \tau_{in}$.)

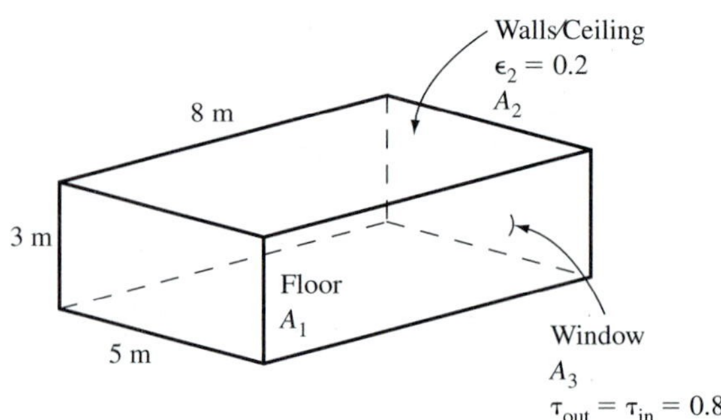

Figure 7–33

7–8 A pizza oven is approximated as a cubic container with upper surface heated, a door that conducts heat out, and all sides and bottom well insulated and reradiating. The inside dimensions of the oven are 1.5 m on a side. When in operation with the door closed, the burner surface is at 500 K and

3 kW of power are used by the heater. Assuming the emissivity of the heater and the inside surface of the door is 0.8.

(a) Determine the equilibrium temperature of the inside surface of the door.

(b) If the door is opened and the room is at 30°C (hot) estimate the equilibrium temperature of the heater assuming 3 kW of power is still used.

7–9 A toaster (see figure 7–34) has two heating elements on either side of a slice of bread when toasting the bread. The element temperature is 450°C, and its emissivity is 0.85 when bread is toasted. If the bread is at 100°C, the bread's emissivity is 0.65, and the surroundings reradiate heat, determine the net heat transfer to the bread.

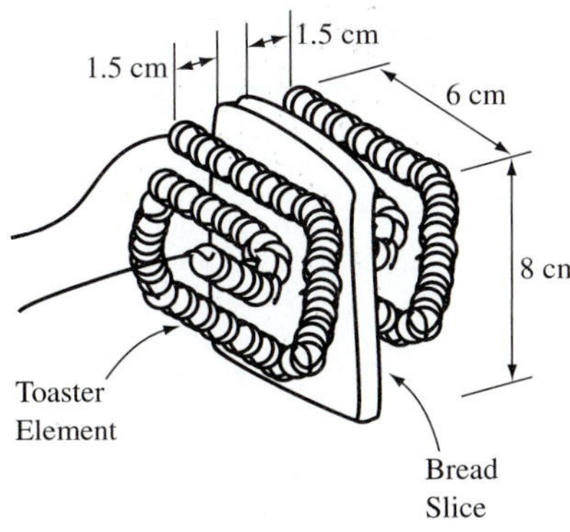

Figure 7–34 Toaster for problems 7–9 and 7–10.

7–10 Determine $\dot{Q}_{net}$ for the toasting bread in problem 7–9 if the surroundings act as a gray surface with $\varepsilon = 0.45$ and $T = 85°C$.

7–11 Consider the radiation heat transfer from an open fireplace that has a fire in it, as shown in figure 7–35. The fire has an effective surface temperature of 600°F and an emissivity of 0.95. The surroundings act as a black body at 60°F. Determine the net heat transfer to a person standing in front of the fireplace and the equilibrium surface temperature facing the fireplace. Neglect radiation from the back side of the person. Use a surface temperature of 80°F for the person, ε_r of 0.8, and shape factor of fire to person of 0.15; $F_{1-2} = 0.15$.

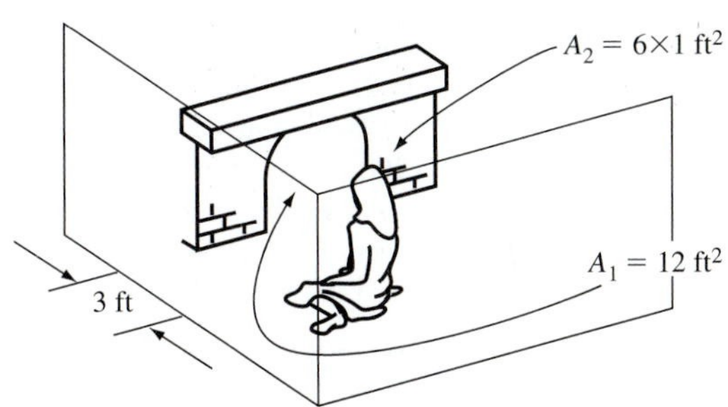

Figure 7–35

7–12 A radiant heater element is made so that it is concentric about a copper tube, as shown in figure 7–36. The heater is 1 cm thick, width OD 10 cm, and ID 8 cm. The copper tubing is 4 cm in diameter, is at 200°C and has an emissivity of 0.05. The heat tubing is at 1000°C, and its emissivity is 0.9 on the inside surface and 0.04 on the outside. If the surroundings are taken as a black body at 300 K, determine the power per unit of length of the heater needed for steady state operation.

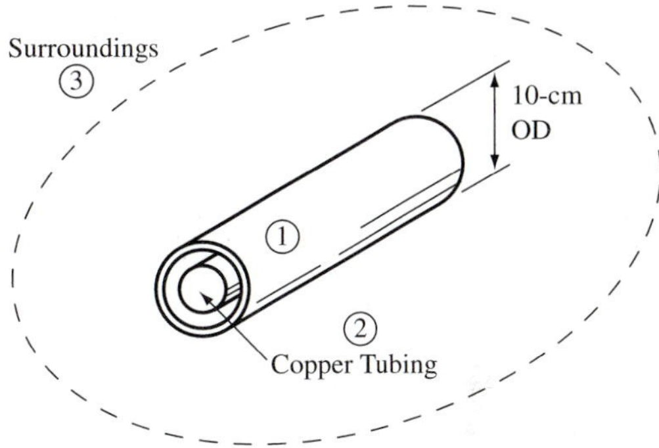

Figure 7–36

7–13 A steam generator is a device that converts water into steam by capturing radiant heat from hot combustion gases. Consider a 15-ft cube steam generator, all sides made up of water tubes and a "fireball" 6 ft in diameter at 4000°F at the center. If the water tubes have a surface temperature of 600°F and an emissivity of 0.7, determine the heat transfer to the water from the combustion gases. Assume the fireball acts as a black body.

7–14 Two parallel disks shown in figure 7–37 have emissivity of 0.6. If one disk is at 200°C and the other one at 30°C,

(a) Determine the net heat transfer between the disks if the surroundings reradiant heat.

(b) Determine the net heat transfer between the two disks if the surroundings are a black body at 20°C.

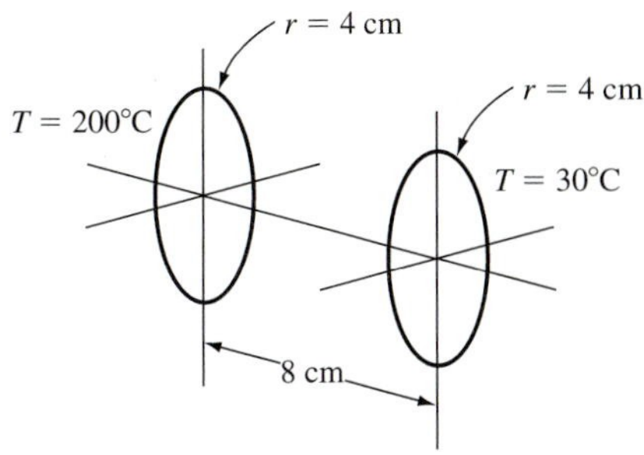

Figure 7–37

Section 7–4

7–15 On a particular foggy day the visibility is $^1/_4$ mile. This means that 95% of the visible light is absorbed after traveling $^1/_4$ mile. What is the extinction coefficient for the atmosphere?

7–16 Determine the effective emissivity of a bank of air 80 cm thick at 20°C, 80% relative humidity, and 1 atm pressure.

7–17 Determine the absorptivity of radiation through a 5-ft-wide plume of exhaust gases at 180°F and 1 atm pressure. The gases are a mixture of air with relative humidity of 15% and carbon dioxide having mole fraction of 20%.

7–18 A 200-ft-high by 60-ft-wide building is on fire and one of its walls is at 1500°F. Air at 100°F and 40% relative humidity is near the building, shown in figure 7–38. Estimate the net heat transfer to the air from this burning wall if the wall's emissivity is taken as 0.85.

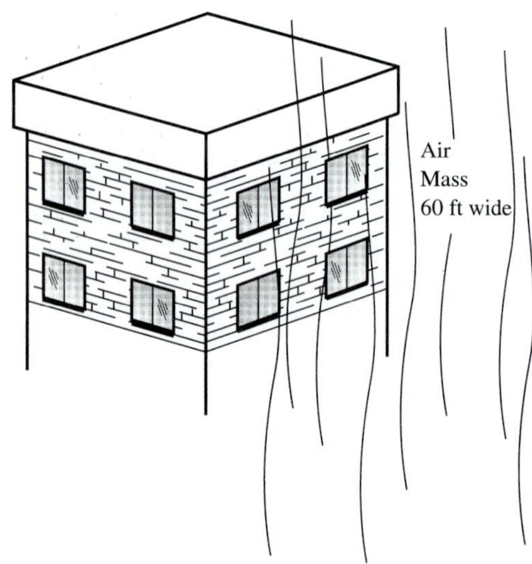

Figure 7–38

Section 7–5

7–19 A mercury-in-glass thermometer reads 20°C when hanging outside in the shade on a calm summer day. Another thermometer hangs in a sunny location and registers 45°C. Assuming a convective heat transfer coefficient of 10 W/m^2 · °C and $\varepsilon_r = 0.2$ for the thermometers, what is the effective surrounding surface temperature in the sun?

7–20 It is desired to study the radiation heat loss (or gain) from the surface of an automobile engine. Assume that engine is a gray opaque convex surface radiating with a heat exchanger for cooling engine coolant (called the radiator) on one side. Assume that the radiator is also a gray opaque flat surface. The engine also radiates with the road surface underneath (a black body), the hood and fender wells (assumes these are one reradiating surface), and the firewall (a black body).

(a) Construct a circuit to describe these processes.

(b) Identify all of the radiosities, black body emitters ($\dot{E}_{bi}$), and resistances in terms of the emissivities, areas, and shape factors.

7–21 For the plan view shown in figure 7–39, determine the mean radiant temperature (MRT) at the location shown.

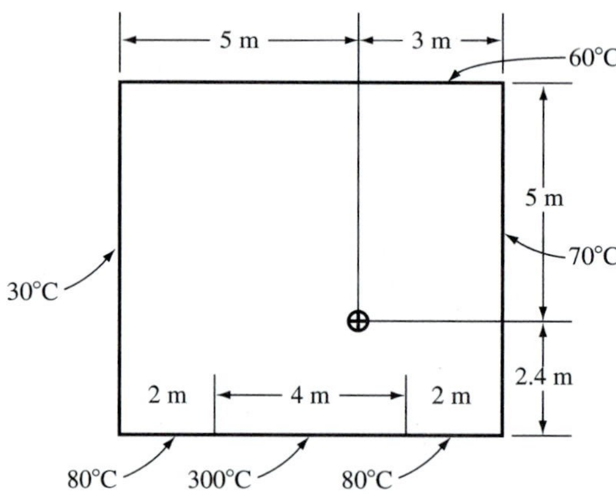

Figure 7–39

7–22 A steam generating unit burns coal with the combustion gases at 900°C, filling the combustion chamber. The combustion chamber is a right circular cylinder 10 m in diameter and 15 m tall. Estimate the heat transfer to the walls if the walls are at 500°C and have emissivity of 0.9.

7–23 An opaque gray surface having reflectivity of 0.5 at 27°C is exposed to irradiation of 1000 W/m^2. Air at 7°C flows over the surface and the convective heat transfer coefficient is 15 W/m^2 · °C. Determine the net heat flux from the surface.

7–24 An opaque horizontal plate is insulated on the backside. The plate irradiation is 2500 W/m^2, of which 500 W/m^2 is reflected. The plate temperature is 227°C and has an emissive power of 1200 W/m^2. Air at 127° and with $h = 12$ W/m^2 · °C flows over the plate. Determine

(a) ε, α, and the radiosity of the plate.

(b) The net heat transfer per unit of area.

7–25 A thermocouple having emissivity of 0.5, total specific heat of 0.34 J · K, and surface area of 0.8 cm^2 is placed in a large heated duct to measure the temperature of a gas flowing through the duct. The duct walls are at 300°C, and they behave as black body radiators. Estimate the value of the convective heat transfer coefficient of the thermocouple in the duct if the indicated temperature is 170°C initially and 290°C 8 min later, when it has just reached equilibrium. Also determine the gas temperature.

7–26 Bricks have been made for centuries by baking them in the direct sunlight. Determine the time required to complete the baking if water is driven off at 95°F and this requires 7000 Btu for each brick. Each brick is 4 in. by 6 in. by 8 in., and the surface emissivity is 0.4.

REFERENCES

1. Oppenheim, A. K., Radiation Analysis by the Network Method, *Trans. ASME,* 65, 725–735, 1956.
2. Hottel, H. C., and R. B. Egbert, *Trans. Am. Inst. Chem. Engr.,* 38, 531–565, 1942.
3. Hottel, H. C., and A. F. Sarofim, *Radiative Transfer,* McGraw-Hill, New York, 1967.
4. Fanger, P. O., *Thermal Comfort,* McGraw-Hill, New York, 1972.
5. Siegel, R., and J. R. Howell, *Thermal Radiation Heat Transfer,* McGraw-Hill, 1981.
6. Sparrow, E. M., and R. D. Cess, *Radiation Heat Transfer,* Hemisphere Publishing, New York, 1978.
7. Tien, C. L., Thermal Radiation Properties of Gases, in J. P. Hartnett and T. F. Irvine, eds., *Advances in Heat Transfer,* vol. 5, Academic Press, New York, 1968.
8. Dunkle, R. V., Radiation Exchange in an Enclosure with a Participating Gas, in W. M. Rosenow and J. P. Hartnett, eds., *Handbook of Heat Transfer,* McGraw-Hill, New York, 1973.

MASS TRANSFER

New Terms

B	permeability	$\dot{N}_A$	molar flow rate per unit area
Bi_m	mass transfer Biot Number	P	Permeance
C_i	molar concentration	Sc	Schmidt Number
Fo_m	mass transfer Fourier Number	Sh	Sherwood Number
Gr_m	mass transfer Grashof Number	$\overline{\mathbf{V}}_N$	local molar average velocity
h_D	convective mass transfer coefficient	$\overline{\mathbf{V}}_m$	local mass average velocity
$\dot{G}$	mass flow rate per unit area		

8–1 THE MECHANISMS OF MASS TRANSFER

Mass transfer is the phenomenon whereby mass moves from one region in space to another region. If we were to identify two control volumes in space, *A* and *B,* as shown in figure 8–1, then mass transfer occurs if some of the mass originally in *A* moves over to *B*. You can probably recall many examples of the occurrence of mass transfer in everyday experience, and in Chapter 1 we discussed some mass-transfer phenomena that are important in engineering analysis and design. We identified eight types of mass transfer and we restate those:

1. Ordinary diffusion resulting from a concentration gradient at the control volume boundary.
2. Thermal diffusion caused by a temperature gradient. This effect is usually insignificant, but some chemical separation processes rely on this effect.
3. Pressure diffusion caused by a hydrostatic pressure gradient. This effect can be important in centrifuging of fluid mixtures.

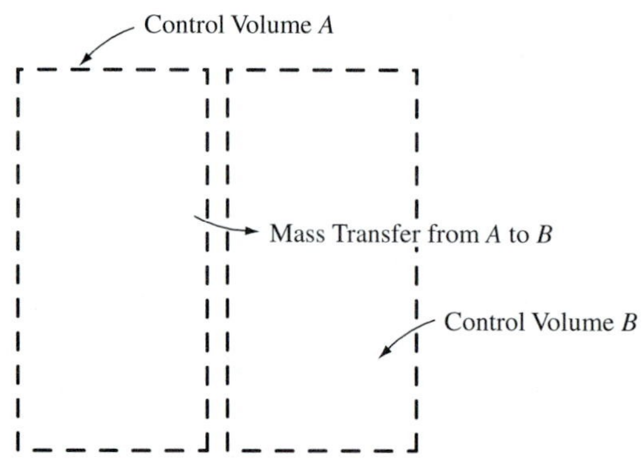

Figure 8–1 The mechanism of mass transfer.

4. Forced diffusion due to external forces such as the separation of ions by electric field strength.
5. Mass transfer in forced convection.
6. Mass transfer in free convection.
7. Turbulent mass transfer occurring due to eddies throughout a fluid.
8. Interphase mass transfer resulting from a nonequilibrium state at an interface.

We will discuss in some detail the mechanisms of diffusion due to concentration gradients, mass diffusion due to forced and free convection, and absorption and adsorption of water, which are types of mass transfer resulting from nonequilibrium states at an interface.

As we will see, mass transfer of a vapor or liquid into a gas is important in drying processes, humidifying, and cooling processes. Mass transfer of a gas or liquid into a liquid or solid is important in dehumidifying, odor control mechanisms, and control of moisture migration from one environment to another. Mass transfer of solids into solids is important in the heat treatment of metals, surface treatment of materials, bonding or gluing processes, and others.

In figure 8–1 the essential concept of mass transfer is indicated, and the driving force or cause for the phenomenon is one of those delineated in the preceding list. This transfer will always result in the mixing of substances because the region into which a mass moves will be occupied by another substance. If the substance in the region and the substance moving into that region are the same, then the process is called *self-diffusion.* For the purposes of this discussion we will normally assume that the mass transfer is not self-diffusion but rather diffusion in which one substance moves into a region occupied by a different substance. The mixture of two different substances is called a *binary mixture,* and the process is called binary diffusion. In addition, we will limit our discussion to diffusion involving substances that do not chemically react with each other.

To treat mass transfer further we need to discuss the concepts of a mixture and the methods for analyzing binary mixtures. This is the subject of the next section.

8–2 ANALYSIS OF MIXTURES

In section 1–3 of Chapter 1 we discussed how the analysis of mixtures requires that the mole or mass fraction of the components of the mixture be determined. The mass fraction for the ith component of the mixtures was defined as

$$\psi_i = \frac{m_i}{m_T} \qquad \textbf{(1–37, repeated)}$$

and the mole fraction as

$$\Omega_i = \frac{N_i}{N_T} \qquad \textbf{(1–40, repeated)}$$

The molar concentration or concentration of the ith component is defined as

$$C_i = \frac{N_i}{V} \qquad \textbf{(8–1)}$$

where V is the volume occupied by the mixture. Also,

$$C_i = \Omega_i C_m \qquad \textbf{(8–2)}$$

where C_m is the total concentration in moles per unit volume of the mixture. For a mixture of n components this is

$$C_m = \sum_{i=1}^{n} C_i \tag{8-3}$$

The mass concentration is its density,

$$\rho_i = \frac{m_i}{V} = \psi_i \rho_m \tag{8-4}$$

where ρ_m is the total density of the mixture. The mixture density is then, for a mixture of n components

$$\rho_m = \sum_{i=1}^{n} \rho_i \tag{8-5}$$

The concentrations and densities are related through the molecular mass, MW_i, by

$$\rho_i = C_i MW_i \tag{8-6}$$

These fractions and concentrations can be interpreted as isotropic and homogeneous properties of a mixture of components in equilibrium; however, most mixtures experiencing mass transfers have not yet achieved equilibrium, so the concentrations may vary and be nonuniform throughout the mixture. In addition, mixtures present a number of problems for modeling; some of which are

1. Chemical reactions.
2. Equations of state for each of components at the mixture conditions.
3. Phase changes during or due to mixing.
4. Capillarity and surface tension effects.

For the purposes of this text we will consider only those mixtures where no chemical reactions occur, where the components are ideal gases, incompressible liquids, or incompressible solids. Phase changes will be considered only for those instances where water vapor condenses from an air–water vapor mixture. Further, surface tensions between various components and capillary action will be ignored. In the following discussions we will consider only ideal gas mixtures and dilute gas/liquid/solid mixtures. Ideal gas mixtures were discussed in Chapter 1, and here we will recall that the most useful model for ideal gas mixtures is the one that assumes that every component of the mixture occupies the full mixture volume. According to this model, each component then has a partial pressure given by

$$p_i = \Omega_i p_T \tag{1-46, repeated}$$

and the density of each component in the mixture is

$$\rho_i = \frac{p_i}{RT} \tag{8-7}$$

The concentrations of each of the components in the mixture is

$$C_i = \frac{p_i}{R_u T} \tag{8-8}$$

EXAMPLE 8–1 An ideal gas mixture, composed of argon and nitrogen, is used as an inert gas in a double-pane window. If the mass fractions of the argon and nitrogen are 0.8 and 0.2, respectively, determine the densities and the concentrations of the two components and of the mixture itself when the gas is at 100 kPa and 10°C.

Solution The densities of the argon and nitrogen can be determined from equation (8–7), but first we need to determine the partial pressures and this requires determining the mole fractions:

$$N_{\text{argon}} = \frac{m_{\text{argon}}}{MW_{\text{argon}}} = \frac{0.8\ \text{kg/kg-mix}}{40\ \text{kg/kg-mol}} = 0.02\ \text{kg-mol/kg-mix}$$

$$N_{\text{nitrogen}} = \frac{0.2}{28} = 0.007\ \text{kg-mol/kg-mix}$$

and

$$N_T = N_{\text{argon}} + N_{\text{nitrogen}} = 0.02 + 0.007 = 0.027\ \text{kg-mol/kg-mix}$$

Then

$$\Omega_{\text{argon}} = \frac{N_{\text{argon}}}{N_T} = \frac{0.02}{0.027} = 0.74$$

$$\Omega_{\text{nitrogen}} \frac{0.007}{0.027} = 0.26$$

The densities are, then,

Answer
$$\rho_{\text{argon}} = \frac{(0.74)(100\ \text{kPa})(40\ \text{kg/kg-mol})}{(8.31\ \text{kJ/kg-mol}\cdot\text{K})(283\ \text{K})} = 1.259\ \text{kg/m}^3$$

Answer
$$\rho_{\text{nitrogen}} = \frac{(0.26)(100\ \text{kPa})(28\ \text{kg/kg-mol})}{(8.31\ \text{kJ/kg-mol}\cdot\text{K})(283\ \text{K})} = 0.310\ \text{kg/m}^3$$

The mixture density is just the sum of the two densities,

Answer
$$\rho = \rho_{\text{argon}} + \rho_{\text{nitrogen}} = 1.569\ \text{kg/m}^3$$

The mixture component concentrations are

Answer
$$C_{\text{argon}} = \frac{\rho_{\text{argon}}}{MW_{\text{argon}}} = 0.0315\ \text{kg-mol/m}^3$$

Answer
$$C_{\text{nitrogen}} = \frac{\rho_{\text{nitrogen}}}{MW_{\text{nitrogen}}} = 0.011\ \text{kg-mol/m}^3$$

and the mixture concentration is

Answer
$$C = C_{\text{argon}} + C_{\text{nitrogen}} = 0.0425\ \text{kg-mol/m}^3$$

Mixtures of solids and many liquids can often be approximated as dilute mixtures. A dilute mixture is one in which one of the mixture components, call it the major component, is mixed with small amounts of other components. As listed earlier, a mixture presents a number of conceptual problems and which volumes and/or pressures describe each of the components is one the most pressing. The concept of the *ideal mixture* or ideal solution can be used to model a dilute mixture. An ideal mixture is one in which the mixture volume is defined as the summation of the volumes of each of the mixture components, as if they were each at the same temperature and pressure of the mixture, usually standard state atmospheric pressure and temperature. That is,

$$V_m = \sum_{n=1}^{n} V_{i,s} \tag{8–9}$$

Thus, the mixture volume can be viewed as the addition of component volumes as diagramed in figure 8–2, except that the mixture components are, finally, occupying the total volume of the mixture for purposes of determining component densities and concentrations. If the amount of the dilute components becomes too great, then the mixture is not a dilute mixture and interactions between the mixture components can affect the mixture volume and other properties.

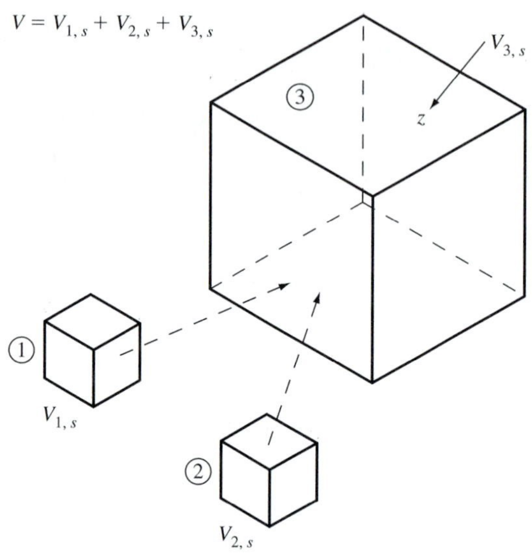

Figure 8–2 The mechanism of mixing in an ideal mixture.

EXAMPLE 8–2 Lead has amalgamated into a cast-iron bar. A particular sample taken from the bar indicates the following conditions: $\Omega_{\text{lead}} = \Omega_{\text{Pb}} = 0.02$, $\Omega_{\text{Fe}} = 0.95$, $\Omega_{\text{carbon}} = 0.03$. Determine the mixture density and molar concentration of the cast iron.

Solution The mixture may be assumed to be a dilute mixture and an ideal mixture. The density of the cast iron can be determined from the mixture mass and the mixture volume. First we determine the mass fractions of the lead, iron, and carbon. Assuming 100 kg of cast iron, we may generate the following table.

Comp.	MW	Ω	ρ (kg/m³)	N (kg-mol)	mass, m (kg)	ψ, mass fract.	$V_{i,s}$ (m³)
Iron	55.8	0.95	7,874	95	5301	0.922	0.673
Carbon	12	0.03	2,000	3	36	0.006	0.018
Lead	207.2	0.02	11,340	2	414.4	0.072	0.037
Total		1.00		100	5751.4	1.000	0.728

In the table the second column lists the molecular weights (masses) as given in the appendix. The third column lists the given mole fractions and the fourth column densities of iron, carbon (graphite), and lead as given in appendix table B–2. The fifth column lists the number of kg-mol of each component in 100 kg-mol of mixture, and the sixth column gives the mass of each component, which is the product of the molecular weight and the number of kg-mol. The seventh column lists the mass fractions, and the eighth column gives the standard state component volumes. The total mixture volume results from the sum of column eight. The mixture density is then

Answer
$$\rho_m = \frac{m_T}{V_m} = \frac{5751.4\ \text{kg}}{0.728\ \text{m}^3} = 7900.27\ \text{kg/m}^3$$

and the mixture concentration is

$$C_m = \frac{\rho_m}{\text{MW}_m}$$

The mixture molecular weight is determined from the following:

$$\text{MW}_m = \sum_{i=1}^{n} \Omega_i \text{MW}_i = \Omega_{\text{iron}}\text{MW}_{\text{iron}} + \Omega_C \text{MW}_C + \Omega_{\text{lead}}\text{MW}_{\text{lead}}$$

$$= 0.95(55.8) + 0.03(12) + 0.02(207.2) = 57.514\ \text{kg/kg-mol}$$

so that the concentration is

Answer
$$C_m = \frac{7900.27\ \text{kg/m}^3}{57.514\ \text{kg/kg-mol}} = 137.36\ \text{kg-mol/m}^3$$

The concentration could also have been determined from the total number of moles (100 kg-mol) and the mixture volume, 0.728 m^3.

One of the major problems of analyzing mass-transfer processes is the determination of mixture and component densities and concentrations, which change with location and time in a mixture. Thus, in general, these terms are not constant or uniform in a mixture. The rates at which mass is transferred through these nonuniform concentrations is the concern of mass-transfer analysis. In particular, if we apply the conservation of mass to a control volume in a mixture, as shown schematically in figure 8–3, we have for the ith component

$$\frac{dm_{i,\text{in}}}{dt} + \frac{dm_{i,\text{gen}}}{dt} = \frac{dm_{i,\text{out}}}{dt} + \frac{dm_{i,\text{accum}}}{dt} \tag{8–10}$$

The mass flow rate into the control volume, $dm_{i,\text{in}}/dt$, may be affected by the mechanisms discussed in the previous section. The mass generation term, $dm_{i,\text{gen}}/dt$, accounts for chemical reactions, absorption, or adsorption. A positive mass generation accounts for the generation of a substance in a chemical reaction; a negative value accounts for the destruction of a substance. In this text we will not consider chemical reactions; however, we will treat absorption and adsorption. The accumulation of mass accounts for the non–steady state conditions that can occur in mass transfer. We will discuss to a limited extent the non–steady state conditions, but the analysis will assume steady state unless otherwise indicated.

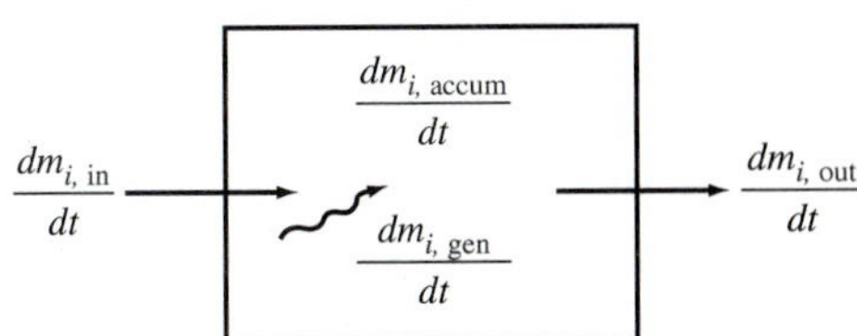

Figure 8–3 Control volume in a mixture.

8–3 DIFFUSION MASS TRANSFER

An isothermal mixture having a nonuniform concentration of components through its volume will tend to a uniform concentration by means of diffusional mass transfer. Consider for instance an isothermal mixture of n components, shown in figure 8–4, and consider the flow of the ith component at a plane Y–Y. The ith component may be moving across the plane Y–Y with a velocity $\mathbf{V}_i$ while the velocity of the another component ($i + 1$, for instance) may be moving in the opposite direction at $\mathbf{V}_{i+1}$. The local average velocity of the mixture is

$$\overline{\mathbf{V}} = \frac{1}{\sum_{i=1}^{n} \rho_i} \sum_{i=1}^{n} \rho_i \mathbf{V}_i = \frac{1}{\rho_m} \sum_{i=1}^{n} \rho_i \mathbf{V}_i \tag{8–11}$$

This local average velocity would be the velocity sensed by placing a Pitot tube in the mixture and measuring the dynamic pressure head. The local average molar velocity at plane Y–Y is

$$\overline{\mathbf{V}}_N = \frac{1}{\sum_{i=1}^{n} C_i} \sum_{i=1}^{n} C_i \mathbf{V}_i = \frac{1}{C_m} \sum_{i=1}^{n} C_i \mathbf{V}_i \tag{8–12}$$

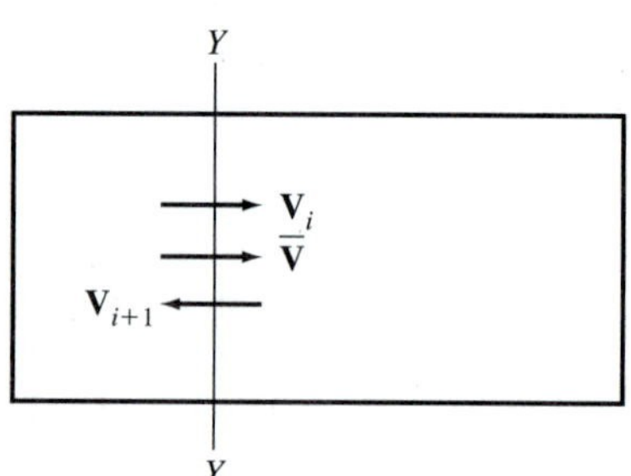

Figure 8–4 Mixture component flows.

The diffusional velocity of the ith component in a mixture can be visualized as the difference between the velocity of the component $\mathbf{V}_i$ and the local average velocity,

$$\mathbf{V}_{i/m} = \mathbf{V}_i - \overline{\mathbf{V}}_m \tag{8–13}$$

or the difference between the velocity of the component and the local average molar velocity,

$$\mathbf{V}_{i/N} = \mathbf{V}_i - \overline{\mathbf{V}}_N \tag{8–14}$$

These velocities are relative velocities and are important concepts because there will tend to be motion of components in both directions across the plane Y–Y or any other plane in the mixture. The mass flow rate per unit area for the ith component is

$$\dot{G}_i = \frac{\dot{m}_i}{A} = \rho_i \mathbf{V}_i \tag{8–15}$$

and the molar flow rate per unit area is

$$\dot{N}_{A,i} = \frac{\dot{N}_i}{A} = C_i \mathbf{V}_i \tag{8–16}$$

Common units for the mass flow rate per unit area are kg/m^2 · s or lbm/ft^2 · s, and for the molar flow rate per unit area, kg-mol/m^2 · s or lbm-mol/ft^2 · s. The mass flow rate of species *i*, or the *i*th mixture component, relative to the mixture motion is

$$\dot{G}_{i,m} = \rho_i(\mathbf{V}_i - \overline{\mathbf{V}}) = \dot{G}_i - \rho_i\overline{\mathbf{V}} \tag{8–17}$$

and the molar flow rate relative to the mixture is

$$\dot{N}_{A,i/N} = C_i(\mathbf{V}_i - \overline{\mathbf{V}}) = \dot{N}_{A,i} - C_i\overline{\mathbf{V}} \tag{8–18}$$

In many mixtures the average velocity may be taken as zero so that the relative mass flow rate is then the same as the mass flow rate and the relative molar flow rate is the same as the molar flow rate. If we consider a binary mixture such that $n = 2$ and denoting the components as A and B, then the molar rate of diffusion of A in the mixture, $\dot{N}_{A,A/N}$ at a plane Y–Y as shown in figure 8–5 can be written

$$\dot{N}_{N,A/N} = -(C_A + C_B)\wp_{AB}\frac{\partial \Omega_A}{\partial x} = -\wp_{AB}\frac{\partial C_A}{\partial x} \tag{8–19}$$

and for component B

$$\dot{N}_{A,B/N} = -(C_A + C_B)\wp_{BA}\frac{\partial \Omega_B}{\partial x} = -\wp_{BA}\frac{\partial C_B}{\partial x} \tag{8–20}$$

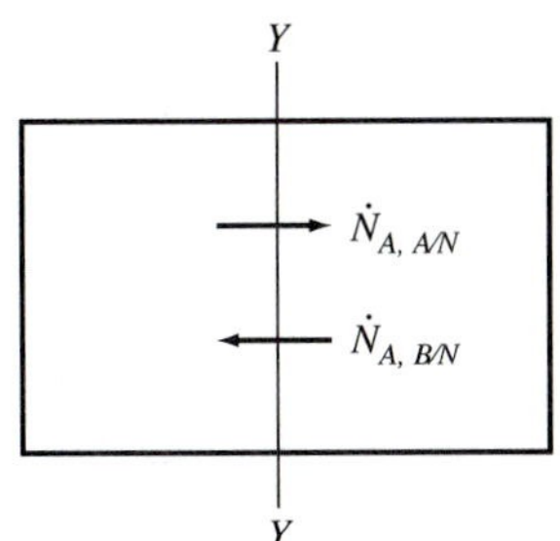

Figure 8–5 Diffusion of A and B in a binary mixture of A and B.

Equations (8–19) and (8–20) are referred to as Fick's Law of Diffusion and represent a definition of the mass diffusivities, $\wp_{AB}$ and $\wp_{BA}$. The mass rate of diffusion, Fick's Law, can also be written for A as

$$\dot{G}_{A/m} = -(\rho_A + \rho_B)\wp_{AB}\frac{\partial \psi_A}{\partial x} = -\wp_{AB}\frac{\partial \rho_A}{\partial x} \tag{8–21}$$

and for B,

$$\dot{G}_{B/m} = -(\rho_A + \rho_B)\wp_{BA}\frac{\partial \psi_B}{\partial x} = -\wp_{BA}\frac{\partial \rho_B}{\partial x} \tag{8–22}$$

The mass diffusivity $\wp_{AB}$ of A into a binary mixture of A and B and the mass diffusivity $\wp_{BA}$ of B into a binary mixture of A and B are equal,

$$\wp_{AB} = \wp_{BA} \tag{8–23}$$

Notice that Fick's Law has a close resemblance to Fourier's Law of Heat Conduction, including the negative sign to account for mass or molar flow rates directed toward lower concentrations or in the direction of negative concentration gradients. The mixture concentration and density are not normally uniform or constant in a mixture, even under steady state conditions, so Fick's Law presents some computational problems. In addition, there is not an abundance of experimental data for mass diffusivities, and this indicates that future work could be done on the experimental determination of various mass diffusivities. Experimental methods, including steady state and unsteady or transient conditions, for the determination of the diffusivity are described by Jost [1] and Bobrova and Rabinovich [2]. In table 8–1 are listed some selected data for mass diffusivity of some binary vapor mixtures, in table 8–2 for some binary liquid mixtures, and in table 8–3 for some vapors, liquids, and solids into another solid.

Table 8–1 Experimental Values for Mass Diffusivity for Some Dilute Vapor Mixtures of Some Organic Compounds at 1 atm Determined by an Evaporation Method

Substance (phase), *A*	**Host Substance, *B***	***T*, K**	**$\wp$, cm²/s**
(sec.)Amyl alcohol	air	303	0.072
Aniline	air	303	0.075
(*n*)Butyl alcohol	air	303	0.088
Chlorobenzene	air	303	0.075
Chloropicrine	air	398	0.088
Cyanogen chloride	air	273	0.111
Ethyl acetate	air	303	0.089
Ethyl alcohol	air	273	0.099
Ethyl ether	air	273	0.0786
Hydrogen cyanide	air	273	0.173
Iodine	air	273	0.0692
Isopropyl alcohol	air	303	0.101
Phosgene	air	273	0.095
Toluene	air	303	0.088
Water (g)	air	273	0.219
Ethyl alcohol	carbon dioxide	273	0.0686
Ethyl ether	carbon dioxide	273	0.0541
Water (g)	carbon dioxide	273	0.1384
Acetone	hydrogen	273	0.361
Benzene	hydrogen	273	0.318
Carbon tetrachloride	hydrogen	273	0.293
Ethyl alcohol	hydrogen	273	0.377
Ethyl ether	hydrogen	273	0.299
Water	hydrogen	273	0.747
Benzene	oxygen	273	0.0797
Carbon tetrachloride	oxygen	273	0.0636
Iodine	nitrogen	273	0.070
Mercury	nitrogen	273	0.1190

It has been difficult to develop analytic expressions for predicting mass diffusivity. The mass diffusivity of an ideal gas into another ideal gas may be approximated by the equation suggested by Bird et al. [3]

$$\wp_{AB} = 0.0018583\frac{\sqrt{T^3\left(\dfrac{1}{\mathrm{MW}_A} + \dfrac{1}{\mathrm{MW}_B}\right)}}{p\sigma_{AB}^2\Omega_{\wp,AB}} \tag{8–24}$$

where the mass diffusivity is in units of cm²/s. The temperature T is in kelvin, the pressure p in atmospheres, and the term σ_{AB} is the effective diameter of the gas molecules in angstrom units (Å). This diameter is taken as

$$\sigma_{AB} = \tfrac{1}{2}(\sigma_A + \sigma_B) \tag{8–25}$$

Table 8–2 Experimental Values for Mass Diffusivity for Some Liquids or Gases in Other Liquids or Gases

Substance, *A*	Phase	Substance *B*	Phase	*T*, K	$\wp_{AB}$, m²/s
Caffeine	liquid	water	liquid	298	0.63×10^{-9}
Ethanol	liquid	water	liquid	298	0.12×10^{-8}
Glucose	liquid	water	liquid	298	0.69×10^{-9}
Glycerol	liquid	water	liquid	298	0.94×10^{-9}
Acetone	gas	water	liquid	298	0.13×10^{-8}
CO_2	gas	water	liquid	298	0.20×10^{-8}
H_2	gas	water	liquid	298	0.63×10^{-8}
N_2	gas	water	liquid	298	0.26×10^{-8}
O_2	gas	water	liquid	298	0.24×10^{-8}
Acetone	gas	air	gas	273	0.11×10^{-4}
Ammonia	gas	air	gas	298	0.28×10^{-4}
Benzene	gas	air	gas	298	0.88×10^{-5}
CO_2	gas	air	gas	298	0.16×10^{-4}
H_2	gas	air	gas	298	0.41×10^{-4}
Naphthalene	gas	air	gas	300	0.62×10^{-5}
O_2	gas	air	gas	298	0.21×10^{-4}
Radon	gas	air	gas	288	0.12×10^{-4}
Water	gas	air	gas	298	0.26×10^{-4}
Hydrogen	gas	carbon dioxide	gas	273	0.55×10^{-4}
Mercury	gas	hydrogen	gas	273	0.53×10^{-4}
Argon	gas	nitrogen	gas	293	0.19×10^{-4}
Cadmium	gas	nitrogen	gas	273	0.17×10^{-4}
CO_2	gas	nitrogen	gas	293	0.16×10^{-4}
Hydrogen	gas	nitrogen	gas	273	0.68×10^{-4}
Mercury	gas	nitrogen	gas	273	0.14×10^{-4}
Oxygen	gas	nitrogen	gas	273	0.18×10^{-4}
Sodium	gas	nitrogen	gas	288	20.4×10^{-4}
CO_2	gas	oxygen	gas	273	0.14×10^{-4}
Hydrogen	gas	oxygen	gas	273	0.70×10^{-4}

where values for the individual molecular diameters, σ_A and σ_B, are given in appendix table A–12–1 for selected ideal gases. The function Ω_{AB} is a dimensionless function of temperature and is listed in table A–12–2 as a function of the parameter $T(k/\varepsilon_{AB})$. The term k/ε_{AB} is defined as

$$\frac{k}{\varepsilon_{AB}} = \sqrt{\left(\frac{k}{\varepsilon_A}\right)\left(\frac{k}{\varepsilon_B}\right)} \tag{8–26}$$

and values for ε/k for individual gases are given in appendix table A–12–1. In table 2–1 collision diameters, Γ, were listed for helium, argon, and nitrogen gas. These collision diameters may be used as a first approximation for the individual molecular diameters. For many gases the collision diameter has a value in the range of 2 Å to 6 Å [3], and a value of $\sqrt{10}$ Å is probably typical. Notice that the form of equation (8–24) suggests that the mass diffusivity is a function of $T^{3/2}/p$ for an ideal gas. Expressions for the mass diffusivity of liquids and solids may be obtained from the literature, and the interested reader may want to see for instance Bird et al. [3], Brodkey and Hershey [4], and Hirschfelder et al. [5]; however, further discussion of analytic expressions for mass diffusivity are beyond the purposes of this text.

Table 8–3 Experimental Values for Mass Diffusivity for Some Substances in Solids

Diffusing substance	Phase	Solid Host Substance	Diffusivity $\wp$ (m^2/s @ 773 K)	Diffusivity $\wp$ (m^2/s @ 1273 K)
Aluminum	solid	copper, Cu	0.13×10^{-33}	
Cadmium	solid	Cu	0.27×10^{-18}	
Carbon	solid	FCC iron	5×10^{-15}	3×10^{-11}
Carbon	solid	BCC iron	1×10^{-12}	2×10^{-9}
Carbon	solid	HCP titanium	3×10^{-16}	2×10^{-11}
Copper	solid	aluminum	4×10^{-14}	1×10^{-10}
Copper	solid	copper	1×10^{-18}	2×10^{-13}
Iron	solid	FCC iron	2×10^{-23}	2×10^{-16}
Iron	solid	BCC iron	1×10^{-20}	3×10^{-14}
Manganese	solid	FCC iron	3×10^{-24}	1×10^{-16}
Nickel	solid	FCC iron	1×10^{-23}	2×10^{-16}
Silver	solid	silver	1×10^{-17}	1×10^{-12}
Zinc	solid	copper	4×10^{-18}	5×10^{-13}
CO_2	gas	rubber	0.11×10^{-9}	
H_2	gas	iron	0.26×10^{-12}	
He	gas	SiO_2 (glass)	0.4×10^{-13}	
N_2	gas	rubber	0.15×10^{-9}	
O_2	gas	rubber	0.21×10^{-9}	

EXAMPLE 8–3 Estimate the mass diffusivity of carbon dioxide, CO_2, in air at atmospheric pressure and 25°C by using equation (8–24) and comparing the value to that given in table 8–1.

Solution We read, from table A–12–1 for carbon dioxide

$$\sigma = 3.996 \text{ Å} \qquad \varepsilon/k = 190.0 \text{ K}$$

and for air

$$\sigma = 3.617 \text{ Å} \qquad \varepsilon/k = 97.0 \text{ K}$$

then, from equation (8–17),

$$\sigma_{AB} = \frac{1}{2}(3.996 + 3.617) = 3.8065 \text{ Å}$$

and from equation (8–26)

$$\frac{k}{\varepsilon_{AB}} = \sqrt{\left(\frac{1}{190.0}\right)\left(\frac{1}{97.0}\right)} = 0.00737$$

Then $T(k/\varepsilon_{AB}) = 2.195$ and from table A–12–2, $\Omega_{AB} = 1.042$. The mass diffusivity can then be computed from equation (8–24):

Answer

$$\wp_{AB} = 0.0018583 \frac{\sqrt{(293 \text{ K})^3\left(\frac{1}{44} + \frac{1}{28.9}\right)}}{(1 \text{ atm})(3.8065)^2(1.042)} = 0.1516 \text{ cm}^2/\text{s}$$

Comparing this value to that listed in table 8–1, $\wp_{AB} = 0.165$ cm^2/s, shows that close agreement can be obtained. Such a circumstance does not always occur, and discrepancies of as much as 30% can be expected when predicting mass diffusivities.

If we consider the diffusion of one ideal gas into another ideal gas, then for Fick's Law the density gradient in an isothermal mixture is

$$\frac{\partial \rho_A}{\partial x} = \frac{1}{R_A T}\frac{\partial p_A}{\partial x} \tag{8–27}$$

and

$$\frac{\partial \rho_B}{\partial x} = \frac{1}{R_B T}\frac{\partial p_B}{\partial x} \tag{8–28}$$

where p_A and p_B are the partial pressures of the ideal gases A and B respectively.

EXAMPLE 8–4 A plume of combustion gases is in contact with the surrounding air as shown in figure 8–6. If the gases are at 300°C and consist of a mixture of 35% CO_2 and 65% air at atmospheric pressure while the air immediately surrounding the plume is at 300°C and has 15% CO_2 measured 10 cm from the plume, estimate the mass transfer of CO_2 into the air and the air that becomes entrained in the plume.

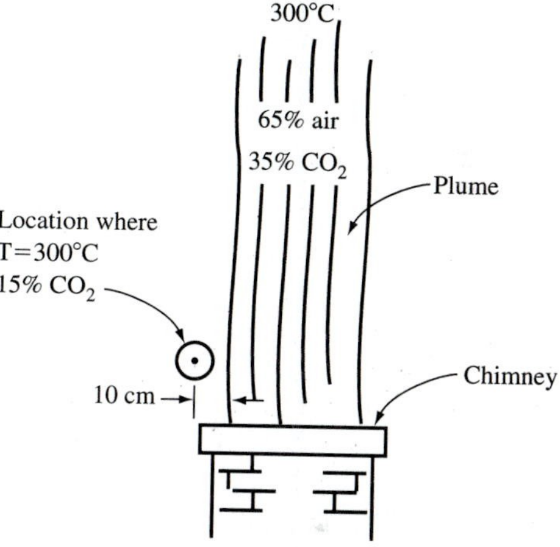

Figure 8–6 Plume of products of combustion.

Solution If we neglect the motion of the plume and the air surrounding the plume, then we may estimate the mass transfer from the plume to the surrounding air from the form of Fick's Law in equations (8–21) and (8–22). For the surrounding air at the location x shown in figure 8–7, we have

$$p_{A,x} = (0.85)(1 \text{ atm}) = 0.85 \text{ atm}$$

and

$$p_{B,x} = (0.15)(1 \text{ atm}) = 0.15 \text{ atm}$$

Figure 8–7 Approximation for concentration gradient of CO_2 at the boundary of a plume of products of combustion.

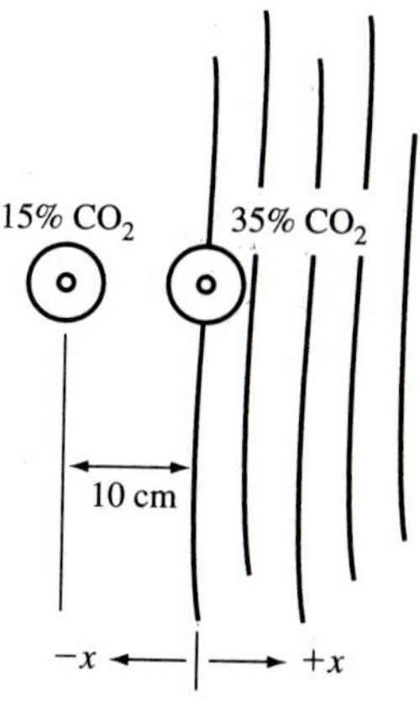

In the plume,

$$p_{A,\text{pl}} = (0.65)(1 \text{ atm}) = 0.65 \text{ atm}$$

and

$$p_{B,\text{pl}} = (0.35)(1 \text{ atm}) = 0.35 \text{ atm}$$

The density gradients are, from equations (8–27) and (8–28),

$$\frac{\partial \rho_A}{\partial x} = \frac{1}{R_A T}\frac{\partial p_A}{\partial x} \approx \frac{1}{(287 \text{ J/kg}\cdot\text{K})(573 \text{ K})}\frac{(p_{A,\text{pl}} - p_{A,x})}{10 \text{ cm}}\frac{1}{(287)(573)}\frac{0.65 - 0.85 \text{ atm}}{10 \text{ cm}}$$

$$= -0.122 \times 10^{-6}\frac{\text{atm}\cdot\text{kg}}{\text{J}\cdot\text{cm}} = -1.228\frac{\text{kg}}{\text{m}^4}$$

and

$$\frac{\partial \rho_B}{\partial x} = \frac{\partial \rho_{CO_2}}{\partial x} \approx \frac{1}{R_B T}\frac{p_{B,x} - p_{B,\text{pl}}}{-10 \text{ cm}} = \frac{1}{(189 \text{ N}\cdot\text{m/kg}\cdot\text{K})(573 \text{ K})}\frac{0.15 - 0.35}{-10 \text{ cm}}$$

$$= 1.8652 \text{ kg/m}^4$$

The mass diffusivity may be estimated from equation (8–24) and from the method demonstrated in example 8–3:

$$\sigma_{AB} = 3.826 \text{ Å}$$

and

$$Tk/\varepsilon_{AB} = (573 \text{ K})(0.00807 \text{ K}^{-1}) = 4.624$$

so that $\Omega_{\wp,AB} = 0.856$ from appendix table A–12–1. Then, from equation (8–24) we have

$$\wp_{AB} = 0.0018583\frac{\sqrt{(573 \text{ K})^3\left(\frac{1}{44} + \frac{1}{28.9}\right)}}{(1 \text{ atm})(3.826)^2(0.856)} = 0.0487 \text{ cm}^2/\text{s}$$

The mass transfers are predicted from equation (8–21), revised with equation (8–27):

$$\dot{G}_{A/m} = -\wp_{AB}\frac{1}{R_A T}\frac{\partial p_A}{\partial x} = -(0.0487 \text{ cm}^2/\text{s})(-1.228 \text{ kg/m}^4) = 5.98 \times 10^{-4} \text{ kg/s}\cdot\text{m}^2$$

Answer

$$= 0.598 \text{ g/s}\cdot\text{m}^2$$

and for the mass transfer of the CO_2 to the surrounding air,

$$\dot{G}_{B/m} = -\wp_{BA}\frac{1}{R_B T}\frac{\partial p_B}{\partial x} = -(0.0487 \text{ cm}^2/\text{s})(1.8652 \text{ kg/m}^4)$$

Answer

$$= -0.908 \text{ g/s}\cdot\text{m}^2$$

The plume is dispersing carbon dioxide, and air is becoming entrained in the plume. The negative sign on the carbon dioxide mass flow rate indicates that its mass flow is opposite to that of the air.

In this last example we estimated the mass transfer from and to a plume of a hot gas mixture. Now we will consider a stagnant isothermal column of a gas through which another gas diffuses, as shown in figure 8–8. A liquid *A* is evaporating and rising into a stagnant column of gas or vapor *B*. The liquid evaporates because its vapor pressure, p_A, in the column is less than its saturation pressure at the liquid and vapor temperature. This action occurs in air drying of clothes, or any other process where water evaporates into still air. It can also describe the action of many other drying processes, such as paint drying where the thinner chemical or carrier evaporates and leaves the pigment on the painted surface.

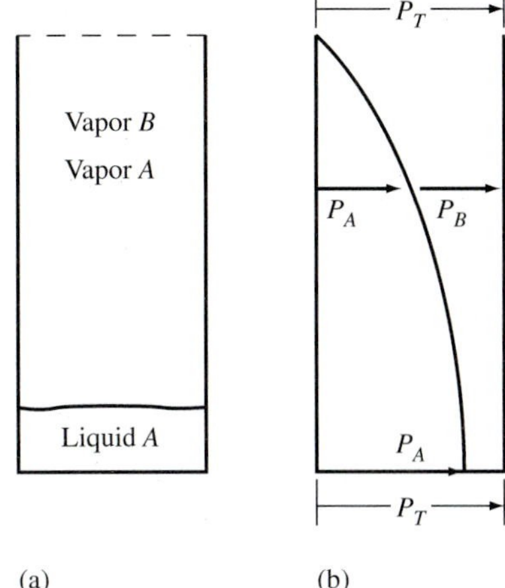

Figure 8–8 Evaporation into a stagnant vapor column.

The vapor pressures of A and B are qualitatively described in figure 8–8b, where the vapor pressure p_A decreases at higher levels in the column while the vapor pressure p_B increases so that at any level we have constant total pressure;

$$p_T = p_A + p_B$$

We assume that the vapor A that reaches the top of the column can flow into a free stream outside of the column. If we consider the motion of the vapors, component B will diffuse down because of the decrease in its vapor pressure and therefore its concentration. Then, the diffusion is

$$\dot{G}_{B/m} = -\wp_{BA}\frac{1}{R_BT}\frac{\partial p_B}{\partial y} \tag{8–29}$$

and in order that the vapor B be stagnant a bulk flow of B and A must be against this downward flow. The vapor B in the bulk flow is

$$\dot{G}_B = \rho_B\overline{\mathbf{V}} \tag{8–30}$$

where $\overline{\mathbf{V}}$ is the local average mass velocity from equation (8–11). Thus, for the stagnant vapor B we have

$$\dot{G}_{B/m} = -\dot{G}_B$$

or, from equation (8–29),

$$\wp_{BA}\frac{1}{R_BT}\frac{\partial p_B}{\partial y} = \rho_B\overline{\mathbf{V}} \tag{8–31}$$

Since

$$\rho_B = \frac{p_B}{R_BT}$$

we get

$$\wp_{BA}\frac{1}{R_BT}\frac{\partial p_B}{\partial y} = \frac{p_B}{R_BT}\overline{\mathbf{V}}$$

The local average mass velocity is then

$$\overline{\mathbf{V}} = \wp_{BA}\frac{1}{p_B}\frac{\partial p_B}{\partial y} \tag{8–32}$$

Also, since $\wp_{AB} = \wp_{BA}$, $p_B = p_T - p_A$, and $\partial p_B/\partial y = -\partial p_A/\partial y$, equation (8–32) is

$$\overline{\mathbf{V}} = \frac{1}{p_A - p_T}\wp_{AB}\frac{\partial p_A}{\partial y} \tag{8–33}$$

The bulk mass flow of vapor A upward through the column is

$$\dot{G}_A = \rho_A\overline{\mathbf{V}} = \frac{p_A}{R_AT}\left(\frac{\wp_{AB}}{p_A - p_T}\right)\frac{\partial p_A}{\partial y} \tag{8–34}$$

The vapor A is also diffusing through the column, driven by a concentration gradient and described by Fick's Law,

$$\dot{G}_{A/m} = -\wp_{AB}\frac{1}{R_AT}\frac{\partial p_A}{\partial y} \tag{8–35}$$

and the total evaporation rate of A through the stagnant column is the sum of the bulk mass flow of vapor A and the mass flow rate due to diffusion, the addition of equations (8–34) and (8–35):

$$\dot{G}_{A,\text{total}} = \dot{G}_A + \dot{G}_{A/m} = \frac{\wp_{AB}}{R_AT}\left(\frac{p_A}{p_A - p_T} - 1\right)\frac{\partial p_A}{\partial y} \tag{8–36}$$

or

$$\dot{G}_{A,\text{total}} = \frac{\wp_{AB}}{R_AT}\left(\frac{p_T}{p_A - p_T}\right)\frac{\partial p_A}{\partial y} \tag{8–37}$$

Equation (8–37) is referred to as *Stefan's Law*. If we neglect any friction or cross-diffusion in the column, then the partial derivative of the pressure with respect to y is an ordinary derivative and Stefan's Law is

$$\dot{G}_{A,\text{total}} = \frac{\wp_{AB}}{R_AT}\left(\frac{p_T}{p_A - p_T}\right)\frac{dp_A}{dy} \tag{8–38}$$

and if this is evaluated between two levels, as indicated in figure 8–9, by integrating equation (8–38) for constant diffusivity and mass flow rate (steady state), we have, per unit area

$$\dot{G}_{A,\text{total}} = \frac{p_T\wp_{AB}(p_{A,1} - p_{A,2})}{R_AT(y_2 - y_1)p_{\text{LM}}} \tag{8–39}$$

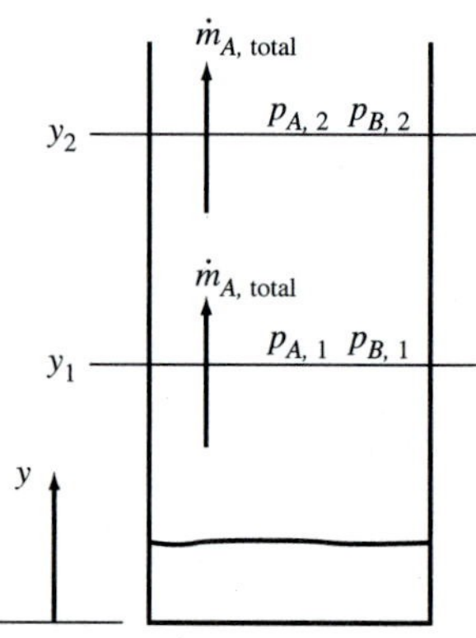

Figure 8–9 Determination of log mean pressure difference.

where p_{LM} is the log mean pressure difference,

$$p_{\mathrm{LM}} = \frac{(p_T - p_{A,2}) - (p_T - p_{A,1})}{\ln\left(\dfrac{p_T - p_{A,2}}{p_T - p_{A,1}}\right)} \tag{8–40}$$

EXAMPLE 8–5 In a papermaking process a 2.5-cm-thick slurry of paper fiber and water is placed on a 5-m-wide wire belt, which moves the slurry through drying boxes. At a particular location shown in figure 8–10 the slurry is 20% dry fiber and 80% water. If there is a 10-cm layer of stagnant air at 30°C over the wet paper slurry and air at 30°C, 30% relative humidity is moving above the stagnant layer, estimate the evaporation rate per unit length of paper. Assume that the stagnant air has 100% relative humidity at the wetted surface between the air and the slurry.

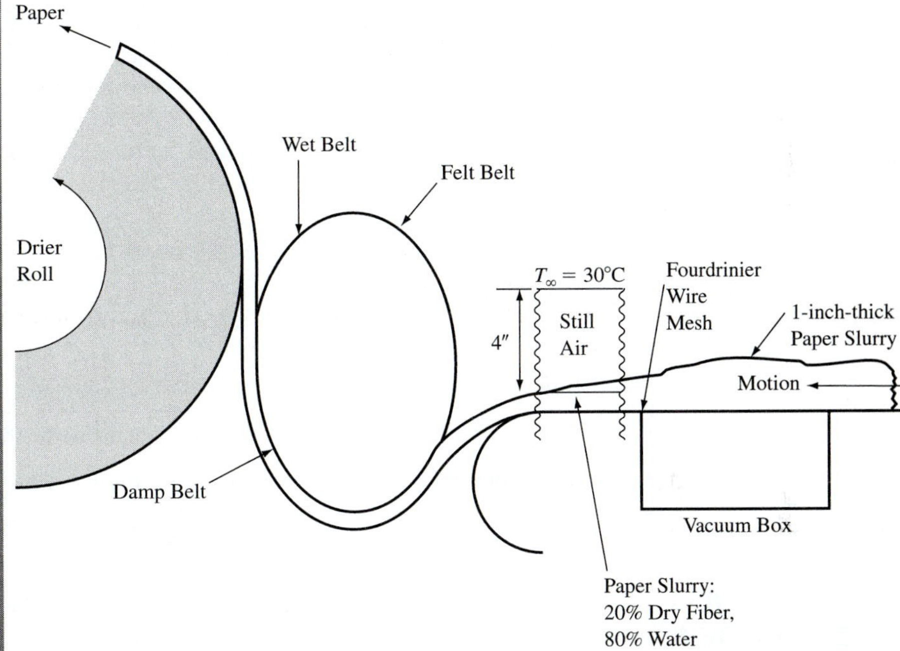

Figure 8–10 Schematic of a papermaking process.

Solution

We have

$$p_{A,1} = \beta p_{V,1} = (1.00)(4.2\ \text{kPa}) = 4200\ \text{N/m}^2$$

At state 2, the outer boundary of the stagnant air layer,

$$p_{A,2} = \beta p_{V,2} = (0.3)(4.2\ \text{kPa}) = 1260\ \text{N/m}^2$$

For water diffusing into air

$$\wp_{AB} = 0.3 \times 10^{-4}\ \text{m}^2/\text{s}$$

Then for the 5-m-wide slurry the evaporation rate of the water may be predicted from equation (8–39) with the revision that we want to determine the evaporation rate over a 5-m-wide slurry. Equation (8–39) is then

$$\dot{G}_{A,\text{total}} = (\text{width})\frac{p_T \wp_{AB}(p_{A,1} - p_{A,2})}{[R_A T(y_2 - y_1) p_{\mathrm{LM}}]}$$

and $$\dot{G}_{A,\text{total}} = (5\ \text{m})\frac{(0.3 \times 10^{-4}\ \text{m}^2/\text{s})(101{,}000\ \text{N/m}^2)(4200 - 1260\ \text{N/m}^2)}{(0.46\ \text{N}\cdot\text{m/g}\cdot\text{K})(303\ \text{K})(0.1\ \text{m})p_{\text{LM}}}$$

The log mean pressure difference is

$$p_{\text{LM}} = \frac{4200 - 1260\ \text{N/m}^2}{\ln\left(\dfrac{101 - 1.26}{101 - 4.2}\right)} = 91{,}578\ \text{N/m}^2$$

giving

Answer $$\dot{G}_{A,\text{total}} = 6.5\ \text{mg/m}^2\cdot\text{s}$$

EXAMPLE 8–6 A wood power pole 1.5 ft in diameter is treated with pentachlorophenol (PCP) such that it has 30,000 parts per million (ppm). The pole is placed in earth to a depth of 4 ft. Estimate the mass transfer of PCP to the ground 3 years later when the ground 1.5 ft from the pole has a concentration of 5000 ppm of PCP. Assume that the wood fiber is impermeable in the earth and that the earth is impermeable in the wood. Assume the diffusivity of PCP into earth is 1.076×10^{-15} ft^2/s and that the earth's density is 125 lbm/ft^3.

Solution The mass flow of PCP may be estimated from Fick's Law of Diffusion in a form of equation (8–22),

$$\dot{m}_{\text{PCP}} = -\wp_{AB}A_{\text{pole}}\rho\frac{\partial\psi_{\text{PCP}}}{\partial x}$$

$$\approx (1.076 \times 10^{-15}\ \text{ft}^2/\text{s})A_{\text{pole}}(125\ \text{lbm/ft}^3)\left(\frac{0.03 - 0.005}{1.5\ \text{ft}}\right)$$

The surface area of the pole in the ground is

$$A_{\text{pole}} = \pi DL = \pi(1.5\ \text{ft})(4\ \text{ft}) = 18.85\ \text{ft}^2$$

so the migration of PCP into the ground over the full area of the pole in the ground is

Answer $$\dot{m}_{\text{PCP}} \approx 42.26 \times 10^{-15}\ \text{lbm/s} = 1.52 \times 10^{-10}\ \text{lbm/hr}$$

8–4 CONVECTION MASS TRANSFER

There are many instances when forced and free convection, as discussed in Chapters 4 and 5, are associated with mass transfer. If the surface along which the convecting fluid interacts, for example, is wetted by a film of fluid such as water, by a solvent carrying paint pigment, or by a solvent carrying a bonding agent in glues, then the water or solvents may evaporate into the convecting fluid. Such evaporation occurs only if the vapor pressure of the water or solvent in the convecting fluid is less than the saturation pressure at the surface temperature. As we saw in section 8–3, a fluid may evaporate through a stagnant gas layer and by diffusion be carried away by some other fluid stream. The reader may recall many everyday examples of how drying processes may be increased by stagnant air changing into a convection air stream; that is, by using fans or blowers, air is moved over a wet surface to increase its drying rate. We can expect that diffusion of a fluid through a stagnant gas is less effective in the evaporative process than forced or free convection. As we saw in convection heat transfer analysis, the complications of convection phenomena are great enough that experimental methods and dimensional analysis often are the most effective methods of analysis. We may define a convective mass coefficient of A into a free stream, $h_{D,A}$ such that

$$\dot{G}_A = h_{D,A}(\rho_{A,S} - \rho_{A,\infty}) \tag{8–41}$$

or

$$\dot{N}_A = h_{D,A}(C_{A,S} - C_{A,\infty}) \quad \textbf{(8–42)}$$

where $\rho_{A,S}$ is the density of component A at a fluid or solid surface interacting with another fluid. The density $\rho_{A,\infty}$ for A occurs within the ambient conditions of the fluid. Using the concentration C_A instead of density gives the molar flow rate $\dot{N}_A$ instead of the mass flow rate $\dot{m}_A$. Comparing equation (8–41) with Fick's Law, equation (8–21),

$$\dot{G}_{A/m} = -\wp_{AB}\frac{\partial \rho_A}{\partial x} \quad \textbf{(8–21, repeated)}$$

indicates that for the approximation

$$\frac{\partial \rho_A}{\partial x} \approx \frac{\Delta \rho_A}{\Delta x}$$

we have, comparing equations (8–21) to (8–41),

$$h_{D,A} = \frac{\wp_{AB}}{\Delta x} \quad \textbf{(8–43)}$$

The units of the convective mass transfer coefficient will be length per unit of time: cm/s, ft/h, and so on.

If the conservation of mass is applied to a control volume involving one-dimensional diffusion (y direction) in a binary mixture of A and B sketched in figure 8–11, in a manner similar to the application of the momentum and conservation of energy in Chapter 4, it can be shown that

$$\frac{\partial \rho_A}{\partial t} + \mathbf{u}\frac{\partial \rho_A}{\partial x} + \mathbf{v}\frac{\partial \rho_A}{\partial y} + \mathbf{w}\frac{\partial \rho_A}{\partial z} = \wp_{AB}\frac{\partial^2 \rho^A}{\partial y^2} \quad \textbf{(8–44)}$$

where **u, v,** and **w** are the velocities in the x, y, and z directions, respectively.

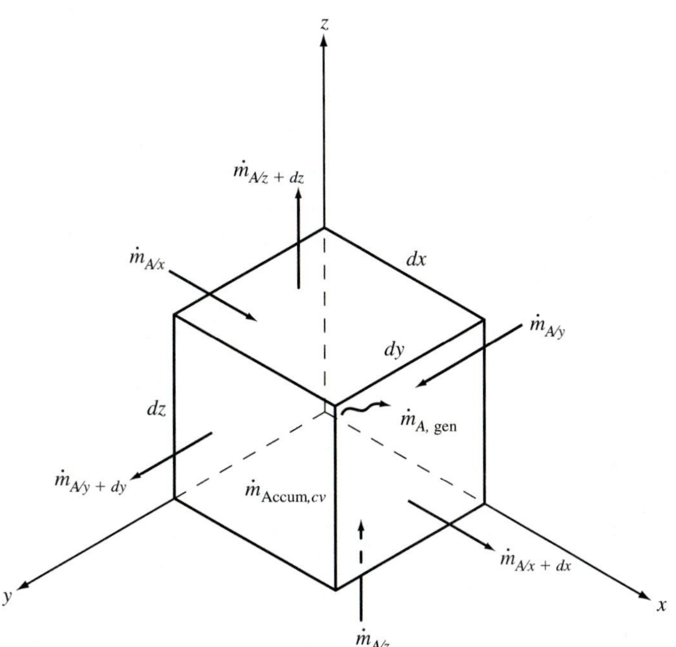

Figure 8–11 Conservation of mass for a control volume in a binary mixture.

For steady state two-dimensional (x and y) conditions this equation reduces to

$$\mathbf{u}\frac{\partial \rho_A}{\partial x} + \mathbf{v}\frac{\partial \rho_A}{\partial y} = \wp_{AB}\frac{\partial^2 \rho_A}{\partial y^2} \qquad \textbf{(8–45)}$$

and if we compare this result to that for the two-dimensional steady state boundary equations of x-direction momentum and energy conservation:

$$\mathbf{u}\frac{\partial \mathbf{u}}{\partial x} + \mathbf{v}\frac{\partial \mathbf{v}}{\partial y} = \nu\frac{\partial^2 \mathbf{u}}{\partial y^2} \quad \text{(neglecting the pressure gradient in the } x \text{ direction):} \qquad \textbf{(4–47, repeated)}$$

$$\mathbf{u}\frac{\partial T}{\partial x} + \mathbf{v}\frac{\partial T}{\partial y} = \alpha\frac{\partial^2 T}{\partial y^2} \quad \text{(revised)} \qquad \textbf{(4–58, repeated)}$$

we can see the close relationship between these three equations.

Recall from Chapter 4 that the Prandtl Number Pr relates the thermal energy to the momentum and if the Prandtl Number has a value near one (1.0), then the momentum and thermal boundary layers are comparable. Here we will relate the diffusion to the momentum through the Schmidt Number, Sc_{AB}, for substance A diffusing into substance B, defined as

$$\text{Sc}_{AB} = \frac{\mu}{\rho\wp_{AB}} = \frac{\nu}{\wp_{AB}} \qquad \textbf{(8–46)}$$

and if the Schmidt Number is near 1.0 the momentum and mass transfer boundary layers are similar. In figure 8–12 is sketched the flow of a fluid past a flat surface with the mass transfer boundary layer indicated. For a Schmidt Number of about 1.0 this boundary layer can be considered to coincide with the momentum boundary layer. Then we may recall the various equations relating the Nusselt Number to the Reynolds and Prandtl Numbers. For instance, the laminar flow ($\text{Re}_L < 5 \times 10^5$) forced convection heat-transfer equation was

$$\text{Nu}_{\text{ave}} = \frac{h_{\text{ave}}L}{\kappa} = 0.664\ \text{Re}_L^{1/2}\ \text{Pr}^{1/3} \qquad \textbf{(4–106, repeated)}$$

Applying similarity to the solution of equation (8–45) compared to equations (4–47) and (4–58), we have

$$\frac{h_{D,A}L}{\wp_{AB}} = 0.664\ \text{Re}_L^{1/2}\ \text{Sc}^{1/3} = \text{Sherwood Number} = \text{Sh} \qquad \textbf{(8–47)}$$

Thus, equation (8–47) may be used to predict the convective mass-transfer coefficient provided that $\text{Re}_L < 5 \times 10^5$ and $\text{Sc} \approx 1.0$.

For the turbulent flow conditions we may use the similarity of solutions again, using equation (4–116), and get

$$\begin{gathered}\text{Re}_L > 5 \times 10^5 \\ \text{Sc} \approx 1.0 \\ \text{Sh} = (0.037\ \text{Re}_L^{0.8} - 871)\text{Sc}^{1/3}\end{gathered} \qquad \textbf{(8–48)}$$

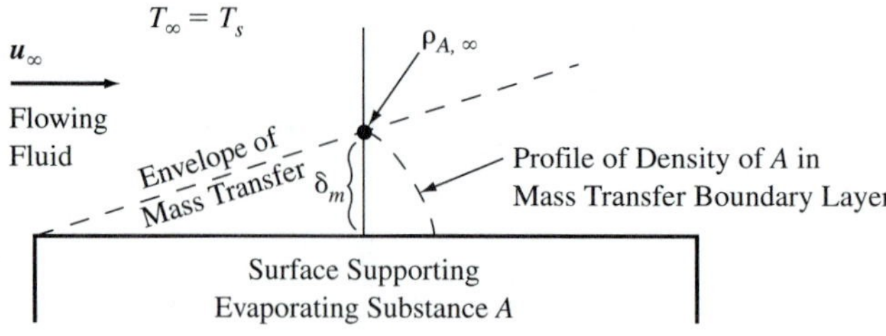

Figure 8–12 Mass transfer boundary layer.

Table 8–4 Analogous Equations for Forced Heat and Mass Transfer

Convection Mass-Transfer Configuration	Heat-Transfer Equation	Mass-Transfer Equation	Regime
Fluid flow over flat surface, average convection	(4–106)	$\text{Sh}_{\text{ave}} = 0.664\ \text{Re}_L^{0.5}\ \text{Sc}^{1/3}$ (8–47)	Laminar flow, $\text{Re} < 5 \times 10^5$, isothermal
Fluid flow over flat surface, local convection	(4–103)	$\text{Sh}_x = 0.332\ \text{Re}_x^{0.5}\ \text{Sh}^{1/3}$	Laminar flow, $\text{Re} < 5 \times 10^5$, isothermal
Fluid flow over flat surface, average convection	(4–116)	$\text{Sh}_{\text{ave}} = (0.037\ \text{Re}_L^{0.8} - 871)\text{Sc}^{1/3}$ (8–48)	Laminar–turbulent flow, isothermal
Fluid flow around round cylinder, average convection	(4–128)	$\text{Sh} = C\ \text{Re}_D^m\ \text{Sc}^{1/3}$	$\text{Re}_D > 0.4$ $\text{Re}_D < 4 \times 10^5$
Fluid flow through tubes or closed channels	(4–151)	$\text{Sh} = 0.023\ \text{Re}_D^{0.8}\ \text{Sc}^{0.4}$	Turbulent flow, $\text{Re}_D > 10{,}000$

Similarity may be applied with caution to other relationships of forced convection heat transfer for analysis of isothermal mass transfer situations.

Analogous Equations for Free Heat and Mass Transfer

Convection Mass-Transfer Configuration	Heat-Transfer Equation	Mass-Transfer Equation	Regime
Vertical flat surface, isothermal	(5–45)	$\text{Sh}_{\text{ave}} = \dfrac{0.825 + 0.387(\text{Gr}_m \cdot \text{Sc})^{1/6}}{\left[1 + \left(\dfrac{0.492}{\text{Sc}}\right)^{9/16}\right]^{8/27}}$	$\text{Gr}_m \cdot \text{Sc} \leq 10^9$
Horizontal flat surface, isothermal	(5–46)	$\text{Sh}_L = 0.54(\text{Gr}_m \cdot \text{Sc})^{1/4}$	$10^4 \leq \text{Gr}_m \cdot \text{Sc} \leq 10^7$
Horizontal flat surface, isothermal	(5–47)	$\text{Sh}_L = 0.15(\text{Gr}_m \cdot \text{Sc})^{1/4}$	$10^7 \leq \text{Gr}_m \cdot \text{Sc} \leq 10^{10}$

Similarity may be applied to other natural or free convection heat-transfer relationships and with caution to isothermal mass-transfer problems with similar geometric conditions.

For the flow of a binary mixture (components A and B) through a tube where the inner surface of the tube is wetted by one of the fluids (A) we may use a form of the Dittus–Boelter Equation (4–151) if $\mathrm{Re}_D > 2300$ and if $\mathrm{Sc} \approx 1.0$,

$$\mathrm{Sh} = 0.023\ \mathrm{Re}_D^{0.8}\ \mathrm{Sc}^{0.4} \tag{8–49}$$

to predict the convective mass-transfer coefficient. As first approximations, other results from convection heat transfer may be used with the Prandtl Number replaced by the Schmidt Number and the Nusselt Number replaced by the Sherwood Number. For instance, the results for forced convection around objects may be used to predict evaporation rates on cylindrical objects.

In free convection involving mass transfer and with isothermal conditions such that no heat transfer occurs the Grashof Number accounting for the buoyancy effects is defined as, after Gebhart et al. [6],

$$\mathrm{Gr}_m = \left(\frac{gL^3}{\nu^2}\right)\left(1 - \frac{\gamma_{m,\infty}}{\gamma_{m,S}}\right) \tag{8–50}$$

where $\gamma_{m,\infty}$ is the specific weight of the mixture at the ambient conditions of the fluid and $\gamma_{m,S}$ is the specific weight at the whetted surface. For a binary mixture of A and B these specific weights are

$$\gamma_{m,\infty} = \gamma_{A,\infty} + \gamma_{B,\infty} \tag{8–51}$$

and

$$\gamma_{m,S} = \gamma_{A,S} + \gamma_{B,S} \tag{8–52}$$

The free convection heat transfer results may be used as first approximations for predicting mass transfers in free convection situations, subject to replacing the Grashof Number Gr by the mass-transfer Grashof Number Gr_m, equation (8–50), the Prandtl Number by the Schmidt Number, and the Nusselt Number by the Sherwood Number. Those situations where both mass and heat transfer occur are beyond our purposes but are treated in advanced texts, such as Gebhart et al. [6]. In table 8–4 are listed the various analogous equations of forced and free convection for heat and mass transfer.

EXAMPLE 8–7

Predict the rate of evaporation of water from the paper in example 8–5 if air at 30°C and 30% relative humidity is blown over the paper surface at 1.5 m/s as shown in figure 8–13.

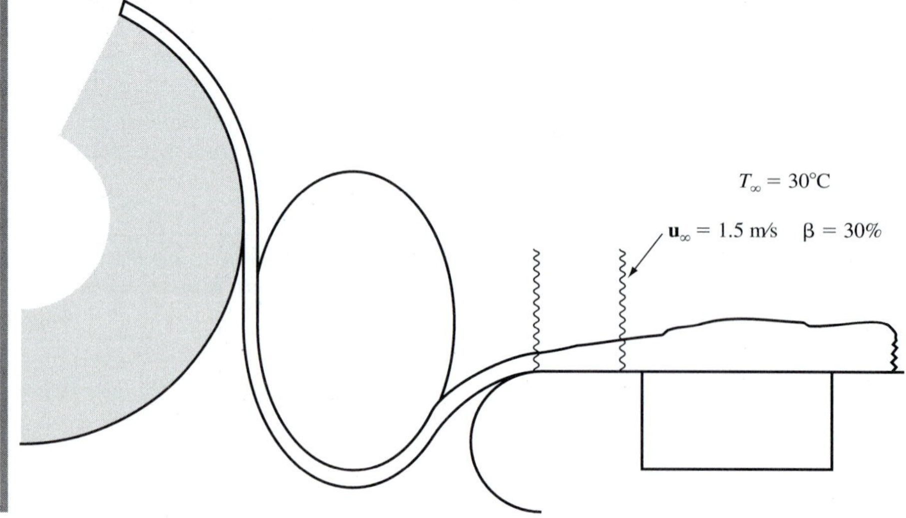

Figure 8–13 Forced convection drying of paper slurry.

Solution

In Example 8–5 we assumed that the evaporation occurred through a stagnant 10-cm-thick air column and was driven by the concentration gradient through the column. Here we will estimate the forced convection mass transfer. The Reynolds number is

$$\text{Re}_L = \frac{\rho_{\text{air}}\mathbf{u}_\infty L}{\mu} = \frac{\mathbf{u}_\infty L}{\nu} = \frac{(1.5 \text{ m/s})(5 \text{ m})}{17 \times 10^{-6} \text{ m}^2/\text{s}} = 4.4 \times 10^5 \qquad \text{(laminar or transition)}$$

We will use equation (8–47) to predict the convective mass-transfer coefficient:

$$\text{Sh}_L = 0.664 \text{ Re}_L^{0.5} \text{ Sc}^{1/3}$$

For water vapor (A) into air (B) we use for the mass diffusivity

$$\wp_{AB} = 0.3 \times 10^{-4} \text{ m}^2/\text{s}$$

so that

$$\text{Sc} = \frac{\nu}{\wp_{AB}} = \frac{17 \times 10^{-6} \text{ m}^2/\text{s}}{0.3 \times 10^{-4} \text{ m}^2/\text{s}} = 0.567$$

which is, as a rough approximation, close to a value of one. Then

$$h_D = \left(\frac{\wp_{AB}}{L}\right)(0.664 \text{ Re}_L^{0.5} \text{ Sc}^{1/3}) = \frac{0.3 \times 10^{-4} \text{ m}^2/\text{s}}{5 \text{ m}}[0.664(4.4 \times 10^5)^{0.5}(0.567)^{1/3}]$$
$$= 21.87 \text{ m/s}$$

The rate of evaporation is then predicted from equation (8–41)

$$\dot{G}_{\text{water}} = h_D(\text{width})(\rho_{A,S} - \rho_{A,\infty})$$

The density of water at the surface will be that of a saturated vapor,

$$\rho_{A,S} = \rho_g = \frac{1}{\nu_g} = \frac{1}{32.9 \text{ m}^3/\text{kg}} = 0.0304 \text{ kg/m}^3$$

The density of the water vapor in the air is

$$\rho_{A,\infty} = \frac{p_{A,\infty}}{R_A T}$$

The vapor pressure of the water vapor in the air is

$$p_{A,\infty} = \beta p_{g@30°\text{C}} = (0.3)(4200 \text{ Pa}) = 1260 \text{ Pa}$$

and the density is then

$$\rho_{A,\infty} = \frac{1260 \text{ N/m}^2}{(460 \text{ N} \cdot \text{m/kg} \cdot \text{K})(303 \text{ K})} = 0.00904 \text{ kg/m}^3$$

thus, we have for the evaporation rate,

Answer

$$\dot{G}_{\text{water}} = (21.87 \times 10^{-4} \text{ m/s})(5 \text{ m})(0.0304 - 0.00904 \text{ kg/m}^3) = 0.23 \text{ g/m} \cdot \text{s}$$

Notice that this evaporation rate is nearly one order of magnitude greater than the evaporation rate predicted by using a stagnant layer of air above the slurry, as assumed in example 8–5.

EXAMPLE 8–8

Isopropyl alcohol at 30°C is rubbed on an arm and then held horizontally in still air. If the arm is 10 cm in diameter, estimate the evaporation rate per unit length of the arm in g/m · s.

Solution

We may use the analogy of free convection heat transfer in the form

$$\text{Sh} = D(\text{Gr}_m \cdot \text{Sc})^n$$

From table 8–3 we read

$$\wp_{AB} = 0.101 \text{ cm}^2/\text{s} = 10.1 \times 10^{-6} \text{ m}^2/\text{s}$$

The Schmidt Number is then

$$\text{Sc} = \frac{\nu}{\wp_{AB}} = \frac{15.89 \times 10^{-6} \text{ m}^2/\text{s}}{10.1 \times 10^{-6} \text{ m}^2/\text{s}} = 1.573$$

The Grashof Number is

$$\text{Gr}_m = \left(\frac{gD^3}{\nu^2}\right)\left(1 - \frac{\gamma_{m,\infty}}{\gamma_{m,s}}\right)$$

The density of isopropyl alcohol may be taken as

$$\rho_{A,S} = 1.44 \text{ kg/m}^3$$

so that the specific weight is

$$\gamma_{A,S} = \rho_{A,S}g = (9.8 \text{ m/s}^2)(1.44 \text{ kg/m}^3) = 14.1 \text{ N/m}^3$$

The specific weight of air,

$$\gamma_{B,\infty} = g\rho_{B,\infty} = \frac{gp_{\text{air}}}{R_{\text{air}}T} = \frac{(9.8 \text{ m/s}^2)(101 \text{ kPa})}{(0.287 \text{ kJ/kg} \cdot \text{K})(303 \text{ K})} = 11.38 \text{ N/m}^3$$

If we assume that the isopropyl alcohol has a mass fraction of 1 (100%) at the arm's surface and is zero (0%) in the air, the mixture specific weights are, from equation (8–42)

$$\gamma_{m,S} = \gamma_{A,S} = 14.1 \text{ N/m}^3$$

and

$$\gamma_{m,\infty} = \gamma_{B,\infty} = 11.38 \text{ N/m}^3$$

Then

$$\text{Gr}_m = \frac{(9.8 \text{ m/s}^2)(0.1 \text{ m})^3}{(15.89 \text{ m}^2/\text{s})^2}\left(1 - \frac{11.38}{14.1}\right) = 7.48 \times 10^6$$

Using an analogy for free convection

$$\text{Gr}_m \cdot \text{Sc} = \text{Ra}_m = 11.78 \times 10^6$$

If we interpret this as an equivalent Rayleigh Number, then we may use equation (5–50) with constants obtained from table 5–2. For a Rayleigh Number of 11.78×10^6, from table 5–2, we read $C = 0.125$ and $n = 0.333$. Then

$$\text{Sh} = \frac{h_D D}{\wp_{AB}} = C\text{Ra}_m^n = 0.125 \text{ Ra}_m^{0.333} = 28.4$$

and

$$h_D = \text{Sh}\left(\frac{\wp_{AB}}{D}\right) = 28.4\left(\frac{10.1 \times 10^{-6} \text{ m}^2/\text{s}}{0.1 \text{ m}}\right) = 0.002868 \text{ m/s}$$

The evaporation rate may then be approximated as

$$\dot{G}_{\text{alcohol}} = h_D(\rho_{m,S} - \rho_{m,\infty}) = (0.002868 \text{ m/s})\left(1.44\frac{\text{kg}}{\text{m}^3} - 0\right) = 4.13\frac{\text{g}}{\text{s} \cdot \text{m}^2}$$

If we assume that evaporation occurs around the arm, then the evaporation rate per length of arm length is

Answer
$$\dot{m}_{\text{alcohol},l} = \pi D G_{\text{alcohol}} = \pi(0.1 \text{ m})(4.13) \text{ g/s} \cdot \text{m}^2 = 1.3 \text{ g/s} \cdot \text{m}$$

8–5 TRANSIENT DIFFUSION

Mass diffusion is seldom a steady state phenomenon; however, due to the lack of reliable information the steady state conditions are used as first approximations to actual conditions. For transient mass transfer one needs to begin with the mass balance of each of the components of the mixture. Thus, we need to write the mass balance for each component A, as in equation (8–44). For one-dimensional transient mass transfer (y direction) in a stationary binary medium (or binary mixture) we have, from equation (8–44), for A

$$\frac{\partial \rho_A}{\partial t} = \wp_{AB} \frac{\partial^2 \rho_A}{\partial y^2} \tag{8–53}$$

and for B

$$\frac{\partial \rho_B}{\partial t} = \wp_{AB} \frac{\partial^2 \rho_B}{\partial y^2} \tag{8–54}$$

If we compare these equations to that for one-dimensional transient conduction heat transfer, equation (4–10),

$$\frac{\partial T}{\partial t} = \alpha \frac{\partial^2 T}{\partial y^2}$$

we can see the similarity of the equations. The mass diffusivity $\wp_{AB}$ is the parameter serving the same purpose as the thermal diffusivity, α. Then the transient mass-transfer situations may be analyzed using methods similar to that used for transient conduction heat transfer. We need to be able to identify boundary and initial conditions so that an analysis can proceed.

The typical situation of transient mass transfer for a binary mixture is shown in the schematic of figure 8–14, where a mixture of A and B has a boundary with another mixture of diluted A or B. The diffusion mechanism will normally be one where the component A transfers to the diluted A mixture from the stronger mixture, as indicated in figure 8–14. A complicating feature of these processes is that the diffusivity $\wp_{AB}$ will normally not be a constant for changes in concentrations or densities so that equations (8–53) and (8–54) will be nonlinear. As first approximations, however, one may use two diffusivities, one for each of the two mixtures, that are each constants.

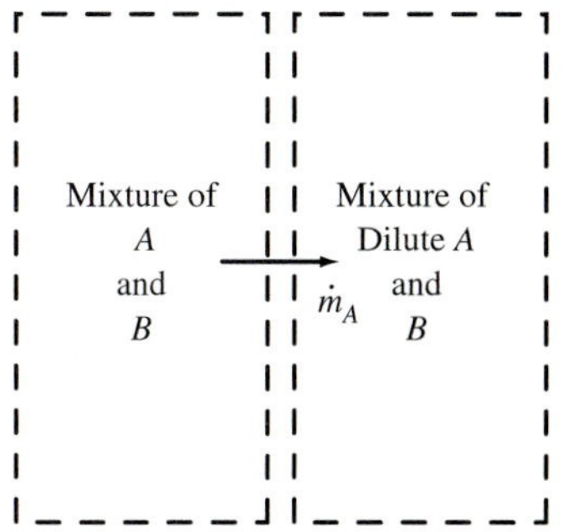

Figure 8–14 Two binary mixtures of A and B.

As a comparison to the transient heat-transfer situations, where the boundary temperature is constant with time, one of the two mixtures may be assumed to be much larger than the other mixture and have an extremely large diffusivity such that its concentrations remain constant with time.

Assuming a constant diffusion rate with time at the boundary is another boundary condition for transient mass transfer, which is analogous to constant heat transfer at a boundary of a transient conduction heat-transfer problem. Further, an analogous problem of the lumped heat capacity for lumped mass transfer can be analyzed if the dilute mixture (or the strong mixture) has a very large diffusivity in relation to the convective mass-transfer between the two mixtures. In the analysis of isothermal transient mass-transfer problems the developments of transient heat transfer may be used if the Biot Number is replaced by the definition for the mass-transfer Biot Number

$$\mathrm{Bi}_m = \frac{h_{D,A}L}{\wp_{\mathrm{self}}} \tag{8–55}$$

and the Fourier Number by the mass-transfer Fourier Number

$$\mathrm{Fo}_m = \wp_{\mathrm{self}}\frac{t}{L^2} \tag{8–56}$$

Notice that the diffusivity is that for self-diffusion or for diffusion through a strong mixture (as opposed to a dilute mixture).

Lumped Mass Capacity System

If we consider a binary mixture, A and B, bounded by another large surrounding binary mixture of A and B, as shown in figure 8–15, such that the Biot Number of the mixture is less than about 0.1, then the density of mixture component A may be described by

$$\rho_A = \rho_{A,\infty} + (\rho_{A,i} - \rho_{A,\infty})e^{-h_D A t/V} \tag{8–57}$$

where h_D is the convective mass transfer coefficient in the surrounding mixture at the boundary, A is the surface of the mixture, and V is its volume. The time is t and the mass-transfer time constant for the mixture is the product

$$t_{\mathrm{tcm}} = \frac{V}{Ah_D} \tag{8–58}$$

This result describes the physical situation where a small region in a binary mixture can be considered as an imbalanced mixture within a large mixture that equilibrates through mass transfer after a period of time. Notice that the lumped mass capacity analysis requires infinite time to equilibrate; however, the density of A is close to that of its density in the surrounding mixture after two or three time constants.

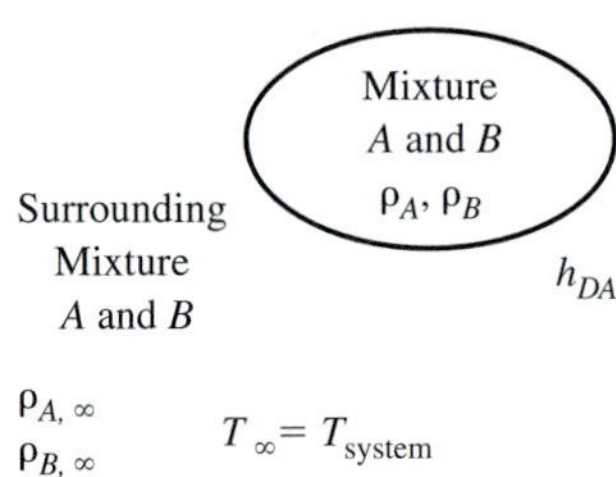

Figure 8–15 Lumped mass capacity system.

EXAMPLE 8–9 Spherical drops of yellow water color are put in water. The drops are 4 mm in diameter and the convective mass-transfer coefficient is expected to be 2 mm/s. Determine the time constant for mixing of the color in water.

Solution Using equation (8–58) and the standard volume and surface areas for a sphere, we obtain

Answer

$$t_{tcm} = \frac{\frac{4}{3}\pi r^3}{4\pi r^2 h_D} = \frac{r}{3h_D} = \frac{2\text{ mm}}{3(2\text{ mm/s})} = 0.333\text{ s}$$

Semi-Infinite Binary Mixtures

The problem of a semi-infinite binary mixture of components A and B has been discussed and illustrated in figure 8–14. Three boundary conditions can be considered that provide an analogy to the transient conduction in a semi-infinite medium, all with the initial condition that $\rho_A = \rho_{A,i}$ for all $x \geq 0$, $t = 0$:

Case I. $\rho_A = \text{constant} = \rho_O$ at $x = 0$, $t > 0$, as delineated in figure 8–16.
Case II. $\dot{G} = C$ (some constant) at $x = 0$, $t > 0$.
Case III. $\dot{G} = h_D(\rho_{A,s} - \rho_{A,\infty})$ at $x = 0$, $t > 0$.

These three cases may be considered as similar to the cases delineated in Chapter 3: case I for constant temperature at the boundary with the density of component A in the semi-infinite region predicted, after the results from heat transfer in Chapter 3, from

$$\rho_A(t) = \rho_C + (\rho_{A,i} - \rho_C)\text{erf}\left(\frac{x}{2\sqrt{\wp_{AB}t}}\right) \tag{8–59}$$

Case II for constant heat transfer at the boundary, and case III for convection boundary condition may be treated with the developments of those given in Chapter 3 for heat transfer.

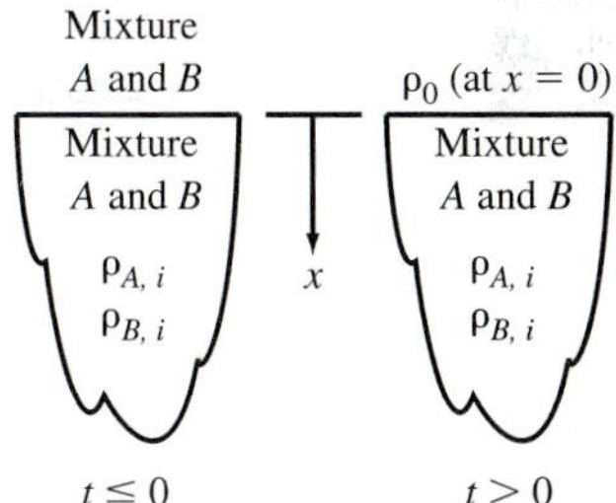

Figure 8–16 Semi-infinite binary mixture with sudden change in density of component A at the boundary.

Infinite Plate, Infinite Cylinder, Sphere with Convection Boundary Condition

The problem of an infinite plate that represents a binary mixture which is then surrounded on its two boundary surfaces by a convective mass transfer as shown in figure 8–17 can be analyzed with the results presented in the Heisler Chart, figure 3–12, using the mass-transfer Biot and Fourier Numbers defined in equations (8–55) and (8–56) in place of the customary Biot and Fourier Numbers.

If the binary mixture volume is better described by an infinite cylinder with convection boundary conditions, the Heisler chart of figure 3–15 can be used to analyze the mass transfer. Finally, if the mixture occupies a spherical region with a convective mass transfer at its surface, then figure 3–17 can be used to analyze the mass transfer through the region.

Figure 8–17 Binary mixture in an infinite plate region.

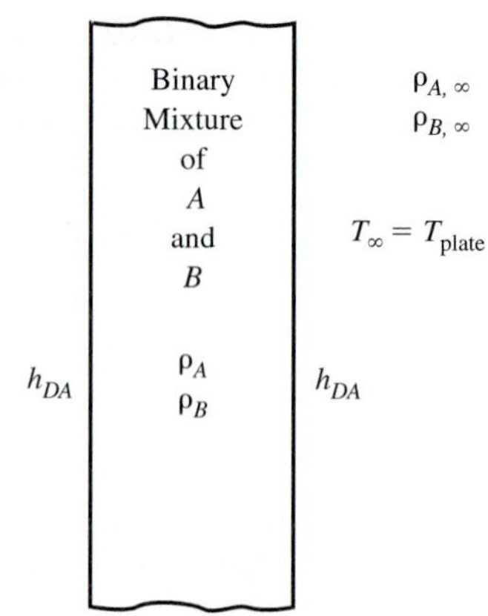

EXAMPLE 8–10 Steel roller bearing rollers 10 mm in diameter are to be carburized at 900°C in a concentrated carbon gas, as shown in figure 8–18. The steel rollers are made of a steel having 0.2% carbon, and the carburization is intended to increase the percentage to 0.5% at a depth of 0.4 mm into the surface. Estimate the time required to accomplish the carburizing process. Assume $\rho_{\text{steel}} = 7850$ kg/m^3 for the steel roller material, and for the carbon gas, $\rho_\infty = 100$ kg/m^3. The diffusivity of carbon in steel may be taken as 2.0×10^{-11} m^2/s.

Figure 8–18 Steel rollers for carburizing.

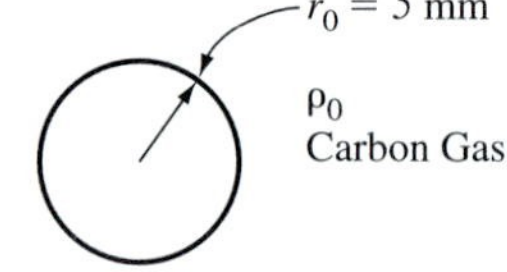

Solution If we assume that carbon concentration at the surface is constant such that the boundary condition is

$$\rho_{\text{carbon}} = 100 \text{ kg/m}^3 \qquad \text{at } x = 0,\ t > 0$$

and that the rollers are large enough to be considered as a semi-infinite medium, then we use case 1 with equation (8–59). For this situation,

$$\begin{aligned} \rho_{A,i} &= (0.2\%)(7850 \text{ kg/m}^3) = 15.7 \text{ kg/m}^3 \\ \rho_C &= 100 \text{ kg/m}^3 \\ \rho_A &= (0.5\%)(7850 \text{ kg/m}^3) = 39.25 \text{ kg/m}^3 \end{aligned}$$

and, from equation (8–59),

$$\rho_A = \rho_{\text{carbon}} + (\rho_{A,i} - \rho_C)\text{erf}(0.0004 \text{ m}/2\sqrt{\wp_{AB}t})$$

or

$$\text{erf}\left(\frac{0.0004 \text{ m}}{2\sqrt{2.0 \times 10^{-11} \text{ m}^2/\text{s}\ t}}\right) = 0.72064$$

From appendix table A–9–1 the argument of the error function is

$$0.0004 \text{ m}/(2\sqrt{(2.0 \times 10^{-11} \text{ m}^2/\text{s})t} = 0.775$$

and the time is

Answer

$$t = 3508.6 \text{ s} = 58.5 \text{ min}$$

8–6 ABSORPTION AND ADSORPTION

We have considered binary mixtures and how one of the mixture components may diffuse across the boundary of the mixture system to change the composition of that mixture. There are increasingly more applications where a component *A* diffuses from a mixture of *A* and *B* to a mixture of *A* and *C*, where we represent *B* and *C* as different components of the separate mixtures. This sort of process is shown schematically in figure 8–19. There are many ways in which a component *A* can leave a binary mixture and associate with another mixture of a different component. We will discuss two of these mechanisms: absorption and adsorption. These two mechanisms are important in many engineering applications such as humidifying or dehumidifying air, removing odors in air, eliminating undesirable gases from products of combustion such as in catalytic converters, or in some chemical separation processes. The mechanism of absorption is used in absorption refrigeration systems, which require significantly less mechanical or electrical power input than do the conventional mechanical vapor compression refrigeration systems. With appropriate design of such absorption systems, they can be operated entirely on heat input such as solar energy and completely eliminate the need for electricity.

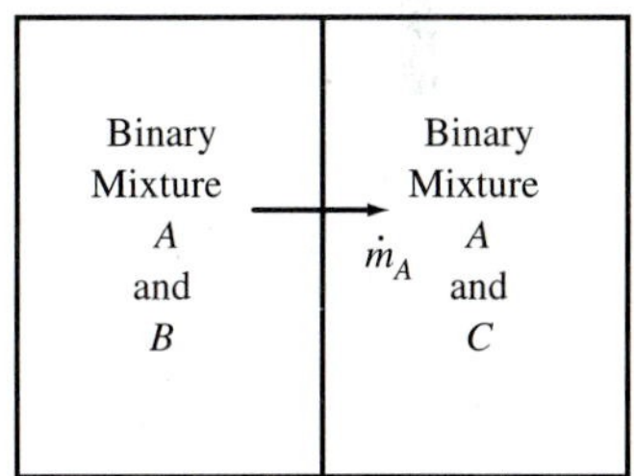

Figure 8–19 Schematic of absorption or adsorption.

Absorption is a mechanism whereby a substance *A* chemically changes a mixture of *A* and *C* when it is absorbed into the mixture. An example of an absorption process is the absorption of water in sodium chloride (salt, NaCl). Sodium chloride is a solid, but when water is absorbed in it, the salt becomes a liquid. Another example is the absorption of oxygen in water to cause the chemical composition of water to change to a higher concentration of oxygen. Adsorption is a mechanism whereby a substance *A* is integrated into another substance *C* without causing it (*C*) to change phase or otherwise change in chemical composition. A sponge adsorbs water, which does not alter the sponge except to increase its total weight or mass. Adsorption is often described as an action of capillary attraction, surface tensions, or weak intermolecular attraction forces that draw molecules onto the adsorbing surfaces.

Materials that absorb or adsorb other materials are called *sorbents,* and a particular type of sorbent that absorbs or adsorbs water is called a *desiccant*. The absorption or adsorption of liquids or vapors may be considered as diffusion processes driven by surface vapor pressure differences. If a liquid or vapor has a condition of lower vapor pressure at the surface of the sorbent than it has outside the sorbent, diffusion of that fluid will occur and drive the fluid into the sorbent volume, to be absorbed and adsorbed. Sometimes the term *sorbed* is used to describe the action if it is not clear whether the process is absorption or adsorption. The absorption and adsorption of solids is much more complex and is beyond the purposes of this text.

A common and important application of sorption is the humidification and/or dehumidification of air. *A* large amount of attention has been devoted to such processes, and much information is available. Most desiccants that rely on absorption to transfer water to or from air are liquids. Lithium chloride and triethylene glycol are two such liquid desiccants. When in a mixture with water they exhibit surface water vapor pressures that are appropriate to conveniently create conditions for water diffusion.

Figures 8–20 and 8–21 show the behavior of surface water vapor pressure of these two desiccants as the vapor pressure is affected by temperature and percent of desiccant (or amount of water in the sorbent). Notice that the surface vapor pressure decreases with increasing temperature and increases with increasing percentage of desiccant. This data is representative of equilibrium conditions so that an analysis of mass transfer requires information regarding how fast the desiccant will reach that equilibrium. For instance, referring to figure 8–20, a 25% lithium chloride mixture with water at 30°C will eventually absorb enough water so that the vapor pressure of the water in the air surrounding the mixture will be 2.1 kPa. The rate at which this occurs depends on the physical configuration of the system and how much surface area can be in contact with the air. If the liquid desiccant is sprayed into an air stream at high velocity then the desiccant can absorb water rapidly.

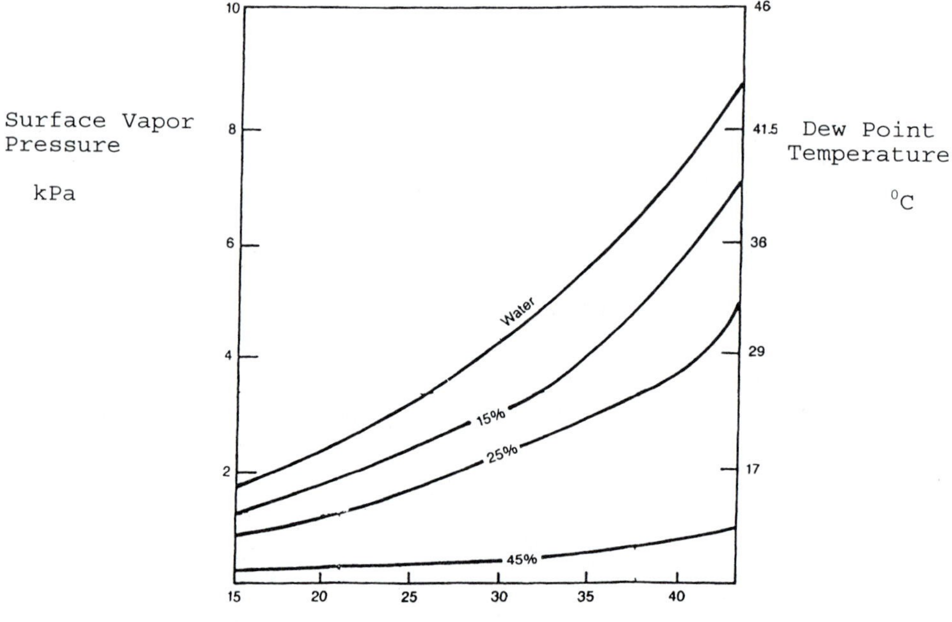

Figure 8–20 Vapor pressure of water–lithium chloride solutions. (Revised from Cyprus Foote [7])

A complicating feature of all absorption or adsorption processes is that the water, when absorbed by the desiccant, will release its latent heat of vaporization. As a consequence of this action the desiccant will increase in temperature. The air can also increase its temperature from this energy release; however, to simplify the analysis, it is usually assumed that the desiccant receives all of this energy.

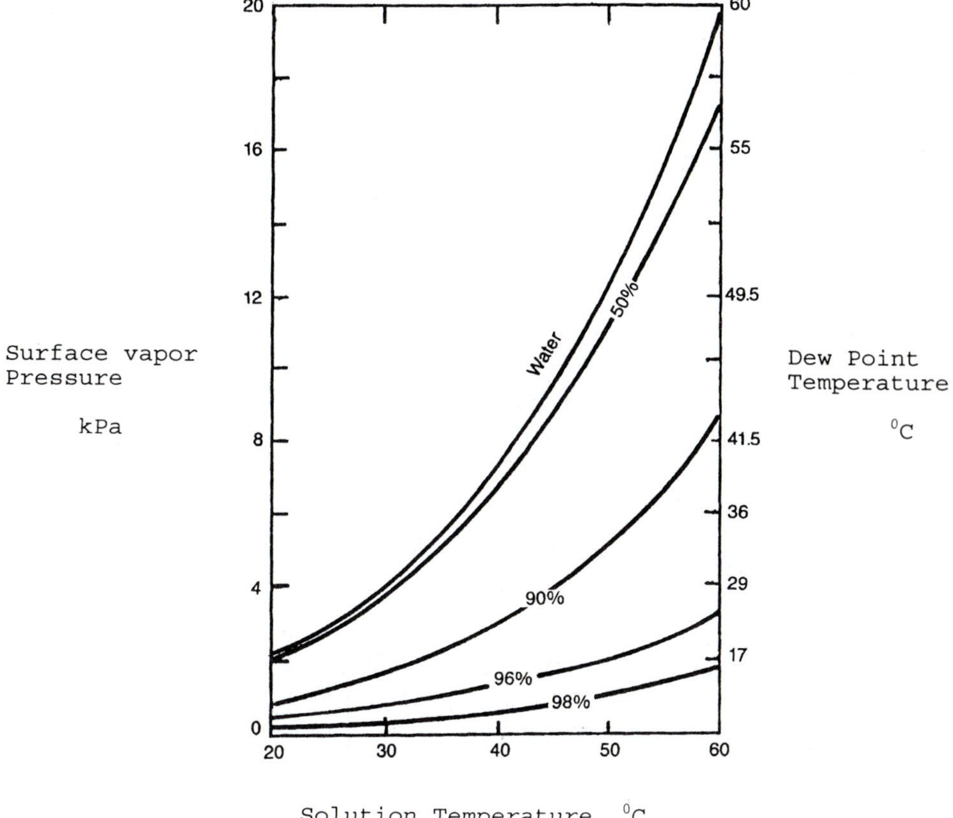

Figure 8–21 Vapor pressure of water–triethelene glycol solutions. (Revised from Dow [8])

If a desiccant is to be used in a continuous process, it needs to have the absorbed (or adsorbed) water removed at some time; otherwise it will become saturated with water and not function as a desiccant. A typical desiccant cycle is shown in figure 8–22, where a "dry" desiccant at state 1 begins to absorb water, following the sorption path from 1 to 2. At state 2 the desiccant is assumed to have reached equilibrium with the surrounding air such that p_{v2} = water vapor pressure in the air = surface vapor pressure of the desiccant. The desiccant is then heated to state 3 without changing the moisture composition in the

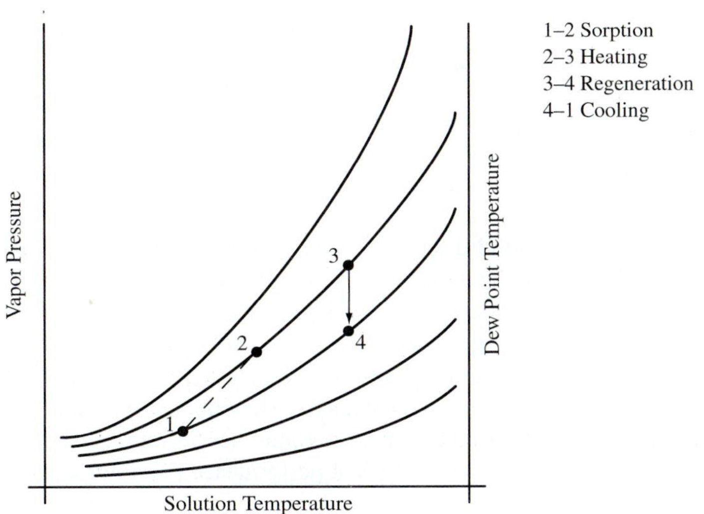

Figure 8–22 Typical desiccant cycle.

desiccant. The vapor pressure at state 3 needs to be high enough that the water can be removed by dry air having a vapor pressure of p_{v4} in a regenerator. The desiccant is then cooled at constant composition to state 1 to continue with another cycle. The desiccant acts as an absorber during process 1–2, and all of the water is driven out of it in the regenerator. Adsorbents that are used as desiccants are usually solids. Some effective solid desiccants are silica gels, zeolites, molecular sieves, activated aluminas, carbon, and some synthetic polymers. Solid desiccants are often formed into circular wheels and rotated to provide a steady state operating condition in which each section of the wheel is continuously passing through various phases of the cycle, described in figure 8–20 for a liquid desiccant.

In figure 8–23 is shown an arrangement of a desiccant wheel. Warm humid air passes through the adsorption portion of the device, where the surface vapor pressure is low enough to adsorb water from the passing air. As the desiccant wheel rotates, it enters the regenerator section, where hot dry air is passed through the desiccant wheel and thereby removes the adsorbed water. The wheel continues to rotate and complete its cycle to dehumidify over hot humid air. Many of these solid desiccants are used to provide dehumidification and air conditioning of buildings. An appropriate application of these systems has been in supermarket dehumidifying, where conventional mechanical vapor compression systems provide air that is too cool and uncomfortable near food storage units [9].

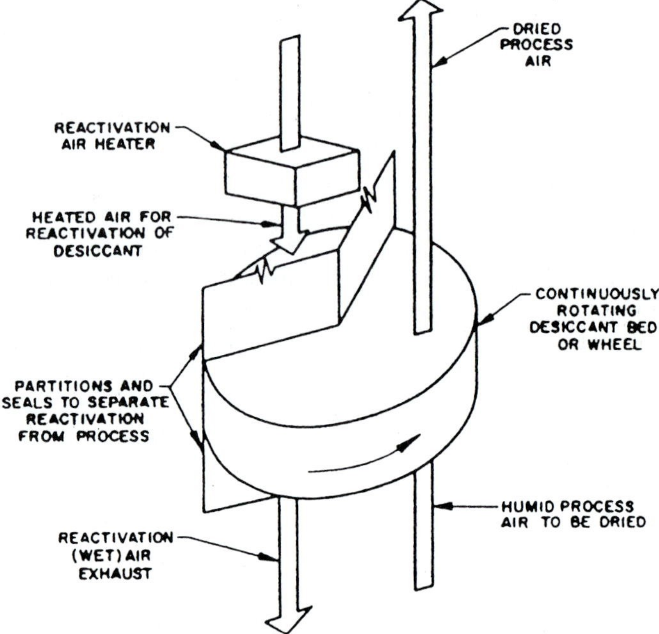

Figure 8–23 Typical rotary solid desiccant dehumidifier. (From ASHRAE [10])

The solid desiccant rotary dehumidifier generally operates at 0.5 to 6 revolutions per hour, with the desiccant adsorbing up to a kilogram of water per kilogram of desiccant during each revolution. Since the latent heat of vaporization of water is released into the desiccant when the water is removed from the air, the desiccant tends to become warmer. However these desiccants act like packed bed heat exchangers and, as we saw in Chapter 4, are effective heat exchangers. The dried air will thus be warmer, and it is difficult to predict the temperatures of these rotary desiccants. Empirical determinations must be used, and figure 8–24 shows a typical performance curve for a rotary solid sorption dehumidifier.

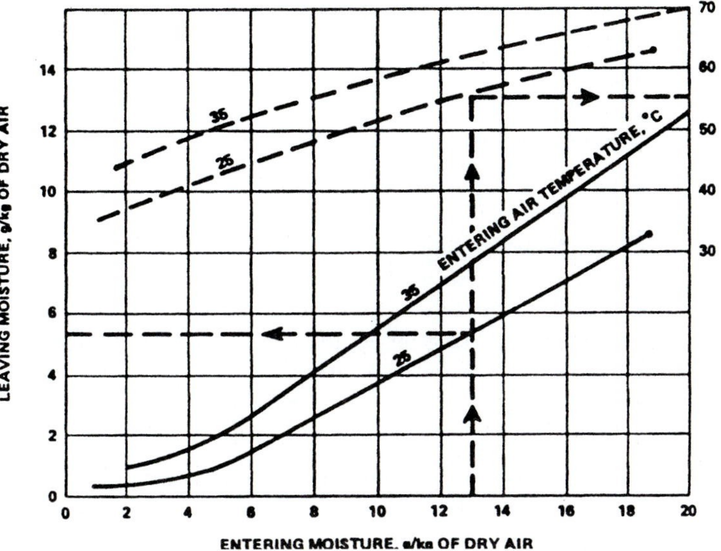

Figure 8–24 Typical performance data for rotary solid sorption dehumidifer. The dashed line indicates incoming air with 13 g/kg moisture, and for 25°C incoming air temperature the leaving air will be at 55°C and with 5.3 g/kg moisture. (From ASHRAE [10])

EXAMPLE 8–11 Three hundred cfm of 90°F 80% relative humidity air is to be conditioned to 80°F, 40% relative humidity. It is proposed to use 40% lithium chloride with water at 64°F to accomplish this process. Estimate the amount of solution needed, based on an assumption that 25% more solution will be sprayed into the air stream, as shown in figure 8–25, to assure that the desiccant properly absorbs the necessary water from the air. Also determine if a heating or a cooling coil is needed in the system so that the conditioned air is at the proper temperature.

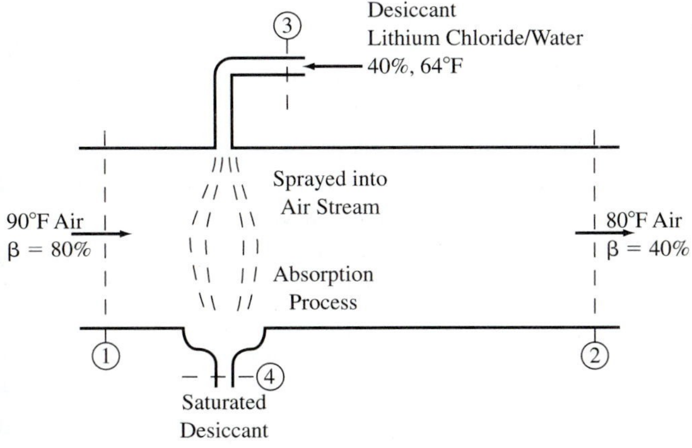

Figure 8–25 Lithium chloride desiccant system to condition air.

Solution From the psychrometric chart C–1 in appendix C we see that the absolute humidity of the entering air is 172 grains/lbm dry air, which is equal to 0.02438 lbm water/lbm dry air. The vapor pressure is 0.7275 psia for the water at 80% relative humidity. The air leaving the conditioner must have a humidity of 61 grains/lbm dry air or 0.0087 lbm water/lbm dry air. The vapor pressure of the conditioned air should be $(0.4)(0.506 \text{ psia}) = 0.2024$ psia. The 40% lithium chloride as given in figure 8–20 has a surface vapor pressure for water of 0.5 kPa $= 0.07252$ psia at 64°F, so the desiccant is capable of absorbing the water from air. If we assume that all of the air is used to heat up the desiccant we may write the energy balances such that, for the desiccant stream,

$$\dot{m}_{\text{des},1}h_1 + \dot{m}_{\text{da}}(\omega_1 - \omega_2)h_{\text{fg}} = \dot{m}_{\text{des},2}h_2$$

and for the air,

$$\dot{m}_{\text{da}}h_1 = \dot{m}_{\text{da}}h_2$$

and this gives

$$h_1 = c_pT_1 + \omega_1 h_{g,1} = h_2 = c_pT_2 + \omega_2 h_{g,2}$$

We can then solve for T_2, the conditioned air temperature if it has no heat transfer:

$$T_2 = T_1 + \frac{1}{c_p}(\omega_1 h_{g,1} - \omega_2 h_{g,2})$$

If an analytic expression is used to describe the relationship between $h_{g,2}$ and T_2, then a direct solution for T_2 can be obtained; otherwise an iterative method must be employed: Assuming $T_2 = 120°\text{F}$ gives $h_{g,2} = 1114$ Btu/lbm from a steam table. Then, using the preceding equation with the values $T_1 = 90°\text{F}$, $h_{g,1} = 1100$ Btu/lbm, and $c_p =$ 0.24 Btu/lbm · °F, gives

$$T_2 = 90°\text{F} + \frac{[(0.2438\ \text{lbm}_{\text{water}}/\text{lbm}_{\text{da}})(1100\ \text{Btu/lbm}_{\text{water}}) - (0.0087)(1131)]}{0.24\ \text{Btu/lbm}\cdot°\text{F}} = 163°\text{F}$$

This result does not agree with the assumed value for T_2, so we try $T_1 = 161°\text{F}$. Here $h_{g,1} = 1131$ Btu/lbm and

$$T_2 = 90 + \frac{1}{0.24}[(0.2438)(1100) - (0.0087)(1131)] = 161°\text{F}$$

This value agrees with its assumed value, so the air temperature will be 161°F leaving the desiccant because of the water removal. The air needs to be cooled, and a cooling coil needs to be included in the device to cool the air to 80°F. The desiccant temperature will increase because of the absorption of the water and its latent heat of vaporization. This absorption process also will decrease the percentage of lithium chloride in the desiccant, and the equilibrium condition will be reached when the vapor pressure of the desiccant is 0.204 psia, or approximately 1.4 kPa.

The mixture specific heat is

$$c_{p,m} = \psi_w c_{p,w} + \omega_{\text{lc}} c_{p,\text{lc}}$$

the subscript notation w refers to the water in the water–lithium chloride mixture, and lc refers to the lithium chloride. For water as a liquid $c_p = 4.186$ kJ/kg · K, and for lithium chloride $c_p = 1.13$ kJ/kg · K. For the 40% lithium chloride mixture we have

$$\begin{aligned} c_{p,m} &= (0.6)(4.186\ \text{kJ/kg}\cdot\text{K}) + (0.4)(1.13\ \text{kJ/kg}\cdot\text{K}) = 2.96\ \text{kJ/kg}\cdot\text{K} \\ &= 0.708\ \text{Btu/lbm}\cdot°\text{F} \end{aligned}$$

The mass balance for the desiccant flow stream is, referring to figure 8–25,

$$\dot{m}_{\text{des},3} + \dot{m}_{\text{da}}(\omega_1 - \omega_2) = \dot{m}_{\text{des},4}$$

which reduces to

$$\dot{m}_{\text{des},4} = \dot{m}_{\text{des},3} + 1.159\ \text{lbm/s}$$

Thus, the mass fraction of desiccant after absorbing water is

$$\psi_{\text{des},4} = \dot{m}_{\text{des},3}\frac{0.4}{\dot{m}_{\text{des},3} + 1.159}$$

Rewriting the energy balance for the desiccant stream,

$$\dot{m}_{\text{des},3}c_{p,3}T_3 + (1.159\text{ lbm/s})(1042\text{ Btu/lbm}) = (\dot{m}_{\text{des},3} + 1.159)c_{p,4}T_4$$

where the specific heats $c_{p,3}$ and $c_{p,4}$ are determined for the mixture. Using these two equations we again iterate to determine the mass flow rate of desiccant mixture, $\dot{m}_{\text{des},3}$, noting that the mass fraction, the exit temperature of the desiccant, T_4, and the specific heats are affected by the desiccant mixture flow rate. We find that the mass flow rate is

Answer $$\dot{m}_{\text{des},3} = 25\text{ lbm/s}$$

the mass fraction is

$$\psi_{\text{des},4} = 0.38 = 38\%$$

and the exit temperature is

Answer $$T_4 = 120°\text{F}$$

8–7 PERMEABILITY

In the discussion of mass transfer we have assumed in many instances that a fluid or vapor diffused, was absorbed, or was adsorbed into another mixture. Here we will analyze how mass may pass through another substance. In figure 8–26 is sketched the transfer of mass through another substance, driven by a concentration gradient. Mass *A* permeates through *B* by diffusion, absorption, or adsorption. There may be other mechanisms that can help to account for the action that results in the migration of *A* through *B*. Mass *A* will usually be a vapor, and that is what we will consider here. Permeation of solids and liquids through other solids is complicated and beyond our purposes.

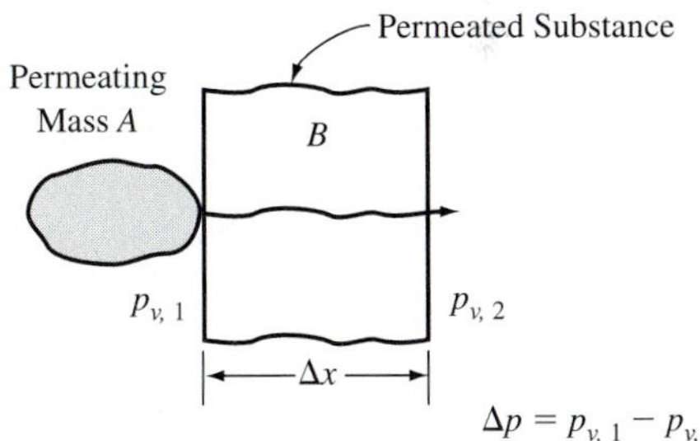

Figure 8–26 Mass transfer by permeating through a substance.

We may write for the mass transfer of *A* through *B*,

$$\dot{m}_A = -B_{AB}A\frac{\Delta p}{\Delta x} \tag{8–60}$$

where $\Delta p/\Delta x$ is the average vapor pressure gradient of substance *A* across the permeated substance, *B*, *A* is the area normal to the mass transfer, and B_{AB} is the permeability of substance *A* through substance *B*. The vapor pressure gradient drives the permeation of *A* through *B* because *A* is assumed to be a vapor. The permeability is a property of substance *B* [1] and is often listed in units of g/[Pa · s · m] (SI units) or grains/hr · ft · in. Hg (English units).

Table 8–5 Perm and Permeability Values for Selected Materials

A. Permeance and Permeability of Materials to Water Vapor

Material	Permeance, perms	Permeability, grains/hr · ft · in. Hg
Concrete (1:2:4 mix)		0.267
Brick (4-in. thick)	0.8	
Concrete block (8 in.)	2.4	
Glazed tile (4 in.)	0.12	
Asbestos board (0.2 in.)	0.54	
Plaster/metal lath	15	
Plaster/wood lath	11	
Plaster/gypsum board	20	
Gypsum board (0.375 in.)	50	
Gypsum sheathing (0.50 in.)		1.67
Insulation board		1.67–4.167
Interior insulating board (0.5 in. thick)	50–90	
Hardboard, standard (0.125 in.)	11	
Hardboard, tempered (0.125 in.)	5	
Built-up roofing, hot mopped	0.0	0.0
Wood, sugar pine		0.03–0.45
Plywood, exterior, glued (0.25 in.)	0.7	
Plywood, interior, glued (0.25 in.)	1.9	
Acrylic, glass-fiber-reinforced sheet (56 mil)	0.12	
Polyester, glass-fiber reinforced (48 mil)	0.05	
Air, still		10.0
Cellular glass		0.0
Corkboard		0.175–0.217
Mineral wool		9.67
Expanded polyurethane (R-11, blown) board stock		0.033–0.133
Expanded polystyrene, extruded		0.10
Expanded polystyrene, bead		0.167–0.483
Unicellular synthetic flexible rubber, foam		0.00167–0.0125
Sand and gravel mix		150
Aluminum foil (1 mil)	0.0	
Aluminum foil (0.35 mil)	0.05	
Polyethylene (2 mil)	0.16	
Polyethylene (4 mil)	0.08	
Polyethylene (6 mil)	0.06	
Polyethylene (6 mil)	0.04	
Polyethylene (10 mil)	0.03	
Polyester (1 mil)	0.7	
Cellulose acetate (125 mil)	0.4	
Polyvinylchloride, unplasticized (2 mil)	0.68	
Polyvinylchloride, plasticized (4 mil)	0.8–1.4	

The transfer of mass A through B may be described through the equation, sometimes referred to as *Darcy's Law*, $\dot{V}$ being a volume flow rate,

$$\dot{V} = \frac{K_{AB}A}{\mu}\frac{\Delta p}{\Delta x} \tag{8–61}$$

In this equation the parameter K_{AB} is also referred to as the permeability by some investigators, and the units of this parameter are found to be length squared, such as m^2 or ft^2. The reader should recognize that much standardization needs to be established in mass

B. Permeance and Resistance by the Dry-Cup Method

Material	Permeance, perms	R_m, 1/perms
Duplex sheet, asphalt laminate/aluminum foil one side	0.002	500
Saturated, coated roll roofing	0.05	20
Kraft paper and asphalt laminate	0.3	3.3
Blanket thermal insulation backup paper, asphalt coated	0.4	2.5
Asphalt-saturated coated vapor barrier paper	0.2–0.3	5.0–3.3
Asphalt-saturated uncoated vapor barrier paper	3.3	0.3
15-lb Asphalt felt	1.0	1.0
15-lb Tar felt	4.0	0.25
Single-Kraft double	31	0.032
Paint, asphalt paint on plywood, 2 coats	0.4*	2.5*
Aluminum varnish on wood, 2 coats	0.3–0.5	3.3–2.0
Paint—exterior, white lead and oil on wood siding, 3 coats	0.3–1.0	3.3–1.0
Paint—exterior, white lead–zinc oxide and oil on wood, 3 coats	0.9	1.1
Styrene–butadiene latex coating (2 oz/sq. ft)	5.5	0.18
Hot melt asphalt (2.0 oz/sq. ft)	0.5	2
Hot melt asphalt (3.5 oz/sq. ft)	0.1	10

*Wet-cup method.

transfer if effective analyses are to be conducted. A discussion of Darcy's Law and permeability may be found in many advanced texts covering transport phenomena, such as Muskat [11].

In the building trades permeation is often analyzed by using the performance parameter called permeance, P, defined as [1]

$$P_{AB} = \frac{B_{AB}}{\Delta x}$$

so that the rate of permeation from equation (8–60) can be written

$$\dot{m}_A = P_{AB} A \Delta p \tag{8–62}$$

The direction of the mass flow is implied in equation (8–62) to be progressing from a high to a low vapor pressure. The units of permeance are $g/Pa \cdot s \cdot m^2$ or grains/in. $Hg \cdot hr \cdot ft^2$. The unit of the grain/in. $Hg \cdot hr \cdot ft^2$ is called the *perm* in the building trade. Thus,

$$1 \frac{\text{grain}}{\text{in. Hg} \cdot \text{hr} \cdot \text{ft}^2} = 1 \text{ perm} \tag{8–63}$$

This concept is used for measuring the moisture migration or mass transfer through a building material [1]. Since there are 7000 grains in 1 lbm, the perm has the following units

$$1 \text{ perm} = 3.968 \times 10^{-8} \frac{\text{lbm}}{\text{in. Hg} \cdot \text{s} \cdot \text{ft}^2} = 5.72 \times 10^{-8} \frac{\text{g}}{\text{Pa} \cdot \text{s} \cdot \text{m}^2}$$

In table 8–5 are listed perm values for selected materials. The dry-cup and wet-cup methods for determining permeability values are described in ASTM Standards [12]. In table 8–6 are presented permeability constants for various gases through some common

Table 8–6 A. Permeability Constants

System, Permeating Substance—Medium being permeated	Temperature °C	Permeability B', $cm^3 \cdot mm/s \cdot cm^2 \cdot cm\ Hg$
Helium—vulcanized neoprene	0	0.0022
	30.4	0.0078
	57.0	0.035
Helium—raw neoprene	21.6	0.0039
	37.6	0.0116
Helium—volcanized rubber, 2% sulfur	19.2	0.0051
	57.0	0.0179
Helium—"Vulcaplas"	50.0	0.000174
	59.0	0.000342
Hydrogen—vulcanized neoprene	17.5	0.0085
	26.9	0.0128
	63.7	0.0534
Hydrogen—German rubber (butadiene–acrylonitrile polymer)	0	0.0032
	20.0	0.0085
Hydrogen—butadiene–methyl methacrylate polymer	20.0	0.023
Hydrogen—polystyrene–butadiene polymer	19.9	0.0084
Hydrogen—chloroprene polymer, pure	31.8	0.0049
	60.8	0.0175
Nitrogen—neoprene	27.1	0.00137
	65.4	0.0106
Nitrogen—German rubber	20.0	0.00061
	59.5	0.0048
Nitrogen—methyl methacrylate polymer	21.2	0.0028
	61.9	0.0132
Nitrogen—polystyrene–butadiene polymer	20.0	0.0029
	64.2	0.0161
Argon—neoprene	36.1	0.0068
	52.2	0.0144
Argon—butadiene–methyl methacrylate polymer	20.0	0.0059
	62.3	0.0395
Argon—polystyrene–butadiene polymer	19.5	0.0109

membrane materials. These data were collected by Barrer [13], and the permeability constant has units of $cm^3 \cdot mm/s \cdot cm^2 \cdot cm$ Hg pressure. Thus, this constant needs to be multiplied by the permeating vapor density in order that the permeability be determined.

In the analysis of permeation, it is sometimes useful to define the mass flow resistance, $R_m = 1/P_{AB}$, in place of the permeance. Thus, equation (8–62) would then be

$$\dot{m}_A = \frac{A}{R_m}\Delta p \tag{8–64}$$

The process of mass transfer through a substance as described by equations (8–60), (8–62), and (8–64) is a complicated one that includes diffusion, sorption, and bulk fluid flow. Since it is so complicated and not easily analyzed, the concept of permeance has been useful for engineering analysis of building materials and structures, filters, semipermeable membranes, and separation processes.

B. Permeability and Diffusivity for Vulcanized Rubber at 298 K

Gas	Permeability, B'_{AB}, cm^3 · mm/s · cm^2 · cm Hg	$\mathscr{D}_{AB}$, cm^2/s
Hydrogen, H_2	0.045×10^{-6}	0.85×10^{-5}
Oxygen, O_2	0.020×10^{-6}	0.21×10^{-5}
Nitrogen, N_2	0.0071×10^{-6}	0.15×10^{-5}
Carbon Dioxide, CO_2	0.132×10^{-6}	0.11×10^{-5}

C. Relative Permeabilities for Various Membranes at 298 K

Membrane	Gas							
	H_2	He	A	N_2	O_2	CO	CO_2	NH_3
Rubber, vulcanized	1.0	0.62		0.16	0.44		2.88	8.0
Rubber, raw	1.0	0.30	0.19	0.11	0.35	0.16	2.50	
Neoprene	1.0	0.61	0.29	0.10				
Chloroprene	1.0	0.22						
Butadiene–methylmethacrylate	1.0		0.26	0.11				
Butadiene–polystyrene	1.0		0.95	0.27				

EXAMPLE 8–12 Ammonia gas at 100 cm Hg pressure is contained in a rubber bladder at 20°C. The rubber bladder is 5 mm thick. Predict the permeation through the rubber by the ammonia per unit of area. Assume the ammonia acts as an ideal gas.

Solution The permeation or mass transfer through the rubber may be predicted from equation (8–60), using permeability data from table 8–6. The ammonia density can be predict from

$$\rho = \frac{p}{RT} = \frac{(100 \text{ cm Hg})(1.333 \text{ kPa/cm Hg})(17 \text{ kg/kg-mol})}{(8.315 \text{ kJ/kg-mol} \cdot \text{K})(298 \text{ K})} = 0.915 \text{ kg/m}^3$$

From table 8–6, part C, ammonia has a relative permeability eight times that of hydrogen gas. From table 8–6, part B, the permeability constant for hydrogen gas through rubber is 0.045×10^{-6} cm^3 · mm/s · cm^2 · cm Hg. Notice that the permeability is given for a volume amount (cm^3), so we must multiply this value by the density of the permeating substance, ammonia. Using equation (8–60) we obtain,

$$\frac{\dot{m}_A}{A} = \dot{G}_A = \rho B'_{AB}\frac{\Delta p}{\Delta x} = \left(0.041 \times 10^{-12}\frac{\text{kg} \cdot \text{mm}}{\text{s} \cdot \text{cm}^2 \cdot \text{cm Hg}}\right)\frac{100 \text{ cm Hg}}{5 \text{ mm}}$$

Answer

$$= 8.235 \times 10^{-10}\frac{\text{kg}}{\text{s} \cdot \text{cm}^2} = 2.9646 \times 10^{-3}\frac{\text{g}}{\text{hr} \cdot \text{cm}^2}$$

EXAMPLE 8–13 Determine the expected moisture migration through the wall of figure 8–27. Also determine the expected direction of the migration.

Figure 8–27 Moisture migration through a wall.

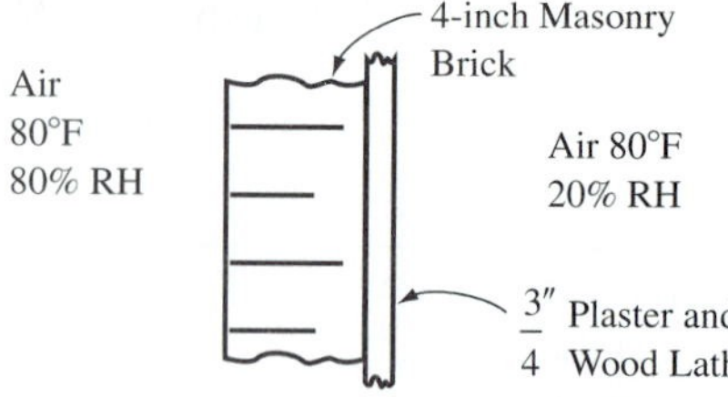

Solution The moisture migration can be predicted from equation (8–64) with the variation that the resistance will be the sum of the resistances of the brick and the plaster in wood lath,

$$\dot{G} = \frac{\Delta p}{\sum R_m}$$

The vapor pressures of the water in the air are

$$p_{vi} = (0.2)(0.506 \text{ psia}) = 0.2024 \text{ psia} = 0.4122 \text{ in. Hg}$$

and

$$p_{vo} = (0.8)(0.506 \text{ psia}) = 0.4048 \text{ psia} = 0.8244 \text{ in. Hg}$$

The resistances are read as the inverse perm values from table 8–6:

$$R_{m,\text{brick}} = 1.25 \text{ hr} \cdot \text{ft}^2 \cdot \text{in. Hg/grain}$$

$$R_{m,\text{plaster/wood}} = 0.067 \text{ hr} \cdot \text{ft}^2 \cdot \text{in. Hg/grain}$$

so that

$$\sum R_m = 1.317 \frac{\text{hr} \cdot \text{ft}^2 \cdot \text{in. Hg}}{\text{grain}}$$

and the moisture migration is

Answer

$$\dot{G} = \frac{0.8244 - 0.4122 \text{ in. Hg}}{1.317 \text{ hr} \cdot \text{ft}^2 \cdot \text{in. Hg/grain}} = 0.313 \text{ grain/h} \cdot \text{ft}^2 = 4.47 \times 10^{-5} \text{ lbm/hr} \cdot \text{ft}^2$$

The moisture migration will proceed from the outside to the inside since the vapor pressure is greater outside than inside.

Vapor barriers or vapor retarders are installed in structures or other enclosures to insure low moisture migration. Often moisture migration through such arrangements occurs along with heat transfer, and the following example demonstrates how the two phenomena of mass and heat transfer may be analyzed.

EXAMPLE 8–14 Consider a flat composite roof shown in figure 8–28. Determine the moisture migration and the heat transfer through the roof and predict whether there will be condensation, and how much can be expected, within the roof section.

Solution We will assume steady state one-dimensional heat and mass transfer. Then using the thermal resistance concept, we can predict the temperature distribution through the roof section. Recalling that the heat transfer is

$$\dot{q}_A = \frac{\Delta T_{\text{overall}}}{\sum R_T} = \frac{\Delta T_i}{R_{T,i}}$$

where ΔT_i is the temperature difference over the ith component of the composite roof and $R_{T,i}$ is its thermal resistance. For instance, the total heat transfer is

$$\dot{q}_A = \frac{25°\text{C} - (-24°\text{C})}{4.8638 \text{ m}^2 \cdot \text{K/W}} = 10.07 \frac{\text{W}}{\text{m}^2}$$

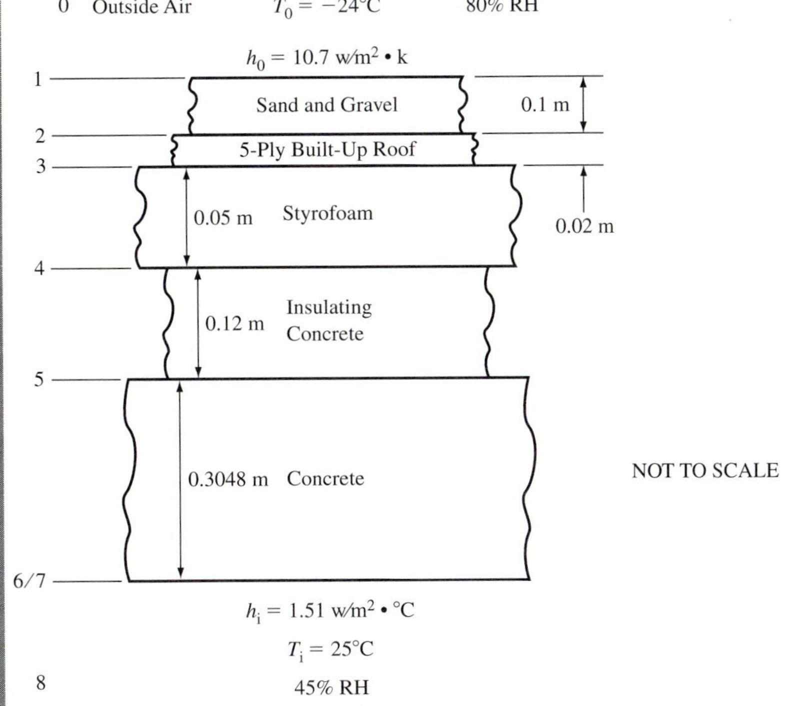

Figure 8–28 Cross section of a flat built-up roof.

Then the temperature difference at the outside convective film is

$$\Delta T_1 = \dot{q}_A R_{T,1} = \left(10.07 \frac{\text{W}}{\text{m}^2}\right)\left(0.1 \frac{\text{m}^2 \cdot \text{K}}{\text{W}}\right) = 1.007°\text{C}$$

and

$$T_1 = T_0 + 1.007°\text{C} = -23°\text{C}$$

Similarly, for the sand and gravel section

$$T_2 = T_1 + \Delta T_2 R_{T,2} = -23°\text{C} + (10.07)(0.25) = -20.5°\text{C}$$

The results of all of these calculations are shown in table 8–7 and in the graph of figure 8–29. The saturation pressure for water vapor in the air through the roof section is tabulated in table 8–8, third column. The various perm values and the mass flow resistances, which are inverse perms, are also given in this table. The total mass flow resistance, $\sum R_m$, is the summation of the resistances, 9.1567 hr · ft² · in. Hg/grain. The expected moisture migration through the roof, assuming no condensation in the roof, is determined from equation (8–64),

$$\dot{G} = \frac{p_i - p_{v,0}}{\sum R_m} = \frac{0.699 - 0.0126 \text{ in. Hg}}{9.1567 \text{ hr} \cdot \text{ft}^2 \cdot \text{in. Hg/grain}} = 0.0746 \frac{\text{grains}}{\text{hr} \cdot \text{ft}^2}$$

The expected vapor pressures of moisture inside the roof section are then predicted from the calculations:

$$p_{v,2} = p_{v,0} + m_a(R_{m,2}) = 0.016 + (0.0746)(0.0067) = 0.0164 \text{ in. Hg}$$
$$p_{v,3} = p_{v,2} + m_a(R_{m,3}) = 0.0164 + (0.0746)(5.0) = 0.3894 \text{ in. Hg}$$
$$p_{v,4} = p_{v,3} + m_a(R_{m,4}) = 0.3894 + (0.0746)(0) = 0.3894 \text{ in. Hg}$$

and so on. The results are given in table 8–8, column 5. The vapor pressure distribution through the roof section is graphically displayed in figure 8–29. From this result we can

Table 8–7 Heat Transfer Analysis Through Roof Section

Station	Description	Δx, m	κ, W/m · K	h, W/m² · K	R_T, m² · K/W	T, °C
0	Outside air	—	—	—	—	−24
1	Outside surface	—	—	10.7	0.1	−23
2	Sand and gravel	0.10	0.4	—	0.25	−20.5
3	5-Ply built-up	0.02	0.062	—	0.323	−17.2
4	Styrofoam	0.05	0.029	—	1.724	0.16
5	Insulating concrete	0.12	0.08	—	1.5	15.27
6	Concrete	0.3048	1.00	—	0.3048	18.34
7	Inside wall surface	—	—	1.51	0.662	25.0
8	Inside air	—	—	—	—	25.0
					$\sum R_T = 4.8638$	

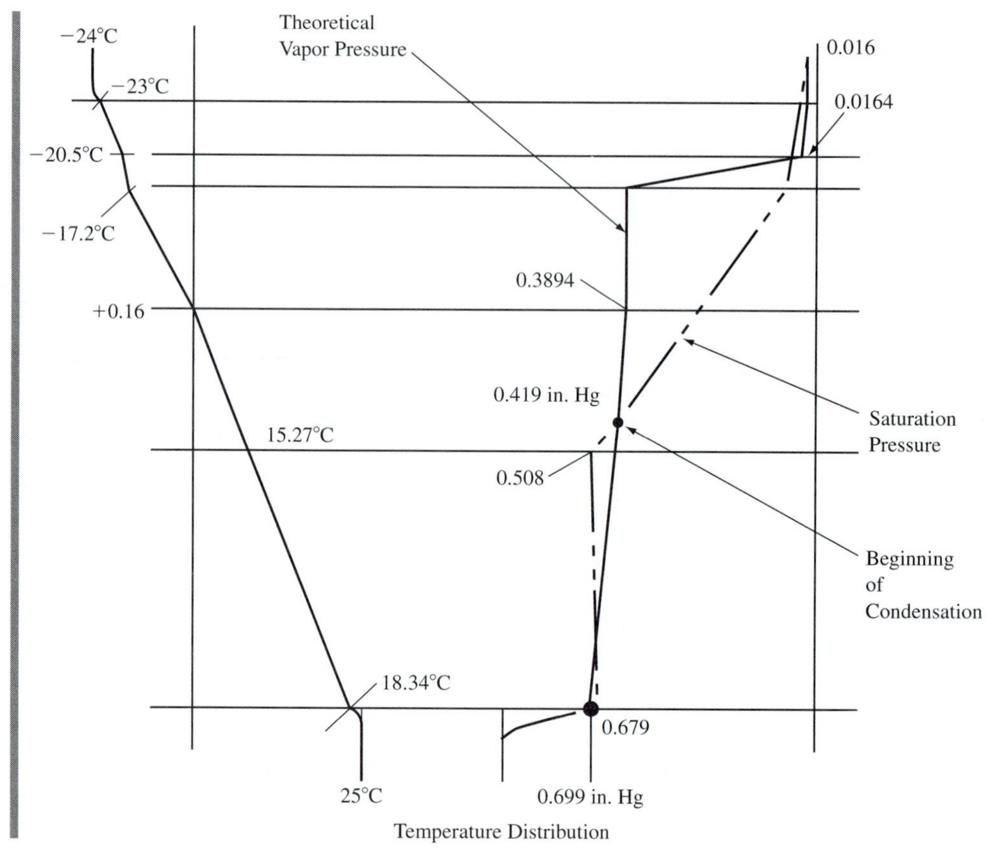

Figure 8–29 Temperature and vapor pressure distribution through roof.

see that there will likely be condensation on the lower surface of the roof (station 6) because the vapor pressure (0.699 in. Hg) is greater than the saturation pressure at the surface temperature, 18.34°C (0.679 in. Hg). The amount of moisture that can be expected to condense at the lower surface of the roof is, from psychrometrics, and assuming the atmospheric pressure is 29 in. Hg,

$$\dot{G} = (0.622)\dot{m}_{air}\frac{\Delta p_v}{p_{v,air}} = (0.622)\left(0.0746\frac{\text{grain}}{\text{hr}\cdot\text{ft}^2}\right)\frac{0.699 - 0.624\text{ in. Hg}}{29 - 0.624\text{ in. Hg}}$$

$$= 8.899 \times 10^{-5}\frac{\text{grain}}{\text{hr}\cdot\text{ft}^2}$$

This is negligible, and the mass flow through the roof is

$$\dot{G} = 0.0746 - 0.00008899 \approx 0.0746\frac{\text{grain}}{\text{hr}\cdot\text{ft}^2}$$

Table 8–8 Mass Transfer Analysis Through Roof Section

Station	Temperature °C	Saturation pressure, in. Hg	R_m, hr · ft² · in. Hg/grain	Vapor Pressure, in. Hg
0	−24	0.0207	0	0.016
1	−23	0.0230	0	0.016
2	−20.5	0.0292	0.0067	0.0164
3	−17.2	0.0400	5.0	0.3894
4	0.16	0.200	0	0.3894
5	15.27	0.508	0.4	0.41924
6	18.34	0.679	3.75	0.699
7	25	0.960	0	0.699
8	25	0.960	0	0.699
			$\sum R_m = 9.1567$	

Since there is another condensation point located inside the insulating concrete near station 5, we see that the moisture migration will exceed the local dew point pressure the greatest near station 3, the lower side of the 5-ply built-up section. The amount of moisture that can be expected to condense between station 5 and station 3 is the difference between the expected total moisture migration from the lower surface up to station 3 and the predicted moisture migration from station 3 to the upper surface. The predicted moisture migration from station 3 to the upper surface is

$$\dot{G}_{3-\text{upper}} = \frac{p_{g,3} - p_{v,0}}{\sum R_m - \sum R_{m,1-3}}$$

where the denominator is the total mass flow resistance from station 5 to the upper surface. Referring to table 8–8, we have

$$\dot{G}_{3-\text{upper}} = \frac{0.040 - 0.016\text{ in. Hg}}{5.4067\text{ hr}\cdot\text{ft}^2\cdot\text{in. Hg/grain}} = 0.0048\frac{\text{grain}}{\text{hr}\cdot\text{ft}^2}$$

The migration from the lower surface to station 3 is

$$\dot{G}_{1-3} = \frac{0.699 - 0.040\text{ in. Hg}}{4.15\text{ hr}\cdot\text{ft}^2\cdot\text{in. Hg/grain}} = 0.1588\frac{\text{grain}}{\text{hr}\cdot\text{ft}^2}$$

and the amount of condensation at station 3 can be approximated as

$$\dot{G}_{1-3} - \dot{G}_{3-\text{upper}} = 0.154\frac{\text{grains}}{\text{hr} \cdot \text{ft}^2}$$

This moisture can accumulate in the styrofoam and diffuse down through the insulating concrete; however, if this condition were to persist, a moisture retarder should be installed near the lower surface of the roof to prevent moisture damage.

8–8 SUMMARY

There are eight mechanisms for mass transfer:

1. Diffusion due to a concentration gradient
2. Thermal diffusion
3. Pressure diffusion
4. Forced diffusion
5. Forced convection diffusion
6. Free convection diffusion
7. Turbulent diffusion
8. Interphase mass transfer

Self-diffusion results when a substance diffuses into a region occupied by its same species.

Fick's Law of Diffusion is

$$\dot{N}_{A/N} = -\wp_{AB}\frac{\partial C_A}{\partial x} \tag{8–19}$$

where $\wp_{AB}$ is the mass diffusivity of A into B.

Stefan's Law describes the diffusion of a substance upward through a stagnant column of vapor,

$$\dot{G}_{A/\text{total}} = \frac{\wp_{AB}}{R_A T}\frac{p_T}{p_A - p_T}\frac{\partial p_A}{\partial y} \tag{8–37}$$

or, for no frictional effects and constant $\wp_{AB}$

$$\dot{G}_{A/\text{total}} = p_T\wp_{AB}\frac{p_{A,1} - p_{A,2}}{R_A T(y_2 - y_1)p_{\text{LM}}} \tag{8–39}$$

where p_{LM} is the log mean pressure difference in the column.

Convective mass transfer is described by

$$\dot{G}_A = h_{D,A}(\rho_{A,S} - \rho_{A,\infty}) \tag{8–41}$$

where $h_{D,A}$ is the convection mass-transfer coefficient

$$h_{D,A} = \frac{\wp_{AB}}{\Delta x} \tag{8–43}$$

The *Schmidt number, Sc*

$$\text{Sc} = \frac{\nu}{\wp_{AB}} \tag{8–46}$$

is analogous to the *Prandtl Number,* and the Sherwood number, Sh

$$\text{Sh} = \frac{h_{D,A}L}{\wp_{\text{self}}}$$

is analogous to the *Nusselt Number* for analyzing forced convection mass transfer. The *mass-transfer Grashof Number, Gr_m*

$$\mathrm{Gr}_m = \frac{gL^3}{\nu^2}\left(1 - \frac{\gamma_{m,\infty}}{\gamma_{m,S}}\right) \tag{8–50}$$

is analogous to the Grashof Number for analyzing free convection. The *mass-transfer Biot Number*, Bi_m

$$\mathrm{Bi}_m = \frac{h_{D,A}L}{\wp_{\text{self}}} \tag{8–55}$$

is analogous to the Bi for transient mass transfer and the *mass-transfer Fourier Number*, Fo_m

$$\mathrm{Fo}_m = \frac{\wp_{\text{self}}t}{L^2} \tag{8–56}$$

is analogous to the Fourier Number for transient mass transfer.

Lumped Mass Capacity can be used if $\mathrm{Bi}_m < 0.1$. Otherwise the analyses of transient heat transfer may be used.

Absorption is the chemical or phase change of a mixture due to a substance transferring into the mixture. *Adsorption* is the absorption of a substance into mixture without chemical or phase changes.

Permeability refers to the transfer of a substance through another mixture.

Darcy's Law

$$\dot{V} = \frac{K_{AB}A}{\mu}\frac{\Delta p}{\Delta x} \tag{8–61}$$

or

$$\dot{m}_A = P_{AB}A\Delta p \tag{8–62}$$

is used to predict the mass transfer through the mixture. P_{AB} is called the *permeance* and is usually given in units of *perm*.

DISCUSSION QUESTIONS

Section 8–1

8–1 How would you describe the term *diffusion*?

8–2 What is binary diffusion?

Section 8–2

8–3 What is the difference between a mass fraction and a mole fraction?

8–4 What is the difference between density and mass concentration?

Section 8–3

8–5 What is meant by local average velocity?

8–6 What is diffusional velocity?

8–7 What is Fick's Law?

8–8 What is mass diffusivity?

8–9 What is Stefan's Law?

8–10 What are the restrictions on Stefan's Law?

Section 8–4

8–11 What does the convective mass transfer coefficient represent?

8–12 What does the Schmidt Number represent?

8–13 What does the Sherwood Number represent?

Section 8–5

8–14 What is meant by self-diffusion?

8–15 When can one use lumped mass capacity analysis?

Section 8–6

8–16 What is the difference between absorption and adsorption?

8–17 What is meant by a sorbent?

8–18 What is meant by a desiccant?

Section 8–7

8–19 What is Darcy's Law?

8–20 What is permeation?

8–21 What is a perm?

PRACTICE PROBLEMS

Section 8–1

8–1 Blood as it enters into the lungs has a high concentration of carbon dioxide (CO_2) and a low concentration of oxygen (O_2). The blood leaves the lungs with a high concentration of O_2 and low CO_2. What are the qualitative concentration gradients of CO_2 and O_2 in the lungs?

Section 8–2

8–2 For a 10-mol mixture of air (8 mol N_2 and 2 mol O_2) at 101 kPa, 20°C, determine the mass concentration and the molar concentration of N_2 and O_2.

8–3 The term parts per million (ppm) is often used to describe small molar concentrations of substances in a mixture. Determine the molar analysis of standard air (79% N_2, 20% O_2, and 1% argon by volume) with 100 ppm of radon gas at 101 kPa, 20°C.

8–4 Using the mass balance (equation 8–10) of the steady state reaction of oxygen and hydrogen to produce water at a rate of 10 mol/min, identify the rates of mass flow for O_2, H_2, and H_2O. Also identify the mass generation rates for the oxygen, hydrogen, and water.

Section 8–3

8–5 Predict the mass diffusivity of benzene, C_6H_6, in air at 20°C and 1 atm pressure using equation (8–24). Compare this value to that from experimental values.

8–6 If the diffusivity of radon gas into nitrogen gas at 20°C, 1 atm pressure is 0.12 cm^2/s, what would you expect the diffusivity to be at 30°C, 2 atm pressure.

8–7 It is found that lead from 2-mm-thick old paint on a certain wood wall has permeated to 1000 ppm at a depth of 2 mm below the outer wood surface. If the paint contains 20% (by mass) lead and was applied 20 years before the lead content in the wood was measured, estimate the diffusivity of the lead in the wood if it is found that the paint has lost 50% of the original amount of the lead. Assume that the lead did not diffuse into the air. Wood may be approximated as cellulose, $C_6H_{10}O_5$, with a molecular weight of 116, lead has a molecular weight of 207, and the paint density may be taken as 1020 kg/m^3 when new.

8–8 Estimate the diffusion of O_2 from dry air into water at 60°F if the O_2 concentration is 0.0003 lbm-mol/ft^3 1 in. below the water surface. (*Note:* The molar concentration of O_2 in air may be taken as 20%.)

8–9 An "empty" steel barrel with its lid off has some gasoline in its bottom, as shown schematically in figure 8–30. If the vapor pressure (saturation pressure) is 1 psia, estimate the evaporation rate of the gasoline and its vapor pressure halfway up in the barrel. Assume the mass diffusivity for gasoline into air is 0.0094 ft^2/s.

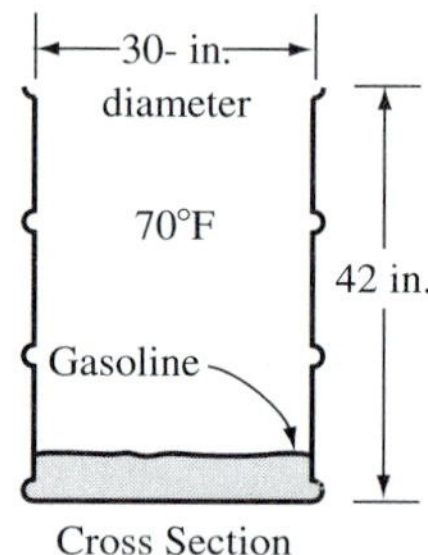

Figure 8–30

8–10 Carburization of iron is a process to increase its strength by diffusing carbon into the metal. If the process is carried out at 500°C, estimate the mass diffusion rate into the iron when the carbon concentration is 2.3% (by volume) at a depth of 2 mm. Assume that the iron is face center cubic (FCC) and the carbon concentration at the surface is 100%.

8–11 Radon gas has a diffusivity of 0.12 cm^2/s into air at 20°C. If the air temperature is raised to 30°C, what would you expect the diffusivity to be?

8–12 A body of water at 80°F has still air at 80°F, 14.7 psia, and 10% relative humidity above it. Estimate the evaporation rate of water if the air has 20% relative humidity 3 ft above the water surface.

8–13 Estimate the mass diffusivity of methane into air at 300 K, assuming ideal gas behavior.

8–14 A copper and an aluminum part are bolted together as shown in figure 8–31. Estimate the rate of diffusion of copper into the aluminum if the copper mass fraction at a particular time is 0.1% at 2 mm into the aluminum and the diffusivity is the same as that at 500°C.

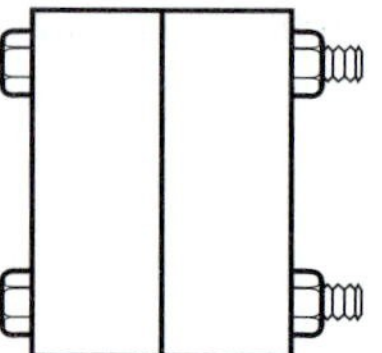

Figure 8–31

Section 8–4

8–15 Determine the Schmidt Number for ethyl alcohol flowing through a copper trough. Assume that the diffusivity of ethyl alcohol into air is 5.0×10^{-6} m²/s and into copper is 3×10^{-16} m²/s.

8–16 Determine the Schmidt Number for benzene evaporating into air at 25°C.

8–17 If steam is used as a cleaning agent for removing caffeine attached to the sides of 2-cm-diameter tubes in a coffee processing plant, estimate the time to remove 1 mm of caffeine from the inside of the tubes if steam at 120°C, 1 atm pressure, is forced through the tubes at 10 m/s. Assume the diffusivity of caffeine into steam is 6×10^{-6} m²/s and caffeine density is 8.0 kg/m³.

8–18 Estimate the rate of evaporation of acetone (C_3H_6O) into 75°F air from an open 4-in.-diameter by 8-in.-tall jar when the acetone covers 2 in. of the bottom of the jar and fresh air passes over the top of the jar. Assume acetone acts as an ideal gas.

8–19 Ammonia is mixed with water in a ratio of 1 part ammonia to 4 parts water (by volume) and used as a cleaning agent for floors. Estimate the evaporation rate of the ammonia per unit of area if the process occurs at 25°C. Assume air is moving over the surface at 5 m/s.

8–20 Acetone is rubbed on a steel roller, 4 cm diameter and 40 cm long to clean it. Air at 25°C is blown over the roller at 5 m/s. If the density of acetone vapor is taken as 2.37 kg/m³, estimate the rate of evaporation of acetone into the air. The viscosity of acetone may be taken as 0.01 g/m · s.

8–21 Dry air at 20°C, 101 kPa, passes through a 2-cm-diameter tube at 5 m/s. If the tube is also at 20°C and has acetone on its inside area, estimate the evaporation rate of acetone into air per unit area. Assume the density of saturated vapor acetone is 2.37 g/m³.

Section 8–5

8–22 Cubes of sugar 1 cm on a side are placed in water at 25°C. Estimate the time before the lumps have dissolved; that is, when the sugar concentration at the lump is 10% by weight. Assume that the diffusivity of sugar into water is 6.9×10^{-7} m²/s and the self-diffusivity of sugar is 1×10^{-3} cm²/s. The density of the sugar may be taken as 1500 kg/m³ and h_D is 6.4×10^{-6} m/s.

8–23 Distilled water at 75°F is suddenly exposed to oxygen gas at 75°F. Determine the time before the oxygen concentration in the water is 0.1% by mass at a point 0.05 in. below the surface of the water.

8–24 Wood boards 1-in. thick have a moisture content of 60% by mass. If the diffusivity of water in wood is taken as 3×10^{-10} ft²/s, estimate the time for the wood to have a moisture content of 50% at its center if the air surrounding the boards was originally dry and h_D is 1×10^{-8} ft/s.

Section 8–6

8–25 A rotary solid sorption dehumidifier is used to dehumidify air at 77°F, 85% relative humidity. If the dehumidifier performance map is like that shown in figure 8–24, estimate the outlet dry air temperature and its relative humidity. Then estimate the rate of water removed from 100 cfm of air.

8–26 Water–90% triethylene glycol is to be used to dry air. If the mixture is added to an air stream at 25°C, determine the amount of mixture required per kilogram of dry air to dehumidify the air if the air stream is at 30°C and 90% relative humidity. Assume that 25% excess mixture is required that theoretically predicted. Also determine the outlet air temperature the relative humidity. The specific heat of triethylene glycol is $2.4 \frac{\text{kJ}}{\text{kg} \cdot \text{k}}$.

Section 8–7

8–27 Determine the moisture migration through the wall shown in figure 8–32.

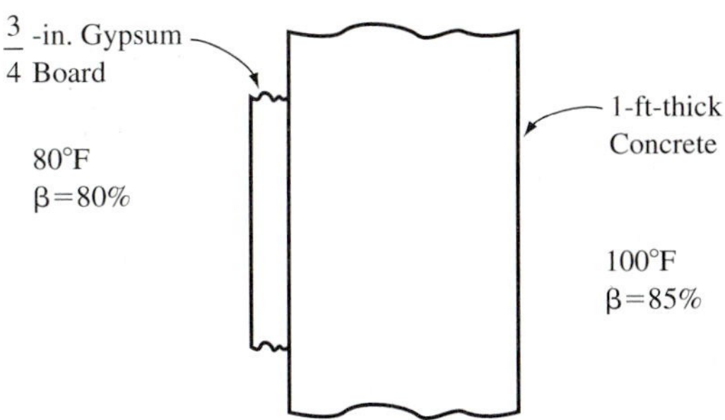

Figure 8–32

8–28 Determine the temperature distribution and the moisture migration through the wall shown in figure 8–33.

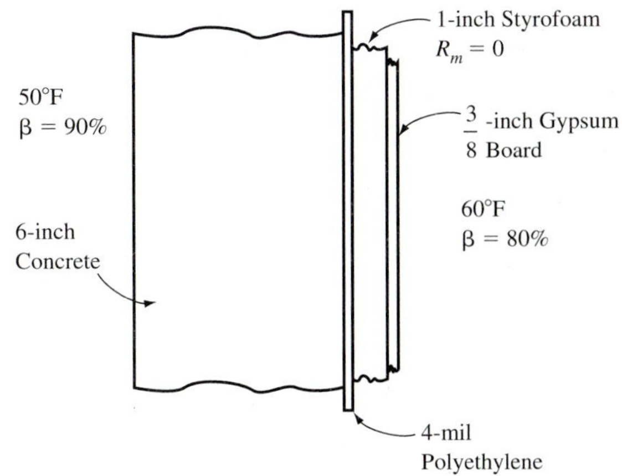

Figure 8–33

REFERENCES

1. Jost, W., *Diffusion in Solids, Liquids, Gases,* Academic Press, New York, 1960.
2. Bobrova, G. I., and G. D. Rabinovich, Determination of the Soret Coefficients and the Concentration Diffusion Coefficients by an Unsteady-State Method, in I. G. Gurevich, I. P. Zhuk, and N. G. Kondrashov eds., *Non-Stationary Heat and Mass Transfer,* N.A.S.A. and N.S.F., U.S. Department of Commerce, Publication TT F-432, TT-67–51368, 1967.
3. Bird, R. B., W. E. Stewart, and E. N. Lightfoot, *Transport Phenomena,* John Wiley & Sons, New York, 1960.
4. Brodkey, R. S., and H. C. Hershey, *Transport Phenomena, A Unified Approach,* McGraw-Hill, New York, 1988.
5. Hirschfelder, J. O., C. F. Curtiss, and R. B. Bird, *Molecular Theory of Gases and Liquids,* John Wiley & Sons, New York, 1966.
6. Gebhart, B., Y. Jaluria, R. L. Mahajan, B. Sammakia, *Buoyancy-Induced Flows and Transport,* Hemisphere Publishing, New York, 1988.
7. *Lithium Chloride Technical Data,* Bulletin 151, Cyprus Foote Mineral Company, Malvern, PA, 1988.
8. *A Guide to Glycols,* Dow Chemical Company, Midland, MI, 1991.
9. ASHRAE 1985, Symposium on Changes in Supermarket Heating, Ventilating, and Air Conditioning Systems, *ASHRAE Transactions,* 91, (1B), 423–468, 1985.
10. *ASHRAE Systems and Equipment Handbook,* chapter 22, American Society of Heating, Refrigerating, and Air Conditioning Engineers, 1996.
11. Muskat, M., *The Flow of Homogeneous Fluids Through Porous Media,* 1st edition, J. W. Edwards, Ann Arbor, MI, 1946.
12. E96-80 ASTM Standards.
13. Barrer, R. M., *Trans. Faraday Soc.,* 35, 628, 1939.

HEAT EXCHANGERS

9

New Terms

C	$[\dot{m}c_p]_{\text{minimum}}/[\dot{m}c_p]_{\text{maximum}}$	NTU	number of thermal units, $UA/[\dot{m}c_p]_{\text{minimum}}$
F	LMTD correction factor	U	overall heat exchanger conductance
F_f	fouling factor	ε_{HX}	heat exchanger effectiveness
LMTD	log mean temperature difference		

9–1 GENERAL CONCEPTS OF HEAT EXCHANGERS

Heat exchangers are devices that transfer thermal energy from one fluid stream to another. Often the two streams are liquids, but there are important applications where the streams are gases or vapors or combinations of liquids and vapors. Heat exchangers are one of the most often used devices in technological systems and have been widely studied and analyzed. They are all around any person who lives in our highly technological society. Most heat exchangers are described by other names: condensers, evaporators, coils, heaters, radiators, coolers, preheaters, reheaters, boilers, steam generators, after heaters, and on and on. In figures 9–1 through 9–6 are shown some arrangements of heat exchangers. They all have in common that thermal energy is transferred from one substance to another. In figure 9–7 is shown schematically the general concept of a heat exchanger. In all heat exchangers there will be some heat loss or gain with the surroundings; however, as a first approximation and for the developments of this text we will assume that these losses or gains are negligible, or zero. Thus, we assume that the heat exchanger will be an adiabatic device.

Applying an energy balance to the adiabatic heat exchanger under steady state conditions gives

$$\dot{m}_c(hn_{c,o} - hn_{c,i}) = \dot{m}_h(hn_{h,i} - hn_{h,o}) \tag{9–1}$$

where the hn's are the enthalpies of the fluids and the subscripts c and h indicate *cold* and *hot* fluids, respectively. In some situations when one of the fluids has a flow rate (relative to the flow rate of the other fluid), its enthalpy change will be nearly zero. Also, if one of the fluids has no flow rate but relies on free convection for heat transfer, then it is representative of a very large, essentially infinite reservoir so that there is no enthalpy change. If one of the fluids is proceeding through a phase change such as in a condenser or a steam generator, then its temperature change will be zero although its enthalpy change will not be zero. In these instances the reader is advised to proceed cautiously with further analysis.

In the analysis of a typical heat exchanger where both fluids exhibit an enthalpy change and have defined flow rates we may identify the internal heat exchange between the two fluid streams as $\dot{Q}_{\text{HX}}$, given by

$$\dot{Q}_{\text{HX}} = \dot{m}_c(hn_{c,o} - hn_{c,i}) = \dot{m}_h(hn_{h,i} - hn_{h,o}) \tag{9–2}$$

Figure 9–1 The double pipe heat exchanger, which is basically one small tube concentrically located inside of a larger tube, is a simple configuration having many applications. The particular heat exchanger shown in this photograph is used as a cooler of reactor effluents. (Reproduced by permission of Kennedy Tank and Manufacturing Co., Inc., Indianapolis, Indiana)

Figure 9–2 The shell and tube heat exchanger is a common configuration for many types of heat exchangers. This photograph shows the tubes, or tube bundles, that are installed inside of a cylindrical shell to make up a typical shell and tube heat exchanger. This particular heat exchanger is used as a hot air cooler. (Reproduced by permission of Kennedy Tank and Manufacturing Co., Inc., Indianapolis, Indiana)

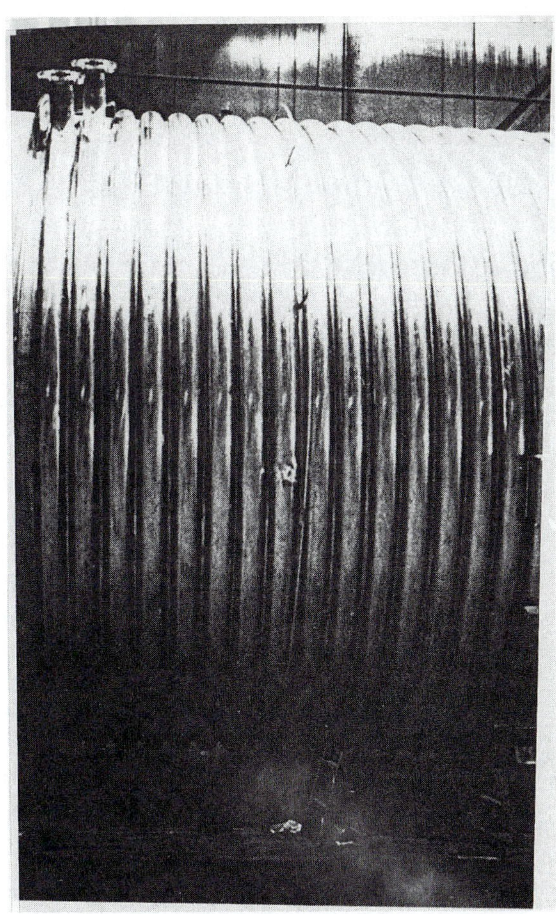

Figure 9–3 This heat exchanger uses a half-tube wrapped around the outer surface of a cylindrical tank to provide for convenient fabrication and maintenance. In addition, this configuration is used for pressure vessels for the pharmaceutical, beverage, paper, and defense industries. One fluid stream passes through the half-tube; another fluid is inside the tank. (Reproduced by permission of Northland Stainless Inc., Tomahawk, WI)

Figure 9–4 Another method for providing an effective heat exchanger is the dimple jacket configuration. Here a cylindrical tank, which could be a pressure vessel, is wrapped with a jacket and attached to the tank with dimple welds. A fluid passes through the cavity between the jacket and outer surface of the tank and another fluid is inside the tank. (Reproduced by permission of Northland Stainless, Inc., Tomahawk, WI)

Figure 9–5 This is a chilled water coil which is used to cool air for central air conditioning of buildings. Cool water (chilled water) or glycol solutions may be passed through the multiple rows of continuous circuit tubes while air is forced over the tube bank. Notice the return bends and the manifold header to connect the rows of tubes. (Reproduced with permission of Aerofin Corporation, Inc., Lynchburg, VA)

This heat exchange can involve conduction, convection, and radiation. It is convenient to define the overall heat exchanger conductance, U, such that

$$\dot{Q}_{\text{HX}} = UA(\Delta T) \tag{9–3}$$

where A is some surface area of the heat exchanger. One of the important tasks of an engineer is to be able to predict the necessary size of a heat exchanger, and this surface area is one of the parameters used to accomplish this task. The temperature difference, ΔT, represents a representative temperature difference between the two fluid streams, which we will see in section 9–3. With an understanding of convection heat transfer augmentation through the use of fins, appropriate material choices, and imaginative flow patterns, the engineer can obtain suitable heat exchanger designs subject to the constraints of size and temperature ranges.

It has become customary since the 1950s to classify heat exchangers into one of two categories:

1. Conventional Heat Exchangers
2. Compact Heat Exchangers

A conventional heat exchanger has a ratio of surface area to volume, called the specific surface, S_m, of less than about 700 m^2/m^3, or 215 ft^2/ft^3. Compact heat exchangers have specific surfaces greater than 700 m^2/m^3. As an example, the common radiator of an automobile water-cooled engine has a specific surface, S_m, of about 1200 m^2/m^3, a refrigerator condenser has an S_m of about 800 m^2/m^3, and the human lungs have an S_m of about 20,000 m^2/m^3.

Figure 9–6 Heat exchangers are used for very low temperature to very high temperature applications. This figure shows a gas turbine regenerator that is designed to recuperate the energy in gas turbine exhaust by passing cool combustion air through the tubes. In this particular application it is to be used to recuperate a portion of the energy from exhaust gases at 650°C before these gases are directed to another heat exchanger to generate steam. (Reproduced by permission of Stahl, Inc., York, PA)

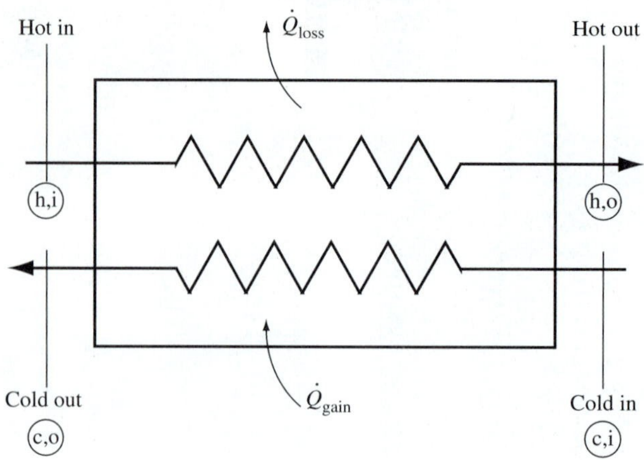

Figure 9–7 General heat exchanger arrangement.

We will analyze the shell and tube heat exchanger because it is an arrangement that is widely used in industrial and chemical processes. The typical shell and tube heat exchanger is sketched in figure 9–8. Such an arrangement is rugged and economical, and requires low maintenance. The shell and tube heat exchanger commonly has a specific surface of between 70 and 500 m^2/m^3.

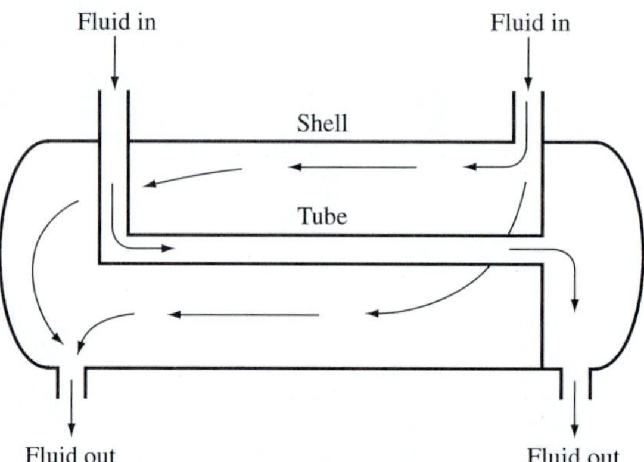

Figure 9–8 Typical shell and tube heat exchanger.

9–2 PARAMETERS IN HEAT EXCHANGERS

In the last section we introduced the overall heat transfer conductance, U. Other parameters that may be useful in describing the heat exchanger performance are its surface area A, its specific surface S_m, the temperature difference ΔT, and the flow patterns. If we consider the single-pass shell and tube heat exchanger shown in figure 9–8 and view a cross section of the tube, as shown in figure 9–9, we can see that the heat transfer may be approximated by the steady state one-dimensional conduction analysis of Chapter 2. Using the thermal resistance concept, we have for the heat exchanger

$$\dot{Q}_{\text{HX}} = \frac{\Delta T_{\text{overall}}}{\sum R_T} \tag{9–4}$$

where $\Delta T_{\text{overall}}$ is the temperature difference between the two bulk fluid temperatures. The total thermal resistance may be written, from Chapter 2,

$$\sum R_T = \frac{1}{h_i \pi D_i L} + \frac{\ln(D_o/D_i)}{2\pi\kappa L} + \frac{1}{h_o \pi D_o L} \tag{9–5}$$

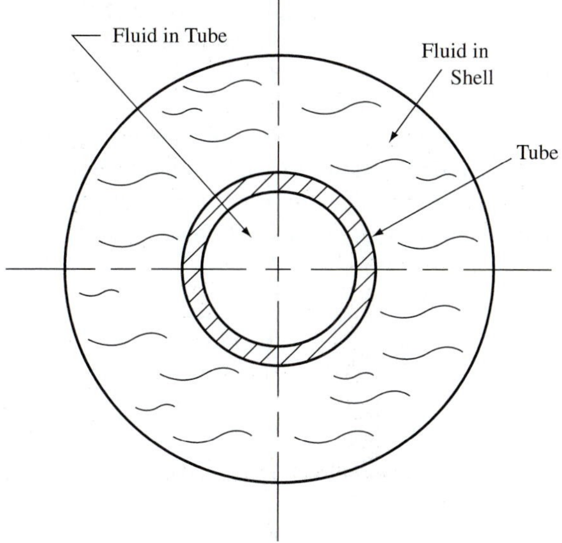

Figure 9–9 Cross section of single tube and shell heat exchanger.

where L is the axial length of the tube and shell. This thermal resistance can then be used to define the overall conduction

$$UA = \frac{1}{\sum R_T} \tag{9–6}$$

It is often advantageous to base the overall UA value on either the inside surface area, A_i, or the outer surface area, A_o. In particular,

$$U_i A_i = \frac{\pi D_i L}{\dfrac{1}{h_i} + \dfrac{D_i \ln(D_o/D_i)}{2\kappa} + \dfrac{D_i}{h_o D_o}} \tag{9–7}$$

where $A_i = \pi D_i L$, the surface of the inside of the tube. Thus, the overall heat exchanger conductance, based on the inside area, is

$$U_i = \frac{1}{\dfrac{1}{h_i} + \dfrac{d_i \ln(D_o/D_i)}{2\kappa} + \dfrac{D_i}{h_o D_o}} \tag{9–8}$$

Similarly, to base the overall heat exchanger conductance on the outside surface area, from equations (9–5) and (9–6) we have

$$U_o A_o = \frac{\pi D_o L}{\dfrac{D_o}{h_i D_i} + \dfrac{D_o \ln(D_o/D_i)}{2\kappa} + \dfrac{1}{h_o}} \tag{9–9}$$

with $A_o = \pi D_o L$. Then the overall heat exchanger conductance based on the outside surface area is

$$U_o = \frac{1}{\dfrac{D_o}{h_i D_i} + \dfrac{D_o \ln(D_o/D_i)}{2\kappa} + \dfrac{1}{h_o}} \tag{9–10}$$

Some typical values for the overall heat exchanger conductance are given in table 9–1 for some general configurations and specific heat exchangers. The reader can see that a large value for the conductance gives a more effective heat exchanger and one that requires less surface area. It can also be seen from equations (9–8) and (9–10) that the value of the overall conductance depends on appropriate material selection (to increase thermal conductivity) and effective flow patterns to cause high convective heat-transfer coefficients. Commercial heat exchangers, particularly very large ones, indicate the overall heat exchanger conductance on the nameplate, which is fastened to the housing of the heat exchanger.

Table 9–1 Some Typical Values for Overall Heat Exchanger Conductance

	U, W/m$^2 \cdot$ °C
Water to oil heat exchangers	60 to 400
Gas to gas heat exchangers	60 to 650
Air condensers	300 to 800
Ammonia condensers and evaporators	800 to 1500
Steam condensers	1500 to 5000

Another factor that needs to be considered in the design and operation of heat exchangers is the potential for the conductance to decrease with time. This could happen if the heat exchanger is operated with fluids that are dirty, contain suspended particles, are chemically reactive, or otherwise may make the convection and/or conduction of heat less effective. The concept of the fouling factor, F_f, has been introduced to account for the degradation of U with particular types of operations. The fouling factor is usually defined such that it can be used in equation (9–8) or (9–10). For a fouling factor based on the outside tube area, for equation (9–10), the corrected overall conductance is then

$$U_{o,\text{corrected}} = \frac{1}{\dfrac{D_o}{h_i D_i} + \dfrac{\ln(D_o/D_i)}{2\kappa} + \dfrac{1}{h_o} + F_f} \tag{9–11}$$

and if the fouling factor is based on the inside tube surface area then it should be included in equation (9–8) similar to that of equation (9–11). Some typical fouling factors listed in table 9–2 for selected applications. The reader interested in further discussion of the various parameters of heat exchangers may want to consult, for instance, Walker [1], Afgan and Schlünder [2], Schlünder [3], and Somerscales and Knudsen [4].

EXAMPLE 9–1 Determine the overall conductance for a single-shell single-tube heat exchanger with the cross-section shown in figure 9–9. The following data pertain to the heat exchanger:

$$\begin{aligned} \text{copper tube, } \kappa &= 400 \text{ W/m} \cdot \text{°C} \\ \text{Tube OD} &= D_o = 6 \text{ cm} \\ \text{Tube ID} &= D_i = 5 \text{ cm} \\ h_o &= h_A = 30 \text{ W/m}^2 \cdot \text{°C} \\ h_i &= h_B = 1220 \text{ W/m}^2 \cdot \text{°C.} \end{aligned}$$

Solution We may determine the overall conductance based on the outside surface area of the tube or the inside surface area. For the inside surface area we use equation (9–8),

Table 9–2 Selected Values for Fouling Factors, F_f

	m² · °C/W	hr · ft²/Btu
Seawater	0.0028	0.0005
Brackish water	0.0011	0.002
Cooling tower water	0.0114	0.002
Well water	0.0057	0.001
Hard water	0.0170	0.003
Muddy or silty water	0.0170	0.003
Transformer oil	0.0057	0.001
Fuel oil	0.0284	0.005
Engine lubricating oil	0.0057	0.001
Vegetable oil	0.0170	0.003
Natural gas	0.0057	0.001
Kerosene	0.0057	0.001
Refrigerant liquids	0.0057	0.001
Compressed air	0.0114	0.002

From reference [5].

Answer

$$U_i = \frac{1}{\dfrac{0.05\ \text{m}}{(30\ \text{W/m}^2\cdot{}^\circ\text{C})} + \dfrac{(0.05\ \text{m})\ln(6/5)}{2(400\ \text{W/m}^2\cdot{}^\circ\text{C})} + \dfrac{1}{1220\ \text{W/m}^2\cdot{}^\circ\text{C}}} = 34.95\frac{\text{W}}{\text{m}^2\cdot{}^\circ\text{C}}$$

For the conductance based on the outside area of the tube we use equation (9–10):

Answer

$$U_o = \frac{1}{\dfrac{1}{30} + \dfrac{(0.06)\ln(6/5)}{2(400)} + \dfrac{0.06}{(1220)(0.05)}} = 29.13\frac{\text{W}}{\text{m}^2\cdot{}^\circ\text{C}}$$

Notice that the overall conduction terms are the same; that is,

$$U_iA_i = U_oA_o$$

9–3 LMTD METHOD OF ANALYSIS

The determination of the heat transfer occurring in a heat exchanger can be made only if the temperature difference between the two fluid streams is known. That is, from equation (9–3) we must know, or be able to predict, the temperature difference ΔT. As we discussed in section 9–2, the overall heat exchanger conductance U and the surface area can be determined for a single-pass single shell and tube heat exchanger, shown in figure 9–7. If we refer to this figure again we see that the fluids are flowing in opposite directions. This arrangement is called a counterflow or opposed flow heat exchanger. If the fluids flow in the same direction, then the heat exchanger is called a parallel flow heat exchanger.

Considering the counterflow heat exchanger of figure 9–8 and repeated in figure 9–10, notice that the temperature distribution in each fluid through the heat exchanger is such that the cold fluid temperature T_c increases as the fluid flows through the heat exchanger. The hot fluid temperature T_h decreases as that fluid passes through the exchanger. As shown in figure 9–10a and b, a differential amount of heat $d\dot{Q}_{\text{HX}}$ is transferred from the hot to the cold fluid over a differential length dx. If we assume that the fluids do not change phase, then the enthalpy changes can be determined by the temperature changes, and from equation (9–3) we write

$$d\dot{Q}_{\text{HX}} = \dot{m}_c c_{p,c} dT_c = -\dot{m}_h c_{p,h} dT_h \qquad \textbf{(9–12)}$$

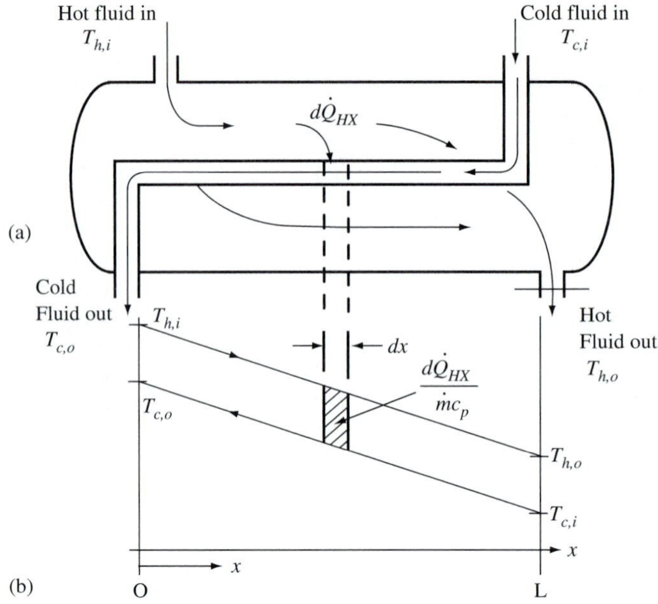

Figure 9–10 Counterflow heat exchanger and fluid temperature distribution.

where the negative sign indicates that the differential heat transfer is leaving the hot fluid. From equation (9–3) we also have that

$$d\dot{Q}_{\mathrm{HX}} = U(\Delta T)dA \tag{9–13}$$

where ΔT is the temperature difference between the hot and cold fluid at the differential element, $\Delta T = T_h - T_c$. Notice that a counterflow heat exchanger may perform in such a way that the outlet temperature of the cold fluid could be warmer than the outlet temperature of the hot fluid.

If we now assume the following for the heat exchanger:

1. Constant specific heats $c_{p,h}$ and $c_{p,c}$.
2. Constant overall conductance U.
3. Adiabatic heat exchanger.
4. No kinetic or potential energy changes.
5. No radiation heat transfer.
6. No heat transfer in the axial direction.

Then we write that, for the counterflow heat exchanger,

$$d(\Delta T) = dT_h + dT_c \tag{9–14}$$

This relationship results from the situation where the hot fluid is flowing in a positive x direction, as indicated in figure 9–10, while the cold fluid is flowing in a negative x direction. For a parallel flow heat exchanger, shown in figure 9–11. The differential temperature difference would be

$$d(\Delta T) = dT_h - dT_c$$

From equation (9–12), we may write, in general,

$$dT_h = -\frac{d\dot{Q}_{\mathrm{HX}}}{\dot{m}_h,c_{p,h}} \quad \text{and} \quad dT_c = \frac{d\dot{Q}_{\mathrm{HX}}}{\dot{m}_c c_{p,c}} \tag{9–15}$$

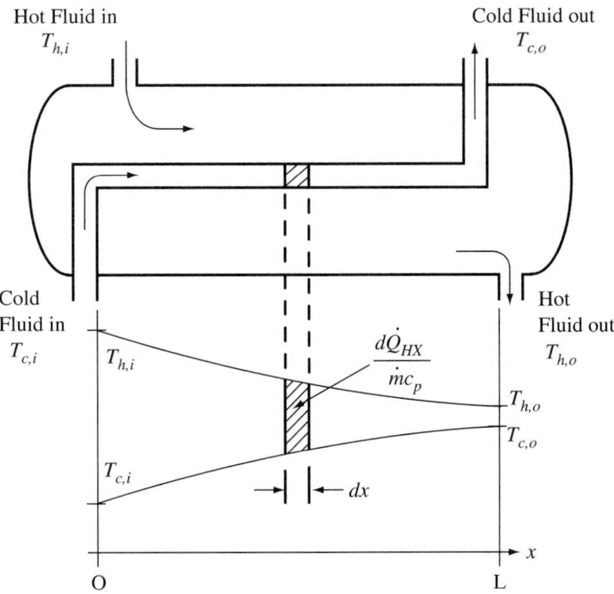

Figure 9–11 Parallel flow heat exchanger.

Substituting equations (9–15) into (9–14) gives, for the counter flow heat exchanger,

$$d(\Delta T) = d\dot{Q}_{\text{HX}}\left(\frac{1}{\dot{m}_c c_{p,c}} - \frac{1}{\dot{m}_h c_{p,h}}\right) \tag{9–16}$$

and substituting equation (9–13) into this result gives

$$d(\Delta T) = U(\Delta T)\left(\frac{1}{\dot{m}_c c_{p,c}} - \frac{1}{\dot{m}_h c_{p,h}}\right)dA$$

or

$$\frac{d(\Delta T)}{\Delta T} = U\left(\frac{1}{\dot{m}_c c_{p,c}} - \frac{1}{\dot{m}_h c_{p,h}}\right)dA \tag{9–17}$$

We can now integrate this over the axial length of the heat exchanger:

$$\int_{\Delta T_1}^{\Delta T_2} \frac{d(\Delta T)}{\Delta T} = U\left(\frac{1}{\dot{m}_c c_{p,c}} - \frac{1}{\dot{m}_h c_{p,h}}\right)\int_A dA \tag{9–18}$$

Since $dA = \pi D dx$ and setting $\Delta T_1 = T_{h,i} - T_{c,o}$ at $x = 0$ and $\Delta T_2 = T_{h,o} - T_{c,i}$ at $x = L$ we then get, after integrating equation (9–18),

$$\ln\left(\frac{\Delta T_2}{\Delta T_1}\right) = U\left(\frac{1}{\dot{m}_c c_{p,c}} - \frac{1}{\dot{m}_h c_{p,h}}\right)\pi DL \tag{9–19}$$

Also, from equation (9–12) we may write that

$$\dot{Q}_{\text{HX}} = \dot{m}_c c_{p,c}(T_{c,o} - T_{c,i}) = \dot{m}_h c_{p,h}(T_{h,i} - T_{h,o})$$

and rearrange these equations to read

$$\frac{1}{\dot{m}_c c_{p,c}} = \frac{T_{c,o} - T_{c,i}}{\dot{Q}_{\text{HX}}} \quad \text{and} \quad \frac{1}{\dot{m}_h c_{p,h}} = \frac{T_{h,i} - T_{h,o}}{\dot{Q}_{\text{HX}}} \tag{9–20}$$

Substituting equations (9–20) into (9–19) gives

$$\ln\left(\frac{\Delta T_2}{\Delta T_1}\right) = \frac{UA}{\dot{Q}_{\mathrm{HX}}}[T_{c,o} - T_{c,i} - T_{h,i} + T_{h,o}] \tag{9–21}$$

where $A = \pi DL$. Since we set $\Delta T_1 = T_{h,i} - T_{c,o}$ and $\Delta T_2 = T_{h,o} - T_{c,i}$, equation (9–21) becomes

$$\ln\left(\frac{\Delta T_2}{\Delta T_1}\right) = \frac{UA}{\dot{Q}_{\mathrm{HX}}}(\Delta T_2 - \Delta T_1)$$

or

$$\dot{Q}_{\mathrm{HX}} = UA\frac{\Delta T_2 - \Delta T_1}{\ln(\Delta T_2/\Delta T_1)} \tag{9–22}$$

It is convenient to define the *log mean temperature difference, LMTD* or ΔT_{LM}, as

$$\Delta T_{\mathrm{LM}} = \frac{\Delta T_2 - \Delta T_1}{\ln(\Delta T_2/\Delta T_1)} \tag{9–23}$$

and then

$$\dot{Q}_{\mathrm{LM}} = UA\Delta T_{\mathrm{LM}} \tag{9–24}$$

describes the heat transfer in a counterflow heat exchanger. This equation may be described as the *LMTD Method.*

For the parallel flow heat exchanger, where the pattern of fluid flow through the heat exchanger follows that shown in figure 9–11, the heat transfer is still described by equation (9–24) with the LMTD, ΔT_{LM}, defined by equation (9–23) with the temperature differences, referring to figure 9–11, given by:

$$\Delta T_2 = T_{h,o} - T_{c,o} \qquad \text{(parallel flow only)} \tag{9–25}$$

and

$$\Delta T_1 = T_{h,i} - T_{c,i} \qquad \text{(parallel flow only)} \tag{9–26}$$

If the temperature differences ΔT_2 and ΔT_1 turn out to be equal, then, whether the heat exchanger is parallel or counter flow, the LMTD becomes undefined from equation (9–23). In such instances the LMTD is replaced by the arithmetic average

$$\Delta T_{\mathrm{LM}} = \frac{1}{2}(\Delta T_1 + \Delta T_2) \tag{9–27}$$

in equation (9–24).

The cross-flow heat exchanger is a heat exchanger in which the flow patterns of the two fluid streams are perpendicular to each other, as diagramed in figure 9–12. The analysis of these devices can be conducted in the same fashion as for the counterflow heat exchangers so that the LMTD for cross flow would be that of equation (9–23).

The use of the LMTD method can be extended for analysis of heat exchangers other than the single-tube, single-shell type by using the LMTD correction factor F defined in the equation

$$\dot{Q}_{\mathrm{HX}} = FUA(\Delta T_{\mathrm{LM}}) \tag{9–28}$$

where the *log mean temperature difference is that for the counterflow heat exchangers, equation (9–23).* For the single-tube, single-pass heat exchanger, which we have just

Figure 9–12 Cross-flow heat exchanger.

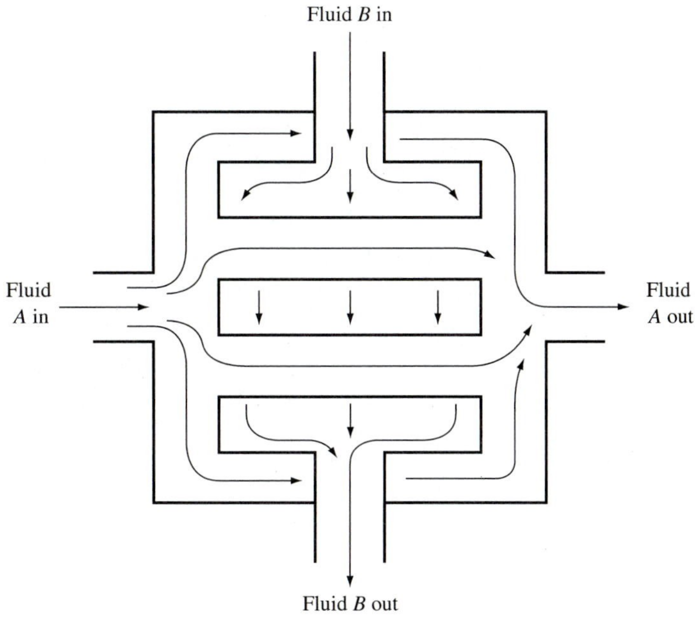

treated and from which we have derived the LMTD factor, the correction factor F is 1.0. For more elaborate devices this factor must be determined. For the one-shell, two-tube-pass heat exchanger shown in figure 9–13, the Tubular Exchange Manufacturers Association (TEMA) has published the appropriate correction factor. In figure 9–14 is shown this correction factor for various combinations of temperature. The correction factors obtained from this chart may be extended for use with multiples of two-tube-pass systems. In figure 9–15, for instance, is shown the configuration for a single-shell, six-tube-pass heat exchanger, which can be analyzed with the correction factor obtained from figure 9–14.

Figure 9–13 One-shell, two-tube-pass heat exchanger.

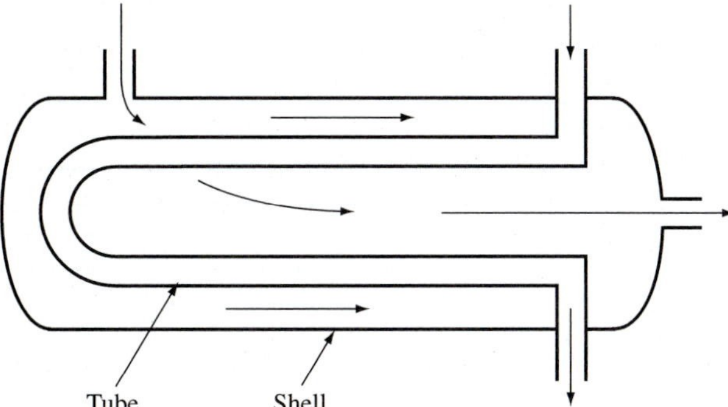

F = LMTD Correction Factor

R=100, 20.0, 15.0, 8.0, 6.0, 4.0, 3.0, 2.5, 2.0, 1.8, 1.6, 1.4, 1.2, 1.0, 0.9, 0.8, 0.7, 0.6, 0.5, 0.4, 0.3, 0.2, 0.1

p = Temperature Efficiency

	LMTD Correction Factor	
	1 Shell Pass	Even Number of Tube Passes
	$P = \dfrac{t_2 - t_1}{T_1 - t_1}$	$R = \dfrac{T_1 - T_2}{t_2 - t_1}$

Figure 9–14 LMTD correction factor for single shell and two-, four-, six-, ... tube-pass heat exchangers. (Reproduced by permission of Tubular Exchange Manufacturers Association, Inc., Tarrytown, New York [5])

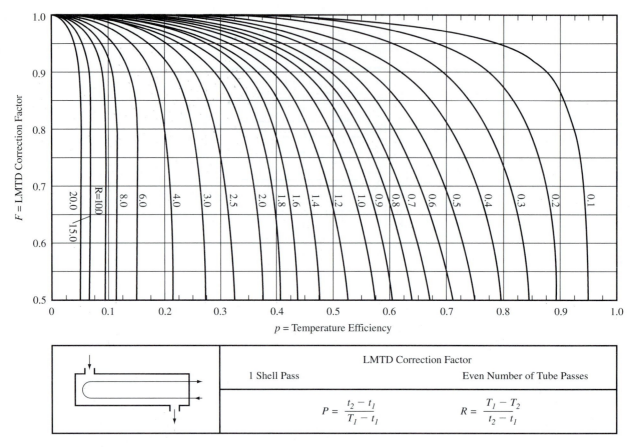

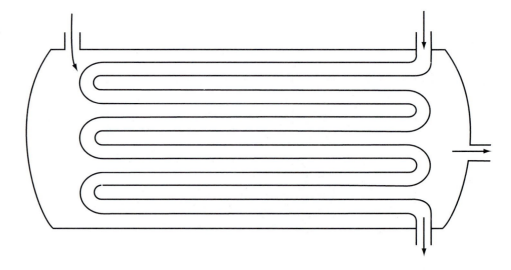

Figure 9–15 Single-shell, six-tube-pass heat exchanger.

Some other configurations for shell and tube heat exchangers and their appropriate correction factors for use with equation (9–28) are given in figures 9–16, 9–17, 9–18, and 9–19. TEMA [5] has provided more extensive information of other heat exchanger arrangements for the interested reader. In figures 9–14, 9–16, and 9–17 the ratios R and P are determined by using temperatures of the fluids. The lower case t's are for the cold fluid and the capital T's are for the hot fluid.

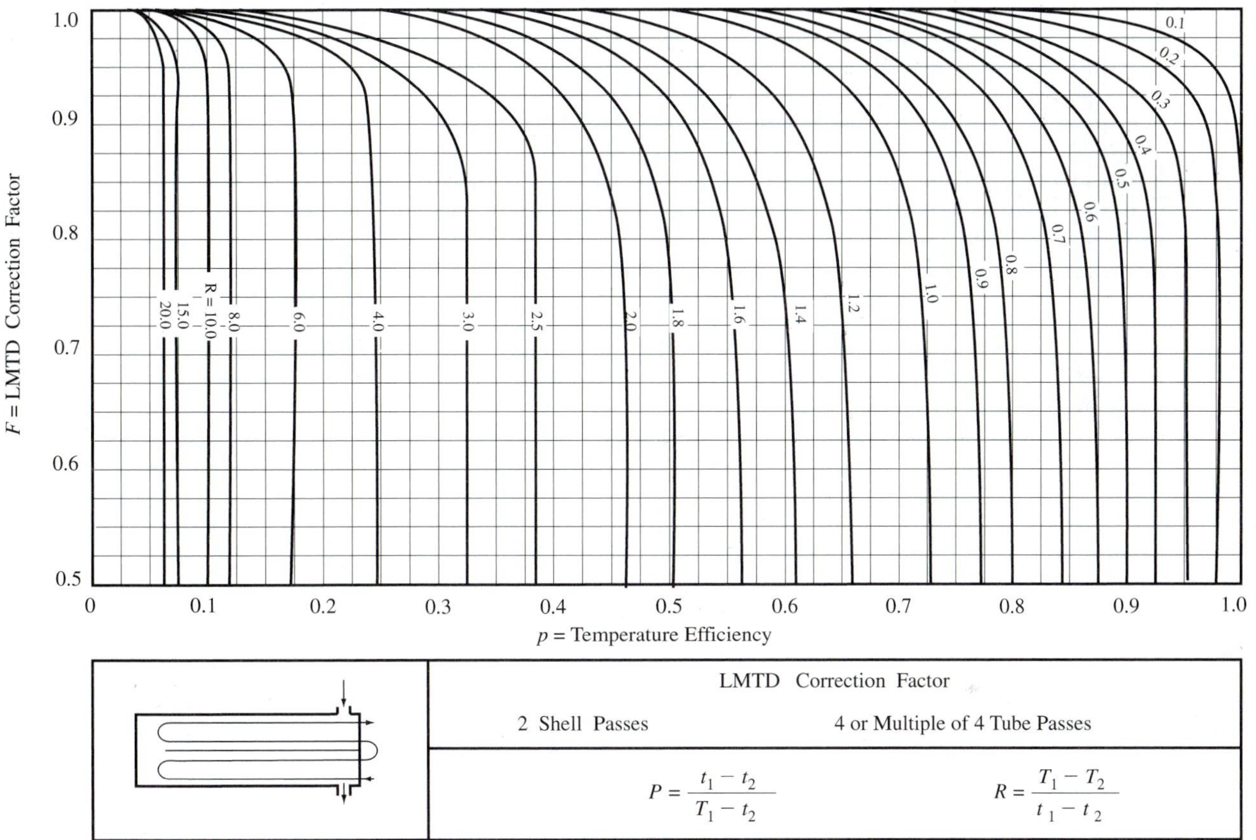

Figure 9–16 LMTD correction factors for two-shell, four or multiples of four–tube-pass heat exchanger. (Reproduced by permission of Tubular Exchanger Manufacturers Association, Inc., Tarrytown, New York [5])

Multiple-Pass Heat Exchanger Analysis

Heat exchangers having multiple passes in a shell may be analyzed using the information provided in figures 9–14, 9–16, and 9–17. The LMTD, ΔT_{LM}, for these arrangements can be determined as if the heat exchanger operated as a counterflow type. Thus, the LMTD is determined from equation (9–23).

Mixed and Unmixed Flow

In the information of figures 9–18 and 9–19 the terms *mixed* and *unmixed* refer to a structural configuration in the heat exchanger such that in an unmixed flow the fluid enters the heat exchanger, is divided to flow through multiple channels, and does not reunite until it leaves the heat exchanger. In a mixed flow heat exchanger the fluid is not directed through multiple channels but is free to mix in the bulk flow throughout the heat exchanger.

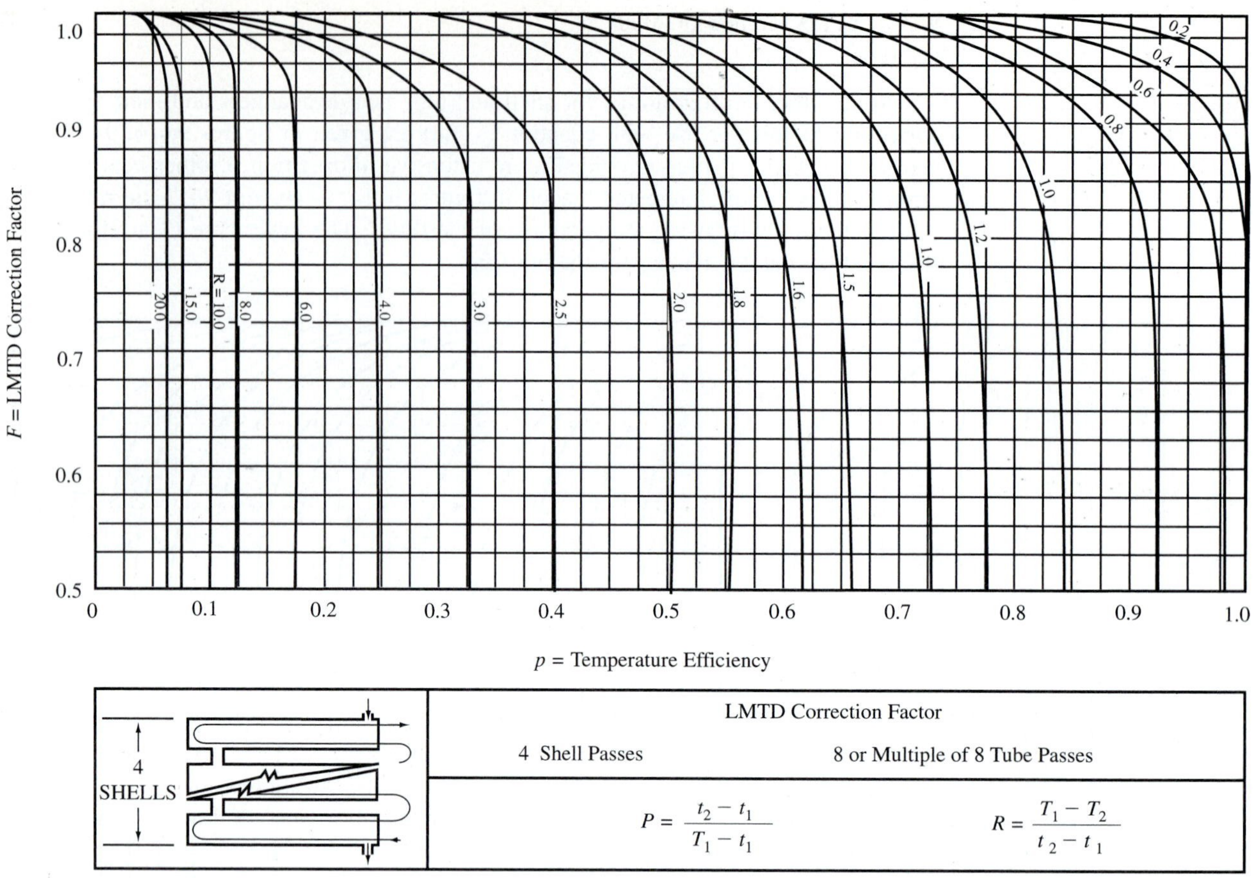

F = LMTD Correction Factor

p = Temperature Efficiency

R = 20.0, 15.0, 10.0, 8.0, 6.0, 4.0, 3.0, 2.5, 2.0, 1.8, 1.6, 1.5, 1.2, 1.0, 0.8, 0.6, 0.4, 0.2

4 SHELLS

LMTD Correction Factor

4 Shell Passes — 8 or Multiple of 8 Tube Passes

$$P = \frac{t_2 - t_1}{T_1 - t_1} \qquad R = \frac{T_1 - T_2}{t_2 - t_1}$$

Figure 9–17 LMTD correction factor for four-shell, eight or multiple of eight–tube-pass heat exchangers. (Reproduced by permission of Tubular Exchanger Manufacturers Association, Inc., Tarrytown, New York [5])

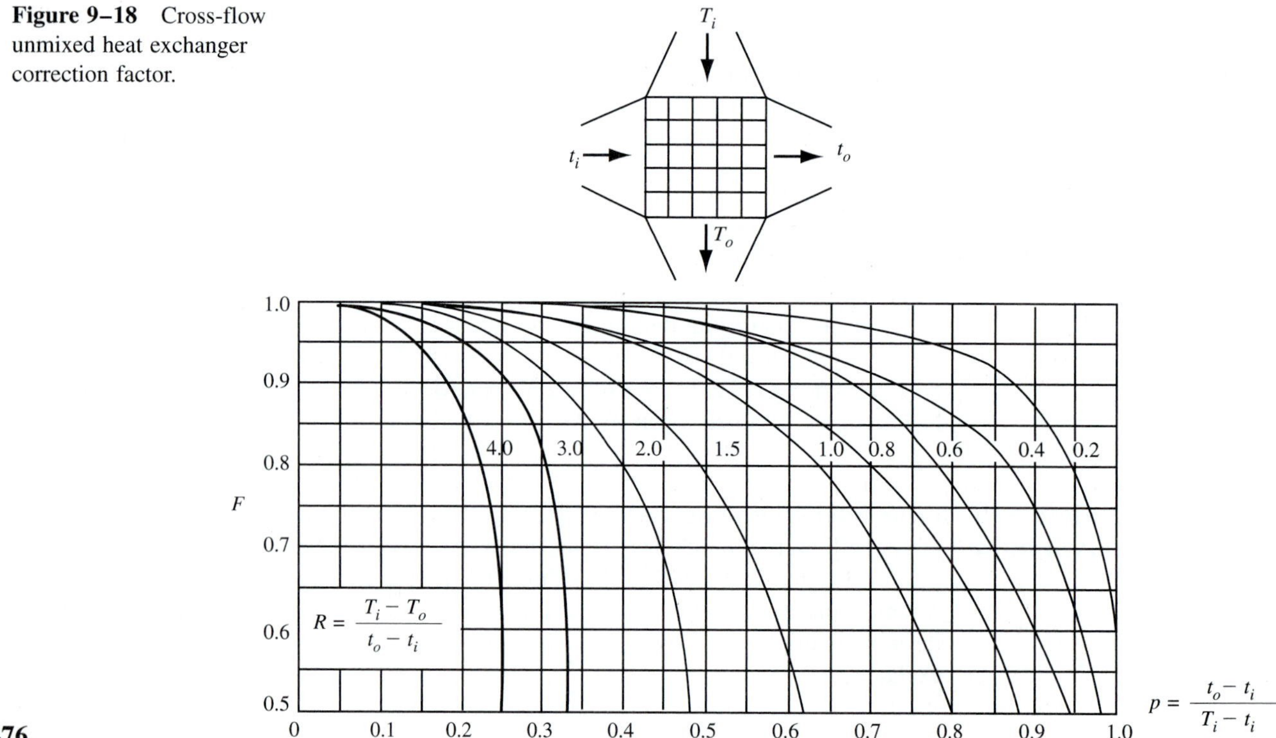

Figure 9–18 Cross-flow unmixed heat exchanger correction factor.

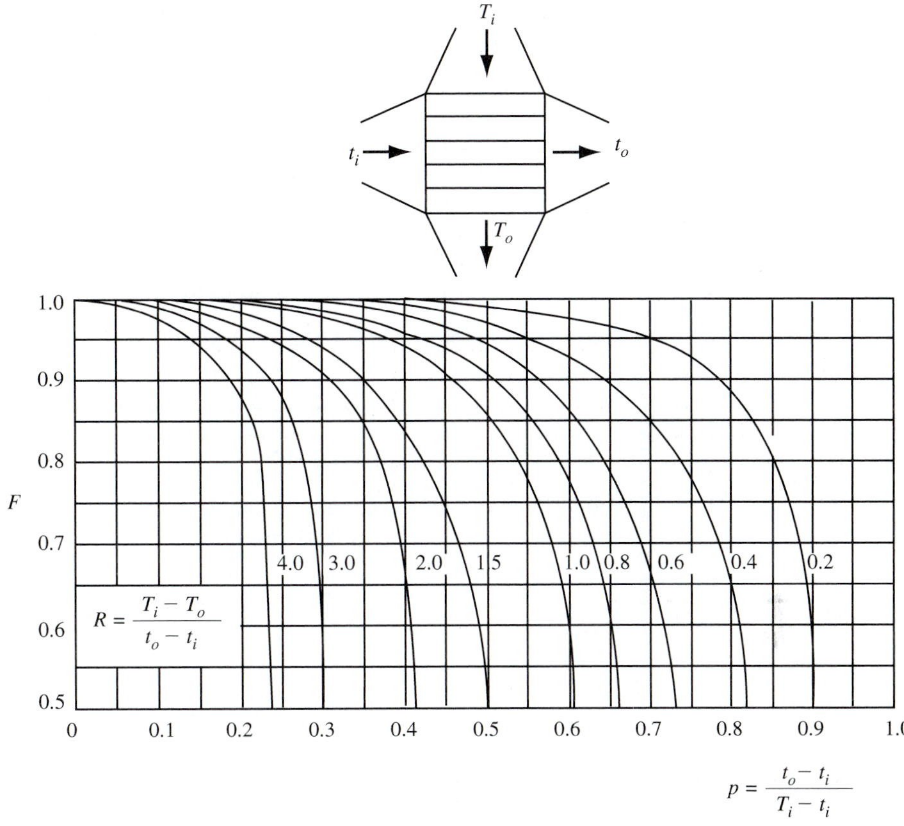

Figure 9–19 Cross-flow, one fluid mixed flow heat exchanger correction factor.

Corrections for Fluid Phase Change

In all of these analyses the fluids are assumed to exhibit temperature changes in the heat exchanger and not have phase changes. For the situation where one of the fluids exhibits a phase change inside the heat exchanger, the correction factor F is 1.0.

EXAMPLE 9–2

A cross-flow heat exchanger is used in a cogeneration system where 100 kg/s of exhaust gases at 300°C are used to heat water from 15°C to 75°C. A schematic of the system is shown in figure 9–20. The overall conductance is 1100 W/m^2 · °C and a fouling factor of

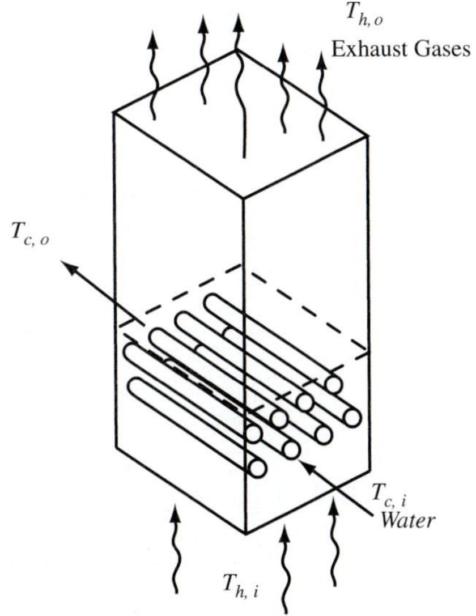

Figure 9–20 Cross-flow cogeneration heat exchanger.

0.0003 $m^2 \cdot °C/W$ is recommended because the gases are unpredictable and may contain significant amounts of soot and other contaminants. Determine the necessary area if 35 kg/s of water is to be heated.

Solution

We will use equation (9–28) to determine the surface area A of the heat exchanger. The overall corrected conductance is, from equation (9–11),

$$U_{\text{corrected}} = \frac{1}{\left(\frac{1}{U} + F_f\right)} = \frac{1}{\left(\frac{1}{1100\ \text{W/m}^2 \cdot °\text{C}} + 0.0003\ \text{m}^2 \cdot °\text{C/W}\right)} = 827 \frac{\text{W}}{\text{m}^2 \cdot °\text{C}}$$

The heat transfer is determined from the energy balance,

$$\dot{Q}_{\text{HX}} = \dot{m}_{\text{water}} c_{p,\text{water}} \Delta T_{\text{water}} = \left(35 \frac{\text{kg}}{\text{s}}\right)\left(4.186 \frac{\text{kJ}}{\text{kg} \cdot °\text{C}}\right)(60°\text{C})$$
$$= 8790.6\ \text{MW}$$

The temperature of the exhaust gases leaving the heat exchanger can be determined from an energy balance of that fluid stream,

$$\dot{Q}_{\text{HX}} = \dot{m}_{\text{gas}} c_{p,\text{gas}} \Delta T_{\text{gas}} - 8790.6\ \text{MW}$$

Assuming a value of 1.007 $kJ/kg \cdot °C$ for the exhaust gas specific heat gives

$$\Delta T_{\text{gas}} = \frac{\dot{Q}_{\text{HX}}}{\dot{m}_{\text{gas}} c_{p,\text{gas}}} = 87.3°\text{C}$$

Then

$$T_{\text{gas},o} = 300°\text{C} - 87.3°\text{C} = 212.7°\text{C}$$

We assume that the heat exchanger allows the exhaust gases to mix but the water is unmixed. Referring to figure 9–19, we have

$$t_i = T_{c,i} - 15°\text{C} \qquad T_i = T_{h,i} = 300°\text{C}$$
$$t_o = T_{c,o} = 75°\text{C} \qquad T_o = T_{h,o} = 212.3°\text{C}$$

Then

$$R = (300 - 212.3)/(75 - 15) = 1.45$$
$$P = (75 - 15)/(300 - 15) = 0.21$$

From figure 9–19, $F \approx 0.975$. For the cross-flow heat exchanger we determine the LMTD from equation (9–23),

$$\Delta T_{\text{LM}} = \frac{212.3 - 15 - 300 + 75}{\ln(197.3/225)} = 210.8°\text{C} = 210.8\ \text{K}$$

The heat exchanger surface area is

Answer

$$A = \frac{\dot{Q}_{\text{HX}}}{U \Delta T_{\text{LM}}} = 51.7\ \text{m}^2$$

EXAMPLE 9–3

A two-shell, eight-tube-pass heat exchanger is to be used to heat 80 lbm/s of oil from 20°F to 110°F by using 100 lbm/s of high-pressure water at 180°F. If the overall conductance is taken as 50 $Btu/hr \cdot ft^2 \cdot °F$, determine the necessary surface area of the heat exchanger.

Solution

We will use equation (9–28) to determine the surface area. The heat exchange between the fluid streams is

$$\dot{Q}_{HX} = [\dot{m}c_p\Delta T]_{oil} = [\dot{m}c_p\Delta T]_{water}$$

From appendix table B–3 we read that the specific heat of unused engine oil is 1.89 kJ/kg · K or 0.45 Btu/lbm · °F, so

$$\dot{Q}_{HX} = \left(80\frac{\text{lbm}}{3}\right)\left(0.45\frac{\text{Btu}}{\text{lbm}\cdot{}^\circ\text{F}}\right)(110 - 20^\circ\text{F}) = 3240\frac{\text{Btu}}{\text{s}}$$

The temperature change of the water can then be determined from the calculation, assuming $c_p = 1.0$ Btu/lbm · °F for water,

$$\Delta T_{water} = \frac{\dot{Q}_{HX}}{[\dot{m}c_p]_{water}} = \frac{3240 \text{ Btu/s}}{(100 \text{ lbm/s})(1.0 \text{ Btu/lbm}\cdot{}^\circ\text{F})} = 32.4^\circ\text{F}$$

and the outlet water temperature,

$$T_{h,o} = 180^\circ\text{F} - 32.4^\circ\text{F} = 147.6^\circ\text{F}$$

Referring to figure 9–17 for a two-shell heat exchanger, we have

$$t_i = T_{c,i} = 20^\circ\text{F} \qquad T_i = T_{h,i} = 180^\circ\text{F}$$
$$t_o = T_{c,o} = 110^\circ\text{F} \qquad T_o = T_{h,o} = 147.6^\circ\text{F}$$

so that

$$R = (180 - 147.6)/90 = 0.36$$

and

$$P = 90/(180 - 20) = 0.56$$

Then the correction factor from figure 9–17 is

$$F \approx 1.0$$

The LMTD is

$$\Delta T_{LX} = \frac{147.6 - 20 - 180 + 110}{\ln\dfrac{127.6}{70}} = 95.9^\circ\text{F}$$

and the surface area is

Answer

$$A = \frac{\dot{Q}_{HX}}{UF\Delta T_{LM}} = 2431.6 \text{ ft}^2$$

9–4 EFFECTIVENESS–NTU METHOD OF ANALYSIS

Heat exchangers can be analyzed by using the LMTD Method; however, if only two of the fluid temperatures are known, an iterative process for the LMTD Method needs to be used to fully analyze and define a heat exchanger. It is then more convenient to use the *Effectiveness–NTU Method* for analysis of heat exchangers. The effectiveness, ε_{HX}, for a heat exchanger is defined as

$$\varepsilon_{HX} = \frac{\text{Actual heat transfer}}{\text{Maximum possible heat transfer}} \tag{9–29}$$

The actual heat transfer is, from equation (9–2)

$$\dot{Q}_{HX,actual} = [\dot{m}c_p\Delta T]_{cold} = [\dot{m}c_p\Delta T]_{hot}$$

and the ideal possible heat transfer is that which would be achieved by changing a fluid's temperature the ideal maximum amount. This ideal maximum temperature change is

$$\Delta T_{\text{ideal}} = T_{h,i} - T_{c,i} \tag{9–30}$$

There is a restriction on this criterion in that the fluid experiencing the ideal maximum temperature change must be that fluid having the minimum mass flow rate/specific heat product; that is,

$$\dot{Q}_{\text{HX,ideal}} = [\dot{m}c_p]_{\text{minimum}} \Delta T_{\text{ideal}} \tag{9–31}$$

Thus, in the analysis of a heat exchanger the mass flow rates and the specific heats must be established and the minimum of the cold or hot fluids is used in equation (9–31).

For the analysis of a counterflow heat exchanger it is useful to recall equation (9–19),

$$\ln\left(\frac{\Delta T_2}{\Delta T_1}\right) = U\left(\frac{1}{\dot{m}_c c_{p,c}} - \frac{1}{\dot{m}_h d_{p,h}}\right)\pi DL$$

and assume that the fluids are such that the cold fluid has minimum mass flow rate/specific heat product,

$$[\dot{m}c_p]_{\text{minimum}} = \dot{m}_c c_{p,c} \qquad \textit{Assumption}$$

Using equation (9–29) we find that the effectiveness for this situation is, for cold fluid minimum $\dot{m}c_p$,

$$\varepsilon_{\text{HX}} = \frac{\dot{m}_c c_{p,c}(T_{c,o} - T_{c,i})}{\dot{m}_c c_{p,c}(T_{h,i} - T_{c,i})} = \frac{T_{c,o} - T_{c,i}}{T_{h,i} - T_{c,i}} \tag{9–32}$$

Also, we define

$$C_{\text{minimum}} = [\dot{m}c_p]_{\text{minimum}} \quad \text{and} \quad C_{\text{maximum}} = [\dot{m}c_p]_{\text{maximum}}$$

$$C = \frac{C_{\text{minimum}}}{C_{\text{maximum}}}$$

and equation (9–19) can then be written

$$\ln\left(\frac{\Delta T_1}{\Delta T_2}\right) = -\frac{UA}{C_{\text{minimum}}}(1 - C) \tag{9–33}$$

or

$$\frac{T_{h,i} - T_{c,o}}{T_{h,o} - T_{c,o}} = e^{-\frac{UA}{C_{\text{minimum}}}(1-C)} \tag{9–34}$$

It is useful now to eliminate the outlet hot fluid temperature from the left side of equation (9–34). Using

$$\dot{m}_c c_{p,c}(T_{c,o} - T_{c,i}) = \dot{m}_h c_{p,h}(T_{h,i} - T_{h,o})$$

we get

$$T_{h,o} = \frac{\dot{m}_c c_{p,c}}{\dot{m}_h c_{p,h}}(T_{c,i} - T_{c,o}) + T_{h,i}$$

Substituting this result into equation (9–34) gives

$$\frac{T_{h,i} - T_{c,o}}{T_{h,i} + C(T_{c,i} - T_{c,o}) - T_{c,i}} = e^{-\frac{UA}{C_{\text{minimum}}}(1-C)}$$

If we now add the term $T_{c,i} - T_{c,i}$ ($= 0$) to the numerator of the left side and rearrange, we have

$$\frac{(T_{h,i} - T_{c,i}) + (T_{c,i} - T_{c,o})}{(T_{h,i} - T_{c,i}) + C(T_{c,i} - T_{c,o})} = e^{-\frac{UA}{C_{\text{minimum}}}(1-C)}$$

Dividing top and bottom of the left side by the term $(T_{h,i} - T_{c,i})$ gives

$$\frac{1 + \dfrac{T_{c,i} - T_{c,o}}{T_{h,i} - T_{c,i}}}{1 + C\left(\dfrac{T_{c,i} - T_{c,o}}{T_{h,i} - T_{c,i}}\right)} = e^{-\frac{UA}{C_{\text{minimum}}}(1-C)} \tag{9–35}$$

The terms $(T_{c,i} - T_{c,o})/(T_{h,i} - T_{c,i})$ in the left side of this equation are equal to $-\varepsilon_{\text{HX}}$ from equation (9–32), so equation (9–35) becomes

$$\frac{1 - \varepsilon_{\text{HX}}}{1 - C\varepsilon_{\text{HX}}} = e^{-\frac{UA}{C_{\text{minimum}}}(1-C)} \tag{9–36}$$

This equation can be rearranged some more to obtain the equation for the heat exchanger effectiveness,

$$\varepsilon_{\text{HX}} = \frac{1 - e^{-\frac{UA}{C_{\text{minimum}}}(1-C)}}{1 - Ce^{-\frac{UA}{C_{\text{minimum}}}(1-C)}} \tag{9–37}$$

Equation (9–37) also results if the hot fluid were assumed to have had the minimum $\dot{m}c_p$ term. Thus, equation (9–37) is the relationship for a counterflow heat exchanger. The term UA/C_{min} is sometimes referred to as the number of thermal units, NTU or just N. If the product $\dot{m}c_p$ is the same for both fluid streams such that

$$[\dot{m}c_p]_{\text{hot}} = [\dot{m}c_p]_{\text{cold}}$$

then $C = 1.0$, and it can be shown that

$$\varepsilon_{\text{HX}} = \frac{\text{NTU}}{\text{NTU} + 1} \tag{9–38}$$

For parallel flow heat exchangers, using a similar analysis to that applied to the counterflow system, the effectiveness is given by the equation

$$\varepsilon_{\text{HX}} = \frac{1}{1 + C}(1 - \text{e}^{-\text{NTU}(1+C)}) \tag{9–39}$$

Effectiveness equations for other heat exchanger configurations are given in table 9–3. In table 9–4 are presented the relationships for determining the NTU's as functions of the effectiveness and the $\dot{m}c_p$ terms. Also, for convenience the graphical results of some of the effectiveness equations are presented in figures 9–21 through 9–27. For heat exchangers where there is a phase change occurring, such as in evaporators or condensers, the fluid's temperature change will be zero ($\Delta T = 0$) and this will imply that $[\dot{m}c_p]_{\text{maximum}} \rightarrow \infty$ and $c \rightarrow 0$. In this situation the heat exchanger effectiveness is

$$\varepsilon_{\text{HX}} = 1 - \text{e}^{-\text{NTU}} \tag{9–40}$$

for all arrangements of heat exchangers.

Table 9–3 ε_{HX}–NTU Relationships

Heat Exchanger Configuration	Relationship (Equation)	Text Equation No.
Concentric tube—parallel flow	$\varepsilon_{HX} = \frac{1}{1+C}(1 - e^{-NTU(1+C)})$	(9–39)
Concentric tube—counter flow	$\varepsilon_{HX} = \frac{1 - e^{\frac{-UA}{C_{minimum}}(1-C)}}{1 - Ce^{\frac{-UA}{C_{minimum}}(1-C)}} = \frac{1 - e^{-NTU(1-C)}}{1 - Ce^{-NTU(1-C)}}$	(9–37)
Shell and tube—1 shell with 2, 4, 6, 8, ... passes	$\varepsilon_{HX,1} = \frac{2}{\left[1 + C + \sqrt{1+C^2}\left(\frac{1 + e^{-NTU\sqrt{1+C^2}}}{1 - e^{-NTU\sqrt{1+C^2}}}\right)\right]}$	(9–41)
Shell and tube—*n* shells with 2*n*, 4*n*, 6*n*, ... passes	$\varepsilon_{HX} = \frac{\left[\left(\frac{1 - \varepsilon_{HX,1}C}{1 - \varepsilon_{HX,1}}\right)^n - 1\right]}{\left[\left(\frac{1 - \varepsilon_{HX,1}C}{1 - \varepsilon_{HX,1}}\right)^n - C\right]}$	(9–42)
Cross flow—single pass unmixed	$\varepsilon_{HX} = 1 - e^{\frac{1}{C}NTU^{0.22}(e^{-C(NTU)^{0.78}} - 1)}$	(9–43)
Cross flow—single pass C_{max} mixed, C_{min} unmixed	$\varepsilon_{HX} - \frac{1}{C}(1 - e^{C(e^{-NTU} - 1})$	(9–44)
Cross flow—single pass C_{min} mixed, C_{max} unmixed	$\varepsilon_{HX} = 1 - e^{\frac{e^{-C(NTU)} - 1}{C}}$	(9–45)
All heat exchangers with $C = 0$	$\varepsilon_{HX} = 1 - e^{-NTU}$	(9–40)

Table 9–4 NTU–ε_{HX} Relationships

Heat Exchanger Configuration	Relationship (Equation)	Text Equation No.
Concentric tube—parallel flow	$NTU = \frac{-\ln[1 - \varepsilon_{HX}(1 + C)]}{1 + C}$	(9–39a)
Concentric tube—Counter flow	$NTU = \frac{1}{1 - C}\ln\left(\frac{\varepsilon_{HX} - 1}{C\varepsilon_{HX} - 1}\right)$	(9–37a)
Shell and Tube—1 shell with 2, 4, 6, 8, ... passes	$NTU = \frac{-1}{\sqrt{1 + C^2}}\ln\left(\frac{\bar{\varepsilon}_{HX} - 1}{\bar{\varepsilon}_{HX} + 1}\right)$, where	(9–41a)
	$\bar{\varepsilon}_{HX} = \frac{1}{\sqrt{1 + C^2}}\left(\frac{2}{\varepsilon_{HX,1}} - (1 + C)\right)$	(9–41b)
Shell and tube—*n* shells with 2*n*, 4*n*, 6*n*, ... passes	Use equations (9–41a) and (9–41b) with $\varepsilon_{HX,1} = \frac{\bar{\varepsilon}_n - 1}{\bar{\varepsilon}_n - C}$ and $\bar{\varepsilon}_n = n\sqrt{\frac{\varepsilon_{HX}C - 1}{\varepsilon_{HX} - 1}}$	(9–42a)
Cross flow—single pass unmixed	$NTU^{0.22}(e^{-CNTU^{0.78}} - 1) = C\ln(1 - \varepsilon_{HX})$ (iterative solution required)	(9–43a)
Cross flow—C_{max} mixed, C_{min} unmixed	$NTU = -\ln\left(1 + \frac{1}{C}\ln(1 - C\varepsilon_{HX})\right)$	(9–44a)
Cross flow—C_{min} mixed, C_{max} unmixed	$NTU = \frac{-1}{C}\ln[C\ln(1 - \varepsilon_{HX}) + 1]$	(9–45a)
All heat exchangers with $C = 0$	$NTU = -\ln(1 - \varepsilon_{HX})$	(9–40a)

Figure 9–21 Effectiveness of concentric-tube, single-pass parallel flow heat exchanger, equation (9–39).

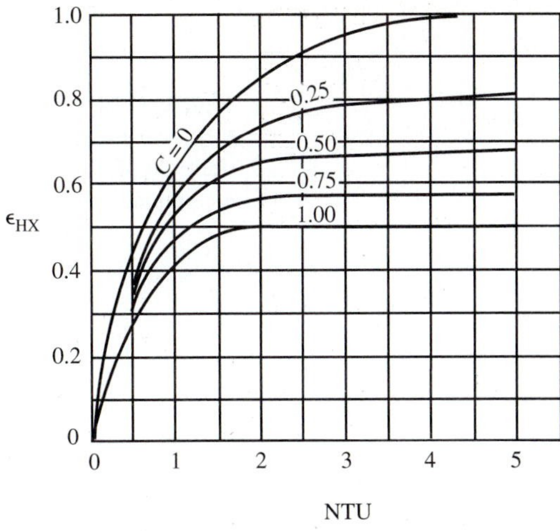

Figure 9–22 Effectiveness of concentric-tube, single-pass counterflow heat exchanger, equation (9–37).

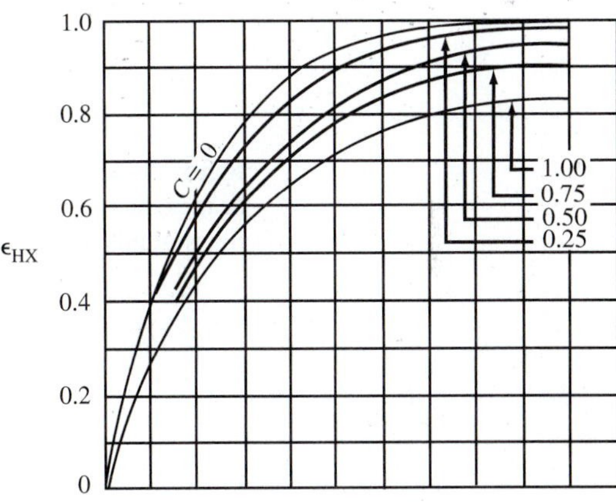

Figure 9–23 Effectiveness of shell and tube heat exchanger with one shell and any multiple of two passes, such as two, four, six, eight, etc., equation (9–41).

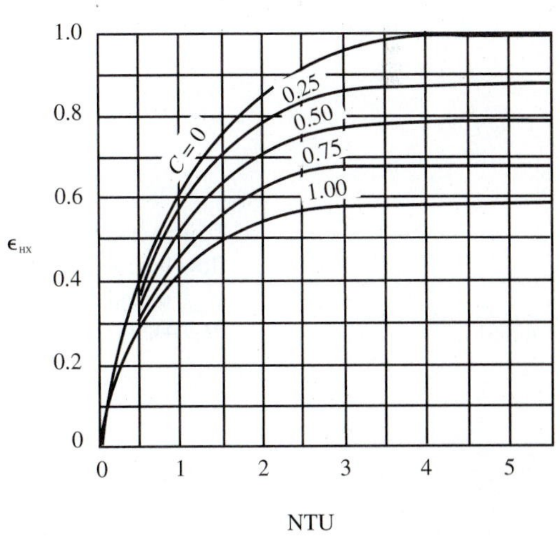

For the effectiveness–NTU relationship of a multi-shell and tube arrangement, given by equation 9–42 of Table 9–3, the calculation for $\varepsilon_{HX,1}$ using equation 9–41 uses an NTU value of the total NTU of the heat exchanger divided by n, the number of shells. Also, for the NTU-ε_{HX} relationship of equation 9–42a the NTU obtained from equation 9–41a needs to be multiplied by n to obtain the total NTU's of the multi-shell heat exchanger.

Figure 9–24 Effectiveness of shell and tube heat exchanger with 2 shells and any multiple of 4 tube passes such as 4, 8, 12, etc., equation (9–42).

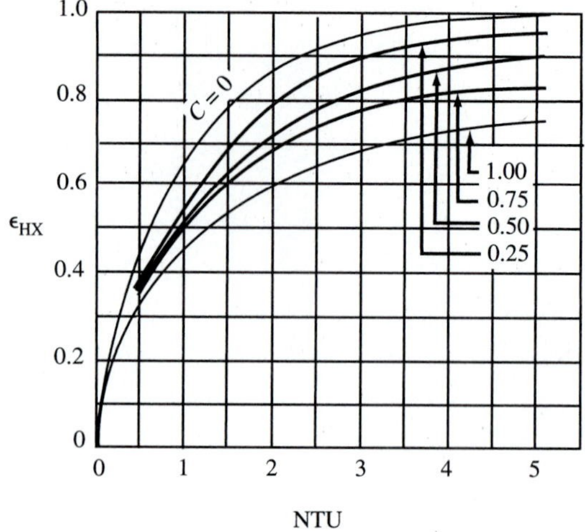

Figure 9–25 Effectiveness of cross flow heat exchanger, single pass with both fluid flows unmixed, equation (9–43).

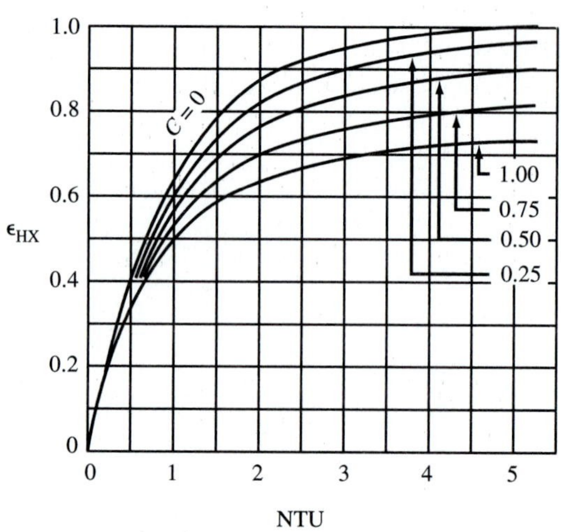

Figure 9–26 Effectiveness of cross flow heat exchanger, single pass, with C_{max} mixed and C_{min} unmixed, equation (9–44).

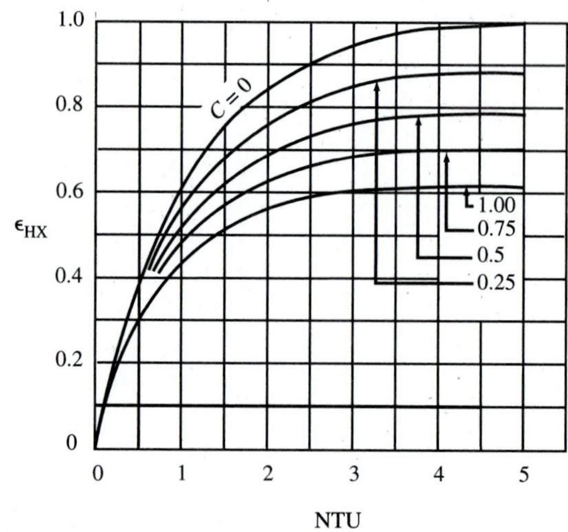

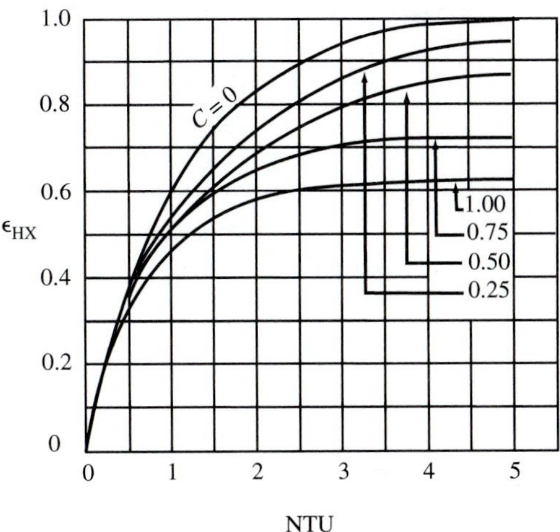

Figure 9–27 Effectiveness of cross flow heat exchanger with C_{min} mixed and C_{max} unmixed, equation (9–45).

EXAMPLE 9–4 A single-shell four-tube-pass heat exchanger is to be used to cool 30 lbm/s of 65°F water by using 40 lbm/s of ammonia at 10°F and 90 psia pressure. The system is intended to use the ammonia as a liquid without it evaporating. Ammonia will evaporate at 51°F and 90 psia. If the heat exchanger's overall conductance is 100 Btu/hr · ft² · °F and its surface area is 800 ft², determine the expected outlet temperature of the water and the ammonia. Does the ammonia evaporate in the heat exchanger?

Solution We determine the specific heat/mass flow rate products:

$$\dot{m}_h c_{p,h} = (30\ \text{lbm/s})(1.0\ \text{Btu/lbm}\cdot{}^\circ\text{F}) = 30\ \text{Btu/s}\cdot{}^\circ\text{F}$$
$$\dot{m}_c c_{p,c} = (40\ \text{lbm/s})(4.8\ \text{kJ/kg}\cdot{}^\circ\text{C})(0.2388\ \text{Btu}\cdot\text{kg}\cdot{}^\circ\text{C/lbm}\cdot{}^\circ\text{F}\cdot\text{kJ}) = 45.85\ \text{Btu/s}\cdot{}^\circ\text{F}$$

The minimum of these two terms is the water or hot fluid, so

$$C = \frac{C_{min}}{C_{max}} = 0.654$$

Also, the number of thermal units, NTU, can be computed:

$$\text{NTU} = \frac{UA}{C_{min}} = \frac{(100\ \text{Btu/ft}^2\cdot{}^\circ\text{F})(800\ \text{ft}^2)}{(30\ \text{Btu/s}\cdot{}^\circ\text{F})} = 0.74$$

From figure 9–25 we read the effectiveness as

$$\varepsilon_{HX} \approx 0.48$$

Then we use the definition of the effectiveness to get

$$\varepsilon_{HX} = \frac{\dot{m}_h c_{p,h}(T_{h,i} - T_{h,o})}{C_{min}(T_{h,i} - T_{c,i})} = 0.48$$

Answer so that $T_{h,o} = 65°\text{F} - 0.48(55°\text{F}) = 38.6°\text{F}$

The ammonia temperature can be determined from the relationship

$$\dot{m}_c c_{p,c}(T_{c,o} - 10°\text{F}) = \dot{m}_h c_{p,h}(65°\text{F} - 38.6°\text{F})$$

Answer or $T_{c,o} = 10°\text{F} + \left(\frac{30}{45.85}\right)(26°\text{F}) = 27.3°\text{F}$

The ammonia will not evaporate.

EXAMPLE 9–5 A condenser in a power plant is constructed as a cross-flow heat exchanger with lake water used to condense steam at 40°C. The lake water is supplied at 700 kg/s and 20°C. The overall heat exchanger conductance is 350 W/m^2 · °C, and its surface area is 3000 m^2. Determine the temperature of the water returning to the lake and the flow rate of the steam if the heat removed from the steam is just that of its heat of condensation, $hn_{fg} = 2406$ kJ/kg.

Solution

Since the condenser involves the phase change of steam, we have

$$C = 0 \quad \text{and} \quad \dot{m}_h c_{p,h} \rightarrow \infty$$

Then
$$[\dot{m}c_p]_{\min} = C_{\min} = (700\ \text{kg/s})(4.18\ \text{kJ/kg} \cdot {}^\circ\text{C}) = 2926\ \text{kW/}^\circ\text{C}$$

The NTU is determined as

$$\text{NTU} = \frac{UA}{C_{\min}} = \frac{(350\ \text{W/m}^2 \cdot {}^\circ\text{C})(3000\ \text{m}^2)}{(2{,}926{,}000\ \text{W/}^\circ\text{C})} = 0.35885$$

Also, the effectiveness is

$$\varepsilon_{\text{HX}} = 1 - e^{-\text{NTU}} = 1 - e^{-0.35885} = 0.30$$

Then we can determine the outlet temperature of the lake water from the calculation

$$T_{w,o} = 20^\circ\text{C} + (0.3)\,(40 - 20^\circ\text{C}) = 26^\circ\text{C}$$

The heat exchange is

$$\dot{Q}_{\text{HX}} = \dot{m}c_p[T_{w,i} - T_{w,o}] = (2926\ \text{kW/}^\circ\text{C})(6^\circ\text{C}) = 17{,}556\ \text{kW}$$

and the flow rate of the steam is

Answer
$$\dot{m}_{\text{steam}} = \frac{\dot{Q}_{\text{HX}}}{h_{f-g}} = \frac{17{,}556\ \text{kJ/s}}{2406\ \text{kJ/kg}} = 7.297\ \text{kg/s}$$

9–5 COMPACT HEAT EXCHANGERS

It has been pointed out by Kays and London [6] that the heat transfer rate or the convection heat-transfer coefficient in heat exchangers is directly proportional to the velocity of the fluids passing through the heat exchanger, while the power required to overcome the frictional resistance inside the flow paths of the heat exchanger is proportional to the velocity squared. Thus, the engineering design of heat exchangers has a restriction on the upper limit to fluid velocity inside the heat exchanger. If the velocity of the fluids becomes too great, then the frictional power becomes excessive. Consequently, it is sometimes more useful to increase the surface area of a heat exchanger rather than attempt to increase the flow velocities to increase capacity. We discussed this earlier with fins and why they are used to increase heat transfer by increasing the surface area as well. By utilizing fins in an appropriate manner, the surface area of heat exchangers can be increased significantly without increasing the volume of the heat exchanger and thereby increasing the specific surface of the heat exchanger.

Heat exchangers having more surface area than the conventional shell and tube type usually have specific surfaces of more than 700 m^2/m^3, and these are then classified as *compact heat exchangers.* Such devices need to be designed with imagination and insight into the actions of conduction and convection heat transfer as well as an appreciation for the need for structural integrity and appropriate materials. In figure 9–28 are shown four specific compact heat exchanger arrangements. Figure 9–28a and b shows the judicious use of fins while figure 9–28c demonstrates an imaginative use of multiple partitions to divide two flow streams into alternate layers of fluid paths. This sort of heat exchanger is called a plate fin because the two fluid streams flow between alternating "plates." Kays and London [6] give an

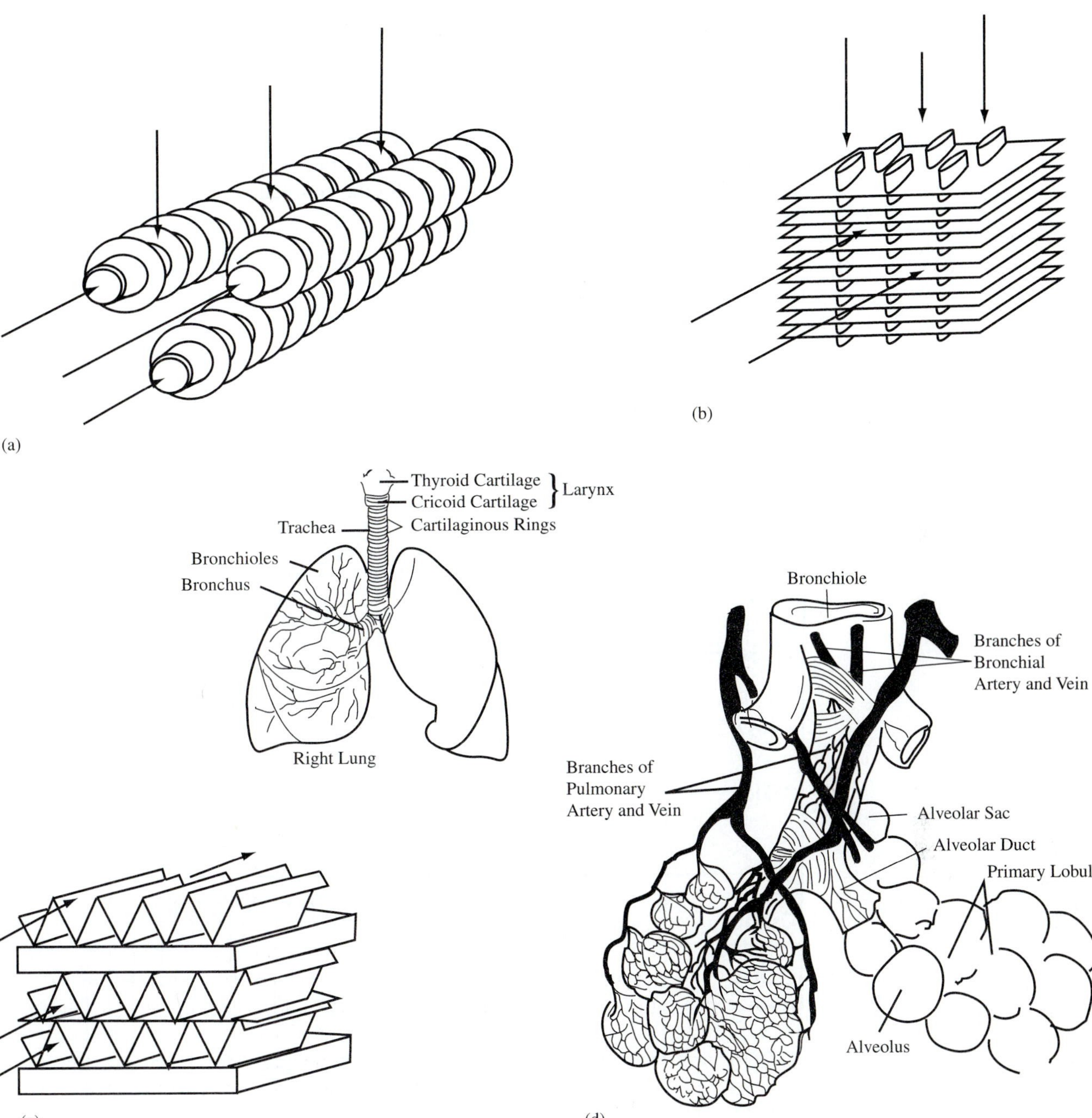

Figure 9–28 Some types of compact heat exchangers.

extensive discussion of the types of compact heat exchangers. Figure 9–28d shows a schematic of the human lungs. As was pointed out in section 9–1, the lungs have a specific surface that is around two orders of magnitude greater than the best compact heat exchanger designs. Thus, the lungs have an effective heat-transfer surface area that is proportionally much greater based on the lung volume than is the heat-transfer surface area of a compact heat exchanger based on its volume. The lungs also are used to transfer mass (oxygen and carbon dioxide), so the physical mechanisms are much more complicated than our simple heat exchangers.

We want to be able to predict the heat exchange for a compact heat exchanger, which may be done with equation (9–3),

$$\dot{Q}_{HX} = UA(\Delta T)_{HX}$$

and in addition we need to be able to predict the pressure drop through the heat exchanger. Since compact heat exchangers generally are constructed to maximize conduction heat transfer, the overall conductance is principally the convection heat transfer coefficients. That is, for a conductance based on the inside area,

$$U_i = \frac{1}{\dfrac{1}{h_i} + \dfrac{A_i}{h_o A_o}} \tag{9–46}$$

Thus, we need to be concerned with the convection heat transfer coefficients and the friction coefficient of the specific compact heat exchanger. Figures 9–29, 9–30, and 9–31 show performance data, taken from Kays and London [6], for compact heat exchangers

Figure 9–29 Finned Flat Tubes. (From [6]; reproduced by permission of William Kays)

Finned flat tubes, surface 9.68 – 0.87.

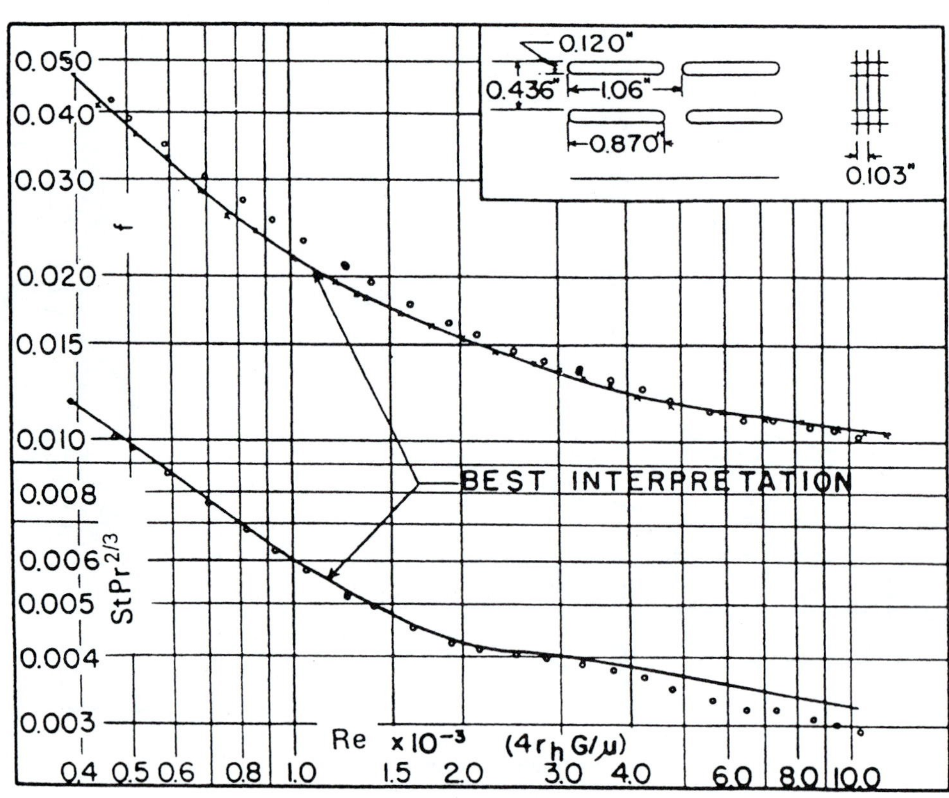

Fin pitch = 9.68 per in = 381 per m

Flow passage hydraulic diameter, $4r_h$ = 0.01180 ft = 3.597 x 10^{-3} m

Fin metal thickness = 0.004 in, copper = 0.102 x 10^{-3} m

Free-flow area/frontal area, σ = 0.697

Total heat transfer area/total volume, α = 229 ft²/ft³ = 751 m²/m³

Fin area/total area = 0.795

Figure 9–30 Finned circular tubes. (From [6]; reproduced by permission of William Kays)

Finned circular tubes, surface CF-9.05–3/4J. *(Data of Jameson.)*

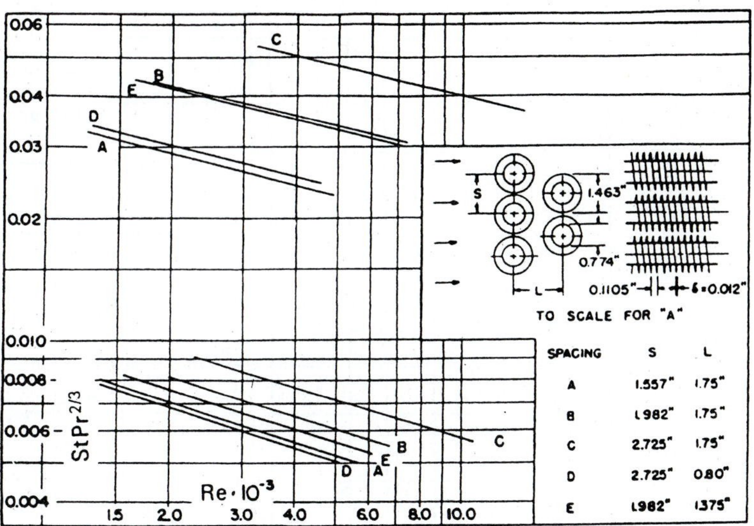

Tube outside diameter = 0.774 in = 19.66×10^{-3} m

Fin pitch = 9.05 per in = 356 per m

Fin thickness = 0.012 in = 0.305×10^{-3} m

Fin area/total area = 0.835

	A	*B*	*C*	*D*	*E*
Flow passage hydraulic diameter, $4r_h$ =	0.01681	0.02685	0.0445	0.01587	0.02108 ft
=	5.131×10^{-3}	8.179×10^{-3}	13.59×10^{-3}	4.846×10^{-3}	6.426×10^{-3} m
Free-flow area/frontal area, σ =	0.455	0.572	0.688	0.537	0.572
Heat transfer area/total volume, α =	108	85.1	61.9	135	108 ft²/ft³
=	354	279	203	443	354 m²/m³

Note: Minimum free-flow area in all cases occurs in the spaces transverse to the flow, except for *D*, in which the minimum area is in the diagonals.

similar to the three mechanical compact heat exchangers shown in figure 9–28. In these the following variables are defined:

$$A_{FT} = \text{Frontal area}$$

$$A_c = \text{Free flow area}$$

$$D_h = \text{Hydraulic diameter (defined in Chapter 4)}$$

$$\dot{G} = \frac{\dot{m}}{A} = \text{Mass flow per unit free flow area}$$

$$A_c/A_{FT} = \sigma = \text{Free flow area/Frontal area}$$

$$\alpha = \frac{A}{V} = \text{Heat transfer area/Total heat exchanger volume}$$

$$\text{Stanton Number} = \text{St} = \frac{h}{\dot{G}c_p}$$

$$\text{Re}_D = \frac{\dot{G}D}{\mu}$$

Strip-fin plate-fin surface 1/8–20.06(D).

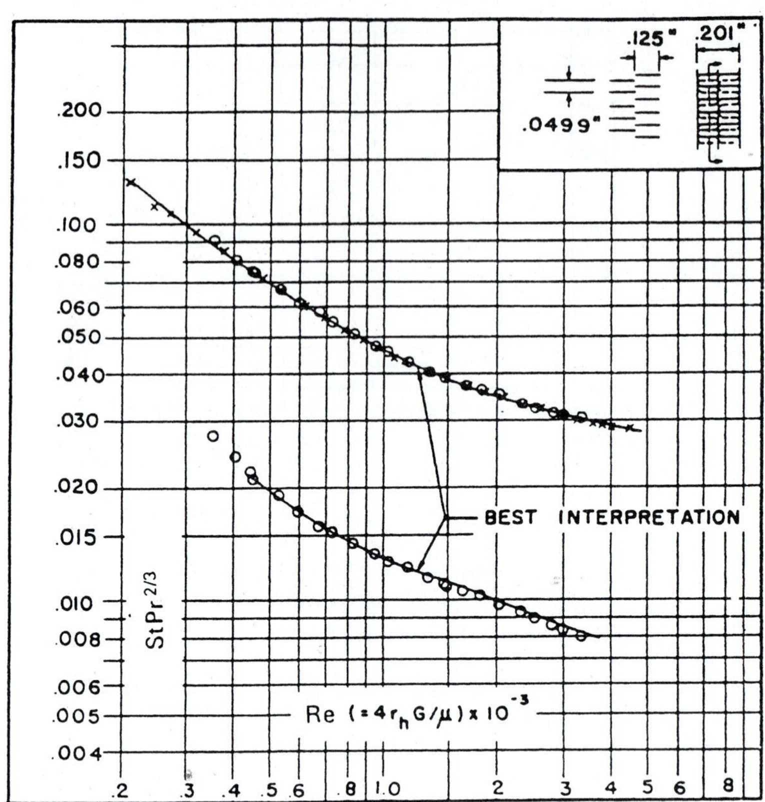

Figure 9–31 Strip-fin plate-fin surface. (From [6]; reproduced by permission of William Kays)

Fin pitch = 20.06 per in = 790 per m

Plate spacing, b = 0.201 in = 5.11×10^{-3} m

Splitter length flow direction = 0.125 in = 3.175×10^{-3} m

Flow passage hydraulic diameter, $4r_h$ 0.004892 ft = 1.491×10^{-3} m

Fin metal thickness = 0.004 in, aluminum = 0.102×10^{-3} m

Splitter metal thickness = 0.006 in = 0.152×10^{-3} m

Total heat transfer area/volume between plates, β = 698 ft^2/ft^3 = 2,290 m^2/m^3

Fin area (including splitter)/total area = 0.843

The fluid properties for analyzing heat exchangers are evaluated at a bulk average temperature, usually estimated as the arithmetic average between the inlet and outlet temperatures. The pressure drop is usually determined from the relationship

$$\Delta p = \frac{\dot{G}\mathbf{V}_1}{2g_c}\left[(1 + \sigma^2)\left(\frac{v_2}{v_1} - 1\right) + f\left(\frac{A}{A_c}\right)\left(\frac{v_m}{v_1}\right)\right] \tag{9–47}$$

where v_1 is the inlet specific volume of the fluid, v_2 its outlet specific volume and v_m the arithmetic mean between the inlet and the outlet volumes. The ratio $\frac{A}{A_c}$ can be written as $\frac{\alpha \mathbf{V}}{\sigma A_{FT}}$ and the ratio $\frac{V}{A_{FT}}$ is approximately equal to r_h. Thus, $\frac{A}{A_c}$ can be written as $\frac{\alpha r_h}{\sigma}$.

EXAMPLE 9–6 An air-to-air heat exchanger is designed with a configuration like that of figure 9–31. Determine the convection heat transfer coefficient, the overall conductance, and the friction factor if air passes through the heat exchanger at 100 ft/s. Assume the average bulk air temperature is 65°F for both fluid streams and they are both at a nominal pressure of 14.7 psia.

Solution The air properties are

$$\rho = \frac{p}{RT} = \frac{14.7 \times 144 \text{ lbf/ft}^2}{(53.3 \text{ ft} \cdot \text{lbf/lbm} \cdot {}^\circ\text{R})(525^\circ\text{R})} = 0.0756 \text{ lbm/ft}^3$$

$$\mu \approx 1.215 \times 10^{-5} \text{ lbm/ft} \cdot \text{s}$$

$$\text{Pr} \approx 0.708$$

$$c_p \approx 0.24 \text{ Btu/lbm} \cdot {}^\circ\text{F}$$

Referring to figure 9–31 we see that the frontal area A_{FT} is 0.0499 in^2/in. The free flow area A_c will then be 0.0499 − 0.004 = 0.0459 in^2/in., and the ratio $A_c/A_{FT} = \sigma$ is then

$$\sigma = 0.0459/0.0499 = 0.920$$

Then the mass flow per unit free flow area is

$$\dot{G} = \frac{\dot{m}}{A_c} = \rho V_f \frac{A_{FT}}{A_c} = \frac{\rho}{\sigma} V_f = \frac{0.0756 \text{ lbm/ft}^3}{0.920}(100 \text{ ft/s}) = 8.22 \frac{\text{lbm}}{\text{s} \cdot \text{ft}^2}$$

The Reynolds Number is

$$\text{Re} = \frac{\dot{G}D}{\mu} = \frac{(0.004892 \text{ ft})(8.22 \text{ lbm/ft}^3)}{1.215 \times 10^{-5} \text{ lbm/ft} \cdot \text{s}} = 3.31 \times 10^3$$

From figure 9–31 we then read

Answer
$$f \approx 0.032$$

From figure 9–31 we also read that

$$\text{St} \cdot \text{Pr}^{2/3} \approx 0.0084$$

where the Stanton Number, St, is

$$\text{St} = \frac{h}{\dot{G}c_p} = 0.0084 \text{ Pr}^{-2/3}$$

so that

$$h = 0.0084 \frac{\dot{G}c_p}{\text{Pr}^{2/3}} = 0.0084\left(\frac{(8.22 \text{ lbm/s} \cdot \text{ft}^2)(0.24 \text{ Btu/lbm} \cdot {}^\circ\text{R})}{(0.708)^{2/3}}\right) 0.02086 \frac{\text{Btu}}{\text{s} \cdot \text{ft}^2 \cdot {}^\circ\text{F}}$$

Answer
$$= 75.1 \frac{\text{Btu}}{\text{h} \cdot \text{ft}^2 \cdot {}^\circ\text{F}}$$

The overall conductance is then determined from equation (9–46),

$$U_o \approx \frac{1}{\dfrac{1}{h_o} + \dfrac{A_o}{h_i A_i}}$$

Since $A_i \approx A_o$ and $h_i \approx h_o (= h)$ we have

Answer
$$U_o \approx \frac{h}{2} = 37.55 \text{ Btu/hr} \cdot \text{ft}^2 \cdot {}^\circ\text{F}$$

9–6 SUMMARY

The general equations for describing a two-fluid heat exchanger are

$$\dot{Q}_{HX} = \dot{m}_c(hn_{c,o} - hn_{c,i}) = \dot{m}_h(hn_{h,i} - hn_{h,o}) \quad \textbf{(9–2)}$$

and

$$\dot{Q}_{HX} = UA\Delta T \quad \textbf{(9–3)}$$

where U is the overall heat exchanger coefficient. Heat exchangers are classified as conventional and compact. The fouling factor is used to quantify the reduction in the effectiveness of heat transfer because of fluid contamination or chemical reactions with the heat exchanger.

The LMTD method is a popular method for analyzing heat exchangers. The log mean temperature difference (LMTD) is often used to describe the temperature distributions in a conventional heat exchanger, defined as

$$\Delta T_{LM} = \text{LMTD} = \frac{\Delta T_2 - \Delta T_1}{\ln(\Delta T_2/\Delta T_1)} \quad \textbf{(9–23)}$$

For counterflow heat exchangers,

$$\Delta T_2 = T_{h,o} - T_{c,i} \quad \text{and} \quad \Delta T_1 = T_{H,i} - T_{c,o}$$

For parallel flow heat exchangers,

$$\Delta T_2 = T_{h,o} - T_{c,o} \quad \text{and} \quad \Delta T_1 = T_{h,i} - T_{c,i}$$

For cases where one or both of the ΔT's is zero, or if the ΔT's are equal, then

$$\Delta T_{LM} = \frac{1}{2}(\Delta T_2 + \Delta T_1) \quad \textbf{(9–27)}$$

The effectiveness–NTU method is used instead of the LMTD Method if only two of the fluid temperatures is known. The heat exchanger effectiveness is defined as

$$\varepsilon_{HX} = \frac{\text{Actual heat transfer}}{\text{Maximum possible heat transfer}} \quad \textbf{(9–29)}$$

For cases where the cold fluid has the minimum $\dot{m}c_p$

$$\varepsilon_{HX} = (T_{c,o} - T_{c,i})/(T_{h,i} - T_{c,i}) \quad \textbf{(9–32)}$$

For cases where the hot fluid has the minimum $\dot{m}c_p$

$$\varepsilon_{HX} = (T_{h,i} - T_{h,o})/(T_{h,i} - T_{c,i})$$

The NTU is a parameter used to analyze conventional heat exchangers. It is defined as

$$\text{NTU} = \frac{UA}{[\dot{m}c_p]_{\text{minimum}}} = \frac{UA}{C_{\min}}$$

where $C_{\min}$ is $[\dot{m}c_p]_{\text{minimum}}$ and is the minimum of the cold or hot fluid. The effectiveness and NTU are functions of each other and can be derived for simple conventional heat exchangers.

Compact heat exchangers are those whose specific surface is 700 m^2/m^3 or more. The use of empirical data is necessary for analyzing the overall conductance and the pressure drop through the heat exchangers. One of the major concerns in the design of compact heat exchangers is the pressure drop in the heat exchanger due to fluid friction. An equation used to predict this pressure drop is

$$\Delta p = \frac{G\mathbf{V}_1}{2g_c}\left[(1 + \sigma^2)\left(\frac{v_2}{v_1} - 1\right) + f\left(\frac{A}{A_c}\right)\left(\frac{v_m}{v_1}\right)\right] \quad \textbf{(9–47)}$$

DISCUSSION QUESTIONS

Section 9–1

9–1 What is the surface of a heat exchanger?

9–2 What is the specific surface of a heat exchanger?

9–3 What is a shell and tube heat exchanger?

Section 9–2

9–4 How is the overall conductance coefficient related to the heat exchanger area?

9–5 What is meant by a fouling factor?

9–6 How would you reduce the fouling factor?

Section 9–3

9–7 What is LMTD?

9–8 What is the difference between parallel and counter flow?

9–9 What is meant by mixed flow? How is it different from unmixed flow?

Section 9–4

9–10 How is heat exchanger efficiency described?

9–11 What is the heat exchanger effectiveness?

9–12 What is meant by NTU?

Section 9–5

9–13 What is a compact heat exchanger? How is it different from a shell and tube heat exchanger?

9–14 Why is pressure drop so important in compact heat exchangers?

9–15 How would you reduce fouling in a compact heat exchanger?

PRACTICE PROBLEMS

Section 9–1

9–1 Determine the mass flow rate of R-134a required for steady state of the cross-flow heat exchanger shown in figure 9–32. Air at 30°C enters at a rate of 3.6 kg/s and leaves at 10°C. The refrigerant enters at −10°C, leaves at 5°C, and has a specific heat of 2.1 kJ/kg · K.

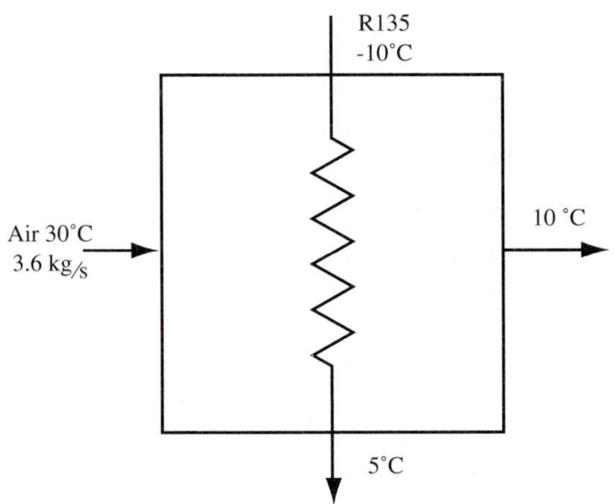

Figure 9–32

Section 9–2

9–2 Determine the overall heat transfer coefficient based on the inside diameter for the tube and shell heat exchanger shown in the schematic cross-sectional view of figure 9–33.

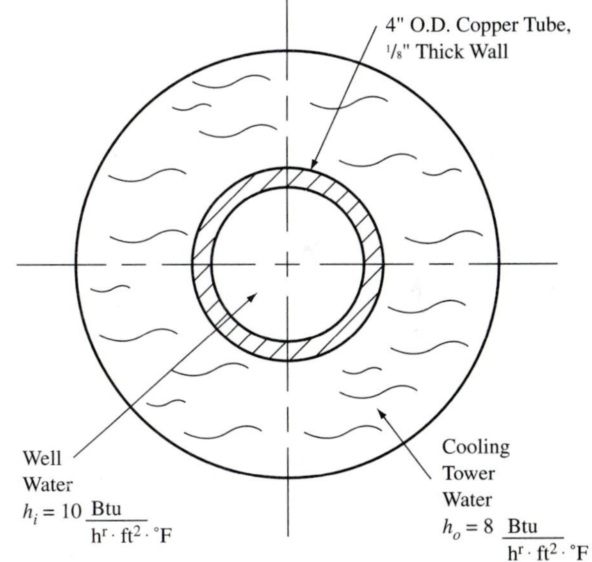

Figure 9–33

9–3 Vegetable oil is boiled by passing steam at 215°F through a 1-in. ID by 1.25-in. OD steel tube. Predict the overall heat transfer coefficient based on the outside area of the tube if the convective heat transfer coefficients are 25 Btu/hr · ft² · °F for steam and 15 Btu/hr · ft² · °F for the vegetable oil. Neglect any fouling factors.

Section 9–3

9–4 A cross-flow heat exchanger is used to heat mixed air from 20°C to 45°C. Water at 75°C enters the tubes and leaves at 45°C. If the overall heat-transfer coefficient is 50 W/m^2 · °C and there is to be 30,000 W of heat exchanged, determine the necessary heat exchanger surface area and the mass flow rate of the water.

9–5 A cross-flow heat exchanger having an overall heat transfer coefficient of 150 W/m^2 · °C is used to heat water from 25°C to 100°C at 3 kg/s. Exhaust gases at 250°C and with a specific heat of 1.0 kJ/kg · °C are used to heat the water, and these gases leave at 130°C. Assume both flows are unmixed.
(a) Determine the mass flow of the exhaust gases.
(b) Determine the surface area of the heat exchanger.

9–6 A counterflow double-pipe heat exchanger is used to heat 0.8 kg/s of water from 40°C to 80°C with oil having a specific heat of 2.1 kJ/kg · K. Assume the maximum oil flow can be 1.0 kg/s and that the oil enters the heat exchanger at 170°C. If the overall heat transfer coefficient is 425 W/m^2 · °C, determine the area and the effectiveness of the heat exchanger. (*Hint:* Use equation 9–29.)

9–7 A two-shell, eight-tube-pass heat exchanger is used to recoup heat from coolant water in a manufacturing plant. The heat exchanger has a rated area of 15 m^2 with an overall heat transfer coefficient of 900 W/m^2 · °C. The device has been used and has a fouling factor of 0.0001 m^2 · °C/W. The coolant water enters at 85°C and leaves at 55°C; the cooling water enters at 15°C and leaves at 45°C.
(a) Determine the mass flow of the coolant water.
(b) Determine the heat exchanger effectiveness (see equation 9–29).

9–8 A single-shell, eight-tube-pass heat exchanger is used in a power plant to preheat 55°F water to 80°F by using 190°F hot water and cooling it to 160°F. If the overall heat exchanger conductance is 400 Btu/hr · ft^2 · °F and 1800 lbm/min of 55°F water is preheated, determine
(a) The mass flow rate of hot water required.
(b) The surface area of the heat exchanger.

9–9 A condenser for a large refrigerator is a four-shell, eight-tube-pass heat exchanger. The condensing fluid is ammonia at 40°C, and the overall heat transfer coefficient is 500 W/m^2 · °C. Ten kg/s of water at 5°C, heated to 30°C, is used to cool the ammonia. Determine the mass flow of ammonia if its heat of vaporization is 1102 kJ/kg. Then determine the surface area of the heat exchanger.

9–10 A two-shell, four-tube-pass heat exchanger has an overall heat exchanger conductance of 350 W/m^2 · °C and is used to cool 10 kg/s of engine oil from 180°C to 80°C. If water at 20°C is used to cool the oil and the water temperature leaving the heat exchanger must be no more than 60°C, determine
(a) The mass flow rate of water required.
(b) The surface area of the heat exchanger.

Section 9–4

9–11 A shell and tube heat exchanger consists of one shell with 100 thin-walled tubes of 15-mm diameter in a double-pass arrangement. The total surface area is 45 m^2, and the overall heat exchanger coefficient is 200 W/m^2 · °C. Water enters the tubes at 15°C and 6.5 kg/s. Hot gases having properties of air at 200°C and 5 kg/s flow through the shell. Determine
(a) The gas outlet temperature.
(b) The water outlet temperature.

9–12 A single-shell, single-tube parallel flow heat exchanger is used to cool oil in a large marine diesel engine. River water at 2 kg/s and 15°C enters the shell while engine oil at 1 kg/s and 140°C enters the tubes. The engine oil has a specific heat of 2.4 kJ/kg · K, the heat exchanger surface area is 7 m^2, and the exchanger's overall heat transfer coefficient is 400 W/m^2 · °C when new.
(a) Determine the outlet oil temperature.
(b) If the river water is found to give a fouling factor of 0.0004 m^2 · °C/W, determine the outlet oil temperature after the system has been well used.

9–13 A single-shell, 16-tube heat exchanger having an overall heat transfer coefficient of 300 W/m^2 · °C is used to cool ethyl alcohol from 80°C to 40°C. Water at 11°C is used as the coolant, and if its flow rate is 300 kg/min determine how the required surface area is affected by the amount of alcohol. Do this by plotting the surface area of the heat exchanger as a function of the alcohol flow rate from 300 kg/min to 500 kg/min.

9–14 A two-shell, eight-tube-pass heat exchanger condenses 3 kg/s of ammonia at 45°C using 65 kg/s of water at 20°C. The heat exchanger has 500 m^2 of surface area and an overall heat exchanger conductance of 300 W/m^2 · °C. If the ammonia enters as saturated vapor, leaves as saturated liquid, and has a heat of vaporization of 1078 kJ/kg, determine
(a) The outlet temperature of the water.
(b) The heat exchanger effectiveness.

Section 9–5

9–15 A compact heat exchanger is proposed to heat tomato juice. The heat exchanger is to be constructed with finned stainless steel tubes as shown in figure 9–30, spacing C. The properties of the tomato juice are $\kappa = 0.6$ W/m · °C, $c_p = 4.1$ kJ/kg · K, $\mu = 0.86$ g/m · s, $\mathbf{V} = 1$ m/s, and $\rho = 890$ kg/m^3. Air at 120°C flows at 10 m/s past the finned tubes. Determine the pressure drop experienced by the air flow, assuming the air temperature average is 100°C. Then determine the convective heat transfer coefficient in the air flow and the overall heat transfer coefficient. Assume the tube wall thickness is 0.05 in.

9–16 An intercooler of an air compressor is to use a finned flat tube heat exchanger as shown in figure 9–29. Compressed air at 100 psia, 130°F passes through the tubes at 100 ft/s, and air at 14.7 psia, 70°F, passes around the tubes at 100 ft/s. Determine the convective heat-transfer coefficient, the friction factor for the cooling air outside the tubes, and the overall convective heat-transfer coefficient for the intercooler. Assume the inside dimensions of the flattened tubes are 0.105 in. by 0.865 in.

9–17 An automobile radiator is approximated by the finned flat tube compact heat exchanger of figure 9–29. Air at 80°F and 14.7 psia flows through the heat radiator at 40 ft/s. If the air temperature leaving the radiator is 100°F and its properties are $\mu = 1.2 \times 10^{-5}$ lbm/ft · s, Pr = 0.708, and $c_p = 0.24$ Btu/lbm · °F, determine
(a) The friction factor for the radiator.
(b) The convective heat-transfer coefficient for the air side.

9–18 A compact heat exchanger uses finned tubes as shown in figure 9–30 with spacing B. Determine the pressure drop and the convective heat transfer coefficient for air passing around the finned tubes at 27°C, 5 m/s and with viscosity of 0.02 g/ms. Assume that the air leaves at

REFERENCES

1. Walker, G., *Industrial Heat Exchangers,* 2nd edition, Hemisphere Publishing, New York, 1990.
2. Afgan, N., and E. U. Schlünder, *Heat Exchangers: Design and Theory Sourcebook,* McGraw-Hill, New York, 1974.
3. Schlünder, E. U., *Heat Exchanger Design Handbook,* Hemisphere Publishing, New York, 1982.
4. Somerscales, E. F. C., and J. G. Knudsen, *Fouling of Heat Transfer Equipment,* Hemisphere Publishing, New York, 1981.
5. *Standards of Tubular Exchange Manufacturers Association,* 6th edition, TEMA, Tarrytown, NY, 1978.
6. Kays, W. M., and A. L. London, *Compact Heat Exchangers,* 3rd edition, McGraw-Hill, New York, 1984.
7. Dunn, P., and D. A. Reay, *Heat Pipes,* Pergamon, Oxford, UK, 1967.
8. Chi, S. Q., *Heat Pipe Theory and Practice,* McGraw-Hill, New York, 1967.
9. Peterson, G. P., L. W. Swanson, and F. M. Gerner, Micro Heat Pipes, Chapter 8 in C-L Tien, A. Majumdar, and F. M. Gerner, eds. Micro Scale Energy Transport, Taylor and Francis, Washington, DC, 1998.

PHASE CHANGE HEAT TRANSFER

10

In this chapter we will consider the processes that are involved when heat transfer occurs with a substance that is experiencing one of its phase changes: evaporation/condensation, melting/freezing, or sublimation. In section 10–1 the phase changes are discussed and a short discussion of bubble formation and behavior in the boiling process is presented. In section 10–2 the theoretical and analytic developments of evaporation or boiling are discussed. Included in these discussions is pool boiling (or natural convection boiling) on horizontal surfaces, along vertical surfaces, around horizontal tubes, and in vertical pipes or tubes. Forced convection boiling in horizontal or slant tubes is also discussed, and various convection heat-transfer relationships are reintroduced for describing the heat transfer in boiling. In section 10–3 condensation is presented, and in section 10–4 some simplified relationships for boiling and condensing of pure substances are given. In section 10–5 melting and freezing phenomena are discussed, and some of the common analytic and empirical methods for describing the heat transfer are presented. In section 10–6 a recapitulation of phase change heat transfer is given along with some design applications.

10–1 THE MECHANISMS OF PHASE CHANGE HEAT TRANSFER

Substances are found in one or more of their three phases: (1) solid, (2) liquid, or (3) vapor. The modes of heat transfer are strongly dependent on the phase of the substances involved in the heat-transfer processes. For substances that are solid, conduction is the predominate mode of heat transfer. For liquids, convection heat transfer predominates, and for vapors convection and radiation are the primary modes of heat transfer. Thus, for heat transfer involving phase changes of the substances, there will be complications beyond those that we have considered up to this point. Then again, phase changes of substances can only occur if heat transfer is involved in the process. The phase change processes that can be studied are

- *Evaporation or boiling*, involving convection heat transfer with radiation at high temperature differences or heat transfer rates.
- *Condensation*, involving convection heat transfer.
- *Melting or thawing*, involving convection and conduction heat transfer.
- *Freezing or fusion,* involving conduction and convection heat transfer.
- *Sublimation,* involving conduction, convection, and radiation heat transfer.

During these processes the temperature and pressure are normally constant or uniform whereas the heats of transformation need to be accounted for in the heat-transfer analyses. These are

- *Heat of vaporization,* hn_{fg}, the heat required to transform a saturated liquid to a saturated vapor.
- *Heat of fusion,* hn_{fs}, the heat required to transform a saturated solid to a saturated liquid.
- *Heat of sublimation,* hn_{sg}, the heat required to transform a saturated solid to a saturated vapor.

Although the pressure and temperature of a pure substance are uniform during a phase change, the density or specific volume of the substance can change dramatically. Thus, we need to consider the specific volume changes

- v_{fg}, volume increase from a saturated liquid to a saturated vapor
- v_{fs}, volume increase (or decrease for water) from a saturated solid to a saturated liquid
- v_{sg}, volume increase from a saturated solid to a saturated vapor

A typical pressure–volume diagram of the evaporization for a pure substance is shown in figure 10–1. Notice that the isotherms (constant temperature) show that pressure, volume, and temperature are independent properties outside the phase change region but that pressure and temperature are constant in the phase change region. For many substances, such as water, the specific volume of the saturated liquid is two or three orders of magnitude less than that of the saturated vapor. This results in a profound effect on boiling heat transfer because convection currents are induced due to the bouyancy of the saturated vapor bubbles created by surface tension effects.

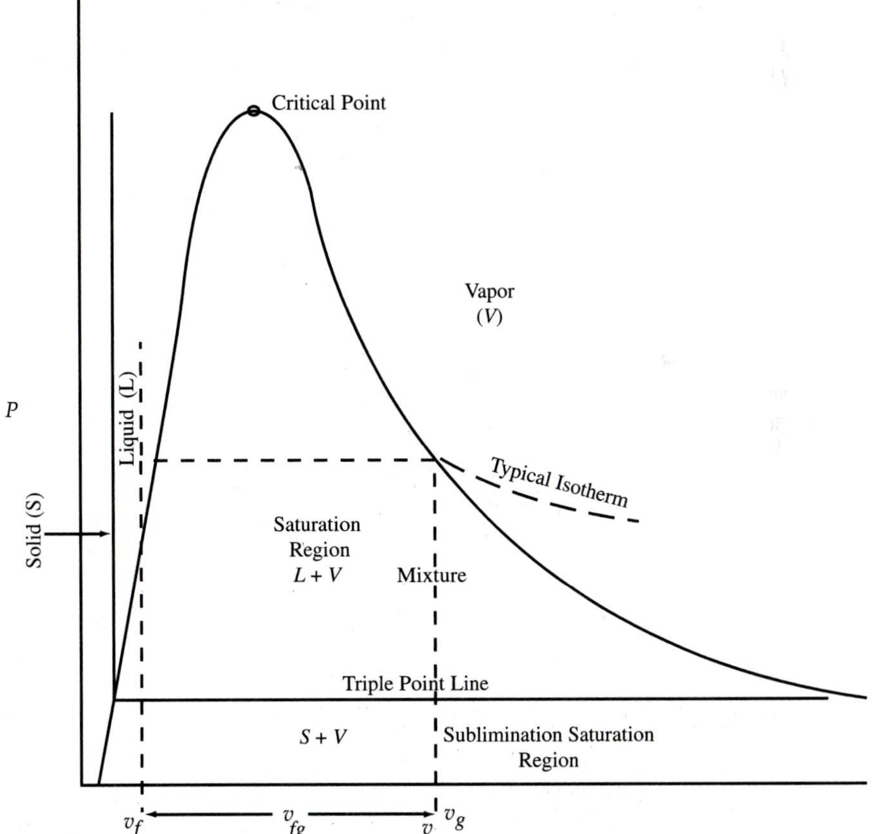

Figure 10–1 A typical p–v diagram for a pure substance in its liquid–vapor phase change region.

EXAMPLE 10–1 Water is boiling in a pot at 100°C. Estimate the bouyant force acting on a bubble that has a spherical diameter of 1 mm.

Solution Bubbles form in a liquid as a result of some of the fluid transforming from a saturated liquid to a saturated vapor. The growth of a bubble often requires an irregularity or microcavity in the surface that can trap the liquid and then heat it to the superheated state. In figure 10–2 is shown a typical sequence of events leading to a bubble in the liquid. The free body diagram of half a bubble is shown in figure 10–3, where the pressure of the vapor in the bubble exceeds the surrounding liquid pressure by the amount expressed in

$$p_g - p_f = \frac{4\sigma_{ST}}{D} \tag{10–1}$$

where σ_{ST} is the surface tension of the liquid/vapor interface. For this problem it would be the surface tension of water at 100°C, which is about 58 mN/m. Thus, the pressure difference between the bubble vapor and the surrounding saturated liquid is

$$p_g - p_f = \frac{4(58 \text{ mN/m})}{0.001 \text{ m}} = 232 \text{ Pa}$$

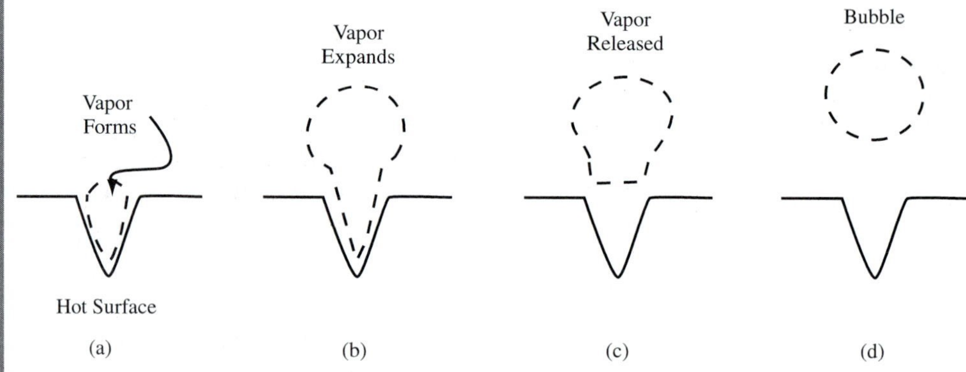

Figure 10–2 Sequence of events in bubble growth during boiling of a pure substance.

Figure 10–3 Free body diagram of a bubble in a saturated liquid.

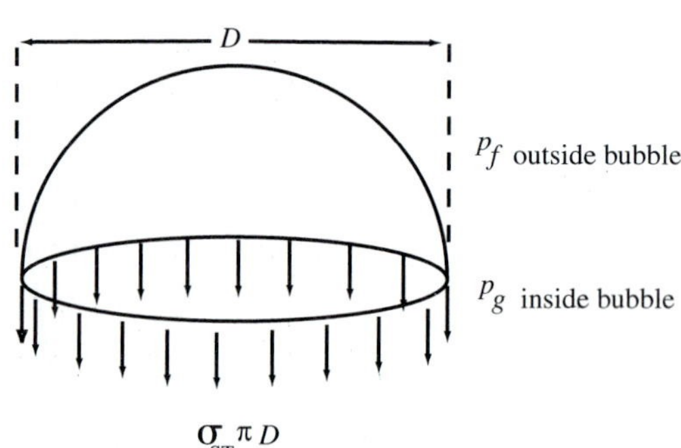

Since the saturation pressure of water at 100°C is 101.31 kPa, the pressure of the vapor in the bubble must be 101.542 kPa. This is slightly higher than the saturation pressure for water at 100°C, so the bubble vapor temperature is slightly higher than 100°C. We will assume it to be saturated vapor at 100°C so that its specific volume is 1.673 m³/kg and its density is then 0.5977 kg/m³. The surrounding liquid's density is

$$\rho_f = \frac{1}{v_f} = \frac{1}{0.0010435} = 958.3 \frac{\text{kg}}{\text{m}^3}$$

Using the concept of buoyancy from Chapter 5, the buoyant force, F_B, is

$$F_B = \gamma_f V = \rho_f g V = (958.3 \text{ kg/m}^3)(9.81 \text{ m/s}^2)(\text{volume of bubble})$$

The bubble volume is

$$V = \tfrac{4}{3}\pi r^3 = \tfrac{4}{3}\pi(0.0005\ \text{m})^3 = 5.236 \times 10^{-10}\ \text{m}^3$$

and the buoyancy is then

$$F_B = 4.922 \times 10^{-6}\ \text{N}$$

The weight of the bubble is just its volume times the vapor density

$$W = \rho_g g V = (0.5977\ \text{kg/m}^3)(9.81\ \text{m/s}^2)(5.236 \times 10^{-10}\ \text{m}^3) = 0.003 \times 10^{-6}\ \text{N}$$

Thus, the buoyant force is about 1700 times greater than the weight, and this means that the bubble will quickly move upward in the liquid, contributing to natural or free convection in the boiling process.

10–2 ANALYSIS OF BOILING HEAT TRANSFER

Boiling, evaporation, and *vaporization* are all terms used to describe the phenomenon whereby a substance experiences the phase change from a liquid or saturated liquid to a vapor or saturated vapor. For a pure substance these events occur such that the temperature does not change if the pressure does not change. As we discussed previously, there is associated with such a phase change an increase in enthalpy, called the heat of vaporization, and a significant decrease in density or increase in specific volume. These phase changes require a heat transfer so that the phase change can occur. It has been convenient to distinguish the boiling phenomena into at least two categories: (1) pool boiling and (2) force convection boiling. We will treat these two types separately.

Pool boiling is characterized as boiling with natural or free convection. It can occur in a pot, kettle, or caldron, where the boiling liquid is contained and heated along a horizontal bottom surface, around heating tubes or cylinders, along vertical or upright sides, and in vertical tubes. The top surface may be open to the environment or closed. The heat transfer involves natural convection, where bubbles of saturated or superheated vapor flow upward through the liquid, thus conveying heat into the boiling substance. In section 10–1 we saw that a vapor bubble has a large buoyant force relative to the bubble's weight and travels rapidly upward. It has been observed that if the bubble temperature is not much greater than that of the saturated liquid surrounding it, the bubble will lose enough heat to return to the saturated liquid state and thus set up the classical natural convection heat transfer we considered in Chapter 5. The liquid will not necessarily boil away. Therefore, the heating surface temperature of the liquid container must be significantly higher than the saturation temperature of the liquid at its pressure. We call this the degrees superheat or superheat temperature,

$$\Delta T_{SH} = T_s - T_g \qquad \textbf{(10–2)}$$

where T_s is the surface temperature and T_g is the saturation temperature.

For water the minimum superheat temperature is about 7°C. That is, unless the heating surface temperature is greater than 7°C, the liquid will not boil away but will experience natural convection. If the heating surface has a temperature greater than this minimum value, the vapor bubbles will not condense to the liquid state but rise and perhaps conglomerate into larger bubbles. If there is a free upper surface to the atmosphere, these larger bubbles will probably be released into the atmosphere. If the upper surface is closed or contained, either the boiling mass will exhibit an increased pressure because of the increase in vapor or such bubbles will be conveyed to another region. The degree to which these bubbles will conglomerate can be qualitatively described as discrete nucleate boiling, where the bubbles are discrete and relatively uniform in size; fully developed nucleate boiling, where the bubbles are becoming quite large; and film boiling, where the bubbles instantaneously become part of a large uniform vapor region. Discrete nucleate boiling occurs at superheat temperatures just above the maximum temperature for natural convection. Fully developed nucleate boiling occurs when the superheat temperature is much above that for discrete nucleate boiling, and film boiling occurs when the superheat temperature is at even higher temperatures.

Figure 10–4 shows the stages of boiling, starting with natural convection (nonboiling) through film boiling for a flat horizontal heating surface below the boiling liquid. Because many liquids are boiled with hot tubes, figure 10–5 shows the equivalent stages of boiling of a liquid surrounding a hot horizontal tube. Of course, for these phase changes to be able to occur, heat needs to transfer to the liquid/vapor. In figure 10–6 is shown the relationship between amount of heat transfer per unit area, in kW/m^2, and the degree of superheat. Notice in the figure that the transition from natural convection to nucleate boiling has an amount of uncertainty. Then, as greater amounts of heat are transferred to the boiling liquid, the degree of superheat increases. At a condition identifed as peak nucleate boiling the amount of vapor in contact with the heating surface becomes great enough that convection heat transfer is retarded. Remember that vapors tend to have lower convective heat-transfer coefficients than liquids, so while the liquid is still boiling, the heating surface temperature continues to rise due to the heat transfer furnished to that surface. But a pseudo-insulation occurs with the vapor at the surface. The curve from point B to C represents an instability in which the thermal inertia of the solid mass below and up to the surface continues to "heat up" even though less outside heat transfer is required to be furnished. This is the curious phenomenon of less heat being required to increase the heating surface temperature. Of course you now know that this is due to

Figure 10–4 Stages of boiling on a flat horizontal surface.

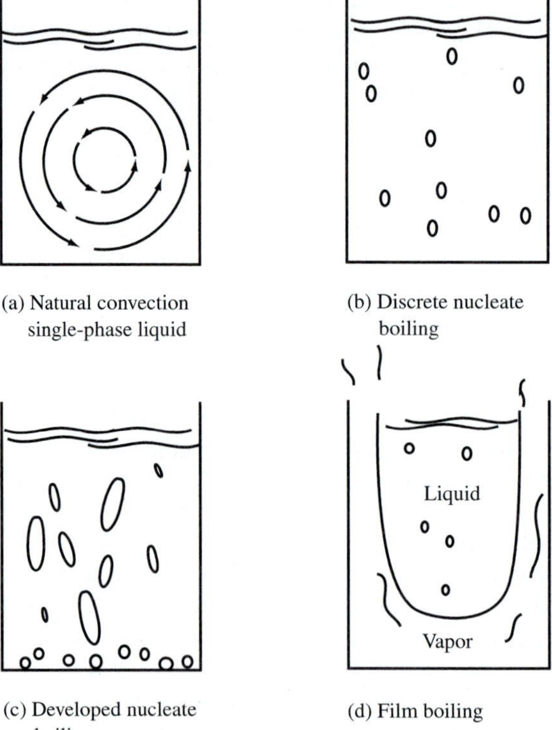

(a) Natural convection single-phase liquid

(b) Discrete nucleate boiling

(c) Developed nucleate boiling

(d) Film boiling

Figure 10–5 Stages of boiling around a horizontal tube.

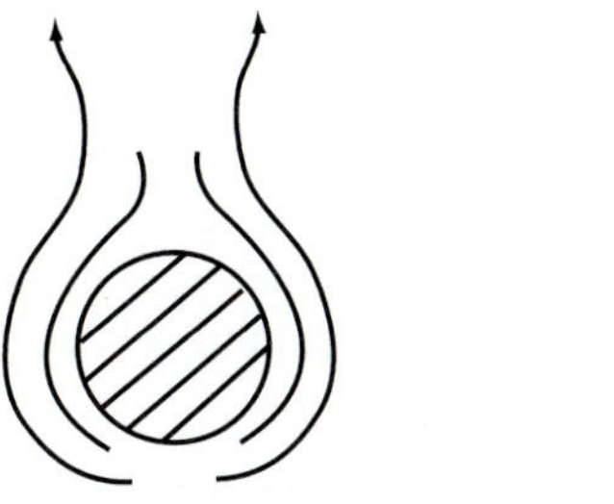

(a) Natural convection single-phase liquid

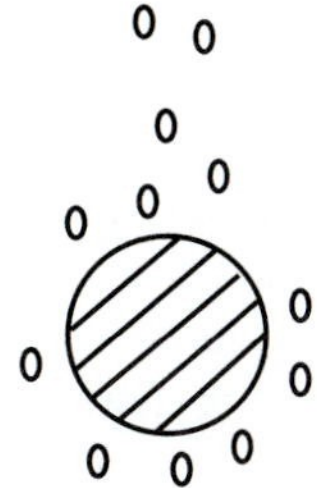

(b) Discrete nucleate boiling

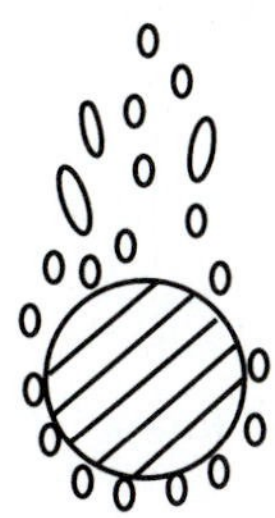

(c) Fully developed nucleate boiling

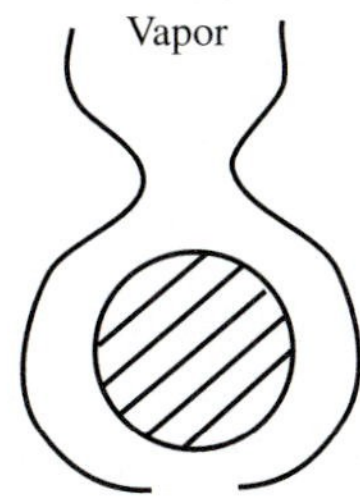

(d) Film boiling

Figure 10–6 Pool boiling curve, water.

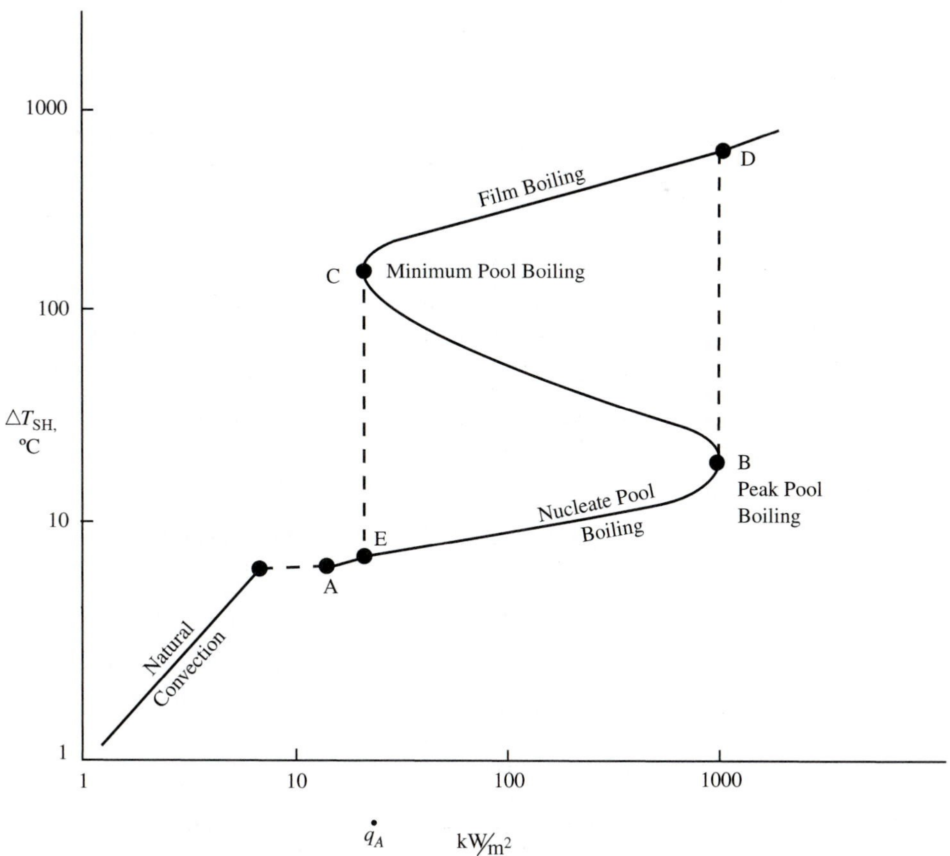

the significant decrease in the convective heat-transfer coefficient as the vapor blankets the heating surface. Film boiling begins at point *C,* and again the degrees of superheat will correspondingly increase with an increase in heat transfer. At some very high superheat, radiation heat transfer becomes a significant mode of heat loss from the surface and to the liquid. Often these temperatures become so high that ionization of the vapor becomes an even more complicating feature of the process. We will not consider that phenomenon here.

Natural Convection

The natural convection heat-transfer relationships developed in Chapter 5 may be used for situations where the heating plate temperature is below that indicated by the minimum superheat. For instance, if a fluid is being heated by a horizontal cylindrical tube, equation (5–52) can be used to predict the Nusselt Number and the convective heat-transfer coefficient,

$$\mathrm{Nu}_D = \left\{0.60 + 0.387\,\mathrm{Ra}_D^{1/6}\left[1 + \left(\frac{0.559}{\mathrm{Pr}_L}\right)^{9/16}\right]^{-8/27}\right\}^2$$

for the range of Rayleigh Numbers

$$10^{-5} \leq \mathrm{Ra}_D \leq 10^{12}$$

Threshhold of Pool Boiling

The temperature at which true boiling, or nucleate boiling, begins can be experimentally observed. Noting, from the analysis of the bubble in example 10–1, that the pressure difference between the inside of the bubble and the outside of the bubble in the liquid can be written

$$p_g - p_f = \frac{4\sigma_{\mathrm{ST}}}{D}$$

and that the saturation pressure and temperature are directly related, we can revise this relationship to read, for the minimum superheat required for boiling,

$$\Delta T_{\mathrm{SH,min}} = \frac{4\sigma_{\mathrm{ST}}}{D}\left(\frac{dT}{dp}\right)_{\mathrm{sat}} \tag{10–3}$$

where the subscript *sat* indicates that the term dT/dp is at the saturation region. Some values for the minimum superheat are listed in table 10–1 for a selection of substances at their saturation or boiling temperature and pressure. Notice that these values range from 1.6 to 24°C. Thus, the surface temperature required for boiling is dependent on the

Table 10–1 Superheat Required for Boiling

Liquid/Surface Configuration	Saturation Pressure, kPa	Saturation Temperature, K	Superheat, ΔT_{SH}
Argon/copper disk	125	87.1	2 to 3
Benzene/copper disk	101	353.1	10 to 11
Ethanol/copper disk	101	351.5	23 to 24
Ethanol/copper tube	101	351.5	6 to 10
Methanol/copper tube	101	338	9 to 10
Nitrogen/copper disk	101	77.2	1.6
n-Propanol/copper tube	101	370.4	8
Water/copper disk	101	373	6 to 8
Water/copper tube	101	373	4 to 8

particular liquid to be boiled. Also, notice that ethanol and water require higher superheat for heating on a horizontal disk than in a tube. This can be attributed to the rougher, more irregular surfaces of a machined flat disk compared to an extruded copper tube and the lack of opportunities to start bubbles on a smooth surface, discussed in section 10–1.

Nucleate Pool Boiling Heat Transfer

Nucleate pool boiling heat transfer has been studied, and Mostinski [1] has suggested the relationship for the convective heat-transfer coefficient,

$$h_{\text{npb}} = 1.2 \times 10^{-8}(\Delta T_{\text{SH}})^{2.33} p_c^{2.3} F_p^{3.33} \qquad (\text{W/m}^2 \cdot {}^\circ\text{C}) \tag{10–4}$$

where p_c is the critical pressure in kPa and

$$F_p = 1.8 p_R^{0.17} + 4 p_R^{1.2} + 10 p_R^{10} \tag{10–5}$$

Here p_R is the reduced pressure, p/p_c. This relationship seems to give values that can be as much as 20% low. A more recent study by Stephan and Abdelsalam [2] resulted in predicting more accurate values for the convective heat-transfer coefficient, with the following relationship for organic fluids,

$$h_{\text{npb}} = 0.0001629 \frac{\kappa_f \rho_g}{\rho_f} \left(\frac{2\sigma_{\text{ST}}}{g(\rho_f + \rho_g)} \right)^{0.274} \left(\frac{\Delta T_{\text{SH}}}{T_g \alpha_f} \right)^{2} \left(\frac{\text{hn}_{fg}}{\alpha^2} \right)^{0.744} \left(\frac{\rho_f}{\rho_f - \rho_g} \right)^{13} \frac{\text{W}}{\text{m}^2 \cdot {}^\circ\text{C}} \tag{10–6}$$

where the parameters all have standard SI units. The surface tension is in mN/m and α_f is the thermal diffusivity of the saturated liquid state in m^2/s.

Nucleate Pool Boiling in a Subcooled Liquid

There are situations where the liquid that is being heated in pool boiling has an ambient temperature below its saturation temperature at the pressure of the liquid and vapor. For these situations the convective heat-transfer relationship is

$$\dot{q}_A = h_{\text{npb,sub}}(T_w - T_\infty) \tag{10–7}$$

where the temperature difference is not the superheat temperature, defined by equation (10–2), but a greater value. It has been observed that the nucleate pool boiling curve, indicated as *A–B* in figure 10–6, will be shifted upward for a given heat transfer as shown in figure 10–7. That is, boiling a subcooled liquid requires a greater temperature difference

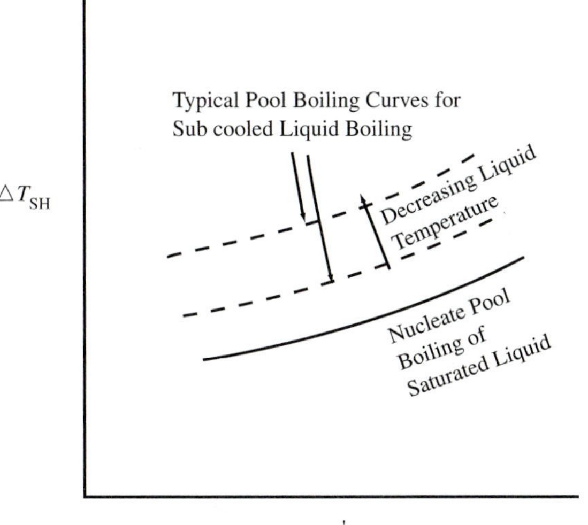

Figure 10–7 Nucleate pool boiling for a subcooled liquid.

between the heating surface and the liquid than does boiling a saturated liquid. Corresponding to this, the convective heat-transfer coefficient, $h_{\text{npb,sub}}$, is less than the nucleate pool boiling value for h_{npb}. In practice, the liquid should become a saturated liquid after a period of time due to the internal heat transfer between the bubbles and the liquid.

Maximum Nucleate Pool Boiling Heat Transfer

The nucleate pool boiling phenomenon becomes unstable at point *B* as indicated in figure 10-6. This represents the maximum heat transfer for a given superheat temperature using nucleate pool boiling of the liquid. It has been of importance in designing boilers and evaporators as well as of scientific interest to study this condition. Tests have been conducted to observe nucleate pool boiling where a horizontal cylinder (actually an electrical resistance heater) was placed in a liquid. The heater was controlled by the amount of power supplied to the wire so that the boiling heat transfer was determined by the amount of electrical power. In figure 10–8 is shown a schematic of such a test apparatus. In experiments conducted by Nukiyama [3] the thermal inertia of the electrical resistance heater wire was great and the power control was not responsive enough, so the boiling curve followed the path indicated by the dashed line in figure 10–6 from *B* to *D*. At *D*, or before *D*, the superheat temperature was great enough that the electrical wire temperature exceeded the melting temperature of the wire material and the wire burnt out. By using a wire material with a higher melting temperature, the curve *B–C–D* was able to be defined. Also, the reverse situation, where film boiling conditions are reduced to those of nucleate boiling, was recorded in the curve *C–E*. Thus, there was observed a hysteresis effect associated with the instability between peak nucleate boiling and film boiling. Subsequently, using better thermal controls for the heater, the unstable cooling curve from *B* to *C* has been observed by, among others, Drew and Mueller [4]. The maximum heat transfer at point *B* for pool boiling on horizontal and vertical surfaces has been approximated by

$$\dot{q}_{A,\text{max nbp}} = 0.149\sqrt{\rho_g}\text{hn}_{fg}[g\sigma_{\text{ST}}(\rho_f - \rho_g)]^{1/4}\ \frac{\text{kW}}{\text{m}^2} \tag{10–8}$$

where all of the parameters in this equation have standard SI units. The surface tension has units of mN/m. For maximum nucleate boiling heat transfer around single horizontal cylinders

$$\dot{q}_{A,\text{max npb}} = 0.131\sqrt{\rho_g}\text{hn}_{fg}[g\sigma_{\text{ST}}(\rho_f - \rho_g)]^{1/4} \tag{10–9}$$

is suggested instead of 10–8. For boiling around tube banks or tube bundles, this equation needs to be revised, and the interested reader is directed to Thome [5] for these details.

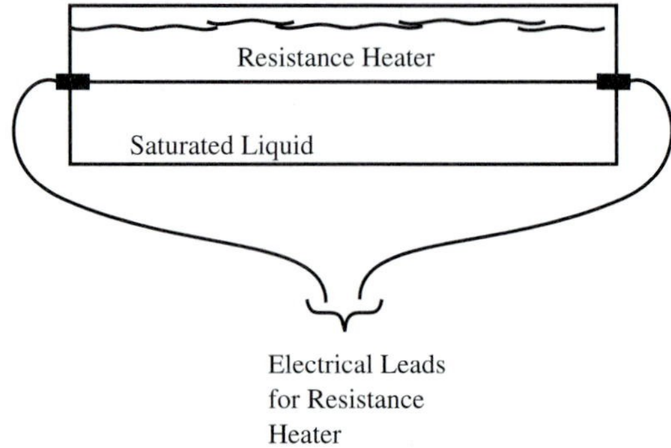

Figure 10–8 Apparatus for determining pool boiling around a cylinder.

EXAMPLE 10–2 Saturated liquid water at 212.4°C is to be boiled without film boiling. What is the maximum temperature of the heating surface allowed?

Solution From figure 10–6 the temperature associated with the maximum nucleate boiling is about 50°C. Using the definition of the superheat temperature we find

$$\Delta T_{SH} = T_w - T_g = 50°C = T_w - 212.4°C$$

and

Answer

$$T_w \leq 262.4°C$$

EXAMPLE 10–3 Methanol is boiled in a container such that the superheat temperature is 10°C when the methanol is a saturated liquid at 50°C. Estimate the convective heat transfer coefficient for this process. The following properties are given for methanol at 50° C:

$$\sigma_{ST} = 20.1 \text{ mN/m} \qquad \text{hn}_{fg} = 1050 \frac{\text{kJ}}{\text{kg}}$$

$$\rho_f = 791.4 \frac{\text{kg}}{\text{m}^3} \qquad \rho_g = 1.03 \frac{\text{kg}}{\text{m}^3}$$

$$c_{p,f} = 2.624 \frac{\text{kJ}}{\text{kg} \cdot \text{K}} \qquad \kappa_f = 0.196 \frac{\text{W}}{\text{m} \cdot \text{K}}$$

Solution The convective heat transfer coefficient may be approximated with equation (10–6). The thermal diffusivity for the saturated liquid is

$$\alpha_f = \frac{\kappa_f}{\rho_f c_{p,f}} = \frac{0.196 \text{ J/s} \cdot \text{m} \cdot \text{K}}{(791.4 \text{ kg/m}^3)(2624 \text{ J/kg} \cdot \text{K})} = 9.44 \times 10^{-8} \text{ m}^2\text{/s}$$

and using equation (10–6) we obtain,

$$h_{npb} = 0.0001629 \frac{0.196(1.03)}{791.4}\left(\frac{2(20.1 \times 10^{-3})}{(9.81)(791.4 + 1.03)}\right)^{0.274}\left(\frac{(10)}{(50 + 273)}\right)^{2}$$
$$\left(\frac{1{,}050{,}000}{(9.44 \times 10^{-8})^2}\right)^{0.744}\left(\frac{791.4}{791.4 - 1.03}\right)^{13} = 8213 \frac{\text{W}}{\text{m}^2 \cdot °\text{C}}$$

Film Boiling

Film boiling, as indicated in the boiling curve of figure 10–6, occurs when the temperature of the heating surface is well above that of the boiling liquid. That is, for very high superheat temperatures, the heating surface is completely covered with saturated vapor or superheated vapor. As a consequence of this condition, convection heat transfer to the liquid must first be transported through the vapor. In addition, however, radiation heat transfer will augment the convection heat transfer so that the heat transfer from the heating surface to the boiling liquid/vapor will be a combination of convection and radiation heat transfer. It is convenient to describe this heat transfer as a psuedo-convection heat transfer with a convection coefficient given by

$$h_{FBT} = h_{FB} + {}^4\!/_3 h_{RAD} \qquad \textbf{(10–10)}$$

In this equation the convective film boiling heat transfer coefficient h_{FB} describes the pure convection heat transfer through the vapor and for film boiling around a horizontal tube, Bromley [6] has suggested the equation

$$\text{Nu}_D = \frac{h_{FB}D}{k_g} = 0.62\left(\frac{(\rho_f - \rho_g)gD^3(\text{hn}_{fg} + 0.34c_{p,g}\Delta T_{SH})}{\nu_g \kappa_g \Delta T_{SH}}\right)^{1/4} \qquad \textbf{(10–11)}$$

The pseudo-convective radiation heat-transfer coefficient, h_{RAD}, is most often defined by

$$h_{\text{RAD}} = \varepsilon_{r,\text{HS}}\sigma\frac{T_w^4 - T_g^4}{\Delta T_{\text{SH}}} \qquad \textbf{(10–12)}$$

which introduces nonlinearity into the computations. Notice that this equation uses the emissivity of the heating surface and σ is the Stefan–Boltzmann constant.

EXAMPLE 10–4 In a particular extrusion of steel wire to 1 cm in diameter the temperature of the wire is found to be 500°C. The wire is cooled by pouring water over the wire as it leaves the extrusion die. Assuming that the water is saturated vapor at 100°C at the surface of the wire, as indicated in figure 10–9, and the wire has an emissivity of 0.8, estimate the convective heat transfer coefficient for this cooling process.

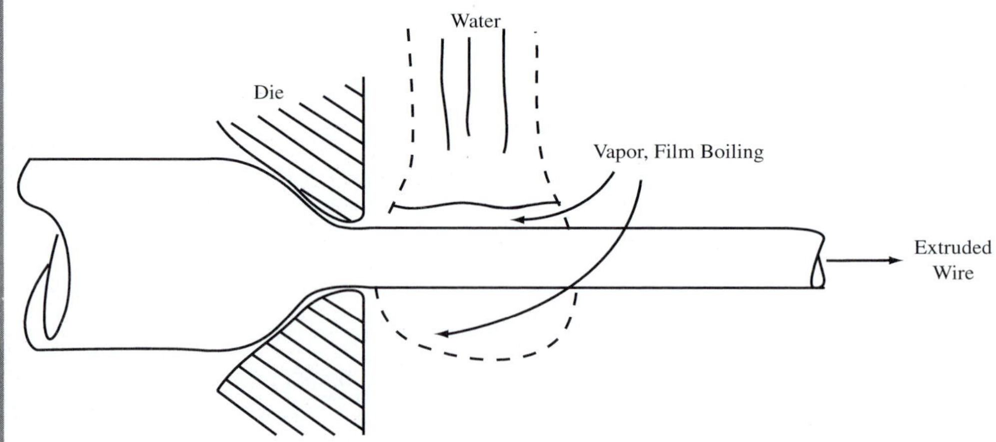

Figure 10–9 Cooling of extruded wire with film boiling of water.

Solution: Since the wire is at 500°C and the water is 100°C, the superheat is 400 K and film boiling is likely to occur. The radiation convective coefficient is, from equation (10–12),

$$h_{\text{RAD}} = \varepsilon_r\sigma(\frac{T_w^4 - T_g^4}{\Delta T_{\text{SH}}}) = (0.8)\left(5.67 \times 10^{-8}\frac{\text{W}}{\text{m}^2 \cdot \text{K}^4}\right)\left(\frac{773^4 - 373^4}{400}\right) = 38.293\,\frac{\text{W}}{\text{m}^2 \cdot \text{K}}$$

Also, using Bromley's Equation (10–11) for film boiling around a tube, we first find from appendix table B–6, $\rho_f = 957.85$ kg/m^3, $\rho_g = 1/v_g = 0.598$ kg/m^3, $hn_{fg} = 2257$ kJ/kg, $D = 0.01$ m, $c_{p,g} = 2.029$ kJ/kg · °C, $\mu_g = 0.012 \times 10^{-3}$ kg/m · s, $\kappa_g \approx 0.025$ W/m · K, and $\nu_g = \mu_g/\rho_g = 12.5 \times 10^{-6}$ m^2/s. Substituting these values into equation (10–11) gives

$$h_{\text{FB}} = 182\,\frac{\text{W}}{\text{m}^2 \cdot {}^\circ\text{C}}$$

Then using equation (10–10),

Answer

$$h_{\text{FBT}} = 182 + \frac{4}{3}(38.293) = 233\,\frac{\text{W}}{\text{m}^2 \cdot {}^\circ\text{C}}$$

Pool Boiling Heat Transfer in Vertical Tubes

Liquids often are boiled in vertical tubes, such as in large steam generating units and in the chemical process industry. Figure 10–10 shows the stages of pool boiling in a vertical tube. Here the fluid is being heated uniformly along the tube so that the tube inside wall temperature will vary as the fluid progresses from a liquid to a saturated liquid, to bubble flow, to slug flow, to annular liquid flow, and to vapor flow. The tube inside wall temperature is indicated in the figure to give you some idea of the behavior of this form of pool boiling. The heat transfer for single-phase liquid flow and single-phase vapor flow can be analyzed with the conventional convective heat-transfer relationships presented in Chapters 4 and 5 for flow through tubes. Although the vertical tube pool boiling process is driven by natural convection or the bouyancy of the lighter vapor bubbles in the liquid, the induced velocity of the upward bubble/vapor flow due to bouyancy can be treated as a forced convection heat-transfer condition. Then the Dittus–Boelter Equation can be used

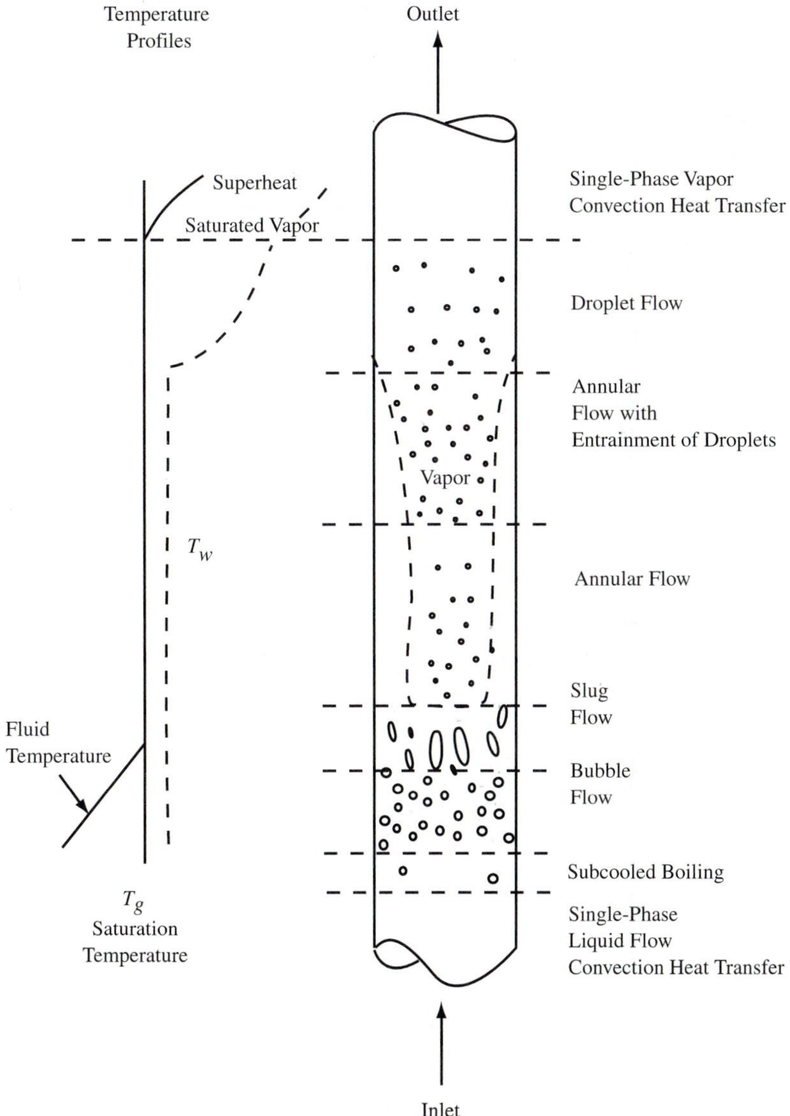

Figure 10–10 Pool boiling in vertical tubes.

if the Reynolds Number is above 10,000 or the single-phase fluid flow is turbulent. The Dittus–Boelter Equation (4–151) is

$$\text{Nu}_D = 0.023\ \text{Re}_D^{0.8}\ \text{Pr}^{0.4}$$

The fluid properties for the liquid need to be those used for single-phase liquid flow and the properties of the saturated vapor those used for single-phase vapor flow.

EXAMPLE 10–5 Estimate the upward bubble velocity for vertical tube boiling of water where bubble flow occurs, as indicated in figure 10–10. The liquid is boiling at 180°C. Assume that the bubbles have a diameter of 1 mm and a drag coefficient of 0.2 when moving through the saturated liquid.

Solution Assuming steady state conditions, the bubbles will rise because the buoyancy is greater than the bubble weight. The balancing force to keep steady state will be due to the drag force. Thus, referring to figure 10–11, we have

$$F_B = W_B + C_f A_F \rho_f \frac{V_B^2}{2}$$

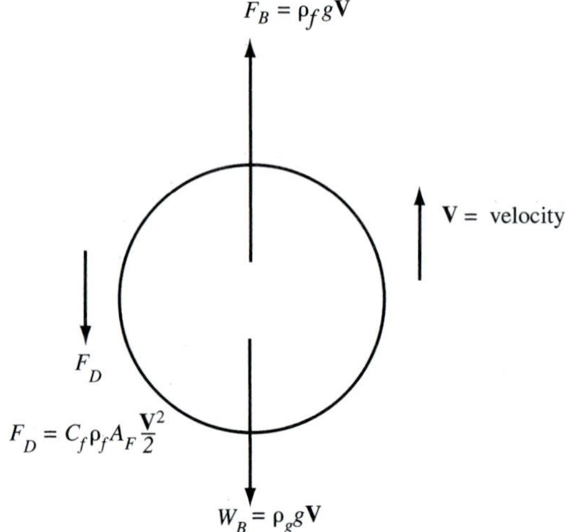

Figure 10–11 Free body diagram of bubble.

The buoyancy is

$$F_B = \rho_f g \frac{4}{3}\pi \frac{D^3}{8} = 4.546 \times 10^{-6}\ \text{N}$$

and the bubble weight is

$$W_B = \rho_g g \frac{4}{3}\pi \frac{D^3}{8} = 0.0265 \times 10^{-6}\ N$$

The frontal area is just the projected area, and the bubble velocity will then be

Answer
$$\mathbf{V} = \sqrt{\frac{2(F_B - W_B)}{C_f A_F \rho_f}} = \sqrt{\frac{2(4.5195 \times 10^{-6}\ \text{N})}{0.2\pi(0.0005\ \text{m})^2(885\ \text{kg/m}^3)}} = 0.255\ \text{m/s}$$

For an analysis of the heat transfer in the two-phase boiling region of the vertical tube, a general method has been outlined by Chen [7], where an equation for the convective heat transfer coefficient is defined as

$$h_{\text{nb,vt}} = h_f F_{\text{TP}} + h_{\text{nb}} S \quad \textbf{(10–13)}$$

and the following successive steps are taken:

1. The Reynolds Number for single-phase liquid flow is determined locally by

$$\text{Re}_{D,x} = \frac{\dot{m}_f D}{A\mu_f} \quad \textbf{(10–14)}$$

Here the mass flow of the liquid will be

$$\dot{m}_f = \dot{m}(1 - x) \quad \textbf{(10–15)}$$

where x is the local quality of the boiling fluid, or mass fraction of vapor to the total mass.

2. Determine the Prandtl Number for the liquid phase, Pr_f.

3. Determine h_f for single-phase liquid convective heat transfer as predicted (e.g., using the Dittus–Boelter Equation).

4. Determine the parameter

$$X_R = \left(\frac{\dot{m}_g}{\dot{m}_f}\right)^{0.9}\left(\frac{v_g}{v_f}\right)^{0.5}\left(\frac{\mu_g}{\mu_f}\right)^{0.1} \quad \textbf{(10–16)}$$

5. Determine the factor F_{TP} as the ratio of the Reynolds Number for two-phase flow to the liquid-only Reynolds Number:

$$F_{\text{TP}} = \frac{\text{Re}_{D,f}}{\text{Re}_{D,\text{TP}}} = 2.35(X_{\text{R}} + 0.213)^{0.736} \qquad \text{for } X_{\text{R}} \geq 0.1$$

$$F_{\text{TP}} = 1.0 \qquad \text{for } X_{\text{R}} > 0.1$$

6. Determine the two-phase Reynolds Number as

$$\text{Re}_{D,\text{TP}} = \frac{\text{Re}_{D,f}}{F_{\text{TP}}}$$

7. Calculate the nucleate pool boiling suppression factor, S, from

$$S = \frac{1}{1 + 2.53 \times 10^{-6}\,\text{Re}_{D,\text{TP}}^{1.17}} \quad \textbf{(10–17)}$$

8. Calculate the nucleate pool boiling convection heat-transfer coefficient from

$$h_{\text{nb}} = 0.00122\left(\frac{\kappa_f^{0.79} c_{p,f}^{0.45} \rho_f^{0.49}}{\sigma_{\text{ST}}^{0.5} \mu_f^{0.29} \text{hn}_{fg}^{0.24} \rho_g^{0.24}}\right)(\Delta T_{\text{SH}})^{0.24}(p_{g,w} - p_g)^{0.75} \quad \textbf{(10–18)}$$

Here all properties are evaluated at the local saturation condition, except $p_{g,w}$, which is the saturation pressure at the wall or tube inside surface temperature.

9. Determine the convective heat transfer coefficient from equation (10–13).

EXAMPLE 10–6 At the particular location in the vertical tube where bubble flow occurs such that the bubble velocities are 0.255 m/s as determined in example 10–5, it is found that the quality of the water is 5%. The tube inside diameter is 2 cm, and the surface temperature is 300°C when the water is 180°C. Detemine the convective heat transfer coefficient using the Chen Method.

Solution The Chen Method defines the convective heat transfer coefficient by equation (10–13). We need to determine the mass flow rates of the saturated vapor and the saturated liquid. Since the quality is 5%, we may write that

$$\frac{m_g}{m_g + m_f} = x = 0.05 = \frac{1}{1 + m_f/m_g} = \frac{1}{1 + V_f \nu_g / V_g \nu_f}$$

Here we know that the specific volumes of the steam at 180°C are

$$\nu_f = 0.00113 \text{ m}^3/\text{kg} \qquad \nu_g = 0.194 \text{ m}^3/\text{kg}$$

and the volume ratio can then be found,

$$\frac{V_f}{V_g} = \frac{v_f}{v_g}\left(\frac{1}{0.05} - 1\right) = 0.111 = \frac{1}{9}$$

Further,

$$\frac{V_T}{V_g} = \frac{V_f + V_g}{V_g} = \frac{10V_f}{9V_f} = \frac{10}{9}$$

and 90% of the volume is bubbles.

Since the tube diameter is 2 cm and the bubbles are 1 mm, at a station in the tube there will be 360 bubbles passing at any time. Therefore the mass flow of vapor will be

$$\dot{m}_g = \frac{\mathbf{V}_B(360 \text{ bubbles})A}{\nu_g} = \frac{(0.255 \text{ m/s})(360)(\pi)(0.005 \text{ m})^2}{0.194 \text{ m}^3/\text{kg}} = 37.2 \times 10^{-5} \text{ kg/s}$$

The mass flow of the saturated liquid is then

$$\dot{m}_f = \dot{m}_g \frac{1 - x}{x} = \left(37.2 \times 10^{-5} \frac{\text{kg}}{\text{s}}\right) \frac{1 - 0.05}{0.05} = 7.06 \times 10^{-3} \frac{\text{kg}}{\text{s}}$$

The Reynolds Number for the saturated liquid is

$$\text{Re}_{D,x} = \frac{\dot{m}_f D}{A\mu_f} = \frac{(7.06 \times 10^{-3} \text{ kg/s})(0.02 \text{ m})}{\pi(0.01 \text{ m})^2(0.00089 \text{ kg/m} \cdot \text{s})} = 505$$

The Prandtl Number for saturated water at 180°C is about 6.21. Then, since the Reynolds Number is below 2000, we assume laminar flow for the fluid. For uniform heat transfer in a tube equation (4–148) may be used,

$$\text{Nu}_D = 4.364 = \frac{h_f D}{\kappa_f} = h_f \frac{0.02 \text{ m}}{0.6 \text{ W/m} \cdot \text{K}}$$

giving

$$h_f = 13.1 \text{ W/m}^2 \cdot \text{K}$$

The parameter X_R is

$$X_R = \left(\frac{0.05}{1 - 0.05}\right)^{0.9} \left(\frac{0.19405}{0.001127}\right)^{0.5} \left(\frac{0.914}{0.857}\right)^{0.1} = 0.933$$

and

$$F_{\text{TP}} = 2.325(0.933 + 0.213)^{0.736} = 2.598$$

so that

$$\text{Re}_{D,\text{TP}} = \frac{\text{Re}_{D,f}}{F_{\text{TP}}} = 194.38$$

The nucleate pool suppression factor is

$$S = \frac{1}{1 + 2.53 \times 10^{-6}\,\mathrm{Re}_{D,\mathrm{TP}}^{1.17}} = 0.4997$$

The nucleate pool boiling convection heat transfer coefficient is

$$h_{\mathrm{nb}} = 0.00122\left(\frac{(0.6)^{0.79}(4.18)^{0.45}(997)^{0.49}}{(50 \times 10^{-3})^{0.5}(0.000857)^{0.29}(2015)^{0.24}(5.15)^{0.24}}\right)$$

$$\times\,(300 - 180)^{0.24}(8590 - 1003)^{0.75} = 437.4\,\frac{\mathrm{W}}{\mathrm{m}^2 \cdot \mathrm{K}}$$

The total convective heat transfer is, from equation (10–13),

Answer

$$h_{\mathrm{nb,vt}} = \left(13.1\,\frac{\mathrm{W}}{\mathrm{m}^2 \cdot \mathrm{K}}\right)(2.598) + \left(437.4\,\frac{\mathrm{W}}{\mathrm{m}^2 \cdot \mathrm{K}}\right)(0.4997) = 252.6\,\frac{\mathrm{W}}{\mathrm{m}^2 \cdot \mathrm{K}}$$

Forced Convection Heat Transfer in Horizontal Tubes

Boiling of liquids inside horizontally placed tubes is common practice in shell and tube heat exchangers as well as evaporators of mechanical refrigeration systems. There are other equally common applications, so this mode of heat transfer has been extensively studied. This particular mode of heat transfer differs from pool boiling because the bouyant forces are normal to the fluid/liquid flow and thus it is possible that the complete spectrum of convection heat transfer/nucleate boiling/film boiling can occur within a fluid-filled tube. In figure 10–12 is shown a typical condition where a liquid enters a tube having an inside surface temperature that is greater than the saturation temperature of the liquid. The liquid is heated by normal single-phase forced convection heat transfer to its saturation temperature and successively progresses through a bubble flow region, a plug flow, a slug flow, a wave flow, and an annular liquid flow, to the single-phase vapor flow.

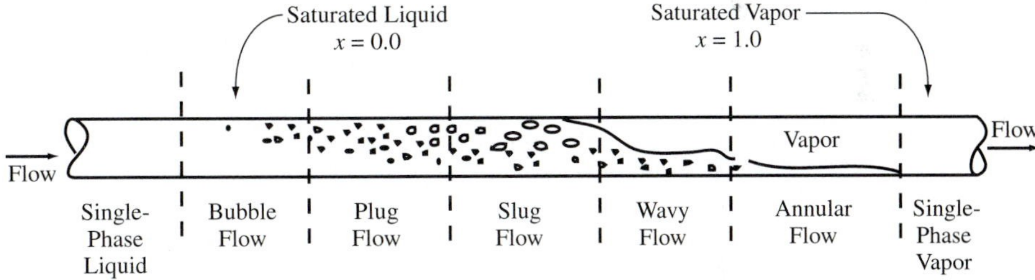

Figure 10–12 Forced convection boiling in a horizontal tube.

Many correlations have been suggested to predict the convective heat-transfer coefficient for horizontal tubes. For the single-phase regions of liquid flow and for vapor flow, the Dittus–Boelter Equation may be used if the flows are turbulent or, as in example 10–6,

$$\mathrm{Nu}_D = 4.364 \qquad \text{for uniform heat transfer} \qquad \textbf{(4–148, repeated)}$$

for laminar flow. For the two-phase region Pierre [8] has suggested the equation

$$h_{\mathrm{FC,ave}} = C\left(\frac{\kappa_f}{D}\right)\left(\mathrm{Re}_{D,f}^2\,\frac{(x_{\mathrm{out}} - x_{\mathrm{in}})\mathrm{hn}_{fg}}{L}\right)^n \frac{\mathrm{W}}{\mathrm{m}^2 \cdot {}^\circ\mathrm{C}} \qquad \textbf{(10–19)}$$

for the average convective heat transfer coefficient over the tube length L where the two-phase region occurs. In equation (10–19) the heat of vaporization, hn_{fg}, needs to have units of J/kg and length L in meters (m). The Reynolds Number is evaluated at the entrance for single-phase liquid flow. The parameters C and n are determined by the

Table 10–2

Quality of Leaving Fluid, %	C	n
Less than 90%	0.0009	0.5
Greater than 90% and superheated	0.0082	0.4

quality of the leaving fluid and are listed in table 10–2. Other more refined and accurate correlations are given in the literature. The interested reader may want to refer to a review by Butterworth and Robertson [9].

EXAMPLE 10–7 An evaporator has 1-cm-diameter tubes placed horizontally. Refrigerant R-134a at $-10°C$ flows at 20 g/s through a tube at a temperature of 0°C. If the R-134a has 20% quality as it enters the evaporator tube and is a saturated vapor leaving, determine the average convective heat-transfer coefficient and the tube length. Use for R-134a that $hn_f = 186.7$ kJ/kg $hn_g = 392.9$ kJ/kg.

Solution The heat transfer to the refrigerant, using an energy balance for the tube, is

$$\dot{Q} = \dot{m}_{\text{R-134a}}(hn_{out} - hn_{in}) = h_{FC,ave}\pi DL(\Delta T_{SH})$$

The following properties are found from appendix table B–3 for R-134a at $-10°C$:

$$\mu_f = 0.33 \text{ g/m} \cdot \text{s} \qquad \kappa_f = 0.098 \text{ W/m} \cdot \text{K}$$

The enthalpies are

$$hn_{out} = hn_g = 392.9 \text{ kJ/kg} \quad \text{and} \quad hn_{in} = x_{in}hn_g + (1 - x_{in})hn_f = 227.94 \text{ kJ/kg}$$

and then

$$\dot{Q} = (0.02 \text{ kg/s})(392.9 - 227.94 \text{ kJ/kg}) = 3.2992 \text{ kW}$$

Also,

$$3.2992 \text{ kW} = h_{FC,ave}\pi(0.01 \text{ m})L(10 \text{ K}) \tag{10–20}$$

Then, for equation (10–19) we have

$$\text{Re}_{D,f} = \frac{4\dot{m}_{\text{R-134a}}}{\pi D\mu_f} = 12732.4$$

giving, for equation (10–19),

$$h_{FC,ave} = (0.0082)\frac{0.098 \text{ W/m} \cdot \text{K}}{0.01 \text{ m}}\left[(12732.4)^2\left(\frac{1.0 - 0.2}{L}\right)(206.3 \text{ kJ/kg})\right]^{0.4}$$

Solving this equation simultaneously with equation (10–20) for $h_{FC,ave}$ and L gives

Answer

$$L = 37.6 \text{ m}$$

and

Answer

$$h_{FC,ave} = 279 \frac{\text{W}}{\text{m} \cdot \text{K}}$$

Since the tube lengths are so large, the tubes probably will need to be in a multipass arrangement.

10–3 CONDENSING HEAT TRANSFER

Condensing is the process where a vapor is cooled by heat transfer to become a saturated liquid. There are at least two types of condensing: *film* or *film-wise condensing* and *drop-wise condensing*. Condensing occurs because of heat transfer from the vapor to a surface that is in contact with the vapor, such as in a tube, around a tube, or in a tank or other container. The surface of the tube or tank needs to be below the saturation temperature of the vapor at its pressure in order for heat transfer to occur from the vapor and thereby allow for condensation. As the condensate or saturated liquid forms, it forms on the surface and normally will accumulate as a layer or film. By gravitational action, the liquid will then flow down the surface as a film. This is the meaning of the term *film-wise condensation,* and it is the most likely type of condensation. If the condensate forms as drops and does not accumulate into a film, this is called drop-wise condensation. We will discuss both of these types of condensation.

Film Condensation on a Vertical Surface

Consider a vertical surface, such as the vertical walls of a tank containing a vapor, in contact with a vapor as shown in figure 10–13, where the surface temperature is less than that of the saturation temperature of the vapor at its pressure. At some location on the surface the vapor will have cooled to its saturation temperature and will lose enough heat to become a saturated liquid. Condensation will begin here, and because the liquid has a greater density than the vapor it will flow down the side of the surface, gathering other particles of saturated liquid and creating a film of liquid flowing down the vertical surface. This situation is similar to the boundary layer ideas of convection heat transfer in Chapters 4 and 5, and it has been considered by many investigators. In particular, Nusselt [10] considered the case where the liquid flow is laminar down the vertical surface, there is no shear stress at the interface between the vapor (which is a uniform saturated vapor) and the condensate, the pressure is the same in the condensate boundary layer and the vapor at the same elevation, heat transfer is by conduction in the condensate boundary layer, and the thermal properties are uniform throughout for the vapor and for the condensate. Figure 10-14a shows this situation. Figure 10–14b shows a differential element within the condensate boundary layer. The momentum equation (4–44b) for this element in the y direction becomes, per unit of element volume,

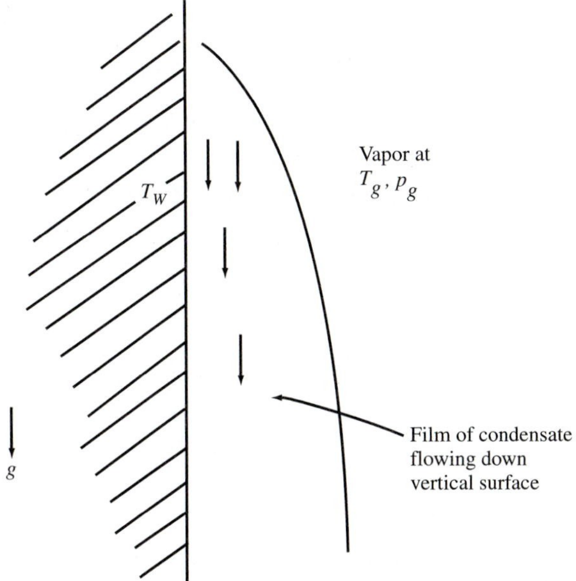

Figure 10–13 Film-wise condensation on a vertical surface.

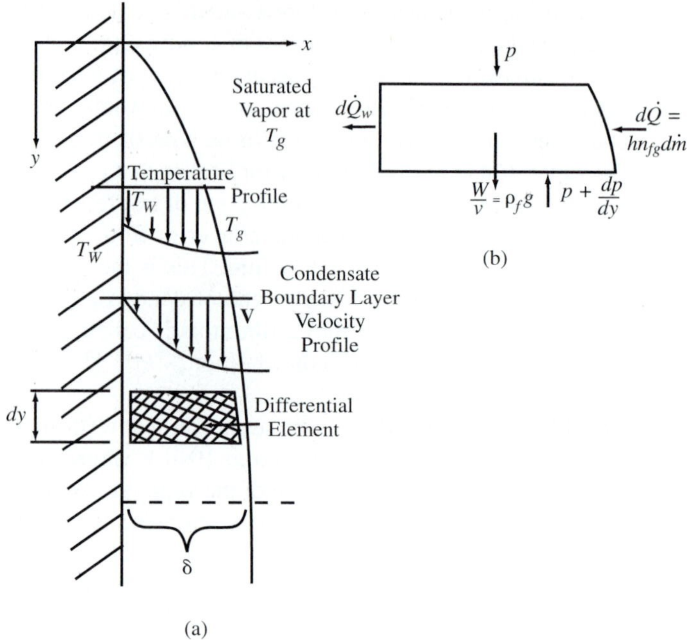

Figure 10–14 Laminar flow film condensation on a vertical surface.

$$\frac{\partial^2 \mathbf{v}}{\partial x^2} = \frac{1}{\mu_f}\frac{dp}{dy} - \frac{\rho_f g}{\mu_f} \tag{10–21}$$

Since the pressure is uniform at any horizontal station, whether in the condensate or the vapor, we may write that

$$\left[\frac{dp}{dy}\right]_{\text{condensate}} = \left[\frac{dp}{dy}\right]_{\text{vapor}} = \rho_g g$$

and substituting this into equation (10–21) gives

$$\frac{\partial^2 \mathbf{v}}{\partial x^2} = -\frac{g}{\mu_f}(\rho_f - \rho_g) \tag{10–22}$$

Assuming that the y-direction velocity, $\mathbf{v}$, can be separated into functions of x and of y, and then integrating at a constant y station, gives

$$\frac{d\mathbf{v}}{dx} = -\frac{g}{\mu_f}(\rho_f - \rho_g)x + C_1 \tag{10–23}$$

Since there is assumed to be no shear stress at the boundary between the condensate and the vapor, this means that

$$\frac{d\mathbf{v}}{dx} = 0 \qquad \text{at } x = \delta$$

Evaluating the constant gives

$$C_1 = \frac{g}{\mu_f}(\rho_f - \rho_g)\delta$$

and equation (10–23) becomes

$$\frac{d\mathbf{v}}{dx} = \frac{g}{\mu_f}(\rho_f - \rho_g)(\delta - x) \tag{10–24}$$

Integrating once more to determine the condensate boundary layer velocity profile gives

$$\mathbf{v} = \frac{g}{\mu_f}(\rho_f - \rho_g)\left(\delta x - \frac{x^2}{2} + C_2\right)$$

Since $\mathbf{v} = 0$ at $x = 0$, $C_2 = 0$, and the velocity profile is

$$\mathbf{v} = \frac{g}{\mu_f}(\rho_f - \rho_g)\left(\delta x - \frac{x^2}{2}\right) \tag{10–25}$$

The condensate flow rate can be determined from

$$\dot{m} = \int_{x=0}^{x=\delta} \rho_f \mathbf{v}(\text{width})dx \tag{10–26}$$

and the mass flow per unit width is

$$\frac{\dot{m}}{\text{width}} = \dot{m}_{\text{WD}} = \int_0^\delta \rho_f \mathbf{v} dx = \int_0^\delta \frac{\rho_f g}{\mu_f}(\rho_f - \rho_g)\left(\delta x - \frac{x^2}{2}\right)dx$$

or, completing the integration,

$$\dot{m}_{\text{WD}} = \frac{\rho_f g}{3\mu_f}(\rho_f - \rho_g)\delta^3 \tag{10–27}$$

The condensate boundary layer δ is a function of y, and to develop the relationship to predict it requires the use of the heat transfers. The heat transfer associated with the condensation of the vapor at the boundary between the condensate and the vapor is

$$d\dot{Q} = \text{hn}_{fg} d\dot{m} \tag{10–28}$$

and the heat transfer associated with the cooling at the vertical surface is

$$d\dot{Q} = \kappa_f \frac{(T_g - T_w)}{\delta}(\text{width})dy \tag{10–29}$$

Equating the heat transfers of equations (10–28) and (10–29), and using the concept of the mass flow per unit width from equation (10–27), we have

$$\text{hn}_{fg} d\dot{m}_{\text{WD}} = \kappa_f \frac{(T_g - T_w)}{\delta} dy \tag{10–30}$$

Further, differentiating equation (10–27) with respect to y gives

$$\frac{d\dot{m}_{\text{WD}}}{dy} = \frac{\rho_f g}{\mu_f}(\rho_f - \rho_g)\delta^2 \frac{d\delta}{dy} \tag{10–31}$$

Combining equations (10–30) and (10–31) gives

$$\delta^3 d\delta = \frac{\kappa_f \mu_f (T_g - T_w)}{\rho_f g(\rho_f - \rho_g)\text{hn}_{fg}} dy \tag{10–32}$$

Noting that the condensate boundary layer is zero at the commencement of condensation and if we set this as $y = 0$ as shown in figure 10–14, then integrating equation (10–32), we get

$$\int_0^{\delta(y)} \delta^3 d\delta = \frac{\delta(y)^4}{4} = \int_0^y \frac{\kappa_f \mu_f (T_g - T_w)}{\rho_f g(\rho_f - \rho_g)\text{hn}_{fg}} dy = \frac{\kappa_f \mu_f (T_g - T_w)}{\rho_f g(\rho_f - \rho_g)\text{hn}_{fg}} y \tag{10–33}$$

and the condensation boundary layer thickness is then

$$\delta(y) = \left(\frac{4\kappa_f\mu_f(T_g - T_w)}{\rho_f g(\rho_f - \rho_g)\mathrm{hn}_{fg}}y\right)^{1/4} \tag{10–34}$$

The local heat transfer can also be written as

$$d\dot{Q} = h_y(T_g - T_w)(\text{width})dy \tag{10–35}$$

and if this is equated to equation (10–29), the following relationship results:

$$h_y = \frac{\kappa_f}{\delta(y)} \tag{10–36}$$

Then substituting equation (10–34) for the condensate boundary layer thickness δ, we have

$$h_y = \left(\frac{\rho_f g(\rho_f - \rho_g)\kappa_f^3\mathrm{hn}_{fg}}{4\mu_f(T_g - T_w)y}\right)^{1/4} \tag{10–37}$$

and the average convective heat-transfer coefficient for a boundary layer L long is

$$h_{\text{ave}} = \frac{1}{L}\int_0^L h_y dy = \frac{4}{3}h_L = 0.943\left(\frac{\rho_f g(\rho_f - \rho_g)\kappa_f^3\mathrm{hn}_{fg}}{\mu_f(T_g - T_w)L}\right)^{1/4} \tag{10–38}$$

The average Nusselt Number for the condensate laminar boundary layer is

$$\mathrm{Nu}_{\text{ave}} = \frac{h_{\text{ave}}L}{\kappa_f} = 0.943\left(\frac{\rho_f g(\rho_f - \rho_g)\mathrm{hn}_{fg}}{\mu_f\kappa_f(T_g - T_w)}L^3\right)^{1/4} \tag{10–39}$$

The total heat transfer per unit width from the condensate to the vertical surface over the height L is

$$\dot{q}_W = h_{\text{ave}}L(T_g - T_w) \tag{10–40}$$

The total rate of condensation per unit width flowing down the vertical surface through a height L is, from equations (10–27) and (10–34) for $y = L$,

$$\dot{m}_{\text{WD,total}} = 2.15\left(\frac{\rho_f g(\rho_f - \rho_g)\kappa_f^3(T_g - T_w)^3}{\mu_f\mathrm{hn}_{fg}^3}L^3\right)^{1/4} \tag{10–41}$$

and by using equation (10–40) (see practice problem 10–14), this can also be written

$$\dot{m}_{\text{WD,total}} = \frac{h_{\text{ave}}L(T_g - T_w)}{\mathrm{hn}_{fg}} \tag{10–42}$$

The Reynolds Number that is most often defined to describe the character of the condensate film flow down a vertical surface is

$$\mathrm{Re}_{\text{cond}} = \frac{4\rho_f\mathbf{v}_{\text{ave}}\delta}{\mu_f} = \frac{4\dot{m}_{\text{WD}}}{\mu_f} \tag{10–43}$$

From experimental observations it has been determined that for

$$\begin{array}{ll} \mathrm{Re}_{\text{cond}} \le 30 & \text{laminar flow occurs} \\ 30 < \mathrm{Re}_{\text{cond}} < 1800 & \text{transition or wavy flow occurs} \\ \mathrm{Re}_{\text{cond}} \ge 1800 & \text{turbulent flow occurs} \end{array} \tag{10–44}$$

Wavy flow is indicated in figure 10–15 as a film of condensation that has an undulating thickness, whereas the laminar boundary layer is regular and the turbulent boundary layer is irregular. The analysis of the transition and turbulent condensate flow leans heavily on

Figure 10–15 Film condensate flow character down a vertical surface.

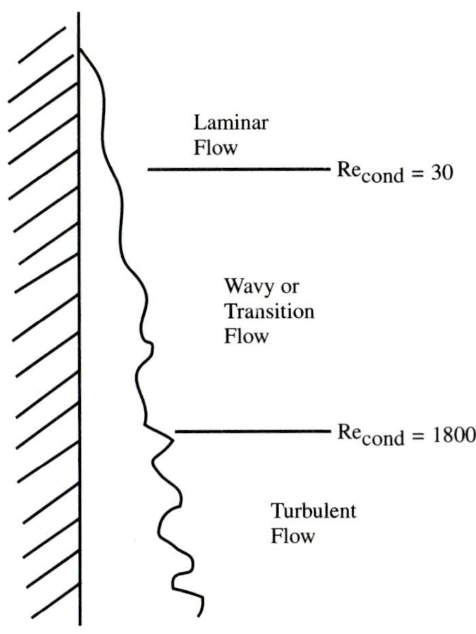

empirical data, as was done with convection heat transfer in Chapters 4 and 5. The average convective heat-transfer coefficient for heat transfer through the transition boundary layer of condensate flow can be evaluated with the relationship suggested by Kutateladze [11],

$$h_{ave} = \frac{\kappa_f}{1.08(Re_{cond}^{1/3})}\left(\left(1 - \frac{\rho_g}{\rho_f}\right)\left(\frac{g}{\nu_f^2}\right)\right)^{1/3} \quad \text{(transition flow)} \qquad \textbf{(10–45)}$$

For turbulent flow of condensate in the film boundary layer, Labuntsov [12] suggests,

$$h_{ave} = \frac{\kappa_f Re_{cond}}{8750 + \dfrac{58(Re_{cond}^{0.75} - 253)}{\sqrt{Pr}}}\left(\frac{g}{\nu_f^2}\right)^{1/3} \quad \text{(turbulent flow)} \qquad \textbf{(10–46)}$$

EXAMPLE 10–8 A dehumidifier is designed with vertical tubes through which "chilled" water passes so as to cool humid air passing around the tubes. For the particular situation shown in figure 10–16, determine the Reynolds Number and whether the condensate is laminar or turbulent flow, the convective heat-transfer coefficient, and the heat transfer from the humid air stream to the chilled water tubes. The condensation rate of 7.0 g/s is uniformly distributed over the 12 vertical tubes. Assume that the tubes act as vertical surfaces and neglect radial effects on the convective heat-transfer coefficient.

Solution The condensation rate of 7.0 g/s is uniformly distributed over the 12 tubes so that the condensation rate per unit width is

$$\dot{m}_{WD} = \frac{\dot{m}}{\text{width}} = \frac{7.0 \text{ g/s}}{2\pi(2 \text{ cm})(12 \text{ tubes})} = 0.04642 \text{ g/cm} \cdot \text{s} = 4.642 \text{ g/m} \cdot \text{s}$$

The Reynolds Number is, from equation (10–43), where μ_f is 1.044 g/m · s from table B.6,

Answer

$$Re_{cond} = \frac{4(4.642 \text{ g/m} \cdot \text{s})}{1.044 \text{ g/m} \cdot \text{s}} = 17.1$$

so the boundary layer flow is laminar. Using equation (10–42) we may predict the convective heat-transfer coefficient.

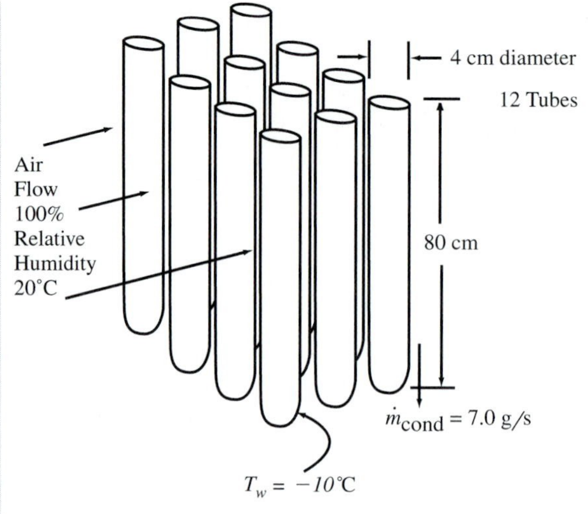

Figure 10–16 Dehumidifier coil as a vertical surface for film condensation.

Answer
$$h_{ave} = \frac{\dot{m}_{WD,total} hn_{fg}}{L(T_g - T_w)} = \frac{(7.0\ \text{g/m}\cdot\text{s})(2454.12\ \text{J/g})}{(0.8\ \text{m})[20°\text{C} - (-10°\text{C})]} = 715.785\ \frac{\text{W}}{\text{m}^2\cdot°\text{C}}$$

and the rate of heat transfer is

Answer
$$\dot{Q} = h_{ave}A(T_g - T_w) = \dot{m}hn_{fg} = (7.0\ \text{g/s})(2454.12\ \text{J/g}) = 17{,}178.84\ \text{W}$$

EXAMPLE 10–9 Refrigerant R-134a condenses at a vertical plate in a compact condenser of a refrigeration system. The vertical height of the plate is 1 ft, its width is 2 ft, and its surface temperature is 80°F when the R-134a is at its saturation temperature of 100°F. Estimate the average convective heat-transfer coefficient and the amount of R-134a that is condensing through film condensation.

The following properties are given for R-134a:

$$\mu_f = 0.355\ \frac{\text{g}}{\text{m}\cdot\text{s}} \qquad \text{at } 100°\text{F} = 311\ \text{K}$$

$$L = 1\ \text{ft} = 0.3048\ \text{m} \qquad \text{Width} = 2\ \text{ft} = 0.6096\ \text{m}$$

$$T_g = 0°\text{F} = 256\ \text{K} \qquad T_w = -20°\text{F} = 245\ \text{K} \qquad hn_{fg} = 186\frac{\text{J}}{\text{g}}$$

$$\kappa_f = 0.081\frac{\text{W}}{\text{m}\cdot\text{K}} \qquad \nu_f = \frac{\mu_f}{\rho_f} \approx 0.166\times 10^{-6}\frac{\text{m}^2}{\text{s}} \qquad \text{Pr} \approx 3.53$$

Solution

In this example the mass flow of condensate needs to be predicted, and without knowing it the Reynolds Number cannot be directly determined. Thus, we need to proceed through an iteration to arrive at an answer. One of the many iterative loops possible is to assume a Reynolds Number so that we may expect whether the flow is laminar, transitional, or turbulent, and predict the convective heat-transfer coefficient, the heat transfer, and then the mass flow of condensate. Finally, the Reynolds Number may then be determined as a check as to whether the character of flow was assumed correctly. Let's assume that the Reynolds Number is 1800 and that the flow is therefore turbulent.

With the initial assumption that the Reynolds Number is 1800, using equation (10–46) we have

$$h_{\text{ave}} = \left(\frac{(0.081\ \text{W/m}\cdot\text{K})(1800)}{8750 + \dfrac{58(1800^{0.75} - 253)}{\sqrt{3.53}}}\right)\left(\frac{9.81\ \text{m/s}^2}{(0.166 \times 10^{-6}\ \text{m}^2\text{/s})^2}\right)^{1/3} = 1091\frac{\text{W}}{\text{m}^2\cdot\text{K}}$$

The heat transfer would be

$$\dot{Q} = h_{\text{ave}}A(T_g - T_w) = \left(1091\frac{\text{W}}{\text{m}^2\cdot\text{K}}\right)(0.3048 \times 0.6096\ \text{m}^2)(11\ \text{K}) = 2230\ \text{W}$$

and this can also be written as

$$\dot{Q} = \dot{m}hn_{fg} = 2230\ \text{W} = \dot{m}\left(186\frac{\text{J}}{\text{g}}\right)$$

The condensate mass flow rate is then

$$\dot{m} = 11.99\frac{\text{g}}{\text{s}}$$

and the mass flow per unit width is

$$\dot{m}_{\text{WD}} = \frac{\dot{m}}{\text{width}} = \frac{11.99\ \text{g/s}}{0.6098\ \text{m}} = 19.67\frac{\text{g}}{\text{m}\cdot\text{s}}$$

The Reynolds Number can now be determined to see if it agrees with the initial assumed value:

$$\text{Re} = \frac{4\dot{m}_{\text{WD}}}{\mu_f} = \frac{4(19.67\ \text{g/m}\cdot\text{s})}{0.355\ \text{g/m}\cdot\text{s}} = 222 \neq 1800$$

Since the computed Reynolds Number is less than the assumed value, we assume a new lower Reynolds Number, say, Re = 310, and recompute new values for the convective heat-transfer coefficient, using equation (10–45) because the flow would be transitional at a Reynolds Number of 310. Recomputing the heat transfer, mass flow rate of condensate, and the mass flow per unit width provides the following results: $h_{\text{ave}} = 1511\ \text{W/m}^2\cdot\text{K}$, $\dot{Q} = 3088$ W, $\dot{m} = 16.6$ g/s, $\dot{m}_{\text{WD}} = 27.23$ g/m · s, and Re ≈ 307. This is close enough to the assumed value so that we may accept these values. The flow is transitional with Re = 310, and

Answer
$$h_{\text{ave}} = 1511\ \text{W/m}^2\cdot\text{K}$$

Answer
$$\dot{m} = 16.6\ \text{g/s} \qquad \text{(mass flow of condensate)}$$

Film Condensation over Horizontal Tubes

Condensers and other devices that are used to condense vapors are often configured with horizontal tubes or pipes to provide the cold surface. Then, too, horizontal cold tubes or water lines are subject to condensation of humidity in the air surrounding these pipes. Figure 10–17 shows some of the types of film condensation that occur on horizontal tubes or pipes. Notice that if tubes are placed directly under one another the film can be continuous or dripping. The heat transfer for condensation is increased if such a continuous film is avoided. This can be accomplished by placing the tubes in a staggered arrangement, similar to the tube bank configuration from Chapter 4. Thus, the dripping condition

Figure 10–17 Film condensation over horizontal tubes.

(a) Single horizontal tube film condensation

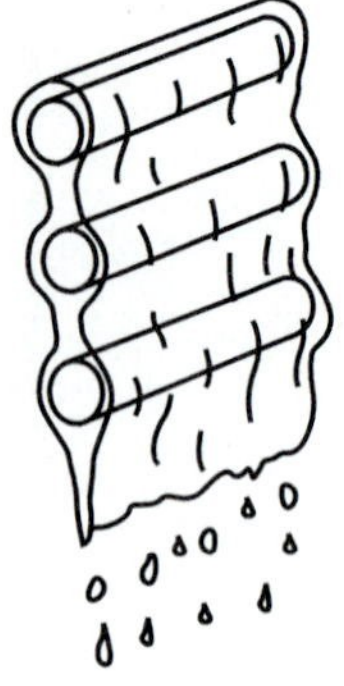

(b) In-line arrangement of film continuous condensation

(c) In-line dripping film condensation

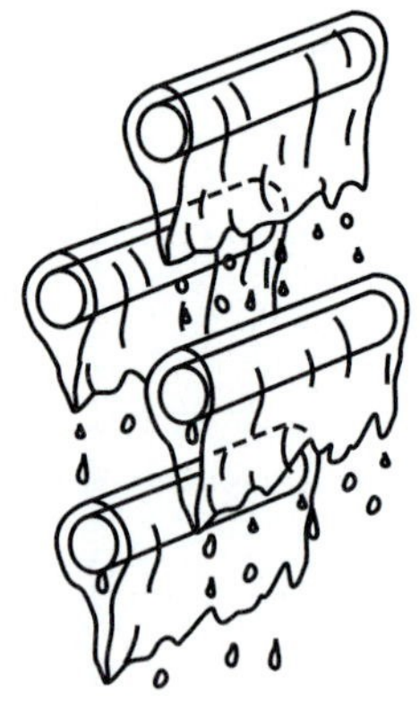

(d) Staggered tube arrangement to augment dripping condensation

shown in figure 10–17c has a greater convective heat-transfer coefficent than the film condensation of figure 10–17b. The convective heat transfer coefficient for film condensation over a single horizontal tube can be predicted from an equation suggested by Nusselt [10],

$$h_{\text{ave}} = 0.725\left(\frac{\rho_f g(\rho_f - \rho_g)\kappa_f^3 \text{hn}_{fg}}{\mu_f(T_g - T_w)D}\right)^{1/4} \tag{10–47}$$

and for film condensation over a single sphere,

$$h_{\text{ave}} = 0.815\left(\frac{\rho_f g(\rho_f - \rho_g)\kappa_f^3 \text{hn}_{fg}}{\mu_f(T_g - T_w)D}\right)^{1/4} \tag{10–48}$$

If a number of horizontal tubes or pipes are arranged vertically in an aligned column as shown in figure 10–17b, the convective heat transfer coefficient may be predicted by a variation of equation (10–47),

$$h_{\text{ave}} = 0.725\left(\frac{\rho_f g(\rho_f - \rho_g)\kappa_f^3 \text{hn}_{fg}}{\mu_f(T_g - T_w)D}\right)^{1/4}\left(\frac{1}{N}\right)^{1/4} \tag{10–49}$$

where N is the number of tubes in the vertically aligned column, or the number of rows of the tubes.

EXAMPLE 10–10 Water at 30°C is a saturated vapor when passing over a vertical set of five 2-cm-diameter tubes as shown in figure 10–18. The tubes are at 10°C and are spaced so that film condensation occurs. Estimate the rate of condensation per unit length of the tubes and compare it to that for a single horizontal tube.

Figure 10–18 Condensation tube bank.

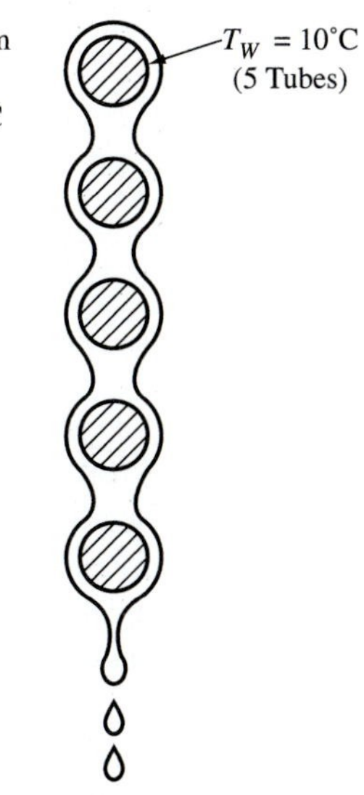

Solution To determine the rate of condensation, we will determine the average convective heat-transfer coefficient, followed by the rate of heat lost by the water, and equate this to the mass flow of condensate times its latent heat of vaporization. The properties of the water are

$$\rho_f = \frac{1}{\nu_f} = \frac{1}{0.001004\ \text{m}^3/\text{kg}} = 996\ \text{kg/m}^3 \qquad \rho = 0.0304\ \text{kg/m}^3$$

$$\kappa_f = 0.21\ \text{W/m}\cdot\text{K} \qquad \mu_f = 0.00085\ \text{kg/m}\cdot\text{s} \qquad \text{hn}_{fg} = 2430.48\ \text{kJ/kg}$$

The average convective heat-transfer coefficient is predicted by equation (10–49),

$$h_{\text{ave}} = 0.725\left(\frac{\left(996\dfrac{\text{kg}}{\text{m}^3}\right)\left(9.81\dfrac{\text{m}}{\text{s}^2}\right)\left(995.07\dfrac{\text{kg}}{\text{m}^3}\right)\left(0.21\dfrac{\text{W}}{\text{m}\cdot\text{K}}\right)^3\left(2430.48\dfrac{\text{kJ}}{\text{kg}}\right)}{(0.00085\ \text{kg/m}\cdot\text{s})(20\ \text{K})(0.02\ \text{m})(\text{N})}\right)^{1/4}$$

$$= 2442\frac{\text{W}}{\text{m}^2\cdot\text{K}}$$

and the heat transfer per unit length of tubes is

$$\dot{q}_L = h_{\text{ave}}\pi D(T_g - T_w) = \left(2442\frac{\text{W}}{\text{m}^2\cdot\text{K}}\right)(5\pi)(0.02\ \text{m})(30\ \text{K} - 10\ \text{K}) = 15{,}344\frac{\text{W}}{\text{m}}$$

This can now be equated as follows

$$\dot{q}_L = 15{,}344\frac{\text{W}}{\text{m}} = \dot{m}_L\text{hn}_{fg} = \dot{m}_L\left(2430\frac{\text{J}}{\text{g}}\right)$$

or

Answer

$$\dot{m}_L = 6.31\frac{\text{g}}{\text{s}}$$

For a single tube,

$$h_{ave} = 3651.75 \frac{W}{m^2 \cdot K}$$

the heat transfer is

$$\dot{q}_L = \left(3651.75 \frac{W}{m^2 \cdot K}\right)(\pi)(0.02\ m)(20\ K) = 4588.9 \frac{W}{m}$$

and the condensate for the single-tube arrangement would be

Answer
$$\dot{m}_L = 1.89 \frac{g}{s}$$

Film Condensation Inside Horizontal Tubes

Vapors are frequently condensed inside horizontal or slightly inclined tubes. The flow of fluid through a tube where condensation occurs is usually assumed to behave as shown in figure 10–19, where the vapor flows in the center area of the tube, gradually diminishing to full liquid flow, and the condensate forms on the outer annular area along the tube surface. Defining the Reynolds Number for the two-phase flow of vapor and liquid as

$$\text{Re}_D = \frac{\rho_g \mathbf{u}_{ave} D}{\mu_g}$$

the laminar condensate flow seems to occur below about 35,000. For laminar condensate flow the convective heat transfer coefficient may be evaluated from the relationship suggested by Chato [13],

$$h_{ave} = 0.555 \left(\frac{\rho_f g(\rho_f - \rho_g)\kappa_f^3 \text{hn}^*_{fg}}{\mu_f(T_g - T_w)D} \right)^{1/4} \quad \textbf{(10–50)}$$

where

$$\text{hn}^*_{fg} = \text{hn}_{fg} + {}^3\!/_8 c_{p,f}(T_g - T_w) \quad \textbf{(10–51)}$$

For turbulent condensate flow, such that $\text{Re}_D > 35{,}000$, the reader is referred to Akers et al. [14] and to Rosenow [15].

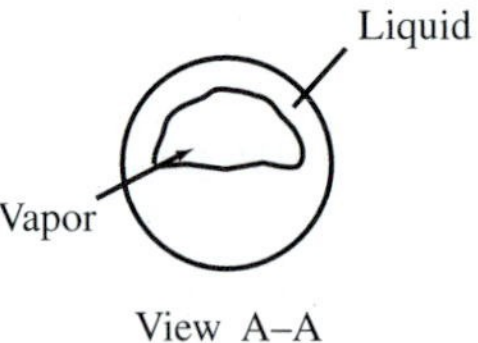

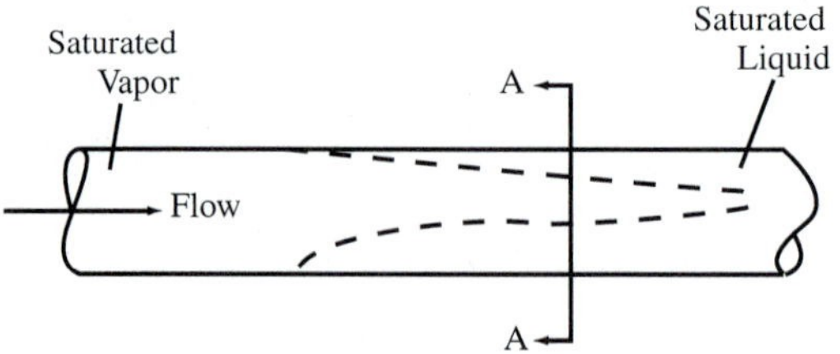

Figure 10–19 Film condensation inside a horizontal tube.

EXAMPLE 10–11 Saturated vapor ammonia at −20°C flows at 5 m/s through a 1-cm-diameter condenser tube that has a surface temperature of −40°C. Determine the expected length of tube needed to fully condense the ammonia.

The properties of the ammonia are

$$\mu_g = 8.5 \times 10^{-3}\ \text{g/m} \cdot \text{s} \qquad \mu_f = 0.214\ \text{g/m} \cdot \text{s} \qquad \kappa_f = 0.514\ \text{W/m} \cdot \text{K}$$

$$\rho_g = \frac{1}{\nu_g} = 1.604\ \text{kg/m}^3 \qquad \rho_f = \frac{1}{\nu_f} = 664.89\ \text{kg/m}^3$$

$$\text{hn}_{fg} = 1329\ \text{kJ/kg} \qquad c_{p,f} = 4.84\ \text{kJ/kg} \cdot \text{K}$$

Solution The Reynolds Number at the entrance is

$$\text{Re}_D = \frac{(1.604\ \text{kg/m}^3)(5\ \text{m/s})(0.01\ \text{m})}{(8.5 \times 10^{-6}\ \text{kg/m} \cdot \text{s})} = 9435$$

so we may use equation (10–50) to predict the average convective heat-transfer coefficient. First we determine the revised heat of vaporization from equation (10–51):

$$\text{hn}^*_{fg} = 1329\ \text{kJ/kg} + {}^3\!/_8(4.84\ \text{kJ/kg} \cdot \text{K})(20\ \text{K}) = 1365.3\ \text{kJ/kg}$$

and the convective heat transfer coefficient is then

$$h_{\text{ave}} = 0.555\left(\frac{\left(664.89\dfrac{\text{kg}}{\text{m}^3}\right)\left(9.81\dfrac{\text{m}}{\text{s}^2}\right)\left(663.286\dfrac{\text{kg}}{\text{m}^3}\right)\left(0.514\dfrac{\text{W}}{\text{m} \cdot \text{K}}\right)^3\left(1365.3\dfrac{\text{kJ}}{\text{kg}}\right)}{(0.214 \times 10^{-3}\ \text{kg/m} \cdot \text{s})(20\ \text{K})(0.01\ \text{m})}\right)^{1/4}$$

$$= 2080.6\frac{\text{W}}{\text{m}^2 \cdot \text{K}}$$

The total heat transfer from the ammonia is

$$\dot{m}_{\text{ammonia}}\text{hn}_{fg} = h_{\text{ave}}(\text{area})(T_g - T_w) = h_{\text{ave}}\pi DL(T_g - T_w)$$

The mass flow of ammonia is

$$\dot{m}_{\text{ammonia}} = \rho_g \mathbf{V}_{\text{ave}}(\pi r^2) = 1.604\frac{\text{kg}}{\text{m}^3}\left(5\frac{\text{m}}{\text{s}}\right)(\pi)(0.005\ \text{m})^2 = 0.6299\frac{\text{g}}{\text{s}}$$

and the tube length required for complete condensation is

Answer

$$L = \frac{\dot{m}_{\text{ammonia}}\text{hn}_{fg}}{h_{\text{ave}}\pi D(\Delta T)} = \frac{\left(0.6299\dfrac{\text{g}}{\text{s}}\right)1329\dfrac{\text{J}}{\text{g}}}{\left(2080.6\dfrac{\text{W}}{\text{m}^2 \cdot \text{K}}\right)\pi(0.01\ \text{m})(20\ \text{K})} = 0.640\ \text{m}$$

Drop-Wise Condensation

Drop-wise condensation has been investigated and developed because it provides a greater condensation rate than film condensation for the same surface area. This occurs because the condensate forms as drops and immediately falls from the surface, allowing the vapor to be in more direct contact with the cold condensing surface. The phenomena of drop-wise condensation requires that the condensing surface be nonwetting to the condensate, and this can usually be achieved by placing a thin film of oil or fatty acids on the condensing surface. It is crucial that the oil be immiscible with the condensation, be thin enough not to hinder heat transfer between the vapor and the condensing surface, and still adhere to the surface as the drops of condensate form and fall from the surface. From extensive empirical information it has been found that drop-wise condensation can be achieved best on very smooth surfaces. The analysis of drop-wise condensation is complicated and beyond the scope of this text.

10–4 SIMPLIFIED RELATIONSHIPS FOR BOILING AND CONDENSING HEAT TRANSFER

The equations developed in sections 10–2 and 10–3 can become more cumbersome and rigorous than is necessary for many preliminary design calculations and estimations for boiling and condensing processes. Here we set down some approximate relationships to give a quick prediction of the convection heat-transfer coefficients involved in specific situations.

Boiling of Water on a Horizontal Surface

Jakob and Hawkins [16] recommend, where $\Delta T = T_w - T_g \leq 14°\text{F} = 7.8\text{ K}$,

$$h_{\text{ave}} = 150(\Delta T)^{1/3} \frac{\text{Btu}}{\text{hr} \cdot \text{ft}^2 \cdot °\text{F}} \qquad (\Delta T \text{ in °F or °R}) \tag{10–52}$$

$$h_{\text{ave}} = 1037(\Delta T)^{1/3} \frac{\text{W}}{\text{m}^2 \cdot \text{K}} \qquad (\Delta T \text{ in °C or K})$$

If $\Delta T > 14°\text{F}$, then

$$h_{\text{ave}} = 0.168(\Delta T)^3 \frac{\text{Btu}}{\text{hr} \cdot \text{ft}^2 \cdot °\text{F}} \tag{10–53}$$

$$h_{\text{ave}} = 5.567(\Delta T)^3 \frac{\text{W}}{\text{m}^2 \cdot \text{K}}$$

Boiling of Water on Vertical Flat Surfaces in a Wide Vessel

Jakob and Hawkins [16] recommend for $\Delta T \leq 7.5°\text{F}$

$$h_{\text{ave}} = 87.43(\Delta T)^{1/7} \frac{\text{Btu}}{\text{hr} \cdot \text{ft}^2 \cdot °\text{F}}$$

$$h_{\text{ave}} = 540(\Delta T)^{1/7} \frac{\text{W}}{\text{m}^2 \cdot \text{K}} \tag{10–54}$$

For $17°\text{F} > \Delta T > 7.5°\text{F}$

$$h_{\text{ave}} = 0.24(\Delta T)^3 \frac{\text{Btu}}{\text{hr} \cdot \text{ft}^2 \cdot °\text{F}}$$

$$h_{\text{ave}} = 7.74(\Delta T)^3 \frac{\text{W}}{\text{m}^2 \cdot \text{K}} \tag{10–55}$$

Water Boiling Inside Vertical Tubes

Jakob [17] recommends for water boiling inside vertical tubes where $\Delta T \leq 14°\text{F}$,

$$h_{\text{ave}} = 187.5(\Delta T)^{1/3} \frac{\text{Btu}}{\text{hr} \cdot \text{ft}^2 \cdot °\text{F}}$$

$$h_{\text{ave}} = 1296(\Delta T)^{1/3} \frac{\text{W}}{\text{m}^2 \cdot \text{K}} \tag{10–56}$$

Pressure Effects on Boiling Water

Pressure effects on the boiling of water may be approximated by determining the average convective heat-transfer coefficient from one of the equations (10–52) through (10–56), which assume a pressure of 1 atm, and using the estimation

$$h_{\text{ave},P} = h_{\text{ave},0}\,p^{0.4} \tag{10–57}$$

Here the pressure p is in atmospheres and $h_{\text{ave},0}$ is determined from one of equations (10–52) through (10–56).

General Expression for Nucleate Boiling

A general expression for nucleate boiling, suggested by Cryder and Finalborgo [18] for many different substances, is

$$\log h_{\text{ave}} = A + 2.5 \log \Delta T + BT_g \qquad \text{Btu/hr} \cdot \text{ft}^2 \cdot {}^\circ\text{F} \tag{10–58}$$

where the constants are listed in table 10–3 for these substances. The temperature and temperature difference in this equation need to be entered as degrees fahrenheit, °F.

Table 10–3 Parameters for Equation (10–58)

Substance	*A*	*B*
Carbon tetrachloride	−2.57	0.012
Glycerol solution (26.3% by weight)	−2.65	0.015
Kerosene	−5.15	0.012
Methanol	−2.23	0.015
Normal butanol	−4.06	0.014
Sodium sulfate solution (10.1% by weight)	−2.62	0.016
Sodium chloride solution (24.2% by weight)	−3.61	0.017
Water	−2.05	0.014

Film Condensation of Ammonia in Horizontal Tubes

For film condensation of ammonia in horizontal tubes of diameter D and for temperature differences between the vapor and the tube surfaces of less than about 15°C, an approximation for the convective heat-transfer coefficient can be made using

$$h_{\text{ave}} = 4000\left(\frac{1}{D\Delta T}\right)^{1/4} \frac{\text{W}}{\text{m}^2 \cdot \text{K}} \tag{10–59}$$

where the diameter is in meters (m) and the temperature difference is K or °C.

Film Condensation of R-134a in Horizontal Tubes

For film condensation of R-134a in horizontal tubes of diameter D (m), the following approximation may be used:

$$h_{\text{ave}} = 983\left(\frac{1}{D\Delta T}\right)^{1/4} \frac{\text{W}}{\text{m}^2 \cdot \text{K}} \tag{10–60}$$

EXAMPLE 10–12

Determine the convective heat-transfer coefficient for methanol boiling at a saturation temperature of 148°F on a horizontal 1-in.-diameter tube having a surface temperature of 158°F. Compare the value for this coefficient using equation (10–58) to that using equation (10–6). Use a value of 1102 kJ/kg for the heat of vaporization of methanol.

Solution

The properties of methanol at 148°F (65°C) are, by interpolating from appendix tables B–3 and B–4,

$$\rho_f = 791 \text{ kg/m}^3 \qquad \rho_g = 1.31 \text{ kg/m}^3$$

$$c_{p,f} = 2.63 \frac{\text{kJ}}{\text{kg} \cdot \text{K}} \qquad \kappa_f = 0.196 \text{ W/m} \cdot \text{K}$$

$$\nu_g = 8.4 \times 10^{-6} \text{ m}^2/\text{s} \qquad \alpha_f = \frac{\kappa_f}{\rho_f c_{p,f}} = 9.4 \times 10^{-8} \frac{\text{m}^2}{\text{s}}$$

From appendix table B–9:

$$\sigma_{\text{ST}} = 2.01 \frac{\text{mN}}{\text{m}}$$

We know that $\Delta T_{SH} = 5.56$ K, $D = 0.0254$ m, and $hn_{fg} = 1{,}102$ kJ/kg. From equation (10–6),

$$h_{ave} = h_{npb} = 0.0001629\frac{\left(0.196\dfrac{\text{W}}{\text{m}\cdot\text{K}}\right)(1.31\ \text{kg/m}^3)}{787\ \text{kg/m}^3}\left(\frac{2\left(20.1\times10^{-3}\dfrac{\text{N}}{\text{m}}\right)}{9.81\dfrac{\text{m}}{\text{s}^2}\left(787+1.31\dfrac{\text{kg}}{\text{m}^3}\right)}\right)^{1/4}$$

$$\left(1102\frac{\text{kJ}}{\text{kg}}\right)^{0.744}\left(\frac{5.56\ \text{K}}{(65+273\ \text{K})\left(9.4\times10^{-8}\dfrac{\text{m}^2}{\text{s}}\right)}\right)^2\left(\frac{787\ \text{kg/m}^3}{787-1.31\ \text{kg/m}^3}\right)^{13}$$

Answer

$$= 1454.3\frac{\text{W}}{\text{m}^2\cdot\text{K}} = 256\frac{\text{Btu}}{\text{hr}\cdot\text{ft}^2\cdot{}^\circ\text{F}}$$

Using equation (10–58),

$$\log h_{ave} = -2.23 + 2.5\log 10 + 0.015(148^\circ\text{F}) = 2.49$$

or

Answer

$$h_{ave} = 309\frac{\text{Btu}}{\text{hr}\cdot\text{ft}^2\cdot{}^\circ\text{F}}$$

These two results are in reasonable agreement.

EXAMPLE 10–13 Compare the convective heat-transfer coefficient values for the condensing of ammonia and for R-134a in horizontal 2-cm-diameter tubes when the vapor temperatures are −10°C and the tube wall surface temperatures are −20°C, using equation (10–50) and the simplified equations (10–59) and (10–60). Use a value of 1296.4 kJ/kg for the heat of vaporization of ammonia at −10°C and $\rho_f = 605$ kg/m³, and $\rho_g = 2.392$ kg/m³. Also, for R-134a use 205.56 kJ/kg for the heat of vaporization, $\rho_f = 1324.5$ kg/m³ and $\rho_g = 10.08$ kg/m³.

Solution For the ammonia at −10°C, interpolating from appendix table B–3,

$$c_{p,f} = 4590\frac{\text{J}}{\text{kg}\cdot\text{K}} \qquad \kappa_f = 0.546\ \text{W/m}\cdot\text{K} \qquad \mu_f = 0.00026\ \text{kg/m}\cdot\text{s}$$

and the average convective heat transfer coefficient, using equation (10–50), is

$$h_{ave} = 0.555\left[\frac{(605)(9.81)(605-2.392)(0.546)^3(1{,}296{,}400+\tfrac{3}{8}(4590)(10))}{(0.00026)(10)(0.02)}\right]^{1/4}$$

Answer

$$= 6112\frac{\text{W}}{\text{m}^2\cdot\text{K}}$$

Using equation (10–59),

Answer

$$h_{ave} = 4000\left(\frac{1}{(0.02)(10)}\right)^{1/4} = 5981\frac{\text{W}}{\text{m}^2\cdot\text{K}}$$

Thus, for ammonia condensation, equation (10–59) provides an estimate that is in substantial agreement with the more accurate prediction by using equation (10–50). For

R-134a, using equation (10–50), and interpolating from appendix table B–3, $c_{p,f} = 1310$ kJ/kg · K, $\kappa_f = 0.098$ W/m · K, and $\mu_f = 0.00034$ kg/m · s. Then,

$$h_{ave} = 0.555\left(\frac{(1324.5)(9.81)(1324.5 - 10.08)(0.098)^3[205{,}560 + \tfrac{3}{8}(1310)](10)}{(0.00034)(10)(0.02)}\right)^{1/4}$$

Answer

$$= 1474\frac{\text{W}}{\text{m}^2 \cdot \text{K}}$$

and using equation (10–60),

Answer

$$h_{ave} = 983\left(\frac{1}{(0.02)(10)}\right)^{1/4} = 1470\frac{\text{W}}{\text{m}^2 \cdot \text{K}}$$

so that there is again good agreement between the two equations.

10–5 EMPIRICAL METHODS AND ANALYSIS OF MELTING AND FREEZING

Freezing and melting are the phase changes associated with the transitions between the solid and the liquid phases or states of substances, and there tend to be many and varied complications that get in the way of any general approaches to analyzing their heat transfers. For instance, two types of solids need to be recognized: crystalline solids and amorphous solids. Crystalline solids are solids that have their atoms or molecules arranged in a regular pattern, or a crystal structure. Many pure substances and elements form into crystalline solids. Water is a good example of this. Of course, being regular does not mean being uniform or predictable, as figure 10–20 alludes to. But although no two snow flakes are alike (?), the formation of sheets of ice can be interpreted as a uniform crystalline structure. Then, too, many metals such as iron, copper, and aluminum form into crystalline solids.

Figure 10–20 The problem of snow flakes. (By permission of Johnny Hart and Creators Syndicate, Inc.)

Amorphous solids are common in mixtures of substances and in many organic compounds. For instance, oils tend to solidify into a solid that has no identifiable crystal structure. Amorphous solids tend to be supercooled liquids in that the molecules or atoms are well mixed and have an irregular pattern. The phase change between the liquid and the solid states occurs in a continuous fashion with no distinct isothermal or constant-temperature phase change. There is also no measureable heat of solidification or heat of fusion, as is associated with the crystalline solid-phase transitions. There is usually an observable "glass transition," where the substance progresses from a liquid with relatively low viscosity to an amorphous solid with a viscosity that is orders of magnitude greater than the liquid viscosity. The glass transition is used to describe this demarcation between the liquid and its amorphous solid states because glass, silicon dioxide and mixtures of other elements, is a prime example of an amorphous solid. Glass has been observed to flow over long time periods, such as years or centuries, and so it has a measureable viscosity, as discussed in Chapter 4. Faced with these complications, and others as well, it is best to consider only some simple restrictive examples in this text.

Melting of a Spherical Lumped System

Consider the situation of a crystalline solid at its saturated solid/liquid temperature and immersed in a liquid bath of its own substance but at a temperature above freezing. Using an analysis similar to that from the lumped heat capacity systems of Chapter 3 and referring to figure 10–21, we may write for an energy balance of the system,

$$\dot{Q} = h_{\text{ave}}A_s(T_\infty - T_s) = \dot{m}_s\text{hn}_{fs} = -\dot{m}_{\text{sys}}\text{hn}_{fs} \tag{10–61}$$

where $\dot{m}_s$ is the rate of melting of the solid, $\dot{m}_{\text{sys}}$ is the rate of change of the system's mass, and hn_{fs} is the heat of fusion or heat of melting. The mass of the system may also be written for a sphere as

$$m_s = \rho_s {}^4\!/_3\pi r^3 \tag{10–62}$$

and the surface area for a sphere is

$$A_s = 4\pi r^2 \tag{10–63}$$

Substituting these into equation (10–61) gives

Figure 10–21 Spherical saturated solid submersed in liquid.

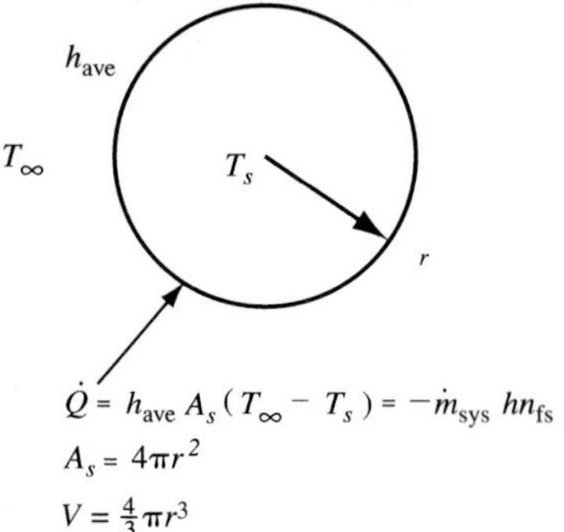

and if we recognize that the radius will be the variable with respect to time, we get

$$4\pi r^2 h_{\text{ave}}(T_\infty - T_s) = -\rho_s hn_{f-s}\frac{4}{3}\pi(3r^2)\frac{dr}{dt} \tag{10–64}$$

Reducing this gives

$$dt = -\frac{\rho_s\text{hn}_{fs}}{h_{\text{ave}}(T_\infty - T_s)}dr$$

so that the time required to melt the sphere from an initial size, r_i, to a final size, r_f, is

$$t = \frac{\rho_s \mathrm{hn}_{fs}}{h_{\mathrm{ave}}(T_\infty - T_s)}(r_i - r_f) \tag{10–65}$$

Notice that the mass of the solid is proportional to the cube of the radius so that the melting time is proportional to the third power of the mass of the system. The time required to completely melt the sphere is predicted by setting the final radius equal to zero in equation (10–65) and obtaining the maximum time as

$$t_{\max} = \frac{\rho_s \mathrm{hn}_{fs} r_i}{h_{\mathrm{ave}}(T_\infty - T_s)} = \frac{\rho_s^{2/3} \mathrm{hn}_{fs}}{h_{\mathrm{ave}}(T_\infty - T_s)}\sqrt[3]{\frac{3m_{\mathrm{sys}}}{4\pi}} \tag{10–66}$$

EXAMPLE 10–14 Estimate the time required to melt 3-cm-diameter spherical taconite pellets in molten iron at 1600°C if the convective heat-transfer coefficient is expected to be 5000 W/m² · K. Assume that the taconite is pure iron having a density of 7270 kg/m³ and heat of fusion of 250 kJ/kg.

Solution Using equation (10–66) we have

$$t = \frac{(7270\ \mathrm{kg/m^3})(250\ \mathrm{kJ/kg})(0.03\ \mathrm{m})}{(5000\ \mathrm{W/m^2 \cdot K})(1600 - 1538°\mathrm{C})} = 175.9\ \mathrm{s} = 2.93\ \mathrm{min}$$

Melting of an Infinite Solid Slab Due to Convection Heating

Consider a solid slab surrounded by two fluids, one on each side. A naturally occurring situation is ice on a body of water, where the ice has a density that is about 10% less than the liquid water and thus will float on the top of the free water surface. Air will be above the upper surface of the ice, as shown in figure 10–22a. Consider, as an approximation, that the ice has constant properties, except that its temperature varies from its saturation value at the lower surface to some higher temperature at its upper surface and that its heat of fusion is much greater than any heat capacity changes in the ice. If we also assume that the liquid water is at the saturation temperature of nearly 0°C (32° F), then the heat balance for the ice slab may be written

$$\dot{q}_{A,\mathrm{conv}} = h_{\mathrm{ave}}(T_\infty - T_U) = \dot{q}_{A,\mathrm{cond}} = \kappa_s \frac{T_U - T_L}{\delta} = -\rho_s \mathrm{hn}_{fs}\frac{d\delta}{dt} \tag{10–67}$$

With some rearrangements this equation can be written

$$\frac{T_\infty - T_L}{\dfrac{1}{h_{\mathrm{ave}}} + \dfrac{\delta}{k_s}} = -\rho_s \mathrm{hn}_{fs}\frac{d\delta}{dt} \tag{10–68}$$

and it can be made dimensionless by defining the terms

$$\bar{\delta} = \frac{h_{\mathrm{ave}}\delta}{\kappa_s} \quad \text{and} \quad \bar{t} = t h_{\mathrm{ave}}^2 \frac{T_\infty - T_L}{\rho_s \mathrm{hn}_{fs}\kappa_s}$$

and substituting these into equation (10–68) to give the form

$$d\bar{t} = -(1 + \bar{\delta})d\bar{\delta} \tag{10–69}$$

Figure 10–22 Melting of an infinite slab due to convection heating.

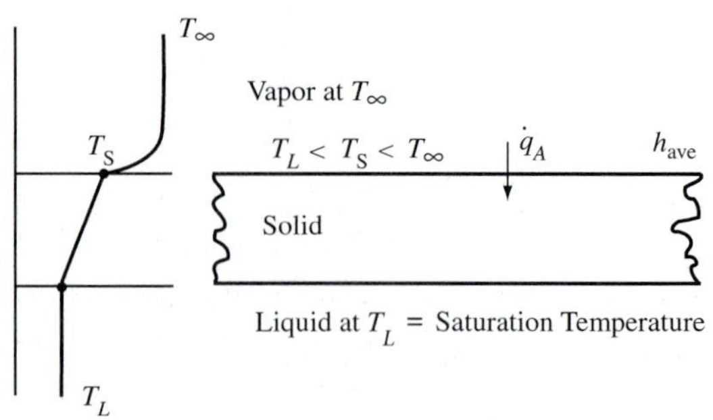

(a) Convection on upper surface, liquid at same temperature as solid on lower surface

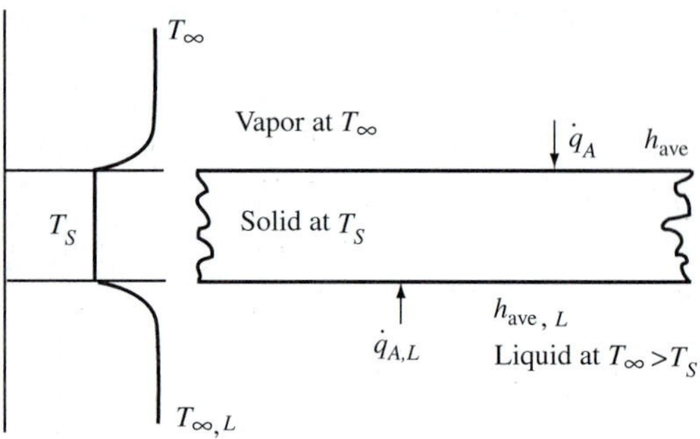

(b) Convection on upper and lower surfaces

If one sets $\bar{\delta}_i$ at $\bar{t} = 0$ and integrates to some time t_{max} such that $\bar{\delta} = \bar{\delta}_f = 0$ when the slab is completely melted, then equation (10–69) becomes

$$\bar{t}_{max} = {}^1\!/_2\,\bar{\delta}_i^2 + \bar{\delta}_i \tag{10–70}$$

If the liquid below the slab is at a higher temperature than the solid slab and a convective heat transfer coefficient $h_{ave,L}$ is identified at the lower surface of the slab, then an energy balance of the slab, assuming here that the slab has a constant temperature, gives

$$h_{ave}(T_\infty - T_s) + h_{ave,L}(T_{\infty,L} - T_s) = -\rho_s \text{hn}_{fs}\frac{d\delta}{dt} \tag{10–71}$$

or, for a particular time period, t, integrating gives

$$\delta_i - \delta_f = \left(\frac{h_{ave}(T_\infty - T_s) + h_{ave,L}(T_{\infty,L} - T_s)}{\rho_s \text{hn}_{fs}}\right)t \tag{10–72}$$

EXAMPLE 10–15 On a warm winter day the air temperature around a frozen pond stays at 50°F from 1:00 P.M. to 3:00 P.M. The ice is 4 in. thick, and the water temperature under the ice has been found to be 33°F. The convective heat-transfer coefficients are estimated as 10 Btu/hr · ft^2 · °F on the upper surface between the air and the ice and 25 Btu/hr · ft^2 · °F between the lower surface of the ice and the water. Estimate the ice thickness at 3:00 P.M. The density of ice may be taken as 57.2 lbm/ft^3 and the heat of fusion as 144 Btu/lbm.

Solution Using equation (10–72) gives

$$\delta_i - \delta_f = 0.0498 \text{ ft} = 0.598 \text{ in.}$$

and the thickness after 2 hr, at 3:00 P.M. is

Answer

$$\delta_f = \delta_i - 0.74 \text{ in.} = 4 - 0.598 = 3.402 \text{ in.}$$

Freezing of Spherical Lumped Systems

The lumped heat capacity system, as an approximation for a freezing spherical mass, growing because it is surrounded by saturated liquid (at the freezing/melting saturation temperature), gives the reverse result of equation (10–65) for a melting sphere. That is, for freezing, we have

$$r_f - r_i = \left(\frac{h_{\text{ave}}(T_s - T_\infty)}{\rho_s \text{hn}_{fs}}\right)t \tag{10–73}$$

where t is the time of freezing.

Freezing of an Infinite Slab due to Convection Heating

The large slab of a crystalline solid subjected to convection cooling on one surface, where the other surface is at some constant saturation temperature and is in contact with a saturated liquid of the same substance as the solid, is indicated in figure 10–23a. For this situation the liquid acts as a heat source and the convective cooling is a heat loss from the slab. The energy balance for the slab gives a heat flow through the solid slab from the liquid at its saturation temperature T_L to the surroundings at T_∞ and

$$\frac{T_L - T_\infty}{\dfrac{\delta}{\kappa_s} + \dfrac{1}{h_{\text{ave}}}} = \rho_s \text{hn}_{fs}\frac{d\delta}{dt} \tag{10–74}$$

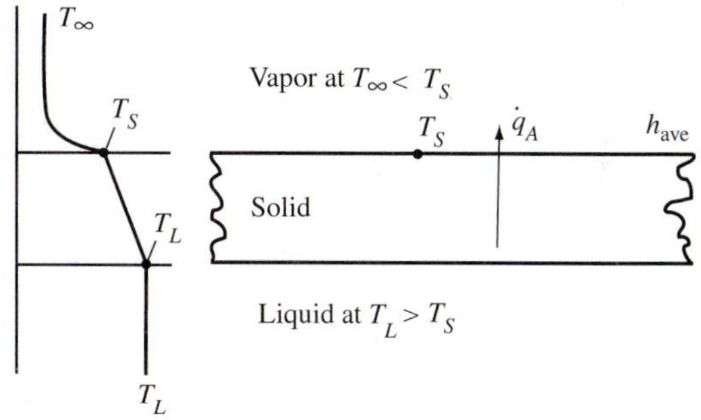

(a) Convection cooling on top surface

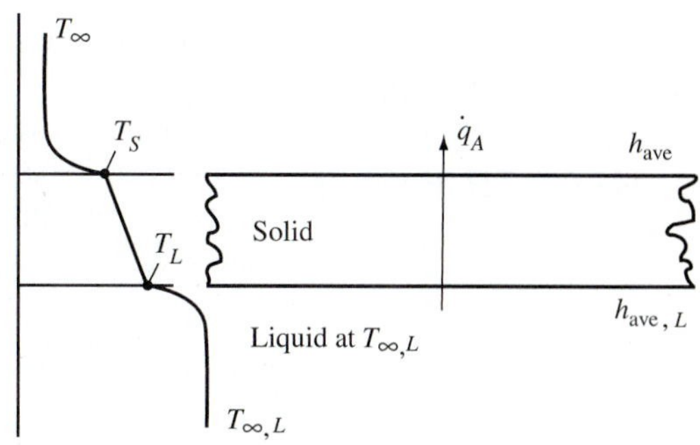

(b) Convection on top and bottom surfaces

Figure 10–23 Freezing of an infinite solid slab due to convection heat transfer.

This is the same equation as (10–68) for melting of an infinite slab, except that the temperature difference is reversed and the slab thickness is increasing with a positive rate of thickness. If this equation is normalized, or made dimensionless, with the parameters

$$\bar{\delta} = \frac{h_{ave}\delta}{\kappa_s} \quad \text{and} \quad \bar{t} = t\text{hn}^2_{f-s}\frac{T_L - T_\infty}{\rho_s \text{hn}_{fs}\kappa_s}$$

then a similar result to equation (10–70) results,

$$\bar{t} = \tfrac{1}{2}(\bar{\delta}_f^2 - \bar{\delta}_i^2) + \bar{\delta}_f - \bar{\delta}_i \tag{10–75}$$

Of course, if one begins with no solid and $\bar{\delta}_i = 0$ then this equation simplifies.

An elaboration of the convective cooling can be made for the case where the liquid is at a higher temperature than the solid and where a convective heat-transfer coefficient, $h_{ave,L}$, is identified for the lower surface as shown in figure 10–23b. London and Seban [19] have developed the analytic expressions for this case where the energy balance is

$$\frac{T_{\infty,L} - T_\infty}{\dfrac{1}{h_{ave}} + \dfrac{\delta}{\kappa_s} + \dfrac{1}{h_{ave,L}}} = \rho_s \text{hn}_{fs}\frac{d\delta}{dt} \tag{10–76}$$

This equation may be made dimensionless by defining the parameters

$$\bar{R} = \frac{h_{ave,L}}{h_{ave}} \quad \text{and} \quad \bar{T} = \frac{T_{\infty,L} - T_L}{T_L - T_\infty}$$

along with

$$\bar{\delta} = \frac{h_{ave}\delta}{\kappa_s} \quad \text{and} \quad \bar{t} = th^2_{ave}\frac{T_L - T_\infty}{\rho_s \text{hn}_{fs}\kappa_s}$$

from the previous development leading to equation (10–75). The resulting equation is

$$\frac{1 + \bar{\delta}}{1 + \overline{RT}(1 + \bar{\delta})}d\bar{\delta} = d\bar{t} \tag{10–77}$$

For the initial condition that $\bar{\delta} = 0$ at $\bar{t} = 0$, equation (10–77) can be integrated to give

$$\bar{t} = \frac{1}{\bar{R}^2\bar{T}^2}\ln\left(1 + \frac{\overline{RT}\bar{\delta}}{1 + \overline{RT}}\right) + \frac{\bar{\delta}}{\overline{RT}} \tag{10–78}$$

and the dimensionless time $\bar{t}$ can then be predicted for the growth of an infinite solid slab to be $\bar{\delta}$ thick.

10–6 APPLICATIONS OF PHASE CHANGE HEAT TRANSFER

The applications of phase change heat transfer are many, and in this textbook a number of them have been demonstrated, including steam generating units where water is boiled, evaporators in refrigeration systems where a refrigerant is vaporized or boiled, and condensers of all sorts that are used to cool vapors to liquids. There are also many applications where melting and freezing heat transfer are involved. The following five examples are given to show how diverse and often highly specialized these applications are.

EXAMPLE 10–16

In the casting of metals, cooling of the molten metal is necessary so that the metal reaches its solid state. In a specialized application where molten iron, cast iron, is poured into a long horizontally rotating cylinder mold that is capped at both ends to prevent the liquid from flowing out as shown in figure 10–24, a concentric hollow cylinder is produced by the centrifugal action of rotation. The common term for this process is *centrifugal casting*

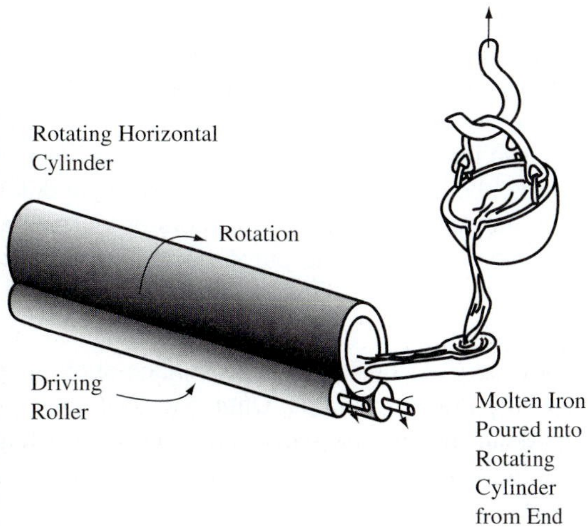

Figure 10–24 A centrifugal casting process.

and the resulting hollow cylinders or *sleeves* are used, for instance, as an insert in an aluminum block for an internal combustion engine as the wearing surfaces for the cylinders and on which the piston rings and pistons are guided. Previous to this application, engine blocks were fabricated of cast iron because of its good wear properties. With the cast-iron sleeve inserted into the aluminum engine block, a significant weight reduction could be achieved because aluminum is about one third as heavy as cast iron. Further, the centrifugal casting of cast iron in this manner produces a sleeve whose inside surface cools slower than the outside surface and, fortuitously, a slower cooling rates tends to produce a better-wearing surface. The heat transfers of this process are indicated in figure 10–24 where a relatively cold mold cools the molten metal in contact with it at a rapid rate by conduction heat transfer. The liquid metal toward the inner regions is cooled by radiation and convection. Since cast iron melts at about 1370°C, radiation will be a significant part of the cooling of the molten metal. A typical cooling curve of the inside surface of the cast iron as it proceeds through its freezing is shown in figure 10–25. That is, the cast iron at the inner surface cools below its freezing temperature (*A*), becoming a subcooled liquid,

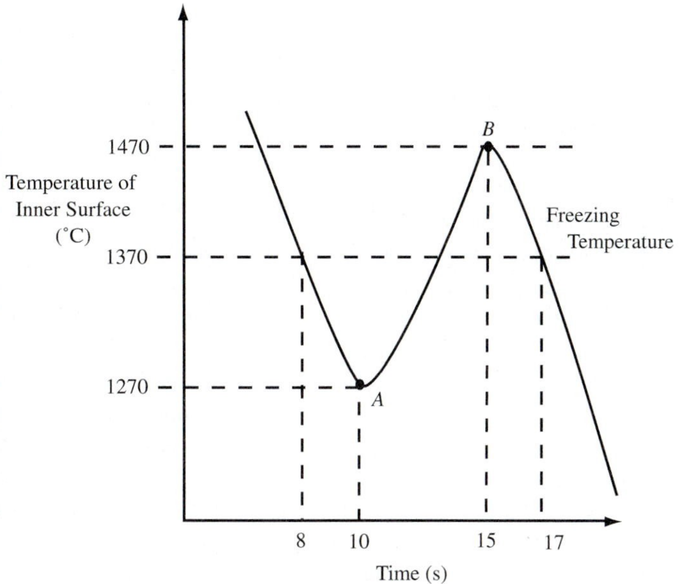

Figure 10–25 A cooling curve for a sleeve.

and because of the heat of fusion liberated in the actual freezing of the cast iron against the mold, the molten cast iron is then heated again to a temperature above the melting temperature (*B*). The radiation and convection heat losses finally drop the temperature below that for melting and solidifying the complete sleeve. Notice that the time period during which the sleeve inside surface temperature drops below freezing and returns to a higher temperature can be significant. Thus, in this example one needs to recognize that surface temperatures are not always good indicators of the phase of material. A complete analysis of this problem is beyond the purposes of this textbook.

EXAMPLE 10–17

Defrosting a window or de-icing a cold surface such as an aircraft wing is often of concern, particularly in cold climates. Often the defrosting is accomplished by warming the surface on which the frost or ice has formed. This situation is shown schematically in figure 10–26, where the under surface is heating and melting a layer of water and the ice has a convection heat transfer on its underside and a convection cooling on its upper side. This problem becomes analogous to the problem of freezing with convection heating as considered in figure 10–23b.

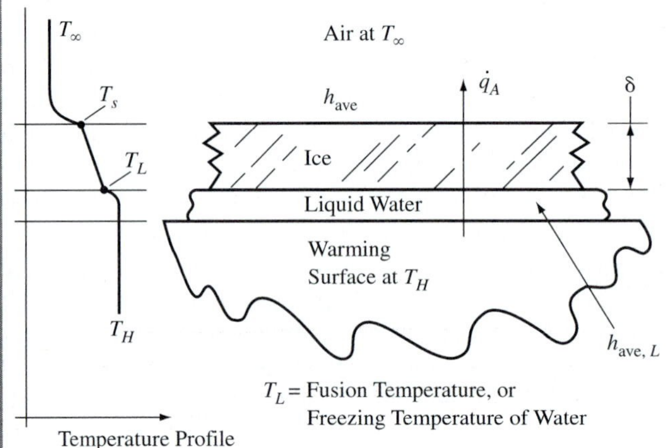

Figure 10–26 The de-icing problem.

There is one subtle difference between these two situations in that the ice is melting or reducing and in figure 10–23b the ice is growing or freezing. The analysis of the freezing problem resulted in equation (10–78) and can be used here with some minor modifications. Thus the dimensionless time $\bar{t}$ predicted by equation (10–78) needs to be interpreted as the time to completely melt or de-ice a surface having ice thickness δ at the beginning of de-icing. Determine the time to remove 1 cm of ice from a flat horizontal surface if the surface is heated to 60°C, the melted ice between the surface and the ice has a convective heat-transfer coefficient of 10 W/m^2 · K, the surrounding air is at −5°C, and the convective heat-transfer coefficient at the ice/air surface is 10 W/m^2 · K.

Solution

The properties of ice are approximated from appendix table B–6 $\rho_s = 916$ kg/m^3, $hn_{fs} = 335$ kJ/kg, and $\kappa_s = 2.2$ W/m · K, and the dimensionless time is

$$\bar{t} = th_{ave}^2 \frac{T_L - T_\infty}{\rho_s hn_{fs} \kappa_s} = t\left(10\frac{\text{W}}{\text{m}^2 \cdot \text{K}}\right)^2 \frac{0 - (-5°\text{C})}{\left(916\frac{\text{kg}}{\text{m}^3}\right)\left(335{,}000\frac{\text{J}}{\text{kg}}\right)\left(2.2\frac{\text{W}}{\text{m} \cdot \text{K}}\right)}$$

$$= t(7.4 \times 10^{-7}) \qquad (t \text{ in seconds})$$

Also,

$$\bar{\delta} = \frac{h_{\text{ave}}\delta}{\kappa_s} = \frac{(10\ \text{W/m}^2\cdot\text{K})(0.01\ \text{m})}{2.2\ \text{W/m}\cdot\text{K}} = 0.0454$$

and

$$\bar{R} = \frac{h_{\text{ave},L}}{h_{\text{ave}}} = 1 \qquad \bar{T} = \frac{T_{\infty,L} - T_L}{T_L - T_\infty} = \frac{60 - 0}{0 - (-5)} = 12$$

Substituting these values into equation (10–78)

$$\bar{t} = \frac{1}{\bar{R}^2\bar{T}^2}\ln\left(1 + \frac{\overline{RT}\,\bar{\delta}}{1 + \overline{RT}}\right) + \frac{\bar{\delta}}{\overline{RT}} = 0.004068$$

and

$$t = 5498\ \text{s} = 1.53\ \text{hr}$$

Thermal storage has become a concept of increasing interest, and materials that are experiencing phase changes are attractive for many of the applications of thermal storage. Thermal storage systems involving phase change materials (PCMs) have been incorporated into solar energy systems, where large amounts of energy, as thermal energy, are available during times when the sun is shining but not when it is cloudy or night. Also, by using concentrating solar collectors, high-temperature thermal energy can be obtained, and by using PCMs the energy can be stored at a relatively high temperature. Further, in the applications of cooling and air conditioning of buildings a high load is placed on air conditioning units during peak times, which is usually during the daylight hours when electricity costs are high. By using ice storage or other PCMs, the "cold" can be stored in the ice during the evening or when electricity rates are low by running the air conditioners or *chillers* to freeze the water. Then the ice can then be used to cool the building during those times when it is heavily used and electric rates are high, thus providing an incentive for cost reductions. As an example the storage of thermal energy or of ice can be accomplished with a version of the shell and tube heat exchanger. In figure 10–27 is shown a model of a portion of such a unit where the tubes convey a cold (or hot) fluid and the heat is transferred from (or to) the PCM surrounding the tubes. In the figure one can see that the PCM will become solid, frozen, around the tubes and at the entrance to the heat exchanger.

Progressing through the heat exchanger to the exit at L, there is less solid and more liquid PCM. As you may expect from the tube bank analyses and the heat exchanger discussions, the arrangement of the tubes affects the performance of the heat exchange between the fluid in the tubes and the surrounding PCM. In figure 10–27 is shown two possible arrangements for these tubes, the square and the staggered patterns. The distances between the tubes also affects the performance. Shamsundar and Srinivasan [20, 21] have developed a solution for the case of this model of a phase change thermal storage unit using the effectiveness–NTU concepts of Chapter 9. The effectiveness is, for the case where the fluid temperature is below the saturation or fusion temperature of the PCM,

$$\varepsilon_{\text{HX}} = \frac{\dot{Q}_{\text{HX}}}{\dot{m}_f c_{p,f}} = \frac{\dot{m}_f c_{p,f}(T_{f,L} - T_{f,0})}{\dot{m}_f c_{p,f}(T_{\text{sat}} - T_{f,0})} = 1 - \frac{T_{\text{sat}} - T_{f,L}}{T_{\text{sat}} - T_{f,0}} \tag{10–79}$$

For a fluid at a higher temperature than the PCM the temperature differences in equation (10–79) should be reversed, although the result is the same. The NTU is defined as

$$\text{NTU} = \frac{\pi D_i h_{\text{ave},i} L}{\dot{m}_f c_{p,f}} \tag{10–80}$$

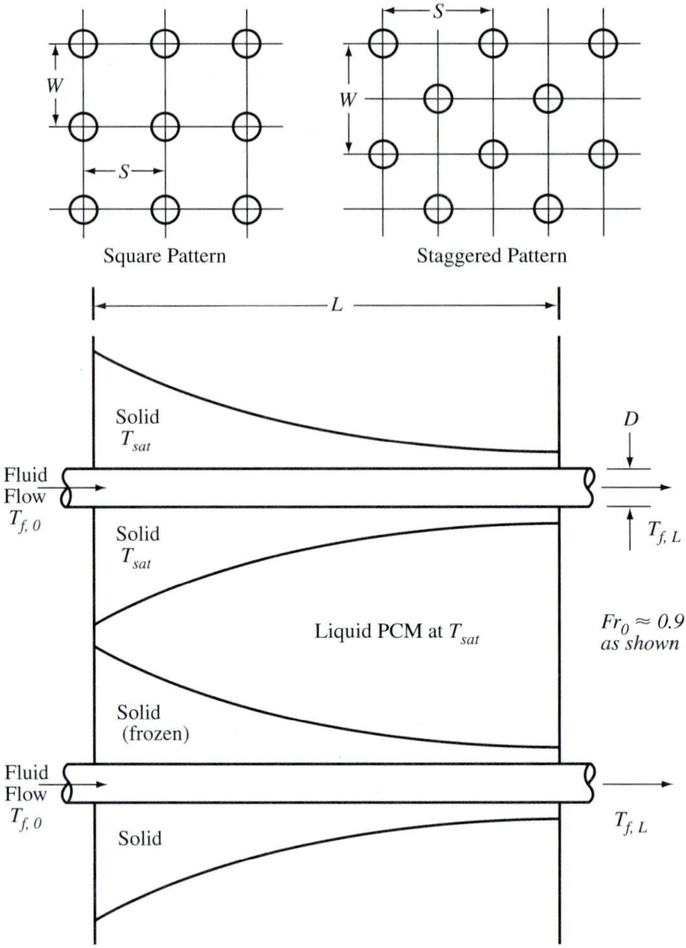

Figure 10–27 A schematic of a phase change thermal storage unit, freezing a PCM.

where D_i is the inside diameter of the tubes and $h_{ave,i}$ is the convective heat transfer coefficient for the fluid. The Biot Number is appropriate for the PCM and is defined as

$$\mathrm{Bi}_{\mathrm{HX}} = \frac{h_i D_0}{\kappa_s} \tag{10–81}$$

where D_0 is the outside diameter of the tubes.

The effectiveness–NTU relationship for this model is

$$\mathrm{NTU} = \ln\left(\frac{1}{1 - \varepsilon_{\mathrm{HX}}}\right) + \mathrm{Bi}_{\mathrm{HX}}\{\bar{f}(\mathrm{Fr}_0) - \bar{f}[\mathrm{Fr}_0(1 - \varepsilon_{\mathrm{HX}})]\} \tag{10–82}$$

Here the parameter Fr_0 is the *frozen fraction* or fraction of the PCM that is solid at the fluid entrance and the functions denoted as $\bar{f}(-)$ are the first frozen fraction functions. They are shown graphically in figure 10–28 for square and staggered tube arrangements. Equation (10–82) is appropriate for any Biot Number, using figure 10–28 for the frozen fraction functions. Notice in figure 10–27 that the frozen fraction at the entrance is indicated as approximately 0.9; however, any fraction from zero (0) to 1 is possible. In figures 10–29, 10–30, and 10–31 are displayed the effectiveness–NTU relationships for Biot Numbers of 1, 10, and 100, respectively.

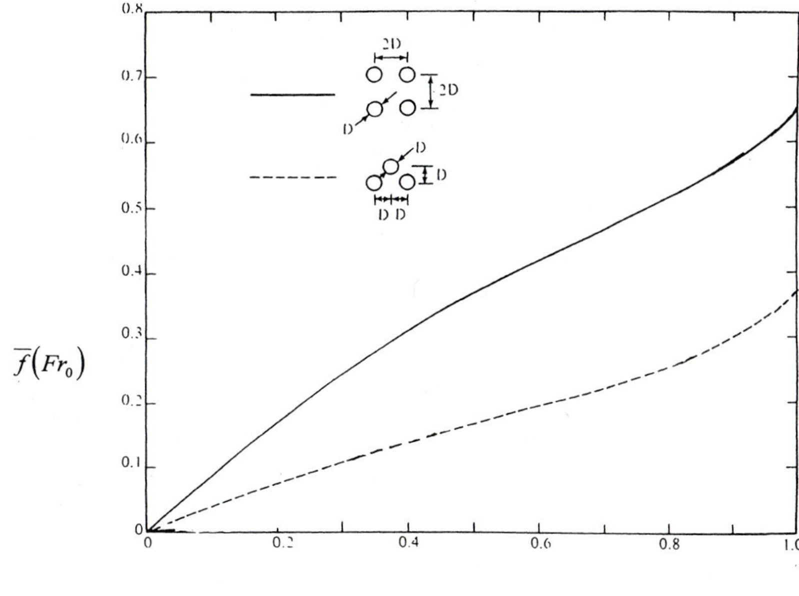

Figure 10–28 First frozen fraction function. (From [21] with permission of the American Society of Mechanical Engineers)

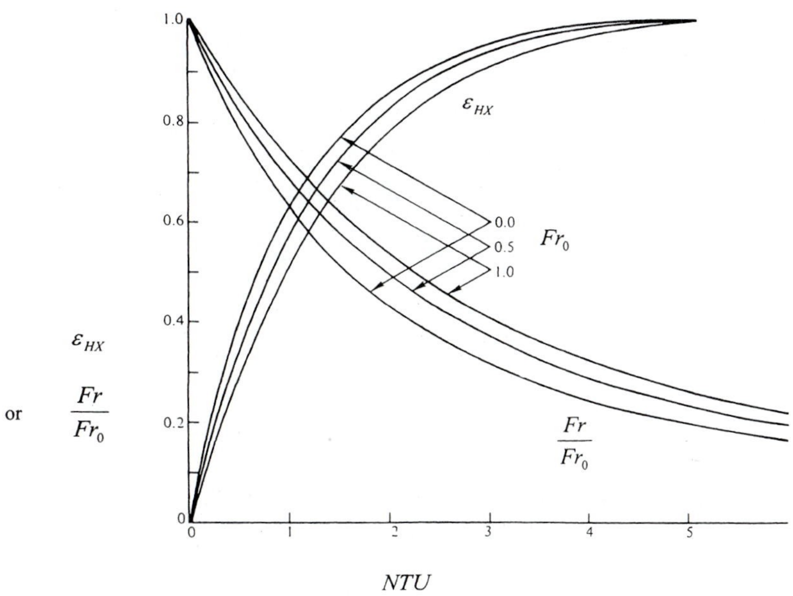

Figure 10–29 Effectiveness–NTU for $Bi_{HX} = 1$ of a PCM Thermal Storage System. (From [21] with permission of the American Society of Mechanical Engineers)

The amount of thermal energy or solid that can be stored in the system requires information about the total frozen fraction of the PCM, Fr, and this is related to the entrance frozen fraction by the equation

$$\text{Fr} = \frac{1}{\text{NTU}}(\text{Fr}_0\varepsilon_{\text{HX}} + \text{Bi}_{\text{HX}}\{\bar{\bar{f}}(\text{Fr}_0) - \bar{\bar{f}}[\text{Fr}_0(1 - \varepsilon_{\text{HX}})]\}) \tag{10–83}$$

The second frozen fraction functions $\bar{\bar{f}}(-)$ are displayed in figure 10–32 for square and staggered tube arrangements. In figures 10–29, 10–30, and 10–31 the ratios Fr/Fr_0 are

Figure 10–30 Effectiveness–NTU relationship for $Bi_{HX} = 10$ of a PCM thermal storage system. (From [21] with permission of the American Society of Mechanical Engineers)

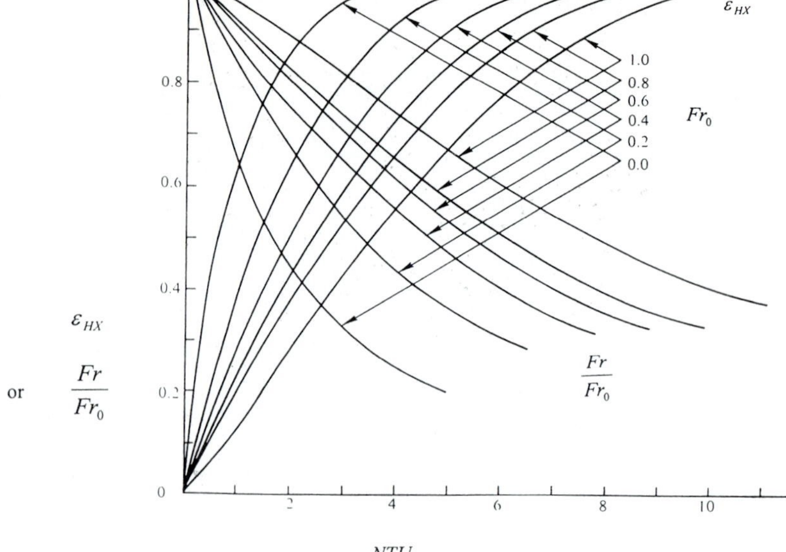

Figure 10–31 Effectiveness–NTU relationship for $Bi_{HX} = 100$ of a PCM thermal storage system. (From [21] with permission of the American Society of Mechanical Engineers)

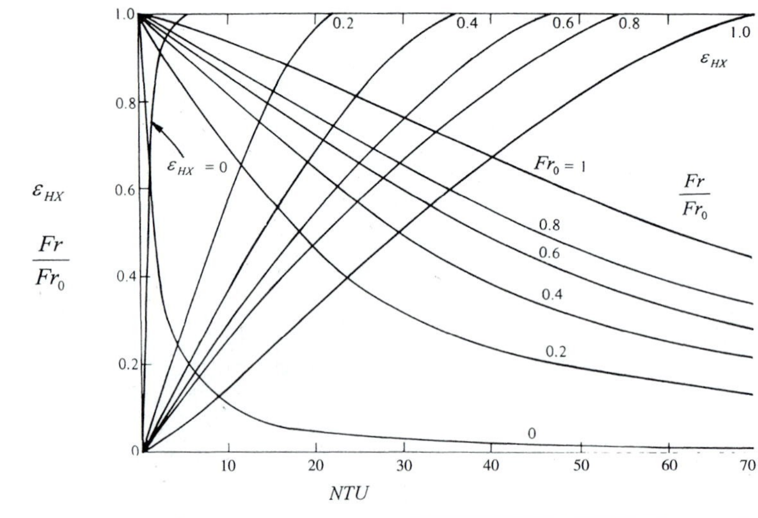

Figure 10–32 Second frozen fraction function. (From [21] with permission of the American Society of Mechanical Engineers)

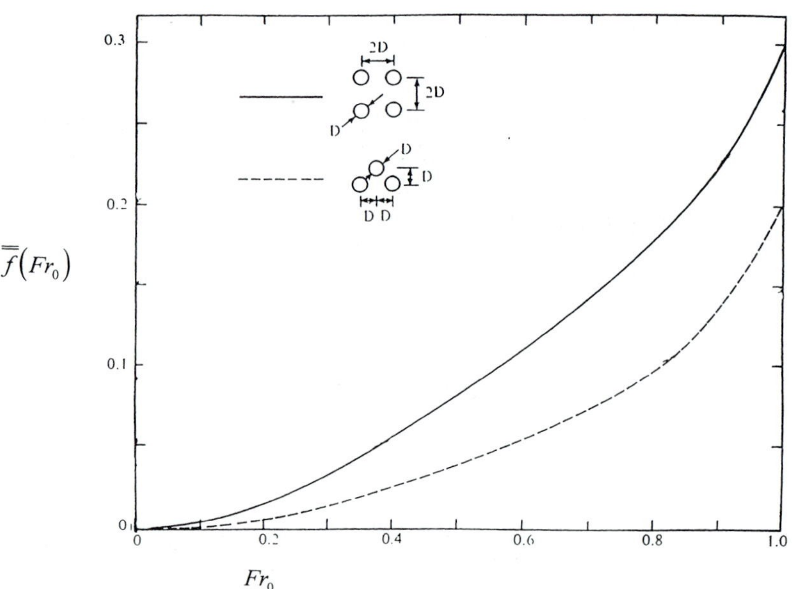

displayed for the Biot Numbers 1, 10, and 100. For other Biot Numbers equation (10–83) is necessary. Since we are here considering transient conditions in the PCM, a dimensionless time parameter is necessary, defined as

$$\bar{t}_{\mathrm{HX}} = \frac{\kappa_s(T_{\mathrm{sat}} - T_{f,0})}{\rho_s \mathrm{hn}_{fs} D_o^2} t \qquad \textbf{(10–84)}$$

and this is also equal to

$$\bar{t}_{\mathrm{HX}} = B_{\mathrm{HX}}\left(\frac{\mathrm{Fr}_0}{\mathrm{Bi}_{\mathrm{HX}}} + \bar{\bar{f}}(\mathrm{Fr}_0)\right) \qquad \textbf{(10–85)}$$

The tube geometry factor is

$$B_{\mathrm{HX}} = \frac{1}{\pi}\left(\frac{SW}{D_o^2} - \frac{\pi}{4}\right) \qquad \textbf{(10–86)}$$

where the terms S and W are indicated in figure 10–27. Combining equations (10–84) and (10–85) gives

$$t = \frac{\rho_s \mathrm{hn}_{fs} D_o^2 B_{\mathrm{HX}}}{\kappa_s(T_{\mathrm{sat}} - T_{f,0})}\left(\frac{\mathrm{Fr}_0}{\mathrm{Bi}_{\mathrm{HX}}} + \bar{\bar{f}}(\mathrm{Fr}_0)\right) \qquad \textbf{(10–87)}$$

EXAMPLE 10–18 Consider an ice storage system using R-134a as the cooling fluid at −5°C entering a bank of 20 tubes. The data obtained for this system are

$$h_{\mathrm{ave},i} = 1000\ \mathrm{W/m^2 \cdot K} \quad D_i = 2\ \mathrm{cm} \quad D_o = 2.2\ \mathrm{cm} \quad L = 8\ \mathrm{m}$$
$$\dot{m}_f c_{p,f} = 126\ \mathrm{W/K} \quad \mathrm{hn}_{fs} = 335\ \mathrm{kJ/kg} \quad \rho_s = 916\ \mathrm{kg/m^3} \quad T_{\mathrm{sat}} = 0°\mathrm{C}$$
$$\kappa_s = 2.2\ \mathrm{W/m \cdot K} \quad \text{Square tube, pitch} = 5\ \mathrm{cm}$$

From these data we have

$$\mathrm{Bi}_{\mathrm{HX}} = \frac{(1000\ \mathrm{W/m^2 \cdot K})(0.022\ \mathrm{m})}{2.2\ \mathrm{W/m \cdot K}} = 10$$

and

$$\mathrm{NTU} = \frac{\pi(0.02\ \mathrm{m})(1000\ \mathrm{W/m^2 \cdot K})(8\ \mathrm{m})}{(126\ \mathrm{W/K})} \approx 4$$

Determine the amount of total frozen fraction after 60 min of cooling, starting with all saturated liquid PCM and total heat transfer at this time.

Solution The dimensionless time is

$$\bar{t}_{\mathrm{HX}} = \frac{(2.2\ \mathrm{W/m \cdot K})[0 - (-5°\mathrm{C})](3600\ \mathrm{s})}{(916\ \mathrm{kg/m^3})(335{,}000\ \mathrm{J/kg})(0.02\ \mathrm{m})^2} = 0.323$$

From the geometry of the square tube arrangement,

$$B_{\mathrm{HX}} = \frac{1}{\pi}\left(\frac{(5\ \mathrm{cm})(5\ \mathrm{cm})}{(2.2\ \mathrm{cm})^2} - \frac{\pi}{4}\right) = 1.394$$

and from equation (10–85),

$$0.323 = 1.394\left(\frac{\mathrm{Fr}_0}{10} + \bar{\bar{f}}(\mathrm{Fr}_0)\right) = \bar{t}_{\mathrm{HX}}$$

We need to determine the entrance frozen fraction after 1 hr, so we need to use this equation and figure 10–32. By iteration, assuming a value for the entrance frozen fraction, we find the following:

$\mathbf{Fr_0}$	$\bar{\bar{f}}(\mathbf{Fr_0})$	$\bar{t}_{\mathbf{HX}}$
0.8	0.185	0.369
0.9	0.148	0.303

By interpolating between these two conditions, the result is $\mathrm{Fr_0} = 0.74$ for $\mathrm{Bi_{HX}} = 10$ and $\bar{t}_{\mathrm{HX}} = 0.323$. Then, for a frozen fraction of 74% (0.74), from figure 10–30 and NTU = 4,

$$\frac{\mathrm{Fr}}{\mathrm{Fr_0}} = 0.64$$

and

Answer
$$\mathrm{Fr} = \frac{\mathrm{Fr}}{\mathrm{Fr_0}}\mathrm{Fr_0} = (0.64)(0.74) = 0.4736 = 47.36\%$$

The effectiveness, from figure 10–30, is

$$0.68 = \frac{\dot{Q}_{\mathrm{HX}}}{\dot{m}_f c_{p,f}(T_{\mathrm{sat}} - T_{f,0})} = \frac{\dot{Q}_{\mathrm{HX}}}{(126\ \mathrm{W/K})(5\ \mathrm{K})}$$

and

$$\dot{Q}_{\mathrm{HX}} = (0.68)(126\ \mathrm{W/K})(5\ \mathrm{K}) = 428.4\ \mathrm{W/tube}$$

For 20 tubes

Answer
$$\dot{Q}_{\mathrm{HX}} = 8.568\ \mathrm{kW}$$

EXAMPLE 10–19 Consider a PCM thermal storage system having a staggered tube arrangement and the following data: $\mathrm{Bi_{HX}} = 20$, NTU = 7, and $\mathrm{Fr_0} = 0.7$. Determine the effectiveness of the heat exchanger.

Solution Using equation (10–82) with the first frozen fraction functions from figure 10–28, we iterate, assuming a value for the effectiveness and interpolating. The first frozen fraction function is about 0.23, for a frozen fraction of 0.7. The following NTU's result from equation (10–82) by assuming effectiveness values:

$\varepsilon_{\mathbf{HX}}$	**NTU**
0.9	6.30
0.95	7.395

By interpolation, we find that

Answer
$$\varepsilon_{\mathrm{HX}} \approx 0.93\ (93\%)$$

Sublimation heat transfer has some applications, one of which is the production of solid carbon dioxide or *dry ice* and the subsequent sublimation back to the vapor state.

EXAMPLE 10–20 Estimate the time for a 10-cm-cube of dry ice, solid carbon dioxide, to sublimate at 1 atm pressure and 20°C and with a convective heat-transfer coefficient of 100 W/m^2 · K at the surface of the cube. The following properties are known for solid carbon dioxide at its saturation freezing temperature of −78.515°C: $\kappa_s = 2$ W/m · K, $\rho_s = 910$ kg/m^3, and $hn_{sg} = 573.4$ kJ/kg.

Solution Using the analogy of the lumped heat capacity, the Biot Number for the cube of carbon dioxide is

$$\text{Bi} = \frac{h_{\text{ave}}L}{\kappa_s} = \frac{h_{\text{ave}}V}{\kappa_s A_s} = \frac{(100\ \text{W/m}^2\cdot\text{K})(0.01\ \text{m})^3}{(2\ \text{W/m}\cdot\text{K})(6)(0.1\ \text{m})^2} = 0.833$$

This is well above the value of 0.1 required for lumped heat capacity, but because the cube is proceeding through sublimation at constant temperature we may use this approximation. The energy balance for the cube may be written as

$$h_{\text{ave}}A_s(T_\infty - T_{\text{sat}}) = -\rho_s hn_{sg}\frac{dV}{dt}$$

The volume V is just the side cubed, l^3, and the surface area A_s is $6l^2$, so

$$h_{\text{ave}}(6l^2)(T_\infty - T_{\text{sat}}) = -\rho_s hn_{sg}\frac{d(l^3)}{dt}$$

and expanding and integrating gives, for the time to completely melt the cube,

$$t = \frac{\rho_s hn_{sg} l}{2h_{\text{ave}}(T_\infty - T_{\text{sat}})}$$

Using the properties given we obtain

$$t = \frac{(910\ \text{kg/m}^3)(573{,}400\ \text{J/kg})(0.1\ \text{m})}{2(100\ \text{W/m}^2\cdot\text{K})(20 + 78.515\ \text{K})} = 2648.3\ \text{s} = 0.74\ \text{hr}$$

Sublimation occurs in other physical situations, such as the sublimation of icebergs and snow; however, there have not been extensive studies on the nature of the heat transfer involved in these process.

10–7 SUMMARY

Bubbles forming in a saturated liquid are due to a pressure difference between the inside and the outside, balanced by the surface tension and given by

$$p_g - p_f = \frac{4\sigma_{\text{ST}}}{D} \tag{10–1}$$

Nucleate pool boiling heat transfer may be analyzed using the relationship for the convective heat-transfer coefficient,

$$h_{\text{npb}} = 1.2 \times 10^{-8}(\Delta T_{\text{SH}})^{2.33}p_c^{2.3}F_p^{3.33}\ (\text{W/m}^2\cdot{}^\circ\text{C}) \tag{10–4}$$

or, for more accurate values for organic fluids,

$$h_{\text{npb}} = 0.0001629\frac{\kappa_f\rho_g}{\rho_f}\left(\frac{2\sigma_{\text{ST}}}{g(\rho_f + \rho_g)}\right)^{0.274}\left(\frac{\Delta T_{\text{SH}}}{T_g}\right)^2\left(\frac{hn_{fg}}{\alpha_f^2}\right)^{0.744}\left(\frac{\rho_f}{\rho_f - \rho_g}\right)^{13}\frac{\text{W}}{\text{m}^2\cdot{}^\circ\text{C}} \tag{10–6}$$

The maximum heat transfer for pool boiling on a flat surface can be approximated by

$$\dot{q}_{\text{A,max npb}} = 0.149\sqrt{\rho_g}\text{hn}_{fg}[g\sigma_{\text{ST}}(\rho_f - \rho_g)]^{1/4}\ \frac{\text{kW}}{\text{m}^2} \tag{10–8}$$

where all of the parameters in this equation have standard SI units. The surface tension has units of mN/m. For maximum nucleate boiling heat transfer around single horizontal cylinders,

$$\dot{q}_{\text{A,max npb}} = 0.131\sqrt{\rho_g}\text{hn}_{fg}[g\sigma_{\text{ST}}(\rho_f - \rho_g)]^{1/4} \tag{10–9}$$

The convective heat transfer coefficient for film boiling on horizontal and vertical surfaces can be predicted by

$$h_{\text{FBT}} = h_{\text{FB}} + {}^{4}\!/_{3}h_{\text{RAD}} \tag{10–10}$$

where the convective film boiling heat transfer coefficient h_{FB} is

$$\text{Nu}_D = \frac{h_{\text{FB}}D}{k_g} = 0.62\left(\frac{(\rho_f - \rho_g)gD^3(\text{hn}_{fg} + 0.34c_{p,g}\Delta T_{\text{SH}})}{\nu_g\kappa_g\Delta T_{\text{SH}}}\right)^{1/4} \tag{10–11}$$

and

$$h_{\text{RAD}} = \varepsilon_{r,\text{HS}}\sigma\frac{T_w^4 - T_g^4}{\Delta T_{\text{SH}}} \tag{10–12}$$

Pool boiling in vertical tubes can be analyzed with the Chen method, where an equation for the convective heat transfer coefficient is defined as

$$h_{\text{nb,vt}} = h_f F + h_{\text{nb}}S \tag{10–13}$$

and the nucleate pool boiling convection heat-transfer coefficient for fluids can be predicted by

$$h_{\text{nb}} = 0.00122\left(\frac{\kappa_f^{0.79}c_{p,f}^{0.45}\rho_f^{0.49}}{\sigma_{\text{ST}}^{0.5}\mu_f^{0.29}\text{hn}_{fg}^{0.24}\rho_g^{0.24}}\right)(\Delta T_{\text{SH}})^{0.24}(p_{g,w} - p_g)^{0.75} \tag{10–19}$$

The mass flow rate of condensate down a vertical surface at a lower temperature than the saturated vapor per unit width is

$$\dot{m}_{\text{WD}} = \frac{\rho_f g}{3\mu_f}(\rho_f - \rho_g)\delta^3 \tag{10–27}$$

The boundary layer thickness is

$$\delta(y) = \left(\frac{4\kappa_f\mu_f(T_g - T_w)}{\rho_f g(\rho_f - \rho_g)\text{hn}_{fg}}y\right)^{1/4} \tag{10–34}$$

and the average convective heat-transfer coefficient for a boundary layer L long is

$$h_{\text{ave}} = \frac{1}{L}\int_0^L h_y\,dy = \frac{4}{3}h_L = 0.943\left(\frac{\rho_f g(\rho_f - \rho_g)\kappa_f^3\text{hn}_{fg}}{\mu_f(T_g - T_w)L}\right)^{1/4} \tag{10–38}$$

For transition flow between laminar and turbulent flow of a liquid, condensing on a vertical surface,

$$h_{\text{ave}} = \frac{\kappa_f}{1.08\text{Re}_{\text{cond}}^{1/3}}\left(1 - \frac{\rho_g}{\rho_f}\right)^{1/3}\left(\frac{g}{\nu_f^2}\right)^{1/3} \quad \text{(transition flow)} \tag{10–45}$$

and for turbulent flow of condensate in the film boundary layer, Labuntsov [12] suggests

$$h_{ave} = \frac{\kappa_f \text{Re}_{cond}}{8750 + \dfrac{58(\text{Re}_{cond}^{0.75} - 253)}{\sqrt{\text{Pr}}}} \left(\frac{g}{\nu_f^2}\right)^{1/3} \quad \text{(turbulent flow)} \tag{10–46}$$

For condensation of a vapor on a horizontal tube, the convective heat-transfer coefficient is

$$h_{ave} = 0.725\left(\frac{\rho_f g(\rho_f - \rho_g)\kappa_f^3 \text{hn}_{fg}}{\mu_f(T_g - T_w)D}\right)^{1/4} \tag{10–47}$$

and for multiple tubes,

$$h_{ave} = 0.725\left(\frac{\rho_f g(\rho_f - \rho_g)\kappa_f^3 \text{hn}_{fg}}{\mu_f(T_g - T_w)D}\right)^{1/4}\left(\frac{1}{N}\right)^{1/4} \tag{10–49}$$

For water boiling on horizontal surfaces, Jakob and Hawkins [16] recommend, where $\Delta T = T_w - T_g \leq 14°\text{F} = 7.8\ \text{K}$, that

$$h_{ave} = 150(\Delta T)^{1/3} \frac{\text{Btu}}{\text{hr} \cdot \text{ft}^2 \cdot °\text{F}} \qquad (\Delta T \text{ in °F or °R}) \tag{10–52}$$

$$h_{ave} = 1037(\Delta T)^{1/3} \frac{\text{W}}{\text{m}^2 \cdot \text{K}} \qquad (\Delta T \text{ in °C or K})$$

For water boiling on a vertical surface and where $17°\text{F} > \Delta T \geq 7.5°\text{F}$,

$$h_{ave} = 0.24(\Delta T)^3 \frac{\text{Btu}}{\text{hr} \cdot \text{ft}^2 \cdot °\text{F}}$$

$$h_{ave} = 7.74(\Delta T)^3 \frac{\text{W}}{\text{m}^2 \cdot \text{K}} \tag{10–55}$$

Jakob [17] recommends for water boiling inside vertical tubes, where $\Delta T \leq 14°\text{F}$,

$$h_{ave} = 187.5(\Delta T)^{1/3} \frac{\text{Btu}}{\text{hr} \cdot \text{ft}^2 \cdot °\text{F}}$$

$$h_{ave} = 1296(\Delta T)^{1/3} \frac{\text{W}}{\text{m}^2 \cdot \text{K}} \tag{10–56}$$

The time needed to melt a spherical solid in its saturated liquid is

$$t_{max} = \frac{\rho_s \text{hn}_{fs} r_i}{h_{ave}(T_\infty - T_s)} = \frac{\rho_s^{2/3} \text{hn}_{fs}}{h_{ave}(T_\infty - T_s)} \sqrt[3]{\frac{3m_{sys}}{4\pi}} \tag{10–66}$$

The time required to melt a solid slab from an initial thickness δ_i to a thickness δ_f with convection heat transfer from above and a saturated liquid below is

$$t = (\delta_i - \delta_f)\left(\frac{h_{ave}(T_\infty - T_s) + h_{ave,L}(T_{\infty,L} - T_s)}{\rho_s \text{hn}_{fs}}\right)^{-1} \tag{10–72}$$

The time required to freeze a saturated liquid to a particular thickness is given by the dimensionless equation

$$\bar{t} = \frac{1}{\overline{R}^2\overline{T}^2}\ln\left(1 + \frac{\overline{RT}\overline{\delta}}{1 + \overline{RT}}\right) + \frac{\overline{\delta}}{\overline{RT}} \tag{10–78}$$

There are many applications for phase change heat transfer. Some of those presented here were

- Boiling and vaporization processes
- Condensation processes
- Freezing of iron, water, and other substances in their vapor states
- Thermal storage systems and ice storage systems
- De-icing and de-frosting processes
- Sublimation

DISCUSSION QUESTIONS

Section 10–1

10–1 Why do you think irregularities in a heating surface provide better conditions for boiling of a saturated liquid than a smooth surface?

10–2 In freezing of a saturated liquid, where would you expect conduction to be greater, in the saturated solid or the saturated liquid? Is this true for water at normal pressures?

10–3 In sublimation the saturation region involves solid and vapor phases. Where would you expect convection heat transfer to be dominant, in the saturated solid or saturated vapor?

Section 10–2

10–4 Is peak nucleate boiling a stable boiling condition?

10–5 Why is film boiling associated with a lower convective heat-transfer coefficient?

10–6 What is meant by the term F_{TP} in tube boiling?

10–7 When would radiation predominate in boiling?

Section 10–3

10–8 What is the difference between film condensation and drop-wise condensation?

10–9 Is film condensation more effective on vertical tubes or on horizontal tubes?

10–10 How do you think condensation would occur inside vertical tubes?

Section 10–4

10–11 What is the degree of accuracy associated with predicted values for the convective heat-transfer coefficient?

10–12 Why should pressure affect the convective heat-transfer coefficient in boiling?

Section 10–5

10–13 What is the difference between amorphous solids and crystalline solids?

10–14 What happens to the water that melts on a frozen pond on a warm day?

10–15 How would you expect pressure to affect melting and freezing processes? Would it increase or decrease the rate of melting or freezing?

Section 10–6

10–16 How would you expect a cooling curve for an amorphous solid through its glass transition to differ from that of freezing iron as shown in figure 10–25?

10–17 Can you think of any refinements to the model for PCM thermal storage as shown schematically in figure 10–27?

10–18 How is sublimation from a solid to a vapor different from the boiling process?

PRACTICE PROBLEMS

Section 10–1

10–1 Determine the expected diameter of water vapor bubbles in liquid water at 200°C if the surface tension is 40 mN/m and the pressure difference between the inside and the outside of the bubble is 100 Pa.

10–2 Estimate the buoyant force of 1-mm-diameter water vapor bubbles in sea water at 10°C. Sea water can be assumed to have a density 10% greater than pure water and the surface tension of water can be assumed to be 65 mN/m.

10–3 Ethyl alcohol bubbles form at 78.5°C, 101 kPa. Assuming that the surface tension is 75 mN/m, and the density of vapor ethyl alcohol is 0.8 kg/m^3 with a vapor pressure of 102 kPa, estimate the buoyant force acting on the bubbles.

Section 10–2

10–4 Determine the heat transfer per unit area for pool boiling water at 200 kPa and where the surface temperature of the container is 135°C. Compare results using equation (10–4) to those using equation (10–6). Assume a surface tension of 45 mN/m for water at 200 kPa.

10–5 Propanol (C_3H_8O) is boiling in a container at 101 kPa and a surface temperature of 120°C. Its saturation temperature is 97.4°C, and its heat of fusion is 697 kJ/kg. Estimate the heat transfer per unit area.

10–6 Estimate the maximum superheat temperature allowable for nucleate pool boiling of methanol. (*Hint:* Use the results of example 10–3, equation 10–8, and the relationship $\dot{q}_{A,\max\text{ npb}} = h_{\text{npb}}\Delta T_{\text{SH}}$.)

10–7 In a chiller unit R-123 evaporates at −20°C in 1-cm-ID tubes. The inside surface temperature of the tubes is estimated as 5°C and film boiling is assumed. Saturated liquid flows in at 20 g/s and is to leave as saturated vapor. Estimate the tube length required. Use $\text{hn}_f = 183.4$ kJ/kg and $\text{hn}_g = 367.9\frac{\text{kJ}}{\text{kg}}$ for R-123.

10–8 Water surrounds horizontal, 3 cm diameter hot tubes in a nuclear steam generator that have a surface temperature of 800°C. The water is at a pressure of 800 kPa, and if film boiling is assumed, estimate the total convective heat-transfer coefficient between the water and the tube surfaces.

10–9 Liquid nitrogen at 77.2 K (its saturation temperature) and 100 kPa is in a well-insulated flask having a surface temperature of 100 K. Estimate the convection heat-transfer coefficient between the liquid nitrogen and the flask surface. The critical pressure of nitrogen is 33.54 atm.

10–10 Ethyl alcohol (ethanol) is to be boiled in a vertical tube, 2 cm inside diameter. The ethanol is to be boiled at a temperature of 78.5°C and a pressure of 3000 kPa when the tube inside surface temperature is 200°C. The ethanol entering the tube may be assumed to be 10% saturated vapor hn_{fg} of 842 J/kg, bubble diameter of 1 mm, and the bubble velocity may be taken as 0.24 m/s. Estimate the nucleate pool boiling convective heat-transfer coefficient.

Section 10–3

10–11 Water is to be condensed at 1 atm pressure in a compact heat exchanger having vertical stainless steel plates at 40°C. Plot the condensation boundary layer thickness flowing down the vertical surface as a function of the distance down the plate, assuming laminar flow. At what point would you assume condensate flow to become turbulent?

10–12 Refer to figure 10–16 as a particular arrangement for a cooling coil of an air conditioning system. If water condensate from the humid air passing around the vertical tubes is 20 g/s evenly distributed over the 12 tubes, determine the Reynolds Number describing the condensate flow down the tubes. Then estimate the convective heat-transfer coefficient and the heat transfer in watts.

10–13 In a chiller unit, ammonia is to be condensed at a saturation temperature of 50°C by passing the ammonia vapor over vertical plates 100 cm high by 150 cm wide and at a surface temperature of 20°C. Estimate the average convective heat-transfer coefficient and the amount of ammonia condensing from one plate surface. Use a value of 1052.8 kJ/kg for hn_{fg} of ammonia.

10–14 Show that the total condensation rate per unit width down a vertical surface as given by equation (10–41) can also be written as

$$\dot{m}_{\text{WD}} = \frac{h_{\text{ave}}L(T_g - T_w)}{\text{hn}_{fg}}$$

10–15 Ethanol is to be condensed around cool 10°C tubes of 1 cm outside diameter and at a saturation temperature of 78.5°C. Plot the heat transfer per unit length on horizontal tube banks arranged as shown in figure 10–17b and as predicted by equation (10–49), as the number of tubes is increased. Use hn_{fg} of 842 J/kg for ethanol. How does this result compare to the heat transfer predicted for a single tube or for a dripping, in-line tube arrangement as shown in figure 10–17c?

10–16 In a condenser of a refrigeration unit, saturated vapor R-134a flows at 0.5 m/s into a 0.5-cm-diameter horizontal tube. The tube inside surface temperature is 30°C and the R-134a is at 80°C. Estimate the convective heat-transfer coefficient and the expected tube length required for complete condensation of the R-134a. Use hn_{fg} of 106.3 kJ/kg, ρ_f of 927.6 kg/m^3, ρ_g of 153.8 kg/m^3, and c_p, k, and μ at 40°C.

Section 10–4

10–17 Using equation (10–52) or (10–53), determine the approximate convective heat transfer for water boiling at atmospheric pressure in a container having a surface temperature of 130°C. Compare this to a value obtained by using equation (10–6).

10–18 Compare the approximate value for the convective heat-transfer coefficient from equation (10–54) or (10–55) for water boiling on vertical flat surfaces at atmospheric pressure and superheat temperature of 40°F to the value predicted by using equation (10–6).

10–19 Determine the approximate value for the convective heat-transfer coefficient for water boiling in vertical tubes at a superheat temperature of 120°C using equation (10–56). Compare this result to that using equation (10–6). Assume the water pressure is 500 kPa.

10–20 Determine the approximate value for the convective heat-transfer coefficient for normal butane boiling with a superheat of 100°F using equation (10–56). The saturation temperature of butane is 243°F at atmospheric pressure.

10–21 Estimate the convective heat-transfer coefficient for condensing of ammonia in horizontal tubes of 1 cm diameter and a tube surface temperature of 40°C below the saturation temperature of the ammonia.

10–22 The average heat transfer per unit area is 100 W/m^2 from saturated vapor R-134a at 50°C flowing at 0.5 kg/s through a 1-cm-diameter horizontal tube having an average surface temperature of 20°C when the R-134a is to be completely condensed. Estimate the tube length required, using hn_{fg} = 205.9 kJ/kg for R-134a at 50°C.

Section 10–5

10–23 For an infinite solid slab subjected to convective heat transfer show that equation (10–68) follows from equation (10–67).

10–24 Lead melts at 327.44°C with a heat of fusion of 23.02 kJ/kg. If a 10-kg solid ingot of lead at its saturation temperature is heated in a container at 500°C having an average convective heat transfer coefficient of 100 W/m^2 · °C, determine the time to completely melt the lead ingot. Assume lumped heat capacity.

10–25 Two kg of molten copper at 1083°C are dropped into water at 40°C. Determine the freezing time for complete solidification of the copper. Assume lumped heat capacity and that the heat of fusion of copper is 204.7 kJ/kg and h_{ave} = 120 W/m^2 · °C.

10–26 For example 10–15, if radiant effects of 95 Btu/hr · ft^2 due to the sun shining are included, determine the amount of decrease in the ice thickness over the 2 hr.

10–27 Water in a pond is at 2°C and air is at −20°C with a convective heat-transfer coefficient of 30 W/m^2 · °C above the water. When the water begins to freeze the lower convective heat-transfer coefficient between the ice and water is 20 W/m^2 · C. Estimate the ice thickness after 8 hr.

10–28 Molten iron is in a large foundry ladle at 1600°C. Air at 150°C is directly above the surface of the molten iron and with a convective heat-transfer coefficient of 300 W/m^2 · C. Iron at atmospheric pressure has a saturation temperature of 1538°C and a heat of fusion of 247.3 kJ/kg. If the convective heat transfer between the molten iron and the freezing iron is taken as 500 W/m^2 · C and assuming that the solid iron remains on the top of the ladle, determine the time to form a 5-mm-thick surface of solid iron in the ladle.

Section 10–6

10–29 Referring to figure 10–25, determine the maximum heating rate to the inner surface of the casting due to the release of heat of fusion in the molten iron.

10–30 Write the energy balance for an ice pond at noon on a sunny day with solar energy of 300 W/m^2 to the ice, air at −8°C and convective heat-transfer coefficient of 10 W/m^2 · °C above the ice, and 1°C water with a convective heat-transfer coefficient of 30 W/m^2 · °C below the ice.

10–31 Estimate the time required to melt a 5-mm sheet of ice on an aircraft wing if the wing surface is heated to 60°C and the surrounding air is −10°C. Assume that the average convective heat-transfer coefficients are 20 W/m^2 · C for the air/ice and 10 W/m^2 · °C for the ice/melted water.

10–32 PCM thermal storage systems can be used to store thermal energy from solar energy collectors. The solar collector heats water and the hot water is then passed through the PCM for storage. Then, energy is extracted by passing cold water through the tubes to heat that water. For a system using a PCM having a phase change at 95°F, with cold water at 65°F passing through the tubes, and Bi = 30, NTU = 8, and Fr_0 = 0.6, determine the outlet water temperature. Assume tubes have a square arrangement with spacing of 2D.

10–33 For example 10–18, determine the total frozen fraction after 60 min of cooling if the tubes are in a staggered arrangement instead of the in-line arrangement.

10–34 Ice or compacted snow sublimes to a saturated vapor in cold climates. Estimate the rate at which ice will sublime per unit area on an iceberg if the ice is −30°C and the air is −10°C. Assume a convective heat transfer coefficient of 20 W/m · °C.

REFERENCES

1. Mostinski, I. L., Application of the Rule of Corresponding States for the Calculation of Heat Transfer and Critical Heat Flux, *Teploenergetika* 4, 66, 1963; English abstract, *Br. Chem. Eng.* 8, 8, 580, 1963.
2. Stephan K., and M. Abdelsalam, Heat Transfer Correlations for Natural Convection Boiling, *Int. J. Heat Mass Transfer,* 23, 73–87, 1980.
3. Nukiyama, S., The Maximum and Minimum Values of Heat Transmitted from Metal to Boiling Water under Atmospheric Pressure, *J. Jpn Soc. Mech. Eng.* 37, 367, 1934.
4. Drew, T. B., and C. Mueller, Boiling, *Trans. AICHE,* 33, 449, 1937.
5. Thome, J. R., *Enhanced Boiling Heat Transfer,* Hemisphere Publishing, New York, 1990.
6. Bromley, A. L., Heat Transfer in Stable Film Boiling, *Chem. Eng. Progr.* 46, 221–227, 1950.
7. Chen, J. C., A Correlation for Boiling Heat Transfer to Saturated Fluids in Convective Flow, *Ind. Eng. Chem. Process Des. Dev.* 5(3), 322–329, 1966.

8. Pierre, B., Flow Resistance with Boiling Refrigerants, *ASHRAE J.,* 6, 58–65, 73–77.

9. Butterworth, D., and J. M. Robertson, Boiling and Flow in Horizontal Tubes, chapter 11 in D. Butterworth and G. F. Hewitt, eds., *Two-Phase Flow and Heat Transfer,* Oxford Univ. Press, London, 1977.

10. Nusselt, W., Die Oberflachen Kondensation des Wasserdampfes, *Z. Ver. Deut. Ing.,* 60, 541, 1916.

11. Kutateladze, S. S., *Fundamentals of Heat Transfer,* Academic Press, New York, 1963.

12. Labuntsov, D. A., Heat Transfer in Film Condensation of Pure Steam on Vertical Surfaces and Horizontal Tubes, *Teploenergetika,* 4, 72, 1957.

13. Chato, J. C., Laminar Condensation Inside Horizontal and Inclined Tubes, *J. ASHRAE,* 4, 52, 1962.

14. Akers, W. W., H. A. Deans, and O. K. Crosser, Condensing Heat Transfer Within Horizontal Tubes, *Chem. Eng. Prog. Symp. Ser.,* 55 (29), 171, 1958.

15. Rosenow, W. M., Film Condensation, *Handbook of Heat Transfer*, Chapter 12A, in W. M. Rosenow and J. P. Hartnett, eds., McGraw-Hill, New York, 1973.

16. Jakob, M., and G. Hawkins, *Elements of Heat Transfer,* 3rd edition, John Wiley & Sons, New York, 1957.

17. Jakob, M., The Influence of Pressure on Heat Transfer in Evaporation, *Proc. Fifth Int. Congr. Applied Mechanics,* p. 561, 1938.

18. Cryder, D. S., and A. C. Finalborgo, *Trans. Am. Inst. Chem. Engrs.* 33, 346–361, 1937.

19. London, A. L., and R. A. Seban, Rate of Ice Formation, *Trans. A.S.M.E.,* 65, 771–778, 1943.

20. Shamsundar, N., and R. Srinivasan, A New Similarity Method for Analysis of Multi-Dimensional Solidification, *J. Heat Transfer,* 101, 585–591, Nov. 1979.

21. Shamsundar, N., and R. Srinivasan, Effectiveness–NTU Charts for Heat Recovery from Latent Storage Units, *J. Solar Eng.,* 102, 263–271, Nov. 1980.

MATHEMATICAL INFORMATION

The information in this appendix section is intended to provide background for the mathematical analyses in the body of the text. More complete and detailed information regarding the various mathematical concepts and operations can be found in many handbooks and textbooks, such as those listed at the end of this section.

A–1 VECTOR OPERATIONS

A scalar is a quantity having magnitude only. In heat transfer temperature, mass, density, volume, and specific heat are scalars. A vector is a quantity having direction and magnitude. Heat transfer is a vector and so are velocity, acceleration, displacement, force, and momentum. A vector may be described in rectilinear Cartesian coordinates by three vectors aligned with the three axes: *x*, *y*, and *z*. That is, the vector **V** may be written as

$$\mathbf{V} = \mathbf{v}_x \oplus \mathbf{v}_y \oplus \mathbf{v}_z = \sum \mathbf{v}_i \qquad \textbf{(A–1–1)}$$

where the sign $\oplus$ is a vector addition. Sometimes $\mathbf{v}_i$ is used to denote a vector without the summation sign for shorthand. The magnitude of a vector is

$$V = \sqrt{\mathbf{v}_x^2 + \mathbf{v}_y^2 + \mathbf{v}_z^2} \qquad \textbf{(A–1–2)}$$

Two vectors **V** and **W** shown in figure A–1 may be multiplied in two ways; by the dot product and by the cross product. The dot product is

$$\mathbf{V} \cdot \mathbf{W} = VW \cos\theta = WV \cos\theta = \mathbf{W} \cdot \mathbf{V} \qquad \textbf{(A–1–3)}$$

The dot product is a scalar quantity and is sometimes referred to as the scalar product of two vectors. The cross product is

$$\mathbf{V} \times \mathbf{W} = \mathbf{n}VW \sin\theta \qquad \textbf{(A–1–4)}$$

Figure A–1

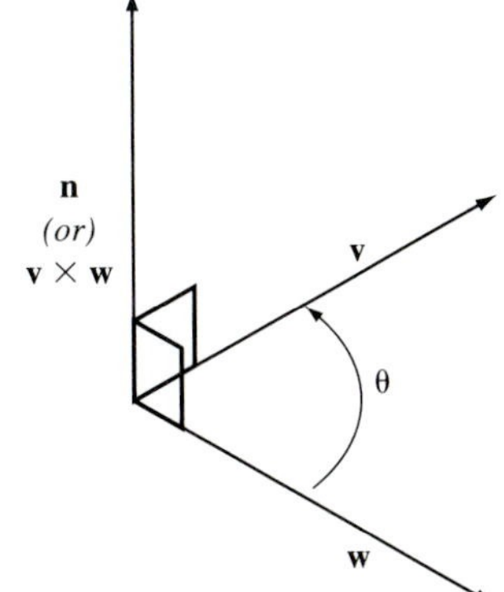

where **n** is a unit vector (of magnitude 1) directed outward from and normal to the plane containing **V** and **W** as shown in figure A–1. Thus, the cross product is a vector directed normal to the plane containing the two vectors

The differential operator for scalars and vectors, called the *nabla* or *del operator*, ∇, is, in Cartesian coordinates,

$$\nabla = \mathbf{n}_x \frac{\partial}{\partial x} \oplus \mathbf{n}_y \frac{\partial}{\partial y} \oplus \mathbf{n}_z \frac{\partial}{\partial z} \tag{A–1–5}$$

where $\mathbf{n}_x$, $\mathbf{n}_y$, and $\mathbf{n}_z$ are unit vectors in the x, y, and z directions. The gradient of temperature, which is an important quantity in conduction heat transfer, or another scalar variable is, in Cartesian coordinates,

$$\nabla T = \mathbf{n}_x \frac{\partial T}{\partial x} \oplus \mathbf{n}_y \frac{\partial T}{\partial y} \oplus \mathbf{n}_z \frac{\partial T}{\partial z} \tag{A–1–6}$$

It can be seen that the gradient of a scalar is a vector. The nabla operator can be used with a vector in two ways. The divergence is the dot product of the nabla operator with the vector,

$$\nabla \cdot \mathbf{V} = \frac{\partial \mathbf{v}_x}{\partial x} + \frac{\partial \mathbf{v}_y}{\partial y} + \frac{\partial \mathbf{v}_z}{\partial z} \tag{A–1–7}$$

which is a scalar product. The cross product of the nabla operator which is a vector, is called the *curl* and is given by

$$\nabla \times \mathbf{V} = \mathbf{n}_x \left(\frac{\partial \mathbf{v}_z}{\partial y} - \frac{\partial \mathbf{v}_y}{\partial z} \right) \oplus \mathbf{n}_y \left(\frac{\partial \mathbf{v}_x}{\partial z} - \frac{\partial \mathbf{v}_z}{\partial x} \right) \oplus \mathbf{n}_z \left(\frac{\partial \mathbf{v}_y}{\partial x} - \frac{\partial \mathbf{v}_x}{\partial y} \right) \tag{A–1–8}$$

A–2 MATRIX OPERATIONS

A matrix is an array of numbers in rows and columns that is a concise method for handling information. In heat transfer, matrices are often used for multidimensional conduction heat-transfer situations and also in radiation heat transfer involving opaque gray surfaces. The matrix A is denoted as $[A]$ or $[a_{ij}]$ and represents the array of numbers

$$\begin{bmatrix} a_{11} & a_{12} & a_{13} & a_{14} \ldots a_{1n} \\ a_{21} & a_{22} & a_{23} & a_{24} \ldots a_{2n} \\ & & \vdots & \\ a_{m1} & a_{m2} & a_{m3} & a_{m4} \ldots a_{mn} \end{bmatrix}$$

The terms a_{11}, a_{12}, and so on are the elements or components of the matrix and can be written a_{ij}. The i represents the row number and j represents the column number.

In heat transfer matrices are used to consider a set of equations containing many variables. A *square matrix* is one where $m = n$, or the number of columns is equal to the number of rows. A special form of square matrix is a *diagonal matrix,* where $a_{ij} = 0$ if i is not equal to j. A unique diagonal matrix is the *identity matrix,* denoted $[I]$, where $a_{ii} = a_{jj} = 1.0$ An example of a diagonal 3×3 matrix is

$$[D] = \begin{bmatrix} d_{11} & 0 & 0 \\ 0 & d_{22} & 0 \\ 0 & 0 & d_{33} \end{bmatrix}$$

and a 3×3 identity matrix is

$$[I] = \begin{bmatrix} 1 & 0 & 0 \\ 0 & 1 & 0 \\ 0 & 0 & 1 \end{bmatrix}$$

A column matrix can be used to represent a vector and is one where $n = 1$. We denote it as $\{B\}$ or $\{b_j\}$.

Matrix Multiplication

The result of multiplying two matrices $[A]$ and $[B]$ is $[C]$, or

$$[A][B] = [C] \tag{A–2–1}$$

or

$$[a_{ij}][b_{ij}] = [c_{ij}] \tag{A–2–2}$$

where

$$c_{ij} = \sum_{k=1}^{m} a_{ik}b_{kj} \tag{A–2–3}$$

Multiplication of two matrices can be defined only if the number of columns of $[A]$, k, is equal to the number of rows of $[B]$, k. For an example, consider a square 3×3 matrix. Here m is 3 and therefore the component c_{11} of the result matrix $[C]$ is, with $i = j = 1$,

$$c_{11} = \sum_{k=1}^{3} a_{1k}b_{k1} = a_{11}b_{11} + a_{12}b_{21} + a_{13}b_{31}$$

For c_{12} we have, since $i = 1$ and $j = 2$,

$$c_{12} = a_{11}b_{12} + a_{12}b_{22} + a_{13}b_{32}$$

Other matrix components can be written using equation (A–2–3).

Matrix Addition and Subtraction

$$[A] \pm [B] = [C] \tag{A–2–4}$$

and the components are $c_{ij} = a_{ij} \pm b_{ij}$

Matrix Multiplied by a Vector (column matrix)

$$[A]\{V\} = \{W\}, \text{ a column matrix}$$

with the components $w_i = a_{ij}v_i$

As an example, for a square 3×3 matrix multiplied by a vector such that $m = n = 3$, the components are

$$\{W\} = \begin{bmatrix} a_{11}v_1 + a_{12}v_1 + a_{13}v_1 \\ a_{21}v_2 + a_{22}v_2 + a_{23}v_2 \\ a_{31}v_3 + a_{32}v_3 + a_{33}v_3 \end{bmatrix}$$

The inverse matrix $[A]^{-1}$, defined only for a square nonsingular matrix, has the property that

$$[A]^{-1}[A] = [I] \qquad \text{(the identity matrix)}$$

For a square matrix multiplied by a vector, we have

$$[A]\{V\} = \{W\}$$

and

$$[A]^{-1}[A]\{V\} = [A]^{-1}\{W\}$$

or

$$\{V\} = [A]^{-1}\{W\} \quad \text{since} \quad [A]^{-1}[A] = [I]$$

A–3 TRIGONOMETRIC RELATIONSHIPS

The following trigonometric identities are often encountered in mathematical analysis involving heat and mass transfer.

$$\sin^2(x) + \cos^2(x) = 1 \quad \textbf{(A–3–1)}$$

$$\sec^2(x) - \tan^2(x) = 1 \quad \textbf{(A–3–2)}$$

$$\csc^2(x) - \cot^2(x) = 1 \quad \textbf{(A–3–3)}$$

$$\sin 2x = 2 \sin x \cos x \quad \textbf{(A–3–4)}$$

$$\cos 2x = \cos^2 x - \sin^2 x = 1 - 2\sin^2 x = 2\cos^2 x - 1 \quad \textbf{(A–3–5)}$$

$$\tan 2x = 2 \tan x/(1 - \tan^2 x) \quad \textbf{(A–3–6)}$$

$$\sin(x + y) = \sin x \cos x \sin y \quad \textbf{(A–3–7)}$$

$$\sin(x - y) = \sin x \cos y - \cos x \sin y \quad \textbf{(A–3–8)}$$

$$\cos(x + y) = \cos x \cos y - \sin x \sin y \quad \textbf{(A–3–9)}$$

$$\cos(x - y) = \cos x \cos y + \sin x \sin y \quad \textbf{(A–3–10)}$$

$$\sin x \cos y = \tfrac{1}{2}[\sin(x + y) + \sin(x - y)] \quad \textbf{(A–3–11)}$$

$$\sin x \sin y = \tfrac{1}{2}[\cos(x - y) - \cos(x + y)] \quad \textbf{(A–3–12)}$$

$$\cos x \cos y = \tfrac{1}{2}[\cos(x + y) + \cos(x - y)] \quad \textbf{(A–3–13)}$$

The values for the trigonometric functions can be computed from the series expansions

$$\sin x = x - \frac{x^3}{3!} + \frac{x^5}{5!} - \frac{x^7}{7!} + \frac{x^9}{9!} - \cdots \quad \textbf{(A–3–14)}$$

$$\cos x = 1 - \frac{x^2}{2!} + \frac{x^4}{4!} - \frac{x^6}{6!} + \frac{x^8}{8!} - \cdots \quad \textbf{(A–3–15)}$$

$$\tan x = x + \frac{x^3}{3!} + \frac{2x^5}{15} + \frac{17x^7}{315} + \cdots \quad \textbf{(A–3–16)}$$

The derivatives of the trigonometric functions are

$$\frac{d}{dx}\sin ax = a \cos ax \quad \textbf{(A–3–17)}$$

$$\frac{d}{dx}\cos ax = -a \sin ax \quad \textbf{(A–3–18)}$$

$$\frac{d}{dx}\tan ax = a \sec^2 ax \quad \textbf{(A–3–19)}$$

A–4 TABLE OF INTEGRALS

The following is a very brief listing of some of the common integrals encountered in heat and mass transfer analysis. Other, more extensive, tables of integrals may be found in the literature, or the mathematical manipulation computer software packages can be used for machine integration.

$$\int [f(x) + g(x)]dx = \int f(x)dx + \int g(x)dx \qquad \int cf(x)dx = c\int f(x)dx$$

$$\int u\,dv = uv - \int v\,du \qquad \int x^n dx = \frac{1}{n+1}x^{n+1} + C \qquad (n \neq -1)$$

$$\int \frac{1}{x}dx = \ln x + C$$

$$\int e^x dx = e^x + c \qquad \int xe^x dx = (x - 1)e^x$$

$$\int \frac{1}{x^2 - a^2} dx = \frac{1}{2a} \ln\left|\frac{x - a}{x + a}\right| \qquad \int \frac{1}{a^2 - x^2} dx = \frac{1}{2a} \ln\left|\frac{a + x}{a - x}\right|$$

$$\int \frac{x}{ax^2 + b} dx = \frac{1}{2a} \ln|ax^2 + b|$$

$$\int \frac{1}{u^4 - a^4} du = \frac{1}{4a^3} \ln\left(\frac{u - a}{u + a}\right) - \frac{1}{2a^3} \arctan\left(\frac{u}{a}\right)$$

$$\int \sin ax\, dx = -\frac{1}{\mathrm{a}} \cos ax$$

$$\int x \sin ax\, dx = \frac{1}{a^2} \sin ax - \frac{1}{a} x \cos ax$$

$$\int \cos ax = \frac{1}{a} \sin ax$$

$$\int \frac{1}{\sin ax} dx = \frac{1}{a} \ln\left|\tan \frac{ax}{2}\right|$$

$$\int \sin^2 ax\, dx = \frac{x}{2} - \frac{\sin 2ax}{4a}$$

$$\int \cos^2 ax\, dx = \frac{x}{2} + \frac{\sin 2ax}{4a}$$

$$\int \tan ax\, dx = -\frac{1}{a} \ln|\cos ax|$$

$$\int e^{ax} dx = \frac{1}{a} e^{ax}$$

$$\int b^{ax} dx = \frac{b^{ax}}{a \ln b}$$

$$\int \ln ax\, dx = x \ln ax - x$$

$$\int \frac{1}{x \ln ax} dx = \ln(\ln ax)$$

$$\int \sinh ax\, dx = \frac{1}{a} \cosh ax$$

$$\int \cosh ax\, dx = \frac{1}{a} \sinh ax$$

$$\int \sinh^2 ax\, dx = \frac{1}{4a} \sinh 2ax - \frac{1}{2} x$$

$$\int \cosh^2 ax\, dx = \frac{1}{4a} \sinh 2ax + \frac{1}{2} x$$

A–5 HYPERBOLIC FUNCTIONS

The hyperbolic functions occur in the analysis of fins in combined conduction/convection heat transfer. The hyperbolic functions of argument x can be defined as

$$\sinh x = {}^1\!/_2(e^x - e^{-x}) \tag{A–5–1}$$

$$\cosh x = {}^1\!/_2(e^x + e^{-x}) \tag{A–5–2}$$

$$\tanh x = \frac{e^x - e^{-x}}{e^x + e^{-x}} = \frac{\sinh x}{\cosh x} \quad \textbf{(A–5–3)}$$

$$\operatorname{csch} x = \frac{1}{\sinh x} \quad \textbf{(A–5–4)}$$

$$\operatorname{sech} x = \frac{1}{\cosh x} \quad \textbf{(A–5–5)}$$

$$\coth x = \frac{1}{\tanh x} \quad \textbf{(A–5–6)}$$

The derivatives of the hyperbolic functions with respect to y are

$$\frac{d}{dy}\sinh x = (\cosh x)\frac{dx}{dy} \quad \textbf{(A–5–7)}$$

$$\frac{d}{dy}\cosh x = (\sinh x)\frac{dx}{dy} \quad \textbf{(A–5–8)}$$

Some integral relationships for hyperbolic sine and hyperbolic cosine are given in section A–4.

The qualitative behavior of the hyperbolic functions is demonstrated graphically in figure A–5, and the values for the hyperbolic functions over a selected domain are given in table A–5. These functions are normally included in a hand calculator and are also included in many mathematical computer software packages. Further information on the hyperbolic functions can also be obtained in any standard mathematical handbook, such as *Handbook of Mathmatical Tables and Formulas* [1].

Figure A–5 The hyperbolic functions.

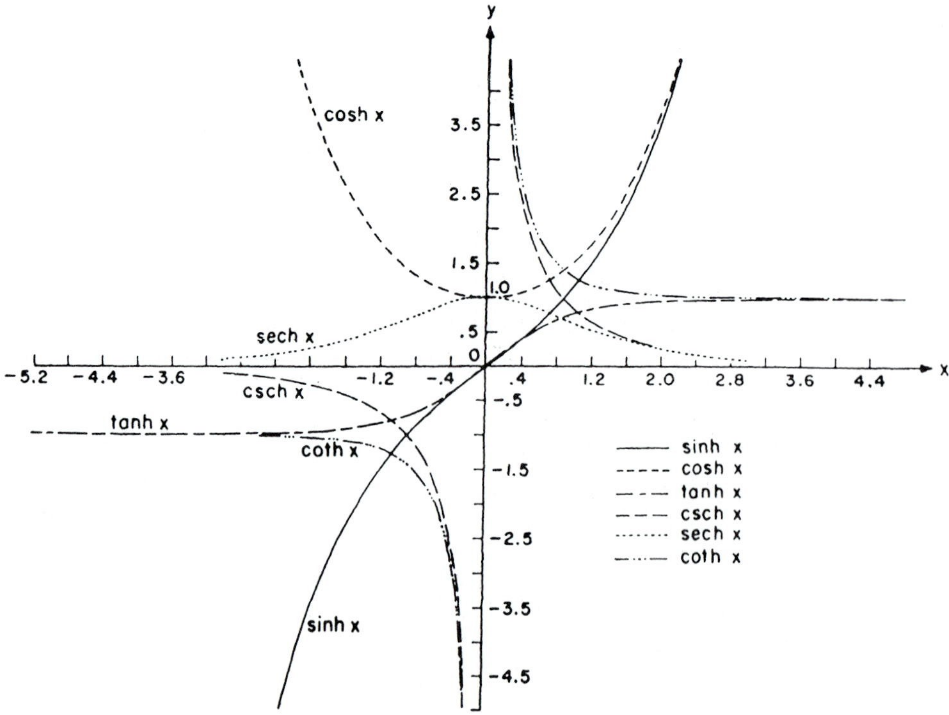

Table A–5 Hyperbolic Functions

x	sin hx	cosh x	tanh x
0.00	0.0000	1.0000	0.00000
0.10	0.1002	1.0050	0.09967
0.20	0.2013	1.0201	0.19738
0.30	0.3045	1.0453	0.29131
0.40	0.4108	1.0811	0.38995
0.50	0.5211	1.1276	0.46212
0.60	0.6367	1.1855	0.53705
0.70	0.7586	1.2552	0.60437
0.80	0.8881	1.3374	0.66404
0.90	1.0265	1.4331	0.71630
1.00	1.1752	1.5431	0.76159
1.10	1.3356	1.6685	0.80050
1.20	1.5095	1.8107	0.83365
1.30	1.6984	1.9709	0.86172
1.40	1.9043	2.1509	0.88535
1.50	2.1293	2.3524	0.90515
1.60	2.3756	2.5775	0.92167
1.70	2.6456	2.8283	0.93541
1.80	2.9422	3.1075	0.94681
1.90	3.2682	3.4177	0.95624
2.00	3.6269	3.7622	0.96403
2.10	4.0219	4.1443	0.97045
2.20	4.4571	4.5679	0.97574
2.30	4.9370	5.0372	0.98010
2.40	5.4662	5.5569	0.98367
2.50	6.0502	6.1323	0.98661
2.60	6.6947	6.7690	0.98903
2.70	7.4063	7.4735	0.99101
2.80	8.1919	8.2527	0.99263
2.90	9.0596	9.1146	0.99396
3.00	10.018	10.068	0.99505
3.50	16.543	16.573	0.99818
4.00	27.290	27.308	0.99933
4.50	45.003	45.014	0.99975
5.00	74.203	74.210	0.99991
6.00	201.71	201.72	0.99999
7.00	548.32	548.32	1.00000
8.00	1,490.5	1,490.5	1.00000
9.00	4,051.5	4,051.5	1.00000
10.00	11,013	11,013	1.00000

A–6 SOME POWER SERIES

Special problems in heat transfer sometimes have mathematical solutions that can be expressed as power series. The following is a brief listing of some common power series that may be encountered in heat or mass transfer analyses.

$$f(x) = f(a) + (x - a)f'(a) + \frac{(x-a)^2}{2!}f''(a) + \cdots + \frac{(x-a)^n}{n!}f^n(a) \quad \text{(Taylor Series)}$$

$$e^x = \sum_{n=0}^{\infty} \frac{x^n}{n!} \qquad (1 - x)^{-1} = \sum_{n=0}^{\infty} x^n \qquad (1 - x)^{-2} = \sum_{n=0}^{\infty} (n - 1)x^n$$

$$(1 + x)^{-1} = \sum_{n=0}^{\infty} (-1)^n x^n$$

$$\ln(x) = \sum_{n=1}^{\infty} \left(\frac{x-1}{x}\right)^n \left(\frac{1}{n}\right)$$

$$\ln(1 + x) = \sum_{n=0}^{\infty} \left(-1\right)^n \left(\frac{1}{n+1}\right) x^{n+1} \qquad \text{for } |x| \le 1 \text{ and } x \neq -1$$

A–7 HARMONIC FUNCTIONS

A group of periodic functions that have many applications to engineering analysis is called the *harmonic functions,* defined as any function of the form

$$F(x) = A_n \sin(nx + \phi) \qquad \textbf{(A–7–1)}$$

where A_n and ϕ are constants. These constants are frequently identified by the terms

A_n = Amplitude of nth harmonic.

n = A positive integer (1, 2, 3, ...) that is used to identify the order of the harmonic function. The *first harmonic* is identified with $n = 1$, the second with $n = 2$, and so on.

ϕ = Phase, or phase shift.

The period is often identified as the location, $x = X_p$, where the function of equation (A–7–1) has completed one cycle. It can also be interpreted as the difference in values of x over one complete cycle. The period, X_p, is defined as

$$X_p = \text{period} = \frac{2\pi}{n} = \frac{360°}{n} \qquad \textbf{(A–7–2)}$$

For the special case where $\phi = \pi/2$, equation (A–7–1) becomes

$$F(x) = A \sin nx \qquad \textbf{(A–7–3)}$$

from the behavior of the sine function (see Appendix A–3). Harmonic functions can also be written as cosine functions. If x is associated with time, t, rather than position, then equation (A–7–1) is usually written as

$$F(t) = A_n \sin(2\pi f t + \phi) \qquad \textbf{(A–7–4)}$$

where f is called the frequency. It is usually expressed in cycles per second (= hertz, Hz). The period from equation (A–7–2) is defined as

$$X_p = T = \frac{1}{f} \qquad \textbf{(A–7–5)}$$

In figure (A–7–1) are shown the harmonic functions for the first harmonic ($n = 1$), the second harmonic ($n = 2$), and the third harmonic ($n = 3$). In general, the nth harmonic will have a period of $2\pi/n$ or $360°/n$ and will have n complete cycles in 2π radians. The terms *even harmonic* and *odd harmonic* are often used to describe those harmonic functions where n is even (2, 4, 6, ...) or n is odd (1, 3, 5, 7, ...) respectively.

When added together, harmonic functions will produce a compound wave form that results from the addition of each of the individual functions. In figure A–7–2 is displayed a graph of the addition of two harmonic functions.

Figure A–7–1 Harmonic functions.

Fundamental or first Harmonic

$y_1(\theta) = A_1 \sin\theta$

$X_{P1} = 360^0 = 2\pi$

$f_1 = 1\dfrac{cycle}{360^0}$

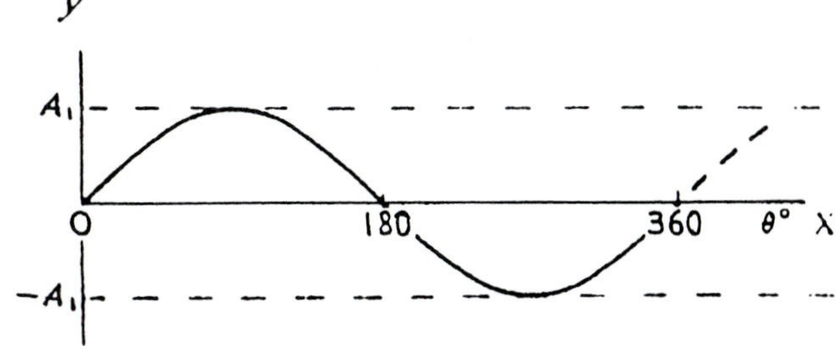

Second Harmonic

$y_2(\theta) = A_2 \sin 2\theta$

$X_{P2} = \dfrac{360^0}{2} = 180^0$

$f_2 = 2\dfrac{cycles}{360^0}$

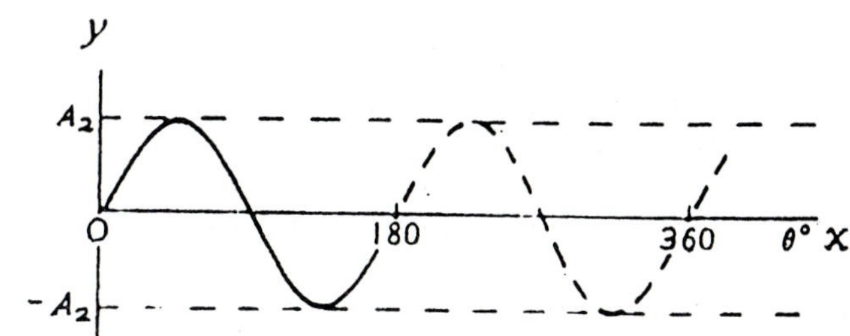

Third Harmonic

$y_3(\theta) = A_3 \sin 3\theta$

$X_{P3} = \dfrac{360^0}{3} = 120^0$

$f_3 = 3\dfrac{cycles}{360^0}$

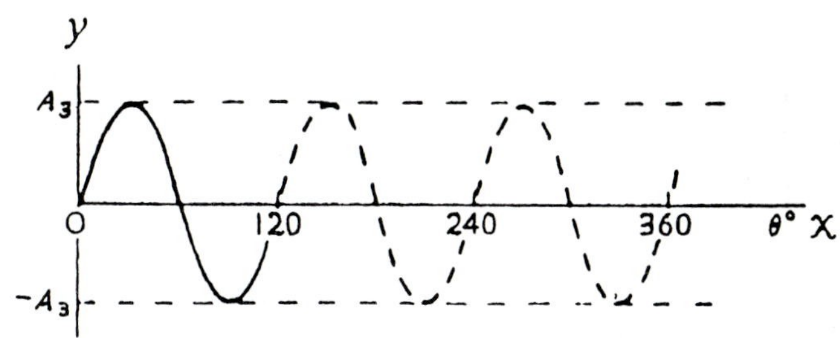

Figure A–7–2 Compound harmonic function.

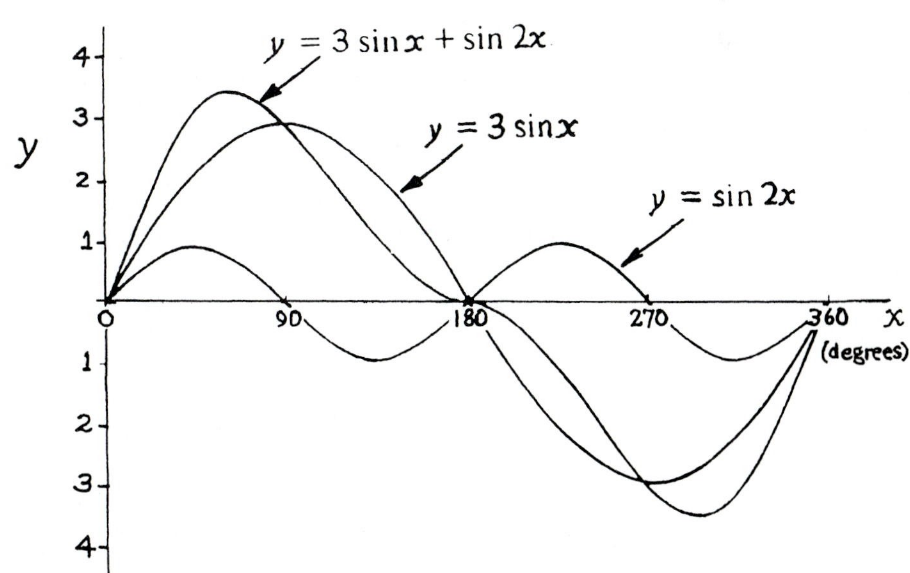

Graph of $y(x) = 3\sin x + \sin 2x$

A–8 FOURIER SERIES

The solution to many differential equations of heat transfer can be written as a series function

$$f(x) = \frac{1}{2}a_0 + \sum_{n=0}^{\infty} a_n \cos nx + \sum_{n=0}^{\infty} b_n \sin nx \qquad \textbf{(A–8–1)}$$

The term or coefficient $\frac{1}{2}a_0$ is introduced in this form for convenience, as we will see. Determining the coefficients a_0, a_n, and b_n is crucial in obtaining the solution to $f(x)$. First we note the following integrals,

$$\int_{-\pi}^{\pi} \cos nx\, dx = 0 \qquad \text{for } n > 0;\ = 2\pi \text{ for } n = 0, \qquad \textbf{(A–8–2)}$$

$$\int_{-\pi}^{\pi} \sin nx\, dx = 0 \qquad \text{for } n > 0;\ = 2\pi \text{ for } n = 0, \qquad \textbf{(A–8–3)}$$

Using equation (A–3–11) we have

$$\int_{-\pi}^{\pi} \sin mx \cos nx\, dx = \frac{1}{2}\int_{-\pi}^{\pi} \sin(m + n)x\, dx + \frac{1}{2}\int_{-\pi}^{\pi} \sin(m - n)x\, dx = 0 \qquad \textbf{(A–8–4)}$$

Using equation (A–3–12) we obtain

$$\int_{-\pi}^{\pi} \sin nx \sin mx\, dx = \frac{1}{2}\int_{-\pi}^{\pi} \cos(m - n)x\, dx - \frac{1}{2}\int_{-\pi}^{\pi} \cos(m + n)x\, dx = 0$$

$$\text{for } m \neq n \text{ and if } m = n = 0;\ = \pi \text{ for } m = n > 0 \qquad \textbf{(A–8–5)}$$

From equation (A–3–13) we have

$$\int_{-\pi}^{\pi} \cos mx \cos nx\, dx = \frac{1}{2}\int_{-\pi}^{\pi} \cos(m + n)x\, dx + \frac{1}{2}\int_{-\pi}^{\pi} \cos(m - n)x\, dx \qquad \textbf{(A–8–6)}$$

The value of equation (A–8–6) is

$$\begin{array}{ll} 0 & \text{if } m \neq 0 \\ \pi & \text{if } m = n > 0 \\ 2\pi & \text{if } m = n = 0 \end{array}$$

If $f(x)$ of equation (A–8–1) can be integrated between $-\pi$ and π, then equation (A–8–1) becomes

$$\int_{-\pi}^{\pi} f(x)dx = \frac{1}{2}a_0\int_{-\pi}^{\pi} dx + \sum_{n=0}^{\infty} a_n \int_{-\pi}^{\pi} \cos nx\, dx + \sum_{n=0}^{\infty} b_n \int_{-\pi}^{\pi} \sin nx\, dx = \pi a_0 \qquad \textbf{(A–8–7)}$$

and then

$$a_0 = \frac{1}{\pi}\int_{-\pi}^{\pi} f(x)dx \qquad \textbf{(A–8–8)}$$

Also, if we revise equation (A–8–1) to read

$$f(x)\cos kx\, dx = \frac{1}{2}a_0 \cos kx\, dx + \sum_{n=0}^{\infty} a_n \cos nx \cos kx\, dx + \sum_{n=0}^{\infty} b_n \sin nx \cos kx\, dx \qquad \textbf{(A–8–9)}$$

where k is an integer. If equation (A–8–9) is integrated over the domain $-\pi$ to π,

$$\int_{-\pi}^{\pi} f(x)\cos kx\,dx = \frac{1}{2}a_0\int_{-\pi}^{\pi}\cos kx\,dx + \sum_{n=0}^{\infty} a_n\int_{-\pi}^{\pi}\cos nx\cos kx\,dx + \sum_{n=0}^{\infty} b_n\int_{-\pi}^{\pi}\sin nx\cos kx\,dx \quad \textbf{(A–8–10)}$$

then the first term on the right side of the equation is zero for $k > 0$, by equation (A–8–2) the second term on the right is nonzero only for $k = n$, and from equation (A–8–4) the third term is zero for all k. The result is

$$\int_{-\pi}^{\pi} f(x)\cos kx\,dx = a_k\pi \quad \textbf{(A–8–11)}$$

for all k. Also, the coefficient is then

$$a_k = \frac{1}{\pi}\int_{-\pi}^{\pi} f(x)\cos kx\,dx \quad \textbf{(A–8–12)}$$

for any k. Notice that for $k = 0$ the first coefficient, c_0, is obtained in agreement with equation (A–8–8). In a similar manner, by multiplying equation (A–8–1) by the term sin $kx\,dx$ and integrating over the domain from $-\pi$ to π, we obtain

$$b_k = \int_{-\pi}^{\pi} f(x)\sin kx\,dx \quad \textbf{(A–8–13)}$$

The series represented by equation (A–8–1) is called a *Fourier Series*, and the coefficients a_k and b_k defined by equations (A–8–12) and (A–8–13) are called the *Fourier Coefficients*.

A–9 ERROR FUNCTION

The error function occurs as a solution to transient conduction heat-transfer problems and the error function of argument z, erf(z), is defined as

$$\text{erf}(z) = \frac{2}{\sqrt{\pi}}\int_0^z e^{-t^2}\,dt \quad \textbf{(A–9–1)}$$

where t is a variable of integration. The error function can be evaluated from the series expansion

$$\text{erf}(z) = \frac{2}{\sqrt{\pi}}\sum_{n=0}^{\infty}(-1)^n\left(\frac{1}{n!(2n+1)}\right)z^{2n+1} \quad \textbf{(A–9–2)}$$

or

$$\text{erf}(z) = \frac{2e^{-z^2}}{\sqrt{\pi}}\sum_{n=0}^{\infty}2^n\left(\frac{1}{1\cdot 3\cdot 5\cdot 7\cdot\ \cdots\ \cdot(2n+1)}\right)z^{2n+1} \quad \textbf{(A–9–3)}$$

The derivative of the error function is

$$\frac{d}{dz}\text{erf}(z) = \frac{2}{\sqrt{\pi}}e^{-z^2} \quad \textbf{(A–9–4)}$$

Values for the error function are given in table A–9.

Table A–9 Error Function

x	**erf** x	x	**erf** x	x	**erf** x	x	**erf** x
0.00	0.00000	0.56	0.57162	1.10	0.88020	1.64	0.97962
0.02	0.02256	0.58	0.58792	1.12	0.88079	1.66	0.98110
0.04	0.04511	0.60	0.60386	1.14	0.89308	1.68	0.98249
0.06	0.06762	0.62	0.61941	1.16	0.89910	1.70	0.98379
0.08	0.09008	0.64	0.63459	1.18	0.90484	1.72	0.98500
0.10	0.11246	0.66	0.64938	1.20	0.91031	1.74	0.98613
0.14	0.15695	0.68	0.66278	1.22	0.91553	1.76	0.98719
0.16	0.17901	0.70	0.67780	1.24	0.92050	1.78	0.98817
0.18	0.20094	0.72	0.69143	1.26	0.92524	1.80	0.98909
0.20	0.22270	0.74	0.70468	1.28	0.92973	1.82	0.98994
0.22	0.24430	0.76	0.71754	1.30	0.93401	1.84	0.99074
0.24	0.26570	0.78	0.73001	1.32	0.93806	1.86	0.99147
0.26	0.28690	0.80	0.74210	1.34	0.94191	1.88	0.99216
0.28	0.20788	0.82	0.75381	1.36	0.94556	1.90	0.99279
0.30	0.32863	0.84	0.76514	1.38	0.94902	1.92	0.99338
0.32	0.34913	0.86	0.77610	1.40	0.95228	1.94	0.99392
0.34	0.36936	0.88	0.78669	1.42	0.95538	1.96	0.99443
0.36	0.38933	0.90	0.79691	1.44	0.95830	1.98	0.99489
0.38	0.30901	0.92	0.80677	1.46	0.96105	2.00	0.995322
0.40	0.42839	0.94	0.81627	1.48	0.96365	2.10	0.997020
0.42	0.44749	0.96	0.82542	1.50	0.96610	2.20	0.998137
0.44	0.46622	0.98	0.83423	1.52	0.96841	2.40	0.999311
0.46	0.48466	1.00	0.84270	1.54	0.97059	2.60	0.999764
0.48	0.50275	1.02	0.85084	1.56	0.97263	2.80	0.999925
0.50	0.52050	1.04	0.85865	1.58	0.97455	3.00	0.999978
0.52	0.53790	1.06	0.86614	1.60	0.97636	3.20	0.999994
0.54	0.55494	1.08	0.87333	1.62	0.97804	3.60	1.000000

A–10 BESSEL FUNCTIONS

The Bessel Functions are special functions that are solutions to the second-order homogeneous differential equation

$$x^2\frac{d^2w(x)}{dx^2} + x\frac{dw(x)}{dx} + (x^2 - \nu^2)w(x) = 0 \qquad \textbf{(A–10–1)}$$

where ν is a real number. The solutions to this equation are the *Bessel Function of the First Kind,* $J_\nu(x)$, the *Bessel Function of the Second Kind,* $Y_\nu(x)$, also called the *Weber Function*, and the *Bessel Functions of the Third Kind,* $H_\nu^1(x)$ and $H_\nu^2(x)$, also called the *Hankel Functions.* The Weber and Hankel Functions are related to the Bessel Function by the definitions

$$Y_\nu(x) = \frac{J_\nu(x)\cos \nu\pi - J_\nu(x)}{\sin\nu\pi} \qquad \textbf{(A–10–2)}$$

$$H_\nu^1(x) = J_\nu(x) + iY_\nu(x) \qquad \textbf{(A–10–3)}$$

$$H_\nu^2(x) = J_\nu(x) - iY_\nu(x) \qquad \textbf{(A–10–4)}$$

The Bessel Function of the First Kind is designated by an order, *n,* so that J_0 is the Bessel Function of zeroth order, J_1 is of the first order and so on. For *n*th order Bessel Function,

$$J_\nu(x) = \left(\frac{x}{2}\right)^\nu \sum_{n=0}^{\infty} \frac{(-x^2/4)^k}{k!\Gamma(\nu + k + 1)} \qquad \textbf{(A–10–5)}$$

where $\Gamma(\nu + k + 1)$ is the *Gamma Function*, defined for an argument x as

$$\Gamma(x) = \int_0^{\infty} t^{x-1} e^{-t} dt \qquad \text{for } x \text{ positive real} \qquad \textbf{(A–10–6)}$$

Frequently the Gamma Function is evaluated from Euler's Formula,

$$\Gamma(x) = \lim_{x \to \infty} \frac{n! n^x}{x(x+1)(x+2)\cdots(x+n)} \qquad \text{for } x \neq 0, -1, -2, -3, \ldots \qquad \textbf{(A–10–7)}$$

where n is a positive integer. Other approximation formulas and relations involving the Gamma Function may be found in *Handbook of Mathematical Functions* [2].

The Gamma Function and its inverse over a limited range and domain are shown in figure A–10–1 and in figure A–10–2 are shown the Bessel and Weber Functions of the zeroth and first order over a limited range and domain. The Bessel Functions of the first kind and of zeroth and first order are tabulated for arguments from 0 to 14.9 in table A–10.

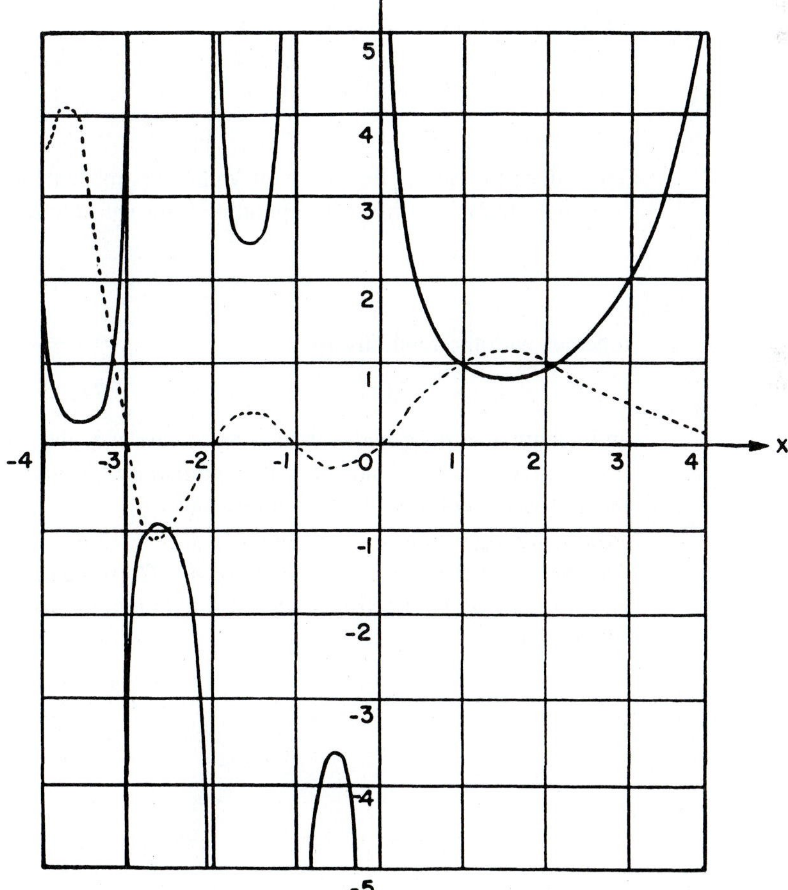

Figure A–10–1 The gamma function.

———, $y = \Gamma(x)$, – – – –, $y = 1/\Gamma(x)$

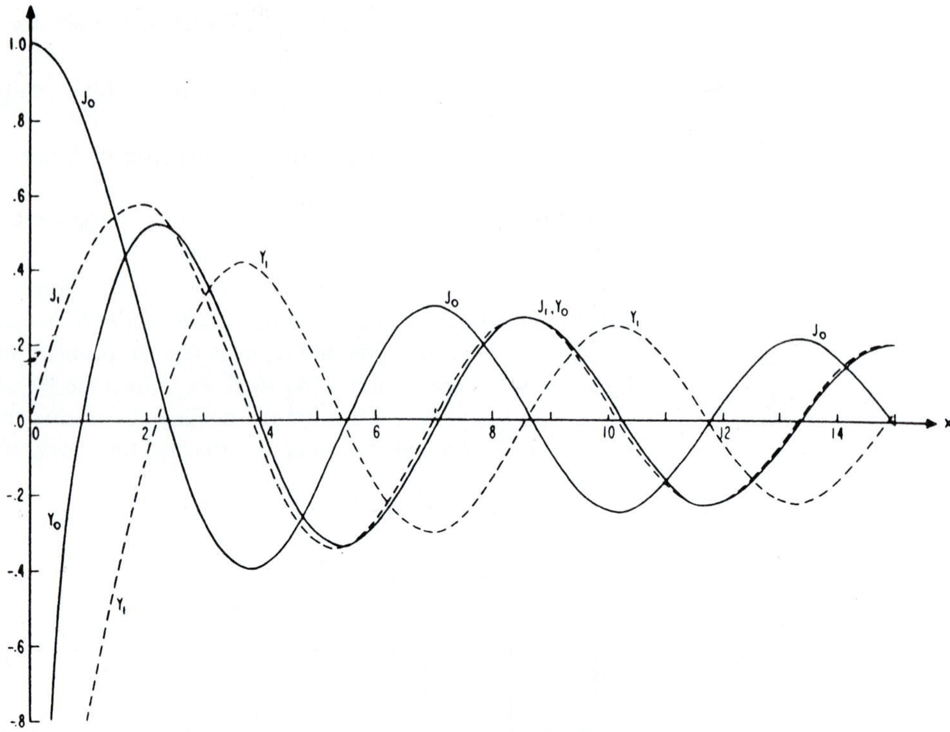

Figure A–10–2 The Bessel Functions, $J_0(x)$, $J_1(x)$, $Y_0(x)$, and $Y_1(x)$.

The values for Bessel Functions of arguments other than those listed in table A–10 may be approximated by linear interpolation; however, the use of the recurrence relation

$$J_{n-1}(x) = \frac{2n}{x}J_n(x) - J_{n-1}(x) \tag{A–10–8}$$

together with the relationship

$$J_0(x) + 2J_2(x) + 2J_4(x) + 2J_6(x) + \cdots = 1.0 \tag{A–10–9}$$

and quantitative information regarding values of higher-order Bessel Functions from, for instance, Abramowitz and Stegun's *Handbook of Mathematical Functions* [2] provides a method of predicting a more accurate value.

Table A–10 Bessel Functions

x	$J_0(x)$	$J_1(x)$	x	$J_0(x)$	$J_1(x)$	x	$J_0(x)$	$J_1(x)$
0.0	1.0000	0.0000	5.0	−0.1776	−0.3276	10.0	−0.2459	0.0435
0.1	0.9975	0.0499	5.1	−0.1443	−0.3371	10.1	−0.2490	0.0184
0.2	0.9900	0.0995	5.2	−0.1103	−0.3432	10.2	−0.2496	−0.0066
0.3	0.9776	0.1483	5.3	−0.0758	−0.3460	10.3	−0.2477	−0.0313
0.4	0.9604	0.1960	5.4	−0.0412	−0.3453	10.4	−0.2434	−0.0555
0.5	0.9385	0.2423	5.5	−0.0068	−0.3414	10.5	−0.2366	−0.0789
0.6	0.9120	0.3867	5.6	0.0270	−0.3343	10.6	−0.2276	−0.1012
0.7	0.8812	0.3290	5.7	0.0599	−0.3241	10.7	−0.2164	−0.1224
0.8	0.8463	0.3688	5.8	0.0917	−0.3110	10.8	−0.2032	−0.1422
0.9	0.8075	0.4059	5.9	0.1220	−0.2951	10.9	−0.1881	−0.1603
1.0	0.7652	0.4401	6.0	0.1506	−0.2767	11.0	−0.1712	−0.1768
1.1	0.7196	0.4709	6.1	0.1773	−0.2559	11.1	−0.1528	−0.1913
1.2	0.6711	0.4983	6.2	0.2017	−0.2329	11.2	−0.1330	−0.2039
1.3	0.6201	0.5220	6.3	0.2238	−0.2081	11.3	−0.1121	−0.2143
1.4	0.5669	0.5419	6.4	0.2433	−0.1816	11.4	−0.0902	−0.2225
1.5	0.5118	0.5579	6.5	0.2601	−0.1538	11.5	−0.0677	−0.2284
1.6	0.4554	0.5699	6.6	0.2740	−0.1250	11.6	−0.0446	−0.2320
1.7	0.3980	0.5778	6.7	0.2851	−0.0953	11.7	−0.0213	−0.2333
1.8	0.3400	0.5815	6.8	0.2931	−0.0652	11.8	0.0020	−0.2323
1.9	0.2818	0.5812	6.9	0.3981	−0.0349	11.9	0.0250	−0.2290
2.0	0.2239	0.5767	7.0	0.3001	−0.0047	12.0	0.0477	−0.2234
2.1	0.1666	0.5683	7.1	0.2991	0.0252	12.1	0.0697	−0.2157
2.2	0.1104	0.5560	7.2	0.2951	0.0543	12.2	0.0908	−0.2060
2.3	0.0555	0.5399	7.3	0.2882	0.0826	12.3	0.1108	−0.1943
2.4	0.0025	0.5202	7.4	0.2786	0.1096	12.4	0.1296	−0.1807
2.5	−0.0484	0.4971	7.5	0.2663	0.1352	12.5	0.1469	−0.1655
2.6	−0.0968	0.4708	7.6	0.2516	0.1592	12.6	0.1626	−0.1487
2.7	−0.1424	0.4416	7.7	0.2346	0.1813	12.7	0.1766	−0.1307
2.8	−0.1850	0.4097	7.8	0.2154	0.2014	12.8	0.1887	−0.1114
2.9	−0.2243	0.3754	7.9	0.1944	0.2192	12.9	0.1988	−0.0912
3.0	−0.2601	0.3391	8.0	0.1717	0.2346	13.0	0.2069	−0.0703
3.1	−0.2921	0.3009	8.1	0.1475	0.2476	13.1	0.2129	−0.0489
3.2	−0.3202	0.2613	8.2	0.1222	0.2580	13.2	0.2167	−0.0271
3.3	−0.3443	0.2207	8.3	0.0960	0.2657	13.3	0.2183	−0.0052
3.4	−0.3643	0.1792	8.4	0.0692	0.2708	13.4	0.2177	0.0166
3.5	−0.3801	0.1374	8.5	0.0419	0.2731	13.5	0.2150	0.0380
3.6	−0.3918	0.0955	8.6	0.0146	0.2728	13.6	0.2101	0.0590
3.7	−0.3992	0.0538	8.7	−0.0125	0.2697	13.7	0.2032	0.0791
3.8	−0.4026	0.0128	8.8	−0.0392	0.2641	13.8	0.1943	0.0984
3.9	−0.4018	−0.0272	8.9	−0.0653	0.2559	13.9	0.1836	0.1165
4.0	−0.3971	−0.0660	9.0	−0.0903	0.2453	14.0	0.1711	0.1334
4.1	−0.3887	−0.1033	9.1	−0.1142	0.2324	14.1	0.1570	0.1488
4.2	−0.3766	−0.1386	9.2	−0.1367	0.2174	14.2	0.1414	0.1626
4.3	−0.3610	−0.1719	9.3	−0.1577	0.2004	14.3	0.1245	0.1747
4.4	−0.3423	−0.2028	9.4	−0.1768	0.1816	14.4	0.1065	0.1850
4.5	−0.3205	−0.2311	9.5	−0.1939	0.1613	14.5	0.0875	0.1934
4.6	−0.2961	−0.2566	9.6	−0.2090	0.1395	14.6	0.0679	0.1999
4.7	−0.2693	−0.2791	9.7	−0.2218	0.1166	14.7	0.0476	0.2043
4.8	−0.2404	−0.2985	9.8	−0.2323	0.0928	14.8	0.0271	0.2066
4.9	−0.2097	−0.3147	9.9	−0.2403	0.0684	14.9	0.0064	0.2069

A–11 ROOTS OF SOME TRANSCENDENTAL EQUATIONS AND ASSOCIATED COEFFICIENTS FOR TRANSIENT CONDUCTION IN INFINITE PLATES, INFINITE CYLINDERS, AND SPHERES

Table A–11–1 First Five Roots of $\xi_n \tan \xi_n = hL/\kappa = Bi$, ξ_n in Radians

	First Five Coefficients $C_n = \dfrac{4 \sin \xi_n}{2\xi_n + \sin 2\xi_n}$									
hL/k	ξ_1	C_1	ξ_2	C_2	ξ_3	C_3	ξ_4	C_4	ξ_5	C_5
0.001	0.0316	1.0002	3.1419	−0.0002	6.2833	0.0000	9.4249	0.0000	12.5665	0.0000
0.002	0.0447	1.0003	3.1422	−0.0004	6.2835	0.0001	9.4250	0.0000	12.5665	0.0000
0.005	0.0707	1.0008	3.1432	−0.0010	6.2840	0.0003	9.4253	−0.0001	12.5668	0.0001
0.01	0.0998	1.0017	3.1448	−0.0020	6.2848	0.0005	9.4258	−0.0002	12.5672	0.0001
0.02	0.141	1.0033	3.1479	−0.0040	6.2864	0.0010	9.4269	−0.0005	12.5680	0.0003
0.05	0.2217	1.0082	3.1574	−0.0100	6.2911	0.0025	9.4301	−0.0011	12.5704	0.0006
0.10	0.3111	1.0160	3.1731	−0.0197	6.2991	0.0050	9.4354	−0.0022	12.5743	0.0013
0.20	0.4328	1.0311	3.2039	−0.0381	6.3148	0.0100	9.4459	−0.0045	12.5823	0.0025
0.50	0.6533	1.0701	3.2923	−0.0873	6.3616	0.0243	9.4775	−0.0111	12.6060	0.0063
1.0	0.8603	1.1191	3.4256	−0.1517	6.4783	0.0466	9.5293	−0.0217	12.6453	0.0123
2.0	1.0769	1.1795	3.6436	−0.2367	6.5783	0.0848	9.6296	−0.0414	12.7223	0.0241
5.0	1.3138	1.2402	4.0336	−0.3442	6.9096	0.1588	9.8938	−0.0876	12.9352	0.0543
10.0	1.4289	1.2620	4.3058	−0.3934	7.2281	0.2104	10.2003	−0.1309	13.2142	0.0881
20.0	1.4961	1.2699	4.4915	−0.4147	7.4954	0.2394	10.5117	−0.1621	13.5420	0.1182
50.0	1.5400	1.2727	4.6202	−0.4227	7.7012	0.2518	10.7832	−0.1779	13.8666	0.1364
100.0	1.5552	1.2731	4.6658	−0.4240	7.7764	0.2539	10.8871	−0.1808	13.9981	0.1401

Table A–11–2 First Five Roots of $\beta_n = J_1(\beta_n) = \frac{hr_0}{\kappa} J_0(\beta_n)$

First Five Coefficients $C_n = \frac{2J_1(\beta_n)}{\beta_n[J_0^2(\beta_n) + J_1^2(\beta_n)]}$

hr_0/κ	β_n	C_1	β_2	C_2	β_3	C_3	β_4	C_4	β_5	C_5
0.001	0.045	1.0002	3.832	0.0000	7.016	0.00027	10.174	−0.0003	13.324	0.0001
0.002	0.063	1.0004	3.832	−0.0004	7.016	0.00027	10.174	−0.0003	13.324	0.0001
0.005	0.100	1.0013	3.833	−0.0013	7.016	0.00027	10.174	−0.0003	13.324	0.0001
0.01	0.141	1.0025	3.834	−0.0032	7.017	0.00121	10.175	−0.0011	13.324	0.0001
0.02	0.200	1.0050	3.837	−0.0068	7.019	0.00310	10.176	−0.0019	13.325	0.0008
0.05	0.314	1.0124	3.845	−0.0167	7.023	0.00689	10.178	−0.0035	13.327	0.0022
0.10	0.442	1.0246	3.858	−0.0333	7.030	0.01351	10.183	−0.0074	13.331	0.0049
0.20	0.617	1.0483	3.884	−0.0658	7.044	0.02672	10.193	−0.0152	13.338	0.0097
0.50	0.941	1.1143	3.959	−0.1573	7.086	0.06591	10.222	−0.0378	13.361	0.0254
1.0	1.256	1.2071	4.079	−0.2904	7.156	0.12908	10.271	−0.0756	13.398	0.0506
2.0	1.599	1.3384	4.292	−0.4936	7.288	0.24201	10.366	−0.1467	13.472	0.0999
5.0	1.990	1.5029	4.713	−0.7978	7.617	0.48396	10.622	−0.3221	13.679	0.2305
10.0	2.180	1.5677	5.034	−0.9580	7.957	0.67492	10.936	−0.5006	13.959	0.3851
20.0	2.288	1.5919	5.257	−1.0321	8.253	0.78977	11.268	−0.6388	14.296	0.5296
50.0	2.357	1.6002	5.411	−1.0574	8.484	0.84018	11.562	−0.7119	14.643	0.6233
100.0	2.381	1.6015	5.465	−1.0567	8.568	0.84916	11.675	−0.7255	14.783	0.6420

Table A–11–3 First Four Roots of $\zeta_n \cos \zeta_n = \left(1 - \frac{hr_0}{\kappa}\right) \sin \zeta_n$

First Four Coefficients $C_n = \frac{4(\sin \zeta_n - \zeta_n \cos \zeta_n)}{2\zeta_n - \sin 2\zeta_n}$

hr_0/k	ζ_1	C_1	ζ_2	C_2	ζ_3	C_3	ζ_4	C_4
0.001	0.055	1.0003	4.494	−0.0012	7.725	−0.0005	10.904	0.0002
0.002	0.077	1.0006	4.494	−0.0012	7.725	−0.0005	10.904	0.0002
0.005	0.122	1.0015	4.495	−0.0033	7.726	0.0015	10.905	−0.0018
0.010	0.173	1.0030	4.496	−0.0053	7.727	0.0035	10.905	−0.0018
0.020	0.242	1.0060	4.498	−0.0094	7.728	0.0055	10.906	−0.0038
0.050	0.385	1.0149	4.504	−0.0216	7.732	0.0135	10.908	−0.0078
0.10	0.542	1.0298	4.516	−0.0461	7.739	0.0277	10.913	−0.0178
0.20	0.759	1.0592	4.538	−0.0904	7.761	0.0717	10.923	−0.0378
0.50	1.166	1.1441	4.604	−0.2207	7.790	0.1294	10.950	−0.0917
1.00	1.571	1.2732	4.712	−0.4237	7.854	0.2547	10.996	−0.1827
2.00	2.030	1.4793	4.913	−0.7670	7.979	0.4905	11.085	−0.3555
5.00	2.569	1.7870	5.354	−1.3733	8.303	0.9311	11.335	−0.8098
10.0	2.836	1.9249	5.717	−1.7379	8.659	1.5144	11.658	−1.3108
20.0	2.986	1.9781	5.978	−1.9162	8.983	1.8247	12.003	−1.7154
50.0	3.079	1.9962	6.158	−1.9850	9.239	1.9669	12.320	−1.9419
100.0	3.110	1.9990	6.220	−1.9960	9.331	1.9914	12.441	−1.9846

A–12 LENNARD–JONES INTERMOLECULAR FORCE PARAMETERS AND MASS DIFFUSION FUNCTION

Table A–12–1

Substance	Molecular Weight, MW	Effective Diameter of Gas, σ (Å)	$\frac{\varepsilon}{\kappa}$ (K)	Substance	Molecular Weight, MW	Effective Diameter of Gas, σ (Å)	$\frac{\varepsilon}{\kappa}$ (K)
H_2	2.016	2.915	38.0	C_2H_4	28.05	4.232	205.00
He	4.003	2.576	10.2	C_2H_6	30.07	4.418	230.00
Ne	20.183	2.789	35.7	C_3H_6	42.08	—	—
Ar	39.944	3.418	124.0	C_3H_8	44.09	5.061	254.00
Kr	83.80	3.498	225.0	n-C_4H_{10}	58.12	—	—
Xe	131.3	4.055	229.0	i-C_4H_{10}	58.12	5.341	313.00
Air	28.97	3.617	97.0	n-C_5H_{12}	72.15	5.769	345.00
N_2	28.02	3.681	91.5	n-C_6H_{14}	86.17	5.909	413.00
O_2	32.00	3.433	113.0	n-C_7H_{16}	100.20	—	—
O_3	48.00	—	—	n-C_8H_{18}	114.22	7.451	320.00
CO	28.01	3.590	110.0	n-C_9H_{20}	128.25	—	—
CO_2	44.01	3.996	190.0	Cyclohexane	84.16	6.093	324.00
NO	30.01	3.470	119.0	C_6H_6	78.11	5.270	440.00
N_2O	44.02	3.879	220.0	CH_4	16.04	3.822	137.00
SO_2	64.07	4.290	252.0	CH_3Cl	50.49	3.375	855.00
F_2	38.00	3.653	112.0	CH_2Cl_2	84.94	4.759	406.00
Cl_2	70.91	4.115	357.0	$CHCl_3$	119.39	5.430	327.00
Br_2	159.83	4.268	520.0	CCl_4	153.84	5.881	327.00
I_2	253.82	4.982	550.0	C_2N_2	52.04	4.380	339.00
CH_4	16.04	3.822	137.0	COS	60.08	4.130	335.00
C_2H_2	26.04	4.221	185.0	CS_2	76.14	4.438	488.00

Table A–12–2 Mass Diffusivity Parameters

$\frac{\kappa T}{\varepsilon}$ or $\frac{\kappa T}{\varepsilon_{AB}}$	Ω_{AB}	$\frac{\kappa T}{\varepsilon_{AB}}$ or $\frac{\kappa T}{\varepsilon_{AB}}$	Ω_{AB}	$\frac{\kappa T}{\varepsilon}$ or $\frac{\kappa T}{\varepsilon_{AB}}$	Ω_{AB}
0.30	2.662	1.65	1.153	4.00	0.8836
0.35	2.476	1.70	1.140	4.10	0.8788
0.40	2.318	1.75	1.128	4.20	0.8740
0.45	2.184	1.80	1.1116	4.30	0.8694
0.50	2.066	1.85	1.105	4.40	0.8652
0.55	1.966	1.90	1.094	4.50	0.8610
0.60	1.877	1.95	1.084	4.60	0.8568
0.65	1.798	2.00	1.075	4.70	0.8530
0.70	1.729	2.10	1.057	4.80	0.8492
0.75	1.667	2.20	1.041	4.90	0.8456
0.80	1.612	2.30	1.026	5.00	0.8422
0.85	1.562	2.40	1.012	6.00	0.8124
0.90	1.517	2.50	0.9996	7.00	0.7896
0.95	1.476	2.60	0.9878	8.00	0.7712
1.00	1.439	2.70	0.9770	9.00	0.7556
1.0	1.406	2.80	0.9672	10.00	0.7424
1.1	1.375	2.90	0.9576	20.00	0.6640
1.1	1.346	3.00	0.9490	30.00	0.6232
1.2	1.320	3.10	0.9406	40.00	0.5960
1.2	1.296	3.20	0.9328	50.00	0.5756
1.3	1.273	3.30	0.9256	60.00	0.5596
1.3	1.253	3.40	0.9186	70.00	0.5464
1.4	1.233	3.50	0.9120	80.00	0.5352
1.45	1.215	3.60	0.9058	90.00	0.5256
1.50	1.198	3.70	0.8998	100.00	0.5170
1.55	1.182	3.80	0.8942		
1.60	1.167	3.90	0.8888		

REFERENCES

1. Burington, R. S., *Handbook of Mathematical Tables and Formulas,* 4th Edition, McGraw-Hill, New York, 1965

2. Abramowitz, M., and I. A. Stegun, *Handbook of Mathematical Functions,* National Bureau of Standards, Applied Mathematics Series, 55, U.S. Government Printing Office, Washington, DC, 1964.

CONVERSION FACTORS AND PROPERTY TABLES

Table B–1

Universal Constants

R_u, Universal gas constant = 1544 ft · lbf/lb · mol · °R
= 8.31 J/g · mol · K
σ, Stefan–Boltzmann constant = 0.174×10^{-8} Btu/hr · ft^2 · °R^4
= 5.67×10^{-8} W/m^2 · K^4
N Avogadro's number = 6.02×10^{23} atoms/g · mol

Conversion Factors

Force

*2000 pounds-force (lbf) = 1 ton-force
2.248×10^{-6} lbf = 1 dyne
8.89644×10^{8} dynes = 1 ton
0.2248 lbf = 1 newton (N)
10^5 dynes = 1 N

Mass

*32.174 pound-mass (lbm) = 1 slug
*16 ounces (oz) = 1 lbm
*2000 lbm = 1 ton
*7000 grains = 1 lbm
2.2046 lbm = 1 kilogram (kg)
0.45359 kg = 1 lbm
*1000 grams (g) = 1 kg
*1000 kg = 1 tonne (metric ton)

Length

*12 inches (in.) = 1 foot (ft)
*3 ft = 1 yard (yd)
*5280 ft = 1 mile (mi)
3.2808 ft = 1 meter (m)
39.37 in. = 1 m
*0.3048 m = 1 ft
*0.0254 m = 1 in.
*2.54 centimeters (cm) = 1 in.
*1000 m = 1 kilometer (km)
*100 cm = 1 m
*1.609 km = 1 mi

Pressure

*1 lbf/in^2 = 144 lbf/ft^2 (psf)
0.0361 psi = 1 in water (at 39.2°F, sea level)
0.491 psi = 1 in mercury (at 39.2°F, sea level)
0.433 psi = 1 ft water (at 39.2°F, sea level)
14.696 psi = 1 atmosphere (atm)
14.504 psi = 1 bar
0.14054 psi = 1 kilopascal (kPa)
1.01325 bar = 1 atm
0.06895 bar = 1 psi
6.895 kPa = 1 psi
*100 kPa = 1 bar
*1000 pascal (Pa) = 1 kPa = 1 kN/m^2
98.0638 Pa = 1 cm water (at 4°C, sea level)
133.3224 Pa = 1 mm mercury (at 4°C, sea level)

Volume

231 in^3 = 1 gallon (gal) (U.S. liquid)
0.1337 ft^3 = 1 gal (U.S. liquid)
3.785 liters (L) = 1 gal (U.S. liquid)
3785.4 cm^3 = 1 gal (U.S. liquid)
0.001 m^3 = 1 L
1000 cm^3 = 1 L
61.02 in^3 = 1 L
0.0353 ft^3 = 1 L

Energy

*1 joule (J) = 1 newton-meter (N · m)
*1 J = 1 watt-second (W · s)
1054 J = 1 British thermal unit (Btu)
1.356 J = 1 foot-pound (ft · lbf)
*4.1868 J = 1 calorie
*1 dyne · cm = 1 erg
*1×10^7 ergs = 1 J
0.2388 calorie = 1 J
252 calories = 1 Btu
3414 Btu = 1 kilowatt-hour (kW · hr)
3.414 Btu = 1 W · hr
2545 Btu = 1 horsepower · hour (hp · hr)
0.9488 Btu = 1 kilojoule (kJ)
778 ft · lbf = 1 Btu
0.737 ft · lbf = 1 J
0.4299 Btu/lbm = 1 kJ/kg
334.5 ft · lbf/lbm = 1 kJ/kg
*2.326 kJ/kg = 1 Btu/lbm
*4.1868 kJ/kg · K = 1 Btu/lbm · °R
0.2388 Btu/lbm · °R = 1 kJ/kg · K

Power

*1 joule/second (J/s) = 1 watt (W)
*1000 W = 1 kW
0.746 kW = 1 horsepower (hp)
1.34 hp = 1 kW
1.41 hp = 1 Btu/s
550 ft · lbf/s = 1 hp
33,000 ft · lbf/min = 1 hp
1.054 kW = 1 Btu/s
3414 Btu/hr = 1 kW
2545 Btu/hr = 1 hp
12,000 Btu/hr = 1 ton (refrigeration)
3.515 kW = 1 ton (refrigeration)

Heat transfer

0.1422 W/m · K = 1 Btu · in./hr · ft^2 · °R
6.9348 Btu · in./hr · ft^2 · °R = 1 W/m · K
519.22 W/m · ft = 1 Btu · in./s · ft^2 · °R
1.73 W/m · K = 1 Btu/hr · ft · °R
0.5779 Btu/hr · ft · °R = 1 W/m · K
1.926 Btu · in./s · ft^2 · °R = 1 kW/m · K
5.678 W/m^2 · K = 1 Btu/hr · ft^2 · °R
0.176 Btu/hr · ft^2 · °R = 1 W/m^2 · K
1 R-value = 1 hr · ft^2 · °F/Btu
1 R-value = 0.176 m^2 · °C/W

*Asterisk indicates exact conversion.

Table B–2 Thermal Properties of Selected Nonmetallic Solids

Material	Temperature, K	Density, kg/m^3	Specific Heat, kJ/kg · K	Thermal Conductivity, κ (W/m · K)	Viscosity, μ (g/m · s)	Isobaric Compressivity $\beta_p \times 10^6$ (K^{-1})
Asbestos	300	2100	1.047	0.156	—	—
Asphalt	300	2115	0.920	0.062	—	600
Brick						
Common	300	1600	0.840	0.70	—	—
Fire	1000	2640	0.960	1.50	—	—
Masonry	300	1700	0.837	0.66	—	—
Carbon, Graphite	300–400	1900–2300	0.711	1.60	—	—
Cardboard	300	700	1.510	0.170	—	—
Clay	300	1460	0.880	1.30	—	24.3
Concrete	300	2300			—	30.0
Reinf.	300	2400	0.840	1.60	—	—
Cork board	300	160	1.90	0.045	—	—
Ebonite	300	—	1.382	—	—	—
Felt Board	300	—	—	0.035	—	—
Food						
Apple	300	*	3.60	0.418	—	—
Beet	300	*	3.77	0.601	—	—
Milk	300	*	3.85	0.550	—	—
Turkey (Cooked)	300	*	3.31	0.500	—	—
Glass	300	2500	0.750	1.40	—	210
Glass Fiber	300	220	—	0.035		—
Glass Wool	300	40	0.70	0.038	—	—
Gypsum board (plaster)	300	817	1.084	0.107	—	—
Ice	273	920	2.04	2.20	—	—
Mineral wool, batts	300	—	—	0.039	—	—
Paper	300	73	1.510	0.096	—	—
Peat	300	275	1.700	0.109	—	—
Polystyrene, Styrofoam	300	37	—	0.029	—	21
Polyurethane foam, rigid	300	40	—	0.023	—	40.5
Rock wool	300	160	—	0.040	—	—
Rubber						
Foam	300	70	—	0.032	—	—
Hard	300	1190	2.01	0.160	—	180
Sand, dry	300	1515	0.80	0.30	—	—
Sawdust	300	215	—	0.059	—	—
Saw shavings	300	215	—	0.059	—	—
Snow						
Loose	273	110	—	0.05	—	—
Packed	273	500	—	0.19	—	—
Soil	300	2050	1.840	0.52	—	—
Sulfur	300	1920	0.736	0.270	—	126
Stone						
Granite	300	2630	0.775	2.79	—	24
Limestone	300	2320	0.810	2.15	—	18
Marble	300	2720	0.920	3.00	—	45
Sandstone	300	2150	0.745	2.90	—	30
Teflon	300	2200	1.00	0.35	—	—

Table B–2 (continued) Thermal Properties of Selected Nonmetallic Solids

Material	Temperature, K	Density, kg/m^3	Specific Heat, kJ/kg · K	Thermal Conductivity, κ (W/m · K)	Viscosity, μ (g/m · s)	Isobaric Compressivity $\beta_p \times 10^6$ (K^{-1})
Tissue, human						
Fat	300	*	—	0.20	—	—
Muscle	300	*	—	0.41	—	—
Skin	300	*	—	0.37	—	—
Wood						
Balsa	300	140	—	0.055	—	—
Pine	300	640	2.805	0.150	—	—
Oak	300	545	2.385	0.170	—	—
Plywood	300	550	1.200	0.120	—	—
Wool	300	90	—	0.036	—	—
Vermiculite, loose	300	80	0.835	0.068	—	—

Thermal Properties of Selected Metallic Solids

Material	Temperature, K	Density, kg/m^3	Specific Heat, kJ/kg · K	Thermal Conductivity, κ (W/m · K)	Viscosity, μ (g/m · s)	Isobaric Compressivity $\beta_p \times 10^6$ (K^{-1})
Aluminum	300	2,700	0.902	236	—	69.9
Brass	300	8,520	0.382	114	—	—
Copper	300	8,900	0.385	400	—	50.1
Iron						
Wrought	300	7,850	0.460	51	—	≈36
Cast	300	7,270	0.420	39	—	≈38
Lead	300	11,340	0.129	36	—	85.5
Silicon	300	2,330	0.691	159	—	7.62
Steel						
Carbon	300	7,800	0.473	43	—	36
Carbon	300	7,854	0.434	60.5	—	35
Stainless	300	7,900	0.477	14	—	33
40% Nickel stainless	300	8,169	0.460	10	—	24
Zinc	300	7,133	0.384	116	—	190

*Density of food varies due to porosity, but a value of 1000 kg/m^3 is probably close.

Table B–3 Thermal Properties of Selected Liquids at Atmospheric Pressure or Where Otherwise Noted

Material	Temperature, T (K)	Density, ρ (kg/m^3)	Specific Heat, c_p (kJ/kg · K)	Thermal Conductivity κ(W/m · K)	Viscosity, μ (g/m · s)	Isobaric Compressivity $\beta_p \times 10^3$ (K^{-1})
Acetone	300	800	2.155	0.160	—	—
Ammonia	240	682	4.46	0.562	0.28	
	300	602	4.80	0.520	0.22	1.93
	460	460	2.45	0.308	0.07	
Benzene	300	895	1.81	0.159	0.65	1.237
Bismuth	600	9,997	0.145	16.4	1.61	0.122
Engine oil (unused)	300	886	1.89	0.145	712.0	0.638
Ethanol	273	790		0.175	1.80	
	300	788	2.395	0.168	1.20	1.120
	373	786		0.149	0.32	
Ethyl glycol	300	1,116	2.385	0.257	21.4	0.6375
Glycerin	300	1,260	2.39	0.287	1490	0.505
Lead	700	10,476	0.157	17.4	2.15	0.120
Lithium	600	498	4.190	48.0	0.57	0.174
Methanol	293	791.4	≈2.63	0.202	0.540	1.120
	373	790	≈2.63	0.184	0.210	
	473	788	≈2.63	0.162	0.072	
Mercury	300	13,546	0.138	—	1.60	0.182
	600	12,816	0.134	14.2	0.84	0.186
Potassium	600	766	0.783	44.0	0.23	0.280
Propanol	300	803.5	2.477	0.154	1.72	
R-12	243	1486.6	0.884	0.0882	0.3594	—
	283 at 0.423 MPa	1362.8	0.948	0.0735	0.232	
	333 at 1.523 MPa	1168.2	1.123	0.0557	0.143	
R-22	223	1409.1	1.092	0.1141	0.280	—
	273 at 0.5 MPa	1284.8	1.161	0.096	0.230	
	323 at 2.0 MPa	1084.7	1.370	0.073	0.130	
R-123	300	1456	1.001	0.0758	0.480	—
	323 at 0.213 MPa	1397.9	1.067	0.0694	0.323	
	353 at 0.490 MPa	1311.4	1.119	0.0607	0.240	
R-134a	247	1374.3	1.268	0.1054	0.4064	—
	273 at 0.300 MPa	1294.0	1.335	0.0934	0.2874	
	313 at 1.02 MPa	1146.5	1.500	0.0750	0.1782	
Sodium	600	871	1.296	76.0	0.320	0.275
	1000	770	1.257	58.0	0.180	0.390
	1500	730	1.260	54.0	0.120	0.450
Water	300	997	4.180	0.608	0.857	

Table B–4 Thermal Properties of Selected Gases at Atmospheric Pressure

Material	Temperature, T (K)	Density, ρ (kg/m^3)	Specific Heat, C_p (kJ/kg · K)	Thermal Conductivity, κ (W/m · K)	Viscosity, μ (g/m · s)	Isobaric Compressivity $\beta_P \times 10^3$ (K^{-1})
Air	250	1.4136	1.007	0.022	0.016	4.00
	300	1.178	1.007	0.026	0.0181	3.33
	350	1.0097	1.010	0.030	0.0204	2.857
	400	0.8835	1.014	0.034	0.023	2.50
Ammonia	200	1.033	2.000	0.0132	0.00689	
	300	0.689	2.096	0.0250	0.01020	
	400	0.540	2.290	0.0374	0.01390	
Carbon dioxide	300	1.81	0.844	0.017	0.0148	
Carbon monoxide	300	1.16	1.040	0.025	0.0182	
Ethanol	300	1.864	1.430	0.0220	0.0090	
	400	1.398	1.765	0.0264	0.0118	
	500	1.120	2.052	0.0381	0.0141	
Helium	200	0.2462	5.193	0.1151	0.0150	
	300	0.1625	5.193	0.152	0.0199	
	400	0.1219	5.193	0.187	0.0243	
	600	0.0836	5.193	0.220	0.0283	
	1000	00.04879	5.193	0.354	0.0446	
Hydrogen	300	0.0817	14.302	0.190	0.0090	
Methane	300	0.655	2.227	0.0341	0.0134	
Methanol	300	1.296	1.378	0.0230	0.0083	
	400	0.972	1.619	0.0261	0.0131	
	500	0.778	1.864	0.0375	0.0165	
Nitrogen	300	1.15	1.040	0.026	0.0176	
Oxygen	300	1.31	0.911	0.027	0.0200	
Propane	300		1.696	0.0182	0.0083	
Propanol	300	2.40	1.917	0.0254	0.0106	
R-12	243	6.289	0.572	0.0070	0.0105	
	283 at 0.423 MPa	24.19	0.658	0.0091	0.0125	
	333 at 1.523 MPa	89.04	0.880	0.0125	0.0160	
R-22	233	4.705	0.606	0.0069	—	
	273 at 0.5 MPa	21.26	0.744	0.0095	0.0118	
	323 at 2.00 MPa	86.13	1.129	0.0124	0.0127	
R-123	300	6.468	0.719	0.0758	—	
	323 at 0.213 MPa	13.02	0.760	0.0119	0.0117	
	353 at 0.490 MPa	29.21	0.819	0.0139	0.0127	
R-134a	243	5.259	0.784	0.0952	0.00979	
	273 at 0.3 MPa	14.42	0.883	0.0118	0.01094	
	313 at 1.02 MPa	50.03	1.120	0.0156	0.0131	

Table B–5 Approximate Values for Emissivities of Surfaces

Surface Material	Temperature, K	Emissivity, ε
Aluminum		
Polished	300–900	0.04–0.06
Oxidized	400–800	0.20–0.33
Brass		
Polished	350	0.09
Dull	300–600	0.22
Copper		
Polished	300–500	0.04–0.05
Oxidized	600–1000	0.5–0.8
Iron		
Cast	300	0.44
Wrought	300–500	0.28
Rusted	300	0.61
Lead		
Polished	300–500	0.06–0.08
Rough	300	0.43
Stainless steel		
Polished	300–1000	0.17–0.3
Oxidized	600–1000	0.5–0.6
Steel		
Polished	300–500	0.08–0.14
Oxidized	300	0.05
Asphalt pavement	300	0.90
Brick (common)	300	0.94
Cloth	300	0.75–0.90
Concrete	300	0.90
Window glass	300	0.92
Ice	273	0.95–0.99
Paint		
Aluminum	300	0.45
Black	300	0.88
Oil (all colors)	300	0.94
White	300	0.90
Red primer	300	0.93
White paper	300	0.93
Sand	300	0.90
Human skin	300	0.95
Snow	273	0.80–0.96
Soil (earth)	300	0.93–0.96
Water	273–373	0.95
Wood		
Beech	300	0.94
Oak	300	0.90

Table B–6 Thermal Properties of Steam
Saturated Liquid–Saturated Vapor

Temperature, °C	Pressure, kPa	Saturated Liquid Density, ρ_f (kg/m^3)	Saturated Vapor Volume v_g (m^3/kg)	hn$_{fg}$, (kJ/kg)	Saturated Liquid Specific Heat, $c_{p,f}$ (kJ/kg · K)	Saturated Vapor Specific Heat $c_{p,g}$ (kJ/kg · K)
0.01	0.6113	1000.00	206.131	2501.35	4.217	1.854
10	1.2276	1000.00	106.376	2477.75	4.191	1.859
20	2.339	998.00	57.7897	2454.12	4.182	1.868
40	7.384	992.06	19.5229	2406.72	4.179	1.887
60	19.941	983.28	7.67071	2358.48	4.186	1.916
80	47.39	971.82	3.40715	2308.77	4.198	1.962
100	101.3	957.85	1.67290	2257.03	4.217	2.029
120	198.5	943.40	0.89186	2202.61	4.238	2.136
140	361.3	925.93	0.50885	2144.75	4.288	2.263
160	617.8	907.44	0.30706	2082.55	4.34	2.40
200	1,553.8	865.05	0.12736	1940.75	4.49	2.84
220	2,317.8	840.34	0.08619	1858.51	4.60	3.15
240	3,344.2	813.67	0.05976	1766.50	4.77	3.54
260	4,688.6	783.70	0.04220	1662.54	4.97	4.06
300	8,581.0	712.25	0.02167	1404.93	5.69	5.89
360	18,651	528.54	0.00694	720.52	14.9	25.4
374.1	22,089	316.96	0.00315	0	—	—

Table B–6 (continued)

Temperature, °C	Pressure, kPa	Saturated Liquid Thermal Conductance, κ_f (W/m · K)	Saturated Vapor Thermal Conductance, κ_g (W/m · K)	Saturated Liquid Viscosity, μ_f (g/m · s)	Saturated Vapor Viscosity, μ_g (g/m · s)	Saturated Liquid Compression, $\beta_p \times 10^3$ (K^{-1})
0.01	0.6113	0.561	0.018	1.750	0.008	−0.068
10	1.2276	0.577	0.019	1.392	0.008	0.087
20	2.339	0.594	0.019	1.044	0.009	0.194
40	7.384	0.627	0.021	0.676	0.010	0.385
60	19.941	0.658	0.022	0.478	0.010	0.523
80	47.39	0.668	0.023	0.358	0.011	0.641
100	101.3	0.679	0.025	0.279	0.012	0.750
120	198.5	0.680	0.027	0.231	0.013	0.857
140	361.3	0.681	0.029	0.196	0.014	1.000
160	617.8	0.680	0.031	0.170	0.014	—
200	1,553.8	0.663	0.037	0.134	0.016	—
220	2,317.8	0.651	0.041	0.122	0.017	—
240	3,344.2	0.638	0.046	0.110	0.017	—
260	4,688.6	0.625	0.052	0.101	0.018	—
300	8,581.0	0.548	0.077	0.088	0.020	—
360	18,651	0.398	0.133	0.065	0.030	—
374.1	22,089	0.238	0.238	0.045	0.045	—

Saturated Solid–Saturated Vapor

Temperature, °C	Pressure, kPa	Saturated Solid Density, ρ_s (kg/m^3)	Saturated Vapor Volume, v_g (m^3/kg)	Heat of Fusion, hn_{sg}, (kJ/kg)	Saturated Solid Specific Heat, $c_{p,s}$ (kJ/kg · K)	Saturated Vapor Specific Heat, $c_{p,g}$ (kJ/kg · K)	Saturated Solid Thermal Conductivity,* κ_s (W/m · K)	Saturated Vapor Thermal Conductivity,* κ_g (W/m · K)
0.01	0.6113	916.76	206.153	2834.7	2.04	1.860	2.250	0.018
−4	0.4376	916.76	283.799	2835.7	2.00	1.859	—	—
−8	0.3102	917.94	394.414	2836.6	1.97	1.859	—	—
−16	0.1510	919.03	785.907	2837.9	1.90	1.857	—	—
−20	0.10355	919.62	1128.11	2838.4	1.87	1.857	—	—
−30	0.03810	920.98	2945.23	2839.0	1.79	1.856	—	—
−40	0.01286	922.42	8366.40	2838.9	1.72	1.855	—	—

* The thermal conductivities of the saturated solid and vapor for water between 0.01°C and −40°C are nearly constant.

Table B–7 Electrical Resistance of Copper Wire

Gauge Number	Diameters (mils)	Area (circular mils)	Ohms per 1,000 ft
0000	460.0	211,600	0.04901
000	409.6	167,800	0.06180
00	364.8	133,100	0.07793
0	324.9	105,500	0.09827
1	289.3	83,690	0.1239
2	257.6	66,370	0.1563
3	229.4	52,640	0.1970
4	204.3	41,740	0.2485
5	181.9	33,100	0.3133
6	162.0	26,250	0.3951
7	144.3	20,820	0.4982
8	128.5	16,510	0.6282
9	114.4	13,090	0.7921
10	101.9	10,380	0.9989
12	80.81	6,530	1.588
14	64.08	4,107	2.525
16	50.82	2,583	4.016
18	40.30	1,624	6.385
20	31.96	1,022	10.15
22	25.35	642.4	16.14
24	20.10	404.0	25.67
26	15.94	254.1	40.81
28	12.64	159.8	64.90
30	10.03	100.5	103.2
32	7.950	63.21	164.1
34	6.305	39.75	260.9
36	5.000	25.00	414.8
38	3.965	15.72	659.6
40	3.145	9.888	1,049

Data from National Bureau of Standards for Bare Cable.

Note: 1 mil = 0.001 in., 1 circular mil = 1 square mil = 1×10^{-6} in^2.

Table B–8 Constants and Coefficients for $\kappa = \kappa_{T0} + \alpha(T - T_0)$

Substance, Solid	Temperature Range, K	κ_{T0}, W/m · K	a, W/m · K^2
Pure copper	273–1273	393	−0.019
Mild steel	273–1273	62.6	−0.015
Brass, 60–40	273–1273	102	0.050
Brass, 90–10	273–1273	100	0.0478
Sodium	200–373	142	−0.01

Table B–9 Surface Tension for Some Selected Materials at Atmospheric Pressure and Interfacing with Air

Material	Temperature, °C	σ_{ST}, mN/m	Material	Temperature, °C	σ_{ST}, mN/m
Acetone, C_3H_6O	0	26.2	Methanol, CH_3OH	0	24.5
	20	23.7		50	20.1
	50	19.9		100	15.7
				200	4.5
Aluminum, Al	800	850	*n*-Octane, C_8H_{18}	−40	27.50
	1600	725		0	23.70
				40	19.78
				100	14.13
				200	5.99
Benzene, C_6H_6	10	30.24	Pentane, C_5H_{12}	−20	20.5
	30	27.49		0	18.2
	60	23.66		40	13.80
Carbon tetrachloride, CCl_4	0	29.5	Propanol	0	25.5
	40	24.5		50	21.2
	100	17.3		100	17.2
	260	1.4			
Copper, Cu	1150	1370	R-12, CCl_2F_2	−60	20.30
	1550	1265		−30	15.91
				−16	13.94
Ethanol, C_2H_5OH	0	24.4	R-22, $CHClF_2$	−100	28.37
	50	20.1		−50	19.70
	100	15.5		−80	16.42
	200	4.3			
Glycerin, $C_2H_5(OH)_3$	25	63.0	R-123, CCl_2HCF_3	−40	23.49
				0	18.43
				50	12.47
				100	6.97
Iron, Fe	1550	1850	R-134a, CF_3CH_2F	−50	19.22
	1650	1790		0	11.71
	1700	1760		50	5.01
				100	0.03
Lead, Pb	350	440	Tin, Sn	300	530
	750	420		450	535
	1000	388		600	525
Mercury, Hg	25	486	Water, H_2O	0	75.50
	200	429		20	72.88
	300	402		40	69.48
				60	66.07
				80	62.69
				100	58.91

CHARTS

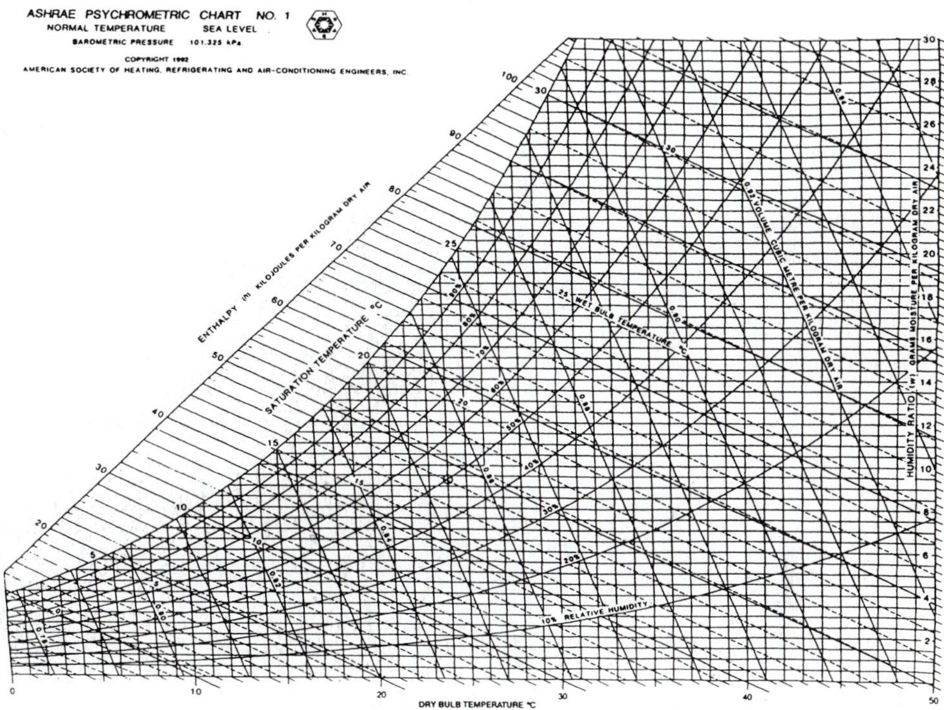

Chart C–1 Psychrometric chart. (From the American Society of Heating, Refrigerating, and Air Conditioning Engineers, *ASHRAE Handbook, 1997 Fundamentals,* ASHRAE, Atlanta, GA, 1997)

Chart C–2 Pressure–enthalpy diagram for water. (From the American Society of Heating, Refrigerating, and Air Conditioning Engineers, *ASHRAE Handbook, 1997 Fundamentals,* ASHRAE, Atlanta, GA, 1997)

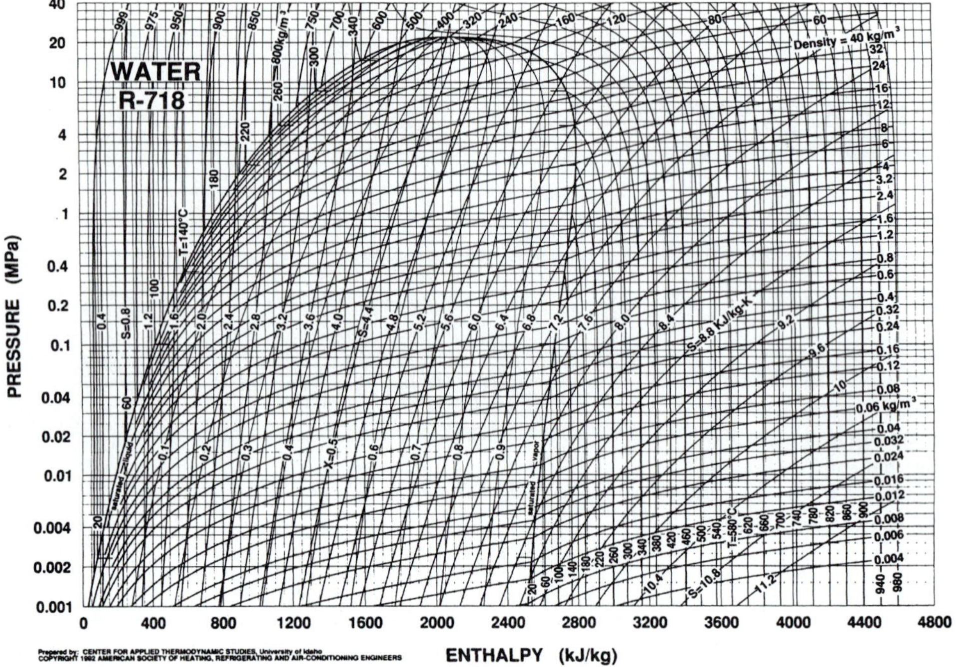

Chart C–3 Pressure–enthalpy diagram for refrigerant-123. (From the American Society of Heating, Refrigerating, and Air Conditioning Engineers, *ASHRAE Handbook, 1997 Fundamentals,* ASHRAE, Atlanta, GA, 1997)

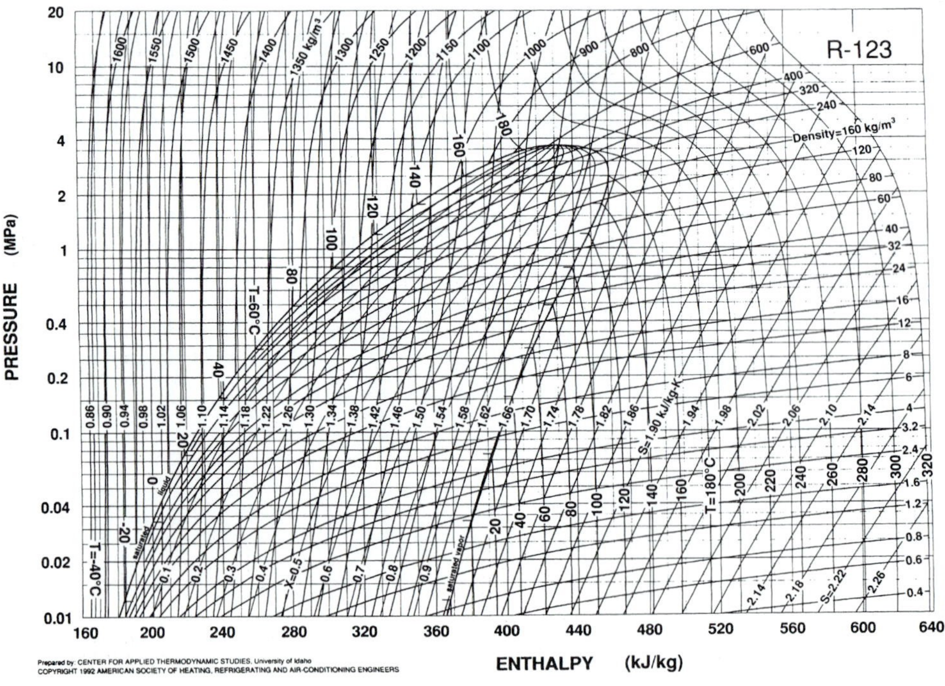

Chart C–4 Pressure–enthalpy diagram for refrigerant-134a. (From the American Society of Heating, Refrigerating, and Air Conditioning Engineers, *ASHRAE Handbook, 1997 Fundamentals,* ASHRAE, Atlanta, GA, 1997)

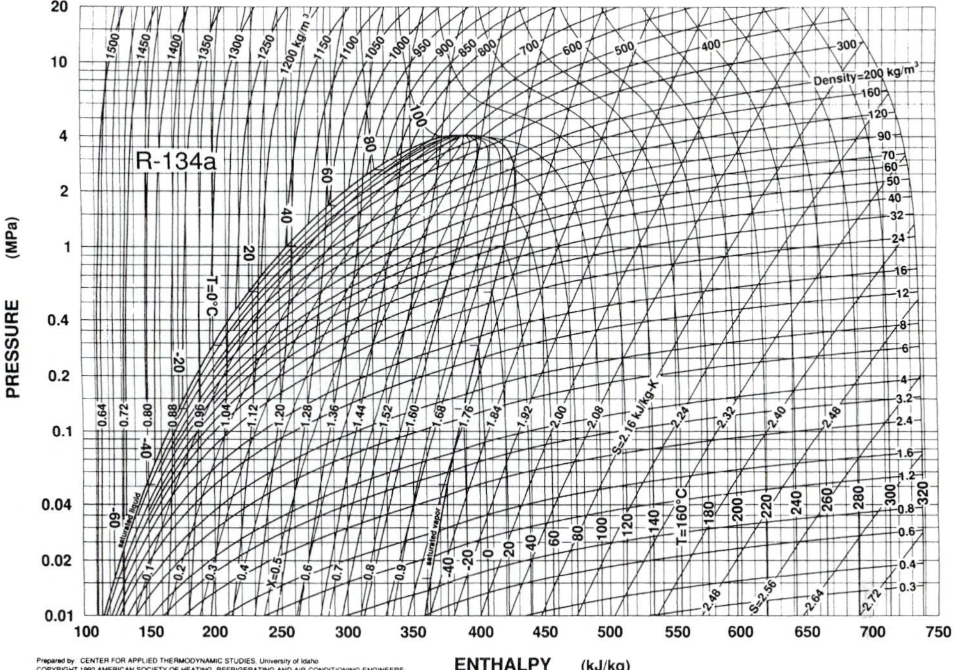

Chart C–5 Pressure–enthalpy diagram for refrigerant-12. (From the American Society of Heating, Refrigerating, and Air Conditioning Engineers, *ASHRAE Handbook, 1997 Fundamentals,* ASHRAE, Atlanta, GA, 1997)

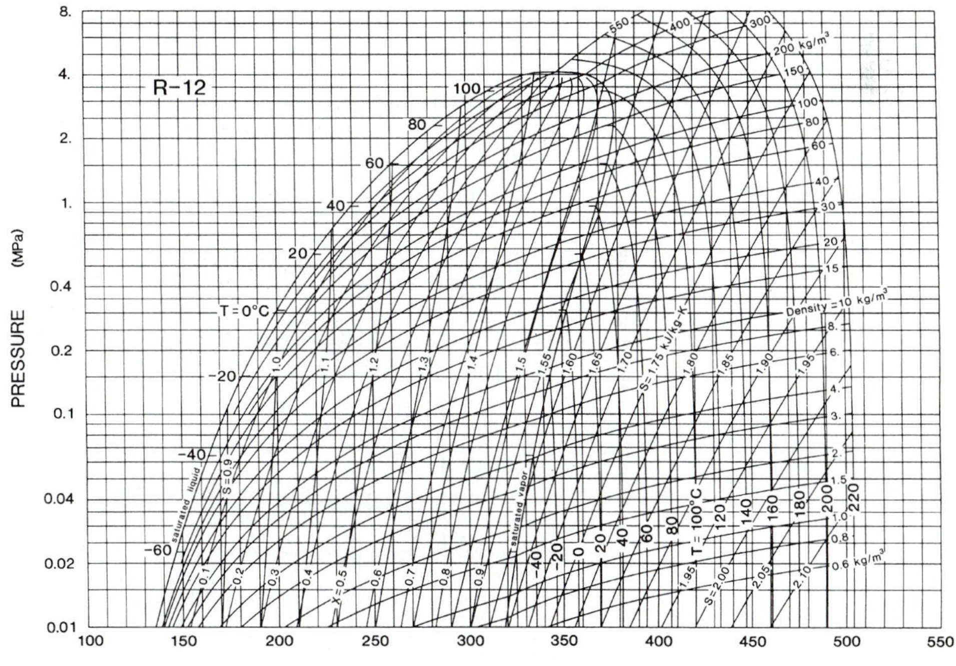

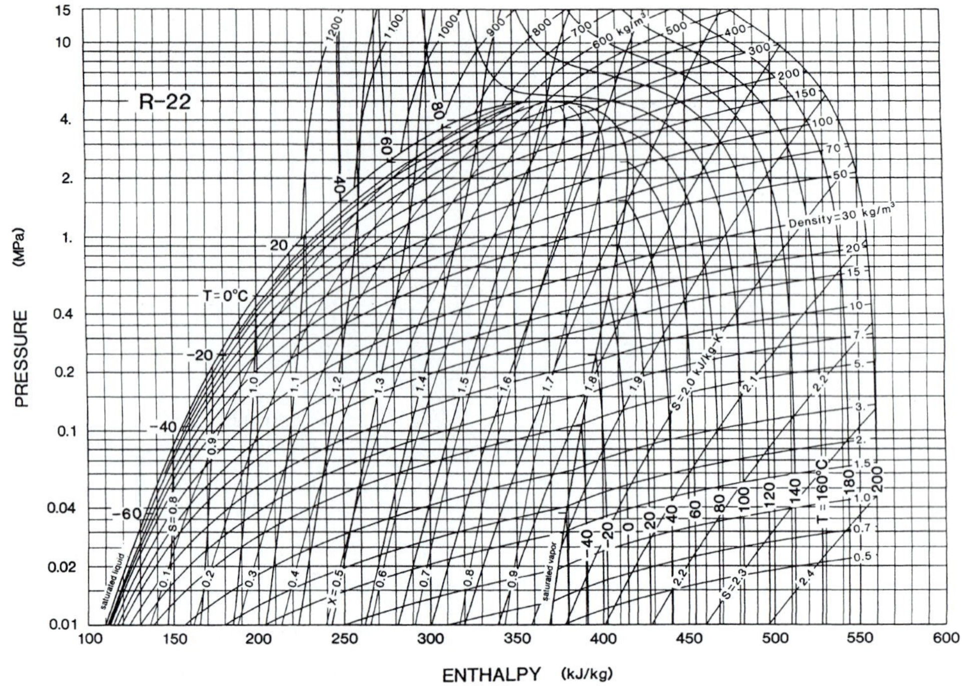

Chart C–6 Pressure–enthalpy diagram for refrigerant-22. (From the American Society of Heating, Refrigerating, and Air Conditioning Engineers, *ASHRAE Handbook, 1997 Fundamentals* ASHRAE, Atlanta, GA, 1997)

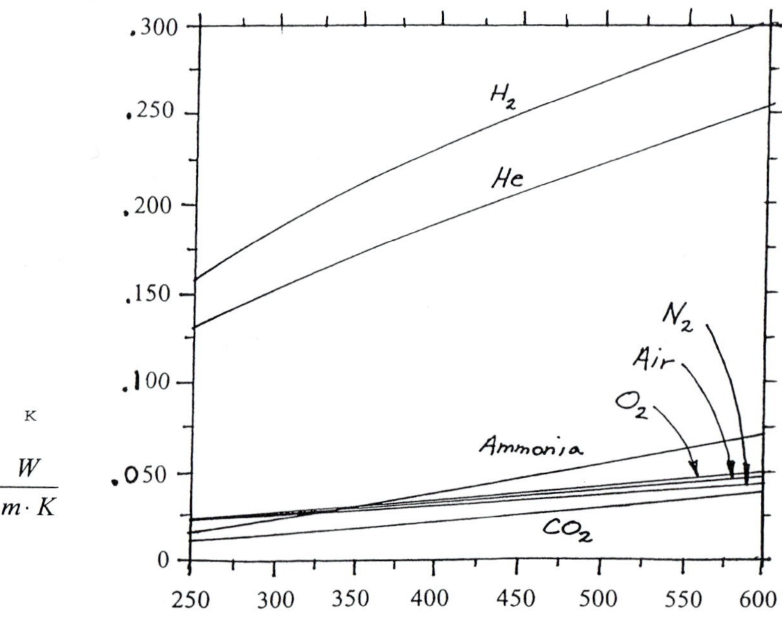

Chart C–7 Thermal conductivity of selected saturated vapors. (Abstracted from *Thermodynamic and Transport Properties* by C. Borgnakke and R. E. Sonntag, 1997, with permission of John Wiley & Sons, Inc.)

Chart C–8 Thermal conductivity of selected saturated liquids. (Abstracted from *Thermodynamic and Transport Properties* by C. Borgnakke and R. E. Sonntag, 1997, with permission of John Wiley & Sons, Inc.)

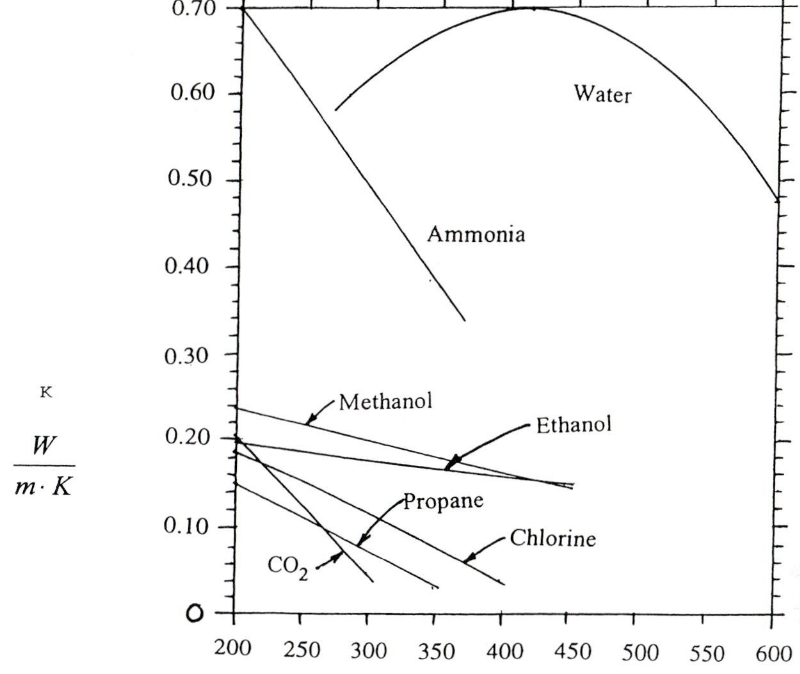

Chart C–9 Viscosity of Selected Saturated Vapors. (Abstracted from *Thermodynamic and Transport Properties* by C. Borgnakke and R. E. Sonntag, 1997, with permission of John Wiley & Sons, Inc.)

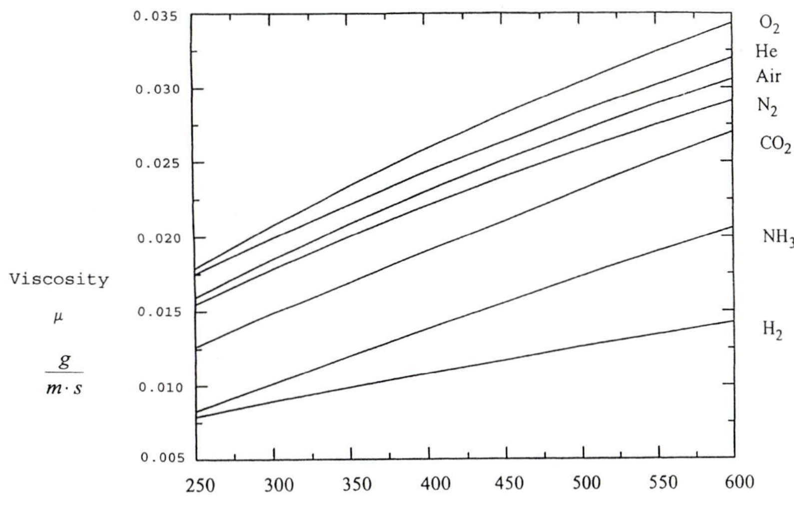

Chart C–10 Viscosity of Selected Saturated Liquids. (Abstracted from *Thermodynamic and Transport Properties* by C. Borgnakke and R. E. Sonntag, 1997, with permission of John Wiley & Sons, Inc.)

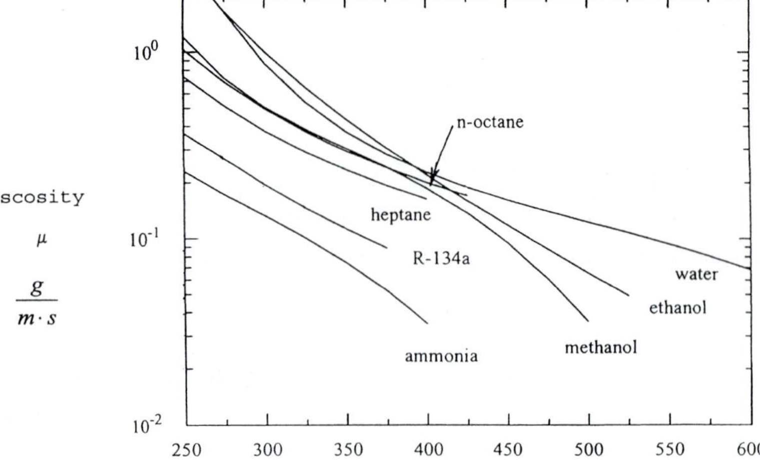

ANSWERS TO SELECTED PROBLEMS

Chapter One

Section 1.3

1–1 $\Delta u = -1165.05$ Btu

1–3 $p_2 = 135$ psia

1–5 $Q = 75.24$ MJ

1–7 $Q = 11.669$ *MW*

1–9 $p_{v1} = 0.64$ psia, $p_{v2} = 0.26$ psia, $\Delta\omega = 0.18$ lbm/lbmda, $T_{dp} = 61$ °F

Section 1.4

1–13 $\dot{q}_A = 323.6$ Btu/hr · ft^2

1–15 $\dot{q}_A = 108{,}183$ Btu/hr · ft^2

1–17 $\dot{q}_l = 19{,}352.2$ Btu/hr · ft

1–19 $\dot{q}_A = 11.85$ Btu/hr · ft^2

1–21 $\dot{q}_{net,A} = 79.2$ W/m^2

1–23 $\dot{Q} = 100.5$ W

Section 1.5

1–25 $\dot{m} = 5.35 \times 10^{-6}$ g/hr

1–27 $\omega = 0.0152$ kg/kgda

Section 1.6

1–29 Two (2) boundary conditions

1–31 Four (4) boundary conditions

Chapter Two

Section 2.1

2–1 $\kappa = 0.152$ W/m-K@300 K

2–5 $\kappa = 227 - 0.02(T - 32°F)$ Btu/hr-ft-°F

2–7 $\kappa = 0.0021$ W/cm-K

Section 2.2

2–11 $\Delta V = (r \sin\theta \cdot \Delta\phi)(\Delta r)(r\Delta\theta)$
$= r^2 \sin\phi \cdot \Delta r \cdot \Delta\theta \cdot \Delta\phi$

Section 2.3

2–13 $\dot{q}_A = 450$ kW/m^2

2–15 $R_V = 2.72$ hr · ft^2 · °F/Btu

2–17 $\dot{q}_l = 12.91$ W/m
$T(r) = 167.07 - 51.385 \ln r$

2–19 $T_i = 80.34°$ C
$R_{TL} = 0.1787$ m · K/W

2–21 $\dot{Q} = 335.9$ Btu/hr

$$T(r) = 1653.4 - \frac{19138.5}{r} \ (r \textit{ in inches})$$

2–23 $R_{TL} = 0.184$ m · K/W
$\dot{q}_l = 162.7$ W/m

2–25 $$T = \left(T_0^{1.001} - \dot{q}_A x\left(\frac{1.001}{a}\right)\right)^{1.001} \quad \textit{for } k = aT^{0.001}$$

$$T = T_0 - \dot{q}_A\frac{x}{a} \quad \textit{for } k = a$$

Section 2.4

2–27 $\dot{Q}_y = 70.67$ Btu/hr

$$T(x, 3\,ft) = 43.68 \sin\frac{\pi x}{L}$$

2–31 $A_0 = A_1 = 0 \quad A_n = 0 \quad$ *for n odd*

$$A_n = \frac{2}{\pi(n+1)} + \frac{2}{\pi(n-1)} \quad \textit{for n even}$$

Section 2.5

2–37 $\dot{Q} = 16.06$ W/m

2–39 $\dot{Q} = 94.6$ W

2–41 $\dot{Q} = 1248.7$ W

2–43 $\dot{q}_l = 8.7725$ W/m

2–45 $\dot{q} \approx 40.1$ W/m

2–47 $\dot{Q} = 21.78$ kW

Section 2.6

2–49 $T_1 = 149.086°F$
$T_7 = 204.22°F$
$T_{12} = 253.53°F$
$T_{17} = 82.177°F$
$T_{20} = 269.157°F$

2–53 $T_1 = 86.375°C$
$T_3 = 83.363°C$
$T_5 = 42.101°C$
$T_8 = 42.161°C$

Section 2.7

2–59 $T_L = 96.148°C$, $\dot{Q}_{fin} = 179.6W$, $\eta_{fin} = 69.8\%$
2–65 $\eta_{fin} = 44\%$, $\dot{Q}_{fin} = 318\ w/fin$
2–67 $\eta_{fin} = 82\%$, $\dot{Q}_{fin} = 291.96$ Btu/hr
2–69 $Q = 70.706\ W$
2–71 $\Delta T = 11°C$
2–73 $\Delta T_{TC} = 3.134°F$
2–75 $\dot{q}_l = 38.17$ W/m, $r_{OC} = 5.2$ cm
2–79 $T(x) = -6147.5x^2 + 1280.5x + 252.5°F$

Chapter Three

Section 3.1

3–1 $\alpha = 9.53 \times 10^{-8}\ m^2/s$
3–3 $\dfrac{dT}{dt} = 265.5°F/s$

Section 3.2

3–7 $t_c = 347.3$ s
3–9 y = 62.7 ft
3–11 $t_c = 4.58$ s, t = 17.9 s

Section 3.3

3–15 erf(0.67) ≈ 0.656
3–17 $\dot{q}_A = 138.565$ W/m², $t = 42\ hrs$
3–21 $\dfrac{\partial T}{\partial t} = 43.9\ °F/hr$
3–23 t = 66 min
3–25 $T_C = 104.34°F$
$T_S = 103.85°F$
3–29 t = 32.56 min
3–31 t = 6.875 min
3–33 t = 1.254 hr

Section 3.4

3–35 $T_C = 396.96°F$,
$T_{corner} \approx 400°F$
3–37 t = 30.78 s
3–39 t = 68 min

Section 3.5

3–41 cooling rate ≈ 30 °C/s
3–43 August 11 (hottest)
February 10 (coldest)
3–45 $T_{max} = 16°\ C$ *August* 29,
$T_{min} = -16°\ C$ *March* 1
3–47 583.5°F

Section 3.6

3–49 Δt ≤ 147.1 s

Chapter Four

Section 4.2

4–1 wp = 8.38 in.
4–3 $\mu = 0.89 \times 10^{-3}\ N \cdot s/m^2$, $v = 8.9 \times 10^{-7}\ m^2/s$
4–5 f = 0.022

Section 4.3

4–7 $Re_L = 160$ (Laminar Flow)
4–9 Pr = 0.00409

Section 4.4

4–11 $\delta_x\ (meters) = 0.00218\sqrt{x}$
4–13 $\tau = \dfrac{-3\mu u_\infty}{9.28}\sqrt{\dfrac{u_\infty}{vx}}$
4–15 **a)** $h_{ave} = 11.928$ W/m² · K (*Laminar*),
$h_{ave} = 19.13$ W/m² · K (*full length*)
b) $h_{ave} = 12{,}003$ W/m² · K (Water)
4–17 $\dot{q}_l = 337.32\ kW/m$

Section 4.5

4–21 h = 87 W/m²-K
4–23 $\dot{Q} = 108\ kW$
4–25 $h_{ave} = 42.3$ W/m² · K

Section 4.6

4–27 $Nu_D = 0.914$
4–29 $\dot{q}_A = 1471$ Btu/hr · ft²
4–31 **a)** $T_0 = 17.69°C$
b) $6.9\ m \le x_F \le 41.4\ m$
4–33 $\dot{q}_A = 20.03$ Btu/hr · ft²
4–35 $T_0 = 88.7°C$

Section 4.7

4–37 $T_{tube} = 5°C$
4–39 $T_0 = 86.9°F$, $\Delta p = 0.0052\ psi$, $p_2 = 14.495\ psi$
4–41 $T_{max} = 135.6°C$

Chapter Five

Section 5.1

5–1 F = 75 lbf

Section 5.2

5–3 $\dot{Q} = 761$ Btu/hr
5–5 $\delta(y) = 3.51 \times 10^{-3}(y)^{1/4}$
5–7 $\dot{Q} = 6644\ W$
5–9 **a)** $L_{crit} = 0.997\ m$
b) $T_w = 36°C$

Section 5.3

5–11 $\dot{q}_A = 415.5\ W/m^2$
5–13 $\dot{Q}_{roof} = 95{,}636\ W$, $\dot{Q}_{total} = 104{,}636\ W$

Section 5.4

5–15 **a)** $\dot{q}_l = 2.83$ W/m

5–17 $\dot{Q} = 1774$ kW

Section 5.5

5–21 $\dot{Q} = 29{,}448$ Btu/hr

Section 5.6/5.7

5–23 $\dot{q}_A = 165.2$ Btu/hr · ft^2

5–25 $\dot{Q} = 4421$ kW

Chapter Six

Section 6.1

6–1 $\lambda_{low} = 0.28\ \mu m$
$\lambda_{high} = 0.49\ \mu m$

6–3 $\omega_{ii} = 0.503$ str
$\omega_i = \omega_{iii} = 0.17777$ str

Section 6.2

6–5 **a)** $\dot{E}_B = 55.4$ W/m^2
b) $\dot{E}_B = 133{,}987$ W/m^2
c) $\dot{E}_B = 6343$ kW/m^2

Section 6.3

6–7 $\tau_r = 0.6$

6–9 $\rho_r = 0.2$

Section 6.4

6–13 $F_{T\text{-}2} = 0.098$

6–15 D = 2.3 cm

6–17 $F_{21} = 0.31$, $F_{23} = 0.69$, $F_{12} = 0.31 \times 10^{-6}$,
$F_{31} = F_{32} \approx 0$

6–19 $F_{31} = 0.32$, $F_{21} = 0.25$, $F_{12} = 0.25$, $F_{13} \approx 0$

Section 6.5

6–21 $\rho_{r,ave} = 0.3805$ $\tau_{r,ave} = 0.4545$ $T = 500\ K$
$\rho_{r,ave} = 0.3085$ $\tau_{r,ave} = 0.6446$ $T = 850\ K$

6–23 **a)** $\dot{E}_{full} = 2.025$ kW/m^2 **b)** $\dot{E}_{1\text{-}2} = 0.375$ kW/m^2

Chapter Seven

Section 7.1

7–1 J = 961.4 Btu/hr-ft^2

Section 7.2

7–3 $\dot{q}_l = 1018$ W/m *for* $\varepsilon_r = 0.28$
$\dot{q}_l = 1677.5$ W/m *for* $\varepsilon_r = 0.8$
$\dot{q}_l = 667.9$ W/m *for* $\varepsilon_r = 0.15$

7–5 $\dot{q}_{net} = 283.2$ W/m^2 (*out*)

Section 7.3

7–7 **b)** $\dot{Q} = 3.421$ kW

7–9 $\dot{Q}_{net} = 113.8$ *W (both sides)*

7–11 $\dot{Q}_2 = 610.7$ Btu/hr

7–13 $\dot{Q} = 7.16 \times 10^7$ Btu/hr

Section 7.4

7–15 $K_\lambda = 11.983\ miles^{-1}$

7–17 $\varepsilon_r = 0.374$

Section 7.5

7–19 $T_S \approx 151°C$

7–21 $T_{MRT} = 91.97°C$

7–23 $\dot{Q}_{net} = 29.635$ W/m^2

7–25 $h_\infty = 21.17$ W/m^2 · K, $T_\infty = 280°C$

Chapter Eight

Section 8.2

8–3 $\Omega_{n_2} = 0.789921$
$\Omega_{O_2} = 0.19998$
$\Omega_{Ar} = 0.009999$
$\Omega_{Radon} = 0.0000999 \ldots$

Section 8.3

8–5 $\wp_{AB} = 0.083\ cm^2/s$ (*eqn*),
$\wp_{AB} = 0.088\ cm^2/s$ (*table*)

8–7 $\wp_{AB} = 4.26 \times 10^{-11}\ cm^2/s$

8–9 $\dot{m}_{AT} = 0.00027\ lbm/s$, $p_2 \approx 0.509\ psia$

8–11 $\wp_{AB} = 0.1262\ cm^2/s$

8–13 $\wp_{AB} = 0.242\ cm^2/s$

Section 8.4

8–15 $Sc_{air} = 0.305$
$Sc_{CU} = 5.07 \times 10^9$

8–17 t = 0.242 s

8–19 $\dot{G}_A = 0.0187\ kg/m^2 \cdot s$

8–21 $\dot{G}_A = 0.0377\ kg/m^2 \cdot s$

Section 8.5

8–23 t = 1.11 s

Section 8.6

8–25 $T_2 \approx 62°C$, $\beta_2 = 5.8\%$, $\dot{m}_W = 125.2\ lbm/\text{min}$

Section 8.7

8–27 $\dot{G} = 0.2158\ grains/hr \cdot ft^2$, $p_2 = 0.835\ in\ Hg$

Chapter Nine

Section 9.1

9–1 $\dot{m}_{134a} = 2.3\ kg/s$

Section 9.2

9–3 $U_0 = 8.537$ Btu/hr · ft² · °F

Section 9.3

9–5 **a)** $\dot{m}_{gas} = 7.8375\ kg/s$
b) $A = 53.44\ m^2$

9–7 **a)** $\dot{m}_{CW} = 3.85\ kg/s$
b) $\varepsilon_{HX} = 42.9\%$

9–9 $\dot{m}_H = 0.948\ kg/s$, $A = 104.7\ m^2$

Section 9.4

9–11 **a)** $T_0 = 52°C$
b) $T_0 = 42.4°C$

Section 9.5

9–15 **a)** $\Delta p = -5.01\ Pa$
b) $h_0 = 64.7$ W/m² · K, $h_i = 4100$ W/m² · K
c) $U = 39.13$ W/m² · K

9–17 **a)** f = 0.0135,
b) h = 12.476 Btu/hr-ft²-°F

Chapter Ten

Section 10.1

10–1 D = 1.6 mm

10–3 $F_{net} = 0.8715 \times 10^{-6}$ N

Section 10.2

10–5 $\dot{q}_A = 1.256 \times 10^6$ W/m²

10–7 L = 0.361 m

10–9 $h_{npb} = 0.066$ W/m² · K

Section 10.3

10–11 Turbulence begins at L = 0.394 m down, where $\delta = 0.173$ mm

10–13 $\dot{m}_{cond} = 2.62\ g/s$
h = 61.3 W/m²-K
$\dot{Q} = 2.759\ kW$

10–15 $\dot{q}_l = 720.234$ W/m *one tube*
$\dot{q}_l = 340.578$ W/m · *tube* *twenty tubes*

Section 10.4

10–17 $h_{ave} = 1.503 \times 10^5$ W/m-K (eqn. 10–53)
$h_{ave} = 0.8707 \times 10^5$ W/m-K (eqn. 10–6)

10–19 $h_{ave} = 6.392 \times 10^3$ W/m-K (eqn. 10–56)
$h_{ave} = 8.892 \times 10^6$ W/m-K (eqn. 10–6)

10–21 $h_{ave} = 5029.7$ W/m²-K

Section 10.5

10–25 t = 9.15 min

10–27 $\delta = 1.95$ mm

Section 10.6

10–29 $\dot{Q}_{max} = 184.47$ kW/m³

10–31 t = 0.436 hr

INDEX